HOMOLOG

HOMOLOGY THEORY

AN INTRODUCTION TO ALGEBRAIC TOPOLOGY

BY

P. J. HILTON

Professor of Pure Mathematics in the University of Birmingham

AND

S. WYLIE

Formerly Fellow of Trinity Hall and Lecturer in Mathematics in the University of Cambridge

CAMBRIDGE

AT THE UNIVERSITY PRESS

1965

PUBLISHED BY
THE SYNDICS OF THE CAMBRIDGE UNIVERSITY PRESS
Bentley House, 200 Euston Road, London, N.W. 1
American Branch: 32 East 57th Street, New York, N.Y. 10022

First printed 1960
Reprinted 1962
1965

Printed in Great Britain at the University Printing House, Cambridge
(Brooke Crutchley, University Printer)

CONTENTS

6 The Fundamental Group and Covering Spaces

PART II. GENERAL HOMOLOGY THEORY

Background to Part II

7 Contrahomology and Maps

8 Singular Homology Theory

GENERAL INTRODUCTION

This book has been written with the intention of providing an introduction to algebraic topology as it is practised today. The reader is not supposed at the outset to possess any knowledge of algebraic topology; indeed, even the reader with no knowledge of analytic topology or abstract algebra is provided, in the Background to Part I, with a synopsis of the facts that are taken for granted in the text. The treatment throughout has been subject to the consideration that, if the book is to serve its purpose, it must provide an account of the basic notions of algebraic topology intelligible to the mathematician inexperienced in the techniques and problems described. However, though the treatment is elementary, we have been more ambitious in our choice of material than is customary in elementary textbooks. It appears to us that the literature is rich in advanced textbooks and adequate in elementary introductory textbooks, but that the two types of book are not very effectively linked. Again, the advanced textbooks themselves fall into two classes which may broadly be described as classical and modern and the rapid shifts of emphasis which the subject has experienced make it difficult always to recognize classical arguments in their modern dress. We have tried to provide the links which we believe the student might find difficulty in providing for himself from a study of the available literature.

Thus, while our beginning is quite elementary, we have been able, by omitting certain topics, particularly those treated canonically in classical works, to reach in later chapters the parts of the subjects which lie in the immediate foreground of present-day research. The book is divided into two parts. The first part, 'Homology Theory of Polyhedra', also includes a chapter on the fundamental group and covering spaces and a chapter introducing the reader to Homological Algebra. With the exception of certain material in the latter chapter, the notions presented are classical—though the style of presentation is intended to be modern. This part is entirely self-contained with one exception (the proof of the Künneth formula for the homology groups of topological products). Homology groups with arbitrary coefficients and the cohomology‡ ring are defined for pairs of simplicial complexes and proved to be topological invariants—indeed, homotopy

‡ In this book called 'contrahomology' for reasons explained below.

invariants—in the case of finite complexes. The axioms of Eilenberg and Steenrod are verified; and the viewpoint of their textbook as expressed in the (Eilenberg–Maclane) theory of *categories* and *functors* is adopted in the sense that we always bring out the naturality (in its technical sense) of the concepts and transformations introduced. We do not actually define the notions of category and functor in Part I (though the reference to Eilenberg and Steenrod appears several times), but we hope that our emphasis on the ideas appropriate to the functorial approach (e.g. covariance and contravariance, canonical isomorphisms) will prepare the reader for the next stage of his battle with the literature of algebraic topology.

We take this opportunity emphatically to reject the idea that, by attempting to be modern and, in particular, by approximating to the Eilenberg–Steenrod approach, we have made the book harder than it need be. On the contrary, the systematic enunciation of homology properties certainly makes it easier to grasp the pattern of homology theory; this is just a special case of a general principle that the systematic introduction of algebraic procedures into any branch of mathematics advances it. This is certainly not to say that the geometric content of algebraic topology should be suppressed; indeed, we have been at pains to give, where appropriate, geometrical significance to the algebraic concepts.

Part II is concerned with the application of homology theory to the study of general topological spaces. For these purposes, and particularly for chapter 7, the reader needs certain facts from homotopy theory; these are provided in the Background to Part II. Chapter 7 applies the simplicial cohomology theory developed in Part I to the study of maps of polyhedra into arbitrary spaces, and is intended to serve as an introduction to obstruction theory. The next two chapters are devoted to the singular homology theory (there is a brief description of Čech cohomology in an appendix to chapter 8). The singular theory is developed in full detail; the axioms are verified, and canonical isomorphisms established (i) between singular simplicial and singular cubical homology, and (ii) between the singular theory for polyhedra and the simplicial theory of Part I. The gap in Part I is filled by showing that the chain complex used in chapter 5 does, in fact, yield the singular homology groups of the topological product. Chapter 9, on the singular cohomology ring, contains a section on the Hopf invariant (of maps of S^{2n-1} in S^n) and closes with an appendix in which the concept of naturality is, at last, defined precisely. It was plainly necessary to do this at some stage since naturality appears as a working tool in many of our arguments, particularly in chapters 8 and 9. We delayed the definition to this late stage in the belief that its general significance would be more easily understood by a reader already possessed of a number of particular examples.

It is held by some that it is unnecessary to present simplicial homology theory since it coincides with singular theory on the category of polyhedra. It has indeed been trenchantly argued that the only two homology theories of importance are the singular theory (appropriate to homotopy theory) and the Čech theory (appropriate to the study of manifolds and algebraic geometry), and that the simplicial theory is, in a sense, the intersection of these two theories. Our case for devoting a considerable part of the book to the simplicial theory of polyhedra is, in the first place, that the book is explicitly an introduction and simplicial theory is, in our view, the best introduction to homology theory; conceptually it is certainly the easiest to comprehend, in that it is geometrically satisfying and that, for a finite complex, all groups involved are finitely generated. Furthermore, in developing simplicial theory, basic notions common to any homology theory (e.g. chain complex and chain map) are made familiar and techniques appropriate to the more powerful homology theories are exemplified in an elementary setting (e.g. the Eilenberg–Maclane technique of acyclic models generalizes the simpler notion of acyclic carriers). We also argue that it is easier to compute by means of simplicial complexes; indeed, some attention is paid in Part I to modifications of simplicial complexes which make computation easier and quicker (pseudodissections and block chains).

The final chapter is concerned with two abstract developments of homology theory, namely, spectral homology theory and the homology theory of groups. These two topics have come into great prominence in recent years, and a knowledge of the basic facts of these theories seems to be part of the essential equipment of the algebraic topologist. Applications of these theories to topology and algebra are indicated.

Each chapter ends with a batch of exercises. Those for the early chapters are meant mainly to give practice and confidence; those for the later chapters are harder and more interesting.

Certain notational innovations have been made and two of them, at least, need some defence. Throughout the book, the prefix 'co-', which traditionally appears before the terms chain, cycle, boundary, homology, to describe the dual notions, is replaced by 'contra-'. We thus adopt the third of three suggestions made by M. M. Postnikov and V. G. Boltyanskii. The preference for 'contra-' over 'co-' is that contra objects transform *contra*variantly. For example, if C is a chain complex and $\mathrm{Hom}(C, G)$ the associated contrachain complex (with values in G), then a chain map $\phi : C \to D$ induces a contrachain map $\phi^{\cdot} : \mathrm{Hom}(D, G) \to \mathrm{Hom}(C, G)$; thus $\phi^{\cdot}$ operates in the direction *opposite* to that of ϕ. Thus, from this point of view, the new terminology is more logical; equally logical would be the replacement of the terms chain, cycle, ..., by *co*chain, *co*cycle, ..., since these objects transform *co*variantly (with respect to maps of topological spaces). The latter replacement is the second of the Postnikov–Boltyanskii suggestions, but, to quote a familiar phrase, the time is clearly not ripe for this innovation! Thus the terminology we have adopted is avowedly a half-way house between the familiar terminology and that demanded by logical consistency. Had it been possible to use the terms cohomology and contrahomology as suggested, the term homology would then have been available to describe a theory (of differential graded groups) of which cohomology and contrahomology are the most important manifestations. Despite our unwillingness to adopt the term cohomology as proposed by Postnikov and Boltyanskii, we have used the word homology at times in the sense indicated above,‡ for example, in the title of the book and the titles of chapters 2 and 8.

Our second major innovation is to write transformations of spaces and of 'co-' objects on the *right* and transformations of 'contra-' objects on the *left* (more precisely, maps in the image category of a category of topological spaces under a covariant (contravariant) functor are written on the right (left)). This device would be inconvenient if we had to deal frequently with mixed functors, but this is not the case in the present work where most functors are either covariant or contravariant; where a mixed object is in question (e.g. $\mathrm{Hom}(A, B)$ as a functor of A and B), no *a priori* rule is given. On the other hand, the device turns out to have distinct advantages. For spaces, for example, if f maps X to Y and g maps Y to Z, the map f *followed* by the map g appears as fg, not gf. Further, reverting to chain maps ϕ and their associated (adjoint) contrachain maps $\phi^{\cdot}$, it is convenient that $(\phi\psi)^{\cdot} = \phi^{\cdot}\psi^{\cdot}$. Again, for the Kronecker product, we have $(c\phi, d) = (c, \phi^{\cdot}d)$, $c \in C$, $d \in \mathrm{Hom}(D, G)$.

‡ The first of the P–B suggestions, namely, that the word 'homology' should be reserved for free chain complexes would preclude this usage.

We accept the statement that contravariant functors reverse arrows in its literal sense when writing out sequences of objects and maps. Thus, for example, we write that

$$C \xrightarrow{\phi} D \xrightarrow{\psi} E$$

induces $$\operatorname{Hom}(C, G) \xleftarrow{\phi^{\cdot}} \operatorname{Hom}(D, G) \xleftarrow{\psi^{\cdot}} \operatorname{Hom}(E, G).$$

Actually, from chapter 3 onwards, we follow E. C. Zeeman in writing $C \pitchfork G$ for Hom (C, G). The reason we adopt this notation is that it brings out better the duality between the tensor product, $\otimes G$, and $\pitchfork G$. In order to have the greatest possible uniformity of notation we write $A * B$ for the torsion product of two abelian groups and, again to emphasize duality, $A \dagger B$ for Ext (A, B).

We have been concerned to try to avoid repeated use of complicated symbols even where the requirements of logical precision appear to demand them. Thus, for example, the p-dimensional chain group, with coefficients in G, of the simplicial pair K, K_0, oriented by the orientation α, may be represented by the cumbersome symbol $C_p^\alpha(K, K_0; G)$. Our general principle is to write the complete symbol in the definitions and then, in subsequent appearances of the same concept, to reproduce only as much of the symbol as is not clearly implied by the context. Thus the symbol mentioned above might well be replaced in subsequent occurrences by C_p. Here in the introduction we merely state the principle—in the text of the book we prepare the reader for each specific abbreviation.

The reader will be familiar with the grammatical flexibility of the symbol '='. In the same way the symbol '$f: A \to B$' should normally be read as 'f, *which is* a transformation from A to B', but it will occasionally have to be read otherwise, in particular as 'f *is* a transformation from A to B'. Similar grammatical ambiguity occurs with the symbol '$a \in A$'.

We have adapted Halmos' notation by marking the end of a proof with ▌, having been impressed by its effectiveness in the book *General Topology* by J. L. Kelley. The appearance of ▌ immediately after the statement of a theorem (or proposition, etc.) indicates either that the proof has immediately preceded the statement or that the proof is left to the reader. As a further guide to the reader we have adopted the devices (i) of starring certain sections or parts of sections; and (ii) of putting some material in small type. If a part is starred the reader may conclude that its contents are somewhat more sophisticated than its place in the book would suggest, and that it may be omitted on first reading. The use of small type, on the other hand, indicates only that the material is not central to the logical development of the subject; thus illustrative remarks and examples are often so treated.

Our numbering system is by triples $\alpha.\beta.\gamma$, where α is the chapter‡ number, β the section number within the chapter, and γ the item number within the section. Even in chapter α, $\alpha.\beta.\gamma$ is *not* abbreviated to $\beta.\gamma$, so that a reference to 5.7, for instance, is a reference to a section and not to an item. The items are numbered consecutively throughout a section, whether they be theorems, propositions, lemmas, corollaries, definitions, examples, or just displayed formulae. The main exception to this rule is that, following Eilenberg and Steenrod, we may index the statement in contrahomology corresponding to statement $\alpha.\beta.\gamma$ in homology by the same triple followed by the affix c; another exception occurs in 5.3 as is there explained.

We have had help of various kinds from many people, most of whom can only be thanked in rather general terms. Our mathematical friends have contributed lavishly in conversation and discussion, and we have only made specific acknowledgements in the text for a fraction of the ideas that we have knowingly adopted; there must be many more that we have adopted thinking them to have been our own. Dr E. C. Zeeman read the manuscript of what we had intended to put in the book and by his imaginative and effective criticisms persuaded us to improve it in numberless ways; in particular, many of the robuster examples are due to him. Mr E. C. Thompson has read the book in the less flexible form presented by galley proofs and has been of the greatest help in eliminating mistakes and obscurities. We are very grateful indeed to both of them.

We are grateful also to many people for the stimulus they have provided. Some have done this by the constructive and occasionally embarrassing interest they have shown in the progress of the book. Others, to whom we are quite as grateful, have stimulated us in another way—those of our friends who have told us that this is not the time to write a book on Homology Theory, at any rate not a book of this kind, and those who have taken a less technical line (our wives, in fact) and have made it clear that this is not the time to write a book of any kind whatsoever.

‡ For the Background to Part I, $\alpha = \text{I}$; for the Background to Part II, $\alpha = \text{II}$.

On the production side we have been extremely lucky. Miss Kerridge, who did the bulk of the typing, contrived to make good-looking copy of some very unpromising handwriting and has earned our warmest thanks. So too have the officers and staff of the Cambridge University Press who have been involved in the book; all of them have shown us unfailing helpfulness and patience.

P.J.H
S.W.

February 1960

Note on the Second Impression

A number of misprints, and two mistakes, have been corrected at the second printing. We are grateful to those who have drawn our attention to these; and, in particular, to Professor W. S. Massey for pointing out that the original statement of 9.4.16 was wrong.

P.J.H.
S.W.

March 1962

PART I

HOMOLOGY THEORY OF POLYHEDRA

BACKGROUND TO PART I

1 Analytic topology

A *topological space* is a set X in which certain subsets, called *open sets*, are distinguished; the collection of open sets satisfies the axioms:

(O 1) the union of any number of open sets is open;

(O 2) the intersection of any finite number of open sets is open;

(O 3) the whole space and the empty set are open.

To prescribe the open sets is to assign a *topology* to the set X. If $\mathscr{U}, \mathscr{V}$ are two topologies on the set X, then $\mathscr{U}$ is *finer* than $\mathscr{V}$ ($\mathscr{V}$ is *coarser* than $\mathscr{U}$) if every set of X which is open in the topology $\mathscr{V}$ is open in the topology $\mathscr{U}$. A set of open sets of X forms a *base* (for the open sets) if every open set of X is a union of sets of the base.

A *closed* subset of the topological space X is the complement of an open set; thus a topology is assigned by prescribing the closed sets and the closed sets must satisfy the axioms:

(C 1) the union of any finite number of closed sets is closed;

(C 2) the intersection of any number of closed sets is closed;

(C 3) the whole space and the empty set are closed.

If X_0 is a subset of the topological space X, the *induced topology* in X_0 is that in which the open sets are the intersections with X_0 of the opens sets of X. Subsets will always be supposed to carry the induced topology. Plainly, if $X_1 \subseteq X_0 \subseteq X$, then X_0 and X induce the same topology in X_1, and an open (closed) subset of an open (closed) subset of X is an open (closed) subset of X.

The *interior* of X_0 is the union of all open subsets of X contained in X_0.

I. 1.1 Proposition. *The interior of X_0 is the largest open set contained in X_0.*

The *closure* of X_0 is the intersection of all closed sets containing X_0.

I. 1.2 Proposition. *The closure of X_0 is the smallest closed set containing X_0.*

If $x \in X$, a *neighbourhood* of x in X is a subset of X containing x in its interior.

I. 1.3 Proposition. *The closure of X_0 is the sets of points $x \in X$ such that every neighbourhood of x meets X_0.*

The *frontier* of X_0 in X is the intersection of the closure of X_0 and the closure of its complement. X_0 is *dense* in X if its closure is X. The sequence $x_1, x_2, \ldots, x_n, \ldots$ of points of X *converges* to $x \in X$ if every neighbourhood of x contains all but a finite number of points of the sequence.

The collection $\{X_i\}$ of subsets of X is a *covering* of X if their union is X. The covering is *open* (*closed*) if each X_i is open (closed). The covering is *finite* (*countable*) if there are finitely many (countably many) sets in the covering. The covering $\{Y_j\}$ is a *subcovering* (*refinement*) of the covering $\{X_i\}$ if each Y_j is (is contained in) an X_i.

A *map* $f : X \to Y$ from the space X to the space Y is a *continuous* function‡ from X to Y; the *continuity* of f is expressed by the property that, if U is any open subset of Y then $f^{-1}(U)$, the set of points of X mapped by f into U, is open in X. Equivalently, f is continuous provided $f^{-1}(F)$ is closed whenever F is closed. If $X_0 \subseteq X$, then $X_0 f$ is the set of points xf, $x \in X_0$, and is called the *f-image* of X_0. If $Y_0 \subseteq Y$, $f^{-1}(Y_0)$ is called the *f-counterimage* of Y_0. A map $f : X \to Y$ determines functions $f_0 : X_0 \to Y$, $f' : X \to Xf$ given by $xf_0 = xf$, $x \in X_0$; $xf' = xf$, $x \in X$. We may write f_0 as $f \mid X_0$, and we say that f_0 is the *restriction* of f to X_0 and that f is an *extension* of f_0 to X, or over X.

I. 1.4 Proposition. *The functions f_0, f' are continuous.*

I. 1.5 Proposition. *If $f : X \to Y, g : Y \to Z$ are maps, then $fg : X \to Z$ is a map.*

Let $\{A_i\}$ be a *finite* covering of X by closed sets and let $f_i : A_i \to Y$ be maps such that $f_i \mid A_i \cap A_j = f_j \mid A_i \cap A_j$. Then we may define a unique function $f : X \to Y$ by $f \mid A_i = f_i$.

I. 1.6 Proposition. *The function $f : X \to Y$ is continuous.*

(For if $F \subseteq Y$ is closed then $f^{-1}F = \bigcup_i f_i^{-1}(F)$; but $f_i^{-1}(F)$ is closed in A_i and hence in X so that, by (C 1), $f^{-1}F$ is closed in X and f is continuous.) A similar result holds for arbitrary coverings by open sets.

If $X_0 \subseteq X$, let $i : X_0 \to X$ be given by $xi = x$, $x \in X_0$. Then i is plainly continuous and we call it an *inclusion* map or *injection*. If

‡ All functions are understood to be single-valued.

there exists a map $p : X \to X_0$ such that $ip = 1$, where 1 is the *identity map* of X_0, then X_0 is a *retract* of X and p is a *retraction*.

The map $f : X \to Y$ is a *homeomorphism* of X *onto* Y (abbreviated to *homeomorphism*) if there is a map $g : Y \to X$ (called the *inverse* of f) such that $fg = 1 : X \to X$ and $gf = 1 : Y \to Y$. We write $f : X \cong Y$ and say that X and Y are *homeomorphic* or *of the same topological type*.

I. 1.7 Proposition. *$X \cong Y$ is an equivalence relation.*

The map $f : X \to Y$ is a *homeomorphism* of X *into* Y if the induced map $f' : X \to Xf$ is a homeomorphism. Plainly $i : X_0 \to X$ is a homeomorphism of X_0 into X; if $f : X \to Y$ is a homeomorphism of X into Y, then X may be *embedded* in Y by identifying x with xf, $x \in X$. The map $f : X \to Y$ is *locally homeomorphic* if each $x \in X$ possesses a neighbourhood U such that f maps U homeomorphically onto a neighbourhood of xf.

Let $f : X \to Y$ be a function from the *space* X *onto the set* Y; then the *identification topology on* Y *determined by* f is the topology in which $Y_0 \subseteq Y$ is closed if and only if $f^{-1}(Y_0)$ is closed. If Y is given the identification topology, f is called an *identification map* or *projection*.

I. 1.8 Proposition. *The identification topology is the finest topology consistent with the continuity of f.*

(Note that this proposition asserts, *a fortiori*, that the identification topology *is* a topology.)

Let R be an equivalence relation on the points of X (thus xRx; xRx' implies $x'Rx$; xRx' and $x'Rx''$ together imply xRx''). Let Y be the set of R-equivalence classes and let $k : X \to Y$ send each point to its equivalence class. If Y is given the identification topology determined by k we may write $Y = X/R$ and say that Y is the *quotient space* of X by the relation R (with the *quotient topology*).

Given two spaces X and Y their *topological product* $X \times Y$ is the set of pairs (x, y), $x \in X$, $y \in Y$, with the topology in which a base of open sets consists of the sets $U \times V$, where U is open in X and V is open in Y. The maps $(x, y) \to x$, $(x, y) \to y$ project $X \times Y$ onto X, Y. If $x_* \in X$, $y_* \in Y$, the maps $x \to (x, y_*)$, $y \to (x_*, y)$ embed X, Y in $X \times Y$; we refer to these maps (by abuse of language) as injections.

The subspace $(X \times y_*) \cup (x_* \times Y)$ of $X \times Y$ is called the *bunch* of X and Y and is written $X \vee Y$ if there is no need to specify the points

x_*, y_*. Then $X \times y_*$, $x_* \times Y$ are subspaces‡ of $X \vee Y$ whose union is $X \vee Y$ and whose intersection is the single point (x_*, y_*). Thus, using the injections defined above, we may think of $X \vee Y$ as the 'union of X and Y with a single common point'.

Clearly the notion of topological product extends to finite collections of spaces (it also extends to infinite collections, but we shall only be concerned with finite products).

A *metric* in the set X is a real-valued function ρ defined on (ordered) pairs of points of X and satisfying

(M 1) for $x, y \in X$, $\rho(x,y) \geqslant 0$;

(M 2) $\rho(x,y) = \rho(y,x)$;

(M 3) $\rho(x,y) = 0$ if and only if $x = y$;

(M 4) (*triangle inequality*) for $x, y, z \in X$, $\rho(x,y) + \rho(y,z) \geqslant \rho(x,z)$.

The *ϵ-ball, centre x*, in the set X (with the metric ρ) is the set, $V(x,\epsilon)$, of points $y \in X$ with $\rho(x,y) < \epsilon$. The *metric topology* in the set X is the topology in which the open sets are unions of sets of the form $V(x,\epsilon)$.

I. 1.9 Proposition. *The metric topology is a topology.*

X is a *metric space* if its topology is given by a metric ρ. The *diameter* of a subset X_0 of a metric space X is $\underset{x, y \in X_0}{\text{l.u.b.}}\ \rho(x,y)$.

The topological space X is *metrizable* if the set X admits a metric such that the metric topology coincides with the given topology.

I. 1.10 Proposition. *A function $f: X \to Y$ from the metric space X to the metric space Y is continuous if and only if, for any $x \in X$ and $\epsilon > 0$, there exists $\delta(x,\epsilon) = \delta > 0$ such that $V(x,\delta)f \subseteq V(xf,\epsilon)$.*

The map f is *uniformly continuous* if δ may be chosen independently of x.

The space X is

Hausdorff if any two distinct points of X possess disjoint neighbourhoods;

compact if every open covering of X has a finite subcovering;

sequentially compact if every sequence of points of X has a convergent subsequence;

locally compact if every point of X has a compact neighbourhood;

connected if X is not the union of two disjoint non-empty closed sets;

separable if X possesses a countable dense subset.

I. 1.11 Proposition. *A metric space is compact if and only if it is sequentially compact.*

‡ They are closed subspaces if the points x_*, y_* are closed sets.

I. 1.12 Proposition. *A compact metric space is separable.*

I. 1.13 Proposition. *A (continuous) map of a compact metric space into a metric space is uniformly continuous.*

I. 1.14 Proposition. *The limit of a convergent sequence in a Hausdorff space is unique.*

I. 1.15 Proposition. *A compact subset of a Hausdorff space is closed.*

I. 1.16 Proposition. *If X is locally compact and Hausdorff then every neighbourhood contains a compact neighbourhood.*

I. 1.17 Proposition. *Let $f: X \to Y$ be a map of the compact space X onto the Hausdorff space Y; then Y has the identification topology determined by f. In particular, f is a homeomorphism if it is* (1, 1).

I. 1.18 Proposition. *A subset of Euclidean space R^n is compact if and only if it is closed and bounded.*

Let $r \geqslant 0$ be a real number and let I_r be the closed interval $0 \leqslant t \leqslant r$ in R^1. A *path* in X is a map $f: I_r \to X$, for some $r \geqslant 0$. The path *starts* at $0f$, its *initial* point, and *ends* at rf, its *final* point. X is *path-connected* if, given any two points $x, y \in X$, there exists a path in X starting at x and ending at y. The *component* (*path-component*) of X containing x is the largest connected (path-connected) subset of X containing x.

I. 1.19 Proposition. *A path-connected space is connected.*

The *Hilbert space* H^∞ is the metric space whose points are infinite sequences of real numbers $\mathbf{u} = (u_1, u_2, \ldots, u_n, \ldots)$ such that $\sum_{n=1}^{\infty} u_n^2$ converges, the metric being

$$\rho(\mathbf{u}, \mathbf{v}) = \left\{ \sum_{n=1}^{\infty} (u_n - v_n)^2 \right\}^{\frac{1}{2}}.$$

I. 1.20 Proposition. *H^∞ is separable.*

(The set of finite sequences of rationals is a countable dense subset.)

Euclidean space R^n (of dimension n) may be embedded in H^∞ by identifying $(u_1, \ldots, u_n)$ with $(u_1, \ldots, u_n, 0, 0, \ldots)$. Then there are inclusions

$$R^1 \subseteq R^2 \subseteq \ldots \subseteq R^n \subseteq R^{n+1} \subseteq \ldots \subseteq H^\infty.$$

The *unit n-sphere* S^n is the subset of R^{n+1} given by $u_1^2 + \ldots + u_{n+1}^2 = 1$. It is the frontier in R^{n+1} of the (closed) ball V^{n+1}, given by

$u_1^2 + \ldots + u_{n+1}^2 \leqslant 1$. An *n-cell* is a homeomorph of V^n; an *n-sphere* is a homeomorph of S^n; the *frontier* of the n-cell E^n is the image of S^{n-1} under a homeomorphism $V^n \to E^n$; the complement in E^n of its frontier is called the *interior of* E^n.

2 Algebra

A *group* G is a collection of elements $g, g', g'', \ldots$ together with a law of composition, written multiplicatively, satisfying the axioms:

(G 1) $(gg')g'' = g(g'g'')$ for all g, g', g'';

(G 2) there exists an element $e \in G$ such that $ge = g$, for all g;

(G 3) to each g there exists $\bar{g} \in G$ such that $g\bar{g} = e$.

The *order* of G is the number (not necessarily finite) of elements in G.

An *abelian group* A is a collection of elements $a, a', a'', \ldots$ together with a law of composition, written additively, satisfying the axioms:

(A 1) $(a+a')+a'' = a+(a'+a'')$ for all a, a', a'';

(A 2) there exists an element $0 \in A$ such that $a+0 = a$, for all a;

(A 3) to each a there exists $(-a) \in A$ such that $a+(-a) = 0$;

(A 4) $a+a' = a'+a$ for all a, a'.

The group G is said to be *commutative* if $gg' = g'g$ for all g, g'. By an abuse of language we may identify a commutative group with the abelian group to which it evidently corresponds. Our main algebraic concern in this book is with abelian groups.

A *subgroup* (more precisely, sub-(abelian group)!) of an abelian group A is a subset of the elements of A which constitutes an abelian group under the law of composition defined in A. The intersection of subgroups is a subgroup.

If B, C are subsets of the abelian group A we write $B+C$ for the subset of A consisting of elements $b+c$, $b \in B$, $c \in C$. We call $B+C$ the *sum* of B and C; this notation evidently extends to any finite collection of subsets of A, or even to an arbitrary collection $\{B_\alpha\}$ provided we select $0 \in B_\alpha$ for all but a finite number of values of α. If each B_α is a subgroup of A, so is their sum.

If A_0 is a subgroup of A, a *coset of A by A_0* is a set of elements $a+A_0$; the *factor group*‡ or *quotient group* A/A_0 is the set of cosets of A by A_0 with the law of composition

$$(a+A_0)+(a'+A_0) = (a+a')+A_0.$$

‡ A strong case exists for calling this the *difference group* and writing it $A - A_0$.

I.2.1 Proposition. *If A_1 is a subgroup of A_0 and A_0 is a subgroup of A, then A_0/A_1 is a subgroup of A/A_1 and*

$$(A/A_1)/(A_0/A_1) \cong A/A_0.$$

I.2.2 Proposition. *If B, C are subgroups of A, $(B+C)/C \cong B/(B \cap C)$.*

A *homomorphism* $\phi : A \to B$ from the abelian group A to the abelian group B is a function satisfying $(a+a')\phi = a\phi + a'\phi$; then $0\phi = 0$ and $(-a)\phi = -a\phi$. The *kernel* of ϕ is the subgroup, $\phi^{-1}(0)$, of A; the *image* of ϕ is the subgroup, $A\phi$, of B; and the *cokernel* of ϕ is the‡ factor group, $B/A\phi$, of B. Then ϕ is a *monomorphism* if its kernel is zero, an *epimorphism* if its cokernel is zero (equivalently, if B is the image of ϕ), and an *isomorphism* if it is monomorphic and epimorphic. We write $0 : A \to B$ for the *zero homomorphism* ($a0 = 0$, all a). An *endomorphism* of A is a homomorphism $A \to A$ and an *automorphism* of A is an isomorphism $A \to A$. We write $\phi : A \cong B$ if ϕ is an isomorphism and we write $1 : A \cong A$ for the *identity automorphism* ($a1 = a$, all a). If A_0 is a subgroup of A, the monomorphism $i : A_0 \to A$, given by $ai = a$, $a \in A_0$, is called the *inclusion* map or *injection.* If $\phi : A \to B$ is a monomorphism then A may be *embedded* in B by identifying a with $a\phi$, all a. If A_0 is a subgroup of A, the epimorphism $p : A \to A/A_0$ which sends each element of A into its coset is called the *projection.*

A sequence of abelian groups and homomorphisms

$$\dots \to A_{n+1} \xrightarrow{\phi_{n+1}} A_n \xrightarrow{\phi_n} A_{n-1} \to \dots,$$

finite or infinite is *exact at A_n* if the kernel of the homomorphism§ ϕ_n is the image of the homomorphism ϕ_{n+1}. The sequence is *exact* if it is exact at A_n for each n.

I.2.3 Proposition. *The sequence $0 \to A_0 \xrightarrow{i} A \xrightarrow{p} A/A_0 \to 0$ is exact.*

The collection of homomorphisms $A \to B$ may be given an abelian group structure by defining the sum $\phi + \phi'$ of homomorphisms ϕ and ϕ' by the rule

$$a(\phi + \phi') = a\phi + a\phi', \ a \in A, \ \phi, \phi' : A \to B.$$

This group is usually written $\mathrm{Hom}(A, B)$; we shall introduce the notation $A \pitchfork B$, due to E. C. Zeeman.

‡ To complete the 'duality', we may define the *co-image* of ϕ as the factor group $A/\phi^{-1}(0)$.

§ In diagrams it is convenient to write '$A \xrightarrow{\phi} B$' for '$\phi : A \to B$'.

If $\phi: A \to B$, $\psi: B \to C$ are homomorphisms, then the transformation $\phi\psi: A \to C$ is again a homomorphism. If $\phi': A \to B$ and $\psi': B \to C$ are also homomorphisms, then

$$(\phi + \phi')\psi = \phi\psi + \phi'\psi$$
$$\phi(\psi + \psi') = \phi\psi + \phi\psi'.$$

Let $\{A_\alpha\}$ be an indexed family of abelian groups; their *direct sum* is the abelian group A defined as follows: an element of A is a collection of elements $\{a_\alpha\}$, $a_\alpha \in A_\alpha$, subject to the restriction that $a_\alpha = 0$ for all but a finite number of values of α; and addition is defined componentwise by $\{a_\alpha\} + \{a'_\alpha\} = \{a_\alpha + a'_\alpha\}$. We write $A = \sum_\alpha A_\alpha$. If the restriction on $\{a_\alpha\}$ is withdrawn the resulting group is called the *direct product* or *unrestricted direct sum* of the groups A_α and written $\prod_\alpha A_\alpha$. If α runs over a *finite* indexing set, say $\alpha = 1, \ldots, k$, then $\sum_\alpha A_\alpha = \prod_\alpha A_\alpha$ and we write either as $\sum_{\alpha=1}^{k} A_\alpha$ or $A_1 \oplus \ldots \oplus A_k$. We may embed A_{α_0} in $\sum_\alpha A_\alpha$ (or in $\prod_\alpha A_\alpha$) by identifying a_{α_0} with $\{a_\alpha\}$, where $a_\alpha = 0$, $\alpha \neq \alpha_0$. By a slight abuse of language we refer to the given embedding as an *injection*; similarly, the epimorphism $\sum_\alpha A_\alpha \to A_{\alpha_0}$ $(\Pi A_\alpha \to A_{\alpha_0})$ given by $\{a_\alpha\} \to a_{\alpha_0}$ will be called a *projection*.

The groups A_α are not at the outset subgroups of $\sum_\alpha A_\alpha$. To be precise, we have defined the *external* direct sums of the groups A_α. If the groups A_α are subgroups of an abelian group A such that A is the sum of the groups A_α and each A_α intersects the sum of the remaining A_α in the zero element alone, then A is called the *internal* direct sum of its subgroups A_α.

I.2.4. Proposition. (i) *If A is the internal direct sum of its subgroups A_α, then* $A \cong \sum_\alpha A_\alpha$; (ii) *for any indexed family $\{A_\alpha\}$, $\sum_\alpha A_\alpha$ is the internal direct sum of the images under injection of the groups A_α.*

If A_0 is a subgroup of A, and A can be expressed as the internal direct sum of A_0 and some other subgroup of A, then A_0 is said to be a *direct factor* of A.

A set $\{a_i\}$ of elements of an abelian group A *generates* the subgroup A_0 of elements of A expressible as $\Sigma n_i a_i$, the n_i being integers of which only a finite number are non-zero; if the set $\{a_i\}$ is enumerable we may write $A_0 = (a_1, a_2, \ldots)$. If $A_0 = A$ we call $\{a_i\}$ a set of *generators* of A. An identity $\Sigma m_i a_i = 0$ in A is then called a *relation* (between the generators a_i). The *order* of $a \in A$ is the order of the subgroup of A generated by a.

The abelian group F is *free* if it possesses a set of elements $\{f_i\}$ such that each element of F is *uniquely* expressible as $\Sigma n_i f_i$. The set $\{f_i\}$ is called a *basis* for F and the *rank* of F is the number (not necessarily finite) of elements in any basis. This definition is justified by

I. 2.5 Proposition. *The rank of F is independent of the choice of basis.*

(This is a classical theorem of linear algebra if the rank is finite. If the rank is infinite it equals the order of F.)

I. 2.6 Proposition. *If F is a free abelian group, $\{f_i\}$ a basis for F and A an arbitrary abelian group, then any function defined on the basis with values in A may be uniquely extended to a homomorphism from F to A.*

(For, given such a function θ, we define $(\sum_i n_i f_i)\phi = \sum_i n_i(f_i\theta)$. Then ϕ is single-valued since the representation $\sum_i n_i f_i$ is unique and is clearly homomorphic. Moreover, ϕ extends θ and is the unique such homomorphism.)

A (free abelian) *presentation* of the abelian group A is an epimorphism $\mu : F \to A$, where F is free.

The symbol '$A = J$' means that A is isomorphic with the additive group of integers; the symbol '$A = J_m$' means that A is isomorphic with the residue classes of integers mod m, where m is a positive integer.

Let G be a group. A subgroup G_0 of G is *normal* or *self-conjugate* if $g^{-1}g_0 g \in G_0$ for all $g_0 \in G_0$, $g \in G$. If G_0 is a normal subgroup of G, the factor group‡ or quotient group is written G/G_0. The *centre* C of G is the subgroup consisting of elements $c \in G$ such that $cg = gc$ for all $g \in G$.

I. 2.7 Proposition. *C is a normal subgroup of G.*

For each $g, g' \in G$ write $[g, g'] = g^{-1}g'^{-1}gg'$ and let $[G, G]$ be the least subgroup of G containing all *commutators* $[g, g']$. Then $[G, G]$ is the *commutator subgroup* or *derived* group of G.

I. 2.8 Proposition. *$[G, G]$ is normal in G and $G/[G, G]$ is commutative. Moreover, if G_0 is normal in G and G/G_0 is commutative then $G_0 \supseteq [G, G]$.*

‡ The definitions of subgroup, coset and factor group are analogous to those for abelian groups.

A *ring* R is an abelian group with a further operation, written multiplicatively, satisfying the axioms:

(R 1) $(rr')r'' = r(r'r'')$ for all $r, r', r'' \in R$;

(R 2) $(r+r')r'' = rr''+r'r''$, $r''(r+r') = r''r+r''r'$, for all $r, r', r'' \in R$.

The ring R is *commutative* if $rr' = r'r$ for all $r, r' \in R$. A *unity element* in the ring R is an element $1 \neq 0$ satisfying $r1 = 1r = r$ for all $r \in R$. The ring R is a *field* if its non-zero elements form a commutative group under multiplication.

A (*two-sided*) *ideal* in the ring R is a subring I with the property that $ra \in I$, $ar \in I$, whenever $a \in I$, $r \in R$. If I is an ideal the projection $R \to R/I$ induces in R/I the structure of a ring, called the *quotient ring* of R by I. If $\phi : R \to S$ is a ring homomorphism, which means that

$$(r+r')\phi = r\phi + r'\phi, \quad (rr')\phi = (r\phi)(r'\phi), \quad r, r' \in R,$$

then the kernel K of ϕ is an ideal in R, the image of ϕ is a subring of S and

I. 2.9 Proposition. *$R/K \cong R\phi$ (as rings).*

A *diagram*

$$\begin{array}{ccc} A & \xrightarrow{\phi} & B \\ \downarrow{\scriptstyle\rho} & & \downarrow{\scriptstyle\psi} \\ C & \xrightarrow{\sigma} & D \end{array}$$

of sets and set transformations is *commutative* if $\rho\sigma = \phi\psi : A \to D$. This terminology will also be applied to more complicated diagrams. The diagrams appearing in this book will most frequently be concerned with abelian groups and homomorphisms.

3 Zorn's lemma

In a few places in the book we need an extension of the familiar Principle of Mathematical Induction. We here give some definitions leading to this extension (Zorn's lemma).

A *partially ordered set* is a set A of elements and an order relation (to be read as 'less than or equal to') $a_1 \prec a_2$, that holds between certain pairs of elements of A. This relation satisfies the conditions

(i) for all $a \in A$, $a \prec a$;

(ii) if $a_1 \prec a_2$, $a_2 \prec a_3$, then $a_1 \prec a_3$.

We remark that we do *not* in general demand

(iii) if $a_1 \prec a_2$, $a_2 \prec a_1$, then $a_1 = a_2$.

A partial ordering $\prec$ satisfying (iii) will be said to be *strict*.‡ Condition (iii) would exclude example (δ) below. If A is a partially ordered set we can introduce an equivalence relation in A by putting $a_1 R a_2$ if and only if $a_1 \prec a_2$ and $a_2 \prec a_1$. If A^* is the set of equivalence classes then the partial ordering in A induces a strict partial ordering in A^*.

A *totally ordered set* is a strictly partially ordered set A such that for each pair $a, a' \in A$ either $a \prec a'$ or $a' \prec a$.

Examples: (α) the real numbers, ordered by $\leqslant$, form a totally ordered set;

(β) the complex numbers, ordered by the relation $a+ib \prec a'+ib'$ if and only if $a \leqslant a'$ and $b \leqslant b'$, form a partially ordered set;

(γ) the set of subsets of a given set, ordered by $\subseteq$, form a partially ordered set;

(δ) the set of coverings of a space (or, indeed, set) X ordered by the relation $\{X_i\} \prec \{Y_j\}$ if and only if $\{Y_j\}$ is a refinement of $\{X_i\}$ (see I.1) is a partially ordered set.

Note that if $\{X_i\}$ is the set of all 2^{-n}-balls in a metric space X and $\{Y_j\}$ is the set of all 3^{-n}-balls, $n=0, 1, 2, \ldots$, then $\{X_i\} \prec \{Y_j\}$ and $\{Y_j\} \prec \{X_i\}$ but $\{X_i\} \neq \{Y_j\}$ in general.

The element b of the partially ordered set A is an upper bound of the subset $A_0 \subseteq A$ if $a_0 \prec b$ for all $a_0 \in A_0$. A is *inductive* if each totally ordered subset of A has an upper bound in A. The element $m \in A$ is *maximal* if for all $a \in A$, $m \prec a$ implies $a \prec m$.

I. 3.1 (Zorn's lemma). *If the (non-empty) partially ordered set A is inductive, then it possesses a maximal element.*

I. 3.2 Proposition. *If A_0 is a totally ordered subset of the inductive set A, then it possesses a maximal upper bound.*

(For the set of upper bounds of A_0 is inductive if A is inductive.)

We regard Zorn's lemma as an axiom of set theory. For further discussion (particularly of its relation to the *well-ordering theorem* and the *axiom of choice*) see J. L. Kelley, *General Topology* ch. 0.

‡ Some authors use the term 'pre-ordering' for a relation satisfying (i) and (ii)

1

THE TOPOLOGY OF POLYHEDRA

1.1 Rectilinear simplexes

Elementary combinatorial topology is concerned with those topological spaces which admit dissection into suitably regular pieces. If we put this the other way round, we require to formulate the concept of a standard space or type of space in a precise manner and then to define the concept of building a space from such pieces or 'bricks'. In this first section we define the bricks and leave the description of the building technique to the next section.

The bricks are called *simplexes*; a simplex is a generalization of an interval (1-simplex), a triangle (2-simplex) or a tetrahedron (3-simplex). In this section we shall be concerned only with subsets of a Euclidean space R^m; the points of R^m can be determined by real m-dimensional position vectors with respect to some chosen Cartesian coordinate system in R^m and we shall use the same letter for a point of R^m and for the vector associated with it. Thus if a, b are points of R^m and λ, μ are real numbers, we may speak of the point $\lambda a + \mu b$. In such a context the symbol 0 represents the origin.

We now proceed with the preliminary notions necessary for the definition of a p-simplex. A set of $(p+1)$ points of R^m, $a^0, a^1, \ldots, a^p$, is said to be a set of *independent* points if the equations

1.1.1 $$\sum_{i=0}^{p} \lambda_i a^i = 0, \quad \sum_{i=0}^{p} \lambda_i = 0,$$

together imply‡ $\lambda_0 = \lambda_1 = \ldots = \lambda_p = 0$. This definition clearly describes a property of the points $a^0, a^1, \ldots, a^p$ themselves and is independent of the choice of origin. A point b is said to be *dependent* on the set $a^0, a^1, \ldots, a^p$ if there exist $\lambda_0, \ldots, \lambda_p$ such that

1.1.2 $$\sum_i \lambda_i a^i = b \quad \text{and} \quad \sum_i \lambda_i = 1.$$

It is then obvious that the set $a^0, a^1, \ldots, a^p, b$ is not independent, so that no point of an independent set is dependent on the others.

‡ This condition is equivalent to the assertion that the vectors $a^i - a^0$, $i = 1, \ldots, p$, are linearly independent.

Moreover we may show

1.1.3 Proposition. *If $a^0, \ldots, a^p$ is an independent set and b is dependent on it, the expression* 1.1.2 *is unique. If $a^0, \ldots, a^p$ is a dependent set and b is dependent on it, the expression* 1.1.2 *is not unique; moreover, if in this case b admits an expression* 1.1.2 *with all $\lambda_i > 0$, it admits more than one such.*

First let the set be independent and suppose

$$b = \sum_i \lambda_i a^i = \sum_i \mu_i a^i, \quad \text{with} \quad \sum_i \lambda_i = \sum_i \mu_i = 1.$$

Then
$$\sum_i (\lambda_i - \mu_i) a^i = 0, \quad \sum_i (\lambda_i - \mu_i) = 0,$$

whence, by 1.1.1, $\lambda_i = \mu_i$, $i = 0, \ldots, p$, and the expression 1.1.2 is unique.

Conversely, suppose the set not to be independent. Then there exist numbers ν_i, $i = 0, \ldots, p$, not all zero, such that $\sum_i \nu_i a^i = 0$, $\sum_i \nu_i = 0$. Then if $b = \sum_i \lambda_i a^i$ we also have

$$b = \sum_i (\lambda_i + \nu_i) a^i \quad \text{and} \quad \sum_i (\lambda_i + \nu_i) = 1,$$

so that the expression for b is not unique. Moreover, if all $\lambda_i > 0$, we may choose the numbers ν_i so small that all $\lambda_i + \nu_i > 0$. ∎

The set of points dependent on $a^0, a^1, \ldots, a^p$ form a (closed) Euclidean subspace of R^m; if the set is independent the subspace is of (Euclidean) dimension p. In the latter case, Proposition 1.1.3 allows us to attach to each point of the subspace a unique set of *coordinates*; namely, we attach to b the coordinates $(\lambda_0, \lambda_1, \ldots, \lambda_p)$, where 1.1.2 holds. The coordinates $(\lambda_0, \lambda_1, \ldots, \lambda_p)$ are called the *barycentric* coordinates of b with respect to the independent set $a^0, a^1, \ldots, a^p$. The name 'barycentric' derives from the fact that b is the centre of mass of masses λ_i placed at the points a^i, $i = 0, \ldots, p$. We recall that $\Sigma \lambda_i = 1$.

Let $a^0, a^1, \ldots, a^p$ be an independent set.

1.1.4 Definition. The *rectilinear p-simplex*, s_p, with *vertices* $a^0, a^1, \ldots, a^p$ is the set of points dependent on $a^0, a^1, \ldots, a^p$ whose barycentric coordinates satisfy $\lambda_i > 0$, $i = 0, \ldots, p$. The simplex s_p with vertices $a^0, a^1, \ldots, a^p$ may be written $(a^0 a^1 \ldots a^p)$; s_p is said to be *spanned* by its vertices.

It is clear that s_p is an open subset in the subspace of R^m consisting of points dependent on $a^0, a^1, \ldots, a^p$. For brevity we may call the latter space the subspace *determined* by $a^0, a^1, \ldots, a^p$ and write it $R(a^0, a^1, \ldots, a^p)$. The closure of s_p, $\bar{s}_p$, is the *closed rectilinear p-simplex* (with vertices $a^0, a^1, \ldots, a^p$) and consists of points of $R(a^0, a^1, \ldots, a^p)$ whose barycentric coordinates satisfy $\lambda_i \geqslant 0$, $i = 0, \ldots, p$. The *frontier* of s_p is written $\dot{s}_p$ and is defined as $\bar{s}_p - s_p$. It is thus the frontier of s_p in the sense of analytic topology if s_p is regarded as a subset of $R(a^0, a^1, \ldots, a^p)$. Then $\dot{s}_p$ consists of those points of $\bar{s}_p$ for which $\lambda_i = 0$ for some i, $0 \leqslant i \leqslant p$. The simplex s_p is said to have *dimension* p; this agrees, of course, with its dimension as a Euclidean space. A *p-simplex* is a simplex of dimension p; we identify the vertex a with the 0-simplex it determines.

1.1.5 Proposition. *The simplex s_p determines its vertices.*

For clearly every point of $\bar{s}_p$ except the vertices is the midpoint of a segment lying in $\bar{s}_p$; this provides a characterization of the vertices. ❚

It follows from this proposition that the dimension of a simplex is a property of the simplex itself.

1.1.6 Proposition. *s_p and $\bar{s}_p$ are convex subsets of R^m.*

Let
$$b = \sum_i \lambda_i a^i, \quad \sum_i \lambda_i = 1, \quad \lambda_i > 0$$
and
$$b' = \sum_i \lambda'_i a^i, \quad \sum_i \lambda'_i = 1, \quad \lambda'_i > 0.$$
Then for $0 < t < 1$,
$$tb + (1-t)\,b' = \sum_i (t\lambda_i + (1-t)\,\lambda'_i)\,a^i$$
and $\sum_i (t\lambda_i + (1-t)\,\lambda'_i) = t\sum_i \lambda_i + (1-t)\sum_i \lambda'_i = 1, \quad t\lambda_i + (1-t)\,\lambda'_i > 0.$

This establishes the convexity of s_p; the proof for $\bar{s}_p$ is almost identical. ❚

If $s_p = (a^0 \ldots a^p)$, any subset of $a^0, a^1, \ldots, a^p$ is also independent and a simplex spanned by a subset of the vertices is called a *face* of s_p. In particular, s_p is a face of itself; the other faces are called *proper* faces of s_p and have dimension less than p. It is sometimes convenient to include the empty set as a subset of the vertices—indeed, of the vertices of any simplex. It is consistent to attribute to the empty simplex the dimension‡ -1. We write $s_q \prec s_p$ if s_q is a face of s_p, so that $s_{-1} \prec s$ for any simplex s.

1.1.7 Proposition. *Each point of $\dot{s}$ belongs to precisely one proper face of s. Conversely, the points of proper faces of s are all points of $\dot{s}$.*

‡ In dimension theory the empty set is given the dimension -1.

For if $s = s_p$ and $a^0, a^1, \ldots, a^p$ are the vertices, then $\sum_i \lambda_i a^i$ is a point of $\dot{s}$ if and only if some λ_i is zero. Suppose $\lambda_{i_1}, \ldots, \lambda_{i_q}$ are non-zero; then $\sum_i \lambda_i a^i$ is a point of the proper face spanned by $a^{i_1}, \ldots, a^{i_q}$. The converse statement is equally obvious. ∎

We close this section with an important observation about the topology of p-simplexes. We have seen that the p-simplex s_p is determined by its vertices. In particular, any map of the vertices of s_p into a Euclidean space R^n may be extended to a linear map of s_p, and the extension is unique. If $b^0, b^1, \ldots, b^p$ is an independent set in R^n spanning the simplex t_p, we obtain a linear map $f : s_p \to t_p$ by the rule

1.1.8 $$(\Sigma \lambda_i a^i) f = \Sigma \lambda_i b^i.$$

Then f is obviously a homeomorphism. Moreover, 1.1.8 may also be regarded as defining a homeomorphism between $\bar{s}_p$ and $\bar{t}_p$. Thus the topological type of closed or open simplexes is entirely determined by their dimensions. But it is important to note that s_p and t_p (equally, $\bar{s}_p$ and $\bar{t}_p$) are more than homeomorphic. For, given an ordering of their vertices, there is a unique *linear* homeomorphism which is order-preserving on the vertices.

1.2 Geometric simplicial complexes

In this section we describe how simplexes may be fitted together to form more interesting configurations. In this way we shall be able to bring combinatorial methods to bear on spaces of widely differing topological types. The configurations in question will be called *finite geometric simplicial complexes*, abbreviated where there can be no ambiguity to *geometric complex* or even just *complex*.

Let R^m be a fixed Euclidean space.

1.2.1 Definition. A *finite geometric simplicial complex* in R^m is a finite collection, K, of simplexes s_p^i of R^m, subject to the conditions:

K 1: If $s_p \in K$ and $s_q \prec s_p$, then $s_q \in K$.
K 2: Distinct simplexes of K are disjoint.

When we adopt the convention that $s_{-1} \prec s$ for all s, the complex is *augmented* and we write K^+ for an augmented complex; that is, K^+ possesses one (-1)-simplex which is a face of every simplex of K. Unless the superscript + appears, K is to be regarded as unaugmented, that is, without a (-1)-simplex; we say that K^+ is obtained by *augmenting* K.

The *dimension* of K is the maximum of the dimensions of its simplexes.

1.2.2 Proposition. *Conditions* **K 1** *and* **K 2** *are equivalent*‡ *to* **K 1** *and* $\bar{\mathbf{K}}$**2**: *Two closed simplexes of K are disjoint or intersect in a closed face of each.* [*Two closed simplexes of* K^+ *intersect in a closed face of each.*] ∎

The reader will have remarked on the convenience, in the statement of $\bar{\mathbf{K}}$**2** of considering the augmented complex K^+. More striking technical advantages of this device reveal themselves in later chapters.

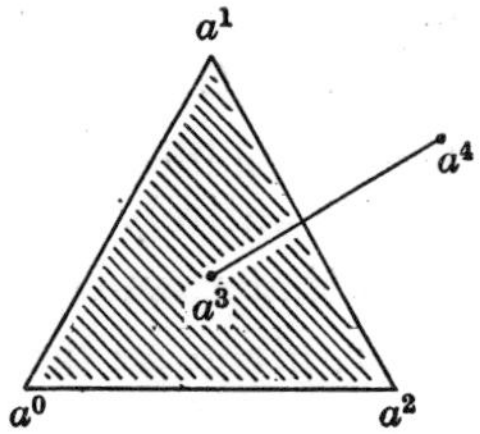

Fig. 1.1

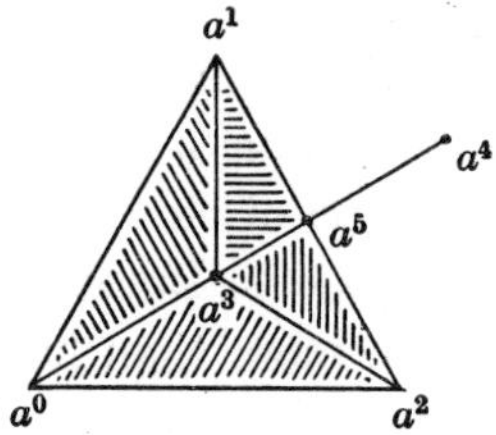

Fig. 1.2

Condition **K 1** ensures that a geometric complex is a union of closed rectilinear simplexes; condition **K 2** ensures that they fit together in a satisfactory way. For instance, in fig. 1.1 we have a collection of simplexes in R^2 satisfying **K 1** but not **K 2**, for the simplex a^3a^4 intersects the simplex $a^0a^1a^2$. On the other hand, we may easily describe a complex whose underlying space coincides with that of fig. 1.1; this is demonstrated in fig. 1.2.

The underlying space of a complex K, that is, the set of points of R^m belonging to some simplex of K, with the topology induced by that in R^m, is called the *polyhedron* of K and written $|K|$; of course, $|K^+| = |K|$. The complex K is also said to be a *dissection* or *triangulation* of the polyhedron $|K|$. Of course, $|K|$ is more than just a topological space; it inherits from R^m a *metric* and an affine structure in each simplex (as described in 1.1). If x is a point of $|K|$, then, by **K 2**, it belongs to just one simplex of K which is called the *carrier* of x. Clearly different complexes may have the same polyhedron; the reader will easily supply examples. Our object essentially is to deter-

‡ The description of a complex in terms of **K1** and $\bar{\mathbf{K}}$**2** may be found in, for example, Pontryagin [4].

mine properties of (or 'functions' of) a complex which are in reality properties of the topological type of the polyhedron.

If K is a complex, a subcollection K_0 of its simplexes is called a *subcomplex* of K if it is a complex. If K_0 is a subcomplex of K, $|K_0|$ is called a *subpolyhedron* of $|K|$. It is, of course, only necessary to verify condition **K 1** for K_0, since **K 2** is automatically satisfied. In an obvious sense we may say that K_0 is a subcomplex if it is *closed*, that is, if it is a union of closed simplexes of K. An important example of a subcomplex is, indeed, the set of all faces of some simplex s of K; then $|K_0| = \bar{s}$. Again, we may take all proper faces of s; then $|K_0| = \dot{s}$. By an abuse of symbols we shall sometimes write $\bar{s}$, $\dot{s}$ for these complexes whose polyhedra are $\bar{s}$, $\dot{s}$. Let K^n be the set of simplexes of K of dimension $\leqslant n$; then K^n is a subcomplex of K, called the *n-section* of K. The empty set of simplexes of K is to be regarded as a subcomplex of K.

1.2.3 Proposition. *If K_0, K_1 are subcomplexes of K, so are $K_0 \cup K_1$, $K_0 \cap K_1$.* ∎

1.2.4 Proposition. *If K, L are complexes in R^m and if*

$$|K| \cap |L| = |M|,$$

where M is a subcomplex of K and of L, then $K \cup L$ is a complex. ∎

We remark that a complex is a collection of simplexes (not of points) so that unions and intersections of complexes are again collections of simplexes. The intersection of two complexes in R^m is certainly a complex, but its polyhedron may not be the intersection of the underlying polyhedra.

We now refer again to the definition of K, but from a different viewpoint. We emphasize in fact that K is known when we know (i) its *vertex set*, that is, the set of points of R^m that are vertices of simplexes of K, and (ii) its *vertex scheme*, that is, the set of subsets of the vertex set which span simplexes of K. Then conditions **K 1** and **K 2** may be interpreted as conditions on the vertex scheme.‡

This point of view is useful in practice, since a complex is often presented as a vertex scheme which is then proved to be the vertex scheme of a complex. Thus we may consider arbitrary vertex schemes

‡ Of course, we also have the condition that the points of a subset in the vertex scheme are independent. We may sometimes identify a complex with its vertex scheme.

in R^m. We say that V is a *vertex scheme on the vertices* $a^1, \ldots, a^N$, where $a^1, \ldots, a^N$ are arbitrary points of R^m, if V is a selection of subsets of $\{a^i\}$, subject only to the condition that every vertex belongs to some selected subset of the scheme. The vertex scheme is said to be *closed* if every subset of a selected set is selected. A vertex scheme V determines a subset $|V|$ of R^m, namely, the set of points $\sum_1^N \lambda_i a^i$, where

(i) $\lambda_i \geqslant 0$, all i;

(ii) $\sum_i \lambda_i = 1$;

(iii) if $\lambda_{i_0}, \ldots, \lambda_{i_p}$ is the complete set of non-zero λ's, then $(a^{i_0}, \ldots, a^{i_p})$ is a selected subset of the scheme.

1.2.5 Proposition. *If each selected subset of V is an independent set, then $|V|$ is the union of the simplexes spanned by these subsets. If V is the vertex scheme of a complex, then V is closed and $|V| = |K|$.* ∎

1.3 Polyhedra

Topology is concerned with properties common to spaces of the same topological type. Thus the proper objects of study in elementary combinatorial topology are spaces homeomorphic to polyhedra. We shall broaden the scope of the term polyhedron to include all such spaces, but temporarily, in order to stress the distinction, we shall refer to the more general object as an *abstract polyhedron*. Notice that a polyhedron, as a subspace of R^m, has more structure than an abstract polyhedron, which is simply to be regarded as a topological space. Of course, the existence of a homeomorphism of the abstract polyhedron P with some polyhedron $|K|$ implies certain facts about the topological type of P; for example, P is metrizable.

If $h : |K| \cong P$ is a homeomorphism, then, for each simplex $s_p \in K$, we can consider the subspace $(s_p)h$ of P. Clearly P is the union of its subspaces $(s_p)h$, each of which will be called a *curvilinear simplex* of P; the set of all curvilinear simplexes $(s_p)h$ forms a *dissection* of P into a *curvilinear complex*. This form of words is often more convenient for discussing P than the alternative, which involves constantly comparing P with $|K|$. For instance, the n-sphere S^n is homeomorphic to $\dot{s}_{n+1}$ (this may be seen by placing the centre of S^n at the barycentre of s_{n+1} and projecting radially); it is often convenient to think of S^n itself as dissected into curvilinear simplexes determined by some $h : \dot{s}_{n+1} \cong S^n$.

1.3.1 Theorem. *An abstract polyhedron is a compact metrizable space.*

Consider the polyhedron $|K|$. Obviously $|K|$ is metrizable. Also $|K|$ is the union of a finite number of closed simplexes; each closed simplex is a bounded closed subset of R^m and hence compact. Thus $|K|$ is itself compact. Since the assertion of the theorem is topological, the truth for polyhedra implies the truth for abstract polyhedra. ∎

We point out that the converse of Theorem 1.3.1 is false. Thus the set of points in R^1 with coordinates $0, 1, \frac{1}{2}, \ldots, 1/n, \ldots$ is a compact metrizable space. But it cannot be a polyhedron, since a complex has only a finite number of vertices and the set obviously contains no 1-simplex.

The last remark in the proof of 1.3.1 exemplifies the fact that it is unhelpful to preserve the distinction between polyhedra and abstract polyhedra when purely topological properties are in question. Thus, henceforward (unless otherwise stated) we drop the terms 'abstract' and 'curvilinear' in discussing polyhedra. Of course, the terms 'vertex' and 'face' may be used in discussing dissections of arbitrary polyhedra.

We shall call the subspace P_0 of the polyhedron P a *subpolyhedron* if, for some dissection K of P, and subcomplex K_0 of K, $|K_0| = P_0$. We remark that this asserts more than that P_0 is a polyhedron.

Let K be a complex, $|K|$ its underlying polyhedron.

1.3.2 Proposition. *A subset $A \subseteq |K|$ is closed if and only if, for each $s_p \in K$, $A \cap \bar{s}_p$ is closed in $\bar{s}_p$.* ∎

It follows that, if K_0 is a subcomplex of K, $|K_0|$ is a closed subset of $|K|$.

1.3.3 Proposition. *Let Y be a topological space and suppose that, for each $s_p^i \in K$ a map $f_p^i : \bar{s}_p^i \to Y$ is defined such that, if $s_q^j \prec s_p^i$, then $f_p^i \mid \bar{s}_q^j = f_q^j$. Then there exists a unique map $f : |K| \to Y$ such that $f \mid \bar{s}_p^i = f_p^i$ for each s_p^i in K.*

Broadly speaking, the proposition asserts that piecewise continuous transformations, defined consistently over the closed simplexes of K, fit together to give a global continuous transformation; it follows at once from I. 1.6. ∎

If s is a simplex of K we define $\mathrm{st}_K(s)$, the *star* of s, to be the set of simplexes of K having s as a face. Notice that $\mathrm{st}_K(s)$ is not in general a subcomplex of K and that if K_0 is a subcomplex of K and $s \in K_0$, then

$\mathrm{st}_{K_0}(s)$ differs in general from‡ $\mathrm{st}_K(s)$. Nevertheless, we shall write simply $\mathrm{st}(s)$ if no ambiguity is possible. Then $|\mathrm{st}(s)|$ is the set of points covered by the simplexes of $\mathrm{st}(s)$.

1.3.4 Theorem. $|\mathrm{st}(s)|$ *is an open subset of* $|K|$.

Consider $K-\mathrm{st}(s)$. If s is not a face of t, then certainly s is not a face of a face of t. Thus the collection $K-\mathrm{st}(s)$ satisfies **K 1** and is a subcomplex of K, so that $|K-\mathrm{st}(s)|$ is compact and hence closed in $|K|$. (We use I.1.15 and the fact that any metric space is Hausdorff.) But since $|K|$ is the union of disjoint simplexes,

$$|K-\mathrm{st}(s)| = |K| - |\mathrm{st}(s)|.$$

Thus $|\mathrm{st}(s)|$ is open in $|K|$. ∎

1.4 Regular subdivision

Let K be a geometric complex in R^m. Then $|K|$ admits the rectilinear dissection K. We now describe a process whereby, for complexes K of positive dimension, further rectilinear dissections $K', K'', \ldots$ of $|K|$ may be found which chop $|K|$ up more and more finely. This process of *regular subdivision* is of fundamental importance in the sequel.

Let V be the vertex scheme of K. We shall define a vertex scheme V' which we shall prove in the next section to be the vertex scheme of a complex K' such that $|K'| = |K|$. Once V' is defined and an example given, readers will have no difficulty in accepting the truth of this statement provisionally, while more immediately accessible facts about K' are discussed. K' is called the *first derived* of K and is said to be obtained from K by *regular subdivision.*

We now define V'. To each simplex $(a^{i_0}, \ldots, a^{i_p})$ of K we define a vertex $b^{i_0 \cdots i_p}$ of V' by

$$b^{i_0 \cdots i_p} = \frac{1}{p+1} \sum_{k=0}^{p} a^{i_k}.$$

Thus $b^{i_0 \cdots i_p}$ is the barycentre, or centre of gravity, of equal masses placed at the vertices $a^{i_0}, \ldots, a^{i_p}$; it follows that the vertices of V' are all distinct. Clearly $b^{i_0 \cdots i_p}$ is symmetric in its superscripts. We next describe the selected subsets of the vertices which are to belong to V'; namely, a subset of the b's is selected if and only if its

‡ Compare $\bar{s}$, $\dot{s}$, which are subcomplexes and do not depend on the complex containing s, provided of course that we confine attention either to augmented or to unaugmented complexes.

elements may be so ordered that the set of superscripts for each vertex is contained in the set of superscripts for the preceding vertex. For instance, if $b^{1,9,3,4,7}$, $b^{4,3}$, $b^{3,9,4}$ are vertices of V', they form a selected subset since they may be reordered as $b^{1,9,3,4,7}$, $b^{3,9,4}$, $b^{4,3}$. On the other hand, $b^{1,3,4,7}$, $b^{4,3}$, $b^{3,9,4}$ is not selected. Certainly V' is closed. We give a diagram for K and K' for a simple case (fig. 1.3).

Granted that K' is a complex and that $|K'| = |K|$, it is clear that we may iterate the process, obtaining $K'', K''', \ldots, K^{(r)}, \ldots$. Notice that we use the same notation as for the derivative of a function in the differential calculus. Notice also that every vertex of K is also a vertex of K'; it is, however, convenient to use different symbols for the same point according to the complex under discussion. By definition $K^{+'} = K'^{+}$.

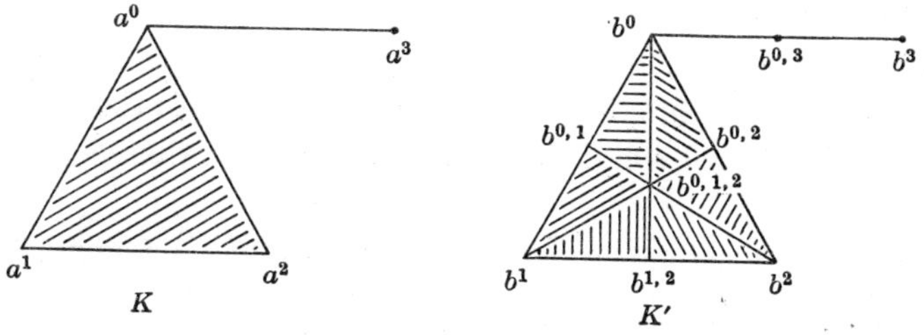

Fig. 1.3

1.4.1 Proposition. *If K_0 is a subcomplex of K, then K_0' is a subcomplex of K'.* ∎

Until Theorem 1.5.3 is proved it would, of course, be more accurate to conclude that K_0' is a closed subscheme of K', in the obvious sense.

1.4.2 Proposition. *If K_0, K_1 are subcomplexes of K such that $K_0 \cup K_1 = K$, then $K_1' \cup K_1' = K'$.*

Certainly $K_0' \cup K_1' \subseteq K'$. Let s' be a simplex of K'. Let $b^{i_0 \cdots i_p}$ be the vertex of s' with the greatest number of superscripts. Then $(a^{i_0}, \ldots, a^{i_p})$ is a simplex of K. Suppose without real loss of generality that it belongs to K_0. Then clearly $s' \in K_0'$. ∎

This proof (and the definition of V') suggests that we call the vertex of $s' \in K'$ with the greatest number of superscripts the *leading vertex* of s'.

We now proceed to show in what sense the process of regular subdivision chops up the polyhedron $|K|$ more and more finely. Let A be a bounded set in R^m. Then the intersection of all convex sets containing

A is a convex set called the *convex cover* of A. It is evident that a closed simplex is the convex cover of its vertices.

1.4.3 Lemma. *The diameter of A is equal to the diameter of its convex cover.*

Let d be the diameter of A and $\tilde{d}$ the diameter of its convex cover $\tilde{A}$. Suppose $a \in A$, $x \in \tilde{A}$. Then‡ $\rho(a, x) \leqslant d$. For the closed ball, with centre a and radius d, is a convex set containing A and hence $\tilde{A}$. It follows that if $x, x' \in \tilde{A}$, then $\rho(x, x') \leqslant d$. For the closed ball, with centre x and radius d, is a convex set containing A, as we have just proved, and hence $\tilde{A}$. Thus $\tilde{d} \leqslant d$; but obviously $d \leqslant \tilde{d}$ so that $d = \tilde{d}$. ∎

1.4.4 Corollary. *The diameter of a simplex is the length of its longest side (one-dimensional face).* ∎

We define the *mesh*, $\mu(K)$, of the complex K to be the maximum of the diameters of its simplexes; by the corollary this is the maximum of the lengths of its 1-simplexes. The main result of this section is

1.4.5 Theorem. *If K has dimension $\leqslant n$, then $\mu(K') \leqslant \dfrac{n}{n+1}\mu(K)$.*

A typical 1-simplex of K' is $(b^{i_0 \dots i_p j_0 \dots j_q}, b^{j_0 \dots j_q})$, where $p+q+1 \leqslant n$. The vector from the first vertex to the second is

$$\frac{1}{q+1}\sum_{s=0}^{q} a^{j_s} - \frac{1}{p+q+2}\left(\sum_{r=0}^{p} a^{i_r} + \sum_{s=0}^{q} a^{j_s}\right)$$

$$= \frac{p+1}{(q+1)(p+q+2)}\sum_{s=0}^{q} a^{j_s} - \frac{1}{p+q+2}\sum_{r=0}^{p} a^{i_r}$$

$$= \frac{p+1}{p+q+2}\left(\frac{1}{q+1}\sum_{s=0}^{q} a^{j_s} - \frac{1}{p+1}\sum_{r=0}^{p} a^{i_r}\right).$$

Now $\dfrac{1}{q+1}\sum_{s=0}^{q} a^{j_s}$ and $\dfrac{1}{p+1}\sum_{r=0}^{p} a^{i_r}$ are both points of the closed simplex of K spanned by the vertices $a^{i_0}, \dots, a^{i_p}, a^{j_0}, \dots, a^{j_q}$. Thus the length of the 1-simplex of K' in question is less than $\dfrac{p+1}{p+q+2}\mu(K)$. But $\dfrac{p+1}{p+q+2} \leqslant \dfrac{p+q+1}{p+q+2} \leqslant \dfrac{n}{n+1}$, since $p+q+1 \leqslant n$. The theorem now follows from the observation following corollary 1.4.4 (it is of course trivial if $\dim K = 0$).

‡ As usual, we use ρ for the metric.

Since $\left(\frac{n}{n+1}\right)^r \to 0$ as $r \to \infty$, and since, as may easily be seen, $\dim K' = \dim K$, we conclude

1.4.6 Corollary. *If K is a geometric complex and ϵ a positive number, then there exists an integer r such that $\mu(K^{(r)}) < \epsilon$.* ∎

This corollary expresses precisely the sense in which regular subdivision yields arbitrarily fine dissections of a polyhedron; the mesh is clearly a satisfactory index of fineness.

1.5 The cone construction

Let W be a vertex scheme in R^m and a a point of R^m which is not a vertex of W. We define aW to be the vertex scheme whose vertices are those of W with a adjoined and whose selected subsets are (i) the selected subsets of W, (ii) a itself and (iii) all subsets consisting of the union of a with a selected subset of W. We call aW the *join* of a to W or the *cone* on W with vertex a. It is closed if W is closed. If W is the vertex scheme of a complex L and aW is the vertex scheme of a complex K we call K the *join* of a to L or the *cone* on L with vertex a and write $K = aL$. Thus, for example, if $|L| \subseteq R^{m-1}$ and R^{m-1} is embedded in R^m, and if $a \in R^m - R^{m-1}$, then aL is a complex and $|aL|$ is the cone (in the ordinary Euclidean sense‡) on the base $|L|$ with vertex a.

1.5.1 Proposition. *Let $a^0, a^1, \ldots, a^p$ be an independent set in R^m and let $s_p = (a^0, a^1, \ldots, a^p)$, $s_{p-1} = (a^1, \ldots, a^p)$. Then $\bar{s}_p = a^0\bar{s}_{p-1}$.* ∎

We now prove a general result.

1.5.2 Theorem. *Let B be a closed bounded convex set in R^m and let $\dot{B}$ be its frontier in the Euclidean space which it spans. If L is a complex such that $|L| = \dot{B}$ and if $a \in B - \dot{B}$, then aL is a complex with polyhedron B.*

We take from the theory of convex sets the fact that every point of $B - a$ is uniquely expressible as $\lambda a + (1-\lambda) f$, where $f \in \dot{B}$ and $0 \leqslant \lambda < 1$, and that every point of this form belongs to $B - a$.

In an obvious notation the selected sets of aL are (i) $\{s\}$, $s \in L$, (ii) a, (iii) $\{as\}$, $s \in L$. All these sets are simplexes. This assertion is trivial in cases (i) and (ii), so we prove only case (iii). Suppose that $s = (a^0 \ldots a^p)$ and that $a, a^0, \ldots, a^p$ is not an independent set; then by

‡ Not, of course, the *infinite* cone.

1.1.3, if $b = \frac{1}{2}a + \frac{1}{2p+2}\sum_0^p a^i$, b has a distinct representation 1.1.2 as $b = \mu a + \sum_0^p \mu_i a^i$, where $\mu + \sum \mu_i = 1$, $0 < \mu < 1$ and all $\mu_i > 0$. Then $b = \frac{1}{2}a + \frac{1}{2}f = \mu a + (1-\mu)g$, where $f, g \in s \subset \dot{B}$. Since the representations 1.1.2 are distinct, either $\mu \neq \frac{1}{2}$ or $f \neq g$; in either case a point of $B - a$ has two distinct representations as $\lambda a + (1-\lambda)f$, with $f \in \dot{B}$. This contradiction proves that the vertices of as are independent.

It remains to show that aL satisfies **K2** for complexes and that $|aL| = B$. Clearly aL satisfies **K2** if, for any distinct $s, t \in L$, as is disjoint from t and from at. These assertions follow from arguments similar to that given above. The assertion that $|aL| = B$ again follows from the representation of points of $B - a$ given at the start of the proof. The reader may supply the details of the argument. ∎

We now prove the result stated in the previous section.

1.5.3 Theorem. *Let K be a geometric complex with vertex scheme V. Then V' is the vertex scheme of a complex K' such that $|K'| = |K|$.*

The proof will be by induction. The theorem is certainly true for complexes of dimension $\leqslant 0$. We assume the theorem true for all complexes of dimension less than n and for complexes of dimension n having fewer than q n-simplexes where $q \geqslant 1$, and shall prove it for an n-dimensional complex K with q n-simplexes.

Let s_n be an n-simplex of K, let K_0 be $\bar{s}_n$, let K_1 be $K - s_n$ and let L be $\dot{s}_n$. Then K_1 is a subcomplex and $K_0 \cap K_1 = L$, $|K_0| \cap |K_1| = |L|$.

By the inductive hypothesis L' is a complex and $|L'| = |L| = \dot{s}_n$. Let b be the barycentre of s_n; then $bL' = K_0'$. Now $\dot{s}_n$ is the frontier of $\bar{s}_n$ in the Euclidean space spanned by $\bar{s}_n$, and $b \in \bar{s}_n - \dot{s}_n$. Thus by Theorem 1.5.2, K_0' is a complex and $|K_0'| = \bar{s}_n = |K_0|$.

Again by the inductive hypothesis K_1' is a complex and $|K_1'| = |K_1|$. By Proposition 1.4.1, L' is a subcomplex of K_0' and K_1'; by what we have proved, $|K_0'| \cap |K_1'| = |K_0| \cap |K_1| = |L| = |L'|$. By Proposition 1.2.4, $K_0' \cup K_1'$ is a complex. By Proposition 1.4.2, $K_0' \cup K_1' = K'$ and thus K' is a complex.‡ Finally, we note that

$$|K'| = |K_0' \cup K_1'| = |K_0'| \cup |K_1'| = |K_0| \cup |K_1| = |K_0 \cup K_1| = |K|$$

and the induction is complete. ∎

Thus far we have defined 'cone' or 'join' as a purely combinatorial notion. It is natural (and, indeed, the source of the terminology) to

‡ Of course, neither Proposition 1.4.1 nor Proposition 1.4.2 depends on the proof of Theorem 1.5.3.

call $|aL|$ the *cone* on $|L|$ with vertex a or the *join* of a to $|L|$ if aL is a complex.

1.5.4 Proposition. *If aL and bL are complexes (lying in some R^m), then*

$$|aL| \cong |bL|.$$

For any point of $|aL| - a$ is uniquely expressible as $\lambda a + (1-\lambda)x$, $0 \leqslant \lambda < 1$, $x \in |L|$, and a homeomorphism $h : |aL| \cong |bL|$ is given by

$$ah = b, \quad (\lambda a + (1-\lambda)x)h = \lambda b + (1-\lambda)x. \;\rrbracket$$

Notice that h is more than a homeomorphism; it maps simplexes to simplexes and is linear in each simplex.

This proposition shows that the choice of vertex plays an unimportant role in the topology of the cone. We may thus speak of 'the cone on L' and 'the cone on $|L|$' when purely topological properties are involved. Thus

1.5.5 Proposition. *The cone on $\dot{s}_n$ is homeomorphic to $\bar{s}_n$.*

This is an easy consequence of Theorems 1.5.2 and 1.5.3; it could also be proved directly and the reader is recommended to do so. ❚

In defining the cone on a polyhedron, the ambient Euclidean space has played a role in imposing a topology on the cone. Indeed, given any subspace of Euclidean space we can define the cone on that space and its topological type is independent of the choice of vertex. Thus, given $X \subseteq R^m$, we may define its *Euclidean cone* $\hat{X}$ and this is realizable in R^{m+1}. However, topology is often concerned with spaces which not only are not subspaces of some Euclidean spaces but are not even homeomorphic to such subspaces. We now proceed to describe how we may construct a cone on an arbitrary topological space.

Let I be the unit interval $0 \leqslant t \leqslant 1$ and let $X \times I$ be the topological product of X with I. We set up an equivalence relation R on the points of $X \times I$ by the simple rule that $(x,t)\,R(x',t')$ if and only if $(x,t) = (x',t')$ or $t = t' = 1$ and we endow the factor set $(X \times I)/R$ with the identification topology; this identification space we call CX, the *cone* on X. The equivalence class $X \times 1$ is called the *vertex* of the cone.

Intuitively, we see that CX has been formed from the *cylinder* $X \times I$ by pinching the top of the cylinder to a point. The topology of CX has been induced by the projection $k : X \times I \rightarrow CX$. Thus the topologies of the cylinder and cone stand in close relation to each other; each, in fact, plays an essential role in the notion of homotopy, as we shall see.

1.5.6 Proposition. *Let W be any space, R an equivalence relation on its points and $k : W \to Z = W/R$ the identification‡ map; if $f : W \to Y$ is a map such that, for all w, w' for which wRw', $wf = w'f$, then f induces a unique map $g : Z \to Y$ such that $kg = f$.*

The condition on f ensures that g exists as a transformation and the fact that k is onto ensures that g is unique. We prove g continuous by observing that, for any closed subset $F \subseteq Y$, $k^{-1}(g^{-1}F) = f^{-1}(F)$, which is closed in W since f is continuous; from the identification topology of Z we infer that $g^{-1}F$ is closed in Z; g is therefore continuous. ∎

We shall use this in the case where W is $X \times I$ and Z is CX. In this case 1.5.6 asserts that, if $f : X \times I \to Y$ has the property that $(X \times 1)f$ is a single point, then f induces a map $g : CX \to Y$ such that $kg = f$.

If X is embedded in Euclidean space, we may clearly regard $\hat{X}$ and CX as two topological spaces on the same set of points. This intuitive notion is made more precise in

1.5.7 Theorem. *The map $f : X \times I \to \hat{X}$ given by $(x, t)f = t.a + (1-t).x$ induces a transformation $g : CX \to \hat{X}$ which is $(1, 1)$ onto and continuous.*

We apply 1.5.6 to the spaces $X \times I$, CX and $\hat{X}$; f satisfies the condition of 1.5.6 and the map g of this theorem is the map g of 1.5.6. Inspection quickly shows that g is $(1, 1)$ onto. ∎

1.5.8 Corollary. *If X is compact $g : CX \cong \hat{X}$.*

See I. 1.17. ∎

If X is compact, we may use the map g to identify CX with $\hat{X}$.

Thus the two definitions of the cone coincide for polyhedra. That they do not coincide in general—in other words, that CX may have more closed sets that $\hat{X}$—is shown by taking for X the set of positive integer points on the line, regarded as embedded in R^2. Then if V_n is the set of points on the ray from a to n within $1/n$ of a, $\bigcup V_n$ is a neighbourhood of a in CX but not in $\hat{X}$§.

1.6 Homotopy

A concept of fundamental importance in algebraic topology is that of homotopic maps and it will recur frequently in subsequent pages. If X, Y are two topological spaces and f_0, f_1 are maps of X into Y, then a *homotopy* from f_0 to f_1 is a continuous chain of maps $f_t : X \to Y$,

‡ See I. 1 for definitions.
§ The authors are indebted to A. H. Stone for this easily explained example.

$0 \leqslant t \leqslant 1$. This intuitive picture of a process of gradually deforming the f_0-image of X into the f_1-image of X is made quite precise by

1.6.1 Definition. We say that $f_0, f_1 : X \to Y$ are *homotopic* if there exists a map $F : X \times I \to Y$ such that $xf_0 = (x, 0)F$, $xf_1 = (x, 1)F$ for all $x \in X$. Then F is said to be a *homotopy* from f_0 to f_1.

The maps $f_t : X \to Y$ mentioned above are then given by $xf_t = (x, t)F$. The definition gives precision to the sense in which f_t depends continuously on x and t. We write $f_0 \simeq f_1$ if f_0 is homotopic to f_1; the symbol $f_0 \overset{F}{\simeq} f_1$ means that F is a homotopy from f_0 to f_1.

Definition 1.6.1 may be relativized. If $X_0 \subseteq X$, we call X, X_0 a *pair* of spaces and we write $f: X, X_0 \to Y, Y_0$ to denote that $f: X \to Y$ such that $X_0 f \subseteq Y_0$. When considering maps $X, X_0 \to Y, Y_0$ the definition of homotopy is obvious: $f_0 \simeq f_1 : X, X_0 \to Y, Y_0$ if there exists $F : X \times I, X_0 \times I \to Y, Y_0$ such that $xf_0 = (x, 0)F$, $xf_1 = (x, 1)F$. Notice that the image of X_0 stays in Y_0 throughout the homotopy. There is another sense in which we may say that a homotopy is relative. Let $f_0, f_1 : X \to Y$ agree on the subset A of X; then we say that f_0 is *homotopic to* f_1 *rel* A if there is a homotopy F from f_0 to f_1 such that, for all $a \in A$ and $t \in I$, $(a, t)F = af_0$ $(= af_1)$. In future we shall not always give the relative form of definitions and assertions if the extension to the relative case is obvious.

1.6.2 Proposition. *The relation $f \simeq g : X \to Y$ is an equivalence relation in the set of maps from X to Y.*

We have to show that $\simeq$ is reflexive, symmetric and transitive.

(*a*) $f \simeq f : X \to Y$ under the homotopy $F : X \times I \to Y$ given by $(x, t)F = xf$.

(*b*) If $f \simeq g : X \to Y$ under the homotopy F, then $g \simeq f$ under the homotopy F', given by $(x, t)F' = (x, 1-t)F$.

(*c*) If $f \simeq g : X \to Y$ under F and $g \simeq h : X \to Y$ under G, then $f \simeq h : X \to Y$ under H, where

$$\begin{aligned}(x, t)H &= (x, 2t)F, & 0 \leqslant t \leqslant \tfrac{1}{2} \\ &= (x, 2t-1)G, & \tfrac{1}{2} \leqslant t \leqslant 1.\end{aligned}$$

We leave to the reader the verification that the homotopies defined above are, indeed, continuous. ∎

Let $1_X : X \to X$ be the identity map; we shall often write this as 1. Then we say that $f : X \to Y$ is a *homotopy equivalence* with *homotopy inverse* $g : Y \to X$ if $fg \simeq 1_X : X \to X$, $gf \simeq 1_Y : Y \to Y$. To justify this terminology, we shall need the following important but elementary lemma.

1.6.3 Lemma. (*a*) *If* $f_0 \simeq f_1 : X \to Y$ *and* $g : Y \to Z$, *then*

$$f_0 g \simeq f_1 g : X \to Z.$$

(*b*) *If* $f_0 \simeq f_1 : X \to Y$ *and* $h : W \to X$, *then* $hf_0 \simeq hf_1 : W \to Y$.

Let $F : X \times I \to Y$ be a homotopy from f_0 to f_1. (*a*) Then $Fg : X \times I \to Z$ is a homotopy from $f_0 g$ to $f_1 g$. (*b*) Then $F' : W \times I \to Y$, given by $(w, t) F' = (wh, t) F$, is a homotopy from hf_0 to hf_1. ∎

The effect of this lemma is that in any composite map we may replace a single map by a homotopic map and obtain a composite map homotopic to the original ($g \simeq g'$ implies $f \ldots g \ldots h \simeq f \ldots g' \ldots h$).

We seek to justify the use of the word 'equivalence' for a map $f : X \to Y$ for which a homotopy inverse $g : Y \to X$ exists with $fg \simeq 1$, $gf \simeq 1$. We write $f : X \simeq Y$ if f is a homotopy equivalence and we write $X \simeq Y$ if there exists $f : X \simeq Y$.

1.6.4 Theorem. $X \simeq Y$ *is an equivalence relation between topological spaces.*

The reflexive and symmetric properties are trivial. Now suppose $f : X \simeq Y$ with homotopy inverse g, and $u : Y \simeq Z$ with homotopy inverse v. Then $fg \simeq 1$, $gf \simeq 1$, $uv \simeq 1$, $vu \simeq 1$. Consider $fu : X \to Z$, $vg : Z \to X$. Then $fuvg \simeq f1g = fg \simeq 1$, and $vgfu \simeq v1u = vu \simeq 1$, by Lemma 1.6.3. Thus we have proved more, namely, that the composition of two homotopy equivalences is a homotopy equivalence with homotopy inverse the composition of the two homotopy inverses. ∎

1.6.5 Definition. We say that X and Y are of the same *homotopy type* if $X \simeq Y$.

Obviously two spaces of the same topological type are of the same homotopy type. On the other hand, two spaces of the same homotopy type may fail to be homeomorphic. For example, let X be any convex bounded subset‡ of R^m and let Y be a point. Let $x_* \in X$, let $f : X \to Y$ be the unique available map and let $g : Y \to X$ be given by $Yg = x_*$. Then $gf = 1$ and $fg : X \to X$ is given by $xfg = x_*$ for all $x \in X$. Define $F : X \times I \to X$ by $(x, t) F = t . x_* + (1 - t) . x$. Then $(x, 0) F = x$,

‡ E.g. a closed simplex.

$(x, 1)F = x_*$, so that $fg \simeq 1$ and $X \simeq Y$. Thus homotopy type is a coarser classification than topological type and invariants of homotopy type, or *homotopy invariants*, are, *a fortiori*, topological invariants The *homology* and *contrahomology groups*, which are the main concern of this book, are homotopy invariants.

An important case in which spaces of the same homotopy type arise is that of a deformation retract. We recall that the subspace X of Y is a retract of Y if there exist a (retracting) map $h : Y \to X$ such that $h \mid X = 1$. We say that X is a (strong) *deformation retract* of Y if there is a homotopy $H : Y \times I \to Y$, rel X, such that

$$(y, 0)H = y,$$

$$(y, 1)H \in X, \quad \text{for all } y \in Y.$$

Then, if $h : Y \to X$ is given by $yh = (y, 1)H$, clearly h retracts Y onto X and H is a homotopy‡ rel X from the identity map to the retracting map h. The homotopy F of the last paragraph was such a homotopy retracting X onto x_*.

It is clear that if X is a deformation retract of Y then $X \simeq Y$. For if $i : X \to Y$ is the embedding then $ih = 1 : X \to X$ and $1 \overset{H}{\simeq} hi : Y \to Y$. Thus, for example, if X is now any space, $X \cong X \times 0 \simeq X \times I$.

Let $f : X \to Y$ be a map. A *nullhomotopy* of f is a homotopy of f to a constant map. We may write $f \simeq 0$ if f is homotopic to a constant map. Now we have seen that the assertion that $f_0 \simeq f_1 : X \to Y$ is equivalent to the assertion that a certain map of $(X \times 0) \cup (X \times 1)$ can be extended to $X \times I$. Similarly, we see

1.6.6 Proposition. *A map $f : X \to Y$ is nullhomotopic if and only if it can be extended to the cone CX. Indeed, a nullhomotopy of f induces a map $CX \to Y$ extending f.*

First let $F : X \times I \to Y$ be a nullhomotopy of f; thus $(x, 0)F = xf$, $(x, 1)F = y_*$, all $x \in X$, where y_* is a given point of Y. Let

$$k : X \times I \to CX$$

be the identification map; then by 1.5.6 F determines a map $g : CX \to Y$ such that $F = kg$. It is this map g that is the extension of f.

‡ X is a *strong* deformation retract of Y if there is a retraction $h : Y \to X$ such that $1 \simeq hi$ rel X, where $i : X \to Y$ is the injection; X is a *weak* deformation retract of Y if, for some retraction h, $1 \simeq hi : Y, X \to Y, X$. We shall use the expression 'deformation retract' to mean 'strong deformation retract'. For polyhedral pairs the two notions are equivalent.

Conversely, if an extension $g : CX \to Y$ exists, then $kg : X \times I \to Y$ is a nullhomotopy of f to the constant map sending X to ag, where a is the vertex of CX. ∎

Up to this point we have allowed the relativizations to remain implicit. For though the relative forms of 1.6.2, 1.6.3, 1.6.4 and 1.6.5 are profoundly important, they are easily derivable from the absolute forms given. However, a curious possibility arises in formulating the relative form of 1.6.6 which deserves explicit mention.

Let $f : X \to Y$ be a map. Then f induces $F : X \times I \to Y \times I$ given by

1.6.7 $$(x, t)\, F = (xf, t).$$

Similarly, f induces $Cf : CX \to CY$. Cf is determined by the commutativity of the diagram

$$\begin{array}{ccc} X \times I & \xrightarrow{F} & Y \times I \\ \downarrow{\scriptstyle k_X} & & \downarrow{\scriptstyle k_Y} \\ CX & \xrightarrow{Cf} & CY \end{array}$$

where k_X, k_Y are the identification maps. Now if f is a homeomorphism (into), so is F. On the other hand, Cf need not be; it is (1, 1) and continuous but CX may have too many closed sets.‡ Thus it would be inaccurate to say that $f : X, X_0 \to Y, Y_0$ is nullhomotopic if and only if it can be extended to CX, CX_0, since CX_0 is not necessarily a subspace of CX. On the other hand, the embedding $X_0 \subseteq X$, as we have said, induces an injection map $CX_0 \to CX$. Let us write $C(X, X_0)$ for the pair consisting of CX and the image in CX of CX_0. Then we do indeed have

1.6.8 Proposition (*relative form of* **1.6.6**). *A map $f : X, X_0 \to Y, Y_0$ is nullhomotopic if and only if it can be extended to $C(X, X_0)$.*

Indeed, a nullhomotopy of f induces a map $C(X, X_0) \to Y, Y_0$ extending f. ∎

There is an important case in which the phenomenon described above cannot occur, namely, when X_0 is closed in X.

1.6.9 Proposition. *If X_0 is a closed subset of X, then CX_0 is a closed subset of CX.*

It will be convenient to write $f : X_0 \subseteq X$ for the embedding. We have a commutative diagram of maps:

$$\begin{array}{ccc} X_0 \times I & \xrightarrow{F} & X \times I \\ \downarrow{\scriptstyle k_0} & & \downarrow{\scriptstyle k} \\ CX_0 & \xrightarrow{Cf} & CX \end{array}$$

Notice first that F embeds $X_0 \times I$ as a closed subset of $X \times I$. We have to show that if Q_0 is a closed subset of CX_0, then $(Q_0)\, Cf$ is a closed subset of CX. This is obviously so if $a \notin Q_0$, where a is the vertex of the cone. Now if $a \in Q_0$, then it is easily verified that

$$k^{-1}((Q_0)\, Cf) = (k_0^{-1}(Q_0))\, F \cup (X \times 1).$$

‡ See Exercise 8.

But $k_0^{-1}(Q_0)$ is closed in $X_0 \times I$, so that $(k_0^{-1}(Q_0))F$ is closed in $X \times I$. Certainly $(X \times 1)$ is closed in $X \times I$, so that $k^{-1}((Q_0)Cf$ is closed in $X \times I$. By definition of the topology in CX, $(Q_0)Cf$ is closed in CX. ∎

Of course if X is embedded in Euclidean space and $X_0 \subseteq X$, then $\hat{X}_0 \subseteq \hat{X}$. Thus the Euclidean cone is distinguished in this respect from the cone that features in homotopy theory. We have seen that the two notions coincide on compact spaces (1.5.8) and Proposition 1.6.9 confirms that the cone construction $X \to CX$ produces no anomalies for a pair $|K|$, $|L|$, where L is a subcomplex of K.

We close this section with a fundamental theorem on maps of polyhedra.

1.6.10 Theorem. (*Homotopy extension theorem for maps of polyhedra*). *Let K be a complex, L a subcomplex, $f_0 : |K| \to X$ a map of $|K|$ into a topological space X and $g_t : |L| \to X$ a homotopy such that $g_0 = f_0 \,|\, |L|$. Then there is a homotopy $f_t : |K| \to X$ of the map f_0 such that $g_t = f_t \,|\, |L|$.*

The theorem asserts that if one map $|L| \to X$ in a homotopy of maps can be extended to $|K|$, then so can the whole homotopy. We prove the theorem by extending g_t step-by-step over the sections‡ of K. In the notation of 1.6.1 we have a map $F' : |K| \times 0 \cup |L| \times I \to X$, and we wish to extend F' to a map of $|K| \times I$. Let us write $K^{(m)}$ for the complex $K^m \cup L$. We may extend F' to $F^{(0)} : (|K| \times 0) \cup (|K^{(0)}| \times I) \to X$ by defining $(a, t) F^{(0)} = af_0$, where a is a vertex of $K - L$. Suppose F' extended to $F^{(m)} : (|K| \times 0) \cup (|K^{(m)}| \times I) \to X$, and let $s_{m+1} \in K - L$. Then $F^{(m)}$ is defined on $(\bar{s} \times 0) \cup (\dot{s} \times I)$. We wish to prove that $F^{(m)}$ may be extended to $\bar{s} \times I$.

Suppose this is granted for the moment. Then $F^{(m)}$ may be extended to $F^{(m+1)} : (|K| \times 0) \cup (|K^{(m+1)}| \times I)$ which is continuous on $\bar{s} \times I$ for each simplex s of $K^{(m+1)}$. Then by an easy extension of Proposition 1.3.3, $F^{(m+1)}$ is continuous. We thus proceed in this way until we reach $K^{(n)}$, where $\dim K = n$. But then $F^{(n)}$ is the required extension.

We now demonstrate the decisive step in the proof. We shall show, in fact, that $(\bar{s} \times 0) \cup (\dot{s} \times I)$ is a retract of $\bar{s} \times I$. That is to say, there exists a map $r : \bar{s} \times I \to (\bar{s} \times 0) \cup (\dot{s} \times I)$ which is the identity on $(\bar{s} \times 0) \cup (\dot{s} \times I)$. Then obviously $rF^{(m)}$ is an extension of $F^{(m)}$ to $\bar{s} \times I$.

Now $\dim s = m + 1$; let s be embedded in R^{m+1} with barycentre b; then $\bar{s} \times I$ is embedded in R^{m+2} in the obvious way. In R^{m+2} let c be the point $(b, 2)$; this is the point whose first $(m+1)$ coordinates are those of b and whose $(m+2)$th is 2. Then radial projection from c retracts $\bar{s} \times I$ onto $(\bar{s} \times 0) \cup (\dot{s} \times I)$ (see fig. 1.4). ∎ We observe the fact

‡ Recall the n-section K^n is the set of all simplexes of K of dimension $\leqslant n$.

(not here needed by us) that this radial projection gives a deformation retraction.

The relativized form of Theorem 1.6.10 is worth noticing. If K_0 is a subcomplex of K such that $|K_0|f_0 \cup |K_0 \cap L|\, g_t \subseteq X_0 \subseteq X$, then we may choose f_t so that $|K_0|f_t \subseteq X_0$.

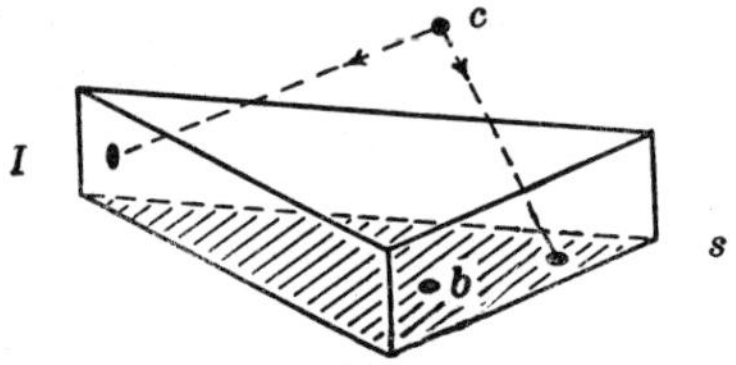

Fig. 1.4

1.7 Simplicial maps

The notions of (continuous) map and homotopy are fundamental in topology. We now describe combinatorial analogues of these notions within the category of simplicial complexes.

Let $s_p = (a^0, \ldots, a^p)$, $t_q = (b^0, \ldots, b^q)$ be two simplexes. A *vertex transformation* v from s_p to t_q is simply a rule assigning to each vertex of s_p a vertex of t_q. We may write this as

$$(a^i)\, v = b^{(i)v}$$

Then v determines a linear map $|v| : \bar{s}_p \to \bar{t}_q$ by the rule

1.7.1 $$(\Sigma \lambda_i a^i)\, |v| = \Sigma \lambda_i b^{(i)v}.$$

We may express 1.7.1 differently by saying that

$$(\lambda_0, \ldots, \lambda_p)\, |v| = (\mu_0, \ldots, \mu_q),$$

where $$\mu_j = \sum_{(i)\, v=j} \lambda_i.$$

The map $|v|$ is obviously continuous; it is said to be induced by v and maps $\bar{s}_p$ onto a closed face of t_q. Let s' be a face of s_p. Then $v \mid s'$ is a vertex transformation v' from s' to t_q and it follows immediately from 1.7.1 that $|v'| = |v| \mid s'$.

Now let K, L be two complexes. We may evidently speak of a vertex transformation from K to L. We say that such a vertex transformation v is *admissible* if, whenever $a^0, \ldots, a^p$ span a simplex of K, then $a^0 v, \ldots, a^p v$ span a simplex of L; note, however, that there may be repetitions among the vertices $a^0 v, \ldots, a^p v$. Such a transformation v determines a map $|v| : |K| \to |L|$ by linearity inside each simplex of K. By 1.3.3 $|v|$ is continuous. We call $|v|$ a *simplicial map* of $|K|$ into $|L|$, since it respects the decompositions of $|K|$ and $|L|$ into simplexes and the affine structure in the simplexes. However, since $|v|$ is determined by v, we shall often speak, by abuse of language, of the *simplicial map* $v : K \to L$. Simplicial maps, then, are special maps of polyhedra; we shall prove later in this chapter that they are not too special, in a sense which will be made precise.

As an example of a simplicial map, consider a complex K and its first derived K'. Each vertex b of K' is the barycentre of some simplex s of K. Let v assign to b an arbitrarily selected vertex of s. We prove that v is admissible. Let s' be a simplex of K' and let b be its leading‡ vertex. Then if b is the barycentre of s, every other vertex of s' is the barycentre of a face of s, so that v assigns to the vertices of s' vertices of s. Thus the v-images of the vertices of s' span a face of s and v is admissible. Thus v induces a simplicial map

$$v : K' \to K,$$

called a *standard* map.§ Of course, v is not uniquely determined by K and K'; one way of choosing v is to order the vertices of K and to assign to b the vertex of s which comes first in the ordering.

1.7.2 Proposition. *If $u : K \to L$ and $v : L \to M$ are simplicial maps, so is $uv : K \to M$.* ∎

Now let $v : K \to L$ be a simplicial map. Then v associates with every simplex s of K a simplex sv of L, namely, the simplex spanned by the v-images of the vertices of s. Thus v induces a vertex transformation v' from K' to L'. For if b is a vertex of K' it is the barycentre of some simplex s of K and we define bv' to be the barycentre of sv.

1.7.3 Proposition. *$v' : K' \to L'$ is a simplicial map.*

Observe first that if $t \prec s$, then $tv \prec sv$. Now let α^0 be the leading vertex of a simplex s'_p of K' and suppose α^0 to be the barycentre of s. Then every other vertex α^i of s'_p is the barycentre of a face s^i of s and we may order the vertices of s'_p so that

$$s^p \prec s^{p-1} \prec \ldots \prec s^1 \prec s.$$

Then
$$s^p v \prec s^{p-1} v \prec \ldots \prec s^1 v \prec sv.$$

‡ See definition following 1.4.2.

§ Some authors use the term 'canonical map', but modern terminology excludes this term if choice must be exercised.

Thus the barycentres of the (not necessarily distinct) simplexes $s^p v, \ldots, sv$ of L span a simplex of L', so that v' is admissible. ∎

We now move towards a combinatorial analogue of homotopy.

1.7.4 Definition. If X is a topological space and L is a complex, we say that two maps $f_0, f_1 : X \to |L|$ are *L-approximate* if, to each $x \in X$, there exists a simplex $t(x) \in L$ such that xf_0 and xf_1 belong to $\overline{t(x)}$.

1.7.5 Proposition. *If $f_0, f_1 : X \to |L|$ are L-approximate, they are homotopic.*

Let $|L| \subseteq R^m$. We define $F : X \times I \to R^m$ by

$$(x, u)\,F = (1-u)\,(xf_0) + u(xf_1).$$

Then F is evidently continuous and $(x, u)\,F \in \overline{t(x)}$. Thus F maps $X \times I$ into $|L|$ and is the desired homotopy. ∎

1.7.6 Corollary. *If $v : K' \to K$ is a standard map, then $|v| \simeq 1$.*

Given $x \in |K'|$, let s' be its carrier‡ and let the leading vertex of s' be the barycentre of s. Then $s' \subseteq s$ and we take $t(x) = s$. Each vertex of s' is mapped by v to a vertex of s so that $x\,|v| \in \bar{s}$. Thus $x, x\,|v|$ belong to $\overline{t(x)}$, so that $|v|$ and 1 are K-approximate. ∎

1.7.7 Corollary. *Let $v : K \to L$ induce $v' : K' \to L'$ as in* 1.7.3; *then* $|v| \simeq |v'|$.

Let $x \in |K'|$ and let s' be its carrier. Taking s again as in 1.7.6, it is clear that $x\,|v|$ and $x\,|v'| \in \overline{sv}$. Thus $|v|$ and $|v'|$ are L-approximate. ∎

There is, as we have seen, a unique simplex s of K containing a given simplex of K'. The same must hold for the rth derived $K^{(r)}$; to each simplex $s^{(r)}$ of $K^{(r)}$, there exists a unique simplex s of K with $s^{(r)} \subseteq s$. We call s the *carrier* of $s^{(r)}$.

1.7.8 Definition. The simplicial maps $v_0 : K \to L$, $v_1 : K^{(r)} \to L$ are called *contiguous* if, to each $s^{(r)} \in K^{(r)}$, there is a simplex t of L such that $sv_0 \prec t$ and $s^{(r)}v_1 \prec t$, where s is the carrier of $s^{(r)}$.

This notion of contiguity is a generalization of the usual notion, which is confined to the case $r = 0$. Let $q > r$ and let $u : K^{(q)} \to K^{(r)}$ be a composition of standard maps. Then we see immediately that v_0 and uv_1 are contiguous if v_0 and v_1 are contiguous. Similarly, if $u : K^{(q)} \to K$ is a composition of standard maps, uv_0 and v_1 are contiguous.

‡ The definition of the carrier of a point may be found on p. 18.

1.7.9 Proposition. *Contiguous maps are homotopic.*

This is an immediate consequence of 1.7.5. ▌ The equivalence relation generated by contiguity is the combinatorial analogue of homotopy to which we have referred.

We continue to prepare the ground for the important results of the next section. We make the fundamental definition.

1.7.10 Definition. If $f : |K| \to |L|$ and the simplicial map $v : K \to L$ have the property that, for each $x \in |K|$, $x\,|v|$ lies in the closure of the carrier of xf, then we say that v is a *simplicial approximation* to f (with respect to K, L).

The definition implies that f and $|v|$ are L-approximate (and hence homotopic), but it asserts more. For instance, let K be a single vertex a^0, L a closed 1-simplex b^0b^1, let $a^0f = b^0$, $a^0v = b^1$. Then v is simplicial and f and $|v|$ are L-approximate; but the closure of the carrier of a^0f is b^0, and therefore v is not a simplicial approximation to f.

1.7.11 Theorem. *If $v : K^{(r)} \to L$ and $w : K^{(q)} \to L$ are simplicial approximations to $f : |K| \to |L|$, then v and w are contiguous.*

There is no loss of generality in taking $q = 0$. Let $s^{(r)}$ be a simplex of $K^{(r)}$ and let $x \in s^{(r)}$. Let s be the carrier of $s^{(r)}$. Now $x\,|v|$ and $x\,|w|$ both lie in the closure of the carrier of xf, say $\bar{t}$. Since v and w are simplicial, $s^{(r)}\,|v|$ and $s\,|w|$ both lie in $\bar{t}$, so that v and w are contiguous. ▌

1.8 The simplicial approximation theorem

We have suggested that, roughly speaking, in studying maps of polyhedra into polyhedra it is sufficient to consider simplicial maps. We shall prove in this section that, if P, Q are polyhedra, then any map $f : P \to Q$ may be approximated arbitrarily closely by a simplicial map of some dissection of P into some dissection of Q.

We now state the main theorem of this section.

1.8.1 Theorem. (*Simplicial approximation theorem*). *Let K, L be two complexes and let $f : |K| \to |L|$ be a map. Then there exists an integer $r \geqslant 0$ and a simplicial map $v : K^{(r)} \to L$ which is a simplicial approximation to f.*

To prove this theorem, we need two lemmas. The first is a well-known result in analytical topology.

1.8.2 Lemma. *If X is a compact metric space, and if $\mathscr{U} = \{U_\lambda\}$ is an open covering of X, then there exists a real number $\delta > 0$ such that each subset of X of diameter $< \delta$ is contained in at least one member of $\mathscr{U}$.*

Let ρ be the metric in X, and, for $x \in X$, $\eta > 0$, let $V(x, \eta)$ be the set of points $y \in X$ with $\rho(x, y) < \eta$.

We argue by contradiction. If the lemma is false, there exists a sequence of subsets $A_1, \ldots, A_n, \ldots$ with A_n of diameter $< \delta_n$ such that $\delta_n \to 0$ and no A_n is contained in a set of the covering $\mathscr{U}$. Let $a_n \in A_n$. Since X is compact metric, by I. 1.11 a subsequence of the sequence $a_1, a_2, \ldots, a_n, \ldots$ converges. We select this subsequence; we may, by a change of notation, suppose that the sequence $\{a_n\}$ itself converges, say $a_n \to a$.

Suppose $a \in U \in \mathscr{U}$. Since U is open, there exists $\eta > 0$ such that the η-ball $V(a, \eta) \subseteq U$. There exist N_1, N_2 such that $\delta_n < \frac{1}{2}\eta$ if $n \geqslant N_1$ and $a_n \in V(a, \frac{1}{2}\eta)$ if $n \geqslant N_2$. Let $N = \max(N_1, N_2)$; then, if $x \in A_N$, we have $\rho(x, a_N) < \delta_N < \frac{1}{2}\eta$, $\rho(a_N, a) < \frac{1}{2}\eta$, so that $\rho(x, a) < \eta$, and $x \in V(a, \eta) \subseteq U$. Thus $A_N \subseteq U$ and we have a contradiction. ▮

Any such number δ will be called a *Lebesgue number* of the covering.

1.8.3 Lemma. *The vertices $a^{i_0}, \ldots, a^{i_p}$ of a complex K span a simplex if and only if their stars have a point in common.*

For if $a^{i_0}, \ldots, a^{i_p}$ are vertices of the simplex s, then $s \in \operatorname{st} a^{i_j}, j = 0, \ldots, p$, so that $s \subseteq \bigcap_j |\operatorname{st} a^{i_j}|$. Conversely, if $x \in \bigcap_j |\operatorname{st} a^{i_j}|$, then s, the carrier of x, is a simplex of each $\operatorname{st} a^{i_j}$, so that each a^{i_j} is a vertex of s. ▮

Notice that, in fact, if $a^{i_0}, \ldots, a^{i_p}$ span the simplex s, then

$$\bigcap_j \operatorname{st} a^{i_j} = \operatorname{st} s, \quad \bigcap_j |\operatorname{st} a^{i_j}| = |\operatorname{st} s|.$$

Before proceeding to the proof of the main theorem, we indicate the line of argument in a simple case. Let K be the closed simplex $\bar{s}$, and let $f : \bar{s} \to |L|$ have a simplicial approximation $v : \bar{s} \to L$. Let $\bar{s}\,|v| = \bar{t}$, $t \in L$. Now, for $x \in s$, $x\,|v|$ lies in the closure of the carrier of xf, so that the carrier of xf belongs to $\operatorname{st} t$. Thus f can only have a simplicial approximation if the f-images of simplexes of K are contained in stars of simplexes of L, and the idea behind the proof is to subdivide K so finely that this property will be ensured. The proof now follows.

Let $\{b^j\}$ be the set of vertices of L. The sets $|\operatorname{st}(b^j)|$ form an open covering of $|L|$, so that the sets $f^{-1}\,|\operatorname{st}(b^j)|$ form an open covering of $|K|$. Let δ be a Lebesgue number of this covering, and let r be chosen such that $\mu(K^{(r)}) < \frac{1}{2}\delta$ (see 1.4.6). If a^i is a vertex of $K^{(r)}$, then any point of $|\mathrm{st}\,(a^i)|$ is distant $< \frac{1}{2}\delta$ from a^i, so that the diameter of $|\mathrm{st}\,(a^i)|$ is less than δ. Thus there exists j such that $|\mathrm{st}\,(a^i)| \subseteq f^{-1}\,|\mathrm{st}\,(b^j)|$. Choose, for each vertex a^i of $K^{(r)}$, a suitable j and define a vertex transformation $v : K^{(r)} \to L$ by $a^i v = b^j$. Then v is admissible; for if $a^0, \ldots, a^p$ span a simplex of $K^{(r)}$, then $\bigcap_i |\mathrm{st}\,(a^i)|$ is non-empty, so that $\bigcap_i (|\mathrm{st}\,(a^i)|f)$ is non-empty. But $\bigcap_i (|\,\mathrm{st}(a^i)|f) \subseteq \bigcap_i |\mathrm{st}\,(a^i v)|$, so that, using 1.8.3, the vertices $(a^i v)$ span a simplex; since v is admissible it determines a simplicial map $v : K^{(r)} \to L$.

Finally, we prove that v is a simplicial approximation to f. Let $x \in |K|$ and let $a^0, \ldots, a^p$ span the carrier of x in $K^{(r)}$. Then $x \in |\mathrm{st}\,(a^i)|$, $i = 0, \ldots, p$, so that $xf \in \bigcap_i (|\mathrm{st}\,(a^i)|f) \subseteq \bigcap_i |\mathrm{st}\,(a^i v)|$. Thus the closure of the carrier of xf in L contains $a^i v$ for $i = 0, \ldots, p$, and hence $x\,|v|$. This completes the proof of the theorem. ∎

We make three remarks about this theorem. First, suppose that, under $f : |K| \to |L|$, $x \in |K|$ is mapped to a vertex b of L. Then b is the closure of the carrier of xf so that $xf = x\,|v|$. Thus we see that a simplicial approximation to f has the property that it agrees with f on counterimages of vertices of L.

Secondly, notice that choices were available at two stages in defining v; namely, we chose a suitable r and then, to each a^i, a suitable b^j. However, Theorem 1.7.11 tells us that any two choices for v are contiguous.

Thirdly, it would be satisfactory to make more precise the degree of 'approximation' possible—to say how close we can get to an arbitrary map by a simplicial map. To do this, we introduce a definition. If f_1, f_2 are maps of a compact space X into a metric space Y (with metric ρ), we may define the *distance* between f_1 and f_2 to be $\underset{x \in X}{\mathrm{l.u.b.}}\, \rho(xf_1, xf_2)$. It may then be verified that this distance is a metric in the set of maps of X into Y. We call the metric space so constructed the *function space* of maps of X into Y and write it Y^X. The compactness of X ensures that the topology of Y^X depends only on the topologies of X and Y and not on the metric in Y. Then we may prove

1.8.4 Theorem. *If K, L are two complexes, the set of all simplicial maps $v : K^{(r)} \to L^{(s)}$, where r, s range over all non-negative integers, is dense in the space $|L|^{|K|}$.*

Let d be the metric in $|L|^{|K|}$. We must prove that, for any $f : |K| \to |L|$ and any $\epsilon > 0$, there exists $v : K^{(r)} \to L^{(s)}$ with $d(f, |v|) < \epsilon$.

Now 1.4.6 enables us to find s so that $\mu(L^{(s)}) < \epsilon$. Let $v : K^{(r)} \to L^{(s)}$ be a simplicial approximation to f. For any $x \in |K|$, xf and $x|v|$ both belong to the closure of some simplex of $L^{(s)}$ so that $\rho(xf, x|v|) < \epsilon$, whence $d(f, |v|) < \epsilon$. ∎

The simplicial approximation theorem (1.8.1) is so formulated as to refer to abstract polyhedra; before Theorem 1.8.4 however could be interpreted for abstract polyhedra, it would be necessary to define the topology of Q^P, for instance, in terms of a metric on a Euclidean homeomorph of Q.

We now prove a second theorem about the proximity of maps in $|L|^{|K|}$.

1.8.5 Theorem. *If K and L are complexes and δ is a Lebesgue number of the covering of $|L|$ by the stars of its vertices, then any two maps $f, g \in |L|^{|K|}$ with $d(f,g) < \frac{1}{3}\delta$ have a common simplicial approximation $v : K^{(r)} \to L$.*

Reference to the proof of Theorem 1.8.1 shows that we would wish to find r so that, for each vertex a^i of $K^{(r)}$, $|\text{st}(a^i)|f$ and $|\text{st}(a^i)|g$ both lie in some $|\text{st}(b^j)|$, where b^j is a vertex of L. For then we shall be able to define a simplicial map v exactly as in Theorem 1.8.1 and conclude that v is a simplicial approximation to f and to g.

The procedure is to shrink the sets of the covering $|\text{st}(b^j)|$ of $|L|$. Precisely, let W^j be the set of points $y \in |L|$ such that

$$\rho(y, |L| - |\text{st}(b^j)|) > \tfrac{1}{3}\delta.$$

Clearly W^j is an open set contained in $|\text{st}(b^j)|$. Moreover, $\{W^j\}$ is a covering; for if $y \in |L|$, then $\overline{V(y, \frac{1}{3}\delta)}$ is a set of diameter $< \delta$ and so is contained in some $|\text{st}(b^j)|$; y must accordingly belong to the corresponding W^j. Since $\{W^j\}$ is a covering and $b^j \notin W^k$ if $j \neq k$, it follows that $b^j \in W^j$.

The covering of $|K|$ by the sets $\{f^{-1}W^j\}$ has a Lebesgue number ϵ and we choose r so that $\mu(K^{(r)}) < \frac{1}{2}\epsilon$. Then, for each vertex a^i of $K^{(r)}$, $|\text{st}(a^i)|f \subseteq W^j$ for some j. But $d(f,g) < \frac{1}{3}\delta$ so that $|\text{st}(a^i)|g \subseteq |\text{st}(b^j)|$. We have thus achieved our object. ∎

To complete our programme of relating continuous maps of polyhedra to combinatorial maps of complexes, we consider homotopic maps and replace the homotopy relation by a relation between simplicial maps which is generated by the contiguity relation.

1.8.6 Theorem. *If $f_0 \simeq f_1 : |K| \to |L|$, there exists a sequence of simplicial maps $v_1, \ldots, v_k$, where $v_i : K^{(r_i)} \to L$, $i = 1, \ldots, k$, such that v_{i-1}*

and v_i are contiguous, v_1 is a simplicial approximation of f_0 and v_k is a simplicial approximation of f_1.

Since $|K| \times I$ is compact metric, by I.1.13 the homotopy $F : |K| \times I \to |L|$ from f_0 to f_1 is uniformly continuous. This implies that, given any $\epsilon > 0$, there exists $\eta > 0$, such that

$$\rho\{(x,t)F, (x,t')F\} < \epsilon$$

for all $x \in |K|$ and all $t, t' \in I$ such that $|t-t'| < \eta$. Hence $d(f_t, f_{t'}) < \epsilon$ if $|t-t'| < \eta$.

Let δ be as in 1.8.5 and take $\epsilon = \frac{1}{3}\delta$. Then, if k is an integer greater than $1/\eta$, the maps $f_0, f_{1/k}, \ldots, f_{i/k}, \ldots, f_1$ have the property that consecutive maps $f_{(i-1)/k}$ and $f_{i/k}$ are distant apart less than ϵ and so (Theorem 1.8.5) have a common simplicial approximation v_i. Moreover, v_i and v_{i+1} are both approximations of $f_{i/k}$ and so are contiguous; and by definition v_1, v_k are simplicial approximations of f_0, f_1 respectively. ∎

1.9 Abstract simplicial complexes

As emphasized in § 3, topology is concerned with abstract polyhedra rather than with concrete realizations as polyhedra of geometric simplicial complexes. Similarly, in our combinatorial approach, we are really only interested in an abstract object, namely, a set of vertices together with a list of subsets or simplexes, and we do not wish constantly to refer to the Euclidean space in which the vertices lie. We now develop this point of view.

1.9.1 Definition. A finite *abstract simplicial complex* is a set of objects, called *vertices*, $a^1, \ldots, a^\alpha$, and a set K of subsets s_p^i of the vertices,‡ where $(p+1)$ is the number of vertices in the *simplex* s_p^i and i is an indexing superscript; the simplexes of K satisfy the condition that any subset of a simplex of K is also a simplex of K. The *dimension* of s_p^i is p and the *dimension* of K is the largest of the dimensions of its simplexes.

1.9.2 Definition. A *realization* of the abstract simplicial complex K is an admissible vertex transformation, v, of K onto a geometric simplicial complex L, such that v possesses an inverse w which is also admissible. We shall also describe the complex L itself as a realization of K.

‡ We may as well assume that every object belongs to some subset, since otherwise it may be discarded from the set.

1.9.3 Proposition. *Every abstract simplicial complex has a realization.*

Let the vertices of the abstract simplicial complex K be $a^1, \ldots, a^\alpha$. Let $b^1, \ldots, b^\alpha$ be vertices of an $(\alpha-1)$-simplex, s, in $R^{\alpha-1}$ and let $v_0 : K \to \bar{s}$ be given by $a^i v_0 = b^i$. Obviously v_0 is admissible. Let L be the subcomplex of $\bar{s}$ consisting of simplexes spanned by those sets of vertices $\{a^i v_0\}$ such that the set $\{a^i\}$ is a simplex of K. Then it is clear that v_0 may be regarded as a vertex transformation $v : K \to L$ which realizes K. ∎

Definition 1.9.2 suggests an obvious definition of *isomorphism* for abstract simplicial complexes.

1.9.4 Definition. Two abstract simplicial complexes K, L are *isomorphic* if there exists an admissible vertex transformation $v : K \to L$ with admissible inverse.

Now any geometric simplicial complex determines, in an obvious way, an abstract simplicial complex of which it is a realization. If we identify, for the moment, a geometric complex with its associated abstract complex, we may say that a realization of K is an isomorphism of K with a geometric complex. To show that it does not matter which realization we choose, it is clearly sufficient to prove

1.9.5 Theorem. *Isomorphic abstract complexes have homeomorphic realizations.*

If K, L are isomorphic abstract complexes and $\tilde{K}$, $\tilde{L}$ are geometric realizations of K, L, then there is an isomorphism $v : \tilde{K} \cong \tilde{L}$; let w be its inverse. Then $|v| : |\tilde{K}| \to |\tilde{L}|$ is a continuous map with inverse $|w|$ and hence $|v|$ is a homeomorphism. ∎

As in the case of geometric complexes, we shall sometimes assume an abstract complex to be *augmented*, in which case the empty set of vertices belongs to the complex.

We shall continue to use the same notation for abstract as for geometric complexes. However, the reader should not think that an abstract complex only arises by 'abstracting' from a geometric complex. The following is an important example of an abstract complex arising in a different way.

Let X be a topological space and let $\mathscr{U} = \{U_\alpha\}$ be a covering of X by a finite collection of non-empty open sets. We define an abstract complex in which the 'vertices' are the members of $\mathscr{U}$ and the simplexes are the subcollections of members of $\mathscr{U}$ with non-empty intersection. This abstract complex, called the *nerve* of the covering $\mathscr{U}$, plays an important role in applying combinatorial methods to general spaces.‡

It follows from 1.9.5 that if P is the polyhedron of some realization of the abstract complex K, then topological properties of P which can be calculated from K are shared by all realizations of K. It is one of the triumphs of combinatorial topology that significant topological properties of a polyhedron can be calculated from the abstract complex associated with an *arbitrary* simplicial dissection of the polyhedron. The description of some of these properties will be the concern of chapter 2 and the proof of their topological invariance is given in chapter 3. Meanwhile, we remark that, though devoid of geometrical content, the notion of an abstract complex admits all the combinatorial concepts developed in this chapter for geometric complexes. For instance, an abstract complex can be regularly subdivided to give an abstract complex K'. The vertices of K' are the simplexes of K and the simplexes of K' are nests of simplexes of K, that is, sequences of simplexes of K each a face of its predecessor. Then it is obvious that if L is a realization of K, L' is a realization of K'. A simplicial map from the abstract complex K to the abstract complex L is just an admissible vertex transformation. The cone on an abstract complex is defined in the obvious way. Thus when we are concerned solely with combinatorial properties of a geometric complex, it is natural to speak only of the associated abstract complex.

We close this section with an important theorem on realization. This theorem is not of direct concern to us in this book, but its fundamental role in dimension theory warrants its inclusion.

1.9.6 Theorem. *An abstract complex K of dimension n has a realization in R^{2n+1}.*

Before we prove this, notice how 'economical' the realization is compared with that of 1.9.3. For K may have arbitrarily many vertices.

In fact, let $a^1, \ldots, a^\alpha$ be the vertices of K. Consider the points $p^1, \ldots, p^\alpha$ of R^{2n+1}, where p^i has coordinates $(i, i^2, \ldots, i^{2n+1})$, $i = 1, \ldots, \alpha$. Any $(2n+2)$ of the points $p^1, \ldots, p^\alpha$ are independent. For if not there

‡ See chapter 8 (Čech theory).

would exist numbers $\lambda_1, \ldots, \lambda_{2n+2}$, not all zero, and some set of mutually distinct integers $i_1, \ldots, i_{2n+2}$ such that

$$\sum_1^{2n+2} \lambda_k = 0,$$

$$\sum_1^{2n+2} \lambda_k i_k = 0,$$

$$\sum_1^{2n+2} \lambda_k i_k^2 = 0,$$

$$\cdots\cdots\cdots\cdots\cdots\cdots$$

$$\sum_1^{2n+2} \lambda_k i_k^{2n+1} = 0.$$

But the determinant of the coefficients in these equations is $\prod_{j<k} (i_j - i_k)$ and is thus non-zero, so that no such numbers $\lambda_1, \ldots, \lambda_{2n+2}$ can exist.

The transformation $v : K \to R^{2n+1}$ given by $a^i v = p^i$ induces a closed vertex scheme V on the vertices $p^1, \ldots, p^\alpha$. Moreover, the selected subsets of V are vertices of simplexes in R^{2n+1} since $(n+1) < (2n+2)$. It remains to prove that condition K 2 for complexes is satisfied. However, this follows immediately from what we have proved; for a common point of two simplexes would be expressible as $\Sigma \lambda_i p^i$ and $\Sigma \mu_i p^i$, where not more than $(n+1)$ of the λ's and not more than $(n+1)$ of the μ's differ from zero and $\Sigma \lambda_i = \Sigma \mu_i = 1$. We should then have, writing $\nu_i = \lambda_i - \mu_i$, that $\Sigma \nu_i p^i = 0$, $\Sigma \nu_i = 0$, and not more than $(2n+2)$ of the ν's differ from zero. Since any $(2n+2)$ of the points $p^1, \ldots, p^\alpha$ are independent this implies that $\nu_i = 0$, for all i; so no point has distinct expressions and the simplexes belonging to v are mutually disjoint. Thus v is a realization of K in R^{2n+1}. ∎

This theorem is best possible in the sense that, for each n, there exists an n-dimensional abstract complex which cannot be realized in R^{2n}; in fact, the n-section of s_{2n+2} constitutes such an example. For $n = 1$, this asserts that the 1-section of a closed 4-simplex cannot be drawn in the plane without crossings, a fact familiar to students of the four-colour problem. A further example if $n = 1$ is the hexagon with its three diagonals. This, too, must cross itself if drawn in the plane; would-be solvers of the famous puzzle about three houses and three public utilities may have reached this conclusion empirically.

Theorem 1.9.6 is an essential step in the proof that a compact

n-dimensional metrizable space can be embedded in R^{2n+1}. The proof uses 1.8.4 and the nerves of coverings. A full account is to be found in Pontryagin [4], chap. 1, §3.

1.10* Infinite complexes

So far we have considered only *finite* complexes, geometric and abstract. For many purposes (e.g. the consideration of non-compact spaces) it is convenient to be able to speak of infinite complexes. The notion of an abstract complex is easily extended.

1.10.1 Definition. An *infinite abstract simplicial complex* K defined on an infinite collection of objects, called vertices, is an infinite set of finite subcollections‡ s_p^i, called simplexes, satisfying the condition that if $s_p^i \in K$ then any subcollection of s_p^i belongs to K.

The subscript p and superscript i have, of course, the same significance as in Definition 1.9.1. Now we should not expect to be able to realize every infinite complex K in Euclidean space—the cardinal of the collection of vertices might exceed that of the continuum—so we apply a more intrinsic procedure and construct a topological space from the complex K.

Let $\{a^i\}$ be the collection of vertices of K, where the superscripts i are drawn from an indexing set I; it is convenient to assume I totally ordered and to display subcollections of the $\{a^i\}$ according to the ordering of their superscripts. We now define a topological space $|K|$ called the *polyhedron* of K. We first describe the points of $|K|$ and then give $|K|$ a topology.

The points of $|K|$ are collections of non-negative real numbers $\{\lambda_i\}$, one number for each $i \in I$, subject to:

(i) The non-zero λ's have as subscripts elements of I which form the superscripts of vertices of a simplex of K;

(ii) $\Sigma\lambda_i = 1$.

Notice that (ii) makes sense in view of (i) and the fact that simplexes of K are *finite* subcollections of vertices.

To the simplex $s_p = (a^{i_0}, \ldots, a^{i_p})$ corresponds a subset $|\bar{s}_p|$ of $|K|$, namely, the set of points $\{\lambda_i\}$ such that $\lambda_i = 0$ for all i other§ than $i_0, \ldots, i_p$. We further define the subset $|s_p|$ to be the set of points $\{\lambda_i\}$ such that $\lambda_i \neq 0$ if and only if i is one of $i_0, \ldots, i_p$. Plainly $|K| = \bigcup \bar{s}_p = \bigcup s_p$. We can define a transformation $t : |\bar{s}_p| \to R^{p+1}$ by $\{\lambda_i\}t = (\lambda_{i_0}, \ldots, \lambda_{i_p})$.

‡ As in the finite case, we may assume that every vertex belongs to some simplex.

§ We may identify the vertex a^{i_0} with the point $\{\lambda_i\}$ for which $\lambda_{i_0} = 1$.

To give $|K|$ a topology, we first topologize $|\bar{s}_p|$ by requiring that t be a homeomorphism into R^{p+1}. We then topologize $|K|$ by specifying its closed sets: $A \subseteq |K|$ is closed if $A \cap |\bar{s}|$ is closed in $|\bar{s}|$ for every simplex s of K. This is clearly a topology.

The space $|K|$ can be defined in terms of the abstract complex K whether K is finite or infinite. The notation $|K|$ is justified by

1.10.2 Proposition. *If K is finite, and $|K|$ is defined as above, then $|K|$ is homeomorphic to any realization of K.* ∎

The topology given to $|\bar{s}_p|$ by the homeomorphism t is, of course, that of a closed p-simplex; $|\bar{s}_p|$ clearly retains this topology as a (closed) subset of $|K|$. We observe that the topology of $|K|$ implies that $\overline{|s|} = |\bar{s}|$.

It is obvious how a simplicial map $v : K \to L$ induces a map $|v| : |K| \to |L|$. The process is equivalent to that whereby, for *finite* complexes, a simplicial map induces a map of a realization of K into a realization of L. If v embeds K in L as a subcomplex, then $|v|$ embeds $|K|$ in $|L|$ homeomorphically as a closed subspace.

1.10.3 Proposition. *$|K|$ is a Hausdorff space.*

Let $\{\lambda_i^1\}$, $\{\lambda_i^2\}$ be two points of $|K|$. If they are distinct, there exists a suffix i_0 such that $\lambda_{i_0}^1 < \lambda_{i_0}^2$. Let A be the subset of $|K|$ consisting of points $\{\lambda_i\}$ with $\lambda_{i_0} \leqslant \frac{1}{2}(\lambda_{i_0}^1 + \lambda_{i_0}^2)$ and let B be the subset of $|K|$ consisting of points $\{\lambda_i\}$ with $\lambda_{i_0} \geqslant \frac{1}{2}(\lambda_{i_0}^1 + \lambda_{i_0}^2)$. Each of A, B is closed since obviously their intersection with any $|\bar{s}|$ is closed. Their complements are disjoint open sets of $|K|$ containing $\{\lambda_i^2\}$ and $\{\lambda_i^1\}$ respectively. ∎

We have shown that $|K|$ is a Hausdorff space with the property that $|\bar{s}_p|$ has in $|K|$ the topology of a closed p-simplex. The topology we have given to $|K|$ is the finest consistent with these requirements. For since $|K|$ is Hausdorff and $|\bar{s}_p|$ is compact, $|\bar{s}_p|$ is closed in $|K|$. Thus a closed set of $|K|$ must intersect each $|\bar{s}_p|$ in a closed set of $|\bar{s}_p|$, so that the topology we have given to $|K|$ may be characterized as having the largest number of closed sets consistent with the two given requirements. For our purposes, the utility of the topology is expressed by the following extension of Proposition 1.3.3.

1.10.4 Proposition. *Let Y be any topological space and K an infinite complex; suppose that for each $s^i \in K$ a map $f^i : |\bar{s}^i| \to Y$ is given such that if $s^j \prec s^i$ then $f^i \mid |\bar{s}^j| = f^j$. Then there is a unique map $f : |K| \to Y$ with $f \mid |\bar{s}^i| = f^i$.*

The proof is that of I. 1.6; but readers should note that 1.10.4 would fail if $|K|$ had been given a strictly coarser topology. ∎

We have observed that, from the point of view of topology, the notations $|K|$ and $|v| : |K| \to |L|$ are consistent extensions of the notations for finite complexes. We may then speak simply of abstract complexes, geometric complexes and polyhedra if our remarks are to apply to finite or infinite complexes. Similarly, we wish to carry over to the infinite case those concepts related to finite complexes which extend in an obvious way.

Thus we may define the *dimension* of an infinite complex in the obvious way, provided we admit the possibility that $\dim K = \infty$. Again, s is the *carrier* of x if $x \in |s|$. We may also extend the notion of st (s) to infinite complexes. Again, $K - \text{st}(s)$ is a subcomplex of K so that $|\text{st}(s)|$ is an open subspace of $|K|$.

1.10.5 Proposition. *If $x \in |s|$ and $t \in \text{st}(s)$, then every neighbourhood of x intersects $|t|$.*

For let U be an open set of $|K|$ which does not intersect $|t|$. Then $|K| - U$ is a closed set containing $|t|$ and hence $\overline{|t|}$, which is the same as $|\bar{t}|$; but $s \prec t$ so that $|s| \subseteq |\bar{t}|$ and $x \in |s|$. Thus $x \in |K| - U$, which proves the assertion. ∎

We now prove some results about the topology of $|K|$, in the first of which we distinguish between finite and infinite complexes.

1.10.6 Theorem. *If K is an infinite complex, $|K|$ is not compact.*

If K is infinite, it certainly has an infinite number of vertices.‡ Then the sets $|\text{st}\, a^i|$, as i runs over the indexing set of the vertices, is an open covering of $|K|$ of which no subcollection is a covering, since to omit $|\text{st}\, a|$ is to uncover $|a|$. ∎

It follows that a polyhedron is the polyhedron of a finite complex if and only if it is compact.

We now pick out an important class of infinite complexes.

1.10.7 Definition. The (infinite) complex K is *locally finite* if, for each $s \in K$, st (s) consists of a finite number of simplexes. (We adopt the obvious convention that finite complexes are locally finite.)

1.10.8 Proposition. *K is locally finite if and only if each vertex belongs only to finitely many simplexes.* ∎

‡ As in 1.1 we identify vertices with 0-simplexes; see the footnote to 1.10.1.

1.10.9 Theorem. *K is locally finite if and only if $|K|$ is locally compact.*

Suppose K locally finite and let $x \in |K|$. We consider s, the carrier of x. Then $\mathrm{st}(s)$ consists of a finite number of simplexes $s^1, \ldots, s^k$ and $|\overline{\mathrm{st}(s)}| = \bigcup_{i=1}^{k} |\bar{s}^i|$, so that $|\overline{\mathrm{st}(s)}|$ is a compact neighbourhood of x. Thus $|K|$ is locally compact.

Conversely, suppose K is not locally finite. Then there exists a vertex a with infinite star. We shall prove that a has no compact neighbourhood. Let V be a neighbourhood, let $\{s^j\}$ be the set of simplexes of $\mathrm{st}(a)$ and let $x^j \in V \cap |s^j|$; we have already observed in 1.10.5 that $V \cap |s^j|$ is not empty. Then $A = \bigcup_j x^j$ is an infinite set of points of V every subset of which is closed in $|K|$ and hence in V. Thus V is not compact. ∎

We may distinguish the polyhedra of locally finite complexes by another criterion, embodied in the following theorem. We omit the proof which is long but not hard.

1.10.10 Theorem. *K is locally finite if and only if $|K|$ is metrizable; if $|K|$ is metrizable it admits the metric $\rho(\{\lambda_i\}, \{\mu_i\}) = \sqrt{[\Sigma(\lambda_i - \mu_i)^2]}$.*

We remark that it follows from this theorem that, if K is not locally finite, the metric topology given by

$$\rho(\{\lambda_i\}, \{\mu_i\}) = \{[\sum_i (\lambda_i - \mu_i)^2]\}^{\frac{1}{2}}$$

is strictly coarser than the topology of $|K|$.

A characterization of a further topological property of $|K|$ in terms of the combinatorial structure of K is given by

1.10.11 Theorem. *$|K|$ is separable if and only if K has at most denumerably many simplexes.* ∎

By defining $|K|$ intrinsically, we avoided the question of realizability. However, if we discuss this question for complexes K with $\dim K = \infty$, we must obviously go beyond Euclidean space.

Let H^∞ be Hilbert space. Then there are the obvious embeddings $R^1 \subset R^2 \subset \ldots \subset R^m \subset R^{m+1} \subset \ldots \subset H^\infty$. We say that K can be *realized* in H^∞ if there is a homeomorphism of $|K|$ into H^∞ such that the image of a closed simplex of $|K|$ is a closed rectilinear simplex. The image of $|K|$ is then a *geometric realization* of K in H^∞.

1.10.12 Theorem. *K can be realized in H^∞ if and only if it is locally finite and has at most denumerably many simplexes.*

Since (see I. 1.20) H^∞ is separable metric, so is any subset of H^∞. If, therefore, K is realizable in H^∞, $|K|$ is separable metric, and we invoke 1.10.10 and 1.10.11.

Now suppose that K is denumerable and locally finite. We define $t : |K| \to H^\infty$ by

$$\{\lambda_i\}t = (\lambda_1, \lambda_2, \ldots).$$

Then t is an isometry with respect to the metric of 1.10.10 and so certainly a homeomorphism. Its linearity shows that the images of closed simplexes are closed rectilinear simplexes in H^∞, and the theorem is proved. ∎

1.10.13 Theorem. *If $|K|$ can be embedded in R^n, K is at most denumerable and locally finite and* $\dim K \leqslant n$*; if K is at most denumerable and locally finite and* $\dim K \leqslant n$*, then it can be realized in* R^{2n+1}.

In the first part of this theorem, we invoke dimension theory (somewhat illicitly, since no proof will be offered) to yield the fact that $\dim K = \dim |K| \leqslant n$. The second part is proved by a minor adaptation of the argument of Theorem 1.9.6. ∎

1.11 Pseudodissections

A major objection to the use of simplicial complexes in computing topological invariants of compact polyhedra is that the dissection of the polyhedron may require an uncomfortably large number of simplexes. Thus the most obvious

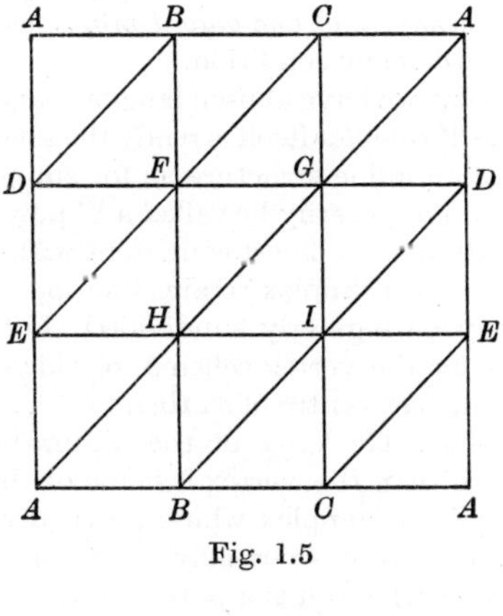

Fig. 1.5

dissection of the torus, as shown in fig. 1.5, requires 9 vertices, 27 edges and 18 triangles; the minimum dissection requires 7 vertices, 21 edges and 14 triangles. Again, the minimum dissection of the real projective plane requires 6 vertices, 15 edges and 10 triangles.

To meet this difficulty, it is natural to try to relax the conditions required of a simplicial complex in such a way as to render calculation more manageable, without, of course, jeopardizing the prospect that significant information about the polyhedron may be obtained from the combinatorial object under study. In this section we shall allow only a mild relaxation of the conditions which will, however, turn out to be of practical advantage. We openly admit that the particular advantages we have in mind are, first, that of facilitating the topological study of projective space and, to a lesser extent, that of simplifying the study of the fundamental group (see chapter 6). We confine attention to finite complexes, but the reader may readily extend the definitions to the infinite case.

The clue to our plan is provided by the remark that we are no longer going to insist that a simplex be determined by its vertices; more than one simplex of a *pseudodissection* may be spanned by a given set of vertices. Thus we should admit, for example, as a pseudodissection of a circle that provided by two points on the circle and the two arcs they determine. This is not a dissection in the sense of § 2 because the intersection of the two closed 1-simplexes is the pair of vertices. Two points of view are possible in describing a pseudodissection (or, indeed, a dissection). On the one hand we may suppose given the space to be dissected and describe how it is broken up into combinatorial pieces; on the other hand, we may describe the rules for putting together the pieces to form spaces suitable for study. We shall take the latter view (as in § 10) and leave the industrious reader to formulate the definition from the former view.

A simplicial complex may be regarded as built up out of disjoint closed simplexes by allowing certain identifications on their frontiers. The identifications permitted are just those induced by a linear homeomorphism of a face of one onto a face of the other. These identifications, are, of course, subject to a consistency condition that ensures that if the face s' of s is identified with the face t' of t, then the faces of s' and t' are themselves appropriately identified.

We now describe a *pseudo-simplicial complex*, or *Ψ-complex* as built up out of disjoint closed simplexes by allowing certain identifications on their frontiers; the identifications permitted are those induced by a *piecewise-linear homeomorphism of a proper subcomplex of one onto a proper subcomplex of the other*, subject to the obvious consistency condition.

In the interests of brevity, we have abused language and talked of complexes instead of polyhedra. The *Ψ-complex* itself is really the set of its (open) simplexes and each simplex carries an affine structure as for simplicial complexes. The space of a Ψ-complex will temporarily be called a Ψ-polyhedron. Its topology is analogous to that of § 10; it is the finest consistent with the requirement that it be Hausdorff and each closed simplex retains its topology.

Now a Ψ-complex may be regularly subdivided. For each simplex has a barycentre and we consider the vertex scheme in which the vertices are the barycentres and, if b^i is the barycentre of s^i, then $(b^0, b^1, \ldots, b^n)$ is a selected set provided $s^0 \succ s^1 \succ \ldots \succ s^n$. The crux of the argument justifying pseudodissections lies in the fact that the vertex scheme of the first derived is the vertex scheme of a simplicial complex whose polyhedron coincides with the original dissected space. The proof is virtually a copy of the argument proving Theorem 1.5.3. Thus a Ψ-polyhedron is a polyhedron so that Ψ-complexes do not involve a widening of the class of spaces under consideration. They are thus simply a convenient combinatorial tool; it is less easy to prove general results about them than about simplicial complexes, but it is easier to make the relevant computations. To demonstrate the gain in simplicity, we give the examples of the projective plane (fig. 1·6) and the torus (fig. 1.7). In fig. 1·6, we have a pseudodissection of the projective plane with 3 vertices A, B, O;

6 edges, 2 running from A to B, 2 from A to O and 2 from B to O; and 4 triangles with vertices A, B, O.

In fig. 1.7 we have a pseudodissection of the torus with 4-vertices A, B, C, O; 12 edges, 4 running from A to O, 2 from B to O, 2 from C to O, 2 from A to B and 2 from A to C, and 8 triangles, 4 with vertices A, B, O, and 4 with vertices A, C, O.

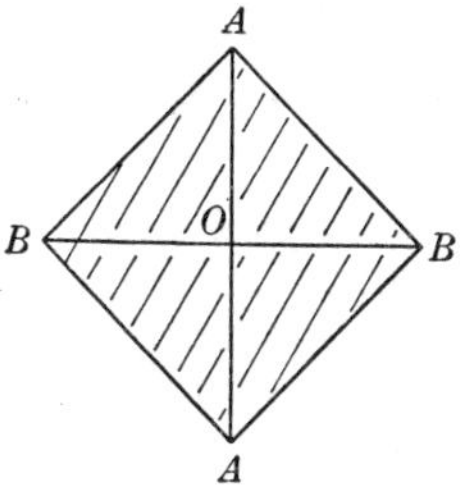

Fig. 1.6

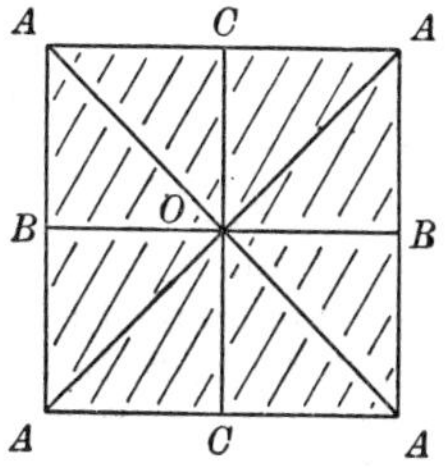

Fig. 1.7

EXERCISES

1. The set $a^0, a^1, \ldots, a^p$ of points of R^m is dependent and A is its convex cover. Prove that, if a point of A has a unique representation as $\sum_i \lambda_i a^i$ ($\sum_i \lambda_i = 1$; all $\lambda_i \geqslant 0$), then at least two of the λ_i are zero. Prove also that any such point lies on the frontier of A regarded as a subspace of $R(a^0, a^1, \ldots, a^p)$.

2. Construct complexes whose polyhedra are
(i) a solid cube,
(ii) a solid octahedron,
(iii) homeomorphic to an annulus,
(iv) homeomorphic to the real projective plane.

3. Prove Proposition 1.2.4.

4. If $s_p = (a^0 a^1 \ldots a^p)$, in the first derived of $\bar{s}_p$ a typical p-simplex is $b^{i_0 i_1 \ldots i_p} b^{i_0 i_1 \ldots i_{p-1}} \ldots b^{i_0}$, where $i_0 i_1 \ldots i_p$ is some permutation of $0\,1 \ldots p$. Prove that a point of $\bar{s}_p$ with barycentric coordinates $(\lambda_0, \lambda_1, \ldots, \lambda_p)$ lies in this p-simplex of the first derived if and only if $\lambda_{i_0} > \lambda_{i_1} > \ldots > \lambda_{i_p} > 0$. Find its barycentric coordinates with respect to this p-simplex in terms of $\lambda_0, \lambda_1, \ldots, \lambda_p$.

Characterize analogously the points belonging to simplexes of lower dimension of the first derived.

5. Supply the details left to the reader of the argument of Theorem 1.5.2.

6. If s_p is the simplex $a^0 a^1 \ldots a^p$ in R^p and t_1 the simplex $b^0 b^1$, which lies in R^{p+1} and is orthogonal to R^p, prove that the space $\bar{s}_p \times \bar{t}_1$ can be triangulated using the vertices $c^{ij} = a^i \times b^j$ $(i = 0, \ldots, p, j = 0, 1)$ if the simplexes are taken to be those spanned by subsets of distinct vertices $c^{i_0 j_0}, \ldots, c^{i_k j_k}$ for which $i_0, \ldots, i_k$ and $j_0, \ldots, j_k$ are both non-decreasing sequences.

Deduce that, if P is a polyhedron, so also is $P \times I$.

7. Generalize the construction of Q. 6 to give a triangulation of $\bar{s}_p \times \bar{t}_q$. Prove that the topological product of two polyhedra is a polyhedron.

8. Let s_1, t_1 be 1-simplexes and $f: s_1 \to \bar{t}_1$ map s_1 homeomorphically onto t_1. Prove that Cf is *not* a homeomorphism of Cs_1 into $C\bar{t}_1$.

9. (i) Prove that, if $f: X \to S^n$ is such that Xf is a proper subset of S^n, then f is nullhomotopic.

(ii) Prove that if $f_0, f_1: X \to S^n$ have the property that for no point x of X do xf_0 and xf_1 form an antipodal pair in S^n, then $f_0 \simeq f_1$.

10. Let $f: A \to B, g: B \to C, h: C \to D$ be maps such that fg and gh are homotopy equivalences. Prove that f, g and h are homotopy equivalences.

11. If $X_0 \simeq X_1$ and $Y_0 \simeq Y_1$, prove that $X_0 \times Y_0 \simeq X_1 \times Y_1$ and that $CX_0, X_0 \simeq CX_1, X_1$.

12. A pair (Y, Z) of spaces is said to have the *homotopy extension property* if, for all spaces X, maps $g_0: Y \to X$ and homotopies $h_t: Z \to X$ such that $h_0 = g_0|Z$, there exists a homotopy $g_t: Y \to X$ with $h_t = g_t|Z$. Prove that, if Z is a closed subset of Y, then (Y, Z) has the homotopy extension property if and only if $(Y \times 0) \cup (Z \times I)$ is a retract of $(Y \times I.)$ Prove that the pair (CZ, Z) has the homotopy extension property.

13. Let $|K|$ be a polyhedron and $|L|$ a subpolyhedron and let $f: |L| \to Z$ be a map. Form the disjoint union $|K| \cup Z$ and let R be the equivalence relation generated by $pRpf$, $p \in |L|$. Let Y be the identification space $(|K| \cup Z)/R$. Show that Z is embedded in Y as a closed subspace and that (Y, Z) has the homotopy extension property.

14. Contiguity is a relation between simplicial maps of $K^{(r)}$ into L, where K, L are fixed and r may vary; prove that this is not an equivalence relation.

15. Let $K = \overline{s_1}$ and let $|L| \subset R^1$ be the segment $[-2, 2]$, the closed 1-simplexes of L being the segments $[-2, -1]$, $[-1, 0]$, $[0, 1]$ and $[1, 2]$. If (λ_0, λ_1) are the barycentric coordinates in $\overline{s_1}$, $f: |K| \to |L|$ is defined by

$$(0, 1)f = 0,$$

$$(\lambda_0, \lambda_1)f = 2\lambda_0 \cos(\pi/\lambda_0^2) \quad \text{for} \quad 0 < \lambda_0 \leqslant 1.$$

Find r and a simplicial map $v: K^{(r)} \to L$ that is a simplicial approximation to f.

16. Prove that the distance between maps of a compact space X into a metric space Y defined in 1.8 is a metric in the set of all maps of X into Y. Prove also that the topology of Y^X depends only on the topology of X and of Y and does not depend on the choice of metric in Y.

17. Prove that, if K is an infinite complex, each compact subset of $|K|$ lies in the polyhedron of a finite subcomplex of K.

18. Prove Theorem 1.10.11.

2

HOMOLOGY THEORY OF A SIMPLICIAL COMPLEX

2.1 Orientation of a simplex

We have now described in detail how we may associate with a polyhedron a combinatorial structure called a simplicial complex. This can be done in many ways and thus the problem of deriving true topological invariants from the structure is non-trivial. We are concerned in this chapter to describe how certain important invariants may be obtained from a simplicial complex, but we shall not prove until the next chapter that they are, in fact, invariants of the underlying polyhedron. We stress (as we did in the Introduction) that these invariants are computable from the complex; they are algebraic in nature and attach to a polyhedron certain abelian groups, its *homology groups*.

We now begin the description of the way in which the homology groups are obtained from a simplicial complex. The first step is to *orient* the simplexes.

We recall that, in Euclidean geometry, two different sets of axes may or may not determine the same orientation of the space. In considering the orientation of a simplex, the role of an ordering of a set of axes is played by an ordering of the vertices. Two orderings of the vertices are said to determine the same orientation of the simplex if and only if an even permutation transforms one ordering into the other; if the permutation is odd, the orientations are said to be opposite.

Suppose that the simplex s_p has been oriented by selecting an ordering of the vertices; we call the pair consisting of s_p and its orientation an oriented simplex and write it as σ_p or $+\sigma_p$. The same simplex s_p with the opposite orientation will be written as $-\sigma_p$. An orientation of s_p can be defined by placing its vertices in an order. If $s_p = (a^0 a^1 \dots a^p)$, the symbol $a^0a^1 \dots a^p$ without brackets means the oriented simplex whose orientation is determined by the given ordering. Thus, for example, if $\sigma_p = a^0a^1 \dots a^p$, then

$$-\sigma_p = a^1a^0 \dots a^p = a^pa^1a^2 \dots a^{p-1}a^0.$$

0-simplexes admit only one orientation; thus the vertex a^0 determines a unique simplex s_0 and a unique oriented simplex σ_0; we write $\sigma_0 = a^0$. In an augmented complex we shall conventionally orient s_{-1} to give σ_{-1}.

A formal device which will prove useful later is to regard the oriented simplex $a^0a^1 \dots a^p$ as the 'product' of its vertices $a^0, a^1, \dots, a^p$. This multiplication is then skew-commutative, since $a^ia^j = -a^ja^i$, and in the augmented case has a unity element, namely, σ_{-1}. It is thus suggested that we should write σ_{-1} as 1; it is also suggested that we regard an expression $a^{i_0} \dots a^{i_p}$ in which a vertex appears more than once as being, in some sense, equal to zero.

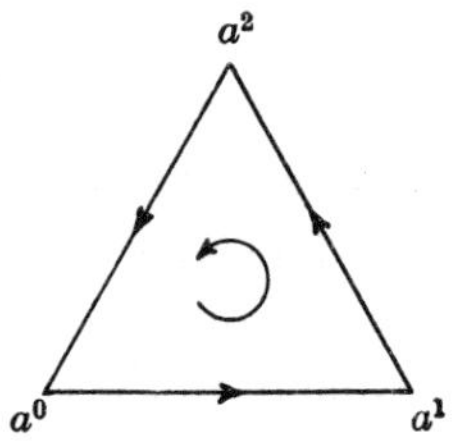

Fig. 2.1

It is natural to think of an oriented 1-simplex a^0a^1 as a segment carrying an arrow from a^0 to a^1; it is also natural to think of an oriented 2-simplex $a^0a^1a^2$ as a triangle with a sense of rotation determined, the sense in which a radius vector turns which leads from a fixed interior point to a point which varies round the frontier from a^0 to a^1, from a^1 to a^2, and from a^2 to a^0 (see fig. 2.1). This suggests a fruitful notion. The rotation in the triangle is in a sense 'coherent' with the orientations a^0a^1, a^1a^2, a^2a^0 of its sides, and this can be generalized.

Let $\sigma_p = a^0a^1 \dots a^p$; we say that the oriented $(p-1)$-face opposite a^i

$$a^0a^1 \dots \widehat{a^i} \dots a^p = a^0a^1 \dots a^{i-1}a^{i+1} \dots a^p$$

is *coherently oriented* with $(-1)^i\sigma_p$; or that $a^0a^1 \dots \widehat{a^i} \dots a^p$ is coherently oriented with σ_p if i is even, and non-coherently if i is odd. The coherent orientation of any $(p-1)$-face is said to be that orientation *induced* by the orientation of σ_p. We shall make further use of this notation; an array of symbols with certain symbols omitted is written by crowning the symbols to be omitted with a $\wedge$.

Before this definition can be accepted, we must show that any even permutation on the ordering $a^0, a^1, \dots, a^p$ gives the same definition of the coherent orientation of a face. It is in fact easy to see that the interchange of any pair of consecutive vertices reverses the induced orientation of each face and hence that any even permutation leaves it unchanged. The standard orientation of σ_{-1} is, by definition, coherent with each σ_0.

The notions of orientation and of induced orientation apply equally well to an abstract simplicial complex. All the results that follow and that are now specifically related to a geometric complex hold for an abstract complex. Nothing further will be said to reinforce this remark; the word 'complex' will be used without prefix when the statement is valid both for an abstract and for a geometric complex. However, unless otherwise stated, a complex will be finite.

Let K be a complex. Then K is said to be *oriented* when each of its simplexes is oriented. A collection α of orientations of its simplexes is called an *orientation* of K, and we may write (K, α) for the complex K with orientation α. However, we shall usually use the symbol K for an oriented complex as well as for the complex before orientation.

An *ordering of* K is a partial ordering of its vertices such that the vertices of each simplex are totally ordered. There is a useful example of such an ordering for K', the first derived of K; we may (partially) order the vertices of K' (which are barycentres $b(s)$ of simplexes of K) by the rule that $b(s) < b(t)$ if and only if $s \succ t$. It is clear from 1.4 that this defines an ordering of K'. The leading vertex of a simplex of K' is the first vertex in this ordering.

If ω is an ordering of K it places the vertices of each simplex of K in an order and so orients K. We may write (K, ω) for the complex K oriented by the ordering ω.

Let K be an oriented complex with (oriented) simplexes σ_p^i. To each pair $(\sigma_p^i, \sigma_{p-1}^j)$, we may assign an integer, called their *incidence number*, $\partial_{j,p}^i$, defined as follows:

$$\begin{aligned} \partial_{j,p}^i &= 0, && \text{if } s_{p-1}^j \text{ is not a face of } s_p^i, \\ &= +1, && \text{if } s_{p-1}^j \text{ is a face of } s_p^i \text{ and is coherently oriented with it,} \\ &= -1, && \text{if } s_{p-1}^j \text{ is a face of } s_p^i \text{ and is not coherently oriented with it.} \end{aligned}$$

The incidence numbers between simplexes of dimension p and $p-1$ form in an obvious way a matrix ∂_p, the element in the ith row and jth column of which is $\partial_{j,p}^i$. The matrices $\partial_1, \partial_2, \ldots$ are called the *incidence matrices* of the oriented complex K.

For example, let K be the closed 2-simplex with vertices a^0, a^1, a^2 and let K be oriented by the ordering $a^0a^1a^2$. Then we have

$$\sigma_2^1 = a^0a^1a^2;\ \sigma_1^1 = a^1a^2,\ \sigma_1^2 = a^0a^2,\ \sigma_1^3 = a^0a^1;\ \sigma_0^1 = a^0,\ \sigma_0^2 = a^1,\ \sigma_0^3 = a^2.$$

Then $$\partial_1 = \begin{pmatrix} 0 & -1 & 1 \\ -1 & 0 & 1 \\ -1 & 1 & 0 \end{pmatrix} \quad \text{and} \quad \partial_2 = (1 \quad -1 \quad 1).$$

We state here and prove in the next section the fundamental theorem on incidence matrices.

2.1.1 Theorem. *If* $p > 1$, *then* $\partial_p \partial_{p-1} = \mathbf{0}$.

Of course, the only difference between the incidence matrices for K and K^+ is the presence in the latter case of the matrix ∂_0, consisting of a column of $+1$'s and satisfying $\partial_1 \partial_0 = \mathbf{0}$. In this chapter the emphasis will be on K, but most statements will have obvious analogues for K^+.

2.2 Chains, cycles and boundaries

Let K be an oriented complex. A *p-chain*, c_p, of K is defined as a formal sum

2.2.1 $$c_p = \sum_i m_i \sigma_p^i,$$

where the summation is over the oriented p-simplexes of K and the coefficients m_i are rational integers. Thus a chain can be visualized as a collection of oriented simplexes taken with certain multiplicities; in the chain 2.2.1, the oriented simplex σ_p^i has multiplicity m_i if $m_i \geqslant 0$, otherwise the oriented simplex $-\sigma_p^i$ has multiplicity $-m_i$.

As an example of a p-chain we may take

2.2.2 $$\sum_j \partial_{j,p+1}^i \sigma_p^j.$$

This is the collection of p-simplexes which are faces of σ_{p+1}^i, each occurring with the induced orientation. The p-chain 2.2.2 is called the *boundary* of σ_{p+1}^i and is written $\sigma_{p+1}^i \partial$.

We may regard σ_p^i as a p-chain; similarly, we may regard $-\sigma_p^i$ as a p-chain. Thus, if $\sigma_p = a^0a^1 \dots a^p$, the symbol $a^1a^0 \dots a^p$ may be regarded as standing for the p-chain, $-\sigma_p$, of K. We then have

2.2.3 Proposition.‡ $\sigma_p \partial = \sum_i (-1)^i (a^0a^1 \dots \widehat{a^i} \dots a^p)$, $p > 0$. ∎

‡ In K^+ we also have $\sigma_0 \partial = 1$.

The set of p-chains of K forms an additive abelian group by the rule

2.2.4 $$\sum_i m_i \sigma_p^i + \sum_i n_i \sigma_p^i = \sum_i (m_i + n_i)\, \sigma_p^i.$$

This group is called the *group of p-chains of K with integer coefficients*; for brevity we may say the *group of p-chains of K* or the *p-chain group of K*. It will be written as $C_p(K)$; similarly, we define $C_p(K^+)$ and $C_p(K) = C_p(K^+)$ for $p \geqslant 0$. When it is clear to which complex, augmented or otherwise, we are referring, the p-chain may simply be written C_p.

The group $C_p(K)$ is evidently a free abelian group with the simplexes σ_p^i constituting a basis. Thus the boundary operator ∂ may be extended uniquely to a homomorphism

$$\partial_p : C_p \to C_{p-1};$$

we write ∂ for ∂_p if no confusion can arise.

Explicitly, ∂_p is given by

$$\Big(\sum_i m_i \sigma_p^i\Big)\, \partial_p = \sum_i m_i (\sigma_p^i \partial) = \sum_i m_i \sum_j \partial_{j,p}^i \sigma_{p-1}^j = \sum_j \Big(\sum_i m_i \partial_{j,p}^i\Big)\, \sigma_{p-1}^j.$$

If $p \leqslant 0$ (or if $p \leqslant -1$ in the augmented case) it is often convenient to take $C_{p-1} = 0$ and ∂_p as the zero homomorphism.

2.2.5 Theorem. *For any p-chain c_p of K, $c_p \partial\partial = 0$.*

Now since ∂ is a homomorphism, it is clearly sufficient to prove this if $c_p = \sigma_p$ a p-simplex. Let $\sigma_p = (a^0 a^1 \ldots a^p)$ and suppose $p > 1$. By Proposition 2.2.3,

$$\begin{aligned} \sigma_p \partial\partial &= \Big(\sum_i (-1)^i (a^0 a^1 \ldots \widehat{a^i} \ldots a^p)\Big)\, \partial \\ &= \sum_i (-1)^i \Big(\sum_{j<i} (-1)^j (a^0 a^1 \ldots \widehat{a^j} \ldots \widehat{a^i} \ldots a^p) \\ &\qquad + \sum_{j>i} (-1)^{j-1} (a^0 a^1 \ldots \widehat{a^i} \ldots \widehat{a^j} \ldots a^p)\Big). \end{aligned}$$

Thus $a^0 a^1 \ldots \widehat{a^k} \ldots \widehat{a^l} \ldots a^p$ appears twice in $\sigma_p \partial\partial$, once with coefficient $(-1)^l (-1)^k$ and once with coefficient $(-1)^k (-1)^{l-1}$. It follows that $\sigma_p \partial\partial = 0$, so that $c_p \partial\partial = 0$. The cases $p \leqslant 1$ are trivial. ∎

Theorem 2.1.1 is an immediate consequence; for the element in the ith row and jth column of $\boldsymbol{\partial}_p \boldsymbol{\partial}_{p-1}$ is just the coefficient of σ_{p-2}^j in $\sigma_p^i \partial\partial$.

We write $Z_p(K)$ or Z_p for the kernel of ∂_p and call an element of Z_p a *p-cycle*.‡ We write $B_p(K)$ or B_p for the image of ∂_{p+1} and call an element of B_p a *p-boundary*.

‡ Recall the convention governing the use of C_p.

2.2.6 Corollary. *B_p is a subgroup of Z_p.* ∎

There is a generalization of the notion of a p-chain which has some important applications. We may replace the coefficients m_i in 2.2.1 by elements g_i of an arbitrary additive abelian group G. The resulting object we call a p-chain of K *with coefficients in* G, and such chains form an additive abelian group $C_p(K; G)$ under the law of addition analogous to 2.2.4. Note that $C_p(K)$ now becomes an abbreviation for $C_p(K; J)$, where J is the additive group of integers. The boundary operator ∂ again induces a homomorphism

$$\partial_p : C_p(K; G) \to C_{p-1}(K; G).$$

We write the formula for ∂_p as

2.2.7 $$(\sum_i g_i \sigma_p^i)\,\partial_p = \sum_j (\sum_i \partial_{j,p}^i g_i)\, \sigma_{p-1}^j.$$

It is important to note that σ_p is not, in general, an element of $C_p(K; G)$; the problem of describing precisely the role played by the simplexes σ_p in $C_p(K; G)$ is purely algebraic and is held over to chapter 5. The reader is advised in the first instance to think in terms of integer coefficients, for which a geometrical picture is more readily available. We continue to write $C_p(K)$, $C_p(K^+)$ for $C_p(K; J)$, $C_p(K^+; J)$; further, we permit ourselves the abbreviation C_p for $C_p(K; G)$ or $C_p(K^+; G)$ when the complex and coefficient group in question are clear; when the symbols $C_p(K)$ or $C_p(K^+)$ appear, it is always to be understood that the coefficient group is J. Similar conventions apply to Z_p and B_p.

We close this section by reinforcing the geometrical picture. If $c_p \in C_p$, we define $\bar{c}_p$, the *carrier* of c_p, to be the subcomplex of K consisting of the closures of those simplexes s_p such that the corresponding oriented simplex σ_p appears in c_p with non-zero coefficient. Thus we might have said that $\bar{c}_p$ is the smallest subcomplex of K of which c_p is a chain.

2.2.8 Proposition. $\overline{c_p^1 \pm c_p^2} \subseteq \overline{c_p^1} \cup \overline{c_p^2}$. ∎

2.2.9 Proposition. $\overline{c_p \partial} \subseteq \bar{c}_p$. ∎

The close relation between the boundary ∂ and the topological frontier is, in fact, provided by

2.2.10 Proposition. $\overline{\sigma_p \partial} = \dot{s}_p$. ∎

Note that we may write $\bar{\sigma}_p$ for $\bar{s}_p$.

It is sometimes convenient to consider not the carrier $\bar{c}_p$ of a chain c_p, which is a subcomplex of K, but the *support* $|c_p|$, which is the polyhedron of its carrier; in fact $|c_p| = |\bar{c}_p|$. The carrier is a set of simplexes, the support is a set of points. The reader is warned that $|\sigma_p|$ and $|s_p|$ are different sets of points; since σ_p is a chain, $|\sigma_p|$ is the set of points of its carrier $\bar{s}_p$, whereas $|s_p|$ is the set of points of the open simplex.

2.3 Homology groups

Let K be a simplicial complex and let $C_p = C_p(K)$; then we have seen that the group of p-cycles, Z_p, is a subgroup of C_p and that the group of p-boundaries, B_p, is a subgroup of Z_p. At present these groups depend on the choice of orientation of K; but we may readily conclude that, up to isomorphism, they are independent of orientation. Precisely, let α, β be two orientations of K; let the simplex s_p of K be oriented as σ_p^α by α and as σ_p^β by β. We shall write $C_p^\alpha(K)$, etc., for $C_p(K, \alpha)$, etc. We may define, for each p,

$$\theta_p^{\alpha,\beta} : C_p^\alpha(K) \to C_p^\beta(K),$$

by

$$\sigma_p^\alpha \theta_p^{\alpha,\beta} = \pm \sigma_p^\beta,$$

the sign being $+$ or $-$ according as α and β impose the same or opposite orientations on s_p. It is then obvious that $\theta_p^{\alpha,\beta}$ is an isomorphism with inverse $\theta_p^{\beta,\alpha}$ and that $\theta^{\alpha,\beta}\theta^{\beta,\gamma} = \theta^{\alpha,\gamma}$. Moreover, $\sigma_p^\alpha \theta_p \partial = \sigma_p^\alpha \partial \theta_{p-1}$ where $\theta = \theta^{\alpha,\beta}$; for, in each chain, the coefficient of σ_{p-1}^β is zero if s_{p-1} is not a face of s_p and otherwise is $+1$ or -1 according as s_{p-1} is oriented by β coherently or non-coherently with the orientation of s_p by α. Thus‡ $\theta_p \partial = \partial \theta_{p-1}$.

It follows immediately that

$$Z_p^\alpha \theta_p \subseteq Z_p^\beta, \quad B_p^\alpha \theta_p \subseteq B_p^\beta.$$

For if $z \in Z_p^\alpha$, then $z\theta_p \partial = z\partial\theta_{p-1} = 0$, so that $z\theta_p \in Z_p$; and if $b \in B_p^\alpha$, then $b = c\partial$, $c \in C_{p+1}^\alpha(K)$, and $b\theta_p = c\partial\theta_p = c\theta_{p+1}\partial$, so that $b\theta_p \in B_p^\beta$. We write our conclusions in a self-evident notation as

2.3.1 Theorem. $\theta_p^{\alpha,\beta} : C_p^\alpha, Z_p^\alpha, B_p^\alpha \cong C_p^\beta, Z_p^\beta, B_p^\beta$. ∎

The reader has probably observed that, algebraically, $\theta_p^{\alpha,\beta}$ may be reinterpreted as a change of basis of a particularly simple kind in the group C_p. Then what we have proved amounts to saying that the

‡ $\theta^{\alpha,\beta}$ is a chain map in the sense of chapter 3.

boundary homomorphism is determined by the group itself and not by the choice of basis. At any rate, orientation stands revealed as a technical device enabling C_p and ∂ to be defined and the groups we study are essentially unaffected by the choice of orientation.

On the other hand, we are primarily interested in the polyhedron $|K|$, and it is evident that the groups C_p, Z_p, B_p are heavily dependent on the choice of K covering the polyhedron. Thus, for example, subdivision of K enlarges all these groups (in the sense of increasing their rank). However, further experiment suggests that variations in K cause strongly correlated variations in Z_p and B_p. For example,

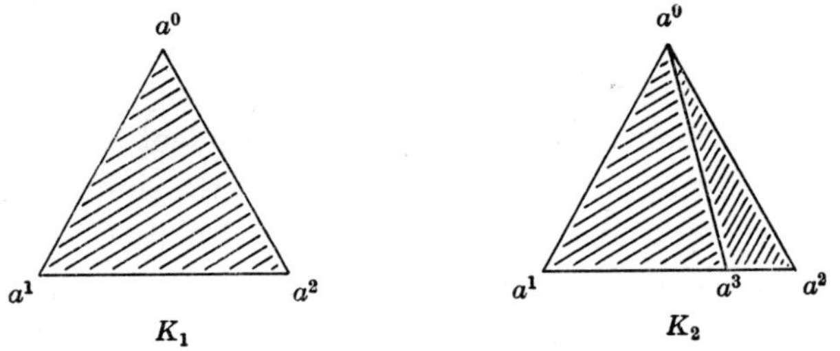

Fig. 2.2

let $|K|$ be the closed 2-simplex spanned by a^0, a^1, a^2 and consider the triangulations K_1 and K_2 of fig. 2.2. Then

$$C_2(K_1) \cong J, \quad Z_2(K_1) = 0, \quad B_2(K_1) = 0;$$
$$C_1(K_1) \cong J \oplus J \oplus J, \quad Z_1(K_1) \cong J, \quad B_1(K_1) \cong J;$$
$$C_0(K_1) \cong J \oplus J \oplus J, \quad Z_0(K_1^+) \cong J \oplus J, \quad B_0(K_1) \cong J \oplus J;$$

whereas

$$C_2(K_2) \cong J \oplus J, \quad Z_2(K_2) = 0, \quad B_2(K_2) = 0;$$
$$C_1(K_2) \cong J \oplus J \oplus J \oplus J \oplus J, \quad Z_1(K_2) \cong J \oplus J, \quad B_1(K_2) \cong J \oplus J;$$
$$C_0(K_2) \cong J \oplus J \oplus J \oplus J, \quad Z_0(K_2^+) \cong J \oplus J \oplus J, \quad B_0(K_2) \cong J \oplus J \oplus J.$$

Thus the insertion of the vertex a^3 and the edge a^0a^3 adds one to the ranks of Z_1 and B_1 and adds one to the ranks of Z_0 and B_0.

This example (and others, which the reader may readily construct) suggests that the significant feature to be derived from the complex K and the boundary ∂ is the factor group§ Z_p/B_p.

2.3.2 Definition. The factor group Z_p/B_p is said to be the pth *homology group* of the oriented complex K and is written $H_p(K)$.

§ Some authors prefer to speak of the *difference group* $Z_p - B_p$.

If the augmented complex K^+ is in question, we write $H_p(K^+)$ for the homology group. Either $H_p(K^+)$ or $H_p(K)$ may be abbreviated to H_p if the context permits. The more precise notation for $H_p(K)$ would be $H_p(K, \alpha) = H_p^\alpha(K)$ if α is the orientation. However, Theorem 2.3.1 immediately implies

2.3.3. Theorem. *$\theta_p^{\alpha,\beta}$ induces an isomorphism*

$$\theta_p^{\alpha,\beta} : H_p^\alpha(K) \cong H_p^\beta(K). \blacksquare$$

Notice that the homology groups, being factor groups of subgroups of finitely generated abelian groups, are themselves finitely generated and abelian. Notice also that we may define $Z_p(K; G)$ as the kernel of $\partial_p : C_p(K; G) \to C_{p-1}(K; G)$ and $B_p(K; G)$ as the image of $C_{p+1}(K; G)$ under ∂_{p+1}, that $B_p(K; G) \subseteq Z_p(K; G)$, and that we may thus define the factor group $H_p(K, G) = Z_p(K; G)/B_p(K; G)$. This is called the *p-th homology group of K with coefficients in G*. It, too, is independent, up to isomorphism, of the orientation of K.

The homology groups, together with variants and generalizations of them, are the main topic of this book. Their importance rests on the fact that they are topological invariants of the underlying polyhedron; their usefulness is, of course, greatly enhanced by the fact that the problem of deciding whether two complexes have isomorphic homology groups is always soluble in a finite number of steps. This last remark is intended to counter the point of view that it would be clearly preferable to define the homology groups in such a way that their invariance is obvious.‡

If c_p^1, c_p^2 are two chains of K whose difference is a boundary, we say that c_p^1 is *homologous* to c_p^2 and write $c_p^1 \sim c_p^2$. Usually the chains in question will be cycles, when $z_p^1 \sim z_p^2$ is equivalent to saying that z_p^1 and z_p^2 belong to the same coset of Z_p mod B_p or that they determine the same element of H_p. This element is called the *homology class* of the cycle z_p^1 (or z_p^2), and is written $\{z_p^1\}$.

The notion of homology is intuitively satisfactory in simple cases. A closed path on a complex represents a 1-cycle and two closed paths which together span a surface are homologous because they combine to form the boundary of the 2-chain represented by the surface. However, the reader should be warned that not every statement about cycles and homology is enriched by reference to a geometrical picture. This is particularly true when we take a general coefficient

‡ Such definitions are given in chapter 8.

group. The reader is advised to consider what changes would be required in subsequent assertions and arguments if the coefficient group were allowed to be arbitrary.

2.4 $H_0(K)$ and connectedness

In this section we give an interpretation of the zero-dimensional homology group, $H_0(K)$. In addition, we shall establish the relation between $H_0(K)$ and $H_0(K^+)$.

We begin with some general remarks. Let K be a complex and let K^+ be the corresponding augmented complex.

2.4.1 Proposition. $C_p(K) = C_p(K^+)$, $p \geqslant 0$; $Z_p(K) = Z_p(K^+)$, $p > 0$; $B_p(K) = B_p(K^+)$, $p \geqslant 0$. ∎

2.4.2 Corollary. $H_p(K) = H_p(K^+)$, $p > 0$. ∎

2.4.3 Theorem. $H_0(K) \cong H_0(K^+) \oplus J$.

Let $a^0, a^1, \ldots, a^m$ be the vertices of K. They may be identified with the elements of a basis for $C_0(K)$ and plainly $C_0(K) = Z_0(K)$. By a change of basis, we may choose the chains $a^0, a^1 - a^0, a^2 - a^0, \ldots, a^m - a^0$ as a basis for $Z_0(K)$. We now show that the chains

$$a^1 - a^0, a^2 - a^0, \ldots, a^m - a^0$$

are a basis for $Z_0(K^+)$. For let $\sum_i n_i a^i \in Z_0(K^+)$; since $a^i \partial = 1$, it follows that $\sum_i n_i = 0$ and $\sum_i n_i a^i = \sum_{i>0} n_i(a^i - a^0)$. Since each $a^i - a^0$ belongs to $Z_0(K^+)$, the conclusion follows. We have $Z_0(K) = J \oplus Z_0(K^+)$, where J is generated by a^0, and since $B_0(K) = B_0(K^+)$, the theorem follows. ∎

Let K be a complex. We may say that the vertex a may be connected to the vertex b in K if there exists a sequence of vertices $a^{i_0}, \ldots, a^{i_k}$ such that $a^{i_0} = a$, $a^{i_k} = b$, and $a^{i_{j-1}} a^{i_j}$ is a simplex of K for $j = 1, \ldots, k$. We write, temporarily, $a \leftrightarrow b$ for this relation which is clearly an equivalence relation on the vertices of K. Let a^0, a^1 be vertices of $s \in K$, and b^0, b^1 vertices of $t \in K$. Then obviously $a^0 \leftrightarrow b^0$ if and only if $a^1 \leftrightarrow b^1$. Thus K is the union of disjoint subcomplexes $K_1, \ldots, K_q$ such that $a \leftrightarrow b$ if and only if a and b belong to the same subcomplex K_r, $1 \leqslant r \leqslant q$. The subcomplexes are called the *components* of K and K is said to be *connected* ‡ if $q = 1$.

‡ The terms are justified: the components of $|K|$, in the sense of analytic topology, are $|K_1|, \ldots, |K_q|$; K is connected if and only if $|K|$ is connected.

2.4.4. Theorem. $H_p(K) = \sum_{r=1}^{q} H_p(K_r)$.

Obviously $C_p(K) = \sum_r C_p(K_r)$ and $C_p(K_r)\partial \subseteq C_{p-1}(K_r)$. The conclusion of the theorem now follows from elementary abelian group theory. ∎

2.4.5 Corollary. $H_p(K^+) = \sum_{r=1}^{q} H_p(K_r^+), \quad p > 0.$ ∎

Finally, we compute $H_0(K)$; in the light of 2.4.4. it is sufficient to compute $H_0(K)$ if K is connected. Then, evidently, for any two vertices a, b of K, $a \sim b$.

2.4.6 Theorem. (i) $H_0(K) = J$ *if and only if* K *is connected.* (ii) $H_0(K^+) = 0$ *if and only if* K *is connected.*

It is sufficient to prove (ii). As in the proof of Theorem 2.4.3 the chains $a^i - a^0$ form a basis for $Z_0(K^+)$; but, if K is connected, $a^i - a^0 \sim 0$ so that $Z_0(K^+) = B_0(K^+)$ and $H_0(K^+) = 0$.

Conversely, suppose $H_0(K^+) = 0$; then $H_0(K) = J$; but if $K_1, \ldots, K_q$ are the components of K, then, by what we have proved,

$$H_0(K) = \sum_r H_0(K_r)$$

and is free abelian on q generators, so that $q = 1$ and K is connected. ∎

2.4.7 Corollary. (i) $H_0(K)$ *is free abelian; its rank is the number of components of* K. (ii) $H_0(K^+)$ *is free abelian; its rank is one less than the number of components of* K. ∎

2.4.8 Corollary. *If* a, b *are vertices of* K, *then* $a \leftrightarrow b$ *if and only if* $a \sim b$.

We have observed that $a \leftrightarrow b$ implies $a \sim b$. Conversely, suppose $a \sim b$. If K is connected, we know that $H_0(K)$ is free cyclic generated by any vertex‡ of K. Thus from Theorem 2.4.4 we infer that two vertices from different components cannot be homologous so that $a \leftrightarrow b$. ∎

The reader may have gained the impression that K is a more convenient gadget than K^+; this will be the case when geometrical interpretations are in question. The advantage of K^+ is to be seen in developing the general theory; in particular, it will prove convenient that connectedness is equivalent to the *vanishing* of $H_0(K^+)$ (2.4.6 (ii)).

‡ Strictly, $H_0(K)$ is generated by the homology class of any vertex, but such abbreviations are usually adopted.

2.5 Some examples and torsion

Since H_p is a finitely generated abelian group, it follows (see the first section of chapter 5) that it is the direct sum of a free abelian group and a finite abelian group. In the simplest examples it generally happens that the finite summand is null and the group is describable by the number of its free generators. A few examples are given; the arguments used in discussing them are not rigorous but they can all be made rigorous by the use of later results. The examples may suggest to the reader that the number of generators of the homology groups is related to the intricacy of the complex.

1. Let K be a connected 1-dimensional complex, or *linear graph*, with α_0 vertices and α_1 edges. By Theorem 2.4.6, $H_0(K)$ is free abelian on 1 generator. Since there are no 2-simplexes, $B_1(K)$ is null and $H_1 = Z_1$. A connected linear graph is said to be a *tree* if it contains no closed (1-dimensional) circuits; for a tree, Z_1 is null and a simple inductive proof shows that $\alpha_0 - \alpha_1 = 1$. If K is not a tree, it is possible to remove an open 1-simplex without disconnecting it; for there will be a closed circuit, any 1-simplex of which may be removed. It is fairly easy to prove that the removal of such a 1-simplex reduces by one the number of free generators of Z_1—which is for any complex a free abelian group. It immediately follows that $H_1(K) = Z_1(K)$ is free abelian on p_1 generators, where $p_1 = \alpha_1 - \alpha_0 + 1$.

2. If $|K|$ is homeomorphic to S^2, then K is connected and $H_0(K) = J$. Intuition correctly suggests that every 1-cycle bounds, because every closed path on the 2-sphere can be deformed to a point; this makes plausible the statement that $H_1(K)$ is null. Since there are no 3-simplexes, $H_2(K) = Z_2(K)$. Now the 2-simplexes admit of an orientation in such a way that $\sum_i \sigma_2^i$ is a cycle; when this is done, it is easy to see that $\sum_i m_i \sigma_2^i$ is a cycle only if $m_i = m_j$ for all i, j. Consequently $H_2(K)$, $= Z_2(K)$, is generated by $\{\sum_i \sigma_2^i\}$ and is free abelian on one generator.

3. If $|K|$ is a *n-cell*—that is, if $|K|$ is homeomorphic to a closed n-simplex—$H_p(K^+)$ is null for all $p \geqslant 0$. Since the n-simplex can be contracted to a point over itself, it is not surprising that every p-cycle bounds for $p > 0$. The device of augmentation now removes the restriction $p > 0$.

4. Let $|K|$ be the *torus*, the topological product of two copies of S^1. As before $H_0(K) = J$. As in the case of the 2-sphere, we can orient the 2-simplexes so that $\sum_i \sigma_2^i$ is a cycle z_2, which generates $Z_2(K)$ freely; therefore $H_2(K) = Z_2(K) = J$. $H_1(K)$ is more interesting. There are 1-cycles z_1^1, z_1^2 in any simplicial decomposition of the torus which approximately traverse the 'long-way round' and the 'short-way round' on the surface (see fig. 2.3). It turns out that $\{z_1^1\}$ and $\{z_1^2\}$ generate $H_1(K)$ freely; any 1-cycle is homologous to some linear combination of these two. This latter statement can be made acceptable by a geometrical argument. Remove somewhere an open 2-simplex, s_2^1. This affects no homology for 1-cycles; for, if $z_1 = c_2\partial$ where $c_2 = \sum_i m_i \sigma_2^i$, then also $z_1 = (c_2 - m_1 z_2)\partial$; and $(c_2 - m_1 z_2)$ is a chain which is not affected by the removal of s_2^1. If the new complex is called K^-, then $H_1(K^-) \cong H_1(K)$. Now K^- can be shrunk over itself onto $|z^1| \cup |z^2| = |L|$, say; since, by Example 1, $H_1(L)$ is free abelian on 2 generators, it is to be expected that $H_1(K)$ is also. If the method of shrinking of K^- onto L is not obvious, imagine K^- cut along $|z_1^1|$ and $|z_1^2|$ and opened out into a square with a 2-simplex missing, bounded by two copies each of $|z_1^1|$

and $|z_1^2|$; it is now clear that this punctured square can be shrunk onto its frontier.

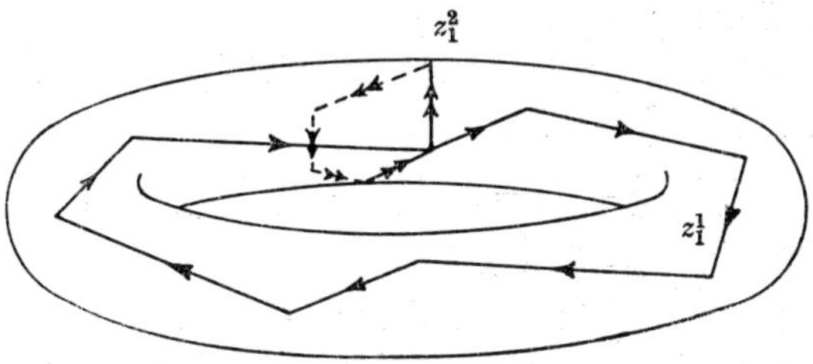

Fig. 2.3

5. Let $|K|$ be the real projective plane, P^2; again $H_0(K) = J$, but the other two groups are a little surprising. Although this is a closed surface, we cannot repeat the argument for $H_2(K)$ that was used for the 2-sphere and the torus; it is not in fact possible so to orient all the 2-simplexes that their sum is a cycle and there are no 2-cycles except the zero 2-cycle. Consequently $H_2(K)$ is null. $H_1(K)$ turns out to be isomorphic to J_2, thus providing our first example of a homology group whose finite summand is not null. To see this, take a 1-cycle, z_1, which (roughly) traverses a projective line and such that $|z_1| \cong S^1$. If c_1 is a 1-chain on K such that $|c_1\partial| \subseteq |z_1|$, then there exists a 1-chain c_1' such that $c_1 \sim c_1'$ and $|c_1'| \subseteq |z_1|$. For, if we cut K along $|z_1|$, it opens out to a 2-cell bounded by two copies of $|z_1|$, each running in the same direction round its frontier; certainly any 1-chain in the 2-cell with boundary on the frontier is homologous to a 1-chain on the frontier. Therefore any 1-cycle of K is homologous to a 1-cycle on $|z_1|$; but, since $|z_1|$ is homeomorphic to S^1, any 1-cycle on $|z_1|$ is equal to mz_1 for some integer m. Every 1-homology class, therefore, of K has an element of the form mz_1. But $2z_1$ belongs to the zero class. This is evident from the representation of K as a 2-cell with two copies of $|z_1|$ on its frontier; the boundary of the obvious 2-chain is the 1-cycle on its frontier and, returning to K itself, we see that $2z_1$ bounds. But $z_1 \not\sim 0$ and $H_1(K)$ therefore has just two elements $\{0\}$ and $\{z_1\}$, the element $\{z_1\}$ being of order 2.

When a complex has non-null finite summands in its homology groups, it is said to possess *torsion*. A cycle not itself bounding, some multiple of which bounds, is said to be a *torsion cycle*, and its homology class (which is an element of finite order in the homology group) is a *torsion class* or torsion element. For some purposes it is convenient to treat torsion cycles as being in some sense boundaries; in that case we shall say that z_p is *weakly homologous to zero* and write $z_p \approx 0$, if there exists an integer $k \neq 0$ such that $kz_p \sim 0$. Thus the symbol '$z_p \approx 0$' means that z_p is either a torsion cycle or a boundary.

2.5.1 Proposition. *The set of cycles that are weakly homologous to zero form a subgroup $\overline{W}_p$ of Z_p.*

(We have chosen the bizarre symbol $\overline{\overline{W}}_p$ rather than W_p to avoid confusion with a group due to be defined in 5.1; the subgroup $\overline{\overline{W}}_p$ makes only a fleeting appearance.)

Suppose $z_p^1 \approx 0$ and $z_p^2 \approx 0$; then there exist k_1, k_2 (neither zero) such that $k_1 z_p^1 \sim 0$ and $k_2 z_p^2 \sim 0$. Therefore

$$k_1 k_2(z_p^1 - z_p^2) = k_2(k_1 z_p^1) - k_1(k_2 z_p^2) \sim 0;$$

and, since $k_1 k_2 \neq 0$, $z_p^1 - z_p^2 \approx 0$. ▌

We shall write $z_p^1 \approx z_p^2$ to mean that $z_p^1 - z_p^2 \approx 0$ or equivalently that z_p^1 and z_p^2 belong to the same coset of Z_p mod $\overline{\overline{W}}_p$. The cosets are elements of the factor group ${}_wH_p = Z_p/\overline{\overline{W}}_p$, called the *weak homology group* of K (of dimension p).

The terminology of 'torsion' is applied to arbitrary abelian groups. If A is an abelian group, the subset of elements of finite order is a subgroup, T, called the *torsion subgroup* of A. Thus the torsion classes of H_p, together with the zero class, form the torsion subgroup of H_p. If $T = A$, A is said to be a *torsion group*.

Since $B_p \subseteq \overline{\overline{W}}_p$, there is a canonical epimorphism‡ $w : H_p \to {}_wH_p$.

2.5.2 Proposition. *The kernel of w is the torsion subgroup of H_p.*

Now $\{z_p\}w = 0$ if and only if $z_p \in \overline{\overline{W}}_p$; and $z_p \in \overline{\overline{W}}_p$ if and only if there exists an integer $k \neq 0$ such that $kz_p \in B_p$. Thus $\{z_p\}w = 0$ if and only if $k\{z_p\} = 0$ for some integer $k \neq 0$. ▌

As remarked at the outset of this section, H_p is the direct sum of a free abelian group, F_p, and a finite abelian group, T_p. It follows that T_p is the torsion subgroup of H_p and, from 2.5.2, that F_p is isomorphic to ${}_wH_p$.

2.6 Contrahomology and the Kronecker product

The concepts leading to the definition of the homology groups correspond reasonably well to natural and familiar geometrical ideas. These concepts are essentially contained in Poincaré's pioneer work in topology at the end of the nineteenth century. We now define groups dual to the homology groups to which full attention was not directed until about 1935, when a number of mathematicians, including Alexander, Čech, Kolmogoroff and Whitney, defined and exploited them. It is easy to see why there should have been this delay; for the intuitive content of these dual groups is a greal deal more elusive.

‡ w is canonical in the sense that it does not depend on a choice of generators for H_p.

After several vicissitudes, the terminology 'cohomology' for these dual groups has become standard. For reasons set out in detail in the Introduction, we will use the term 'contrahomology', but we wish to establish firmly at the outset that the change is purely terminological.

The starting point for contrahomology is the notion of a *p-contrachain*, c^p, of an oriented complex K, with values in an additive abelian group G. It is a function defined on the oriented simplexes of K, with values in G. We may then write

$$(\sigma_p^i)\, c^p = g^i,$$

g^i being the value of the contrachain c^p on the oriented simplex σ_p^i.

The reader may have remarked that we could have assigned the coefficient g^i to the simplex σ_p^i and thus determined a p-chain. Thus a $(1, 1)$ correspondence may be established between p-contrachains and elements of $C_p(K; G)$. This does not justify us in concealing the distinction between the two notions, for the correspondence depends on a choice of basis for $C_p(K)$. The situation in vector-space theory is entirely analogous. Although a finite-dimensional vector space is isomorphic to its dual, the elements of the dual are linear functionals and it is essential that the two spaces be not confused; for example, a map from one vector space to another induces a *contravariant* map of the dual spaces. A further decisive argument favouring the preservation of the distinction between chains and contrachains is to be found in the study of infinite complexes (2.11).

The reader will also remark that we do not single out the group J as a favoured value group for contrachains. In fact, the notion of contrachain provides a clear case for singling out the group J as coefficient group for chains. For every contrachain c^p determines a unique homomorphism $C_p(K) \to G$, given by

2.6.1 $$\Sigma m_i \sigma_p^i \to \Sigma m_i((\sigma_p^i)\, c^p),$$

and conversely, every such homomorphism determines a unique contrachain. We identify c^p with the homomorphism 2.6.1. If A is any group, then, as in I. 2, the set of homomorphisms of A into G, written Hom (A, G), admits a group operation given by

$$a(\phi + \psi) = a\phi + a\psi; \quad a \in A;\ \phi, \psi \in \mathrm{Hom}\,(A, G).$$

We give Hom $(C(K), G)$ this group structure and call the resulting group the *group of p-contrachains* of K with values in G, and write it

$C^p(K; G)$, with the abbreviations $C^p(K)$ or C^p if no ambiguity is to be feared.‡ Notice that, if $c^p, d^p \in C^p(K; G)$, then

2.6.2 $$(\sigma_p)(c^p + d^p) = (\sigma_p)c^p + (\sigma_p)d^p.$$

We shall often prefer to use the notation of 'inner product' suggested by vector-space theory and write (c_p, d^p) instead of $(c_p)d^p$, where $c_p \in C_p(K)$, $d^p \in C^p(K; G)$. Then $(c_p, d^p) \in G$ and we may call (c_p, d^p) the *Kronecker index* (or *product*) of c_p and d^p; it is, of course, bilinear. Let δ^i_j be the Kronecker symbol ($\delta^i_j = 0$, if $i \neq j$; $\delta^i_i = 1$) and let $c^p \in C^p(K; J)$ be the contrachain given by

2.6.3 $$(\sigma^i_p, c^p) = \delta^i_j,$$

all i. We shall write σ^p_j for this contrachain§; clearly $C^p(K; J)$ is free abelian and the contrachains σ^p_j form a basis, adjoint to the basis of $C_p(K)$ consisting of the p-simplexes.

A p-contrachain is defined by giving its value on each p-simplex. Given a $(p-1)$-contrachain d^{p-1}, we define a p-contrachain $\delta^{p-1}d^{p-1}$, which we abbreviate to δd^{p-1}, by the rule

2.6.4 $$(\sigma_p, \delta d^{p-1}) = (\sigma_p \partial, d^{p-1}).$$

We call δd^{p-1} the *contraboundary* of d^{p-1}; then δ^{p-1} is adjoint to ∂_p and raises dimension by 1. It is a homomorphism, since

$$\begin{aligned}(\sigma_p, \delta d^{p-1}_1 + \delta d^{p-1}_2) &= (\sigma_p, \delta d^{p-1}_1) + (\sigma_p, \delta d^{p-1}_2)\\ &= (\sigma_p \partial, d^{p-1}_1) + (\sigma_p \partial, d^{p-1}_2)\\ &= (\sigma_p \partial, d^{p-1}_1 + d^{p-1}_2)\\ &= (\sigma_p, \delta(d^{p-1}_1 + d^{p-1}_2)).\end{aligned}$$

Thus δ is a left homomorphism

$$\delta : C^{p-1}(K; G) \to C^p(K; G),$$

defined for each p, and we have

2.6.5 Theorem. $\delta\delta = 0$.

First note that it follows by linearity from 2.6.4 that

2.6.6 $$(c_p, \delta d^{p-1}) = (c_p \partial, d^{p-1}).$$

Then $$(\sigma_p, \delta\delta d^{p-2}) = (\sigma_p \partial, \delta d^{p-2}) = (\sigma_p \partial\partial, d^{p-2}) = (0, d^{p-2}) = 0. \blacksquare$$

‡ Recall that, on the contrary, if we write $C_p(K)$, the coefficient group is J.

§ Notice the notation; in 'contra' theory, the dimension affix is superscript and indexing symbol subscript.

We can now proceed as in 2.3. We define *contracycles* as contrachains with vanishing contraboundaries; they form a group $Z^p(K;G)$ of which $\delta C^{p-1}(K;G) = B^p(K;G)$, the group of contraboundaries, is a subgroup. The factor group $H^p(K;G) = Z^p/B^p$ is the *p-th contrahomology group* of K with values in G. Two contracycles z_1^p and z_2^p determining the same contrahomology class are called *contrahomologous* and we write $z_1^p \sim z_2^p$. The contrahomology class of z^p may be written as $\{z^p\}$.

2.6.7 Proposition. *A p-contrachain is a contracycle if and only if its Kronecker index with every p-boundary is zero.*

If d^p is a contracycle, and b_p a boundary, then $b_p = c_{p+1}\partial$ and

$$(b_p, d^p) = (c_{p+1}\partial, d^p) = (c_{p+1}, \delta d^p) = (c_{p+1}, 0) = 0.$$

Conversely, if d^p is a contrachain such that $(b_p, d^p) = 0$ for every $b_p \in B_p$, then for each σ_{p+1},

$$(\sigma_{p+1}, \delta d^p) = (\sigma_{p+1}\partial, d^p) = 0 \quad \text{and} \quad \delta d^p = 0. \;\blacksquare$$

2.6.8 Corollary. *If G is* torsion-free (*i.e. its torsion subgroup is zero*), *then a p-contrachain with values in G is a contracycle if and only if its Kronecker index with every element of $\overline{W}_p$ is zero.*

The proof is an obvious modification of the proof of 2.6.7. ∎ We remark that if $G=J$, then there is a perfect symmetry between the roles played by C_p and C^p in the Kronecker index, so that 2.6.7 and 2.6.8 will then have duals in which the roles of C_p and C^p are interchanged.

2.6.9 Theorem. *A Kronecker product between $H_p(K)$ and $H^p(K;G)$ is given by*

$$(\{z_p\}, \{z^p\}) = (z_p, z^p).$$

The essential point here is that the definition of $(\{z_p\}, \{z^p\})$ is, in fact, independent of the choice of representatives. For

$$\begin{aligned}(z_p + c_{p+1}\partial, z^p + \delta c^{p-1}) &= (z_p, z^p) + (c_{p+1}\partial, z^p + \delta c^{p-1}) + (z_p, \delta c^{p-1}) \\ &= (z_p, z^p) + (c_{p+1}, \delta(z^p + \delta c^{p-1})) + (z_p\partial, c^{p-1}) \\ &= (z_p, z^p) + (c_{p+1}, 0) + (0, c^{p-1}) \\ &= (z_p, z^p). \;\blacksquare\end{aligned}$$

We may re-express 2.6.8 by saying that if G if torsion-free, then Z^p is the *annihilator* of $\overline{W}_p$, in the sense that $(\overline{W}_p, c^p) = 0$ if and only if $c^p \in Z^p$. It may also be shown that $\overline{W}_p$ is the annihilator of Z^p, but this is deferred to chapter 5. We may define $\overline{W}^p(K; G)$ and ${}_wH^p(K; G)$ for any value group G in the obvious way and we see immediately

2.6.10 Proposition. *If G is torsion-free, then a Kronecker product may be defined between ${}_wH_p$ and ${}_wH^p(K; G)$.* ▮

2.7 Contrahomology examples

Contrahomology theory has analogues in other branches of mathematics. For instance, if F is a field, the chain group with coefficients in F is a vector space over F and the contrachain group with values in F is the set of linear functionals on this vector space, in fact its dual vector space. Much of the algebra of homology and contrahomology smacks of the theory of vector spaces. When, however, the coefficient or value group is not a field, difficulties arise relating to torsion, which are not found in the theory of vector spaces.

There is another parallel to contrahomology theory which is of considerable importance and which gives a fruitful mental picture for a contrachain. Let $|K|$ be a line segment chopped into 1-simplexes and 0-simplexes in some way, and let us consider it as embedded in the x-axis of a Cartesian plane, covering $a \leqslant x \leqslant b$. Any integrable real function defined over $[a, b]$ attaches to each oriented subinterval a real number, its integral over the interval; in particular, it attaches a real number to each oriented 1-simplex of K. So the integral of the function establishes a homomorphism of $C_1(K)$ into the real numbers, and therefore is a 1-contrachain of K with values in the real numbers.

In general, we may profitably think of a p-contrachain as associated with an integral which has values in some abelian group G and which is defined over the p-chains of K. This suggests a relationship with the theory of differential forms on a manifold,‡ on which there is an operation analogous to δ. The theory of differential forms and de Rham's generalization, the theory of currents, are defined only over manifolds and for integrals having values in a field; any reader, however, who is familiar with either of these theories will find contrahomology theory natural and easy.

There is a familiar example of differential forms and their 'contraboundaries'. We shall not attempt a careful treatment but offer the example to relate the notion of contraboundary to something rather well known.

Starting from R^3 we construct an infinite complex. Its 0-simplexes are the points of R^3; its 1-simplexes are the differentiable arcs of R^3; its 2-simplexes are the differentiable images of s_2 in R^3 that have differentiable edges; its 3-simplexes are the differentiable images of s_3 in R^3 that have differentiable faces. The reader is asked to overlook bizarre properties of this so-called complex, since the present purpose is purely illustrative. We consider contrachains of this complex with values in the real numbers.

A 0-contrachain is clearly a real function defined over R^3; we shall say a real function ϕ determines a contrachain $c^0(\phi)$. A 1-contrachain assigns a real number to each differentiable arc; a vector field $\mathbf{v}$ over R^3 determines such a 1-contrachain $c^1(\mathbf{v})$ by the rule

2.7.1 $$(\sigma_1, (c^1\mathbf{v})) = \int_{\sigma_1} \mathbf{v},$$

‡ See, for instance, Hodge, *Harmonic Integrals*, p. 78.

where the right-hand side means the line integral of the vector field $\mathbf{v}$ along the arc σ_1. A 2-contrachain $c^2(\mathbf{w})$ can also be determined by a vector field $\mathbf{w}$ over R^3 by the rule

2.7.2
$$(\sigma_2, c^2(\mathbf{w})) = \int_{\sigma_2} \mathbf{w},$$

where the right-hand side means the integral over the surface σ_2 of the normal component of $\mathbf{w}$. A 3-contrachain $c^3(\psi)$ can be determined by a real function ψ by the rule

2.7.3
$$(\sigma_3, c^3(\psi)) = \int_{\sigma_3} \psi,$$

the integral being the ordinary volume integral.

We now find what the contraboundaries are for the contrachains associated with differentiable functions ϕ and differentiable vector fields $\mathbf{v}$, $\mathbf{w}$.

The contraboundary of $c^0(\phi)$ is defined by

$$(\sigma_1, \delta c^0(\phi)) = (\sigma_1 \partial, c^0(\phi)) = (p_1)\,\phi - (p_0)\,\phi,$$

where σ_1 leads from the point p_0 to the point p_1. But it is known that the gradient of ϕ, $\nabla\phi$, is a vector field such that

$$\int_{\sigma_1} \nabla\phi = (p_1)\,\phi - (p_0)\,\phi;$$

therefore, since, for each σ_1, $(\sigma_1, \delta c^0(\phi)) = (\sigma_1, c^1(\nabla\phi))$, we know that

2.7.4
$$\delta c^0(\phi) = c^1(\nabla\phi) = c^1(\operatorname{grad}\phi).$$

In words, the contraboundary of the 0-contrachain determined by ϕ is the 1-contrachain determined by $\nabla\phi$.

Consider now a 1-contrachain determined by the vector field $\mathbf{v}$. Its contraboundary satisfies

$$(\sigma_2, \delta c^1(\mathbf{v})) = (\sigma_2 \partial, c^1(\mathbf{v})) = \int_{\sigma_2 \partial} \mathbf{v}.$$

But by Stokes's theorem we know that

$$\int_{\sigma_2 \partial} \mathbf{v} = \int_{\sigma_2} \nabla \times \mathbf{v} = (\sigma_2, c^2(\nabla \times \mathbf{v})).$$

Therefore

2.7.5
$$\delta c^1(\mathbf{v}) = c^2(\nabla \times \mathbf{v}) = c^2(\operatorname{curl}\mathbf{v});$$

or the contraboundary of the 1-contrachain determined by $\mathbf{v}$ is the 2-contrachain determined by its curl, $\nabla \times \mathbf{v}$.

Finally, consider a 2-contrachain determined by the vector field $\mathbf{w}$. Its contraboundary satisfies

$$(\sigma_3, \delta c^2(\mathbf{w})) = (\sigma_3 \partial, c^2(\mathbf{w})) = \int_{\sigma_3 \partial} \mathbf{w}.$$

But Gauss's theorem asserts that

$$\int_{\sigma_3 \partial} \mathbf{w} = \int_{\sigma_3} \nabla . \mathbf{w} = (\sigma_3, c^3(\nabla . \mathbf{w})),$$

so that

2.7.6
$$\delta c^2(\mathbf{w}) = c^3(\nabla . \mathbf{w}) = c^3(\operatorname{div}\mathbf{w});$$

or the contraboundary of the 2-contrachain determined by $\mathbf{w}$ is the 3-contrachain determined by its divergence $\nabla . \mathbf{w}$.

It will be noticed that the law $\delta\delta = 0$ corresponds to well-known properties in the theory of vector fields.

From this example the reader may see that the contraboundary operator corresponds to very natural notions in differential geometry. The point of view suggested by this example may be made quite precise; for, by using differential forms on a differentiable manifold as 'contrachains' and exterior differentiation as the 'contraboundary operator', it has been shown by de Rham that we obtain contrahomology groups isomorphic to those arising from a triangulation of the manifold. In view of the correspondence indicated by 2.7.4–2.7.6, the relation 2.6.4 is often called a generalized Stokes's formula.

We now give a further example in which contra-theory appears in a natural way in a geometrical context. Again the treatment will be descriptive, but the arguments may be made quite rigorous by using the material of subsequent chapters.‡

Let K be an oriented 2-dimensional complex, let S^1 be an oriented circle and let $y_* \in S^1$. We shall investigate the homotopy classes of maps of $|K|$ in S^1, giving arguments towards the conclusion that there is a (1, 1) correspondence between these classes and the elements of $H^1(K; J)$. We start by considering homotopy classes of maps of $|K^1|$ in S^1, K^n being the n-section of K, and lead up to the preliminary result that there is a (1, 1) correspondence between these classes and the elements of $C^1(K; J)/B^1(K; J)$, the group of contrahomology classes of 1-contrachains.

An easy application of the Homotopy Extension Theorem (1.6.10) shows that any class of maps of $|K^1|$ in S^1 contains a map $f : |K^1|, |K^0| \to S^1, y_*$. For each such map we define a 1-contrachain c_f^1 of K^1 (and so also of K) by the rule that, for each $\sigma_1^i \in K$, (σ_1^i, c_f^1) is the number of times that the f-image of σ_1^i winds round S^1 in the positive sense. If $f \simeq g : |K^1|, |K^0| \to S^1, y_*$ then clearly§ $c_f^1 = c_g^1$ and conversely. If, on the other hand $f, g : |K^1|, |K^0| \to S^1, y_*$ are homotopic as maps of $|K^1|$ in S^1, the image of $|K^0|$ being allowed to wander over S^1 during the homotopy F, then we define a 0-contrachain $c_{f,g}^0$ by the rule that, for each vertex a of K, $(a, c_{f,g}^0)$ is the number of times the image of a traverses S^1 in the positive sense during the homotopy F. We now assert that

2.7.7 $$\delta c_{f,g}^0 = c_g^1 - c_f^1.$$

For, if σ_1 is the simplex ab, we must establish that

$$(b, c_{f,g}^0) - (a, c_{f,g}^0) = (\sigma_1, c_g^1) - (\sigma_1, c_f^1);$$

and this equation only asserts the highly credible fact that the increase, in passing from f to g, in the number of times σ_1 is wound round S^1 is the number of times its front end goes round minus the number of times its rear end goes round.

From 2.7.7 we infer that $c_f^1 \sim c_g^1$; thus to each homotopy class of maps of $|K^1|$ in S^1 is associated an element of $C^1(K; J)/B^1(K; J)$. Further, to each $c^1 \in C^1(K; J)$ there is a map

$$f : |K^1|, |K^0| \to S^1, y_*$$

such that $c_f^1 = c^1$: we just prescribe that f winds σ_1^i round S^1 (σ_1^i, c^1) times. Also if $c^1 \sim d^1$ and $f, g : |K^1|, |K^0| \to S^1, y_*$ are such that $c_f^1 = c^1$, $c_g^1 = d^1$, then we can use a 0-contrachain whose contraboundary is $d^1 - c^1$ to determine a homotopy between f and g. These facts collected together establish the first step in the

‡ In particular, see chapter 7.

§ The *proof* is by reference to a property of homology.

argument, namely, that classes of maps of $|K^1|$ in S^1 are in (1, 1) correspondence with the elements of $C^1(K; J)/B^1(K; J)$.

We are interested, however, in classes of maps not of $|K^1|$ but of $|K|$ in S^1. Any such class, by 1.6.10, has an element $f : |K|, |K^0| \to S^1, y_*$, and such an element, on restriction to $|K^1|$, determines a contrachain c^1. We next show that $g : |K^1|, |K^0| \to S^1, y_*$ may be extended to a map of $|K|$ into S^1, if and only if c^1_g is a contracycle. By 1.3.3, g can be extended to $|K|$ if and only if it can be extended to each $\bar{s}_2 \in K$; by 1.5.5, 1.5.8 and 1.6.6, $g \mid \dot{s}_2$ can be extended to $\bar{s}_2$ if and only if it is nullhomotopic. Now a map $\dot{s}_2 \to S^1$ is nullhomotopic if and only if its degree‡ is zero and the degree of $g \mid \dot{s}_2$ is, in fact, $\pm(\sigma_2 \partial, c^1_g) = \pm(\sigma_2, \delta c^1_g)$. It follows that g may be extended to a map of $|K|$ if and only if c^1_g is a contracycle.

The results so far obtained establish a transformation from the homotopy classes of $|K|$ in S^1 onto the set of elements of $Z^1(K; J)/B^1(K; J) = H^1(K; J)$. This transformation is, in fact, (1, 1); for, if $f, g : |K| \to S^1$ have the property that $c^1_f \sim c^1_g$, our earlier argument shows that $f \mid |K^1| \simeq g \mid |K^1|$, and a special property of S^1 (namely, that all maps of S^2 in S^1 are nullhomotopic) implies that therefore $f \simeq g$. This completes the skeleton of a proof that, if K is 2-dimensional, the homotopy classes of maps of $|K|$ in S^1 are in (1, 1) correspondence with the elements of $H^1(K; J)$.

This statement holds, in fact, for complexes K of arbitrary dimension (since the *higher homotopy groups* of S^1 vanish) and the correspondence may be enriched to an algebraic isomorphism by giving to the set of homotopy classes the group structure induced by the group structure of S^1 as the multiplicative group of complex numbers with absolute value 1.

2.8 Relative homology and contrahomology

The theories we have described may be called the *absolute* homology and contrahomology theory of a complex, in contrast with the *relative* theory due to Lefschetz. In the latter we consider a *pair* consisting of a complex K and a subcomplex K_0.

Let c_p be a chain of K_0; in an obvious sense c_p may be regarded as a chain of K. However, it is frequently important to keep the two notions separate; thus, for example, the statement that 'c_p is a boundary' has a different meaning according to whether we regard c_p as a chain of K_0 or of K. When we wish to make this distinction we write $c_p \lambda$ for the chain of K determined by c_p. Thus precisely, $c_p \lambda$ is the chain of K which has the same coefficients as c_p for the simplexes of K_0 and has coefficient zero for the simplexes of K not in K_0. Thus λ is a monomorphism

2.8.1 $$\lambda : C_p(K_0) \to C_p(K)$$

whose image is a direct factor in $C_p(K)$.

Let us write ∂^0 for the boundary operator in K_0.

‡ See, for instance, Lefschetz [2], p. 124; a precise definition of degree can be found in 5.8.

2.8.2 Theorem. $\lambda\partial = \partial^0\lambda$.

This follows immediately from the fact that a subcomplex is closed; if s is a simplex of K_0, then all the faces of s belong to K_0. Thus for any $\sigma \in K_0$, $\sigma\partial$ is a chain of K_0 and the conclusion follows. ∎

Theorem 2.8.2, trivial as it may seem, is fundamental. It establishes the fact that λ is, in the terminology of chapter 3, a chain map. It follows immediately, as is intuitively obvious, that λ sends cycles to cycles and boundaries to boundaries and thus induces a homomorphism

2.8.3 $$\lambda_* : H_p(K_0) \to H_p(K).$$

This homomorphism λ_* is not, in general, monomorphic; thus we have reaped a reward for our apparent pedantry in distinguishing between $C_p(K_0)$ and $C_p(K_0)\lambda$; the reader is welcome to try to express the homomorphism λ_* without bringing in λ.

Notice that all that has been said so far holds for arbitrary coefficient groups; we shall not stress this but leave to the reader the formal changes in foregoing and subsequent statements required by this generalization.

Let

2.8.4 $$\beta : C_p(K) \to C_p(K_0)$$

be the homomorphism which assigns to $c_p \in C_p(K)$, the chain $c_p\beta$ of $C_p(K_0)$ whose coefficients for the simplexes of K_0 are precisely those of c_p. Then, clearly, for $d_p \in C_p(K_0)$, $d_p\lambda\beta = d_p$, or

2.8.5 $$\lambda\beta = 1.$$

On the other hand, $\beta\partial^0 \neq \partial\beta$, in general. However, from 2.8.2 and 2.8.5 we deduce that

2.8.6 $$\partial^0 = \lambda\partial\beta.$$

It is also fruitful to consider the set of simplexes of K not in K_0; this set covers an open subset of $|K|$ and we may refer to it as the *open subcomplex*‡ $K \bmod K_0$. A p-chain of $K \bmod K_0$ is a linear combination of the p-simplexes of K not lying in K_0. Such a p-chain obviously determines a p-chain of K, namely, the p-chain with the same coefficients off K_0 and coefficient 0 on the simplexes of K_0; and a p-chain of K determines a p-chain of $K \bmod K_0$, namely, the

‡ Notice that $K \bmod K_0 = K^+ \bmod K_0^+$.

p-chain with the same coefficients off§ K_0. Thus we have homomorphisms

2.8.7 $$\alpha : C_p(K \bmod K_0) \to C_p(K),$$

2.8.8 $$\mu : C_p(K) \to C_p(K \bmod K_0),$$

with

2.8.9 $$\alpha\mu = 1.$$

2.8.10 Proposition. $C_p(K) = C_p(K_0)\,\lambda \oplus C_p(K \bmod K_0)\,\alpha$ *and*

$$\lambda\mu = 0,\ \alpha\beta = 0,\ \beta\lambda + \mu\alpha = 1. \;]$$

We have to turn $K \bmod K_0$ into a 'complex' by defining a boundary operator; this will require some modification of ∂, since a simplex of $K \bmod K_0$ may have faces in K_0. The definition we make is

2.8.11 $$\hat{\partial} = \alpha\,\partial\mu : C_p(K \bmod K_0) \to C_{p-1}(K \bmod K_0).$$

Thus, conceptually, if $c_p \in C_p(K \bmod K_0)$, $c_p\hat{\partial}$ is the part of its boundary lying off K_0.

2.8.12 Theorem. $\hat{\partial}\hat{\partial} = 0$.

For $$\hat{\partial}\hat{\partial} = \alpha\,\partial\mu\alpha\,\partial\mu = \alpha\partial(1-\beta\lambda)\,\partial\mu = \alpha\,\partial\partial\mu - \alpha\,\partial\beta\lambda\,\partial\mu$$
$$= \alpha\,\partial\partial\mu - \alpha\,\partial\beta\,\partial^0\lambda\mu = 0. \;]$$

The reader is advised to provide himself with a less formal proof.

2.8.13 Theorem. $\mu\hat{\partial} = \partial\mu$.

For $$\mu\hat{\partial} = \mu\alpha\,\partial\mu = (1-\beta\lambda)\,\partial\mu - \partial\mu - \beta\,\partial^0\lambda\mu = \partial\mu. \;]$$

Thus μ commutes with the boundary operators and so sends cycles of K to cycles of $K \bmod K_0$, boundaries to boundaries and thus induces

2.8.14 $$\mu_* : H_p(K) \to H_p(K \bmod K_0).$$

On the other hand, $\alpha\partial \neq \hat{\partial}\alpha$, in general. Consider now

2.8.15 $$\nu = \alpha\,\partial\beta : C_p(K \bmod K_0) \to C_{p-1}(K_0).$$

Thus ν operates as follows: let $c_p \in C_p(K \bmod K_0)$; regard c_p as a chain of K and take its boundary; restrict the boundary to the simplexes of K_0.

§ We express the fact that a simplex belongs to $K \bmod K_0$ by saying it is *off* K_0.

2.8.16 Theorem. $\nu\partial^0 = -\hat{\partial}\nu.$

For
$$\nu\partial^0 = \alpha\partial\beta\partial^0 = \alpha\partial\beta\lambda\partial\beta = \alpha\partial(1-\mu\alpha)\partial\beta$$
$$= -\alpha\partial\mu\alpha\partial\beta = -\hat{\partial}\nu. \ \rbrack$$

The reader is again advised to arm himself with a less formal proof. The minus sign in the relation‡ we have proved clearly does not affect the validity of the conclusion we seek, namely, that ν sends cycles to cycles and boundaries to boundaries and so induces

2.8.17 $$\nu_* : H_p(K \bmod K_0) \to H_{p-1}(K_0).$$

We may sometimes write ∂_* for ν_*.

We remark that, if $c_p \in C_p(K \bmod K_0)$ and $c'_p\mu = c_p$, then $c'_p - c_p\alpha = c'_p\beta\lambda$, so that $c'_p\partial\beta - c_p\nu = c'_p\beta\lambda\partial\beta = c'_p\beta\partial^0$. Thus, to obtain $\{z_p\}\nu_*$, we may extend z_p *in any way* to a chain of K, apply $\partial\beta$, and take the resulting homology class. The homomorphism α enables us to pick a convenient chain c'_p such that $c'_p\mu = c_p$. We may sometimes regard $c_p\alpha$ itself as a chain of $K \bmod K_0$, thereby identifying $C_p(K \bmod K_0)$ with $C_p(K \bmod K_0)\alpha$; we may also sometimes identify $C_p(K_0)$ with $C_p(K_0)\lambda$.

The homomorphisms λ_*, μ_*, ν_* play a fundamental role in homology theory. The following example may help the reader to visualize μ_* and ν_*.

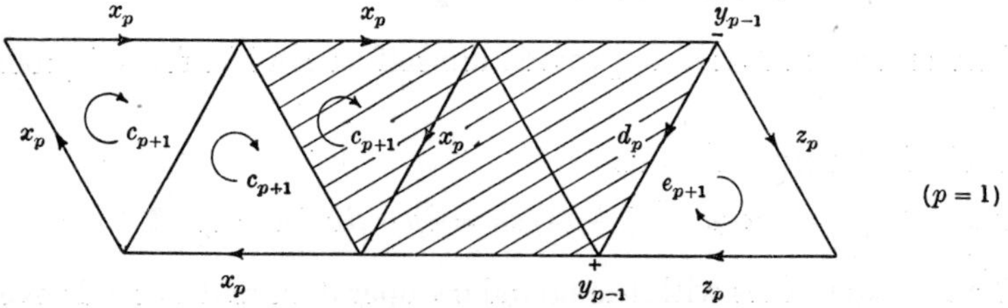

Fig. 2.4

Let K consist of six closed 2-simplexes (as in fig. 2.4) and let K_0 be the shaded subcomplex.

Let $x_p \in Z_p(K)$. Then $x_p\mu$ is a relative cycle of $K \bmod K_0$. If x_p bounds c_{p+1}, then $x_p\mu$ (x_p cut down to $K \bmod K_0$) bounds $c_{p+1}\mu$ in $K \bmod K_0$. Thus the homology class of $x_p\mu$ depends only on that of x_p.

Let $z_p \in Z_p(K \bmod K_0)$; as a chain of K it has a boundary $y_{p-1} = z_p\partial$ which lies in K_0 and is, indeed, a cycle of K_0. If z_p bounds in $K \bmod K_0$, say $z_p = e_{p+1}\hat{\partial}$, then $z_p - e_{p+1}\partial$ is a chain of K_0 and $z_p\partial = (z_p - e_{p+1}\partial)\partial$

‡ According to the usual definition, ν is a chain map lowering dimension by 1.

so that y_{p-1} bounds in K_0; more precisely, $y_{p-1} = d_p \partial^0$, where $d_p = z_p - e_{p+1}\partial$. Thus the homology class of y_{p-1} depends only on that of z_p and $z_p \to y_{p-1}$ induces ν_*. In describing the mechanics of the example we have, in effect, discussed the general case, since the figure simply identifies the chains described with certain chains of the complex we have drawn. Indeed, it is possible to formulate the homomorphism ν_* quite precisely along these lines, but we have preferred to present it as being induced by an explicit map of the chain group.

The whole discussion of this section may now be repeated with modifications for contrahomology. We shall describe how the modifications are to be made and state the results. Corresponding results in the two theories will be indicated, as in Eilenberg and Steenrod [14], by giving them the same index number with the 'contra-' statement carrying the affix *c*.

Let A, B be abelian groups. Let $A^{\cdot} = \mathrm{Hom}(A, G)$, $B^{\cdot} = \mathrm{Hom}(B, G)$, where G is a fixed abelian group, and let $\phi : A \to B$ be a homomorphism. Then a homomorphism $\phi^{\cdot} : B^{\cdot} \to A^{\cdot}$ is given by

2.8.18 $$(a, \phi^{\cdot} b^{\cdot}) = (a\phi, b^{\cdot}), \quad a \in A,\ b^{\cdot} \in B^{\cdot}.$$

Here we have written $(a, \phi^{\cdot} b^{\cdot})$ for $a(\phi^{\cdot} b^{\cdot})$ and $(a\phi, b^{\cdot})$ for $(a\phi)\, b^{\cdot}$ in view of our primary interest in the case in which A and B are chain groups, so that $A^{\cdot}$ and $B^{\cdot}$ are contrachain groups. Notice also that we write $\phi^{\cdot}$ as a *left* operator. We say that $\phi^{\cdot}$ is *adjoint* to ϕ, and will restrict the use of the notation $\phi^{\cdot}$ to the adjoint of ϕ; however, we normally continue to write δ for the contraboundary although $\delta = \partial^{\cdot}$.

2.8.19 Proposition. *If* $\phi : A \to B$, $\psi : B \to C$, *then* $(\phi\psi)^{\cdot} = \phi^{\cdot}\psi^{\cdot}$.

For if $a \in A$, $c^{\cdot} \in C^{\cdot}$, then

$$(a, (\phi\psi)^{\cdot} c^{\cdot}) = (a\phi\psi, c^{\cdot}) = (a\phi, \psi^{\cdot} c^{\cdot}) = (a, \phi^{\cdot}\psi^{\cdot} c^{\cdot}). \ \rule{0.4em}{0.8em}$$

2.8.20 Proposition. *If* $\phi = \{\phi_p\}$ *is a set of homomorphisms,* $\phi_p : C_p \to D_p$, *where* $\{C_p\}$, $\{D_p\}$ *are chain groups, and if* $\phi\partial = \partial\phi$, *where* ∂ *is the boundary operator in* C_p *or* D_p, *then* $\phi^{\cdot}\delta = \delta\phi^{\cdot}$.

For $\phi^{\cdot}\delta = \phi^{\cdot}\partial^{\cdot} = (\phi\partial)^{\cdot} = (\partial\phi)^{\cdot} = \partial^{\cdot}\phi^{\cdot} = \delta\phi^{\cdot}$. $\rule{0.4em}{0.8em}$

This proposition asserts that the adjoint of a chain map is a contrachain map (see chapter 3). Here we develop the ideas of chain and contrachain maps only as far as is required to discuss the homology and contrahomology of a pair (K, K_0).

It is now clear that we may translate all the results 2.8.1–2.8.17

into contrahomology. We recall that $C^p(K; G) = \mathrm{Hom}\,(C_p(K), G)$, $C^p(K_0; G) = \mathrm{Hom}\,(C_p(K_0), G)$; similarly we define

$$C^p(K \bmod K_0; G) = \mathrm{Hom}\,(C_p(K \bmod K_0), G).$$

We shall write ϕ^p for $\phi_p^{\cdot}$, δ_0 for $\partial^{0\cdot}$, $\hat{\delta}$ for $\hat{\partial}^{\cdot}$. Thus δ_0 is the contraboundary operator

$$\delta_0 : C^p(K_0; G) \to C^{p+1}(K_0; G)$$

and $\hat{\delta}$ is the contraboundary operator

2.8.11 c $$\hat{\delta} : C^p(K \bmod K_0; G) \to C^{p+1}(K \bmod K_0; G).$$

We have left homomorphisms

2.8.1 c $$\lambda^{\cdot} : C^p(K; G) \to C^p(K_0; G),$$

2.8.4 c $$\beta^{\cdot} : C^p(K_0; G) \to C^p(K; G),$$

2.8.7 c $$\alpha^{\cdot} : C^p(K; G) \to C^p(K \bmod K_0; G),$$

2.8.8 c $$\mu^{\cdot} : C^p(K \bmod K_0, G) \to C^p(K; G),$$

satisfying

2.8.5 c $$\lambda^{\cdot}\beta^{\cdot} = 1,$$

2.8.9 c $$\alpha^{\cdot}\mu^{\cdot} = 1,$$

2.8.10 c $$C^p(K; G) = \beta^{\cdot}C^p(K_0; G) \oplus \mu^{\cdot}C^p(K \bmod K_0; G),$$

$$\lambda^{\cdot}\mu^{\cdot} = 0, \quad \alpha^{\cdot}\beta^{\cdot} = 0, \quad \beta^{\cdot}\lambda^{\cdot} + \mu^{\cdot}\alpha^{\cdot} = 1.$$

The only point requiring comment is the direct sum decomposition of $C^p(K; G)$. This may, of course, easily be verified directly; however, we remark that if B is a direct factor of A, then $\mathrm{Hom}\,(B, G)$ may be embedded as a direct factor of $\mathrm{Hom}\,(A, G)$, the embedding being by composition with the projection $A \to B$.

2.8.2 c Theorem. $\lambda^{\cdot}\delta = \delta_0\lambda^{\cdot}$. ∎

2.8.13 c Theorem. $\mu^{\cdot}\hat{\delta} = \delta\mu^{\cdot}$. ∎

These results are consequences of Proposition 2.8.20. Then $\lambda^{\cdot}$ induces

2.8.3 c $$\lambda^* : H^p(K; G) \to H^p(K_0; G),$$

and $\mu^{\cdot}$ induces

2.8.14 c $$\mu^* : H^p(K \bmod K_0; G) \to H^p(K; G).$$

We have a homomorphism

2.8.15c $\quad \nu^{\cdot} = \alpha^{\cdot}\delta\beta^{\cdot} : C^p(K_0; G) \to C^{p+1}(K \bmod K_0; G).$

2.8.16c Theorem. $\nu^{\cdot}\delta_0 = -\hat{\delta}\nu^{\cdot}$. ❳

Thus $\nu^{\cdot}$ induces $\nu^* : H^p(K_0; G) \to H^{p+1}(K \bmod K_0; G)$. Just as for homology, we remark that, to obtain $\nu^*\{z^p\}$, we may extend z^p *in any way* to a contrachain of K, apply $\alpha^{\cdot}\delta$, and take the resulting contrahomology class.

We shall not usually be so scrupulous in repeating auxiliary results and concepts in contrahomology. However, we wish to emphasize the adjoint property at this stage; we advise the reader to think out *ab initio* the definitions of the various adjoint homomorphisms introduced; we point out, for example, that $\lambda^{\cdot}$ restricts a contrachain of K to the simplexes of K_0 and $\mu^{\cdot}$ extends a contrachain from $K \bmod K_0$ to K by giving it the value 0 on the simplexes of K_0. We may sometimes identify $C^p(K \bmod K_0; G)$ with its image under $\mu^{\cdot}$, and $C^p(K_0; G)$ with its image under $\beta^{\cdot}$. It is also worth noting that an open subcomplex is, in a sense, a more convenient notion for contrahomology than a closed subcomplex. This is indicated by 2.8.13c; the contraboundary of a contrachain of $K \bmod K_0$ is unaffected if the contrachain is regarded as belonging to K.

We close this section by proving the *Excision Theorem*‡ for complexes. Let $K_0 \cup K_1 = K$ and let the open subcomplex $K \bmod K_1$ be removed or *excised* from K. Then $K_1 \bmod (K_0 \cap K_1) = K \bmod K_0$, whence

2.8.21 $$C_p(K_1 \bmod (K_0 \cap K_1)) = C_p(K \bmod K_0).$$

2.8.22 Theorem. *If* K_0, K_1 *are subcomplexes of* K *such that* $K_0 \cup K_1 = K$, *then* $H_p(K_1 \bmod (K_0 \cap K_1)) = H_p(K \bmod K_0)$.

In view of 2.8.21 it is only necessary to show that the boundary operators in $K_1 \bmod (K_0 \cap K_1)$ and $K \bmod K_0$ coincide. The operators are $\alpha^1\partial^1\beta^1$ and $\alpha\partial\beta$ where we have a diagram

$$\begin{array}{ccc}
C_p(K_1 \bmod (K_0 \cap K_1)) & = & C_p(K \bmod K_0) \\
\downarrow{\scriptstyle \alpha^1} & & \downarrow{\scriptstyle \alpha} \\
C_p(K_1) & \xrightarrow{\lambda} & C_p(K) \\
\downarrow{\scriptstyle \partial^1} & & \downarrow{\scriptstyle \partial} \\
C_{p-1}(K_1) & \xrightarrow{\lambda} & C_{p-1}(K) \\
\downarrow{\scriptstyle \beta^1} & & \downarrow{\scriptstyle \beta} \\
C_{p-1}(K_1 \bmod (K_0 \cap K_1)) & = & C_{p-1}(K \bmod K_0)
\end{array}$$

‡ See Eilenberg and Steenrod [14], pp. 11, 31 et seq. See also chapter 8 of this book.

Now $\alpha^1\lambda = \alpha$, since the effect of either on a chain c_p of $K \bmod K_0$ is to regard c_p as a chain of K having coefficient 0 on the simplexes of K_0; $\lambda\partial = \partial^1\lambda$, since K_1 is a closed subcomplex of K; and $\lambda\beta = \beta^1$, since the effect of either on a chain c_{p-1} of K_1 is to restrict c_{p-1} to $K \bmod K_0$. We deduce that $\alpha^1\partial^1\beta^1 = \alpha\partial\beta$. ∎

2.8.22c Theorem. $H^p(K_1 \bmod (K_0 \cap K_1); G) = H^p(K \bmod K_0; G)$. ∎

2.9 The exact sequences

The homomorphisms λ_*, μ_*, ν_* (2.8.3, 2.8.14, 2.8.17) are defined for all integers p. Thus we may describe a *sequence* of groups and homomorphisms

2.9.1 $$\ldots \xrightarrow{\nu_*} H_p(K_0) \xrightarrow{\lambda_*} H_p(K) \xrightarrow{\mu_*} H_p(K \bmod K_0) \xrightarrow{\nu_*} H_{p-1}(K_0) \xrightarrow{\lambda_*} \ldots.$$

This sequence plays a fundamental role in homology theory and is called the *homology sequence* of the pair (K, K_0). It terminates in the unaugmented case with

2.9.2 $$\ldots H_0(K_0) \xrightarrow{\lambda_*} H_0(K) \xrightarrow{\mu_*} H_0(K \bmod K_0) \to 0,$$

and in the augmented case with

2.9.3 $$\ldots H_0(K_0^+) \xrightarrow{\lambda_*} H_0(K^+) \xrightarrow{\mu_*} H_0(K \bmod K_0)$$
$$\xrightarrow{\nu_*} H_{-1}(K_0^+) \xrightarrow{\lambda_*} H_{-1}(K^+) \to 0;$$

that is to say, all subsequent terms are zero. In fact, unless K_0 is the empty complex, both sequences terminate at $H_0(K \bmod K_0)$. For if K_0 is not the empty complex, then K is also not the empty complex and σ_{-1} bounds in K_0^+ and K^+, so that $H_{-1}(K_0^+) = H_{-1}(K^+) = 0$.

We shall establish the crucial property of the sequence 2.9.1, namely, that it is exact. We recall from I. 2 that a sequence of groups and homomorphisms
$$\ldots \to G_{i+1} \xrightarrow{\phi_{i+1}} G_i \xrightarrow{\phi_i} G_{i-1} \to \ldots$$
is exact at G_i if the kernel of ϕ_i coincides with the image of ϕ_{i+1}; and that the sequence is exact if it is exact at each G_i (other than the first and last).

Before proving that 2.9.1 is exact, we draw attention to certain interesting consequences of exactness.

2.9.4 Proposition. *Let* $G_{i+1} \overset{\phi_{i+1}}{\to} G_i \overset{\phi_i}{\to} G_{i-1} \overset{\phi_{i-1}}{\to} G_{i-2}$ *be exact. Then the following statements are equivalent:* (i) ϕ_{i+1} *is epimorphic*; (ii) $\phi_i = 0$; (iii) ϕ_{i-1} *is monomorphic.*

For ϕ_{i+1} is epimorphic if and only if $\phi_i^{-1}(0) = G_i$, that is, $\phi_i = 0$. Further, $\phi_i = 0$ if and only if $\phi_{i-1}^{-1}(0) = 0$, that is, ϕ_{i-1} is monomorphic. ∎

2.9.5 Corollary. *Let* $G_{i+2} \overset{\phi_{i+2}}{\to} G_{i+1} \overset{\phi_{i+1}}{\to} G_i \overset{\phi_i}{\to} G_{i-1} \overset{\phi_{i-1}}{\to} G_{i-2}$ *be exact and let* $G_i = 0$. *Then* ϕ_{i-1} *is monomorphic and* ϕ_{i+2} *is epimorphic.*

For $\phi_i = 0$ and $\phi_{i+1} = 0$. ∎

2.9.6 Corollary. *Let* $0 \to A \overset{\phi}{\to} B \to 0$ *be exact. Then* ϕ *is an isomorphism.* ∎

2.9.7 Theorem. *The homology sequence*‡ 2.9.1 *is exact.*

Before proving this we remark that this theorem is really a special case of a much more general result proved in§ chapter 5. However, we give a proof here for completeness (and immediate availability). The proof will be expressed in informal language, in the sense that the identifications referred to in the previous section will generally be made.‖

To prove exactness at G_i in $\ldots \to G_{i+1} \overset{\phi_{i+1}}{\to} G_i \overset{\phi_i}{\to} G_{i-1} \to \ldots$ one has to prove two inclusions, namely,

$$G_{i+1}\phi_{i+1} \subseteq \phi_i^{-1}(0) \quad \text{and} \quad \phi_i^{-1}(0) \subseteq G_{i+1}\phi_{i+1}.$$

The former is often easy and the latter usually harder. To prove the exactness of 2.9.1, we first divide the proof into three parts: (i) exactness at $H_p(K_0)$, (ii) exactness at $H_p(K)$, and (iii) exactness at $H_p(K \bmod K_0)$. Then each part subdivides into two parts as shown above.

(i) *Exactness at* $H_p(K_0)$. We have

$$H_{p+1}(K \bmod K_0) \overset{\nu_*}{\to} H_p(K_0) \overset{\lambda_*}{\to} H_p(K).$$

Let $z_{p+1} \in Z_{p+1}(K \bmod K_0)$; then $\{z_{p+1}\}\nu_*$ is the homology class of

‡ This assertion holds in both the augmented and the unaugmented cases.

§ Theorem 5.5.1.

‖ We use λ, α to identify chains of K_0, $K \bmod K_0$, with chains of K; we also identify ∂^0 with ∂.

$z_{p+1}\partial$ in K_0, and $\{z_{p+1}\}\nu_*\lambda_*$ is the homology class of $z_{p+1}\partial$ in K. Thus $\{z_{p+1}\}\nu_*\lambda_* = 0$ and

$$H_{p+1}(K \bmod K_0)\,\nu_* \subseteq \lambda_*^{-1}(0).$$

Conversely, suppose $\{z_p\} \in H_p(K_0)$ and $\{z_p\}\lambda_* = 0$. Then $z_p\lambda$ is a boundary, $z_p\lambda = c_{p+1}\partial$ and c_{p+1} is a cycle of $K \bmod K_0$ such that $\{c_{p+1}\}\nu_* = \{z_p\}$. Thus

$$\lambda_*^{-1}(0) \subseteq H_{p+1}(K \bmod K_0)\,\nu_*.$$

(ii) *Exactness at* $H_p(K)$. We have

$$H_p(K_0) \xrightarrow{\lambda_*} H_p(K) \xrightarrow{\mu_*} H_p(K \bmod K_0).$$

Since $\lambda\mu = 0$, it is obvious that $\lambda_*\mu_* = 0$, so that

$$H_p(K_0)\,\lambda_* \subseteq \mu_*^{-1}(0).$$

Conversely, suppose $\{z_p\} \in H_p(K)$ and $\{z_p\}\mu_* = 0$. Then

$$z_p\mu = c_{p+1}\hat{\partial}, \quad c_{p+1} \in C_{p+1}(K \bmod K_0).$$

Then $z_p - c_{p+1}\partial \in C_p(K_0)$, indeed $\in Z_p(K_0)$. It follows that

$$\{z_p\} = \{z_p - c_{p+1}\partial\} \in H_p(K_0)\,\lambda_*,$$

so that
$$\mu_*^{-1}(0) \subseteq H_p(K_0)\,\lambda_*.$$

(iii) *Exactness at* $H_p(K \bmod K_0)$. We have

$$H_p(K) \xrightarrow{\mu_*} H_p(K \bmod K_0) \xrightarrow{\nu_*} H_{p-1}(K_0).$$

Let $\{z_p\} \in H_p(K)$. Then $\{z_p\}\mu_*\nu_*$ is the homology class in K_0 of $z_p\partial = 0$. Thus

$$H_p(K)\,\mu_* \subseteq \nu_*^{-1}(0).$$

Conversely, suppose $\{y_p\} \in H_p(K \bmod K_0)$ and $\{y_p\}\nu_* = 0$. Then $y_p\partial$ is a boundary in K_0, $y_p\partial = c_p\partial$, $c_p \in C_p(K_0)$. Then $y_p - c_p \in Z_p(K)$ and $\{y_p - c_p\}\mu_* = \{y_p\}$. Thus

$$\nu_*^{-1}(0) \subseteq H_p(K)\,\mu_*. \;\blacksquare$$

The reader will observe that the dimension suffix p is unrestricted in the argument, so that the exactness holds right down to the bottom dimension. Since the sequence is of interest only if K_0 is non-empty, we may think of it as terminating $\ldots \to H_0(K_0^+) \xrightarrow{\lambda_*} H_0(K^+) \xrightarrow{\mu_*} H_0(K \bmod K_0) \to 0$ in the augmented case.

2.9.7c Theorem. *The sequence*

$$\ldots \leftarrow H^p(K_0; G) \overset{\lambda^*}{\leftarrow} H^p(K; G) \overset{\mu^*}{\leftarrow} H^p(K \bmod K_0; G) \overset{\nu^*}{\leftarrow} H^{p-1}(K_0; G) \overset{\lambda^*}{\leftarrow} \ldots$$

is exact.]

The reader is strongly advised to provide himself with a proof. We remark that $H^{-1}(K^+; G) = 0$ unless K is empty. For if c^{-1} is the (-1)-contrachain determined by $(\sigma_{-1}, c^{-1}) = g$, then $(a, \delta c^{-1}) = g$ for each vertex a, so that c^{-1} is a contracycle if and only if $g = 0$. Thus $Z^{-1}(K^+; G) = 0$. It follows that if neither K nor K_0 is empty, then the sequence commences

$$\ldots \leftarrow H^0(K_0; G) \overset{\lambda^*}{\leftarrow} H^0(K; G) \overset{\mu^*}{\leftarrow} H^0(K \bmod K_0; G) \leftarrow 0$$

or $$\ldots \leftarrow H^0(K^+; G) \overset{\lambda^*}{\leftarrow} H^0(K^+; G) \overset{\mu^*}{\leftarrow} H^0(K \bmod K_0; G) \leftarrow 0.$$

Theorem 2.9.7 constitutes an axiom in the axiomatic homology theory of Eilenberg and Steenrod [14]. There and in most books the relative homology groups are written

$$H_p(K, K_0) \quad \text{(similarly, } H^p(K, K_0; G)\text{)},$$

and we shall generally adopt this shortened form in the future. We remark that the homology sequence can be defined for an arbitrary coefficient group G and the proof of exactness is formally unaffected.

2.10 Homology groups of certain complexes

In this section we compute some homology and contrahomology groups. The exact homology sequence is one of the tools we use in the computations.

Let K_0 be a complex and let $K = \hat{K}_0$ be the cone on K_0 with vertex a. If K_0 is oriented we may extend the orientation to K by agreeing that $aa^0 \ldots a^p$ is an oriented simplex of K if $a^0 \ldots a^p$ is an oriented simplex of K_0. We write

2.10.1 $$\sigma = a\tau,$$

where $\tau = a^0 \ldots a^p$, $\sigma = aa^0 \ldots a^p$. Then the notation 2.10.1 may be extended in an obvious way to the chain group of K with coefficients in an abelian group G. We write

2.10.2 $$c_{p+1} = ad_p,$$

where $d_p = \Sigma g_i \tau^i_p$, $\tau^i_p \in K_0$ and $c_{p+1} = \Sigma g_i(a\tau^i_p) \in C_{p+1}(K; G)$. Then from 2.10.1 we have

2.10.3 $$\sigma\partial = \tau - a(\tau\partial),$$

and, more generally,

2.10.4 $$(ad_p)\partial = d_p - a(d_p\partial).$$

2.10.5 Theorem. *If K is a cone, then $H_p(K^+; G)$ and $H^p(K^+; G)$ are zero for all p.*

Now any $c_p \in C_p(K^+; G)$ is uniquely expressible as

2.10.6 $$c_p = ad_{p-1} + e_p,$$

$d_{p-1} \in C_{p-1}(K_0^+; G)$, $e_p \in C_p(K_0^+; G)$. If $p = 0$, $d_{p-1} = g\sigma_{-1}$ for some $g \in G$ and $a\sigma_{-1} = a$ (hence the notation $\sigma_{-1} = 1$). If $p = -1$, $d_{p-1} = 0$.

Then if c_p is a cycle,

$$0 = c_p\partial = (ad_{p-1})\partial + e_p\partial = d_{p-1} - a(d_{p-1}\partial) + e_p\partial.$$

By the uniqueness of 2.10.6 we deduce that $d_{p-1} + e_p\partial = 0$ (and $d_{p-1}\partial = 0$). Thus $(ae_p)\partial = e_p - a(e_p\partial) = e_p + ad_{p-1} = c_p$, and c_p is a boundary. This proves that $H_p(K^+; G) = 0$, all p.

The proof that $H^p(K^+; G) = 0$ is analogous. Given $d^p \in C^p(K_0^+; G)$ we define $ad^p \in C^{p+1}(K^+; G)$ by

2.10.2c $$(\sigma_{p+1}, ad^p) = 0,\ \sigma_{p+1} \in K_0^+,$$
$$(a\tau_p, ad^p) = (\tau_p, d^p),\quad \tau_p \in K_0^+.$$

Then we verify that‡ $\delta(ad^p) = -a(\delta_0 d^p)$,

2.10.4c $$\delta d^p = \delta_0 d^p + ad^p,\quad d^p \in C^p(K_0^+; G);$$

here we identify d^p with $\beta^{\cdot} d^p \in C^p(K^+; G)$. There is a decomposition

2.10.6c $$c^p = ad^{p-1} + e^p,$$

$d^{p-1} \in C^{p-1}(K_0^+; G)$, $e^p \in C^p(K_0^+; G)$, and the proof proceeds as in the homology case. The details are left to the reader. ∎

We have stated this result in terms of the augmented complex. It could, of course, have been stated, but rather less conveniently, for the unaugmented complex.

2.10.7 Corollary. *If K is a cone on K_0,*

$$\nu_* : H_p(K, K_0; G) \cong H_{p-1}(K_0^+; G),$$
$$\nu^* : H^{p-1}(K_0^+; G) \cong H^p(K, K_0; G).$$

This follows from the exact sequences and Corollary 2.9.6. ∎

‡ Recall that δ_0 is the contraboundary in K_0.

2.10.8 Corollary. *If L is a complex, L_0 a subcomplex and if a is a vertex independent of L, then*

$$H_p(L, L_0; G) \cong H_p(L^+ \cup aL_0; G), \quad H^p(L, L_0; G) \cong H^p(L^+ \cup aL_0; G).$$

We may think of $L \cup aL_0$ as arising from L by erecting a cone on L_0; if L is realized in R^m, we may take $a \in R^{m+1} - R^m$.

Put $L \cup aL_0 = K$, $aL_0 = K_0$. Now $\operatorname{st}(a)$ is an open subcomplex of K lying in K_0. By the Excision Theorem 2.8.22 it may be removed from K and K_0 without affecting the homology of K mod K_0. Now $K - \operatorname{st}(a) = L$, $K_0 - \operatorname{st}(a) = L_0$, so that

$$H_p(K, K_0) = H_p(L, L_0).$$

But $H_p(K, K_0) \cong H_p(K^+)$—we apply 2.9.6 to the exact homology sequence of the pair K, K_0. It follows that

$$H_p(L, L_0) \cong H_p(K^+) = H_p(L^+ \cup aL_0).$$

The proof for contrahomology is similar. ▌

In the proof of this corollary we have not specified the coefficient or value group; this will be our custom if it is arbitrary, and this fact has been made clear by the enunciation. Of course, Theorem 2.8.22 holds for arbitrary coefficient groups.

Let K, L be two complexes; we may extend the notion of *join* to define the join of K and L, assuming the vertices of K are distinct from those of L. We may also extend, in an obvious way, the device of orienting aL from an orientation of L to provide ourselves with an orientation of the join KL, based on orientations of K and L. The reader will easily provide precise definitions.

2.10.9 Theorem.

$$H_p(\dot{s}_q L^+; G) \cong H_{p-q}(L^+; G), \quad H^p(\dot{s}_q L^+; G) \cong H^{p-q}(L^+; G).$$

Let s_r, $0 \leqslant r \leqslant q$, be the simplex $b^0 b^1 \dots b^r$. We remark first that $\bar{s}_{q-1} \cup b^q \dot{s}_{q-1} = \dot{s}_q$, so that

$$(\bar{s}_{q-1} \cup b^q \dot{s}_{q-1}) L^+ = \dot{s}_q L^+.$$

Thus we may apply 2.10.8; the L, L_0, K, K_0 of that corollary being replaced here by $\bar{s}_{q-1}L$, $\dot{s}_{q-1}L$, $\dot{s}_q L$, $b^q \dot{s}_{q-1}L$. We conclude that

$$H_p(\dot{s}_q L^+) \cong H_p(\bar{s}_{q-1}L, \dot{s}_{q-1}L).$$

Now $\bar{s}_{q-1}L$ is obviously a cone, so that we deduce from the homology sequence of the pair $(\bar{s}_{q-1}L^+, \dot{s}_{q-1}L^+)$ that

$$H_p(\bar{s}_{q-1}L, \dot{s}_{q-1}L) \cong H_{p-1}(\dot{s}_{q-1}L^+),$$

whence

2.10.10 $$H_p(\dot{s}_q L^+) \cong H_{p-1}(\dot{s}_{q-1} L^+).$$

We may repeat the argument with $(q-1)$ replacing q; continuing in this way we eventually prove that $H_p(\dot{s}_q L^+) \cong H_{p-q}(\dot{s}_0 L^+)$. But $\dot{s}_0$ is the empty complex so that $\dot{s}_0 L^+ = L^+$ and the homology part of the theorem is proved. The contrahomology part is proved similarly. ❙

2.10.11 Corollary. $H_p(\dot{s}_q^+) = 0, \quad p \neq q-1; \quad H_{q-1}(\dot{s}_q^+) = J;$

$$H^p(\dot{s}_q^+) = 0, \; p \neq q-1; \quad H^{q-1}(\dot{s}_q^+) = J.$$

Put $L =$ empty complex; then

$$H_{-1}(L^+) = H^{-1}(L^+) = J, \quad H_n(L^+) = H^n(L^+) = 0, \quad n \neq -1. \;❙$$

If s_q is a simplex of K, then $\operatorname{st} s_q$ is, in a generalized sense, the join of s_q with some augmented subcomplex, K_0^+, of K^+. We call K_0 the *linked complex* of s_q in K. Then $\overline{\operatorname{st}(s_q)} = \bar{s}_q K_0$; and, if $\operatorname{st}(s_q)^{\boldsymbol{\cdot}}$ means $\overline{\operatorname{st}(s_q)} - \operatorname{st}(s_q)$, then $\operatorname{st}(s_q)^{\boldsymbol{\cdot}} = \dot{s}_q K_0$. Significant local properties‡ of $|K|$ can be defined in terms of the relative homology group

$$H_p(\overline{\operatorname{st}(s_q)}, \operatorname{st}(s_q)^{\boldsymbol{\cdot}}).$$

Since $\bar{s}_q K_0$ is a cone, we infer from the homology sequence and Theorem 2.10.9

2.10.12 Corollary. $H_p(\overline{\operatorname{st}(s_q)}, \operatorname{st}(s_q)^{\boldsymbol{\cdot}}) \cong H_{p-q-1}(K_0^+)$. ❙

Thus local homology properties 'near s_q' are determined by the absolute homology of the linked complex.

2.11 Homology and contrahomology in infinite complexes

If K is an infinite complex, we define the *finite p-chains* or *p-chains* of K with coefficients in G to be finite linear combinations of the oriented p-simplexes of K with coefficients in G. We write the group $C_p(K; G)$, abbreviating $C_p(K; J)$ to $C_p(K)$. Thus $C_p(K)$ is, as in the case of a finite complex, the free abelian group freely generated by the oriented p-simplexes of K.

It is clear that 2.2.7 furnishes a boundary operator

$$\partial : C_p(K; G) \to C_{p-1}(K; G)$$

‡ See 3.7.

such that $\partial\partial = 0$. We may thus define $Z_p(K; G)$, $B_p(K; G)$, $H_p(K; G)$ just as for finite complexes. We call $H_p(K; G)$ the *p-th homology group of K* with coefficients in G.

It is now natural to define $C^p(K; G)$ as $\mathrm{Hom}\,(C_p(K); G)$. We then call $C^p(K; G)$ the group of *p-contrachains of K* with values in G. It is important to observe that the elements of $C^p(K; G)$ are *infinite* contrachains, in the sense that, in general, they take non-zero values on infinitely many p-simplexes. However, we see that C_p and C^p stand in the required adjoint relationship to each other; we define

$$\delta : C^p(K; G) \to C^{p+1}(K; G)$$

as the adjoint of ∂; that is, to satisfy

$$(c_{p+1}, \delta c^p) = (c_{p+1}\partial, c^p), \quad c_{p+1} \in C_{p+1}(K), \quad c^p \in C^p(K; G);$$

and we obtain groups $Z^p(K; G)$, $B^p(K; G)$, $H^p(K; G)$. We call $H^p(K; G)$ the *p-th contrahomology group of K* with values in G.

All the developments of this chapter apply to this homology and contrahomology theory; we can form Kronecker products between integral homology classes and contrahomology classes and define weak homology and contrahomology classes; we may relativize the theory and we have exact homology and contrahomology sequences. We may consider the augmented K^+ as well as K.

It is natural to ask whether infinite chains yield a homology theory. Rather than consider this in the restricted context of simplicial complexes, we prefer to consider in the next section a more general class of objects, called *cell complexes*. The results of that section apply to simplicial complexes and the homology theory just defined for an infinite simplicial complex coincides with the homology theory 'of the first kind' in the terminology of the next section; one of the results of the next section implies that for a general infinite simplicial complex there is another set of homology and contrahomology groups, those 'of the second kind', distinct from those defined in this section.

2.12* Abstract cell complexes‡

We have taken the view hitherto that a complex describes a topological space called its polyhedron. In the case of an infinite simplicial complex, however, the complex was defined first and a geometrical object constructed from it. We now introduce a generalization of an (abstract) simplicial complex, namely, an *abstract cell complex*,

‡ For a further treatment see S. Wylie, 'Algebraic Geometry and Topology' (Lefschetz symposium), pp. 389–99.

with which we do not associate a topological space. Nonetheless, of course, the motive for considering such a structure lies in the application to geometrical situations and, in particular, those in which the relatum of a cell is not a simplex. For example, the space may be divided up into cubes instead of simplexes, or, more generally, into blocks or chunks of the space whose closures contain blocks of lower dimension themselves corresponding to cells. Actually, it is more convenient to cut out an intermediate step present in the passage from simplicial complexes to homology theory and to regard the cells of an abstract cell complex as generalizing the oriented simplexes of an *oriented* simplicial complex. We now give the precise definition.

An *abstract cell complex* is a collection, K, of objects, ξ, called *abstract cells*. To each cell ξ is assigned an integer p called its *dimension*, and we write ξ_p if the dimension is to be explicit. An order relation $\prec$ is given in K (we say ξ is a *face* of ξ' if $\xi \prec \xi'$); and to each pair of cells ξ, ξ' of consecutive dimension there is defined an integer $[\xi, \xi']$ called the *incidence number*. The objects, relation and integers satisfy the following axioms.

AC 1: $\prec$ is a strict partial ordering (see I. 3).

AC 2: $\xi_p \prec \xi'_q$ and $\xi_p \neq \xi'_q$ imply $p < q$.

AC 3: For each pair ξ, ξ'' there exists only a finite number of cells ξ' with $\xi \prec \xi' \prec \xi''$.

AC 4: $[\xi_p, \xi'_{p+1}] \neq 0$ implies $\xi_p \prec \xi'_{p+1}$.

AC 5: For each pair ξ_{p-1}, ξ''_{p+1}, $\sum_{\xi'_p} [\xi_{p-1}, \xi'_p][\xi'_p, \xi''_{p+1}] = 0$.

Notice that AC 3 and AC 4 ensure that the sum in AC 5 is finite. It turns out that, to obtain a homology theory on an infinite complex, care must be taken in choosing the chain group. Axioms involving the facing relation $\prec$ are relevant to that choice and AC 5 is then designed to ensure that '$\partial\partial = 0$'. If AC 5 were strengthened to ensure the finiteness of the sum, we could then *define* a relation $\prec'$ by saying that $\xi_p \prec' \xi'_{p+1}$ if $[\xi_p, \xi'_{p+1}] \neq 0$ and thus generate an order relation $\prec'$ by transitivity. However, it would not necessarily coincide with the given facing relation since we allow $[\xi_p, \xi'_{p+1}] = 0$ even where ξ_p is a face of ξ'_{p+1}.

The set of axioms we have given is intermediate between those of Tucker ('An abstract approach to manifolds', *Ann. Math.* **34** (1933), 191) and those of Lefschetz (*Algebraic Topology*, ch. 3). The novel feature is axiom AC 3, the axiom of *intercept finiteness*; it is less restrictive than Tucker's condition of overall finiteness, but our axioms are more restrictive than those of Lefschetz. However, they do not appear to exclude any interesting cases; star- and closure-finite complexes (Lefschetz) are included.

AC 2 appear to us to play no part in the subsequent discussion; it is included simply on the grounds that it is valid for all cell complexes of geometrical interest.

A *subcomplex* of an abstract cell complex K is a subcollection K_0 of its cells such that for each $\xi, \xi'' \in K_0$, $\xi \prec \xi' \prec \xi''$ implies $\xi' \in K_0$. Manifestly a subcomplex satisfies AC 1–5.

Given $\xi \in K$, the *star* of ξ, $\mathrm{st}\,\xi$, is the set of all cells ξ' with $\xi \prec \xi'$; and the *closure* of ξ, $\mathrm{cl}\,\xi$, is the set of cells ξ' with $\xi' \prec \xi$.

2.12.1 Proposition. $\mathrm{st}\,\xi$ *and* $\mathrm{cl}\,\xi$ *are subcomplexes.*]

We say that K_0 is an *open* subcomplex of K if, for each $\xi \in K_0$, $\mathrm{st}\,\xi \subseteq K_0$. Similarly, K_0 is a *closed* subcomplex of K if, for each $\xi \in K_0$, $\mathrm{cl}\,\xi \subseteq K_0$. Then certainly $\mathrm{st}\,\xi$ is open and $\mathrm{cl}\,\xi$ is closed. The complement of a closed (open) subcomplex is open (closed).

2.12.2 Proposition. *The intersection of any set of subcomplexes is a subcomplex.*]

2.12.3 Proposition. *The union and intersection of any set of open (closed) subcomplexes are open (closed) subcomplexes.*]

The reader will notice that the term subcomplex, applied to a cell complex, is wider than the same term applied to a simplicial complex. The subcomplexes of the latter are all closed. The subcomplex $\mathrm{st}\,\xi \cap \mathrm{cl}\,\xi'$ is, in general, neither open nor closed; if $\xi \prec \xi'$ this is the *intercept* determined by ξ, ξ' (Tucker). AC 3 ensures that all intercepts are finite. Of course, $\mathrm{st}\,\xi \cap \mathrm{cl}\,\xi'$ is empty if ξ is not a face of ξ'.

A cell complex K is said to be *star-finite* if, for each $\xi \in K$, $\mathrm{st}\,\xi$ is a finite complex; similarly, it is *closure-finite* if $\mathrm{cl}\,\xi$ is always a finite complex. If it is both star- and closure-finite it is called *locally finite.* A simplicial complex is always closure-finite, but not necessarily star-finite; this fact causes a certain loss of duality in the theory of infinite simplicial complexes.

Let L be a simplicial complex and L_0 a subcomplex. Then the oriented simplexes of $L \bmod L_0$ form a cell complex. It is open and closure-finite, but it is not a simplicial complex, in general. We now given an example of a cell complex which is not closure-finite.

In the Euclidean plane with coordinates x, y, consider the infinite strip $0 < y < 1$; with this, positively oriented, we associate a 2-cell ξ_2. With each integer n we associate a 1-cell ξ_1^n corresponding to the segment $n < x < n+1$, $y = 0$ oriented from n to $n+1$, and a 0-cell ξ_0^n, corresponding to the point $(n, 0)$. Incidence relations and numbers are those suggested by the geometric picture; the complex K corresponds to the strip $0 \leqslant y < 1$. Then K is star-finite but not closure-finite, since $K = \mathrm{cl}\,\xi_2$ is infinite.

We now turn to the question of constructing homology theories for K. We define $C_p(K; G)$ to be the group of linear combinations of the p-cells with coefficients in G with no finiteness assumption (this is a direct product in the group theoretic sense); and we define $C^p(K; G)$ to be the group of functions defined on the p-cells of K with values in G. Also we define the boundary of a cell by

2.12.4
$$\xi_p \partial = \sum_{\xi'_{p-1} \in K} [\xi'_{p-1}, \xi_p]\, \xi'_{p-1},$$

so that ∂ is a function from the p-cells of K to $C_{p-1}(K; J)$.

In general, we cannot extend ∂ to operate on $C_p(K; G)$ because we get involved in infinite sums in G. Moreover, in the absence of closure-finiteness, we cannot restrict attention to finite chains since the boundary of a finite chain would generally be infinite. Thus we may proceed in two different ways: we may enlarge the group of finite chains so that it is closed‡ under ∂, or we may reduce the group of infinite chains so that it always admits ∂.

We define the *carrier* $\bar{c}_p$ of the chain c_p as $\bigcup \operatorname{cl} \xi_p$, the union being taken over all ξ_p with non-zero coefficient in c_p. By 2.12.3, $\bar{c}_p$ is a closed subcomplex of K. By axiom AC 4 it follows that

$$\overline{\xi\partial} \subseteq \operatorname{cl} \xi,$$

for each $\xi \in K$.

2.12.5 Definition. c_p is a *chain of the first kind* or a *globally finite chain* if $\bar{c}_p$ has finite intersection with all open closure-finite subcomplexes of K.

Since $\overline{c^1_p - c^2_p} \subseteq \overline{c^1_p} \cup \overline{c^2_p}$, it is clear that the chains of the first kind form a group $C^{(1)}_p(K; G)$. Also if K is closure-finite then a globally finite c_p must be such that $\bar{c}_p$ is finite, so that the group $C^{(1)}_p(K; G)$ is just the group of finite chains. An infinite simplicial complex is closure-finite and $C^{(1)}_p(K; G)$ then coincides with the chain group of 2.11.

2.12.6 Definition. c_p is a *chain of the second kind* or a *locally finite chain* if $\bar{c}_p$ is star-finite.

Equivalently, c_p is locally finite if $\bar{c}_p \cap \operatorname{st} \xi$ is finite (or vacuous) for every $\xi \in K$. Again it is clear that the chains of the second kind form a group $C^{(2)}_p(K; G)$. Also if K is star-finite every c_p is locally finite so that $C^{(2)}_p(K; G) = C_p(K; G)$. If K is an infinite simplicial complex, c_p is a chain of the second kind if and only if $|c_p|$ is locally compact.

‡ We think of the 'group of chains' as $\sum_p C_p(K; G)$; thus we may speak precisely of its being closed under ∂.

2.12.7 Proposition. $C_p^{(1)}(K; G) \subseteq C_p^{(2)}(K; G)$.

For any c_p, $\bar{c}_p \cap \operatorname{st} \xi$ is vacuous if $\xi \notin \bar{c}_p$. Suppose $c_p \in C_p^{(1)}(K; G)$ and $\xi \in \bar{c}_p$. Now, by axiom AC 3, $\operatorname{st} \xi$ is itself closure-finite, so that $\bar{c}_p \cap \operatorname{st} \xi$ is finite, and $c_p \in C_p^{(2)}(K; G)$. ∎

Obviously $c_p \partial$ is defined for $c_p \in C_p^{(2)}(K; G)$, since no ξ_{p-1} can be a face of infinitely many $\xi_p \in \bar{c}_p$. Moreover

$$\overline{c_p \partial} \subseteq \bar{c}_p.$$

From this it follows that both $C_p^{(1)}$ and $C_p^{(2)}$ are closed under ∂. Now if K_0 is an open closure-finite subcomplex of K it follows immediately that $\operatorname{cl} \xi \cap K_0$ is finite for all $\xi \in K$. Thus $\xi_p \partial \in C_{p-1}^{(1)}(K; J)$ and we may reapply ∂. The coefficient of ξ''_{p-2} in $\xi_p \partial\partial$ is $\sum_{\xi'_{p-1}} [\xi''_{p-2}, \xi'_{p-1}][\xi'_{p-1}, \xi_p] = 0$, so that $\xi_p \partial\partial = 0$. More generally,

2.12.8 Theorem. *The boundary operator ∂ is a homomorphism*

$$\partial : C_p^{(2)}(K; G), C_p^{(1)}(K; G) \to C_{p-1}^{(2)}(K; G), C_{p-1}^{(1)}(K; G)$$

and $\partial\partial = 0$. ∎

We can now define cycles, boundaries and homology groups of the two kinds; the homology groups are written $H_p^{(1)}(K; G)$, $H_p^{(2)}(K; G)$.

The contrachains of the two kinds are now defined in a way completely dual to that for chains. The *carrier* $\underline{c}^p$ of a contrachain $c^p \in C^p(K; G)$ is $\bigcup \operatorname{st} \xi_p$, the union being taken over all ξ_p with $(\xi_p, c^p) \neq 0$.

2.12.5c Definition. c^p is a *contrachain of the first kind* or a *locally finite contrachain* if $\underline{c}^p$ is closure-finite.

2.12.6c Definition. c^p is a *contrachain of the second kind* or a *globally finite contrachain* if $\underline{c}^p$ has finite intersection with all closed star-finite subcomplexes of K.

We write $C^p_{(1)}(K; G)$, $C^p_{(2)}(K; G)$ for the contrachain groups of the two kinds and we have

$$C^p(K; G) \supseteq C^p_{(1)}(K; G) \supseteq C^p_{(2)}(K; G).$$

If K is closure-finite, $C^p(K; G) = C^p_{(1)}(K; G)$; and if K is star-finite, $C^p_{(2)}(K; G)$ is the group of finite contrachains. If K is an infinite simplicial complex, $C^p_{(1)}(K; G)$ is then the contrachain group of 2.11; $C^p_{(2)}(K; G)$ has no easy description.

We may define a Kronecker index between $C_p^{(1)}(K; J)$ and $C^p_{(1)}(K; G)$. For if $c_p = \sum_\alpha m_\alpha \xi_p^\alpha$, the definition $(c_p, c^p) = \sum_\alpha m_\alpha (\xi_p^\alpha, c^p)$ makes sense provided $\bar{c}_p \cap \underline{c}^p$ is finite. If $c^p \in C^p_{(1)}$ then $\underline{c}^p$ is an open closure-finite subcomplex, so that $\bar{c}_p \cap \underline{c}^p$ is finite if $c_p \in C^{(1)}_p$. In fact, $C^p_{(1)}(K; J)$ *is the set of all p-contrachains for which a Kronecker index is defined with* $C^{(1)}_p(K; J)$, *and conversely*. Similar statements hold, of course, for chains and contrachains of the second kind. In particular $(\xi_{p+1}\partial, c^p)$ is defined if $c^p \in C^p_{(1)}(K; G)$ and we define δc^p by

$$(\xi_{p+1}\partial, c^p) = (\xi_{p+1}, \delta c^p).$$

We easily prove that $\underline{\delta c}^p \subseteq \underline{c}^p$, whence it follows that

$$\delta C^p_{(2)}(K; G) \subseteq C^{p+1}_{(2)}(K; G).$$

Thus we can define contracycles, contraboundaries and contrahomology groups of the two kinds; the contrahomology groups are written $H^p_{(1)}(K; G)$, $H^p_{(2)}(K; G)$ and Kronecker products are defined between integral homology groups and contrahomology groups of the same kind.

EXERCISES

1. If K is a finite complex with incidence matrices $\boldsymbol{\partial}_1, \boldsymbol{\partial}_2 \dots$, we may associate with an integral p-chain $c_p = \sum_i m_i \sigma^i_p$ the row-vector $m = (m_1, m_2, \dots)$, and with the integral p-contrachain d^p that takes for each i the value n_i on σ^i_p the column-vector $n = \begin{pmatrix} n_1 \\ n_2 \\ \vdots \end{pmatrix}$. Express the boundary and contraboundary operations in terms of matrices and use these expressions to prove that

$$(c_p, \delta d^p) = (c_p \partial, d^p).$$

2. Consider what obstacles there are to defining chains, boundaries, ... and contrachains, contraboundaries, ..., using a group rather than an abelian group for coefficients and values.

3. Establish satisfactory extensions of the definition of the Kronecker product

(i) for chains and contrachains whose coefficients and values are taken from a ring R,

(ii) for chains with coefficients in J_p and contrachains with values in J_q, the product being an element of J_r, where $r = (p, q)$ the H.C.F. of p and q.

4. The complex K consists of the ten closed 2-simplexes $a^0a^1a^3$, $a^0a^3a^5$, $a^0a^2a^5$, $a^0a^2a^4$, $a^0a^1a^4$, $a^1a^2a^3$, $a^2a^3a^4$, $a^3a^4a^5$, $a^1a^4a^5$, $a^1a^2a^5$; the subcomplex K_0 consists of the three closed 1-simplexes a^1a^3, a^3a^5, a^1a^5. Verify that K, K_0 is a triangulation of the pair P^2, P^1.

Calculate the groups and homomorphisms arising in the exact sequences (homology and contrahomology) for the pair K, K_0 with integer coefficients and values.

Prove that the induced sequences of homomorphisms of weak homology and contrahomology groups are not exact.

5. Give examples of a complex K and abelian group G

(i) such that, for some p, $H_p(K; J) = 0$ and $H_p(K; G) \neq 0$;
(ii) such that, for some p, $H^p(K; G) \not\cong \operatorname{Hom}(H_p(K; J); G)$.

[The complex K of Q. 4 with suitable G provides an example in each case.]

6. Triangulate the following subsets of R^3, each of which is the union of two pieces $|K_1|$ and $|K_2|$, and calculate the homology groups of the triangulations:

(i) $|K_1| = \{(u_1, u_2, u_3) \mid |u_1| + |u_2| = 1 \quad \text{and} \quad u_3 = 0\}$,

$|K_2| = \{(u_1, u_2, u_3) \mid |u_1| \leqslant 1 \quad \text{and} \quad u_2 = u_3 = 0\}$;

(ii) $|K_1| = \{(u_1, u_2, u_3) \mid |u_1| + |u_2| + |u_3| = 1\}$,

$|K_2| = \{(u_1, u_2, u_3) \mid |u_1| \leqslant 1 \quad \text{and} \quad u_2 = u_3 = 0\}$;

(iii) $|K_1| = \{(u_1, u_2, u_3) \mid |u_1| + |u_2| + |u_3| = 1\}$,

$|K_2| = \{(u_1, u_2, u_3) \mid |u_1| + |u_2| \leqslant 1 \quad \text{and} \quad u_3 = 0\}$.

7. Subcomplexes K_0, K_1, K_2 of K are such that $K_1 \cup K_2 = K$ and $K_1 \cap K_2 = K_0$. The symbols λ^{0i}, μ^{0i}, ν^{0i} have the meanings of λ, μ, ν in 2.8 but referring to the pair K_i, K_0; the symbols λ^i, μ^i, ν^i refer similarly to the pair K, K_i; the symbol ι stands for the identity relating $C_p(K \bmod K_1)$ and $C_p(K_2 \bmod K_0)$ used in 2.8.22. Homomorphisms ϕ, χ, ψ are defined

$$C_p(K_0) \xrightarrow{\phi} C_p(K_1) \oplus C_p(K_2) \xrightarrow{\chi} C_p(K) \xrightarrow{\psi} C_{p-1}(K_0)$$

as follows:

$$c^0\phi = (c^0\lambda^{01}, \; -c^0\lambda^{02}), \quad \text{all} \quad c^0 \in C_p(K^0),$$

$$(c^1, c^2)\,\chi = c^1\lambda^1 + c^2\lambda^2, \quad \text{all} \quad c^1 \in C_p(K^1),\; c^2 \in C_p(K^2),$$

$$c\psi = c\mu^1\iota\nu^{02}, \quad \text{all} \quad c \in C_p(K).$$

Prove that ϕ, χ, ψ induce homomorphisms ϕ_*, χ_*, ψ_* such that

$$\ldots \xrightarrow{\psi_*} H_p(K_0) \xrightarrow{\phi_*} H_p(K_1) \oplus H_p(K_2) \xrightarrow{\chi_*} H_p(K) \xrightarrow{\psi_*} H_{p-1}(K_0) \xrightarrow{\phi_*} \ldots$$

is exact. [This sequence is called the *Mayer-Vietoris sequence of the triad* $(K; K_1, K_2)$.]

Use this sequence to confirm the calculations of Q. 6.

8. If A, B are groups (not necessarily commutative), then a product can be defined in the set $T(A, B)$ of all transformations $t: A \to B$ as follows:

$$at_1t_2 = (at_1)(at_2), \quad \text{all} \quad a \in A.$$

Prove that the set $T(A, B)$ forms a group under this product and that the subset $\operatorname{Hom}(A, B)$ of homomorphisms from A to B does not in general form a subgroup. (Compare Q. 2.)

9. Give a formal proof of 2.9.7c, the exactness of the contrahomology sequence.

10. Supply the details left to the reader at the end of the proof of 2.10.5.

11. The complex K_1 has subcomplexes K_2, K_3 such that $K_2 \supset K_3$. The symbols α^{ij}, λ^{ij}, μ^{ij}, ν^{ij} have the meanings of 2.8 referred to the pair K_i, K_j. Homomorphisms ζ, η, θ are defined

$$C_p(K_2, K_3) \xrightarrow{\zeta} C_p(K_1, K_3) \xrightarrow{\eta} C_p(K_1, K_2) \xrightarrow{\theta} C_{p-1}(K_2, K_3)$$

by

$$\zeta = \alpha^{23}\lambda^{12}\mu^{13},$$
$$\eta = \alpha^{13}\mu^{12},$$
$$\theta = \nu^{12}\mu^{23}.$$

Prove that ζ, η, θ induce homology homomorphisms ζ_*, η_*, θ_* such that

$$\dots \xrightarrow{\theta_*} H_p(K_2, K_3) \xrightarrow{\zeta_*} H_p(K_1, K_3) \xrightarrow{\eta_*} H_p(K_1, K_2) \xrightarrow{\theta_*} H_p(K_2, K_3) \xrightarrow{\zeta_*} \dots$$

is exact.

[This sequence is called the *homology sequence of the triple* $(K_1; K_2; K_3)$.]

12. Establish the isomorphism $\gamma: C_n(K) \cong H_n(K^n, K^{n-1})$, where K^p is as usual the p-section of K.

Prove that $\partial\gamma = \gamma\theta_*$, where θ_* is the homomorphism of Q. 11 for the triple $(K^n; K^{n-1}; K^{n-2})$.

What are the corresponding facts in contrahomology?

13. The cell complexes M, N have cells $\{\xi\}$, $\{\eta\}$ respectively. Two cell complexes $M \cdot N$ and $M \times N$, their *join* and *product*, are defined as follows. The cells both of $M \cdot N$ and of $M \times N$ are the pairs (ξ, η) and in both $(\xi, \eta) \prec (\xi', \eta')$ if and only if $\xi \prec \xi'$ and $\eta \prec \eta'$. In $M \cdot N$ the dimension of (ξ_p, η_q) is $p+q+1$; in $M \times N$ it is $p+q$. The incidence numbers are given by

$$[(\xi_p, \eta), (\xi'_{p+1}, \eta)] = [\xi_p, \xi'_{p+1}]$$

and

$$[(\xi_p, \eta_q), (\xi_p, \eta'_{q+1})] = (-1)^d [\eta_q, \eta'_{q+1}],$$

where

$$d = p+1 \quad \text{for} \quad M \cdot N$$
$$= p \quad \text{for} \quad M \times N.$$

Verify that $M \cdot N$ and $M \times N$ both satisfy the axioms for cell-complexes and that

$$H^{(i)}_{r+1}(M \cdot N) \cong H^{(i)}_r(M \times N), \quad i = 1, 2.$$

Prove also that, if $H^{(i)}_r(N) = 0$ for all r, then

$$H^{(i)}_r(M \cdot N) = 0 \quad \text{and} \quad H^{(i)}_r(M \times N) = 0$$

for all r.

14. Any oriented simplicial complex K determines in an obvious way two cell complexes $M(K)$, $M(K^+)$ such that

$$H_r(K) \cong H^{(1)}_r(M(K)) \quad \text{and} \quad H_r(K^+) \cong H^{(1)}_r(M(K^+))$$

for all r. Prove that if K, L are oriented simplicial complexes, $M(K^+) \cdot M(L^+)$ is isomorphic to $M(K^+ \cdot L^+)$.

15. If K, a^0 and L, b^0 are oriented simplicial complexes together with a vertex of each, prove that $M(K^+) \times M(L^+)$ has a subcomplex

$$M_1 = (M(K^+) \times M(b^{0+})) \cup (M(a^{0+}) \times M(L^+))$$

and that $H^{(1)}_r(M_1) = 0$ for all r. Regarding $M(K) \times M(L)$ as a subcomplex of $M(K^+) \times M(L^+)$, verify that $M_0 = M_1 \cap (M(K) \times M(L))$ is isomorphic with $M(K \vee L)$, where $K \vee L$ is obtained from $K \cup L$ by identifying a^0 and b^0.

Prove that

$$H_{r+1}(K^+ \cdot L^+) = H^{(1)}_r(M(K) \times M(L), M_0).$$

3

CHAIN COMPLEXES

3.1 Chain and contrachain complexes

An oriented simplicial complex‡ determines, for each dimension p, a chain group C_p and a boundary homomorphism $\partial : C_p \to C_{p-1}$. From these data the homology and contrahomology groups may be obtained. We now propose to confine attention to these purely algebraical concepts and accordingly define

3.1.1 Definition. A *chain complex* $C. = \{C_p, \partial_p\}$, is a collection of abelian groups C_p, one for each integer p, and of (right) homomorphisms $\partial_p : C_p \to C_{p-1}$ such that $\partial_p \partial_{p-1} = 0$, for each p.

Dually, we define

3.1.1*c* Definition. A *contrachain complex*, $C^{\cdot} = \{C^p, \delta^p\}$ is a collection of abelian groups C^p, one for each integer p, and of (left) homomorphisms $\delta^p : C^p \to C^{p+1}$ such that $\delta^{p+1}\delta^p = 0$, for each p.

We shall generally write C for $C.$, and shall often write ∂, δ for ∂_p, δ^p.

It is clear how we may define the *homology groups* $H_p(C)$ of the chain complex C; if Z_p or $Z_p(C)$, the *p-th cycle group*, is the kernel of ∂_p and B_p or $B_p(C)$, the *p-th boundary group*, is the image of ∂_{p+1}, then B_p is a subgroup of the abelian group Z_p and $H_p(C)$ is the factor group Z_p/B_p. The definition of $H^p(C^{\cdot})$ is analogous; if the context makes it clear that a contrachain complex is in question, we may suppress the superscript dot, so that we may write $H^p(C)$. The reader will notice that if K is an oriented complex, and G a coefficient or value group, then we have (2.2 and 2.6) a chain complex $C(K; G)$ and a contrachain complex $C^{\cdot}(K; G)$; and

3.1.2 $H_p(K; G) = H_p(C(K; G)), \quad H^p(K; G) = H^p(C^{\cdot}(K; G)).$

A *subcomplex* (more precisely, *chain subcomplex*) of the chain complex C is a collection§ $C^0_{\cdot}$ of subgroups $C^0_p \subseteq C_p$ with $C^0_p\partial \subseteq C^0_{p-1}$.

‡ Abstract cell complexes (2.12) also determine chain and contrachain groups.

§ We shall write C^0 for $C^0_{\cdot}$ if no confusion with 0-dimensional contrachains is to be feared.

Thus $C^0_{\cdot}$ is itself a chain complex; if we wish to regard it apart from C, we may write ∂^0 for its boundary operator. Given a chain complex C and subcomplex $C^0_{\cdot}$ we may consider the collection, $\hat{C}_{\cdot}$, of groups $\hat{C}_p = C_p/C^0_p$ and homomorphisms $\hat{\partial}_p : \hat{C}_p \to \hat{C}_{p-1}$, given by

$$[c_p]\hat{\partial}_p = [c_p \partial_p],$$

where $[c_p]$ is the coset of $C_p \operatorname{mod} C^0_p$ containing c_p. Then $\hat{\partial}_p$ is a well-defined homomorphism since $C^0_p \partial \subseteq C^0_{p-1}$ (that is, C^0 is *stable* for ∂), and $\hat{\partial}_p \hat{\partial}_{p-1} = 0$. The chain complex $\hat{C} = \{\hat{C}_p, \hat{\partial}_p\}$ is called the *factor complex* or *quotient complex* of $C \operatorname{mod} C^0$ and we may write $\hat{C} = C/C^0$.

Analogous notions are assigned to contrachain complexes but there is a reversal of notation. If $\hat{C}^{\cdot}$ is a subcomplex of $C^{\cdot}$ we write $\hat{\delta}$ for the contraboundary in $\hat{C}^{\cdot}$ and δ_0 for the contraboundary in the factor complex $C^{\cdot}_0 = C^{\cdot}/\hat{C}^{\cdot}$.

If K_0 is a subcomplex of K, then $C(K_0; G)$ is a chain subcomplex of $C(K; G)$ and the factor complex $C(K; G)/C(K_0; G)$ is isomorphic with the chain complex $C(K \operatorname{mod} K_0; G)$. The situation is less obvious in contrahomology; the contrachain complex $C^{\cdot}(K \operatorname{mod} K_0; G)$ is a contrachain subcomplex of $C^{\cdot}(K; G)$ and the factor complex

$$C^{\cdot}(K; G)/C^{\cdot}(K \operatorname{mod} K_0; G)$$

is isomorphic with the contrachain complex $C^{\cdot}(K_0; G)$. This apparent reversal explains the reversal of notation and will be clarified later in this section.

As in the case of any algebraic structure, we are concerned with transformations which are structure-preserving. The structure carried by a chain (or contrachain) complex consists of (i) the dimension index of the constituent groups, (ii) the group structure of the groups, and (iii) the boundary homomorphisms.

3.1.3 Definition. A *chain map* ϕ from the chain complex C to the chain complex D is a set of (right) homomorphisms $\phi_p : C_p \to D_p$ such that‡ $\phi_p \partial_p = \partial_p \phi_{p-1}$.

An analogous definition holds for a *contrachain map*; the homomorphisms ϕ^p are then written on the left. We often write ϕ for ϕ_p or for ϕ^p.

3.1.4 Proposition. *The kernel of a chain map $\phi : C \to D$ is a subcomplex of C; the image of ϕ is a subcomplex of D.*

‡ Where no confusion is to be feared we use the symbol ∂ for the boundary homomorphism in any chain complex.

The *kernel* of ϕ is the collection, C^0_p, of kernels of the homomorphisms $\phi_p : C_p \to D_p$. If $c \in C^0_p$, then $c\partial\phi = c\phi\partial = 0$, so that $c\partial \in C^0_{p-1}$ and C^0 is a subcomplex of C.

The *image* of ϕ is the collection, D^0_p, of images of the homomorphisms $\phi_p : C_p \to D_p$. If $d \in D^0_p$, then $d = c\phi$, for some $c \in C_p$ and

$$d\partial = c\phi\partial = c\partial\phi,$$

so that $d\partial \in D^0_{p-1}$, and D^0 is a subcomplex of D. ∎

The *cokernel* of ϕ is the factor complex $D/D^0 = D/C\phi$.

Let C^0 be a subcomplex of C. Then there is an *injection* chain map $i : C^0 \to C$ which embeds each C^0_p in C_p. Also there is a *projection* chain map $j : C \to \hat{C}$ which projects each C_p onto C_p/C^0_p.

We observe that the sequence

3.1.5 $$0 \to C^0 \overset{i}{\to} C \overset{j}{\to} \hat{C} \to 0$$

is exact in the sense that $0 \to C^0_p \overset{i_p}{\to} C_p \overset{j_p}{\to} \hat{C}_p \to 0$ is exact for each p.

If the chain map $\phi : C \to D$ maps C^0 to D^0 we shall write

3.1.6 $$\phi : C, C^0 \to D, D^0.$$

3.1.7 Proposition. *A chain map $\phi : C, C^0 \to D, D^0$ induces a chain map $\hat{\phi} : \hat{C} \to \hat{D}$ by the rule $[c]\hat{\phi} = [c\phi]$, $c \in C$.*

First we observe that $[c\phi]$ depends only on $[c]$, since $C^0\phi \subseteq D^0$. Now let $c \in C_p$, so that $[c] \in \hat{C}_p$. Then

$$[c]\hat{\partial}\hat{\phi} = [c\partial]\hat{\phi} = [c\partial\phi] = [c\phi\partial] = [c\phi]\hat{\partial} = [c]\hat{\phi}\hat{\partial},$$

and $\hat{\phi}$ is a chain map. ∎ We may rewrite this into a commutative diagram of chain maps

3.1.8 $$\begin{array}{ccccccccc} 0 & \longrightarrow & C^0 & \overset{i}{\longrightarrow} & C & \overset{j}{\longrightarrow} & \hat{C} & \longrightarrow & 0 \\ & & \downarrow{\scriptstyle \phi^0} & & \downarrow{\scriptstyle \phi} & & \downarrow{\scriptstyle \hat{\phi}} & & \\ 0 & \longrightarrow & D^0 & \overset{i}{\longrightarrow} & D & \overset{j}{\longrightarrow} & \hat{D} & \longrightarrow & 0 \end{array}$$

where ϕ^0 is the chain map induced by ϕ.

Let Z_p, B_p be the groups of p-cycles, p-boundaries of the chain complex C. In an obvious sense‡ $Z = \{Z_p, 0\}$ is a subcomplex of C and $B = \{B_p, 0\}$ is a subcomplex of Z. The factor complex will be written $H_*(C)$; its groups are the homology groups $H_p(C)$. A chain map $H_*(C) \to H_*(D)$ is just a set of homomorphisms $H_p(C) \to H_p(D)$.

‡ Here 0 stands for the zero homomorphism.

3.1.9 Theorem. *A chain map $\phi : C \to D$ induces a homomorphism*

$$\phi_* : H_*(C) \to H_*(D).$$

If $\phi : C \to D$, $\psi : D \to E$ are chain maps then $\phi\psi$ is a chain map and $(\phi\psi)_* = \phi_*\psi_*$.

Let $z \in Z(C)$ and let $\{z\}$ be its homology class. Now $z\phi$ is a cycle, since $z\phi\partial = z\partial\phi = 0$; we define $\{z\}\phi_* = \{z\phi\}$, and leave to the reader the verification that a homomorphism $\phi_* : H_*(C) \to H_*(D)$ is thus defined. The crux, as in the special cases 2.3.1, 2.8.3 and 2.8.14, is that ϕ maps cycles to cycles and boundaries to boundaries and hence induces a homomorphism ϕ_* of the quotient groups.

Now $\phi\psi$ is a chain map since $\phi_p\psi_p$ is a homomorphism $C_p \to E_p$ and $\phi\psi\partial = \phi\partial\psi = \partial\phi\psi$. Moreover,

$$\{z\}\phi_*\psi_* = \{z\phi\}\psi_* = \{z\phi\psi\} = \{z\}(\phi\psi)_*,$$

completing the proof. ∎

A special case is afforded by the chain maps i and j. We have homomorphisms $i_* : H_*(C^0) \to H_*(C)$, $j_* : H_*(C) \to H_*(\hat{C})$ which coincide with λ_*, μ_* in 2.8 when $C = C(K; G)$, $C^0 = C(K_0; G)$. A general theorem in chapter 5 enables us to deduce an exact homology sequence from the exact sequence 3.1.5. All these statements have their analogues in contrahomology.

Just as the contrachain groups of a complex may be regarded as the groups of homomorphisms of the integral chain groups into the value group, so from any chain complex C and value group G the *associated or adjoint contrachain* complex may be defined. For abelian groups A, B we write‡ $A \pitchfork B$ for the group Hom (A, B); we also write $C \pitchfork G$ for the associated contrachain complex; $(C \pitchfork G)^p = C_p \pitchfork G$ and δ^p is given by

3.1.10 $\quad (c_{p+1}, \delta^p c^p) = (c_{p+1}\partial_{p+1}, c^p), \quad c_{p+1} \in C_{p+1}, \; c^p \in C_p \pitchfork G;$

we use the Kronecker index notation as in chapter 2. As in the special case discussed there, δ^p is the adjoint of ∂_{p+1} and we have $\delta^{p+1}\delta^p = 0$, so that $C \pitchfork G$ is a contrachain complex. We also take over the notion of adjoint maps introduced there; if $\phi : C \to D$ is a chain map, then $\phi^{\cdot} : D \pitchfork G \to C \pitchfork G$, given by

$$(c_p, \phi^{\cdot} d^p) = (c_p\phi, d^p), \quad c_p \in C_p, \; d^p \in D_p \pitchfork G,$$

‡ This notation is due to E. C. Zeeman, *J. Lond. Math. Soc.* **30** (1955), 195.

is a contrachain map, as the reader may verify, called the *adjoint* of ϕ; and $(\phi\psi)^{\cdot} = \phi^{\cdot}\psi^{\cdot}$. We may write $C^{\cdot}$ for $C \pitchfork G$ if no confusion may arise. Then $\phi^{\cdot}$, and so also ϕ, induces $\phi^* : H^*(D^{\cdot}) \to H^*(C^{\cdot})$.

There are algebraic advantages in regarding C as the direct sum ΣC_p together with an endomorphism ∂ lowering dimension (or degree) by 1 and satisfying $\partial\partial = 0$. Then an element of C is a *chain* or finite sum Σc_p, where $c_p \in C_p$, and is called homogeneous if it belongs to some C_p. C is then also called a *differential graded group*.‡ The cycle, boundary and homology groups are also graded in that each is the direct sum of subgroups of elements of a given dimension. Equivalently, a contrachain complex may be regarded as a differential graded group, wherein the differential raises dimension by 1. However, the reader should notice that $C \pitchfork G$ is *not* the group of homomorphisms of the group C into the group G, since such a homomorphism would be an element of the unrestricted direct sum, or direct product, $\Pi C_p \pitchfork G$, and so would be an infinite sum. It is consistent with the functorial idiom that we regard $C \to C \pitchfork G$ as a contravariant functor from the category of chain complexes to the category of contrachain complexes‖; if we were to regard C as a group we should change its category and thus change the functor $\pitchfork G$.

Consider the exact sequence 3.1.5

$$0 \to C^0 \xrightarrow{i} C \xrightarrow{j} \hat{C} \to 0.$$

Its analogue in contrahomology is the exact sequence

3.1.5c $$0 \leftarrow D_0^{\cdot} \xleftarrow{i} D^{\cdot} \xleftarrow{j} \hat{D}^{\cdot} \leftarrow 0.$$

On the other hand, 3.1.5 yields, by passing to the adjoint, a sequence

3.1.11 $$0 \leftarrow C^{0\cdot} \xleftarrow{i^{\cdot\cdot}} C^{\cdot} \xleftarrow{j^{\cdot\cdot}} \hat{C}^{\cdot} \leftarrow 0,$$

where $C^{0\cdot} = C^0 \pitchfork G$, $C^{\cdot} = C \pitchfork G$, $\hat{C}^{\cdot} = \hat{C} \pitchfork G$. Notice that, 3.1.11 is not always exact at $C^{0\cdot}$, though, as may easily be proved, it is exact at $C^{\cdot}$ and $\hat{C}^{\cdot}$. For example, if $C_p = J$, $C_p^0 = 2J \subseteq J$ and $G = J$, then not every homomorphism $2J \to J$ can be extended to a homomorphism $J \to J$, so that i^p is not epimorphic. However, if C_p^0 is a direct factor in C_p for each p—the situation we meet when K_0 is a subcomplex of K, $C_p = C_p(K)$, $C_p^0 = C_p(K_0)$—then plainly 3.1.11 is exact at $C^{0\cdot}$. Thus

3.1.12 Proposition. *The sequence* 3.1.11 *is exact at* $\hat{C}^{\cdot}$ *and* $C^{\cdot}$; *if* C^0 *is a direct factor in* C *it is exact at* $C^{0\cdot}$. ∎

We may accordingly always regard $\hat{C}^{\cdot}$ as a subcomplex of $C^{\cdot}$; and the factor complex as $C^{0\cdot}$ if C^0 is a direct factor in C. This explains the notation $\hat{D}^{\cdot}$ for a *subcomplex* of $D^{\cdot}$ and $D_0^{\cdot}$ for the *factor complex* in 3.1.5c.

‡ This terminology is used in chapter 10.
‖ See Eilenberg and Steenrod [14].

We may express the final assertion of 3.1.12 by saying that, under the given hypothesis, $C^{0\cdot} \cong C_0^{\cdot} = C^{\cdot}/\hat{C}^{\cdot}$, and 3.1.11 then essentially coincides with 3.1.5c for the pair $(C^{\cdot}, \hat{C}^{\cdot})$. Moreover, if $\phi : C, C^0 \to D, D^0$ is a chain map inducing $\hat{\phi} : \hat{C} \to \hat{D}$, then $\hat{\phi}^{\cdot} : \hat{D}^{\cdot} \to \hat{C}^{\cdot}$ is obtained by restriction from $\phi^{\cdot} : D^{\cdot} \to C^{\cdot}$ and the factor map $\phi_0^{\cdot} : D^{\cdot}/\hat{D}^{\cdot} \to C^{\cdot}/\hat{C}^{\cdot}$ coincides, when C^0, D^0 are direct factors in C, D, with the map $\phi^{0\cdot} : D^{0\cdot} \to C^{0\cdot}$ adjoint to ϕ^0 of 3.1.8. We shall appeal repeatedly to these facts in the later sections of this chapter.

3.2 Examples of chain complexes and chain maps

Let K be an oriented complex and let α be its orientation. Then we have defined chain groups $C_p^{\alpha}(K; G)$ and contrachain groups

$$C_{\alpha}^{p}(K; G) = C_p^{\alpha}(K) ⋔ G.$$

The resulting homology and contrahomology groups are those of the complex K. If the orientation of K is induced by an ordering ω, we write $C_p^{\omega}(K; G)$, $C_{\omega}^{p}(K; G)$ for the chain and contrachain groups. The chain complex of which $C_p^{\alpha}(K; G)$ is the pth chain group will be written $C^{\alpha}(K; G)$; the contrachain complex will be written $C_{\alpha}^{\cdot}(K; G)$ or $C^{\alpha}(K) ⋔ G$ if we wish to stress that it is adjoint to $C^{\alpha}(K)$. For two orientations α, β there are chain isomorphisms (2.3.1)

$$\theta^{\alpha\beta} : C^{\alpha}(K; G) \to C^{\beta}(K; G).$$

We now describe a different method for constructing a chain complex from a simplicial complex which is of fundamental theoretical importance and which will be proved to give rise to the same homology groups. Let K be a simplicial complex (not oriented). A finite ordered array $[a^{i_0} \dots a^{i_p}]$ of vertices of K is *admissible* if and only if the vertices span a simplex of K. Notice that repetitions are allowed in admissible arrays; the array $[a^3a^1a^8a^8a^3]$ is admissible if the vertices a^1, a^3, a^8 span a triangle of K. We now define a chain complex $C^{\Omega}(K; G)$, the *total* chain complex of K with coefficients in G. $C_p^{\Omega}(K; G)$ is the group of all linear combinations of admissible arrays of length $(p+1)$ with coefficients in G and $\partial_p : C_p^{\Omega}(K; G) \to C_{p-1}^{\Omega}(K; G)$ is defined by linearity from the rule‡

3.2.1 $$[a^{i_0} \dots a^{i_p}]\,\partial_p = \sum_{k=0}^{p} (-1)^k [a^{i_0} \dots \hat{a}^{i_k} \dots a^{i_p}].$$

‡ We may also define $C^{\Omega}(K^+)$; $C_{-1}^{\Omega}(K^+)$ is generated by the empty array and $[a^{i_0}]\,\partial^+ = [\ ]$.

The nub of the proof of Theorem 2.2.5 shows that $\partial_p \partial_{p-1} = 0$. The contrachain complex associated with $C^{\Omega}(K)$ is written $C^{\cdot}_{\Omega}(K; G)$ or $C^{\Omega}(K) \pitchfork G$.

Let $\theta^{\alpha}_p : C^{\Omega}_p(K) \to C^{\alpha}_p(K)$ be the homomorphism given by

$$[a^{i_0} \dots a^{i_p}]\, \theta^{\alpha}_p = 0, \quad \text{if the vertices } a^{i_0}, \dots, a^{i_p} \text{ are not all distinct,}$$
$$= a^{i_0} \dots a^{i_p}, \text{ otherwise.}$$

3.2.2 Proposition. *θ^{α} is a chain map.*

We have to prove that $\theta^{\alpha}_p \partial = \partial \theta^{\alpha}_{p-1}$. Now clearly this holds on $[a^{i_0} \dots a^{i_p}]$ if the vertices are all distinct; moreover, $\theta^{\alpha}_p \partial = \partial \theta^{\alpha}_{p-1} = 0$ if the array contains fewer than p distinct vertices. It remains to consider the case when the array contains exactly p distinct vertices; suppose, for instance, that $a^{i_j} = a^{i_k}$. Then $[a^{i_0} \dots a^{i_p}]\, \theta^{\alpha}_p = 0$ and

$$[a^{i_0} \dots a^{i_p}]\, \partial = (-1)^j [a^{i_0} \dots \hat{a}^{i_j} \dots a^{i_p}] + (-1)^k [a^{i_0} \dots \hat{a}^{i_k} \dots a^{i_p}] + c,$$

where c is a sum (with coefficients ± 1) of arrays with repeated vertices. Thus $c\theta^{\alpha}_{p-1} = 0$ and

$$[a^{i_0} \dots a^{i_p}]\, \partial \theta^{\alpha}_{p-1} = (-1)^j a^{i_0} \dots \hat{a}^{i_j} \dots a^{i_p} + (-1)^k a^{i_0} \dots \hat{a}^{i_k} \dots a^{i_p};$$

but $$a^{i_0} \dots \hat{a}^{i_k} \dots a^{i_p} = (-1)^{k-j-1} a^{i_0} \dots \hat{a}^{i_j} \dots a^{i_p},$$

so that $[a^{i_0} \dots a^{i_p}]\, \partial \theta^{\alpha}_{p-1} = 0$ and the proposition is proved in every case. ▌

We shall say that the oriented complex K is an *ordered complex* if its orientation is induced by an ordering of its vertices; if K is an ordered complex the vertices of a simplex of K will always be written in the given order.

Let K be an ordered complex with ordering ω. Let

$$\bar{\theta}^{\omega}_p : C^{\omega}_p(K) \to C^{\Omega}_p(K)$$

be given by $$a^{i_0} \dots a^{i_p}\, \bar{\theta}^{\omega}_p = [a^{i_0} \dots a^{i_p}].$$

3.2.3 Proposition. *$\bar{\theta}^{\omega}$ is a chain map, and $\bar{\theta}^{\omega}\theta^{\omega} = 1$.* ▌

It follows from this and Theorem 3.1.9 that

3.2.4 $$\bar{\theta}^{\omega}_* \theta^{\omega}_* = 1 : H^{\omega}_*(K; G) \cong H^{\omega}_*(K; G).$$

One of the main results of this chapter is that

$$\theta^{\omega}_* \bar{\theta}^{\omega}_* = 1 : H^{\Omega}_*(K; G) \cong H^{\Omega}_*(K; G),$$

even though $\theta^{\omega}\bar{\theta}^{\omega} \neq 1$. The reader will note that the definitions of $C^{\Omega}(K)$, θ^{α}, $\bar{\theta}^{\omega}$, and the assertions made about them relativize in the obvious way; we shall not be more explicit. Also we have similar statements for the associated contrachain complex $C^{\cdot}_{\Omega}(K; G)$ and the adjoint contrachain maps $\theta^{\cdot}_{\alpha}$, $\theta^{\cdot}_{\omega}$. In particular,

3.2.4 c $$\bar{\theta}^*_{\omega}\theta^*_{\omega} = 1 : H^*_{\omega}(K; G) \cong H^*_{\omega}(K; G).$$

We again leave the relativization to the reader.

Let $v : K \to L$ be a simplicial map from the oriented complex K to the oriented complex L. Then v induces a chain map

$$\phi : C(K) \to C(L)$$

in the following way. Let $a^0 \ldots a^p$ be an oriented p-simplex of K. Then the vertices $a^0v, \ldots, a^pv$ span a simplex of L. We define

3.2.5 $$\begin{aligned} a^0 \ldots a^p\phi_p &= a^0v \ldots a^pv, \quad \text{if} \quad a^0v, \ldots, a^pv \text{ are all distinct} \\ &= 0, \quad \text{otherwise.} \end{aligned}$$

Recall that $a^0v \ldots a^pv$ is $\pm$(the oriented simplex of L spanned by $a^0v, \ldots, a^pv$). Clearly $a^0 \ldots a^p\phi_p$ depends only on the orientation of the simplex $(a^0 \ldots a^p)$ and not on the particular ordering of the vertices.

We extend ϕ_p by linearity to a homomorphism $C_p(K) \to C_p(L)$. That $\phi\partial = \partial\phi$ is now proved precisely as in the proof that θ^{α} is a chain map (Proposition 3.2.2). In fact, the connexion with θ^{α} is very close; for v also induces a chain map

$$\phi^{\Omega} : C^{\Omega}(K) \to C^{\Omega}(L),$$

by the rule

3.2.6 $$[a^{i_0} \ldots a^{i_p}]\phi^{\Omega} = [a^{i_0}v \ldots a^{i_p}v],$$

and we have $$\theta\phi = \phi^{\Omega}\theta.$$

Here we have written θ briefly for $\theta^{\alpha}(\theta^{\beta})$ if K (L) is oriented by $\alpha(\beta)$.

3.2.7 Proposition. *If $v : K \to L$, $w : L \to M$ are simplicial maps inducing the chain maps ϕ, ϕ^{Ω}, ψ, ψ^{Ω}, then vw induces $\phi\psi$, $\phi^{\Omega}\psi^{\Omega}$.* ∎

The expressions for θ and ϕ may be simplified by adopting the convention that if $a^{i_0}, \ldots, a^{i_p}$ are not distinct vertices of K, then $a^{i_0} \ldots a^{i_p}$ is the zero element of $C^{\alpha}_p(K)$. Thus we may write

$$[a^{i_0} \ldots a^{i_p}]\,\theta = a^{i_0} \ldots a^{i_p}, \quad a^0 \ldots a^p\,\phi = a^0v \ldots a^pv.$$

As an example of this convention, we note that, if a is a vertex of the oriented simplex σ, then $a(\sigma\partial) = \sigma$ as a chain. For if $\sigma = a^0 \ldots a^p$ and $a = a^j$, then

$$a(\sigma\partial) = \sum_{i=0}^{p} (-1)^i a^j a^0 \ldots \widehat{a^i} \ldots a^p = (-1)^j a^j a^0 \ldots \widehat{a^j} \ldots a^p = a^0 \ldots a^p.$$

We revert to this example in Theorem 3.2.10.

If v is a simplicial map and $\phi : C(K) \to C(L)$ is the associated chain map, then ϕ induces‡ $\phi_* : H_*(K) \to H_*(L)$. We may write v_* for ϕ_*. Similarly, v induces v^* (or ϕ^*) : $H^*(L; G) \to H^*(K; G)$.

3.2.8 Proposition. *If* $v : K \to L$, $w : L \to M$ *are simplicial maps, then*

$$(vw)_* = v_* w_*, \quad (vw)^* = v^* w^*. \,\rule{0.5em}{1em}$$

The homomorphisms of homology and contrahomology groups induced by a simplicial map play a vital role in the theory.§ For example, if v is a standard simplicial map from K', the first derived, to K, then v induces $v_* : H_*(K') \to H_*(K)$, $v^* : H^*(K; G) \to H^*(K'; G)$. If v induces the chain map ϕ, we call ϕ a *standard* chain map. We shall see that the homomorphism v_* (or ϕ_*) is independent of the choice of v and is indeed an isomorphism.

A step toward this end is to define a chain map $\chi : C(K) \to C(K')$ which, however, is not associated with a simplicial map. Nevertheless, the chain map χ has an intuitive geometric content; it replaces each $\sigma_p \in K$ by the p-chain of K' consisting of the (suitably oriented) p-simplexes of K' lying in s_p (the simplex underlying σ_p). It is most convenient to define χ inductively in the augmented case and we may begin by defining χ in dimension -1 by

$$\sigma_{-1}\chi_{-1} = \sigma_{-1}.$$

We now suppose $\chi_q : C_q(K^+) \to C_q(K'^+)$ defined in dimensions $q < p$ in such a way that $\chi_q \partial = \partial \chi_{q-1}$ and $\overline{\sigma_q \chi_q} \subseteq \bar{s}'_q$ for each σ_q. We define χ_p by prescribing it on each $\sigma_p \in K$ and extending by linearity over $C_p(K)$. Thus

3.2.9 $$\sigma_p \chi_p = b(\sigma_p \partial \chi_{p-1}),$$

where b is the barycentre of s_p. Since $\overline{\sigma_p \partial \chi_{p-1}} \subseteq \dot{s}'_p$, it follows that

‡ Of course, we may regard ϕ_* or v_* as mapping $H_*(K; G)$ to $H_*(L; G)$, for an arbitrary coefficient group G.

§ Recall that every continuous map is 'near' a simplicial map (1.8.1, 1.8.4).

$b(\sigma_p \partial \chi_{p-1})$ is a chain of K' and that its carrier is contained in $\bar{s}'_p$. Moreover,

$$\begin{aligned}\sigma_p \chi_p \partial &= \sigma_p \partial \chi_{p-1} - b(\sigma_p \partial \chi_{p-1} \partial) \quad (2.10.4)\\ &= \sigma_p \partial \chi_{p-1} - b(\sigma_p \partial\partial \chi_{p-2}), \quad \text{by the inductive hypothesis}\\ &= \sigma_p \partial \chi_{p-1}.\end{aligned}$$

Thus $\chi_p \partial = \partial \chi_{p-1}$ and the induction is complete. Note that if a is a vertex, then

$$a\chi_0 = b,$$

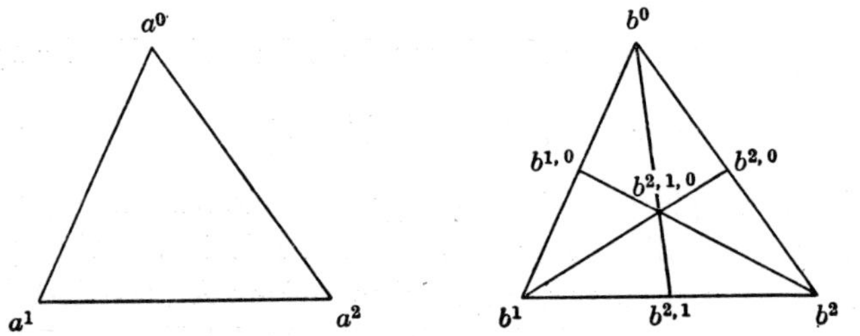

Fig. 3.1

where b is the vertex of K' lying at a. In the unaugmented case we use this to start the induction. As an example consider fig. 3.1. Then

$$\begin{aligned}1\chi &= 1, \quad \text{writing 1 for } \sigma_{-1};\\ a^i\chi &= b^i, \quad i = 0, 1, 2;\\ a^0a^1\chi &= b^{1,0}(a^0a^1\partial\chi) = b^{1,0}b^1 - b^{1,0}b^0,\\ a^1a^2\chi &= b^{2,1}b^2 - b^{2,1}b^1,\\ a^0a^2\chi &= b^{2,0}b^2 - b^{2,0}b^0;\\ a^0a^1a^2\chi &= b^{2,1,0}(a^0a^1a^2\partial\chi)\\ &= b^{2,1,0}b^{2,1}b^2 - b^{2,1,0}b^{2,1}b^1 - b^{2,1,0}b^{2,0}b^2\\ &\quad + b^{2,1,0}b^{2,0}b^0 + b^{2,1,0}b^{1,0}b^1 - b^{2,1,0}b^{1,0}b^0.\end{aligned}$$

The reader will note that each triangle in K' appears with its expected orientation.

3.2.10 Theorem. *Let $v : K' \to K$ be a standard simplicial map inducing $\phi : C(K') \to C(K)$. Then $\chi\phi = 1$.*

This is proved by induction, being trivial in dimensions -1 and 0. Suppose then that $\chi_{p-1}\phi_{p-1} = 1$ and let $\sigma_p \in K$. Then

$$\sigma_p \chi\phi = (b(\sigma_p \partial\chi))\,\phi = bv(\sigma_p \partial\chi\phi).$$

By the definition of v, bv is some vertex a of s_p, and $\sigma_p \partial \chi \phi = \sigma_p \partial$ since $\chi_{p-1} \phi_{p-1} = 1$. Thus $\sigma_p \chi \phi = a(\sigma_p \partial) = \sigma_p$ and $\chi_p \phi_p = 1$. ∎

The reader should note that $\phi_p \chi_p \neq 1$ in dimensions $p > 0$. Nonetheless, we shall prove that $\phi_* \chi_* = 1$.

No difficulty arises in relativizing the definitions of ϕ and χ and obtaining the corresponding results; also we may use general coefficients for homology and we may obtain corresponding results for contrachains. We briefly review the notations which have been introduced and will be employed in the relative case. If K_0 is a subcomplex of K, and L_0 is a subcomplex of L, let

$$\phi : C(K; G), C(K_0; G) \to C(L; G), C(L_0; G)$$

be a chain map. Then ϕ induces, by restriction

$$\phi^0 : C(K_0; G) \to C(L_0; G),$$

and the factor map $\hat{\phi} : C(K, K_0; G) \to C(L, L_0; G)$. The adjoint maps of ϕ, ϕ^0, $\hat{\phi}$ will be written $\phi^{\cdot}$, $\phi^{0\cdot}$, $\hat{\phi}^{\cdot}$. Then (Proposition 3.1.12 et seq.) $\hat{\phi}^{\cdot}$ is induced by $\phi^{\cdot}$ by restriction and $\phi^{0\cdot} = \phi^{\cdot}_0$ is the factor map. Notice also that $C^{\Omega}(K_0)$ is a subcomplex (indeed, a direct factor) in $C^{\Omega}(K)$ so that the same notions and similar notations apply also to the total complex.

3.3 Chain and contrachain homotopy

We have seen how a chain map induces homology homomorphisms. Since the chain group is an auxiliary concept and our real interest lies in the homology groups it is obviously desirable to discover a useful criterion for two chain maps to induce the same homology homomorphisms.

If C, D are chain complexes and $\phi, \psi : C \to D$ are chain maps such that $\phi_* = \psi_* : H_*(C) \to H_*(D)$, then for any $z \in Z_p(C)$, there exists $d \in D_{p+1}$ with $z\phi - z\psi = d\partial$. We now describe an equivalence relation on the set of chain maps $C \to D$; this has the property that, provided ϕ, ψ belong to the same equivalence class, there is a systematic procedure for associating such a $(p+1)$-chain d of D with a p-cycle z of C.

3.3.1 Definition. A *chain homotopy*, Δ, between the chain maps $\phi, \psi : C \to D$ is a set of homomorphisms‡ $\Delta_p : C_p \to D_{p+1}$ such that $\phi - \psi = \partial\Delta + \Delta\partial$. If such a chain homotopy Δ exists, ϕ and ψ are said to be *chain homotopic* and we write $\phi \overset{\Delta}{\simeq} \psi$, or simply $\phi \simeq \psi$.

‡ Notice that Δ itself is *not* a chain map.

A chain homotopy has some formal resemblance to a homotopy. A homotopy F connecting maps f, g of a simplex s is a map of $s \times I$. Now $(s \times I)^{\cdot} = (s \times 1) \cup (s \times 0) \cup (\dot{s} \times I)$ and the F images of $(s \times 1)$ and $(s \times 0)$ are loosely speaking the f and g images of s, while the F image of $(\dot{s} \times I)$ is the homotopy of $\dot{s}$. This situation is a geometrical parallel of the statement that

$$(\sigma\Delta)\,\partial = \sigma\phi - \sigma\psi - (\sigma\partial)\,\Delta.$$

It is obvious that chain homotopy is an equivalence relation between chain maps; it is also obvious that $\phi_* = \psi_*$ if $\phi \simeq \psi$. For if $z \in Z_p(C)$ then $z\partial = 0$ and $z(\phi - \psi) = (z\Delta)\,\partial$.

The following example shows that we may have $\phi_* = \psi_*$ without $\phi \simeq \psi$, even when the chain groups are free abelian. Suppose $C_0 = J$, generated by a_0, $C_1 = J$, generated by a_1, $D_0 = J$, generated by b_0, $D_1 = J \oplus J$, generated by b_1, b_1', with all other groups zero. The boundary homomorphisms are given by $a_1\partial = 2a_0$, $b_1\partial = 2b_0$, $b_1'\partial = b_0$. Let $\phi : C \to D$ be given by $a_0\phi = b_0$, $a_1\phi = b_1$. Then ϕ is a chain map and clearly $\phi_* = 0$. If $\phi = \partial\Delta + \Delta\partial$, then

$$b_1 = a_1\phi = a_1\partial\Delta = 2a_0\Delta,$$

which is a contradiction, and $\phi \not\simeq 0$. Thus the condition $\phi \simeq \psi$ is sufficient but not necessary for the conclusion $\phi_* = \psi_*$.

Suppose ϕ, ψ are chain maps $C \to D$ each mapping C^0 into D^0. If also $\phi - \psi = \partial\Delta + \Delta\partial$, where Δ maps C_p^0 into D_{p+1}^0, we say that Δ is a chain homotopy between $\phi, \psi : C, C^0 \to D, D^0$. In that case, as is readily verified, Δ induces a chain homotopy $\hat{\Delta} : \hat{C} \to \hat{D}$ between $\hat{\phi}$ and $\hat{\psi}$. We thus have

3.3.2 Theorem. *If $\phi \simeq \psi : C, C^0 \to D, D^0$, then*

$$\hat{\phi}_* = \hat{\psi}_* : H_*(\hat{C}) \to H_*(\hat{D}). \;\rule{0.5em}{1em}$$

The dual definitions for contrachain homotopy are evident. Now let $\Delta^{\cdot}$ be the set of homomorphisms $\Delta^p : D^p \to C^{p-1}$, where $D^p = D_p \pitchfork G$, $C^{p-1} = C_{p-1} \pitchfork G$ and Δ^p is adjoint to $\Delta_{p-1} : C_{p-1} \to D_p$.

3.3.3 Theorem. *If $\phi \overset{\Delta}{\simeq} \psi$, then $\phi^{\cdot} \overset{\Delta^{\cdot}}{\simeq} \psi^{\cdot}$.*

We have $\phi_p - \psi_p - \partial_p \Delta_{p-1} - \Delta_p \partial_{p+1} = 0$. Taking adjoints,

$$\phi^p - \psi^p - \delta^{p-1}\Delta^p - \Delta^{p+1}\delta^p = 0,$$

so that $\Delta^{\cdot}$ is a contrachain homotopy between $\phi^{\cdot}$ and $\psi^{\cdot}$. ▌

3.3.4 Corollary *If $\phi \simeq \psi : C, C^0 \to D, D^0$, then*

$$\hat{\phi}^* = \hat{\psi}^* : H^*(\hat{D}^{\cdot}) \to H^*(\hat{C}^{\cdot}).$$

For if Δ is a chain homotopy between ϕ and ψ then $\hat{\Delta}^{\cdot}$ is a contrachain homotopy between $\hat{\phi}^{\cdot}$ and $\hat{\psi}^{\cdot}$ ▮

3.3.5 Theorem. *If $\phi \simeq \psi : C(K), C(K_0) \to C(L), C(L_0)$, then*

$$\hat{\phi}_* = \hat{\psi}_* : H_*(K, K_0; G) \to H_*(L, L_0; G)$$

and

$$\hat{\phi}^* = \hat{\psi}^* : H^*(L, L_0; G) \to H^*(K, K_0; G). ▮$$

We have permitted ourselves to state the result in homology for arbitrary coefficients. It should be clear to the reader that a chain map $\phi : C(K) \to C(L)$ induces in an obvious way a chain map $\phi(G) : C(K; G) \to C(L; G)$ and similarly a chain homotopy Δ between ϕ and ψ induces a chain homotopy $\Delta(G)$ between $\phi(G)$ and $\psi(G)$. We may write ϕ, Δ for $\phi(G), \Delta(G)$ if no confusion is possible. Such obvious generalizations will not be given in detail.

Let $\phi : C \to D$, $\psi : D \to C$ be chain maps such that $\phi\psi \simeq 1 : C \to C$ and $\psi\phi \simeq 1 : D \to D$. Then ϕ is said to be a *chain equivalence* with *chain inverse* ψ. We also say that C and D are *chain equivalent*.

To prove that we really do get an equivalence relation between chain complexes, we proceed as for homotopy (Lemma 1.6.3, Theorem 1.6.4).

3.3.6 Lemma. *If $\phi_1 \simeq \phi_2 : C \to D$, $\theta : D \to E$, $\eta : F \to C$ are chain maps, then $\phi_1\theta \simeq \phi_2\theta$, $\eta\phi_1 \simeq \eta\phi_2$.*

For if Δ is a chain homotopy between ϕ_1 and ϕ_2, then $\Delta\theta$ is a chain homotopy between $\phi_1\theta$ and $\phi_2\theta$, and $\eta\Delta$ is a chain homotopy between $\eta\phi_1$ and $\eta\phi_2$. ▮

3.3.7 Theorem. *The relation of chain equivalence between chain complexes is an equivalence relation.*

The proof imitates that of Theorem 1.6.4. ▮ We note in particular that the composition of two chain equivalences is a chain equivalence. We write $\phi : C \simeq D$ if ϕ is a chain equivalence; if we do not wish to stress the map ϕ, we write $C \simeq D$.

3.3.8 Theorem. *If $\phi : C \simeq D$, then*

$$\phi_* : H_*(C) \cong H_*(D), \quad \phi^* : H^*(D \pitchfork G) \cong H^*(C \pitchfork G).$$

Let $\psi : D \simeq C$ be a chain inverse of ϕ. Then $\phi_*\psi_* = (\phi\psi)_* = 1 : H_*(C) \cong H_*(C)$ and similarly $\psi_*\phi_* = 1 : H_*(D) \cong H_*(D)$. This means that ϕ_* is an isomorphism with inverse ψ_*.

If $\phi : C \cong D$ with chain inverse ψ, then, with the obvious definition of contrachain equivalence, $\phi^{\cdot} : D \pitchfork G \simeq C \pitchfork G$ with contrachain inverse $\psi^{\cdot}$. Thus, as in the first part, $\phi^* : H^*(D \pitchfork G) \cong H^*(C \pitchfork G)$. ∎

We leave to the reader the relativization of this theorem.

3.4 Acyclic Carriers

We have seen that, to prove that two chain maps induce the same homology homomorphisms, it is sufficient to demonstrate the existence of a chain homotopy between them. In this section we describe how the existence of such a chain homotopy may be assured without actually carrying out the tedious construction. Our main theorem is restricted to a subclass of chain complexes broad enough to include chain groups of complexes with integer coefficients; thus our results will have useful geometrical corollaries.

We say that a chain complex C is *free* if for each p, C_p is free abelian. We say that C is *geometric* if (i) it is free, and (ii) for all $p < -1$, $C_p = 0$.

Then certainly $C(K)$ and $C(K^+)$ are geometric and also $C(K \bmod K_0)$; so too are $C^\Omega(K)$, $C^\Omega(K^+)$ and $C^\Omega(K \bmod K_0)$. These examples are responsible for the use of the word 'geometric'.

Given a geometric chain complex C, we may select a basis $\{\gamma_p^i\}$ for each C_p. When we speak of a geometric chain complex we shall always suppose such a selection to have been made. We shall write $\gamma_{p-1} \prec \gamma_p$ if γ_{p-1} appears with non-zero coefficient in $\gamma_p \partial$.

A chain complex D is said to be *acyclic* if $H_p(D) = 0$, for all p. Equivalently D is acyclic if every cycle of D bounds in D. Thus (2.10.5) the augmented chain complex of a cone is acyclic. We now make our main definitions.

3.4.1 Definition. An *algebraic carrier function* E from the geometric chain complex C to the chain complex D is a function defined on the basis elements $\{\gamma_p^i\}$ of C whose values are subcomplexes of D such that, if $\gamma_{p-1} \prec \gamma_p$, then $E(\gamma_{p-1}) \subseteq E(\gamma_p)$. E is said to be *acyclic* if each $E(\gamma)$ is acyclic.

3.4.2 Definition. The chain map $\phi : C \to D$ is said to be *carried by* the algebraic carrier function E if, for each γ, $\gamma\phi$ is a chain of $E(\gamma)$. We shall talk of E as a *carrier* for brevity. The main algebraic theorem is

3.4.3 Theorem. *If the chain map ϕ from the geometric chain complex C to the chain complex D is carried by an acyclic carrier, then $\phi \simeq 0$.*

We must find $\Delta : C_p \to D_{p+1}$ such that $\phi = \partial\Delta + \Delta\partial$. We define Δ inductively starting from $\Delta_p = 0$, $p < -1$. Now suppose Δ_q defined for $q < p$ so that $\phi = \partial\Delta + \Delta\partial$ and also $\gamma_q \Delta_q \in E(\gamma_q)$ for all γ_q; we shall define Δ_p. For this it is sufficient to specify Δ_p on each γ_p. Now $\gamma_p \partial\Delta \in E(\gamma_p)$; for $\gamma_p \partial\Delta \in \bigcup_{\gamma_{p-1} \prec \gamma_p} E(\gamma_{p-1})$ and $E(\gamma_{p-1}) \subseteq E(\gamma_p)$ whenever $\gamma_{p-1} \prec \gamma_p$. Thus $\gamma_p \phi - \gamma_p \partial\Delta \in E(\gamma_p)$; moreover, $\gamma_p \phi - \gamma_p \partial\Delta$ is actually a cycle. For

$$(\gamma_p \phi - \gamma_p \partial\Delta)\,\partial = \gamma_p \phi\partial - \gamma_p \partial\Delta\partial = \gamma_p \partial\phi - \gamma_p \partial\Delta\partial = \gamma_p \partial(\phi - \Delta\partial).$$

Now $\phi = \Delta\partial + \partial\Delta$ in dimension $(p-1)$. Thus

$$\gamma_p \partial(\phi - \Delta\partial) = \gamma_p \partial(\partial\Delta) = 0.$$

Since $\gamma_p \phi - \gamma_p \partial\Delta$ is a cycle and $E(\gamma_p)$ is acyclic, it follows that there exists a $(p+1)$-chain of $E(\gamma_p)$, say d, such that $\gamma_p \phi - \gamma_p \partial\Delta = d\partial$. We put $\gamma_p \Delta = d$. Then $\gamma_p \phi = \gamma_p(\partial\Delta + \Delta\partial)$ and $\gamma_p \Delta \in E(\gamma_p)$. We proceed in this way for each $\gamma_p \in C_p$ and thus complete the induction. ❚
Notice that Δ is far from unique; we exercise a choice for each γ.

It is clear that this proof depends heavily on the restrictions placed on C; we could not start the induction if there were no 'bottom dimension' and the right to describe a homomorphism of C_p in terms of its effect on a set of generators depends on the freedom of the set.

3.4.4 Corollary. *If the chain maps $\phi, \psi : C \to D$ are both carried by the acyclic carrier E, then $\phi \simeq \psi$.* ❚

We make a significant generalization of this trivial result. We shall say that a chain map $\phi : C \to D$ is carried by E *in dimension* p if $\gamma_p \phi \in E(\gamma_p)$ for all γ_p.

3.4.5 Corollary. *If $\phi, \psi : C \to D$ agree in dimensions $< m$ and have a common acyclic carrier in dimensions $\geqslant m$, then $\phi \simeq \psi$.*

For, if ϕ, ψ agree in dimensions $< m$ and are carried by E in dimensions $\geqslant m$, then $\phi - \psi$ is a chain map carried by E in all dimensions. ❚

We shall use 3.4.5 with $m = 0$, for we remark that all the chain maps of chain complexes of augmented simplicial complexes which we shall meet will agree in dimensions < 0, mapping σ_{-1} to σ_{-1}. Notice that 3.4.4 can be compared with Proposition 1.7.9 on contiguous maps; the analogy will be pressed more closely later in this section.

We have remarked that Theorem 3.4.3 is applicable to $C(K)$ and $C(K \bmod K_0)$; it is also, of course, applicable to $C^\Omega(K)$ and $C^\Omega(K \bmod K_0)$. Since we wish to regard $C^\Omega(K)$ as a 'geometric' object, we shall carry over to it the language used for describing K itself. Thus the admissible arrays $[a^{i_0} \ldots a^{i_p}]$ will be called *simplexes* and the *faces* of $[a^{i_0} \ldots a^{i_p}]$ are those arrays obtained by suppressing some of the vertices without, of course, changing the order. When taking this point of view, we may write K^Ω for $C^\Omega(K)$, and it is then clear what is meant by a *subcomplex* of K^Ω. The simplexes of K^Ω constitute the chosen basis for the chain complex $C^\Omega(K)$.

We introduce one further device to connect the geometry and the algebra. In practice we shall probably recognize a *simplicial* subcomplex (of L) and not a *chain* subcomplex (of $C(L)$) as carrying a certain chain, and we therefore introduce the notion of a *geometric carrier function*.

Our object will be to use 3.4.3 to show that if $\phi, \psi : C(K) \to C(L)$ are two (suitably restricted) chain maps carried by the same *acyclic* geometric carrier function then $\phi \simeq \psi$. We make use of augmentation in that we may describe K as *acyclic* if and only if $H_p(K^+) = 0$, all p. Then K is acyclic if and only if $C(K^+)$ is acyclic. Note that $C(K)$ is not acyclic unless K is empty.

3.4.6 Definition. A *geometric carrier function* P from the complex K to the complex L is a function defined on the simplexes of K whose values are subcomplexes of L such that, if $s_{p-1} \prec s_p$, then

$$P(s_{p-1}) \subseteq P(s_p).$$

P is said to be *acyclic* if $P(s)$ is acyclic for all $s \in K$.

3.4.7 Lemma. *A geometric carrier function P from K to L determines an algebraic carrier function $E^+(P) = E^+$ from $C(K^+)$ to $C(L^+)$ by the rule*

$$E^+(\sigma_p) = C(P(s_p)^+), \quad p \geqslant 0,$$

$$E^+(\sigma_{-1}) = 0.$$

Moreover, E^+ is acyclic if P is acyclic. ∎

3.4.8 Definition. The chain map $\phi : C(K) \to C(L)$ is said to be *carried* by the geometric carrier function P if, for each $s_p \in K$, $\sigma_p \phi$ is a chain of $P(s_p)$.

A chain map $\phi^+ : C(K^+) \to C(L^+)$ determines by restriction a chain map $\phi : C(K) \to C(L)$. Conversely, if $\phi : C(K) \to C(L)$ is the restriction of some ϕ^+, we shall say that ϕ is *augmentable* (to ϕ^+); we remark that if ϕ is augmentable (and K is non-empty) then ϕ^+ is determined by ϕ, since $\sigma_{-1}\phi^+ = \sigma_0 \partial \phi^+ = \sigma_0 \phi \partial$.

3.4.9 Lemma. *If* $\phi : C(K) \to C(L)$ *is augmentable and is carried by* P, *then* ϕ^+ *is carried by* $E^+(P)$ *in dimensions* $\geqslant 0$. ∎

We shall write $E(P)$ for $E^+(P)$ restricted to simplexes of K, and we shall say that ϕ is carried by $E(P)$ if ϕ^+ is carried by $E^+(P)$ in dimension $\geqslant 0$. Notice that the *values* of $E(P)$ are *augmented* complexes. The essential fact we need is (cf. Corollary 3.4.5)

3.4.10 Proposition. *If* $\phi, \psi : C(K) \to C(L)$ *are augmentable and are both carried by* $E(P)$ *and if* $\sigma_{-1}\phi^+ = \sigma_{-1}\psi^+$, *then* $\phi^+ - \psi^+$ *is carried by* $E^+(P)$. ∎

This suggests that we impose a restriction on chain maps $C(K) \to C(L)$ to ensure that $\sigma_{-1}\phi^+ = \sigma_{-1}\psi^+$. We describe a chain map $\phi : C(K) \to C(L)$ as *proper* if it is augmentable with $\sigma_{-1}\phi^+ = \sigma_{-1}$. That this is a reasonable notion in geometrically significant situations is ensured by the following composite lemma:

3.4.11 Lemma. (i) *If* $\phi : C(K) \to C(L)$ *is induced by a simplicial map* K *to* L, *then* ϕ *is proper;* (ii) $\chi : C(K) \to C(K')$ *is proper;* (iii) $\theta^\alpha : C^\Omega(K) \to C^\alpha(K)$ *is proper;* (iv) $\bar{\theta}^\omega : C^\omega(K) \to C^\Omega(K)$ *is proper;* (v) *the composite of two proper chain maps is proper.* ∎

We remark that in (iii) and (iv) we have carried out our declared intention to regard K^Ω as a 'simplicial' complex to which the notions associated with a geometric carrier function are appropriate.

3.4.12 Corollary. *If* $\phi, \psi : C(K) \to C(L)$ *are proper chain maps carried by* $E(P)$ *then* $\phi^+ - \psi^+$ *is carried by* $E^+(P)$. ∎

We may evidently relativize the notions of algebraic and geometric carrier functions; thus we say that P is a carrier from the pair K, K_0 to the pair L, L_0 if P is a carrier from K to L and $P(s) \subseteq L_0$ whenever $s \in K_0$, and similarly for the algebraic carrier E. We now sum up the discussion in this section in the following theorem:

3.4.13 Theorem. *If* $\phi, \psi : C(K), C(K_0) \to C(L), C(L_0)$ *are proper chain maps carried by the acyclic carrier* P, *then* $\phi \simeq \psi$ *and*

$$\hat{\phi}_* = \hat{\psi}_* : H_*(K, K_0; G) \to H_*(L, L_0; G),$$

$$\hat{\phi}^* = \hat{\psi}^* : H^*(L, L_0; G) \to H^*(K, K_0; G).$$

Since ϕ, ψ are carried by P, they are carried by $E(P)$ (3.4.9). Since ϕ, ψ are proper, $\phi^+ - \psi^+$ is carried by $E^+(P)$ (3.4.12). Since P is acyclic, $E^+(P)$ is acyclic (3.4.7) so that $\phi^+ - \psi^+ \simeq 0$ (3.4.3). Moreover, the chain homotopy constructed in the proof of 3.4.3 will map $C(K_0^+)$ into $C(L_0^+)$ and σ_{-1} to 0. Thus by restricting to dimensions $\geqslant 0$, $\phi \simeq \psi : C(K), C(K_0) \to C(L), C(L_0)$. The remaining assertions simply repeat Theorem 3.3.5. ∎

Clearly the augmentable chain maps $C(K) \to C(L)$ form a subgroup of the additive group of all chain maps $C(K) \to C(L)$; on the other hand the proper chain maps do not form a subgroup and, in particular, the zero chain map, though augmentable, is not proper. Thus 3.4.13 does not warrant the false conclusion that every simplicial map is chain-homotopic to zero!

It is worth noting that we may generalize Theorem 3.4.13 by requiring simply that $\sigma_{-1}\phi^+ = \sigma_{-1}\psi^+$ (i.e. replacing 3.4.12 by 3.4.10). We may define the *index* of ϕ as that integer m such that $\sigma_{-1}\phi^+ = m\sigma_{-1}$; then the conclusions of Theorem 3.4.13 follow if we replace the requirement that ϕ, ψ be proper by the requirement that they be augmentable with the same index.

Let $\epsilon_K : C_0(K) \to J$ be the homomorphism given by $(\Sigma n_i \sigma_0^i)\,\epsilon = \Sigma n_i$; then $\phi : C(K) \to C(L)$ is augmentable with index m if and only if $\phi\epsilon_L = m\epsilon_K$. It is very convenient for some purposes (e.g. 4.4.3) to regard ϵ_K itself as the augmentation of the complex K. This point of view will appear to advantage in chapter 10; for the purposes of this chapter we have preferred the alternative viewpoint.

Theorem 3.4.13 has an analogue for total complexes which requires no explicit proof; namely,

3.4.14 Theorem. *If $\phi, \psi : C^\Omega(K), C^\Omega(K_0) \to C^\Omega(L), C^\Omega(L_0)$ are proper chain maps carried by the acyclic carrier P, then $\phi \simeq \psi$ and*

$$\hat{\phi}_* = \hat{\psi}_* : H^\Omega_*(K, K_0; G) \to H^\Omega_*(L, L_0; G),$$

$$\hat{\phi}^* = \hat{\psi}^* : H^*_\Omega(L, L_0; G) \to H^*_\Omega(K, K_0; G). \;\blacksquare$$

Theorems 3.4.13 and 3.4.14 in a sense define our problem; wherever we wish to establish that two chain maps induce the same homology and contrahomology homomorphisms, we shall in fact find a common acyclic (geometric) carrier. This technique will yield a rich harvest in the next section‡. Meanwhile, however, we give one immediate application of the technique.

We shall say that the simplicial maps $v, w : K, K_0 \to L, L_0$ are contiguous if they are contiguous as simplicial maps $K \to L$ and $v \mid K_0$, $w \mid K_0$ are contiguous simplicial maps $K_0 \to L_0$.

3.4.15 Theorem. *If $v, w : K, K_0 \to L, L_0$ are contiguous and induce $\phi, \psi : C(K), C(K_0) \to C(L), C(L_0)$, then $\phi \simeq \psi$.*

If $s \in K$, let $P(s)$ be the closure of the simplex of L of lowest dimension of which sv and sw are faces. Then $P(s) \subseteq L_0$ if $s \in K_0$ and P is

‡ A generalization of the technique is applied extensively in chapters 8 and 9.

obviously a carrier function for ϕ and ψ. Moreover, it is acyclic (2.10.5). Finally, ϕ and ψ are proper (3.4.11 (i)). Thus Theorem 3.4.13 may be applied. ▌ It follows that contiguous maps induce the same homology and contrahomology homomorphisms. In particular, then, this must be true of any two standard maps from the first derived K' to K.

3.5 Chain equivalences in simplicial complexes

In this section we use Theorems 3.4.13 and 3.4.14 to demonstrate two fundamental chain equivalences.

Let (K, ω) be an ordered simplicial complex and K_0 an ordered subcomplex. We recall that there are chain maps

$$\theta = \theta^\omega : C^\Omega(K), C^\Omega(K_0) \to C^\omega(K), C^\omega(K_0),$$
$$\bar{\theta} = \bar{\theta}^\omega : C^\omega(K), C^\omega(K_0) \to C^\Omega(K), C^\Omega(K_0)$$

such that $\bar{\theta}\theta = 1$ (3.2.2 and 3.2.3). We prove

3.5.1 Theorem. $\theta\bar{\theta} \simeq 1 : C^\Omega(K), C^\Omega(K_0) \to C^\Omega(K), C^\Omega(K_0)$.

We use the geometrical language foreshadowed in the previous section. Given any 'simplex' $s^\Omega = [a^{i_0} \dots a^{i_p}]$, let $P(s^\Omega)$ be the set of all arrays $[a^{j_0} \dots a^{j_q}]$, where each j_r is one of the indices $i_0, \dots, i_p$. Obviously all such arrays are 'simplexes' and $P(s^\Omega)$ is a subcomplex of K^Ω. Further, if $s^\Omega \prec t^\Omega$ then $P(s^\Omega) \subseteq P(t^\Omega)$ and $P(s^\Omega) \subseteq K_0^\Omega$ if $s^\Omega \in K_0^\Omega$. Thus P is a carrier function. It obviously carries the identity map and it also carries $\theta\bar{\theta}$; for $s^\Omega\theta\bar{\theta} = 0$ if s^Ω has repeated vertices and otherwise is $(\pm)$ a simplex obtained by permuting the vertices of s^Ω. Thus to apply Theorem 3.4.14 it remains to establish that P is acyclic.

We must show that $P(s^\Omega)$ is acyclic; the proof follows closely that of 2.10.5. For any $t^\Omega = [a^{j_0} \dots a^{j_q}]$ of $P(s^\Omega)^+$, let

$$at^\Omega = [aa^{j_0} \dots a^{j_q}] \in P(s^\Omega)^+,$$

where $a = a^{i_0}$. We extend this definition by linearity to yield a mapping $c \to ac$ from $C_q(P(s^\Omega)^+)$ to $C_{q+1}(P(s^\Omega)^+)$, for all q. Then $(ac)\partial = c - a(c\partial)$. Thus, if z is a cycle of $P(s^\Omega)^+$, $z\partial = 0$ and $(az)\partial = z$, so that z is a boundary. Every cycle of $P(s^\Omega)^+$ bounds in $P(s^\Omega)^+$ and $P(s^\Omega)$ is acyclic. ▌

3.5.2 Corollary. $\theta^\omega : C^\Omega(K), C^\Omega(K_0) \simeq C^\omega(K), C^\omega(K_0)$ *with chain inverse* $\bar{\theta}^\omega$. ▌

3.5.3 Corollary. (i) θ^ω *induces* $\hat{\theta}^\omega_* : H^\Omega_*(K, K_0; G) \cong H^\omega_*(K, K_0; G)$, *and the inverse of* $\hat{\theta}^\omega_*$ *is* $\hat{\bar{\theta}}^\omega_*$; (ii) θ^ω *induces* $\hat{\theta}^*_\omega : H^*_\omega(K, K_0; G) \cong H^*_\Omega(K, K_0; G)$, *and the inverse of* $\hat{\theta}^*_\omega$ *is* $\hat{\bar{\theta}}^*_\omega$.

It is only necessary to observe that Theorem 3.5.1 as stated implies that it also holds when the chains are taken with coefficients in G. ∎

3.5.4 Theorem. *Let (K, α) be an oriented simplicial complex and let K_0 be a subcomplex. Then $\theta^\alpha : C^\Omega(K), C^\Omega(K_0) \to C^\alpha(K), C^\alpha(K_0)$ is a chain equivalence, and so $\hat{\theta}^\alpha_*$, $\hat{\theta}^*_\alpha$ are isomorphisms.*

Let ω be an ordering of K. Then (2.3.1)

$$\theta^{\alpha\omega} : C^\alpha(K), C^\alpha(K_0) \to C^\omega(K), C^\omega(K_0)$$

is an isomorphic chain map with inverse $\theta^{\omega\alpha}$. Moreover,

$$\theta^\alpha = \theta^\omega \theta^{\omega\alpha} : C^\Omega(K), C^\Omega(K_0) \to C^\alpha(K), C^\alpha(K_0).$$

Now we have proved (3.5.2) that θ^ω is a chain equivalence with chain inverse $\bar{\theta}^\omega$. It follows that θ^α is a chain equivalence with chain inverse $\theta^{\alpha\omega}\bar{\theta}^\omega$. ∎

This theorem justifies the importance we attach to the total complex. Though $C^\Omega(K)$ is uncomfortably large for computation it is theoretically more satisfactory than $C(K)$, as it is defined independently of an orientation or ordering in K. Its technical advantages will be revealed in the next chapter and in chapter 8.

Now let K be an oriented simplicial complex and K_0 a subcomplex. Let (K', K'_0) be the first derived of (K, K_0), let $v : K', K'_0 \to K, K_0$ be a standard simplicial map inducing $\phi : C(K'), C(K'_0) \to C(K), C(K_0)$ and let $\chi : C(K), C(K_0) \to C(K'), C(K'_0)$ be the subdivision chain map (3.2.9) so that $\chi\phi = 1$ (3.2.10).

3.5.5 Theorem. $\phi\chi \simeq 1 : C(K'), C(K'_0) \to C(K'), C(K'_0)$.

Let s' be a simplex of K' and let b be the leading vertex of s'. Then b is the barycentre of a simplex s of K ‡. We set $P(s') = \bar{s}'$. We have referred to s as the *carrier* of s' (see 1.7). Then if $t' \prec s'$, the carrier of t' is a face of the carrier of s' so that $P(t') \subseteq P(s')$. Also if $s' \in K'_0$, then $P(s') \subseteq K'_0$. Thus P is a carrier from K', K'_0 to K', K'_0; P obviously carries the identity and it also carries $\phi\chi$. For v maps s' onto a face of its carrier, s, and χ maps a chain of $\bar{s}$ onto a chain of $\bar{s}'$. Thus it only remains to prove that P is acyclic. Now $\bar{s}'$ is just the cone with vertex b and base $\dot{s}'$. Thus $P(s')$ is acyclic (2.10.5) and Theorem 3.4.13 (and Lemma 3.4.11) may be applied. ∎

‡ The notation is misleading: K' is the first derived of K but s' is only a simplex of the first derived of s. Thus $\bar{s}'$ should not be confused with the closure of s'.

3.5.6 Corollary. *If $\phi : C(K'), C(K'_0) \to C(K), C(K_0)$ is a standard chain map then $\phi : C(K'), C(K'_0) \simeq C(K), C(K_0)$ with chain inverse χ.* ∎

3.5.7 Corollary. (i) *χ induces $\hat{\chi}_* : H_*(K, K_0; G) \cong H_*(K', K'_0; G)$ and the inverse of $\hat{\chi}_*$ is $\hat{\phi}_*$, where ϕ is a standard chain map*

$$C(K'), C(K'_0) \to C(K), C(K_0);$$

(ii) *χ induces $\hat{\chi}^* : H^*(K', K'_0); G) \cong H^*(K, K_0; G)$ and the inverse of $\hat{\chi}^*$ is $\hat{\phi}^*$.* ∎

This result is fundamental in homology theory‡. It establishes the existence of a natural isomorphism between the homology groups of a pair and those of the first derived. The sense in which $\hat{\chi}_*$ is natural is as follows. We saw (1.7.3) that a simplicial map $v : K, K_0 \to L, L_0$ induces a simplicial map $v' : K', K'_0 \to L', L'_0$. Then v, v' induce homomorphisms§

$$\hat{\phi}_* : H_*(K, K_0; G) \to H_*(L, L_0; G), \quad \hat{\phi}'_* : H_*(K', K'_0; G) \to H_*(L', L'_0; G).$$

We may prove

3.5.8
$$\hat{\chi}_* \hat{\phi}'_* = \hat{\phi}_* \hat{\chi}_*,$$

so that $\hat{\chi}_*$ is, as it were, an isomorphism of the system consisting of homology groups and induced homomorphisms. This is an example of a natural equivalence in the sense of Eilenberg and Steenrod [14]. To establish 3.5.8, we prove

3.5.9 Proposition. $\chi\phi' = \phi\chi$.

This is proved by induction on dimension. It is trivial in dimensions -1 or 0. Let $s \in K$ with barycentre b; then

$$\begin{aligned} \sigma\chi\phi' &= (b(\sigma\partial\chi))\,\phi' = bv'(\sigma\partial\chi\phi') \\ &= bv'(\sigma\partial\phi\chi), \quad \text{by the inductive hypothesis,} \\ &= bv'(\sigma\phi\partial\chi). \end{aligned}$$

Now if s degenerates under v then $\sigma\phi = 0$, so that $\sigma\phi\chi = 0$ and $\sigma\chi\phi' = 0$. If s does not degenerate under v then bv' is the barycentre

‡ The reader had probably been thinking in terms of finite simplicial complexes; this was intended, but the arguments and results of this section apply to infinite complexes.

§ ϕ is of course no longer a standard chain map.

of sv (by definition of v') and $\sigma\phi$ is sv after orientation‡. Thus, by definition of χ,
$$bv'(\sigma\phi\,\partial\chi) = \sigma\phi\chi$$
and the induction is complete. ∎

It will be our object to show that homology groups and, in a sense, induced homomorphisms of homology groups depend only on the underlying polyhedra and maps; 3.5.7 and 3.5.8 are the first major strides in this direction. Meanwhile we use 3.2.10 to improve 3.4.15.

Let $K^{(r)}$, $K_0^{(r)}$ be the rth derived of K, K_0. Then we define by composition a subdivision chain map $\chi^{(r)} : C(K), C(K_0) \to C(K^{(r)}), C(K_0^{(r)})$ and a standard chain map $\theta^{(r)} : C(K^{(r)}), C(K_0^{(r)}) \to C(K), C(K_0)$ and we have

3.5.10
$$\theta^{(r)}\chi^{(r)} \simeq 1, \chi^{(r)}\theta^{(r)} = 1.$$

Now suppose $v : K, K_0 \to L, L_0$ and $w : K^{(r)}, K_0^{(r)} \to L, L_0$ are contiguous (Definition 1.7.8). Then if $u^{(r)} : K^{(r)}, K_0^{(r)} \to K, K_0$ is a standard simplicial map, $u^{(r)}v$ and w are contiguous. Let v induce
$$\phi : C(K), C(K_0) \to C(L), C(L_0)$$
and let w induce
$$\psi : C(K^{(r)}), C(K_0^{(r)}) \to C(L), C(L_0).$$

3.5.11 Theorem. $\theta^{(r)}\phi \simeq \psi$, $\phi \simeq \chi^{(r)}\psi$.

For let $u^{(r)}$ induce $\tilde{\theta}^{(r)}$. Then $\tilde{\theta}^{(r)} \simeq \theta^{(r)}$, each being a standard chain map, and $u^{(r)}v$ induces $\tilde{\theta}^{(r)}\phi$. By 3.4.15, $\tilde{\theta}^{(r)}\phi \simeq \psi$ and, by 3.3.6 $\theta^{(r)}\phi \simeq \tilde{\theta}^{(r)}\phi$. Thus $\theta^{(r)}\phi \simeq \psi$. Applying $\chi^{(r)}$ on the left, and using 3.5.10, we get $\phi \simeq \chi^{(r)}\psi$. ∎

3.6 Continuous maps of polyhedra and the main theorems

In this section we prove that a continuous map from one (compact) polyhedron to another induces a homomorphism of the homology groups of the associated complexes and that the homomorphism is an isomorphism if the given map is a homotopy equivalence. This is a considerably strengthened form of the statement that the homology groups are topologically invariant, for a homeomorphism is, *a fortiori*, a homotopy equivalence. However, it is of great importance to have the isomorphism between the groups in a natural form, that is, induced by a continuous map, and we therefore wish to stress that the association of a homomorphism of homology (and contrahomology) groups

‡ It is irrelevant to this argument whether or not the orientation of sv in L is that of $\sigma\phi$.

with a given continuous map is in itself a fundamental fact of homology theory.

We shall deal with the relative case since no greater difficulty is experienced in this case and the absolute case may then be deduced from it by suppressing the subcomplex. However, for the first time in this chapter we must confine attention to finite complexes, since we wish to invoke the results of 1.6–1.8 in particular.

Suppose given two pairs K, K_0 and L, L_0 and a map

$$f : |K|, |K_0| \to |L|, |L_0|.$$

By the simplicial approximation theorem (1.8.1), there exists an integer r and a simplicial approximation of f, say $v : K^{(r)} \to L$. This implies that, for each $x \in |K|$, $x|v|$ belongs to the closure of the carrier of xf so that, if $x \in |K_0|$, $x|v| \in |L_0|$ and v is a map

$$v : K^{(r)}, K_0^{(r)} \to L, L_0.$$

Then v induces $\phi : C(K^{(r)}), C(K_0^{(r)}) \to C(L), C(L_0)$. Let

$$\chi^{(r)} : C(K), C(K_0) \to C(K^{(r)}), C(K_0^{(r)})$$

be the subdivision chain map. Then $\chi^{(r)}\phi$ is a chain map

$$\chi^{(r)}\phi : C(K), C(K_0) \to C(L), C(L_0).$$

We wish to define the homology homomorphism‡ induced by f as $(\chi^{(r)}\phi)_*$. Before we may do this, however, we have to show that $(\chi^{(r)}\phi)_*$ depends only on f. Suppose then that $w : K^{(s)}, K_0^{(s)} \to L, L_0$ is a second simplicial approximation to f, and let w induce

$$\psi : C(K^{(s)}), C(K_0^{(s)}) \to C(L), C(L_0).$$

We must show that $$\chi^{(s)}\psi \simeq \chi^{(r)}\phi.$$

We may assume without real loss of generality that $s \geqslant r$. Let $\chi^{(s-r)} : K^{(r)}, K_0^{(r)} \to K^{(s)}, K_0^{(s)}$ be the subdivision chain map. Then $\chi^{(s)} = \chi^{(r)}\chi^{(s-r)}$. By Theorem 1.7.11, v and w are contiguous so that, by 3.5.11, $\phi \simeq \chi^{(s-r)}\psi$. Applying $\chi^{(r)}$ to each side gives the result. We may now define the induced homomorphism.

3.6.1 Definition. We call $\hat{\chi}_*^{(r)}\hat{\phi}_* : H_*(K, K_0; G) \to H_*(L, L_0; G)$ and $\hat{\chi}^{(r)*}\hat{\phi}^* : H^*(L, L_0; G) \to H^*(K, K_0; G)$ the *homomorphisms‡ induced by* f. We write f_* for $\hat{\chi}^{(r)}\hat{\phi}_*$, f^* for $\hat{\chi}^{(r)*}\hat{\phi}^*$.

With a view to an application in 5.8 we set on record the following obvious consequence of 3.5.10 and 3.6.1.

‡ This is of course the homomorphism induced by f for the triangulations K, K_0 and L, L_0. The same map f also induces a homology homomorphism for any two simplicial pairs covering the polyhedra $|K|, |K_0|$ and $|L|, |L_0|$.

3.6.2 Proposition. *Let $f : |K|, |K_0| \to |L|, |L_0|$ be a map and let $v : K^{(q)} \to L^{(r)}$ be a simplicial approximation to f; then $f_* \hat{\chi}_*^{(r)} = \hat{\chi}_*^{(q)} \hat{\phi}_*$.*

Let $w : L^{(r)} \to L$ be a standard map inducing $\hat{\psi}_*$; then vw is an approximation to f, so that $f_* = \hat{\chi}_*^{(q)}(\hat{\phi}\hat{\psi})_* = \hat{\chi}_*^{(q)}\hat{\phi}_*\hat{\psi}_*$. Apply $\hat{\chi}_*^{(r)}$ on the right to get $f_*\hat{\chi}_*^{(r)} = \hat{\chi}_*^{(q)}\hat{\phi}_*$. ∎

We now prove the two basic properties about the induced homomorphism.

3.6.3 Theorem. *If $f_0 \simeq f_1 : |K|, |K_0| \to |L|, |L_0|$, then $f_{0*} = f_{1*}, f_0^* = f_1^*$.*

The enunciation implies that there exists a homotopy

$$f_t : |K|, |K_0| \to |L|, |L_0|$$

connecting f_0 and f_1. By Theorem 1.8.6 (relativized in the obvious way) there exists a sequence of simplicial maps

$$v_i : K^{(r_i)}, K_0^{(r_i)} \to L, L_0, \quad 1 \leqslant i \leqslant k,$$

such that v_1 is a simplicial approximation to f_0, v_k is a simplicial approximation to f_1, and v_{i-1}, v_i are contiguous. If

$$\chi^{(r)} : K, K_0 \to K^{(r)}, K_0^{(r)}$$

is the subdivision chain map we have‡ $f_{0*} = \hat{\chi}_*^{(r_1)} v_{1*}$, $f_{1*} = \hat{\chi}_*^{(r_k)} v_{k*}$, and, since v_{i-1}, v_i are contiguous,

$$\hat{\chi}_*^{(r_{i-1})} v_{i-1*} = \hat{\chi}_*^{(r_i)} v_{i*}, \quad i = 2, \ldots, k,$$

as proved in Theorem 3.5.11. Thus $f_{0*} = f_{1*}$ and the assertion $f_0^* = f_1^*$ follows similarly. ∎

3.6.4 Theorem. *If $f : |K|, |K_0| \to |L|, |L_0|$, $g : |L|, |L_0| \to |M|, |M_0|$ are maps, then $(fg)_* = f_* g_*$, $(fg)^* = f^* g^*$.*

Let $v : K^{(r)}, K_0^{(r)} \to L, L_0$ be a simplicial approximation to f and let $w : L^{(s)}, L_0^{(s)} \to M, M_0$ be a simplicial approximation to g. As described in Proposition 1.7.3, v induces $v^{(s)} : K^{(r+s)}, K_0^{(r+s)} \to L^{(s)}, L_0^{(s)}$ and, by iterating Proposition 3.5.9,

3.6.5 $$\chi^{(s)}\phi^{(s)} = \phi\chi^{(s)},$$

where $\chi^{(s)}$ refers to each of the subdivision chain maps,

$$K^{(r)}, K_0^{(r)} \to K^{(r+s)}, K_0^{(r+s)} \quad \text{and} \quad L, L_0 \to L^{(s)}, L_0^{(s)}$$

and $\phi, \phi^{(s)}$ are the chain maps induced by $v, v^{(s)}$. Moreover,§ $|v^{(s)}| \simeq |v| \simeq f$, so that $fg \simeq |v^{(s)}| g \simeq |v^{(s)}|\,|w| = |v^{(s)}w|$. Now, trivially,

$$v^{(s)}w : K^{(r+s)}, K_0^{(r+s)} \to M, M_0$$

is a simplicial approximation to $|v^{(s)}w|$.

‡ By convention, v_* is the relative homology homomorphism $\hat{\phi}_*$, where ϕ is the chain map induced by v. § See 1.7.7.

Thus
$$
\begin{aligned}
(fg)_* &= |v^{(s)}w|_*, && \text{by theorem 3.6.3,}\\
&= \hat{\chi}_*^{(r+s)}(v^{(s)}w)_*, && \text{by definition,}\\
&= \hat{\chi}_*^{(r)}\hat{\chi}_*^{(s)}v_*^{(s)}w_*, && \text{by 3.2.8,}\\
&= \hat{\chi}_*^{(r)}v_*\hat{\chi}_*^{(s)}w_*, && \text{by 3.6.5,}\\
&= f_*g_*, && \text{by definition.}
\end{aligned}
$$

A similar argument establishes that $(fg)^* = f^*g^*$. ∎

The following diagram may help the reader to understand the proof.

$$
\begin{array}{ccccc}
K, K_0 & & & & \\
\downarrow \chi^{(r)} & & & & \\
K^{(r)}, K_0^{(r)} & \xrightarrow{\ v\ } & L, L_0 & & \\
\downarrow \chi^{(s)} & & \downarrow \chi^{(s)} & & \\
K^{(r+s)}, K_0^{(r+s)} & \xrightarrow{\ v^{(s)}\ } & L^{(s)}, L_0^{(s)} & \xrightarrow{\ w\ } & M, M_0
\end{array}
$$

Definition 3.6.1 expresses the *covariance* of homology and the *contravariance* of contrahomology. Theorem 3.6.4, together with the evident fact that an identity map induces an identity automorphism, says that each of homology, contrahomology is a *functor*. That is to say, in the language of Eilenberg and Steenrod‡ [14], homology and contrahomology are both functors from the *category* of triangulated pairs and continuous maps to the category of abelian groups and homomorphisms; but whereas f_* goes in the *same* direction as f, f^* goes in the *opposite* direction. Also, the image, under the homology functor, of the map f followed by the map g is the homomorphism f_* followed by the homomorphism g_*, whereas the image under the contrahomology functor is the homomorphism g^* followed by the homomorphism f^*. These facts are the basis of our use of the term '*contra*homology' where other authors (excluding Postnikov and Boltyanskii) use the term 'cohomology', and for our writing covariant operators on the right, contravariant operators on the left. Unfortunately, the use of 'cohomology' for 'homology', though logically unassailable, would lead to terrifying confusion.

It is interesting to trace the proof of Theorem 3.6.4. The starting-off point is the observation that the composition of simplicial maps is simplicial (1.7.2). Next we observe (3.2.7) that the passage from simplicial maps to chain maps is functorial and then we observe (3.1.9) that the passage from chain maps to homology homomorphisms is functorial. The simplicial approximation theorem is invoked to effect a passage from continuous maps to simplicial maps and so to define the induced homomorphism in terms of the 'natural equivalence' χ and the covariant functor from simplicial maps to homology homomorphisms.

Theorems 3.6.3 and 3.6.4 enable us to prove our main theorem.

‡ More precisely, the language of Eilenberg and Maclane, but we refer the reader to the description in chapter 4 of [14].

3.6.6 Theorem.‡ *If $f : |K|, |K_0| \to |L|, |L_0|$ is a homotopy equivalence, then*

$$f_* : H_*(K, K_0; G) \cong H_*(L, L_0; G),$$
$$f^* : H^*(L, L_0; G) \cong H^*(K, K_0; G).$$

For let g be a homotopy inverse of f, where $g : |L|, |L_0| \to |K|, |K_0|$. Then $fg \simeq 1 : |K|, |K_0| \to |K|, |K_0|$. Since 1 is simplicial it is obvious that $1_* = 1$. Thus $1 = (fg)_* = f_* g_*$ by 3.6.3, 3.6.4. Similarly, $g_* f_* = 1$. This means that

$$f_* : H_*(K, K_0; G) \cong H_*(L, L_0; G)$$

with inverse g_*. ∎

We may extend the statements of this section so that the algebraic objects we associate with the pair K, K_0 are not just the homology and contrahomology groups, but the homology and contrahomology *sequences*.

If $f : |K|, |K_0| \to |L|, |L_0|$ is a map, then f not only induces

$$f_* : H_*(K, K_0; G) \to H_*(L, L_0; G)$$

but also

$$f_* : H_*(K; G) \to H_*(L; G) \quad \text{and} \quad f_* : H_*(K_0; G) \to H_*(L_0; G).$$

More precisely, we may refer to these three homomorphisms as $\hat{f}_*, f_*$ and f^0_*. We thus get a diagram

3.6.7
$$\begin{array}{ccccccccc} \dots \to & H_{p+1}(K, K_0; G) & \xrightarrow{\nu_*} & H_p(K_0; G) & \xrightarrow{\lambda_*} & H_p(K; G) & \xrightarrow{\mu_*} & H_p(K, K_0; G) & \to \dots \\ & \downarrow \hat{f}_* & & \downarrow f^0_* & & \downarrow f_* & & \downarrow \hat{f}_* & \\ \dots \to & H_{p+1}(L, L_0; G) & \xrightarrow{\nu_*} & H_p(L_0; G) & \xrightarrow{\lambda_*} & H_p(L; G) & \xrightarrow{\mu_*} & H_p(L, L_0; G) & \to \dots \end{array}$$

3.6.8 Theorem. *Commutativity holds in* 3.6.7 *; that is,*

$$\hat{f}_* \nu_* = \nu_* f^0_*, \quad f^0_* \lambda_* = \lambda_* f_* \quad \text{and} \quad f_* \mu_* = \mu_* \hat{f}_*.$$

We first observe that the assertion is quite trivial if f is a simplicial map, or, indeed, if the vertical homomorphisms are induced by any chain map $\phi : C(K), C(K_0) \to C(L), C(L_0)$. For then the commutativities hold at the chain level; the first expresses the fact that ϕ is a chain map ($\phi\partial = \partial\phi$) and the other two are self-evident.

We now return to the general case: let $v : K^{(r)}, K_0^{(r)} \to L, L_0$ be a

‡ We emphasize that we are stating and proving this theorem only for compact polyhedra. The result is in fact true for non-compact polyhedra and will be proved in chapter 8.

simplicial approximation to f and let χ be the subdivision chain map $C(K), C(K_0) \to C(K^{(r)}), C(K_0^{(r)})$. We have the diagram

$$\begin{array}{ccccccccc}
\ldots \to H_{p+1}(K, K_0; G) & \xrightarrow{\nu_*} & H_p(K_0; G) & \xrightarrow{\lambda_*} & H_p(K; G) & \xrightarrow{\mu_*} & H_p(K, K_0; G) & \to \ldots \\
\downarrow \hat{\chi}_* & & \downarrow \chi_*^0 & & \downarrow \chi_* & & \downarrow \hat{\chi}_* & \\
\ldots \to H_{p+1}(K^{(r)}, K_0^{(r)}; G) & \xrightarrow{\nu_*} & H_p(K_0^{(r)}; G) & \xrightarrow{\lambda_*} & H_p(K^{(r)}; G) & \xrightarrow{\mu_*} & H_p(K^{(r)}, K_0^{(r)}; G) & \to \ldots \\
\downarrow \hat{v}_* & & \downarrow v_*^0 & & \downarrow v_* & & \downarrow \hat{v}_* & \\
\ldots \to H_{p+1}(L, L_0; G) & \xrightarrow{\nu_*} & H_p(L_0; G) & \xrightarrow{\lambda_*} & H_p(L; G) & \xrightarrow{\mu_*} & H_p(L, L_0; G) & \to \ldots
\end{array}$$

Then we have the following facts: $\hat{f}_* = \hat{\chi}_* \hat{v}_*$, $f_*^0 = \chi_*^0 v_*^0$, $f_* = \chi_* v_*$ and commutativity holds round each square. We immediately infer, for example, that $\hat{f}_* \nu_* = \hat{\chi}_* \hat{v}_* \nu_* = \hat{\chi}_* \nu_* v_*^0 = \nu_* \chi_*^0 v_*^0 = \nu_* f_*^0$; the other two commutativities follow similarly. A similar statement holds in contrahomology. ∎

We shall use the single symbol f_* for the triple of homomorphisms $(\hat{f}_*, f_*, f_*^0)$ and refer to it as the *homomorphism of the homology sequence* of (K, K_0) into that of (L, L_0) induced by f. Similarly, we use f^* for the homomorphism of the contrahomology sequence of (L, L_0) into that of (K, K_0).

3.6.9 Theorem. *If* $f_0 \simeq f_1 : |K|, |K_0| \to |L|, |L_0|$, *then* $f_{0*} = f_{1*}$, $f_0^* = f_1^*$ *as homomorphisms of exact sequences.* ∎

3.6.10 Theorem. *If*

$$f : |K|, |K_0| \to |L|, |L_0|, \quad g : |L|, |L_0| \to |M|, |M_0|$$

are maps, then $(fg)_* = f_* g_*$, $(fg)^* = f^* g^*$ *as homomorphisms of exact sequences.* ∎

3.6.11 Theorem. *If* $f : |K|, |K_0| \simeq |L|, |L_0|$, *then* f_* *is an isomorphism of the homology sequence of* (K, K_0) *onto that of* (L, L_0) *and* f^* *is an isomorphism of the contrahomology sequence of* (L, L_0) *onto that of* (K, K_0). ∎

We draw attention to the fact that Theorems 3.6.3, 3.6.4, 3.6.6, 3.6.9, 3.6.10, 3.6.11 all have valid analogues if we replace the homology of the oriented complex by the homology of the total complex. In fact we could have taken the view from the outset that the groups $H_*^{\Omega}(K, K_0; G)$, $H_{\Omega}^*(K, K_0; G)$ were the natural homology objects to associate with the pair (K, K_0) and proved their topological invariance by a procedure parallel to that adopted in proving Theorem 3.6.6. We should then have invoked, for computational reasons, the chain equivalence $\theta : C^\Omega(K, K_0) \simeq C(K, K_0)$ and obtained homology and contrahomology isomorphisms. We did not adopt this point of view because $C(K, K_0)$ is a more intuitive construct than $C^\Omega(K, K_0)$, but we recognize that $C^\Omega(K, K_0)$ is logically satisfactory and, in fact, lends itself more readily to the generalization to singular homology theory carried out in chapter 8.

The process of copying the arguments and results for the total complex is mechanical. We shall only state explicitly the main results.

3.6.12 Theorem. *A map* $f : |K|, |K_0| \to |L|, |L_0|$ *induces homomorphisms*

$$f^\Omega_* : H^\Omega_*(K, K_0; G) \to H^\Omega_*(L, L_0; G),$$

$$f^*_\Omega : H^*_\Omega(L, L_0; G) \to H^*_\Omega(K, K_0; G).$$

The homomorphisms depend only on the homotopy class of f and $(fg)^\Omega_* = f^\Omega_* g^\Omega_*$, $(fg)^*_\Omega = f^*_\Omega g^*_\Omega$. *Commutativity holds in the diagrams*

$$\begin{array}{ccc} H^\Omega_*(K, K_0; G) & \xrightarrow{f^\Omega_*} & H^\Omega_*(L, L_0; G) \\ \downarrow{\scriptstyle \theta_*} & & \downarrow{\scriptstyle \theta_*} \\ H_*(K, K_0; G) & \xrightarrow{f_*} & H_*(L, L_0; G) \end{array}$$

$$\begin{array}{ccc} H^*_\Omega(K, K_0; G) & \xleftarrow{f^*_\Omega} & H^*_\Omega(L, L_0; G) \\ \uparrow{\scriptstyle \theta^*} & & \uparrow{\scriptstyle \theta^*} \\ H^*(K, K_0; G) & \xleftarrow{f^*} & H^*(L, L_0; G) \end{array}$$

Indeed, one could *define* f^Ω_*, f^*_Ω so that commutativity holds in the diagrams, since θ_*, θ^* are isomorphisms and the assertions of the theorem would follow. Alternatively, from the direct definition of f^Ω_*, f^*_Ω along the lines of Definition 3.6.1 we readily establish the commutativity of the diagrams in the light of 3.2.7. ❙

3.6.13 Corollary. *If f is a homotopy equivalence then f^Ω_*, f^*_Ω are isomorphisms.* ❙

In the course of this section we have effectively proved all but one of the Eilenberg–Steenrod axioms‡ for a homology theory for the polyhedra of pairs of complexes, called by Eilenberg and Steenrod *triangulable pairs*; we may also describe these as *polyhedral pairs*. With such a pair (P, P_0) and an abelian group G we have associated

‡ [14], p. 10; see also chapter 8 of this book.

groups $H_*(P, P_0; G)$, $H^*(P, P_0; G)$ in a *canonical* way. That is to say, for any oriented pair of complexes K, K_0 with $|K| = P$, $|K_0| = P_0$, we have groups $H_*(K, K_0; G)$, $H^*(K, K_0; G)$; and if also $|L| = P$, $|L_0| = P_0$, then the identity map $1 : |K|, |K_0| \to |L|, |L_0|$ induces definite isomorphisms

$$H_*(K, K_0; G) \cong H_*(L, L_0; G), \quad H^*(L, L_0; G) \cong H^*(K, K_0; G).$$

We might regard an element of $H_*(P, P_0)$ as a collection of elements $\{\xi, \eta, \ldots\}$, $\xi \in H_*(K, K_0)$, $\eta \in H_*(L, L_0) \ldots$, any two elements in the collection corresponding under the isomorphism induced by the identity map of P, P_0.

The Eilenberg–Steenrod axiom which remains to be verified is the Excision Axiom. This axiom was proved for pairs of complexes (K, K_0) as Theorem 2.8.22. We prove the axiom for triangulable pairs in the following strong form:

3.6.14 Theorem. *Let $f : |K|, |K_0| \to |L|, |L_0|$ be a map of compact triangulable pairs such that f maps $|K| - |K_0|$ to $|L| - |L_0|$ and is a homeomorphism of their closures. Then f_* and f^* are isomorphisms.*

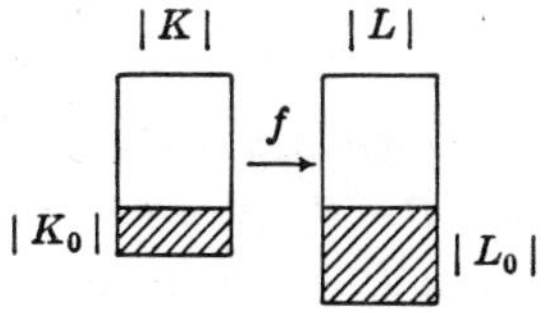

Fig. 3.2

We first prove a lemma. Let M be the smallest subcomplex of K containing $K - K_0$. Then $M \cup K_0 = K$ and

3.6.15 Lemma. $|M| = \overline{|K| - |K_0|}$.

For $|K| - |K_0| = \bigcup_{s \in K - K_0} s$ and $|M| = \bigcup_{s \in K - K_0} \bar{s}$, so that $|M|$ is closed. But

$$\bigcup s \subseteq \bigcup \bar{s} \subseteq \overline{\bigcup s},$$

so that $\bigcup \bar{s} = \overline{\bigcup s}$, or $|M| = \overline{|K| - |K_0|}$. ∎

We now revert to the theorem. Let N be the smallest subcomplex of L containing $L - L_0$ and let $M_0 = M \cap K_0$, $N_0 = N \cap L_0$. By the lemma, f maps $|M|$ homeomorphically onto $|N|$. Moreover, $|M_0| f = |N_0|$. For $|M_0| = |M| \cap |K_0|$, $|N_0| = |N| \cap |L_0|$. Thus certainly $|M_0| f \subseteq |N_0|$. On the other hand, if $y \in |N_0|$, then $y \in |N|$ so that $y = xf$, $x \in |M|$. Also $x \notin |K| - |K_0|$ since f maps $|K| - |K_0|$ into $|L| - |L_0|$ and $y \notin |L| - |L_0|$. Thus $y = xf$ with $x \in |M| \cap |K_0| = |M_0|$, so that, as asserted, $|M_0|f = |N_0|$. Thus if $g : |M|, |M_0| \to |N|, |N_0|$ is given by $xg = xf, x \in |M|$, g is a homeomorphism. Let $i : M, M_0 \to K, K_0$, $j : N, N_0 \to L, L_0$ be the inclusion (simplicial) maps. Then $gj = if$, by definition of g. Since $M \cup K_0 = K$, $N \cup L_0 = L$, the Excision theorem for complexes (Theorem 2.8.22) shows that

$$i_* : H_*(M, M_0; G) \cong H_*(K, K_0; G)$$

and

$$j_* : H_*(N, N_0; G) \cong H_*(L, L_0; G).$$

Also $g_* : H_*(M, M_0; G) \cong H_*(N, N_0; G)$, since g is a homeomorphism (Theorem 3.6.6). Since $g_* j_* = i_* f_*$, it follows that f_* is an isomorphism. A similar argument shows that f^* is an isomorphism. ▌ Notice that it is not assumed that f maps $|K|$ homeomorphically into $|L|$, so that the theorem is more general than an 'excision theorem'; but the excision aspect of the theorem is brought out in the corollary.

3.6.16 Corollary. *Let (P, P_0) be a triangulable pair and let U be an open set in P which is a subset of P_0 and such that $(P - U, P_0 - U)$ is triangulable. Then the injection*

$$i : P - U, P_0 - U \to P, P_0$$

induces homology and contrahomology isomorphisms.

For $(P - U) - (P_0 - U) = P - P_0$. Moreover $P - U$ is closed in P so that the closure of $P - P_0$ in $P - U$ coincides with its closure in P. Thus i satisfies the conditions imposed on f in Theorem 3.6.14. ▌ This result is more general than Theorem 2.8.22, in that it is not required that $P - U$ be triangulated as a subcomplex of a triangulation of P.

We close with a remark about augmentation. By $H_*(|K|, |K_0|)$ we shall understand the group $H_*(K, K_0)$, and it is irrelevant whether we augment or not. If $|K_0| = \varnothing$, the empty set, then K_0 is the empty complex and $H_*(K, K_0) \cong H_*(K)$. The group $H_*(K^+)$ is obtained from $H_*(K^+, K_0)$ with K_0 empty. If P is a polyhedron with complex K, we may write $H_*(P)$ for $H_*(K)$ and $\tilde{H}_*(P)$ for $H_*(K^+)$. $\tilde{H}_*(P)$ is called the *reduced* homology group. Similar conventions hold in contrahomology.

3.7 Local homology groups at a point of a polyhedron

Theorem 3.6.6. establishes that the homology groups are invariants of topological type; we may thus regard them as expressing a 'global' property of a topological space. We now describe certain homology groups which relate to the 'local' nature of a space; we shall leave to the reader the description of the corresponding contrahomology theory.

We start by defining the *local homology groups* $H_*^K(a)$ of a vertex a of a complex K, namely,

3.7.1 $$H_*^K(a) = H_*(\overline{\text{st}(a)}, \text{st}(a)^{\cdot}).$$

Equivalently (from 3.6.16 or 2.8.22) if $K_0(a) = K - \text{st}(a)$, then

3.7.2 $$H_*^K(a) = H_*(K, K_0(a)).$$

It is immediately clear that if L is a subcomplex of K containing st a, then the injection $i : L \to K$ induces the identity automorphism

3.7.3 $$i_* : H_*^L(a) \cong H_*^K(a).$$

Our main aim in this section is to prove a polyhedral form of this result; if x is a point of a (compact) polyhedron P, and Q is a compact subset of P which contains x in its interior and is also a polyhedron (but not necessarily a subpolyhedron of P), then we shall define $H_*^P(x)$, $H_*^Q(x)$ in such a way that the injection induces an isomorphism between these groups. This will fully justify us in regarding $H_*^P(x)$ as a local invariant of P at x. Certainly we may take x to be a vertex of triangulations of P and Q, and this will be done in what follows. We first make some preparatory definitions.

We consider the collection of pairs $(|K|, a)$, where $|K|$ is a (compact) polyhedron and a is a vertex of K. A map‡ $f : |L|, b \to |K|, a$ is an *l-map* if $f^{-1}a = b$. Clearly if $g : |M|, c \to |L|, b$ is also an l-map, then so is gf. A homotopy $F : |L| \times I, b \times I \to |K|, a$ between l-maps f_0 and f_1 is an *l-homotopy* if $F^{-1}a = b \times I$; we then write $f_0 \underset{l}{\overset{F}{\simeq}} f_1$. Clearly if F is an l-homotopy and $g : |M|, c \to |L|, b$ is an l-map, then $(g \times 1) F$ is an l-homotopy; also if $h : |K|, a \to |N|, d$ is an l-map, then Fh is an l-homotopy. Finally we remark that if $f_0 \underset{l}{\overset{F}{\simeq}} f_1$ and $f_1 \underset{l}{\overset{G}{\simeq}} f_2$, then $f_0 \underset{l}{\overset{H}{\simeq}} f_2$, where H is defined as in 1.6.2 (*c*).

We now prove

3.7.4 Proposition. *Let $f : |L|, b \to |K|, a$ be an l-map. Then $f \underset{l}{\simeq} f'$, where $f' : |L|, |L_0(b)| \to |K|, |K_0(a)|$. Moreover, if also $f \underset{l}{\simeq} f''$ with $f'' : |L|, |L_0(b)| \to |K|, |K_0(a)|$, then $f' \simeq f'' : |L|, |L_0(b)| \to |K|, |K_0(a)|$.*

‡ We consider maps $|L| \to |K|$ rather than $|K| \to |L|$ because our main interest is in the case in which $|L|$ is a subset of $|K|$.

Since $a \notin |L_0(b)|f$ and $|K|$ is compact we may subdivide K to $K^{(r)}$, say, so that‡ $|L_0(b)|f \subseteq |K_0^{(r)}(a)|$. Thus f may be regarded as a map $f: |L|, |L_0(b)| \to |K^{(r)}|, |K_0^{(r)}(a)|$. Let $v: K^{(r)}, a \to K, a$ be a standard simplicial map such that $v^{-1}a = a$; for example, we may order the vertices of K with a last and map each vertex of $K^{(r)}$ to the leading vertex of its carrier in K. Then clearly $|v|$ is an l-map and moreover $|v| \underset{l}{\simeq} 1$. Thus $f \underset{l}{\simeq} f|v|$; since plainly v maps $K_0^{(r)}(a)$ to $K_0(a)$ we may take $f' = f|v| : |L|, |L_0(b)| \to |K|, |K_0(a)|$.

We describe such a map f' as an *l-approximation* to f; we must prove that any two l-approximations f', f'' are homotopic. Certainly $f' \underset{l}{\simeq} f'': |L|, b \to |K|, a$; let F be an l-homotopy. Then $a \notin (|L_0(b)| \times I)F$. Thus there is an open set U containing a contained in st(a), and disjoint from $(|L_0(b)| \times I)F$ and there is a map $p: |K|, a \to |K|, a$ which projects $|K| - U$ radially from a onto $|K_0(a)|$. Then

$$p \simeq 1: |K|, |K_0(a)| \to |K|, |K_0(a)|,$$

so that $$f' \simeq f'p \overset{Fp}{\simeq} f''p \simeq f'' : |L|, |L_0(b)| \to |K|, |K_0(a)|. \;\blacksquare$$

It follows from 3.7.4 that we may define $f_*: H_*^L(b) \to H_*^K(a)$ unambiguously by $f_* = f'_*$, where f' is an l-approximation to f. We immediately draw three obvious conclusions about this definition. First, if f is, in fact, an injection of a subcomplex containing st(a) (with $b = a$), then f_* coincides with the isomorphism i_* of 3.7.3. Secondly, if $g: |M|, c \to |L|, b$ is also an l-map, then

3.7.5 $$(gf)_* = g_* f_*.$$

For if g' is an l-approximation to g and f' is an l-approximation to f, then $g'f'$ is an l-approximation to gf, so that

$$(gf)_* = (g'f')_* = g'_* f'_* = g_* f_*.$$

Thirdly, let $i: |K^{(r)}|, a \to |K|, a$ be the identity map. Then i_* is an isomorphism. For if $j: |K|, a \to |K^{(r)}|, a$ is the identity map then $ij = 1: |K^{(r)}|, a \to |K^{(r)}|, a$ and $ji = 1: |K|, a \to |K|, a$, so that, by 3.7.5, $i_* j_* = 1, j_* i_* = 1$ and i_* is an isomorphism.

We are now ready to prove

3.7.6 Theorem. *Let $f: |L|, b \to |K|, a$ be a homeomorphism of $|L|$ into $|K|$ such that a is interior to $|L|f$. Then $f_*: H_*^L(b) \to H_*^K(a)$ is an isomorphism.*

‡ $K_0^{(r)}(a) = K^{(r)} - \text{st}_{(r)}(a)$, where $\text{st}_{(r)}$ is the star in $K^{(r)}$.

Subdivide K to $K^{(r)}$, say, so that

$$|L_0(b)|f \subseteq |K_0^{(r)}(a)| \quad \text{and} \quad |\overline{\mathrm{st}_{(r)}(a)}| \subseteq |L|f;$$

the latter condition will be fulfilled for sufficiently large r since a is interior to $|L|f$. Put $M = \overline{\mathrm{st}_{(r)}(a)}$. Then there is a homeomorphism $g : |M|, a \to |L|, b$ of $|M|$ into $|L|$ such that $gf = j$, where

$$j : |M|, a \to |K|, a$$

is the injection. Since M is a subcomplex of $K^{(r)}$ containing $\mathrm{st}_{(r)}a$, we deduce from 3.7.3 and the first and third remarks preceding the theorem that j_* is isomorphic. Thus, by 3.7.5, $g_*f_* : H_*^M(a) \to H_*^K(a)$ is isomorphic, so that g_* is monomorphic (and f_* epimorphic). Now b is interior to $|M|g$ since $\mathrm{st}_{(r)}(a) \subseteq M$. Thus we may repeat the argument with g replacing f, and M, L replacing L, K, to prove that g_* is epimorphic. This establishes that g_*, and hence f_*, is isomorphic. ∎

This is the polyhedral form of 3.7.3; we deduce immediately

3.7.7 Corollary. *The local homology groups at a point of a polyhedron are topological invariants.* ∎

We shall say that the (compact) polyhedron P is *locally n-Euclidean* at $a \in P$ if there exists a homeomorphism $h : V^n, O \to P, a$, of the n-dimensional ball V^n onto a closed neighbourhood of a.

3.7.8 Proposition. *If P is locally n-Euclidean at a, then $H_n^P(a) = J$, $H_m^P(a) = 0$, $m \neq n$.*

By 3.7.6. it is sufficient to prove that $H_n^{V^n}(O) = J$, $H_m^{V^n}(O) = 0$, $m \neq n$. We triangulate V^n as the join of the origin O to a triangulation of S^{n-1} and have $H_m^{V^n}(O) = H_m(V^n, S^{n-1}) \cong H_m(\bar{s}_n, \dot{s}_n)$, whence by 2.10.7 and 2.10.11 the conclusion follows. ∎

We may apply 3.7.8 to any n-cell E^n, taking a as an interior point (see I. 1); we may also take for P any topological n-sphere and for a any point of P. Indeed, we may define a *topological n-manifold* as a polyhedron which is locally n-Euclidean at each of its points and 3.7.8 is valid for any point a of such a space. Real projective n-space P^n (see 3.9) is a topological n-manifold.

3.8 Simplex blocks

In this section we revert to a problem raised in 1.11. Though theoretically satisfactory, the homology theory of polyhedra based on simplicial complexes may be unpractical from the standpoint of computation. In 1.11 we introduced *pseudodissections*; and in this section we prove a theorem which has as one consequence the fact that the homology groups of a polyhedron may be calculated from a pseudodissection of the polyhedron. We present the absolute theory explicitly to avoid notational complications in the main argument, but the reader will be able to supply the appropriate relativizations.

To exemplify the technique of *simplex blocks,* consider the torus. As pointed out in 1.11, a simplicial dissection of the torus requires, at best, 7 vertices, 21 edges and 14 triangles. However, a torus may be described quite simply as a rectangle with opposite edges identified. It is an attractive suggestion that we might be able to calculate the homology groups from such a representation, with one 2-element, the rectangle, two 1-elements, the pairs of opposite edges, and one 0-element, to which the four corners are identified. Notice that this way of breaking up the torus into blocks is considerably more economical than the pseudodissection proposed in 1.11.

We now show that the homology groups of a torus may indeed be computed as suggested. We consider a simplicial complex K, and collect together suitable sets of simplexes of K and weld them into *blocks*. We then show that the homology of $|K|$ may be calculated from chain groups based on these blocks.

Let e be a collection of simplexes of K (not necessarily forming a subcomplex). We write $\bar{e}$ for the smallest subcomplex of K containing e and $\dot{e}$ for $\overline{\bar{e}-e}$. Clearly if $s \in e$ and $t \prec s$, then either $t \in e$ or $t \in \dot{e}$. Also $\bar{e}-e \subseteq \bar{e}$, so $\dot{e} \subseteq \bar{e}$; thus $\bar{e} = e \cup \dot{e}$.

3.8.1 Definition. A collection, e_p, of simplexes of K is said to form a *block of dimension* p or p*-block* if (i) it contains no simplex of dimension $> p$, and (ii) $H_q(\bar{e}_p, \dot{e}_p) = 0$, $q \neq p$, $H_p(\bar{e}_p, \dot{e}_p) = J$.

By condition (i), $H_p(\bar{e}_p, \dot{e}_p) = Z_p(\bar{e}_p, \dot{e}_p)$, since $C_{p+1}(\bar{e}_p, \dot{e}_p) = 0$. Thus, by (ii), $Z_p(\bar{e}_p, \dot{e}_p) = J$. We *orient* e_p by choosing a generator ϵ_p of $Z_p(\bar{e}_p, \dot{e}_p)$; this will be a linear combination of the p-simplexes of e_p. We may regard ϵ_p as a chain of K or of $\bar{e}_p$; we remark that $\epsilon_p \partial$ is a chain of $\dot{e}_p$ so that $\overline{\epsilon_p \partial} \subseteq \dot{e}_p$.

3.8.2 Definition. The set $\{e_p^i\}$ of simplex blocks of K is said to form a *block dissection* if (i) each $s \in K$ belongs to precisely one block of the set, $e(s)$, and (ii) $\dot{e}_p$ is a union of blocks of dimension $< p$.

Let $\{e_p^i\}$ form a block dissection. We observe that $\dot{e}_p^i \cap e_p^j = \varnothing$, and that, if $i \neq j$, $\bar{e}_p^i \cap e_p^j = \varnothing$. Let $B^{(q)} = \bigcup_{p \leqslant q} e_p^i$, and let A be a union of q-blocks. Then

3.8.3 Proposition. *$B^{(q)} - A$ is a subcomplex of K.*

Let $s \in B^{(q)} - A$; then $s \in e_p$ for some e_p in $B^{(q)} - A$ and $p \leqslant q$. If $t \prec s$, then either $e(t) = e_p$ or $t \in \dot{e}_p$. In the latter case, $e(t)$ is an r-block with $r < p$, so that, in either case, $t \in B^{(q)} - A$.]

3.8.4 Theorem. *If $c_p \in C_p(K)$ and $\overline{c_p \partial} \subseteq B^{(p-1)}$, there exists a block chain $\sum_i m_i e_p^i$ homologous‡ to c_p.*

Suppose that $\bar{c}_p \subseteq B^{(q)}$, $q > p$. We first prove that there is a chain d_p of K such that $c_p \sim d_p$ and $\bar{d}_p \subseteq B^{(p)}$; that is to say, we show that we can slip c_p off the blocks of dimension $> p$. It is clearly sufficient

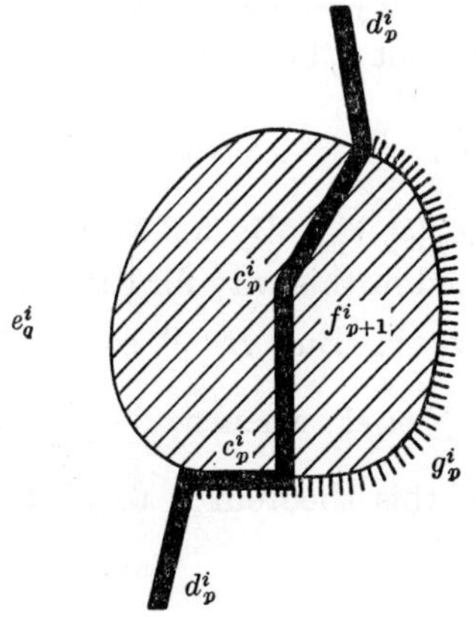

Fig. 3.3

for the proof to demonstrate the existence of a chain f_p of K such that $c_p \sim f_p$ and $\bar{f}_p \subseteq B^{(q-1)}$. For then $f_p \partial = c_p \partial$, so that $\overline{f_p \partial} \subseteq B^{p-1}$ and, if $q - 1 > p$, we reapply the argument with f_p in place of c_p.

Let e_q^i be any q-block (see Fig. 3.3, in which e_q^i is the whole region and $q = 2$, $p = 1$); we proceed to slip c_p off e_q^i. We may write

3.8.5 $$c_p = c_p^i + d_p^i, \quad \bar{c}_p^i \subseteq \bar{e}_q^i, \quad \bar{d}_p^i \subseteq B^{(q)} - e_q^i.$$

Then $\overline{c_p^i \partial} \subseteq \bar{e}_q^i$; on the other hand,

$$\overline{c_p^i \partial} \subseteq \overline{c_p \partial} \cup \overline{d_p^i \partial} \subseteq B^{(p-1)} \cup (B^{(q)} - e_q^i) = B^{(q)} - e_q^i.$$

Thus $\overline{c_p^i \partial} \subseteq \bar{e}_q^i \cap (B^{(q)} - e_q^i) = \dot{e}_q^i$. Then $c_p^i \in Z_p(\bar{e}_q^i, \dot{e}_q^i)$; since $p \neq q$, it follows from 3.8.1 (ii) that

$$c_p^i = f_{p+1}^i \partial + g_p^i, \quad \text{where} \quad f_{p+1}^i \in C_{p+1}(\bar{e}_q^i),\ g_p^i \in C_p(\dot{e}_q^i).$$

‡ Note that this theorem asserts a homology between chains which are not necessarily cycles.

Then $c_p \sim c_p - f^i_{p+1}\partial = d^i_p + g^i_p$, and so

$$\overline{c_p - f^i_{p+1}\partial} \subseteq (B^{(q)} - e^i_q) \cup \dot{e}^i_q = B^{(q)} - e^i_q.$$

Thus c_p has been slipped off e^i_q.

Now, if $i \neq j$, $\bar{e}^i_q \cap e^j_q = \emptyset$, so that $f^i_{p+1}\partial$ has coefficient zero on each p-simplex of e^j_q, since $\overline{f^i_{p+1}\partial} \subseteq \bar{e}^i_q$. It follows that we may find $f^i_{p+1}\partial$, for each i, in such a way that $\overline{c_p - \sum_i f^i_{p+1}\partial} \subseteq B^{(q-1)}$. We take

$$f_p = c_p - \sum_i f^i_{p+1}\partial.$$

Thus we have effectively shown that $c_p \sim d_p$ with $\bar{d}_p \subseteq B^{(p)}$. We proceed again as above: let e^i_p be any p-block; we may write (see 3.8.5)

3.8.6 $$d_p = k^i_p + l^i_p, \quad \bar{k}^i_p \subseteq \bar{e}^i_p, \quad \bar{l}^i_p \subseteq B^{(p)} - e^i_p,$$

and, as before, $k^i_p \in Z_p(\bar{e}^i_p, \dot{e}^i_p)$. By 3.8.1 (ii), $k^i_p = m_i \epsilon^i_p$ for some m_i. Clearly we may proceed in this way for each i and achieve the result that $\overline{d_p - \sum_i m_i \epsilon^i_p} \subseteq B^{(p-1)}$. But $B^{(p-1)}$ contains no p-simplexes by 3.8.1 (i), so that $d_p = \sum_i m_i \epsilon^i_p$, whence $c_p \sim \sum_i m_i \epsilon^i_p$. ∎

The significance of this theorem is to be found in the following corollary:

3.8.7 Corollary. (i) *Every p-cycle of K is homologous to a p-cycle of the form $\sum_i m_i \epsilon^i_p$.* (ii) *If $\sum_i m_i \epsilon^i_p$ bounds in K, then it bounds a $(p+1)$-chain of the form $\sum_i n_i \epsilon^i_{p+1}$.* (iii) *$\epsilon^i_p \partial$ is of the form $\sum_i m^i_j \epsilon^j_{p-1}$.*

Let $z_p \in Z_p(K)$; then $z_p\partial = 0$ and, *a fortiori*, $\overline{z_p\partial} \subseteq B^{(p-1)}$. Thus (i) follows from Theorem 3.8.4. To prove (ii), let $\sum_i m_i \epsilon^i_p = c_{p+1}\partial$, $c_{p+1} \in C_{p+1}(K)$; then $\overline{c_{p+1}\partial} \subseteq B^{(p)}$. Theorem 3.8.4 assures us that $c_{p+1} \sim \sum_i n_i \epsilon^i_{p+1}$ for some n_i; but then $\sum_i m_i \epsilon^i_p = c_{p+1}\partial = (\sum_i n_i \epsilon^i_{p+1})\partial$, proving (ii). To prove (iii), we observe that $\overline{\epsilon^i_p\partial} \subseteq \dot{e}^i_p \subseteq B^{(p-1)}$; thus we may apply the last part of the proof of 3.8.4 with $(p-1)$ replacing p and $\epsilon^i_p\partial$ replacing d_p. ∎

We define the chain complex $C(B(K))$ of the block dissection $B(K)$ by taking C_p to be the free abelian group generated by $\epsilon^1_p, \epsilon^2_p, \ldots$ and the boundary ∂^B to be given by 3.8.7 (iii), $\epsilon^i_p\partial^B = \sum_j m^i_j \epsilon^j_{p-1}$. Since $\epsilon^i_p\partial^B = \epsilon^i_p\partial$, it follows, of course, that $\partial^B\partial^B = 0$; we write ∂ for ∂^B.

Since each chain of $C(B(K))$ may be regarded as a chain of $C(K)$, we have an injection chain map

$$i : C(B(K)) \to C(K),$$

inducing $$i_* : H_*(B(K)) \to H_*(K).$$

3.8.8 Theorem.‡ $i_* : H_*(B(K)) \to H_*(K)$ *is an isomorphism.*

For 3.8.7 (i) asserts that i_* is an epimorphism and 3.8.7 (ii) that i_* is a monomorphism. ∎

It follows, in fact, that

3.8.9 $i_* : H_*(B(K); G) \cong H_*(K; G), \quad i^* : H^*(K; G) \cong H^*(B(K); G);$

this can be proved in the same way or deduced from Theorem 3.10.1.

Before proceeding to general conclusions from Theorem 3.8.8, let us compute the homology of a torus. Consider the simplicial dissection of fig. 3.4.

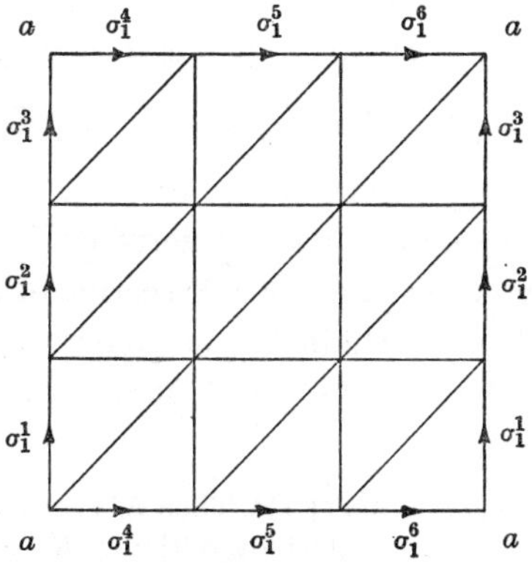

Fig. 3.4

We consider the following blocks:

$$e_0 = a, \quad e_1^1 = \bar{s}_1^1 \cup \bar{s}_1^2 \cup \bar{s}_1^3 - a,$$

$$e_1^2 = \bar{s}_1^4 \cup \bar{s}_1^5 \cup \bar{s}_1^6 - a, \quad e_2 = K - (e_0 \cup e_1^1 \cup e_1^2).$$

Then $$\bar{e}_0 = a, \quad \dot{e}_0 = \varnothing, \quad \bar{e}_1^1 = \bar{s}_1^1 \cup \bar{s}_1^2 \cup \bar{s}_1^3,$$

$$\dot{e}_1^1 = a, \quad \bar{e}_1^2 = \bar{s}_1^4 \cup \bar{s}_1^5 \cup \bar{s}_1^6, \quad \dot{e}_1^2 = a, \quad \bar{e}_2 = K, \quad \dot{e}_2 = \bigcup_{i=1}^{6} \bar{s}_1^i.$$

Obviously conditions 3.8.1 (i), 3.8.2 (i, ii) are satisfied and it remains only to verify 3.8.1 (ii)—that is, of course, the only condition one expects to give trouble in verifying that one has a block dissection.

‡ A more sophisticated proof is given in 10.9.

Clearly $H_0(a, \emptyset) = J$, $H_p(a, \emptyset) = 0$, $p \neq 0$. Also $\bar{e}_1^1$ is the boundary of a 2-simplex (not, of course, a simplex of K); $\sigma_1^1 + \sigma_1^2 + \sigma_1^3$ is a cycle of $\bar{e}_1^1 \bmod a$ which generates $H_1(\bar{e}_1^1, a) = J$ and $H_p(\bar{e}_1^1, a) = 0$, $p \neq 1$. We treat $\bar{e}_1^2$ similarly. Finally, consider $K \bmod \bigcup_{i=1}^{6} \bar{s}_1^i$. Let L be the simplicial complex of a rectangle subdivided isomorphically to the subdivision of fig. 3.4 (but without identifications) and let L_0 be the subdivision of its frontier. Then there is an obvious simplicial map $L, L_0 \to K, \bigcup \bar{s}_1^i$, which is an isomorphism of $L \bmod L_0$ onto $K \bmod \bigcup \bar{s}_1^i$. Thus $H_p(\bar{e}_2, \dot{e}_2) \cong H_p(L, L_0)$; but $(|L|, |L_0|)$ is homeomorphic with $(\bar{s}_2, \dot{s}_2)$ and we know (2.10) that $H_p(\bar{s}_2, \dot{s}_2) = H_{p-1}(\dot{s}_2^+) = \begin{cases} 0, & p \neq 2 \\ J, & p = 2 \end{cases}$. Thus e_2 is a 2-block and we have verified that we have constructed a block dissection. We orient the blocks to obtain block chains ϵ_2, ϵ_1^1, ϵ_1^2, ϵ_0. Computing from the simplicial complex K, we see that $\epsilon_2 \partial = 0$, $\epsilon_1^1 \partial = \epsilon_1^2 \partial = 0$. Thus

$$H_2(B(K)) = J, \quad H_1(B(K)) \cong J \oplus J, \quad H_0(B(K)) = J,$$

whence
$$H_2(K) = J, \quad H_1(K) \cong J \oplus J, \quad H_0(K) = J.$$

This gives the homology groups of the torus.

Let K be a complex and K' its first derived. Then the simplexes of K furnish, in an obvious way, a block dissection of K'. The block chain complex $C(B(K'))$ has the chains $\sigma\chi$ as generators so that there is a chain isomorphism $\psi : C(K) \to C(B(K'))$ such that $\psi i' = \chi$, i' being the chain map of 3.8.8.

Now let K be a *pseudodissection* (1.11). Then K' is again a simplicial complex and the simplexes of K furnish a block dissection of K'. For each $s_p \in K$, $(\bar{s}_p, \dot{s}_p)$ is the pair consisting of a closed p-simplex and its frontier so that, by 3.5.7, $\sigma_p\chi$ generates $H_p(\bar{s}'_p, \dot{s}'_p)$, and $H_q(\bar{s}'_p, \dot{s}'_p) = 0$, $q \neq p$, where χ is the subdivision chain map $C(K) \to C(K')$. Thus $C(B(K'))$ is generated by the block chains $\sigma_p\chi$ and we deduce as above that there is a chain isomorphism

3.8.10 $$\psi : C(K) \cong C(B(K')),$$

such that $$\psi i' = \chi.$$

3.8.11 Theorem. *If K is a pseudodissection, $H_*(|K|)$ may be computed from the chain complex $C(K)$.*

For $H_*(K) \overset{\psi_*}{\cong} H_*(B(K')) \overset{i'_*}{\cong} H_*(K') = H_*(|K'|) = H_*|K|$. ∎ We observe that this is an alternative proof that $H_*(K) \underset{\chi_*}{\cong} H_*(K')$ for any simplicial complex K. Theorem 3.8.11 justifies the introduction of pseudodissections, since it shows that we may proceed as for a simplicial dissection and calculate homology groups from chains based on simplexes of the pseudodissection; the groups we get will in fact be the homology groups of the polyhedron.

We may generalize Theorem 3.8.11, using for example Theorem 3.10.1, to give

3.8.12 Corollary. *If K is a pseudodissection, $H_*(|K|; G) \cong H_*(K; G)$ and $H^*(|K|; G) = H^*(K; G)$.*

We suggest to the reader the exercise of computing the homology groups of the real projective plane from the pseudodissection of 1.11. The homology groups of the real projective space of n dimensions are computed in the next section.

Just as for a simplicial complex and its first derived, so for a pseudodissection K and its first derived K' we may define a standard simplicial map

$$u : K' \to K,$$

and

$$\phi\chi = 1 : C(K') \to C(K'),$$

where ϕ is the standard chain map $C(K') \to C(K)$ induced by u.

Since $\chi_* : H_*(K) \cong H_*(K')$, it follows that

3.8.13

$$u_*(= \phi_*) : H_*(K') \cong H_*(K).$$

We may deduce from Theorem 3.10.1 and 3.8.13

3.8.14 Theorem. *If K is a pseudodissection and ϕ is a standard chain map from $C(K')$ to $C(K)$, then ϕ and χ are chain equivalences.* ∎

The reader may readily supply himself with a more direct proof, by using the method of acyclic carriers.

3.9 Homology of real projective spaces

The real projective space P^n may be defined as the image of the n-sphere S^n under identification of all pairs of antipodal points. In this section we apply the techniques of the previous section to compute the homology and contrahomology groups of P^n. More precisely, we combine the techniques of pseudodissection and block dissection to obtain convenient chain groups for the computations.

To facilitate the pseudodissection of P^n we replace S^n by a suitable homeomorph, namely, the subset Σ^n of R^{n+1} consisting of points $(u_1, \ldots, u_{n+1})$ with $\sum_1^{n+1} |u_r| = 1$. A homeomorphism $h : \Sigma^n \to S^n$ is given by radial projection from the origin and antipodal points of Σ^n are mapped by h to antipodal points of S^n. Thus a homeomorph of P^n is obtained by identifying antipodal points of Σ^n; we shall henceforth use the symbol P^n to refer to this homeomorph. Let $k : \Sigma^n \to P^n$ be the identification map.

We may readily construct a geometric complex to cover Σ^n. Let a^r_+, a^r_- be the points of Σ^n lying on the rth coordinate axis, a^r_+ being, of course, on the positive axis. Then the points $a^r_\pm$, $r = 1, \ldots, n+1$, are the vertices and we define a complex $K^{(n)}$ covering Σ^n inductively by $K^{(0)} = \{a^1_+, a^1_-\}$, $K^{(n)} =$ join of $\{a^{n+1}_+, a^{n+1}_-\}$ with $K^{(n-1)}$. We order $K^{(n)}$ by the rule that $a^r_\pm < a^s_\pm$ if $r < s$, so that an ordered simplex of $K^{(n)}$ is of the form $a^{i_0}_\pm a^{i_1}_\pm \ldots a^{i_p}_\pm$ with $1 \leqslant i_0 < i_1 < \ldots < i_p \leqslant n+1$ and any such set of vertices determines an ordered simplex of $K^{(n)}$. This triangulation has the property that the antipodal set to the simplex $a^{i_0}_{\epsilon_0} \ldots a^{i_p}_{\epsilon_p}$ where each ϵ_i is a sign $\pm$, is the simplex $a^{i_0}_{-\epsilon_0} \ldots a^{i_p}_{-\epsilon_p}$. It follows that the identification map $p : \Sigma^n \to P^n$ induces a pseudodissection $L^{(n)}$ of P^n; the image of each simplex of $K^{(n)}$ is a simplex of $L^{(n)}$. $L^{(n)}$ is not a simplicial complex if $n \geqslant 1$, since, for instance, $a_+^1 a_+^2$ and $a_-^1 a_+^2$ give rise to distinct 1-simplexes of $L^{(n)}$ with the same vertices. We note that $L^{(n)} \supseteq L^{(n-1)} \ldots \supseteq L^{(0)}$; $L^{(n)}$ is, moreover, ordered in the obvious way.

We may describe a simplex of $L^{(n)}$ by either of the two simplexes of $K^{(n)}$ which determine it; to show that the simplex is to be regarded as belonging to $L^{(n)}$ we replace the symbol a_+^r by α_+^r. Thus the ordered simplexes of $L^{(n)}$ are $\alpha_{\epsilon_0}^{i_0} \alpha_{\epsilon_1}^{i_1} \ldots \alpha_{\epsilon_p}^{i_p}$, where $1 \leqslant i_0 < i_1 < \ldots < i_p \leqslant n+1$, and

3.9.1 $$\alpha_{\epsilon_0}^{i_0} \alpha_{\epsilon_1}^{i_1} \ldots \alpha_{\epsilon_p}^{i_p} = \alpha_{-\epsilon_0}^{i_0} \alpha_{-\epsilon_1}^{i_1} \ldots \alpha_{-\epsilon_p}^{i_p}.$$

We shall generally write the simplex 3.9.1 with $\epsilon_p = +$.

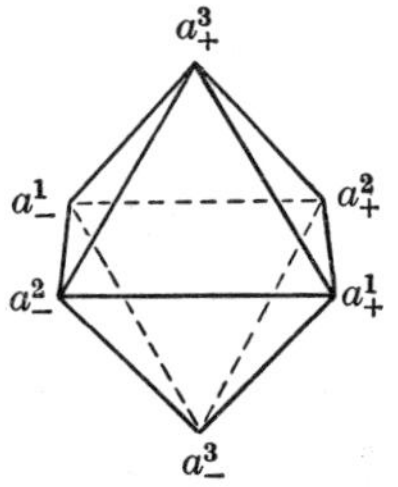

Fig. 3.5

To compute the homology of P^n we make a block dissection of $L^{(n)}$. The reader will have no difficulty in applying the method of block dissection to pseudo-dissections. The justification may be sought either in adapting the proof of Theorem 3.8.4 to pseudodissections or by passing to the first derived, which is a simplicial complex. We leave the details to the reader.

For each $r \leqslant n$, we define a single r-block, e_r, namely, the union of all (ordered) simplexes $\alpha_{\epsilon_0}^{i_0} \ldots \alpha_{\epsilon_p}^{i_p}$ with $i_p = r+1$. Then $\bar{e}_r$ is the union of all simplexes $\alpha_{\epsilon_0}^{i_0} \ldots \alpha_{\epsilon_p}^{i_p}$ with $i_p \leqslant r+1$, and‡ $\dot{e}_r = \bar{e}_{r-1}$. The only substantial point which arises in showing that this describes a block dissection is, as usual, the homology condition. Consider the subcomplex E^r of $K^{(n)}$ consisting of simplexes $a_{\epsilon_0}^{i_0} \ldots a_{\epsilon_p}^{i_p}$, with $i_p = r+1$, $\epsilon_p = +$, together with their faces. Then the identification map k induces a simplicial map $k_r : E^r \to \bar{e}_r$. Moreover, if $\dot{E}^r$ is the subcomplex of E^r consisting of those simplexes of E^r not containing a_+^{r+1}, then k_r maps $\dot{E}^r$ to $\dot{e}_r$, and k_r is an isomorphism of $E^r - \dot{E}^r$ with $\bar{e}_r - \dot{e}_r$. It follows that

$$H_p(\bar{e}_r, \dot{e}_r) \cong H_p(E^r, \dot{E}^r).$$

But $E^r = a_+^{r+1} \dot{E}^r$ and $\dot{E}^r = K^{(r-1)}$. Since $K^{(r-1)}$ is a complex covering Σ^{r-1}, which is homeomorphic to S^{r-1}, the known homology of S^{r-1} guarantees the correct homology properties for the block e_r. It remains to orient the blocks by choosing generators ζ_r of $H_r(\bar{e}_r, \dot{e}_r)$. We can do this by choosing generators of $H_r(E^r, \dot{E}^r)$ and applying k_r. In fact, it turns out that we may take

3.9.2 $$\zeta_r = \sum_{\epsilon} \epsilon_1 \ldots \epsilon_r \alpha_{\epsilon_1}^1 \ldots \alpha_{\epsilon_r}^r \alpha_+^{r+1},$$

where the summation is over all choices of $\epsilon_1, \ldots, \epsilon_r$. Then $|\zeta_r| \subseteq \bar{e}_r$, and

$$\zeta_r \partial = \sum_{i=1}^{r} (-1)^{i-1} \sum_{\epsilon} \epsilon_1 \ldots \epsilon_r \alpha_{\epsilon_1}^1 \ldots \widehat{\alpha_{\epsilon_i}^i} \ldots \alpha_{\epsilon_r}^r \alpha_+^{r+1} + (-1)^r \sum_{\epsilon} \epsilon_1 \ldots \epsilon_r \alpha_{\epsilon_1}^1 \ldots \alpha_{\epsilon_r}^r.$$

‡ Notice that, in fact, $\bar{e}^r = L^{(r)}$.

The first term of this expression is zero; for each i, we obtain zero by summing with respect to ϵ_i. Thus

$$\zeta\partial = (-1)^r \sum_{\epsilon} \epsilon_1 \ldots \epsilon_r \alpha^1_{\epsilon_1} \ldots \alpha^r_{\epsilon_r}.$$

We consider the coefficient of $\alpha^1_{\epsilon_1} \ldots \alpha^{r-1}_{\epsilon_{r-1}} \alpha^r_+$ in this expression; by virtue of 3.9.1, this is

$$(-1)^r \{\epsilon_1 \ldots \epsilon_{r-1} + (-\epsilon_1) \ldots (-\epsilon_{r-1})(-1)\} = (-1\ \ (1+(-1)^r)\,\epsilon_1 \ldots \epsilon_{r-1}.$$

We conclude that

3.9.3
$$\zeta_r\partial = 0, \qquad r \text{ odd},$$
$$= 2\zeta_{r-1}, \qquad r \text{ even}.$$

In any case $|\zeta_r\partial| \subseteq \bar{e}_{r-1} = \dot{e}_r$, so that ζ_r is a cycle of $\bar{e}_r \bmod e_r$. Since

$$Z_r(\bar{e}_r, \dot{e}_r) = H_r(\bar{e}_r, \dot{e}_r) = J$$

and since ζ_r may not be divided by any integer $\neq \pm 1$, it follows that ζ_r is a suitable oriented block chain.

It is now easy to compute the homology groups of P^n. For $0 < 2q \leqslant n$, $C_{2q}(B(L^{(n)}))$ is generated by ζ_{2q} and $\zeta_{2q}\partial = 2\zeta_{2q-1}$. Thus $Z_{2q} = 0$, so that $H_{2q}(P^n) = 0$. For $2q-1 < n$, C_{2q-1} is generated by ζ_{2q-1} and $\zeta_{2q-1}\partial = 0$; thus $Z_{2q-1} = J$, generated by ζ_{2q-1} and B_{2q-1} is generated by $2\zeta_{2q-1}$. Thus $H_{2q-1}(P^n) = J_2$. Obviously $H_0(P^n) = J$, since P^n is connected, and finally $H_n(P^n) = J$ if n is odd, generated by the class of ζ_n. We collect these results in

3.9.4 Theorem. *If P^n is real projective space, then*

$$H_0(P^n) = J; \quad H_{2q}(P^n) = 0, \quad q > 0;$$
$$H_{2q-1}(P^n) = J_2, \quad 0 < 2q-1 < n;$$
$$H_n(P^n) = J \quad \textit{if } n \textit{ is odd.} \ \blacksquare$$

We remark on the geometrical significance of the block chains ζ_r. In fact, the carrier of ζ_r is $\bar{e}_r$ itself and $\bar{e}_r$ is precisely $L^{(r)}$, the pseudodissection of P^r lying in P^n as a subspace. Thus we may think intuitively of ζ_r as being the projective subspace of dimension r, so that the generator of $H_{2q-1}(P^n)$ is the class of $P^{2q-1} \subseteq P^n$. This notion may be made quite precise: we have picked out an increasing sequence $P^0 \subset P^1 \subset \ldots \subset P^n$ and P^n is an orientable manifold ‡ if n is odd. If ζ_m is the basic cycle on P^m, m odd, then $H_{2q-1}(P^n)$ is generated by the class of the basic cycle ζ_{2q-1} of $P^{2q-1} \subseteq P^n$.

From Theorem 3.9.4 and the universal coefficient theorems of chapter 5 or by elementary ad hoc reasoning, we may deduce

3.9.5 Theorem. $H^0(P^n) = J; \qquad H^{2q}(P^n) = J_2, \quad 0 < 2q \leqslant n;$

$$H^{2q-1}(P^n) = 0, \quad 0 < 2q-1 < n;$$
$$H^n(P^n) = J \quad \textit{if } n \textit{ is odd.} \ \blacksquare$$

3.9.6 Theorem. *For $0 \leqslant p \leqslant n$, $H_p(P^n; J_2) = J_2$ and $H^p(P^n; J_2) = J_2$.*

In this case, ζ_r is a cycle for each r and the block contrachain (with values in J_2) which takes the non-zero value on ζ_r is a contracycle for each r. $\blacksquare$

‡ See, for example, Seifert and Threlfall [6].

3.10* Appendix on chain equivalence

We saw (Theorem 3.3.8) that a chain equivalence from the chain complex C to the chain complex D induces homology isomorphisms. In this appendix we prove a partial converse which is frequently useful.

3.10.1 Theorem. *If C, D are geometric chain complexes‡ and if $\phi : C \to D$ induces homology isomorphisms, then ϕ is a chain equivalence.*

A consequence of this theorem is that, once we know that ϕ is a chain equivalence, we can deduce that ϕ induces isomorphisms of homology groups with arbitrary coefficients and contrahomology groups with arbitrary values. We also know that the isomorphism ϕ_*^{-1} from $H_*(D)$ to $H_*(C)$ is induced by a chain map. Having motivated the theorem, we now motivate the proof.

J. H. C. Whitehead, in studying the homotopy type problem, introduced the notion of the *mapping cylinder*§ of a given map $f : X \to Y$. It is a 'cylinder', Z, with X at one end and Y at the other connected by segments from points of X to their f-images in Y. Whitehead proved that if X and Y are polyhedra and f induces isomorphisms of the *homotopy groups* of X with those of Y, then f is a homotopy equivalence; his procedure was to point out that the relative homotopy groups of Z mod X were null and thence to construct a deformation of Z into X. The embedding of Y in Z being a homotopy equivalence, this deformation gives rise to a map $Y \to X$ which is a homotopy inverse of f.

The proof of Theorem 3.10.1 imitates Whitehead's procedure. We construct a chain complex E which is a mapping cylinder for ϕ and contains C and prove that E/C is acyclic. We apply Theorem 3.4.3 to obtain a chain deformation connecting the identity and zero maps of E/C to itself and from this we extract by restriction a chain inverse of ϕ. Of course, the logic of the proof does not require the introduction of E; we could proceed directly to $\hat{E} = E/C$. But we have preferred to introduce the chain complex E in the hope that the reader will be helped in seeing his way through the proof.

We define $E = \{E_p, \partial_p^E\}$ as follows. $E_p = C_p \oplus C_{p-1} \oplus D_p$; a typical element will be written as $e_p = (c_p, c_{p-1}, d_p)$, $c_p \in C_p$, $c_{p-1} \in C_{p-1}$, $d_p \in D_p$. We define ∂^E by

3.10.2 $$(c_p, c_{p-1}, d_p)\,\partial^E = (c_p\partial - c_{p-1}, -c_{p-1}\partial, d_p\partial + c_{p-1}\phi).$$

‡ For definition, see 3.4. § See 9.5 for a formal definition.

Then
$$(c_p, 0, 0)\,\partial^E = (c_p\,\partial, 0, 0),$$
$$(0, 0, d_p)\,\partial^E = (0, 0, d_p\,\partial),$$
$$(0, c_{p-1}, 0)\,\partial^E = (-c_{p-1}, -c_{p-1}\partial, c_{p-1}\phi).$$

The element $(0, c_{p-1}, 0)$ may be thought of as a 'membrane' connecting c_{p-1} with $c_{p-1}\phi$.

We now observe that $(c_p, c_{p-1}, d_p)\,\partial^E\partial^E = (c_{p-2}, c_{p-3}, d_{p-2})$, where
$$c_{p-2} = (c_p\,\partial - c_{p-1})\,\partial + c_{p-1}\partial = 0,$$
$$c_{p-3} = -(-c_{p-1}\partial)\,\partial = 0,$$
$$d_{p-2} = (d_p\,\partial + c_{p-1}\phi)\,\partial - c_{p-1}\partial\phi = 0.$$

Thus $\partial^E\partial^E = 0$ and E is a chain complex. Further, the map $j : C \to E$ given by $c_p j = (c_p, 0, 0)$ embeds C in E; we identify C with Cj and regard the factor complex $\hat{E} = E/C$ as consisting of chain groups
$$\hat{E}_p = C_{p-1} \oplus D_p$$
with boundary $\hat{\partial}$ given by

3.10.3 $$(c_{p-1}, d_p)\,\hat{\partial} = (-c_{p-1}\partial, d_p\,\partial + c_{p-1}\phi).$$

Then $\hat{E}$ is a geometric chain complex and the map $k : D \to \hat{E}$ given by $d_p k = (0, d_p)$ embed D in $\hat{E}$; we identify D with $Dk \subseteq \hat{E}$.

We now prove that $\hat{E}$ *is acyclic*. Let (c_{p-1}, d_p) be a p-cycle of $\hat{E}$, so that $c_{p-1}\partial = 0$, $d_p\,\partial + c_{p-1}\phi = 0$. Thus c_{p-1} is a cycle and $c_{p-1}\phi$ is a boundary. Since $\phi_* : H_*(C) \cong H_*(D)$ and so, in particular, ϕ_* is monomorphic, it follows that c_{p-1} is a boundary, say $c_{p-1} = -c_p\,\partial$. Then $d_p - c_p\phi$ is a cycle, since $(d_p - c_p\phi)\,\partial = d_p\,\partial + c_{p-1}\phi = 0$. Since ϕ_* is epimorphic, there exists a p-cycle z_p of C and a $(p+1)$-chain d_{p+1} of D such that
$$z_p\phi = d_p - c_p\phi - d_{p+1}\partial.$$

Then $$(c_p + z_p, d_{p+1})\,\hat{\partial} = (-c_p\,\partial, d_{p+1}\partial + c_p\phi + z_p\phi) = (c_{p-1}, d_p),$$

and (c_{p-1}, d_p) is a boundary. Thus every cycle is a boundary and $\hat{E}$ is acyclic.

We now apply Theorem 3.4.3 to the identity map $1 : \hat{E} \to \hat{E}$; the acyclic carrier function is simply that which assigns the value $\hat{E}$ to every generator of $\hat{E}$. We deduce that there exists a chain homotopy $\Delta = \{\Delta_p\}$, $\Delta_p : \hat{E}_p \to \hat{E}_{p+1}$, such that

3.10.4 $$\Delta\hat{\partial} + \hat{\partial}\Delta = 1.$$

Let maps $\psi_p : D_p \to C_p$, $\chi_p : D_p \to D_{p+1}$ be given by

$$d_p \Delta_p = (d_p \psi_p, d_p \chi_p).$$

Obviously‡ ψ and χ are homomorphic. Applying 3.10.4 we have

$$d_p = d_p \Delta\hat{\partial} + d_p \hat{\partial}\Delta = (d_p \psi, d_p \chi)\hat{\partial} + (d_p \partial\psi, d_p \partial\chi)$$
$$= (-d_p \psi\partial, d_p \chi\partial + d_p \psi\phi) + (d_p \partial\psi, d_p \partial\chi),$$

whence $\quad -d_p \psi\partial + d_p \partial\psi = 0, \quad d_p \chi\partial + d_p \psi\phi + d_p \partial\chi = d_p.$

Thus

3.10.5 $$\psi\partial = \partial\psi,$$

3.10.6 $$\chi\partial + \partial\chi = 1 - \psi\phi.$$

This shows that $\psi : D \to C$ is a chain map and $\psi\phi \simeq 1$. It remains only to show that $\phi\psi \simeq 1$. To achieve this we define homomorphisms $\lambda_p : C_p \to C_{p+1}$, $\mu_p : C_p \to D_{p+2}$ by

$$(c_p, 0)\Delta_{p+1} = (c_p \lambda_p, c_p \mu_p).$$

Applying 3.10.4, we have

$$(c_p, 0) = (c_p, 0)\Delta\hat{\partial} + (c_p, 0)\hat{\partial}\Delta = (c_p \lambda_p, c_p \mu_p)\hat{\partial} + (-c_p \partial, c_p \phi)\Delta$$
$$= (-c_p \lambda\partial, c_p \mu\partial + c_p \lambda\phi) - (c_p \partial, 0)\Delta + (0, c_p \phi)\Delta$$
$$= (-c_p \lambda\partial, c_p \mu\partial + c_p \lambda\phi) - (c_p \partial\lambda, c_p \partial\mu) + (c_p \phi\psi, c_p \phi\chi).$$

Confining attention to the first component, we deduce that

3.10.7 $$\lambda\partial + \partial\lambda = \phi\psi - 1,$$

whence $\phi\psi \simeq 1$, and the proof is complete. ∎

EXERCISES

1. Supply a proof for Proposition 3.1.12.

2. K is a complex consisting of a single vertex; prove that $C_p^\Omega(K) = J$ for all $p \geqslant 0$. Determine $\partial_p : C_p^\Omega(K) \to C_{p-1}^\Omega(K)$ for p odd and for p even; hence calculate directly $H_p^\Omega(K)$ for all p.

3. Prove that

$$(a^{j_0} a^{j_1} \dots a^{j_p})\chi = \sum_{i_0, \dots, i_p} \epsilon_{i_0 \dots i_p}^{j_0 \dots j_p} b^{i_0 i_1 \dots i_p} b^{i_1 \dots i_p} \dots b^{i_p},$$

where $\epsilon_{i_0 \dots i_p}^{j_0 \dots j_p}$ is zero unless $i_0 \dots i_p$ is a permutation of $j_0 \dots j_p$ and is otherwise equal to the sign of that permutation.

Using this as a definition for χ, give a direct proof of Theorem 3.2.10.

‡ We drop suffixes, as usual.

4. Let $s = (a^0 a^1 \ldots a^p)$ be a p-simplex and $\{F^0, F^1, \ldots, F^p\}$ a covering of $\bar{s}$ by closed sets with the property that, if $a^{i_0} a^{i_1} \ldots a^{i_q}$ is any face of s, then it is contained in the union of the sets $F^{i_0}, F^{i_1}, \ldots, F^{i_q}$. Use 3.2.10 to show that $\bigcap_{i=0}^{p} F^i$ is not empty. (This is *Sperner's Lemma*: it leads quickly to the proof that the dimension of $|K^n|$ is n. Argue by contradiction, using the fact that, however finely $\bar{s}$ is subdivided, any standard map from the subdivision back to $\bar{s}$ must map some p-simplex onto the whole of s.)

5. Find an abelian group G, complexes K and L, and simplicial maps $v, w: K \to L$ such that

$$v_* = w_*: H_*(K) \to H_*(L)$$

but that

$$v_* \neq w_*: H_*(K; G) \to H_*(L; G).$$

[An example can be constructed using for K the complex of Ch. 2, Exercises, Q. 4 and for L a copy of K with vertices $b^0, \ldots, b^5$ to which is adjoined the closed simplex $b^1 b^3 b^5$.]

6. Let ω be an ordering of the complex K and let K^{Ω}_{ω} be the subcomplex of K^{Ω} consisting of those arrays whose vertices appear in the order prescribed by ω. Show that the inclusion map $C^{\omega}(K) \to C(K^{\Omega}_{\omega})$ is a chain equivalence.

7. The pair of spaces S^2, S^1 is mapped to the pair P^2, P^1 by the map f that identifies antipodal point-pairs of S^2. Calculate the homomorphism f_* of the homology sequence of S^2, S^1 into that of P^2, P^1, verifying the commutativity of the diagram.

8. For any complex K its suspension SK is the join of K to a pair of points. Prove that there are isomorphisms

$$\sigma^K_*: H_p(K^+) \cong H_{p+1}(SK),$$

such that for any map $f: |K| \to |L|$,

$$\sigma^K_*(Sf)_* = f_* \sigma^L_*,$$

where $Sf: |SK| \to |SL|$ is the map induced by f.

9. K is a complex and $\{e^i_p\}$ a block dissection of K; L is a subcomplex which is the union of blocks. Prove that the homology and contrahomology sequences for the pair K, L may be calculated from $C(B(K))$, $C(B(L))$ and $C(B(K))/C(B(L))$.

10. Apply Q. 9 above to Q. 6 in the Exercises on Chapter 2.

11. The 3-dimensional torus is obtained from a 3-dimensional cube by identifying opposite faces; a point on one face is identified with the point nearest to it on the opposite face. Find a block-dissection for the 3-dimensional torus and calculate its homology groups.

12. Let D be a chain complex and C a subcomplex of D. Construct the chain complex $\hat{E}$ as in 3.10 using the inclusion map for ϕ. Prove that $\hat{E} \cong D/C$.

4

THE CONTRAHOMOLOGY RING FOR POLYHEDRA

4.1 Definition of the ring for a complex

Hitherto we have developed side by side the dual theories of homology and contrahomology, but have given little justification for introducing contrahomology. In chapter 7 we shall show, certainly, how the obstruction to extending a partial map of a polyhedron is naturally expressible as a contrahomology class; on the other hand, in chapter 5 we shall show that the integral homology groups of a finite complex determine the homology and contrahomology groups with arbitrary coefficients. The reader therefore may be disposed to think that everything could be done in terms of homology alone.

In this section, however, we enrich the concept of the contrahomology group by introducing a multiplication of contrahomology classes (provided the value group is a ring), which is topologically invariant and is preserved under the contrahomology homomorphisms induced by continuous maps. Under this multiplication the contrahomology classes form a ring and this ring cannot be deduced from the integral homology groups. The homology classes of a general complex appear not to have an invariant multiplicative structure; in the case of a complex whose polyhedron is an orientable manifold (locally Euclidean) the homology classes form the *intersection ring*, and a strict duality can then be established between the homology and contrahomology rings; in general, however, the contrahomology multiplication ‡ provides a structural element not available from the

‡ There are also other invariant contrahomology operations that play a vital role in the problem of classifying homotopy types. Basic among these are the Steenrod pth powers (p prime)

$$\mathscr{S}^k_p : H^n(K; J_p) \to H^{n+2k(p-1)}(K; J_p),$$

the Pontryagin pth powers (p prime)

$$\mathscr{P}_p : H^n(K; J_{pm}) \to H^{pn}(K; J_{p^2m}), \quad \text{for any integer } m,$$

and the Bockstein operations (see chapter 5). The literature on contrahomology operations is tabulated in [19]; in these lectures Steenrod gives an excellent description of their role in algebraic topology.

homology groups alone. We shall not be concerned with the intersection ring, preferring to deal with the more general contrahomology product.

For the definition of the contrahomology ring of a complex K we shall use the total complex K^Ω and regard the contrahomology group of K as being $H^*_\Omega(K) = \sum_p H^p_\Omega(K)$. The reader may refer to 3.5.4 and what follows for the justification of the use of the total complex. The product in $H^*(K)$ is defined by means of a product, called the *cup product*, introduced into $C_\Omega(K; R)$, where R is an arbitrary ring,‡ which we shall frequently suppress from the notation.

We recall that an element $c \in C^p_\Omega(K; R)$ is a function with values in R defined over the admissible arrays $[a^{i_0} \dots a^{i_p}]$. We define the cup product $c \cup d$, where $d \in C^q_\Omega(K)$, to be the element of $C^{p+q}_\Omega(K)$ for which

4.1.1 $$[a^{i_0} \dots a^{i_{p+q}}]\,(c \cup d) = [a^{i_0} \dots a^{i_p}]\,c \times [a^{i_p} \dots a^{i_{p+q}}]\,d,$$

for an arbitrary admissible array $[a^{i_0} \dots a^{i_{p+q}}]$. This definition makes sense because R is a ring ($\times$ stands for the product in R) and because a subarray of an admissible array is admissible. Notice that the vertex a^{i_p} appears twice on the right-hand side of 4.1.1, forming a hinge between the two subarrays.

We first analyse the purely algebraic properties of this multiplication of contrachains.

4.1.2 Theorem. *The cup product of contrachains of K^Ω with values in R is associative and distributive over addition. If R has a left (right, two-sided) unity element, so has the ring of contrachains.*

The associativity and distributivity follow at once from the associativity and distributivity of R. If 1 is a left unity element of R, let e^0 be the contrachain with value 1 on each vertex of K; clearly $e^0 \cup d^q = d^q$ for any contrachain d^q. A similar argument holds for a right or two-sided unity element. ∎

When the cup product is introduced into $C^{\cdot}_\Omega(K) = \sum_{p=0}^{\infty} C^p_\Omega(K)$, this contrachain complex is a *graded ring*; that is, a multiplication is defined in $C^{\cdot}_\Omega(K)$ under which it is a ring, such that $C^p \cup C^q \subseteq C^{p+q}$.

4.1.3 Theorem. *If $c^p \in C^p_\Omega(K)$ and $d^q \in C^q_\Omega(K)$ then*

$$\delta(c \cup d) = (\delta c) \cup d + (-1)^p\, c \cup (\delta d).$$

‡ In applications R is usually commutative.

This, is, of course, proved by mere computation. Take an arbitrary $s^{\Omega}_{p+q+1} = [a^{i_0} \dots a^{i_{p+q+1}}]$; then

$$[a^{i_0} \dots a^{i_{p+q+1}}](\delta c) \cup d = [a^{i_0} \dots a^{i_{p+1}}]\, \delta c \times [a^{i_{p+1}} \dots a^{i_{p+q+1}}]\, d$$

$$= \sum_{r=0}^{p+1} (-1)^r [a^{i_0} \dots \widehat{a^{i_r}} \dots a^{i_{p+1}}]\, c \times [a^{i_{p+1}} \dots a^{i_{p+q+1}}]\, d$$

$$= \sum_{r=0}^{p} (-1)^r [a^{i_0} \dots \widehat{a^{i_r}} \dots a^{i_{p+q+1}}]\, c \cup d$$

$$+ (-1)^{p+1} [a^{i_0} \dots a^{i_p}]\, c \times [a^{i_{p+1}} \dots a^{i_{p+q+1}}]\, d.$$

Further

$$[a^{i_0} \dots a^{i_{p+q+1}}]\, c \cup (\delta d) = [a^{i_0} \dots a^{i_p}]\, c \times [a^{i_p} \dots a^{i_{p+q+1}}]\, \delta d$$

$$= [a^{i_0} \dots a^{i_p}]\, c$$

$$\times \sum_{r=p}^{p+q+1} (-1)^{r-p} [a^{i_p} \dots \widehat{a^{i_r}} \dots a^{i_{p+q+1}}]\, d$$

$$= [a^{i_0} \dots a^{i_p}]\, c \times [a^{i_{p+1}} \dots a^{i_{p+q+1}}]\, d$$

$$+ (-1)^p \sum_{r=p+1}^{p+q+1} (-1)^r [a^{i_0} \dots \widehat{a^{i_r}} \dots a^{i_{p+q+1}}]\, c \cup d.$$

Consequently

$$[a^{i_0} \dots a^{i_{p+q+1}}]\, ((\delta c) \cup d + (-1)^p c \cup (\delta d))$$

$$= \sum_{r=0}^{p+q+1} (-1)^r [a^{i_0} \dots \widehat{a^{i_r}} \dots a^{i_{p+q+1}}]\, c \cup d$$

$$= [a^{i_0} \dots a^{i_{p+q+1}}]\, \delta(c \cup d),$$

or $$(s^{\Omega}, (\delta c) \cup d + (-1)^p c \cup (\delta d)) = (s^{\Omega}, \delta(c \cup d)). \blacksquare$$

A contrachain complex in which a product is defined in terms of which it is a graded ring, satisfying the formula of 4.1.3, will be called a *contrachain ring*. From that theorem we see that the product of two contracycles is a contracycle. Moreover, if x^p is a contraboundary and z a contracycle, then $x^p \cup z$ is a contraboundary: for, if $x^p = \delta c^{p-1}$, $\delta(c \cup z) = (\delta c) \cup z + (-1)^{p-1} c \cup (\delta z) = x \cup z$. Similarly, if x is a contracycle and z a contraboundary, $x \cup z$ is a contraboundary.

4.1.4 Corollary. *The set $Z^{\cdot}_{\Omega}(K)$ of contracycles is a subring of $C^{\cdot}_{\Omega}$; the set $B^{\cdot}_{\Omega}(K)$ of contraboundaries is a two-sided ideal in $Z^{\cdot}_{\Omega}$. The quotient ring $R^*(K) = Z^{\cdot}_{\Omega}/B^{\cdot}_{\Omega}$ is a graded ring; if the value ring R has a left (right, two-sided) unity element, so has $R^*(K; R)$.*

In the light of I. 2, only the last statement remains to be proved; it depends on the evident fact that the contrachain e^0 defined above is a contracycle. ∎

The ring $R^*(K)$ is called the *contrahomology ring of* K; we shall write the product in R^* by juxtaposition.

Of course qua graded group $R^*(K) = H^*_\Omega(K)$. The reader may have noticed that the symbol $R^*(K)$ includes no hieroglyph to refer to the total complex; we may, however, write $R^*_\Omega(K)$ for $R^*(K)$ when we want to stress a distinction from some other ring defined from K.

It is easy to see that serious difficulty would arise in any attempt to extend these definitions to provide a product for the augmented K^+. In that case e^0 is a contraboundary and, if R has a two-sided unity element, B^0 is not an ideal; for any ideal containing the unity element is the whole ring.

We have not imposed the restriction on K that it should be finite; everything is plainly valid for contrachains of an infinite complex.

As we have observed before, it is easier to compute using a smaller complex than K^Ω. We now indicate how to define a contrahomology ring $R^*_\omega(K)$, where ω is an ordering of the vertices of K. This is isomorphic to $R^*_\Omega(K)$ and arises when $C^{\cdot}_\omega(K)$, which has fewer generators in each dimension than has $C^{\cdot}_\Omega(K)$, is given a cup-product structure. Clearly we can do this in a manner analogous to that adopted for $C^{\cdot}_\Omega(K)$; moreover, the analogues of 4.1.2–4.1.4 hold, so that a contrahomology ring $R^*_\omega(K)$ is indeed defined.

4.1.5 Proposition. *The contrahomology isomorphism*

$$\bar{\theta}^*_\omega : H^*_\Omega(K) \to H^*_\omega(K)$$

is a ring isomorphism between $R^*_\Omega(K)$ *and* $R^*_\omega(K)$.

Indeed, the contrachain map $\bar{\theta}^{\cdot}_\omega : C^{\cdot}_\Omega(K) \to C^{\cdot}_\omega(K)$ (see 3.2.3) is plainly a homomorphism between contrachain rings. By restriction we therefore have

$$\bar{\theta}^{\cdot}_\omega : Z^{\cdot}_\Omega(K), B^{\cdot}_\Omega(K) \to Z^{\cdot}_\omega(K), B^{\cdot}_\omega(K),$$

the homomorphism being a ring homomorphism; since from 3.5.3 $\bar{\theta}^*_\omega$ is isomorphic as a homomorphism between groups, it is a ring isomorphism. ∎

4.1.6 Corollary. $\theta^*_\omega : H^*_\omega(K) \to H^*_\Omega(K)$ *is a ring isomorphism.*

For it is the inverse to $\bar{\theta}^*_\omega$. ∎

Observe that $\theta^{\cdot}_\omega$ is not in general a contrachain *ring* homomorphism.

Let ω, ω' be two orderings of the vertices of K and let

$$\theta^{\omega\omega'} : C^\omega(K) \to C^{\omega'}(K)$$

be the chain isomorphism of 2.3.1.

4.1.7 Corollary. *$\theta^*_{\omega\omega'}: H^*_{\omega'}(K) \to H^*_{\omega}(K)$ is a ring isomorphism between $R^*_{\omega'}(K)$ and $R^*_{\omega}(K)$.*

For $\theta^{\omega}\theta^{\omega\omega'} = \theta^{\omega'}$. ∎

In view of 4.1.5–4.1.7, we are entitled to use any ordering ω to calculate the contrahomology ring for K.

4.1.8 Theorem. *If the value ring R is commutative, $R^*(K; R)$ is skew-commutative; that is, for $\xi \in H^p_{\Omega}(K)$, $\eta \in H^q_{\Omega}(K)$, $\xi\eta = (-1)^{pq}\eta\xi$.*

In the light of 4.1.5 it is clearly sufficient to prove that, if ω is any ordering, then $R^*_{\omega}(K; R)$ is skew-commutative. Let $\bar{\omega}$ be the reverse ordering to ω and let $\theta = \theta^{\bar{\omega}\omega}$. Thus θ is given by

$$a^0 \dots a^r \theta = \epsilon_r a^r \dots a^0,$$

where $\epsilon_r = (-1)^{\frac{1}{2}r(r+1)}$. We prove that, if $c \in C^p_{\omega}(K; R)$, $d \in C^q_{\omega}(K; R)$, then

4.1.9 $$\theta^{\cdot}(c \cup d) = (-1)^{pq}(\theta^{\cdot}d \cup \theta^{\cdot}c).$$

For
$$\begin{aligned}(a^0 \dots a^{p+q}, \theta^{\cdot}(c \cup d)) &= (a^0 \dots a^{p+q}\theta, c \cup d)\\ &= \epsilon_{p+q}(a^{p+q} \dots a^0, c \cup d)\\ &= \epsilon_{p+q}(a^{p+q} \dots a^q, c) \times (a^q \dots a^0, d)\\ &= \epsilon_{p+q}\epsilon_p\epsilon_q(a^0 \dots a^q\theta, d)\\ &\quad \times (a^q \dots a^{p+q}\theta, c) \quad \text{since } R \text{ is commutative}\\ &= (-1)^{pq}(a^0 \dots a^q, \theta^{\cdot}d) \times (a^q \dots a^{p+q}, \theta^{\cdot}c)\\ &= (a^0 \dots a^{p+q}, (-1)^{pq}(\theta^{\cdot}d \cup \theta^{\cdot}c)).\end{aligned}$$

From 4.1.9 we deduce that, if $\xi \in H^p_{\omega}(K)$, $\eta \in H^q_{\omega}(K)$, then

$$\theta^*\xi\eta = (-1)^{pq}(\theta^*\eta)(\theta^*\xi).$$

But by 4.1.7 θ^* is a ring isomorphism, so that $\xi\eta = (-1)^{pq}\eta\xi$, and 4.1.8 immediately follows. ∎

To sum up, we have associated with a simplicial complex K and a ring R a graded contrahomology ring $R^*(K; R)$; if R has a unity element, so has $R^*(K; R)$, and, if R is commutative, $R^*(K; R)$ is skew-commutative.

We remark finally that we may generalize the notion of a value ring R. If A, B are any abelian groups, let $\gamma: A \times B \to G$ be a bilinear transformation or *pairing* from the set product $A \times B$ to the abelian group G; this means that

$$(a + a', b)\gamma = (a, b)\gamma + (a', b)\gamma \quad \text{and} \quad (a, b + b')\gamma = (a, b)\gamma + (a, b')\gamma.$$

Under these circumstances we can define a bilinear transformation from $H^*(K; A) \times H^*(K; B)$ to $H^*(K; G)$, starting from a bilinear transformation from $C^\cdot(K; A) \times C^\cdot(K; B)$ to $C^\cdot(K; G)$ analogous to the cup product. If $A = B$, it is meaningful to assert that γ is commutative and, if so, a theorem analogous to 4.1.8 holds.

4.2 Relativization, induced homomorphisms and topological invariance

We recall the abstract algebraic notion of a contrachain ring, mentioned in the previous section.

4.2.1 Definition. A *contrachain ring* is a contrachain complex $C^\cdot = \Sigma C^p$ in which a multiplication (cup product) is defined such that (i) $C^\cdot$ is a ring; (ii) $C^p \cup C^q \subseteq C^{p+q}$; and (iii) if $c \in C^p$, $d \in C^\cdot$, then

$$\delta(c \cup d) = (\delta c) \cup d + (-1)^p c \cup (\delta d).$$

As we have already observed, condition (iii) ensures that $Z^\cdot$ is a contrachain subring of which $B^\cdot$ is an ideal; the cup product in $C^\cdot$ therefore induces a multiplication in H^*, which may then be written as R^*. R^* is then a ring for which $H^p \,.\, H^q \subseteq H^{p+q}$.

Suppose now that $\hat{C}^\cdot$ is a subcomplex of $C^\cdot$ which is also an ideal. Then $C_0^\cdot = C^\cdot/\hat{C}^\cdot$ is also a contrachain ring with a cup product induced by that of $C^\cdot$; for the conditions (i), (ii), (iii) hold in $C_0^\cdot$, as they hold in $C^\cdot$. We may describe this situation by saying that we have an exact sequence of contrachain rings and ring homomorphisms

4.2.2 $$0 \leftarrow C_0^\cdot \overset{\lambda^\cdot}{\leftarrow} C^\cdot \overset{\mu^\cdot}{\leftarrow} \hat{C}^\cdot \leftarrow 0.$$

We refer to this sequence as the *cup-product structure for the pair* $C^\cdot, \hat{C}^\cdot$. If $D^\cdot, \hat{D}^\cdot$ is another pair, consisting of a contrachain ring and a sub complex which is an ideal, we shall say that a contrachain map $\phi^\cdot : C^\cdot, \hat{C}^\cdot \to D^\cdot, \hat{D}^\cdot$ *preserves the cup-product structure* if the contrachain maps induced by $\phi^\cdot$, namely,

$$\phi_0^\cdot : C_0^\cdot \to D_0^\cdot, \quad \phi^\cdot : C^\cdot \to D^\cdot \quad \text{and} \quad \hat{\phi}^\cdot : \hat{C}^\cdot \to \hat{D}^\cdot,$$

are all ring homomorphisms.

4.2.3 Proposition. *If the contrachain map* $\phi^\cdot : C^\cdot, \hat{C}^\cdot \to D^\cdot, \hat{D}^\cdot$ *is a ring homomorphism as a map of* $C^\cdot$ *into* $D^\cdot$, *then it preserves the cup-product structure.* ▌

There is a highly important example of a contrachain ring and an ideal. Let K_0 be a subcomplex of K; then $C_\Omega^\cdot(K, K_0)$, interpreted to mean those elements of $C^{\bullet}_{\Omega}(K)$ which take the value zero on all admissible arrays of vertices of K_0, forms a subcomplex of $C^{\bullet}_{\Omega}(K)$ which is an ideal, as the reader will easily see. The quotient contrachain ring $C^{\bullet}_{\Omega}(K)/C^{\bullet}_{\Omega}(K, K_0)$ is isomorphic with $C^{\bullet}_{\Omega}(K_0)$, not only as a contrachain complex but also as a contrachain ring.

In the next chapter we shall show that any sequence 4.2.2 gives rise to an exact sequence

4.2.4 $$\ldots \xleftarrow{\mu^{p+1}} \hat{H}^{p+1} \xleftarrow{\nu^{p+1}} H^p_0 \xleftarrow{\lambda^p} H^p \xleftarrow{\mu^p} \hat{H}^p \xleftarrow{\nu^p} \ldots,$$

where $\hat{H}^p = H^p(\hat{C}^{\bullet})$, etc. The homomorphisms $\{\lambda^p\}$ combine to give $\lambda^* : H^* \to H^*_0$, which is the contrahomology homomorphism induced by $\lambda^{\bullet}$; since $\lambda^{\bullet}$ is a ring homomorphism, so is λ^*. Similarly $\{\mu^p\}$ give $\mu^* : \hat{H}^* \to H^*$ which is also a ring homomorphism. The composite homomorphism ν^* is not a ring homomorphism: $\nu^*(H^p_0 . H^q_0) \subseteq \hat{H}^{p+q+1}$, whereas $\nu^* H^p_0 . \nu^* H^q_0 \subseteq \hat{H}^{p+1} . \hat{H}^{q+1} \subseteq \hat{H}^{p+q+2}$. From 4.2.4 we can therefore extract a short exact sequence of contrahomology rings and ring homomorphisms:

4.2.5 $$R^*_0 \xleftarrow{\lambda^*} R^* \xleftarrow{\mu^*} \hat{R}^*.$$

We can define a multiplication or bilinear pairing from R^*_0, R^* to R^*_0 by the rule $\xi_0 \eta = \xi_0(\lambda^* \eta)$. Then λ^* is an *operator* homomorphism in the sense that it preserves the operation of multiplication on the right by an element of R^*; for

$$\lambda^*(\xi\eta) = (\lambda^*\xi)\,\eta, \quad \xi, \eta \in R.$$

We can also define a bilinear pairing from $\hat{R}^*$, R^* to $\hat{R}^*$. For, where $\hat{C}^{\bullet}$ is an ideal in $C^{\bullet}$, $\hat{B}^{\bullet} \cup Z^{\bullet} \subseteq \hat{B}^{\bullet}$ and $\hat{Z}^{\bullet} \cup B^{\bullet} \subseteq \hat{B}^{\bullet}$; thus a bilinear pairing is given by the rule

$$\{\hat{x}\}\{y\} = \{\hat{x} \cup y\},$$

where $\hat{x} \in \hat{Z}^{\bullet}$, $y \in Z^{\bullet}$, so that $\hat{x} \cup y \in \hat{Z}^{\bullet}$. Then μ^* is an operator homomorphism, since

$$\mu^*(\hat{\xi}\eta) = (\mu^*\hat{\xi})\,\eta,$$

where $\hat{\xi} \in \hat{R}^*$, $\eta \in R^*$. We observe also that

$$\hat{\xi}\hat{\eta} = \hat{\xi}(\mu^*\hat{\eta}).$$

For any $\eta \in H^q$, multiplication on the right by η provides an endomorphism, that we shall designate‡ by $\cup_\eta$, in H^*_0, H^* and $\hat{H}^*$. We

‡ We consider $\cup_\eta$ to be a left homomorphism as it operates on contrahomology elements, but admit that $\cup_{\eta\zeta} = \cup_\zeta \cup_\eta$. If we had defined $\cup_\eta$ by multiplication on the left, the diagram of 4.2.7 would have been partly skew-commutative.

shall refer to the sequence 4.2.5 together with these endomorphisms $\cup_\eta$ of the contrahomology groups as the *product structure* of the pair $C^\cdot, \hat{C}^\cdot$. If a distinction is to be stressed between this product structure and the cup-product structure, we shall refer to the product structure as the *contrahomology product structure.* If $D^\cdot$, $\hat{D}^\cdot$ is another pair, consisting of a contrachain ring and a subcomplex which is an ideal, we shall say that a contrachain map $\phi^\cdot : C^\cdot, \hat{C}^\cdot \to D^\cdot, \hat{D}^\cdot$ *preserves the (contrahomology) product structure* if ϕ_0^*, ϕ^* and $\hat{\phi}^*$ are all ring homomorphisms and the bilinear pairings are also preserved; that is $\hat{\phi}^* \cup_\eta = \cup_{\phi^*\eta} \hat{\phi}^*$ and $\phi_0^* \cup_\eta = \cup_{\phi^*\eta} \phi_0^*$.

4.2.6 Proposition. *If $\phi^\cdot : C^\cdot, \hat{C}^\cdot \to D^\cdot, \hat{D}^\cdot$ preserves the cup-product structure, it preserves the contrahomology product structure.* ∎

All this applies in particular to contrachain rings of a complex K and a subcomplex K_0.

4.2.7 Theorem. *$\cup_\eta$ is an endomorphism of the contrahomology sequence of the simplicial pair K, K_0. In other words the diagram*

$$\begin{array}{ccccccccc}
\xleftarrow{\mu^*} & \hat{H}^{p+1} & \xleftarrow{\nu^*} & H_0^p & \xleftarrow{\lambda^*} & H^p & \xleftarrow{\mu^*} & \hat{H}^p & \xleftarrow{\nu^*} \\
 & \downarrow{\scriptstyle \cup_\eta} & & \downarrow{\scriptstyle \cup_\eta} & & \downarrow{\scriptstyle \cup_\eta} & & \downarrow{\scriptstyle \cup_\eta} & \\
\xleftarrow{\mu^*} & \hat{H}^{p+q+1} & \xleftarrow{\nu^*} & H_0^{p+q} & \xleftarrow{\lambda^*} & H^{p+q} & \xleftarrow{\mu^*} & \hat{H}^{p+q} & \xleftarrow{\nu^*}
\end{array}$$

is commutative, where $\hat{H}^* = H_\Omega^*(K, K_0)$, $H^* = H_\Omega^*(K)$, $H_0^* = H_\Omega^*(K_0)$.

We remark that a similar theorem holds in the more general algebraic case.

Commutativity of the middle and right-hand squares simply follows from the fact that λ^* and μ^* are operator homomorphisms. To prove that the left-hand square is commutative, we recall the remark following 2.8.16*c* about the definition of ν^*. If $\xi_0 \in H_0^p$ and x_0 represents ξ_0, we may take any extension of x_0, that is, any $c \in C^p$ such that $c \mid K_0 = x_0$, and form δc; this contracycle lies in $\hat{C}^{p+1}$ and its class, as an element of $\hat{H}^{p+1}$, does not depend on the extension c chosen. We have then to prove that $\nu^*(\xi_0 \eta) = (\nu^* \xi_0)\,\eta$. Let y represent η and let $y \mid K_0 = d_0$; then $\xi_0 \eta$ is represented by $x_0 \cup d_0$, from the definition of ξ_0 and η; one extension of $x_0 \cup d_0$ is $c \cup y$ and $\nu^*(\xi_0 \eta)$ is therefore represented by $\delta(c \cup y) = (\delta c) \cup y$, since y is a contracycle. But δc represents $\nu^* \xi_0$ and $\delta c \cup y$ represents $(\nu^* \xi_0)\,\eta$. ∎

We now study the behaviour of the product under homomorphisms induced by maps of polyhedra. Consider first a simplicial map $v : K, K_0 \to L, L_0$, inducing

$$\phi_\Omega^{\cdot} : C_\Omega^{\cdot}(L), C_\Omega^{\cdot}(L_0) \to C_\Omega^{\cdot}(K), C_\Omega^{\cdot}(K_0).$$

4.2.8 Proposition. *The contrachain map $\phi_\Omega^{\cdot}$ induced by a simplicial map from K, K_0 to L, L_0 preserves the cup product and therefore preserves the contrahomology product structure.* ∎

4.2.9 Proposition. *If $\phi_\Omega^{\cdot} : C_\Omega^{\cdot}(K), C_\Omega^{\cdot}(K_0) \to C_\Omega^{\cdot}(K^{(r)}), C_\Omega^{\cdot}(K_0^{(r)})$ is induced by a standard simplicial map from $K^{(r)}$ to K, then $\phi_\Omega^{\cdot}$ has a contrachain inverse $\chi_\Omega^{\cdot}$ which preserves the contrahomology product structure.*

The reader is referred to the remarks that follow 3.6.11; the chain map $\chi^\Omega : C^\Omega(K) \to C^\Omega(K')$ can be defined very much as in 3.2.9, and it can be proved to be a chain inverse to the chain map induced by a standard map. In this proposition we consider the adjoint contrachain map to the composition of r such χ^Ω's. Since ϕ_Ω^* is an isomorphism and preserves the product, its inverse χ_Ω^* must also preserve the product. Notice that we do not assert that $\chi_\Omega^{\cdot}$ preserves the cup product. ∎

4.2.10 Theorem. *For any map $f : |K|, |K_0| \to |L|, |L_0|$ the induced contrahomology homomorphisms $f_0^*, f^*, \hat{f}^*$ preserve the product structure.*

We are now adopting the point of view that f^*, for instance, is a homomorphism $H_\Omega^*(L) \to H_\Omega^*(K)$. It is defined in terms of a simplicial approximation $u : K^{(r)} \to L$; if u induces $\psi_\Omega^{\cdot} : C_\Omega^{\cdot}(L) \to C_\Omega^{\cdot}(K^{(r)})$, then we consider $f^{\cdot} = \chi_\Omega^{\cdot}\psi_\Omega^{\cdot} : C_\Omega^{\cdot}(L), C_\Omega^{\cdot}(L_0) \to C_\Omega^{\cdot}(K), C_\Omega^{\cdot}(K_0)$. The homomorphisms $f_0^*, f^*, \hat{f}^*$ are those induced by $f^{\cdot}$. But $\psi_\Omega^{\cdot}$ preserves the cup-product by 4.2.8, and by 4.2.9 $\chi_\Omega^{\cdot}$ preserves the contrahomology product; therefore $f^{\cdot}$ also preserves the product. ∎

4.2.11 Corollary. *Any map $f : |K| \to |L|$ induces a ring homomorphism $f^* : R^*(L) \to R^*(K)$.* ∎

4.2.12 Theorem. *The product structure of a polyhedral pair is an invariant of homotopy type.*

For, if $f : |K|, |K_0| \to |L|, |L_0|$ is a homotopy equivalence, f^* is an additive isomorphism of the contrahomology sequence of L, L_0 onto that of K, K_0; but since f^* preserve the product, the whole product structure of the pair L, L_0 is carried over isomorphically to that of the pair K, K_0. ∎

4.2.13 Corollary. *The product structure of a polyhedral pair is a topological invariant.* ∎

In view of 4.2.13 we may write $R^*(|K|)$ and $R^*(|K|, |K_0|)$ when we wish to regard the ring as a property of the underlying polyhedron or polyhedral pair.

We can use 4.2.12 to prove homotopically inequivalent certain pairs of spaces which have isomorphic homology and contrahomology groups. Suppose, for example, that $|K| = S^m \times S^n$, with m, n different positive integers, and that $|L| = S^m \vee S^n \vee S^{m+n}$ (see I. 1 for definition of $\vee$). For K and for L the homology and contrahomology groups are null except in dimension $0, m, n, m+n$, where the group is in all cases J. If, however, ξ generates $H^m(K)$ and η generates $H^n(K)$, $\xi\eta$ is a generator of $H^{m+n}(K)$; this follows from results in chapter 9. On the other hand, in $R^*(L)$, $H^m(L).H^n(L) = 0$. The condition $m \neq n$ was only imposed to simplify the description of the argument; the conclusion that $K| \not\simeq |L|$ remains valid if $m = n$.

4.3 Calculations, examples and applications

For a calculation it is clearly easier to use an ordered complex K than to use K^Ω; in the same way we shall normally prefer to use as small a complex K covering a given polyhedron as we can. We now show that the contrahomology ring of a polyhedron can be calculated from an ordered pseudodissection K (see 1.11).

Let K be a Ψ-complex and let ω be an ordering of its vertices; we shall define a cup product in $C_\omega(K)$. If $c \in C^p$ and $d \in C^q$ and σ_{p+q} is an ordered simplex with vertices (in order) $a^0, a^1, \ldots, a^{p+q}$, we define $\theta_{(p)}\sigma$ to be that ordered face of s_{p+q} spanned by $a^0, a^1, \ldots, a^p$ and $\psi_{(q)}\sigma$ to be that ordered face spanned by $a^p, \ldots, a^{p+q}$. Notice that, although a set of vertices in a Ψ-complex may span more than one simplex, yet they can span at most one face of any given simplex. We now define $c \cup d$ by the rule

$$(\sigma_{p+q}, c \cup d) = (\theta_{(p)}\sigma, c) \times (\psi_{(q)}\sigma, d).$$

This rule is strictly parallel to 4.1.1. It is easy to verify that, in terms of $\cup$ so defined, $C_\omega(K)$ is a contrachain ring and we define $R^*(K)$ to be its derived contrahomology ring.

Now K', the first derived of K, is a simplicial complex and we can define a pseudo-simplicial order-preserving standard map $v : K' \to K$ as follows: each vertex of K' is a barycentre of an ordered simplex s of K and it is mapped by v into the leading vertex of s; each simplex of K' is mapped linearly into that face of its carrier in K spanned by the images of its vertices. This induces a chain map $\phi : C^{\omega'}(K') \to C^\omega(K)$, which is a chain inverse to $\chi : C^\omega(K) \to C^{\omega'}(K')$ as defined before the proof of 3.8.11; here ω' is the (partial) ordering of K' in 2.1. The reader will have no difficulty in verifying that $\phi^{\cdot}$ preserves the cup product. By 3.8.11 χ^* is an additive isomorphism and its inverse ϕ^* preserves the product; therefore χ^* is an isomorphism between $R^*_{\omega'}(K')$ and $R^*_\omega(K)$. Since, by 4.1.5 and 4.2.13, $R^*_{\omega'}(K') \cong R^*(|K'|)$ and $|K| = |K'|$, we have proved that $R^*_\omega(K) \cong R^*(|K|)$.

We apply this result to calculate the integral contrahomology ring of a real projective space P^n, using the ordered pseudodissection K of 3.9. We recall that $H^0(P^n) = J$, $H^{2q}(P^n) = J_2$ for $0 < 2q \leqslant n$ and $H^n(P^n) = J$ if n is odd; other contrahomology groups of P^n are null. The unity element e^0 of the contrachain ring of K belongs, say, to $\eta^0 \in H^0(P^n)$; then η^0 generates $H^0(P^n)$ and is the unity element in $R^*(P^n)$. Let the non-zero element of $H^{2q}(P^n)$ be ξ^{2q} and let a generator of $H^n(P^n)$, if n is odd, be η^n. Since $H^r(P^n)$ is null for $r > n$, clearly $\xi^{2q}\eta^n = 0$ and $\eta^n\eta^n = 0$. The necessary information about the product structure of $R^*(P^n)$ is contained in

4.3.1 Theorem. *If* $2(q+r) \leqslant n$, $\xi^{2q}\xi^{2r} = \xi^{2q+2r}$.

The main calculation will be based on the definition, for each p, $1 \leqslant p \leqslant n$, of a particular contrachain c_p. It is defined by

$$\begin{aligned}(\alpha^{i_0}_{\epsilon_0} \dots \alpha^{i_p}_{\epsilon_p}, c^p) &= 0, \quad \text{unless the signs } \epsilon_0, \epsilon_1, \dots, \epsilon_p \text{ alternate,} \\ &= 1, \quad \text{if the signs alternate.}\end{aligned}$$

If $p > n$, we take $c^p = 0$, perforce, and we define c^0 to be e^0. A moment's thought shows that, for all $p, q \geqslant 0$, for which $p + q \leqslant n$,

4.3.2
$$c^p \cup c^q = c^{p+q}.$$

We next calculate δc^1;

$$(\alpha^{i_0}_{\epsilon_0}\alpha^{i_1}_{\epsilon_1}\alpha^{i_2}_{+}, \delta c^1) = (\alpha^{i_1}_{\epsilon_1}\alpha^{i_2}_{+}, c^1) - (\alpha^{i_0}_{\epsilon_0}\alpha^{i_2}_{+}, c^1) + (\alpha^{i_0}_{\epsilon_0}\alpha^{i_1}_{\epsilon_1}, c^1).$$

A scrutiny of the possible values for ϵ_0, ϵ_1 shows that the right-hand side is zero unless $\epsilon_0 = +$, $\epsilon_1 = -$. Thus

$$\delta c^1 = 2c^2,$$

and hence $\delta c^2 = 0$. Applying 4.3.2 we can now prove inductively that

$$\begin{aligned}\delta c^p &= 0, \quad &&\text{if } p \text{ is even,} \\ &= 2c^{p+1}, \quad &&\text{if } p \text{ is odd and } p+1 \leqslant n.\end{aligned}$$

We shall next prove that $\{c^{2q}\}$ is the non-zero element of H^{2q}, for $0 < 2q \leqslant n$ and that $\{c^n\}$ generates H^n, in case n is odd.

Certainly in each of these cases c^p is a contracycle. We now recall from 3.9.2 the chain $\zeta_p = \sum\limits_{\epsilon} \epsilon_1 \dots \epsilon_p \alpha^1_{\epsilon_1} \dots \alpha^p_{\epsilon_p}\alpha^{p+1}_{+}$; it was there proved that

$$\begin{aligned}\zeta_p \partial &= 2\zeta_{p-1}, \quad &&\text{if } p \text{ is even and } p \leqslant n, \\ &= 0, \quad &&\text{if } p \text{ is odd.}\end{aligned}$$

For any $p \leqslant n$, $(\zeta_p, c^p) = (-1)^{[\frac{1}{2}(p+1)]}$, where $[x]$ means the greatest integer $\leqslant x$. This follows from the fact that c^p takes the value zero on all but one of the simplexes occurring in ζ_p, namely, that in which $\epsilon_1, \dots, \epsilon_p$, $+$ alternate; for this the number of $-$'s is $[\frac{1}{2}(p+1)]$.

The fact that $(\zeta_p, c^p) = \pm 1$, for $1 \leqslant p \leqslant n$, excludes the possibility that c^p is a contraboundary. For, if $c^p = \delta d^{p-1}$, $(\zeta_p, c^p) = (\zeta_p, \delta d^{p-1}) = (\zeta_p \partial, d^{p-1})$; but $(\zeta_p \partial, d^{p-1})$ is an even number; this contradiction shows that, in the cases where c^p is a contracycle, $\{c^p\}$ is non-zero. Therefore for $0 < 2q \leqslant n$, $\{c^{2q}\}$ is the non-zero element of H^{2q}.

If n is odd and if $\{c^n\}$ were not a generator of H^n, then for some $\eta \in H^n$ and integer $k > 1$, $\{c^n\} = k\eta$. But $(\{\zeta_n\}, \{c^n\}) = (\zeta_n, c^n) = \pm 1$; and, on the other hand, $(\{\zeta_n\}, \{c^n\}) = (\{\zeta_n\}, k\eta)$ which is divisible by k. This contradiction shows that $\{c^n\}$ is a generator of H^n, when n is odd.

If we write ξ^{2q} for $\{c^{2q}\}$ where $0 < 2q \leqslant n$, the equation $c^{2q} \cup c^{2r} = c^{2q+2r}$ implies that $\xi^{2q}\xi^{2r} = \xi^{2q+2r}$, for $2q + 2r \leqslant n$. ∎

4.3.3 Theorem. $R^*(P^n; J_2) \cong J_2[x]/[x^{n+1}]$.

The right-hand side means the polynomial ring in a single variable x and with coefficients in J_2, reduced modulo the ideal generated by x^{n+1}.

We may define ζ_p, c^p as above, interpreting all integer coefficients and values as elements of J_2. Then ζ_p is a cycle and c^p a contracycle for all p; $H_p = J_2$ for $0 \leqslant p \leqslant n$ and the above proof, modified to the case of J_2, proves that $\{c^p\}$ is the non-zero element of H^p for $0 \leqslant p \leqslant n$. Writing $\{c^1\} = \xi^1$ as ξ and $\{c^0\}$ as 1, the relation $c^p = (c^1)^p$ shows that $\xi^p = (\xi)^p$. Any element of R^* is therefore a polynomial in ξ of degree $\leqslant n$ with coefficients in J_2; the additive structure of R^* is that of such polynomials. The multiplicative structure is that of a polynomial ring except that all powers of ξ above the nth vanish. The isomorphism of the theorem is deducible from the homomorphic mapping

$$\phi : J_2[x] \to R^*,$$

given by $\phi(x) = \xi$. This is epimorphic and has kernel $[x^{n+1}]$; the result is immediate. ∎

We may use these results to establish a point of technical interest. It will be proved in the next chapter that the integral contrahomology groups of a compact polyhedron determine its contrahomology groups with an arbitrary value group. We now establish that the corresponding result for contrahomology rings is not true.‡

4.3.4 Example.§ Let $|K| = P^3$ and $|L| = P^2 \vee S^3$; then $R^*(K; J) \cong R^*(L; J)$, but $R^*(K; J_2) \not\cong R^*(L; J_2)$.

We shall use some easily proved facts about $\vee$. If $|L| = |L_1| \vee |L_2|$, then for $p > 0$, $H^p(L)$ may be identified with $H^p(L_1) \oplus H^p(L_2)$; any element $\xi \in H^p(L)$ is uniquely expressible as $\xi_1 + \xi_2$, where $\xi_i \in H^p(L_i)$. If $\eta \in H^q(L)$, $q > 0$, and $\eta = \eta_1 + \eta_2$ then $\xi\eta = \xi_1\eta_1 + \xi_2\eta_2$; in other words, $H^p(L_1).H^q(L_2) = 0$ for $p, q > 0$. Returning to the example, $R^*(K)$ has generators η^0, ξ^2, ξ^3 as in 4.3.1; ξ^2 is of order 2 and ξ^3 is of infinite order; for the multiplicative structure η^0 is the unity element and products otherwise are zero. $R^*(L)$ has generators ϵ^0, ζ^2, θ^3; ζ^2 is of order 2 and θ^3 is of infinite order; ϵ^0 is the unity element and products otherwise are zero. Plainly therefore $R^*(K) \cong R^*(L)$. On the other hand, $R^*(K, J_2)$ has generators ξ^0, ξ^1, ξ^2, ξ^3 and $\xi^1\xi^2 = \xi^3$; whereas $R^*(L; J_2)$ has generators ζ^0, ζ^1, ζ^2, θ^3 and $\zeta^1\zeta^2 = 0$. Therefore $R^*(K; J_2) \not\cong R^*(L; J_2)$.

We can make an application of the facts proved about the rings of real projective spaces which has surprising and attractive consequences.

4.3.5 Theorem. *If $m > n$ and if α is the non-zero element of $H_1(P^m)$, then for any map $f : P^m \to P^n$, $\alpha f_* = 0$.*

The cases $n = 0, 1$ are trivial, so we suppose that $n \geqslant 2$. Let ξ, η be the non-zero elements of $H^1(P^m; J_2)$, $H^1(P^n; J_2)$. Since $f^* : R^*(P^n; J_2) \to R^*(P^m; J_2)$ is homomorphic and $\eta^m = 0$, $(f^*\eta)^m = 0$, whence $f^*\eta \neq \xi$ and so $f^*\eta = 0$. Therefore $(\alpha f_*, \eta) = (\alpha, f^*\eta) = 0$, where the Kronecker products are between elements of an integral homology group and of a J_2-contrahomology group, the value lying in J_2.

If β is the non-zero element of $H_1(P^n)$, and if z, c are representatives of β, η, $(\beta, \eta) = (z, c) = 1$; for we may take for z the cycle ζ_1 of the proof of 4.3.1 and

‡ For complexes without torsion the anomaly exemplified in 4.3.4 cannot occur. See 5.2.24.

§ We owe this example to E. C. Zeeman.

for c the J_2-contracycle c^1 of the proof of 4.3.3. But $H_1(P^n)$ has but two elements, 0 and β; we have proved that $\alpha f_* \neq \beta$, so that $\alpha f_* = 0$. ∎

A map $f : S^m \to S^n$ is said to be *antipodal* if each pair of antipodal points of S^m is mapped onto a pair of antipodal points of S^n.

4.3.6 Theorem. *If $f : S^m \to S^n$ is an antipodal map, $m \leqslant n$.*

Let Q, R with vertices $a^i_\pm$, $b^j_\pm$ be the dissections of S^m, S^n as in 3.9; let K, L be the pseudodissections of P^m, P^n described in 3.9 and let $q : Q \to K$, $r : R \to L$ be the pseudo-simplicial maps relating these dissections: then $|q| : S^m \to P^m$, $|r| : S^n \to P^n$ are the identification maps which identify all antipodal pairs. Since f is an antipodal map, there is a map $g : P^m \to P^n$ such that $|q|\, g = f\, |r|$. We shall assume, without losing generality, that $a^1_+ f = b^1_+$.

By applying the simplicial approximation theorem, valid for Ψ-complexes, we deduce that g has an approximation $v : K^{(s)} \to L$. Then v can be lifted to an antipodal simplicial map $u : Q^{(s)} \to R$ such that $q^{(s)}v = ur$; for each vertex a of $Q^{(s)}$, au is that vertex of $r^{-1}(aq^{(s)}v)$ whose star contains af. Since v is an approximation to g, $a^1_+ u = b^1_+$.

Let us assume first that $n \geqslant 2$. In Q we define a 1-chain $c = a^1_+ a^2_+ - a^1_- a^2_+$ and in R we define $d = b^1_+ b^2_+ - b^1_- b^2_+$. Since $a^1_+ u = b^1_+$ and u is antipodal, $a^1_- u = b^1_-$ and hence‡ $c\chi^{(s)}u - d$ is a 1-cycle on $R = S^n$ and therefore, as $n \geqslant 2$, $c\chi^{(s)}u \sim d$. Now $cq = \zeta_1$, the 1-cycle of K referred to above whose homology class is α; so by 3.6.5 $\{c\chi^{(s)}q^{(s)}\} = \{cq\chi^{(s)}\} = \alpha\chi^{(s)}_*$. Hence

$$\beta = \{dr\} = \{c\chi^{(s)}ur\} = \{c\chi^{(s)}q^{(s)}v\} = \alpha\chi^{(s)}_* v_* = \alpha\, |v|_*.$$

But $|v|$ is a map from P^m to P^n and we can now deduce from 4.3.5 that $m \leqslant n$.

If $n = 0$, R is disconnected and the conclusion that $m = 0$ is immediate. If $n = 1$ and $m > 0$, let $c_- = a^1_- a^2_- - a^1_+ a^2_-$, the chain antipodal to c, and similarly let $d_- = b^1_- b^2_- - b^1_+ b^2_-$. Then $c\chi^{(s)}u - d$ is, as above, a cycle and so, for some integer k, $c\chi^{(s)}u - d = k(d + d_-)$; for $R = S^1$ and so $Z_1(R) = J$, generated by $d + d_-$. Since $\chi^{(s)}$ and u are antipodal, $c_-\chi^{(s)}u - d_- = k(d_- + d)$, so that

$$(c + c_-)\,\chi^{(s)}\,u = (2k+1)\,(d + d_-).$$

But $(2k+1)\,(d+d_-) \nsim 0$, therefore $c + c_- \nsim 0$ and $m = 1$. ∎

4.3.7 Corollary (*Borsuk-Ulam*). *For any $f : S^m \to R^m$, there is a pair of antipodal points $x, -x \in S^m$ such that $(x)f = (-x)f$.*

We regard S^m as being the unit sphere in R^{m+1} and treat points of a Euclidean space as position vectors; it is for that reason that we write the point antipodal to x as $-x$. Suppose now that the corollary is not true; for all $x \in S^m$, $(x)f - (-x)f$ is a non-zero vector in R^m; let $(x)g$ be the unit vector in the direction of $(x)f - (-x)f$. Then g is a map from S^m to S^{m-1}, since the unit vectors of R^m determine points of S^{m-1}. On the other hand, $(-x)g = -(x)g$, so that g is an antipodal map from S^m to S^{m-1}. This contradiction to 4.3.6 establishes 4.3.7. ∎

4.3.8 Corollary. *If $A_1, \ldots, A_m$ are bounded measurable regions of R^m, there is a prime ($(m-1)$-dimensional Euclidean subspace of R^m) which bisects each of the regions.*

Let the measure of A_i be μ_i. Let $h : R^{m+1}, O \to R^{m+1}, p$ be a translation, where $p \notin R^m$. For each point $x \in S^m$, let π^m_x be the prime through p orthogonal to Ox

‡ We allow ourselves the use of u, q, r to stand for induced chain maps.

since p does not lie in R^m, π_x^m does not coincide with R^m. We define $u_i(x)$, $1 \leqslant i \leqslant m$, to be the measure of that part of A_i lying on the same side of π_x^m as does xh. Under these definitions $u_i(x)$ is a continuous real-valued function on S^m, for each i. We may therefore define $f: S^m \to R^m$ by the rule

$$xf = (u_1(x), \ldots, u_m(x)).$$

Now if x, $-x$ are antipodal points of S^m, $\pi_x^m = \pi_{-x}^m$; on the other hand, xh and $(-x)h$ lie on opposite sides of π_x^m; consequently $u_i(x) + u_i(-x) = \mu_i$. By 4.3.7, there is a point $x_0 \in S^m$ such that $(x_0)f = (-x_0)f$ so that $u_i(x_0) = u_i(-x_0)$ for all i, $1 \leqslant i \leqslant m$. This implies that, for all i, $u_i(x_0) = \frac{1}{2}\mu_i$; this π^m is not parallel to R^m and intersects it in a prime which bisects each of the regions. ∎

This corollary in the case of $m = 3$ has enjoyed a certain vogue under the name of The Ham Sandwich Theorem; the three measurable regions are the three constituents of a sandwich and the bisecting plane is the plane of a knife-cut.

4.4* The cap product

In this section we describe a product between homology and contrahomology classes which is in a sense adjoint to the contrahomology product. We do not place emphasis on the role of this product and we confine ourselves, therefore, to discussing it, in outline, in the absolute case and with integral coefficients.

Let K be a simplicial complex and let $[a^{i_0} \ldots a^{i_{p+q}}]$ be an admissible array and $d \in C_\Omega^p(K)$. We associate with this pair an element

$$[a^{i_0} \ldots a^{i_{p+q}}] \cap d \in C_q^\Omega(K)$$

by the rule‡

4.4.1 $$[a^{i_0} \ldots a^{i_{p+q}}] \cap d = ([a^{i_0} \ldots a^{i_p}], d)\, [a^{i_p} \ldots a^{i_{p+q}}].$$

We then extend 4.4.1 linearly to give a bilinear pairing of $C_{p+q}^\Omega(K)$ and $C_\Omega^p(K)$ to $C_q^\Omega(K)$. The product $c_{p+q} \cap d^p$ is called the *cap product* of c and d. If $r < p$ and $c \in C_r^\Omega$ and $d \in C_\Omega^p$, we define $c \cap d$ to be zero.

This definition can be extended to give a product between $C_{p+q}^\Omega(K; G)$ and $C_\Omega^p(K; H)$ if a bilinear pairing is given from G and H to some abelian group Q; the cap product of chains and contrachains then lies in $C_p^\Omega(K; Q)$. The most useful special cases of this are (i) $H = J$, $Q = G$; the pairing is the natural pairing of an integer with an element of G; and (ii) $G = H = Q = R$, a ring, the pairing being given by multiplication in R.

4.4.2 Theorem. *If $c \in C_{p+q+r}^\Omega(K)$, $d \in C_\Omega^p(K)$, $e \in C_\Omega^q(K)$, then*

$$(c \cap d) \cap e = c \cap (d \cup e).$$

‡ There is another definition current, which we distinguish by inverting the order of the members, namely,

$$d \cap [a^{i_0} \ldots a^{i_{p+q}}] = ([a^{i_q} \ldots a^{i_{p+q}}], d).[a^{i_0} \ldots a^{i_q}].$$

Results analogous to those of this section hold for this form of the cap-product, with some changes of sign; the variant form of 4.4.2 reads: $c \cap (d \cap e) = (c \cup d) \cap e$, where c, d are contrachains and e a chain.

This is the main formula connecting $\frown$ and $\smile$. We need only verify in case $c = [a^{i_0} \dots a^{i_{p+q+r}}]$ and appeal to linearity. Now

$$
\begin{aligned}
&([a^{i_0} \dots a^{i_{p+q+r}}] \frown d) \frown e \\
&\qquad = ([a^{i_0} \dots a^{i_p}], d) . [a^{i_p} \dots a^{i_{p+q+r}}] \frown e \\
&\qquad = ([a^{i_0} \dots a^{i_p}], d) \times ([a^{i_p} \dots a^{i_{p+q}}], e) . [a^{i_{p+q}} \dots a^{i_{p+q+r}}] \\
&\qquad = ([a^{i_0} \dots a^{i_{p+q}}], d \smile e) . [a^{i_{p+q}} \dots a^{i_{p+q+r}}] \\
&\qquad = [a^{i_0} \dots a^{i_{p+q+r}}] \frown (d \smile e). \ \blacksquare
\end{aligned}
$$

We recall from p. 112 the 'augmentation' $\epsilon : C_0^\Omega(K) \to J$ which associates with each 0-chain the sum of its coefficients.‡ We use ϵ to relate the cap product with the Kronecker product.

4.4.3 Proposition. *If $c \in C_p^\Omega(K)$ and $d \in C_\Omega^p(K)$, then $(c \frown d)\,\epsilon = (c, d)$.*

We verify for $c = [a^{i_0} \dots a^{i_p}]$ and appeal to linearity. ∎

4.4.4 Corollary. *If $c \in C_{p+q}^\Omega(K)$, $d \in C_\Omega^p(K)$, $e \in C_\Omega^q(K)$,*

$$(c \frown d, e) = (c, d \smile e). \ \blacksquare$$

4.4.5 Theorem. *If $c \in C_{p+q}^\Omega(K)$, $d \in C_\Omega^p(K)$,*

$$(c \frown d)\,\partial = (-1)^p\,(c\partial \frown d - c \frown \delta d). \ \blacksquare$$

This theorem may be proved by direct computation; it may also be less laboriously proved from 4.4.4 using the fact that a chain is necessarily zero if its Kronecker product with every contrachain is zero; it is not hard to prove that the Kronecker products of the two sides of the equation 4.4.5 with an arbitrary contrachain are equal, using 4.1.3. The reader is recommended to try this as a pleasant exercise. The less elegant proof is the more easily extended to cases of general coefficients.

From 4.4.5 we infer that the cap product induces a bilinear product (also written with $\frown$) between elements of $H_{p+q}^\Omega(K) = H_{p+q}(K)$ and of $H^p(K)$ with values in $H_q(K)$. If

$$\alpha \in H_{p+q+r}(K), \quad \xi \in H^p(K), \quad \eta \in H^q(K),$$

then from 4.4.2 $(\alpha \frown \xi) \frown \eta = \alpha \frown (\xi\eta)$; we may define $\epsilon_* : H_0(K) \to J$ as the homomorphism induced by ϵ and deduce from 4.4.3 that, if $\alpha \in H_p(K)$ and $\xi \in H^p(K)$, $(\alpha \frown \xi)\,\epsilon_* = (\alpha, \xi)$.

‡ ϵ is related to ∂_0^+ by the rule $c_0 \partial_0^+ = c_0\,\epsilon[\ ]$, $c_0 \in C_0^\Omega(K)$.

Precisely similar definitions can be made for cap products in an ordered complex and similar results can be proved by indistinguishable methods. The cap products in K^ω and K^Ω are related by

4.4.6 Proposition. *If* $c \in C^\omega_{p+q}(K)$, $d \in C^p_\Omega(K)$,

$$c\bar{\theta}^\omega \cap d = (c \cap \bar{\theta}^\cdot_\omega d)\,\bar{\theta}^\omega.$$

Let $a^0 \dots a^{p+q}$ be an ordered simplex of K^ω; then

$$\begin{aligned} a^0 \dots a^{p+q}\bar{\theta}^\omega \cap d &= [a^0 \dots a^{p+q}] \cap d \\ &= ([a^0 \dots a^p], d)\,.\,[a^p \dots a^{p+q}] \\ &= (a^0 \dots a^p\bar{\theta}^\omega, d)\,.\,a^p \dots a^{p+q}\,\bar{\theta}^\omega \\ &= (a^0 \dots a^{p+q} \cap \bar{\theta}^\cdot_\omega d)\,\bar{\theta}^\omega. \; \rrbracket \end{aligned}$$

We can express the cap product in K^ω in terms of that in K^Ω by

4.4.7 Corollary. *If* $\alpha \in H^\omega_*(K)$, $\eta \in H^*_\omega(K)$

$$\alpha \cap \eta = (\alpha\bar{\theta}^\omega_* \cap \theta^*_\omega \eta)\,\theta^\omega_*.$$

From 4.4.6 we see that, for $\xi \in H^*_\Omega(K)$, $(\alpha\bar{\theta}^\omega_* \cap \xi) = (\alpha \cap \bar{\theta}^*_\omega \xi)\,\bar{\theta}^\omega_*$. In this replace ξ by $\theta^*_\omega \eta$, where θ^*_ω is the inverse of $\bar{\theta}^*_\omega$.]

We consider next how the cap product behaves under simplicial maps.

4.4.8 Proposition. *If* $v : K \to L$ *is a simplicial map inducing* $\phi : C^\Omega(K) \to C^\Omega(L)$, *then, if* $c \in C^\Omega(K)$ *and* $d \in C^\cdot_\Omega(L)$, $c\phi \cap d = (c \cap \phi^\cdot d)\,\phi$.

The proof is formally the same as that of 4.4.6.]

4.4.9 Corollary. *If* $\alpha \in H_*(K)$ *and* $\eta \in H^*(L)$, *then*

$$\alpha\phi_* \cap \eta = (\alpha \cap \phi^*\eta)\,\phi_*. \; \rrbracket$$

From this we can deduce the main theorem.

4.4.10 Theorem. *For any* $f : |K| \to |L|$, *if* $\alpha \in H_*(|K|)$ *and* $\eta \in H^*(|L|)$, *then*

$$\alpha f_* \cap \eta = (\alpha \cap f^*\eta) f_*.$$

The technique of proof is that of the corresponding theorem 4.2.10 for contrahomology products.]

4.4.11 Corollary.‡ *The cap product is an invariant of homotopy type; precisely, if* $f : |K| \to |L|$ *is a homotopy equivalence with homotopy inverse* $g : |L| \to |K|$, *then for any* $\alpha \in H_*(|K|)$ *and* $\xi \in H^*(|K|)$

$$(\alpha \frown \xi) f_* = \alpha f_* \frown g^* \xi.$$

This follows from 4.4.10 on substituting $g^*\xi$ for η and using the fact that g^* is the inverse of f^*. ▌ The topological invariance of the cap product follows of course at once; this implies that the product does not depend on the particular triangulation chosen.

We finish this chapter with some remarks on an application of the cap product to the study of manifolds. A complex K is said to be a (*homology*)-*n-manifold* if, for each p-simplex of K, its linked complex (see 2.10.11–2.10.12) has the homology groups of a sphere of dimension $(n-p-1)$. It is an *orientable* n-manifold if its n-simplexes can be so oriented that their sum is a cycle z_n; in that case $\alpha = \{z_n\}$ generates $H_n(K) = J$.

An important property of such manifolds§, which can be proved by using two mutually dual block dissections of K', is that for any abelian group G

4.4.12 $$H^p(K; G) \cong H_{n-p}(K; G).$$

The cap product may be conveniently used to give an explicit isomorphism between these groups; in fact, if $\eta \in H^p(K; G)$, the correspondence $\eta \to \alpha \frown \eta$ provides such an isomorphism.

From 4.4.12, which we do not propose to prove, and from results in 5.2, follows

4.4.13 Theorem (*Poincaré Duality Theorem*). *If K is an orientable* (*homology*)-*n-manifold*, ${}_wH_p(K) \cong {}_wH_{n-p}(K)$ *and* $T_{p-1}(K) \cong T_{n-p}(K)$. ▌

The isomorphisms 4.4.12 and the product structure in $H^*(K; R)$ provide a product structure in $H_*(K; R)$, where K is an orientable manifold and R is a ring. If $\beta \in H_r(K; R)$, $\gamma \in H_s(K; R)$, their *intersection class* $\beta . \gamma$ belongs to $H_{r+s-n}(K; R)$. If $\beta = \alpha \frown \xi$, $\gamma = \alpha \frown \eta$, then $\beta . \gamma$ is defined to be $\alpha \frown (\xi \smile \eta)$. The notion of the intersection of two cycles in a manifold is intuitively much more immediate than that of the cup product of two contracycles, and it has a longer history‖.

‡ The formulation of this result is rendered awkward by the fact that the cap product is a mixed product, compounded of covariant homology and contravariant contrahomology. The cup product, on the other hand, gives a pure contravariant product.

We may, however, formulate the cap product as a (right) homomorphism

$$\frown : H_{p+q}(K) \to H^p(K) \pitchfork H_q(K).$$

It is then a covariant operation (or functor in the terminology of Eilenberg and Steenrod). In this formulation $((\alpha)\frown)(\xi) = \alpha \frown \xi$.

§ See, for instance, Seifert and Threlfall [6], chapter 10.

‖ See Lefschetz, *Topology*, chapter 4, for a treatment that owes nothing to contrahomology.

EXERCISES

1. Compute the contrahomology ring over the integers of the torus $S^1 \times S^1$. (Use a pseudodissection.)

2. Use the remark at the end of 4.2 to show that $S^m \vee S^n$ is not a retract of $S^m \times S^n$ if m, n are positive.

3. Show that if P, Q are connected polyhedra with a single common point then, for any value ring R,

$$\tilde{R}^*(P \vee Q; R) \cong \tilde{R}^*(P; R) \oplus \tilde{R}^*(Q; R).$$

[Here $\tilde{R}^*$ is the reduced contrahomology ring.]

4. Show that if P, P_0 is a polyhedral pair and if P is contractible then cup products of elements of $R^*(P, P_0)$ of positive dimension all vanish. Use the Excision theorem to deduce that if SK is the suspension of K then all cup products of elements of $R^*(SK)$ of positive dimension vanish.

5. Let K be a simplicial complex and K_0, K_1 subcomplexes of K. Define a cup product of elements of $H^p(K, K_0; R)$ and $H^q(K, K_1; R)$ to elements of $H^{p+q}(K, K_0 \cup K_1; R)$ and prove its invariance.

6. The polyhedron P is said to be of *category* $\leqq n$ (written cat $P \leqslant n$) if, in some triangulation, it is covered by n subcomplexes $K_1, \ldots, K_n$ such that each $|K_i|$ is contractible in P. Show that if $\operatorname{cat} P \leqslant n$ then all n-fold cup products of elements of positive dimension are zero. [Hint: use the product defined in Q. 5.]

7. Show that if K is a connected n-dimensional complex then $\operatorname{cat}|K| \leqslant n+1$. Deduce, using the result of Q. 6, that $\operatorname{cat} P^n = n+1$, where P^n is projective n-space.

8. Generalize 4.2.7 to the contrahomology sequence of a triple.

9. Prove Theorem 4.4.5.

10. Let L be an orientable n manifold and let $f: |K| \to |L|$ be a map of the polyhedron $|K|$ into $|L|$. Show that if f_* maps $H_n(K)$ on to $H_n(L)$ then it maps $H_r(K)$ onto $H_r(L)$ for all r.

5

ABELIAN GROUPS AND HOMOLOGICAL ALGEBRA

In this chapter we bring together the results in the theory of abelian groups‡ which are relevant to homology theory. Certain of our results will be applicable to the homology theory of finite complexes and may be regarded as classical. Those parts of abelian group theory which are applicable to infinite complexes are of a more recondite nature and constitute, with material in chapter 10, an introduction to Homological Algebra (see [13], [17]). The classical results are intended to illuminate the results of previous chapters; the rest of this chapter should rather be regarded as preparatory for chapters 8, 9 and 10.

All groups in this chapter are abelian and will be written additively. Thus we may suppress the word 'abelian'; we shall call G an *fg-group*, or say that G is *fg*, if G is a finitely generated abelian group.

5.1 Standard bases for chain complexes

In this section our main objective is to obtain *standard bases* for the chain groups of a *geometric fg-complex*. In this chapter a chain complex C is said to be *geometric* (with bottom dimension m) if each C_p is free and, for some integer m, $C_p = 0$ for all $p < m$. This slightly generalizes the definition of 3.4. A chain complex C is *fg* if each C_p is *fg*. Thus our results will apply to the chain complex $C(K)$ of a finite oriented complex K, and to $C(K^+)$. The significant feature of a standard basis is that the homology groups may be read off immediately, and in a particularly convenient form.

We first obtain the fundamental theorem on *fg*-groups.

5.1.1 Theorem. *Let F be a free fg-group of rank§ n and R a subgroup. Then R is a free fg group of rank $m \leqslant n$, and we may choose bases*

‡ The reader is referred to [20], chapter 6, for the basic facts about abelian groups.

§ Recall that the *rank* of F is the number of elements in any basis; the null group is, by convention, a free *fg*-group of rank 0.

$(a_1, \ldots, a_n)$ for F, $(b_1, \ldots, b_m)$ for R, such that there exist integers $h_1, \ldots, h_m$ with $h_j \mid h_{j+1}$ $(j = 1, \ldots, m-1)$ and $b_i = h_i a_i$ $(i = 1, \ldots, m)$.

We start by defining the *height* of an element of F and establishing some of its properties.

If $g_1, \ldots, g_n$ is any basis for F and a is a non-zero element of F, a is uniquely expressible as $\sum_1^n h_i g_i$ and we call $h_1, \ldots, h_n$ the *coordinates* of a with respect to the basis. We define the *height*, $h(a)$, of a to be the least positive coordinate of a as the basis ranges over the set of all bases; since a is non-zero there is a basis with respect to which at least one coordinate of a is positive. Suppose now that $g_1, \ldots, g_n$ is a basis with respect to which the first coordinate of a is $h(a)$; then $a = h(a)\, g_1 + \sum_2^n h_i g_i$. Putting $h_i = q_i h(a) + r_i$, with $0 \leqslant r_i < h(a)$ $(i = 2, \ldots, n)$, we may write $a = h(a)\, g_a + \sum_2^n r_i g_i$, where $g_a = g_1 + \sum_2^n q_i g_i$. Now $g_a, g_2, \ldots, g_n$ form a basis and the minimal property of $h(a)$ implies that $r_i = 0$ $(i = 2, \ldots, n)$. This shows that, for each $a \in F$, there is an element g_a belonging to a basis for F such that

5.1.2 $$a = h(a)\, g_a.$$

We now prove that for any basis $h(a)$ is the greatest common divisor (g.c.d.) of the set of coordinates of a with respect to that basis. Since g_a is expressible in terms of this basis and $a = h(a)\, g_a$, certainly $h(a)$ divides each coordinate and so also their g.c.d. On the other hand, if h divides each coordinate, $a = hg$; we express g in terms of a basis containing g_a and deduce at once that $h \mid h(a)$. The g.c.d. therefore divides $h(a)$ and is accordingly equal to it.

Let R be a non-null subgroup of F; we define $h(R)$ to be the least of the heights of non-zero elements of R, and prove that $h(R)$ divides the height of each non-zero element of R. Suppose, then, that $b_1 \in R$ and that $h(b_1) = h_1 = h(R)$; let $b_1 = h_1 a_1$ and let $a_1, g_2, \ldots, g_n$ be a basis for F. Let $c \in R$ and let $c = l_1 a_1 + \sum_2^n m_i g_i$. For any q, put $c - qb_1 = c_q$; since $c_q \in R$, $c_q = 0$ or $h(c_q) \geqslant h_1$. This implies that the non-zero first coordinates of c_q are all in absolute value $\geqslant h_1$ and that therefore $h_1 \mid l_1$; if $l_1 = q_1 h_1$, then $c_{q_1 - 1} = h_1 a_1 + \sum_2^n m_i g_i$. The g.c.d. of these coordinates $\leqslant h_1$ and, by the minimal property of $h(R)$, the g.c.d. $= h_1$; so h_1 divides m_i $(i = 2, \ldots, n)$ and therefore $h(c)$ is divisible by h_1.

We now make the inductive hypothesis that the theorem has been proved for free *fg*-groups of rank $(n-1)$. Certainly it holds for free *fg*-groups of rank 0. We suppose R non-null (since, if R is null, we have nothing to prove) and that $b_1, h_1, a_1, g_2, \ldots, g_n$ are as above. Then, if A_1 is the subgroup generated by a_1 and F' is the subgroup generated by $g_2, \ldots, g_n$, $F = A_1 \oplus F'$ and $R = h_1 A_1 \oplus R'$ where $R' = R \cap F'$. Then F' is a free *fg*-group of rank $(n-1)$ and R' is a subgroup. From the inductive hypothesis we know that there are bases‡ $(b_2, \ldots, b_m)$ for R', $(a_2, \ldots, a_n)$ for F' where $m - 1 < n - 1$ and integers $h_2, \ldots, h_m$ such that $h_j \mid h_{j+1}$ $(j = 2, \ldots, m-1)$ and $b_j = h_j a_j$ $(j = 2, \ldots, m)$. But h_2 is the height of $b_2 \in R$, so that h_1 divides h_2; further, $(a_1, a_2, \ldots, a_n)$ is a basis for F and $(b_1, b_2, \ldots, b_m)$ is a basis for R; the truth therefore of the theorem for free *fg*-groups of rank $(n-1)$ implies its truth for rank n. ▌

‡ Perhaps $m = 1$ and $R' = 0$, in which case the required consequence is immediate.

5.1.3 Theorem. *Let G be an fg-group. Then G is expressible as*

5.1.4 $$G = A \oplus J_{k_1} \oplus \ldots \oplus J_{k_s}, \quad k_i \neq 1,$$

where A is free of rank ρ and $k_i \mid k_{i+1}$, $i = 1, \ldots, s-1$. Moreover, the numbers ρ and $k_1, k_2, \ldots, k_s$ are determined by G.

The number ρ is called the *rank* of G and the numbers $k_1, \ldots, k_s$ are called the *invariant factors*. Although the rank and invariant factors are determined by G, the subgroups occurring in the decomposition are not in general unique.

This theorem is Theorem 5 of [20], chapter 6. We give a proof based on Theorem 5.1.1. Let G be generated by the set $(g_1, \ldots, g_n)$. We may present G by means of the free fg-group F, with basis $(\alpha_1, \ldots, \alpha_n)$ and the epimorphism $\phi : F \to G$ determined by $\alpha_i \phi = g_i$, $i = 1, \ldots, n$. Let $R = \phi^{-1}(0)$, so that $F/R \cong G$. Choose bases for F, R in accordance with Theorem 5.1.1. Then

$$F/R \cong A \oplus J_{k_1} \oplus \ldots \oplus J_{k_s},$$

where A is free of rank $n - m$ and $k_1, \ldots, k_s$ are those of $h_1, \ldots, h_m$ that are greater than 1; precisely, if $h_1 = \ldots = h_{m-s} = 1$, $h_{m-s+1} > 1$, then $k_i = h_{m-s+i}$, $i = 1, \ldots, s$.

It remains to characterize the quantities $\rho, k_1, k_2, \ldots, k_s$ in terms of G itself. For any (abelian) group G let T be the torsion subgroup of G, that is, the subset of elements of finite order. Then, if G is decomposed as in 5.1.4, $T = J_{k_1} \oplus \ldots \oplus J_{k_s}$. For obviously $J_{k_1} \oplus \ldots \oplus J_{k_s} \subseteq T$; on the other hand, let $a + b \in G$ be of finite order q, where $a \in A$, $b \in J_{k_1} \oplus \ldots \oplus J_{k_s}$. Then $qk_s a = qk_s(a+b) - qk_s b = 0$. Since A is free this implies that $a = 0$ and $a + b \in J_{k_1} \oplus \ldots \oplus J_{k_s}$. Thus

5.1.5 $$T = J_{k_1} \oplus \ldots \oplus J_{k_s}.$$

It follows that $G/T \cong A$ so that G/T is free and ρ is characterized as the rank of G/T. We now characterize $k_1, \ldots, k_s$ in terms of T. We characterize k_s as the *exponent* of T; that is, the smallest positive integer n such that $nT = 0$. Now suppose $k_{r+1}, \ldots, k_s$ characterized in terms of T. Since for any n the order of nJ_m is $m/(m, n)$, where (m, n) is the greatest common divisor of m and n, it follows from (5.1.5) that the order of nT is

$$\frac{k_1}{(k_1, n)} \frac{k_2}{(k_2, n)} \cdots \frac{k_s}{(k_s, n)}.$$

We characterize k_r (assuming $k_{r+1}, \ldots, k_n$ to have been characterized) as the smallest integer n such that the order of nT is, in fact,

$$\frac{k_{r+1}}{(k_{r+1}, n)} \frac{k_{r+2}}{(k_{r+2}, n)} \cdots \frac{k_s}{(k_s, n)}.$$

It is clear that this does characterize k_r, and the proof is complete. ▮

It is often convenient to choose a somewhat different decomposition for a finite abelian group from that given by Theorem 5.1.3. Let T be any (abelian) torsion group (see p. 66). Then the subset of T consisting of elements whose order is a power of a given prime p is called its *p-primary component.*

5.1.6 Lemma. *A torsion group is the direct sum of its p-primary components.*

[20], chapter 6, theorem 1—Ledermann's proof goes over without change to torsion groups. ▮

A *p-group* is a group the order of each of whose elements is a power of p, where p is a prime.

5.1.7 Lemma. *A finite abelian p-group, $T_{(p)}$, is expressible as the direct sum of cyclic groups; thus, $T_{(p)} = J_{p^{m_1}} \oplus \ldots \oplus J_{p^{m_k}}$. The orders $p^{m_1}, \ldots, p^{m_k}$ are uniquely determined by $T_{(p)}$.*

This is an immediate consequence of 5.1.3 (see also [20], chap. 6, theorems 2 and 3). ∎ From 5.1.6 and 5.1.7 we deduce an alternative to Theorem 5.1.3, namely,

5.1.8 Theorem. *Every fg-group is expressible as the direct sum of a free abelian group and of cyclic groups of prime-power order. The rank and the prime-power orders are determined by the group.* ∎

Here, as in 5.1.4, the subgroups in the decomposition are not in general unique, although the rank and prime-powers are determined by the group. The prime-power orders of this theorem are called the *elementary divisors* of the given group. It is obvious how the elementary divisors of a given group may be obtained by factorizing the invariant factors into primes. The advantage of working with elementary divisors rather than with invariant factors rests on the fact that the p-primary component of a finite abelian group is a *fully invariant* subgroup in the sense that any homomorphism of finite abelian groups maps p-primary component to p-primary component. The p-primary component has played an increasingly important role in studying the relation between homotopy and homology.

Before proceeding to the application of these theorems to homology groups, we give three further results of importance.

5.1.9 Theorem. *Let $\phi : A \to B$, $\psi : B \to A$ be homomorphisms such that $\phi\psi = 1 : A \to A$. Then ϕ is a monomorphism, ψ is an epimorphism and $B = A\phi \oplus \psi^{-1}(0)$. Moreover,* $\operatorname{coker}\phi \cong \ker\psi$.

Let $a\phi = 0$; then $a = a\phi\psi = 0$, so that ϕ is monomorphic. Also $a = (a\phi)\psi$, so that ψ is epimorphic. Consider $b - b\psi\phi$; then $(b - b\psi\phi)\psi = b\psi - b\psi = 0$, so that $b = b\psi\phi + b_0$, where $b_0 \in \psi^{-1}(0)$. This shows that $B = A\phi + \psi^{-1}(0)$. Now if $a\phi = b_0 \in \psi^{-1}(0)$,

$$a = a\phi\psi = b_0\psi = 0,$$

so that $b_0 = a\phi = 0$ and the sum is direct. We immediately deduce that $B/A\phi \cong \psi^{-1}(0)$ or $\operatorname{coker}\phi \cong \ker\psi$. ∎

We may express the conclusion of 5.1.9 in the following useful form:

5.1.9′ Theorem *If $0 \to A \xrightarrow{\lambda} B \xrightarrow{\mu} C \to 0$ is exact, the following three statements are equivalent:*

(i) *there exists $\theta : B \to \lambda\theta = A$ with* 1;

(ii) *there exists $\eta : C \to B$ with $\eta\mu = 1$;*

(iii) $B = A\lambda \oplus B'$, *where $\mu \mid B' : B' \cong C$.*

From 5.1.9 clearly (i) $\Longleftrightarrow$ (iii) $\Longleftrightarrow$ (ii). ∎

5.1.10 Theorem. *Let* $0 \to H \xrightarrow{\lambda} G \xrightarrow{\mu} F \to 0$ *be exact and let* F *be free; then there exists a monomorphism* $\kappa : F \to G$ *such that* $\kappa\mu = 1$ *and*

$$G = H\lambda \oplus F\kappa.$$

Let $\{f_\alpha\}$ be a basis for F; for each basis element f_α choose $g_\alpha \in G$ such that $g_\alpha \mu = f_\alpha$. The association $f_\alpha \to g_\alpha$ induces a homomorphism $\kappa : F \to G$ and it is clear that $\kappa\mu = 1$. From 5.1.9′ we deduce the theorem. ∎

5.1.11 Theorem. *A subgroup of an fg-group is an fg-group of no greater rank.*‡

Let G be an *fg*-group and $G_0 \subseteq G$. Let $\phi : F \to G$ be a presentation of G by means§ of the free *fg*-group F and let $F_0 = \phi^{-1}G_0$. Then F_0 is *fg* (5.1.1) and $F_0\phi = G_0$ so that the ϕ-images of a set of generators of F_0 form a set of generators of G_0, and G_0 is *fg*.

Let T, T_0 be the torsion subgroups of G, G_0. Then $T_0 = G_0 \cap T$. Thus, by I.2.2, $G_0/T_0 = G_0/(G_0 \cap T) \cong (G_0 + T)/T \subseteq G/T$. Since the rank of G is the rank of the free abelian group G/T, and similarly for G_0, it follows that the rank of $G_0 \leqslant$ rank of G. ∎

We now turn our attention to homology groups with integer coefficients.

5.1.12 Definition. Let K be a finite simplicial complex. Then the *n-th Betti number* of $|K|$ is the rank of $H_n(K)$ and the *n-th torsion coefficients* of $|K|$ are the invariant factors of $H_n(K)$.

5.1.13 Theorem. *The homology groups of* $|K|$ *are completely determined by its Betti numbers and torsion coefficients.* ∎

Theorems 5.1.1 and 5.1.3 do not indicate, by their proofs, how the Betti numbers and torsion coefficients are to be obtained. Suppose we wish to choose bases for the groups $Z_n(K)$, $B_n(K)$ in accordance with Theorem 5.1.1. If $(z_1, \ldots, z_r)$, $(b_1, \ldots, b_s)$ are bases for these groups, then they determine an integer-valued matrix $\mathbf{P} = (p_{ij})$, where $b_i = \sum_{j=1}^{r} p_{ij} z_j$, $(i = 1, \ldots, s)$, and the practical problem is to reduce the matrix $\mathbf{P}$ to diagonal form by unimodular transformations in such a way that the invariant factors appear on the diagonal. Of course, it is not necessary to start with $Z_n(K)$ and $B_n(K)$; it may be

‡ A subgroup of a finitely generated *non-abelian* group may not even be finitely generated.

§ See the proof of 5.1.3; ϕ is an epimorphism.

more convenient to replace them with a free *fg*-group F and a subgroup R such that $F/R \cong H_n(|K|)$.

5.1.14 *Example.* The abelian group G is generated by g_1, g_2, g_3 subject to the relations $2g_1+2g_2+3g_3 = 0$, $3g_1-6g_3 = 0$. Let $F = (\gamma_1, \gamma_2, \gamma_3)$, $R = (\rho_1, \rho_2)$, where $\rho_1 = 2\gamma_1+2\gamma_2+3\gamma_3$, $\rho_2 = 3\gamma_1-6\gamma_3$; then $G \cong F/R$.

We transform the relation matrix $\mathbf{P}$ to diagonal form and make the corresponding changes in the bases for F, R explicitly. We thus find the rank and invariant factors of G and at the same time a set of generators for G arising from the decomposition 5.1.4.

$$\begin{array}{lll}
\begin{pmatrix} 2 & 2 & 3 \\ 3 & 0 & -6 \end{pmatrix} & (\gamma_1, \gamma_2, \gamma_3) & (\rho_1, \rho_2) \\
\downarrow & \downarrow & \downarrow \\
\begin{pmatrix} 2 & 2 & 1 \\ 3 & 0 & -6 \end{pmatrix} & (\gamma_1, \gamma_2+\gamma_3, \gamma_3) & (\rho_1, \rho_2) \\
\downarrow & \downarrow & \downarrow \\
\begin{pmatrix} 1 & 2 & 2 \\ -6 & 3 & 0 \end{pmatrix} & (\gamma_3, \gamma_1, \gamma_2+\gamma_3) & (\rho_1, \rho_2) \\
\downarrow & \downarrow & \downarrow \\
\begin{pmatrix} 1 & 0 & 0 \\ -6 & 15 & 12 \end{pmatrix} & (2\gamma_1+2\gamma_2+3\gamma_3, \gamma_1, \gamma_2+\gamma_3) & (\rho_1, \rho_2) \\
\downarrow & \downarrow & \downarrow \\
\begin{pmatrix} 1 & 0 & 0 \\ 0 & 15 & 12 \end{pmatrix} & (2\gamma_1+2\gamma_2+3\gamma_3, \gamma_1, \gamma_2+\gamma_3) & (\rho_1, \rho_2+6\rho_1) \\
\downarrow & \downarrow & \downarrow \\
\begin{pmatrix} 1 & 0 & 0 \\ 0 & 3 & 12 \end{pmatrix} & (2\gamma_1+2\gamma_2+3\gamma_3, \gamma_1, \gamma_1+\gamma_2+\gamma_3) & (\rho_1, \rho_2+6\rho_1) \\
\downarrow & \downarrow & \downarrow \\
\begin{pmatrix} 1 & 0 & 0 \\ 0 & 3 & 0 \end{pmatrix} & (2\gamma_1+2\gamma_2+3\gamma_3, 5\gamma_1+4\gamma_2+4\gamma_3, \gamma_1+\gamma_2+\gamma_3) & (\rho_1, \rho_2+6\rho_1)
\end{array}$$

The first step produced an equivalent matrix containing as an element the greatest common divisor of the elements of the original matrix; in general, this process may involve several steps. The second step placed this element in the leading position; the next two steps cleared the rest of the first row and column. The fifth step produced an equivalent matrix containing as an element the greatest common divisor of the elements other than the leading element; the sixth step cleared the rest of the second row.

It follows that $G = J \oplus J_3$, and that we may take J to be generated by $g_1+g_2+g_3$ and J_3 to be generated by $5g_1+4g_2+4g_3$. The reader is reminded that the generators are not unique nor even is the subgroup J.

We are now ready to prove the theorem on standard‡ bases for geometric *fg*-complexes. Let C be a geometric *fg*-complex with bottom dimension m and top dimension§ n (i.e. $C_p = 0$ if $p < m$ or $p > n$).

‡ The word 'canonical' appears instead of 'standard' in the literature, but we prefer to reserve the word 'canonical' for situations where no choices are involved.

§ We allow $n = \infty$.

5.1.15 Theorem. *Each chain group C_p may be expressed as*

$$C_p = Z_p \oplus Y_p,$$

where Z_p is the group of p-cycles. Further, Z_p, Y_p may be expressed as

$$Z_p = A_p \oplus V_p \oplus W_p,$$
$$Y_p = D_p \oplus E_p,$$

where $D_p \partial = V_{p-1}$, $E_p \partial \subseteq W_{p-1}$. Further $V_n = 0$, $W_n = 0$, $D_m = 0$, $E_m = 0$, and we may choose standard bases

$$V_p = (v_p^1, \ldots, v_p^{\beta_p}),$$
$$W_p = (w_p^1, \ldots, w_p^{\gamma_p}),$$
$$D_p = (d_p^1, \ldots, d_p^{\beta_{p-1}}),$$
$$E_p = (e_p^1, \ldots, e_p^{\gamma_{p-1}}),$$

such that the boundary homomorphism ∂_p is given by

$$d_p^i \partial = v_{p-1}^i \qquad (i = 1, \ldots, \beta_{p-1});$$
$$e_p^i \partial = k_{p-1}^i w_{p-1}^i \qquad (i = 1, \ldots, \gamma_{p-1}),$$

where each $k_{p-1}^i > 1$ and $k_{p-1}^i \mid k_{p-1}^{i+1}$ $(i = 1, \ldots, \gamma_{p-1} - 1)$.

Now $C_m = Z_m$, since $C_{m-1} = 0$, and $C_{m+1}\partial$ is a subgroup of Z_m. Thus, by Theorem 5.1.1, there is a basis $(a_m^1, \ldots, a_m^{\rho_m}, v_m^1, \ldots, v_m^{\beta_m}, w_m^1, \ldots, w_m^{\gamma_m})$ for Z_m such that $(v_m^1, \ldots, v_m^{\beta_m}, k_m^1 w_m^1, \ldots, k_m^{\gamma_m} w_m^{\gamma_m})$ is a basis for $C_{m+1}\partial$. We choose this basis for $C_m = Z_m$, so that $A_m = (a_m^1, \ldots, a_m^{\rho_m})$ and call it a standard basis for C_m. Now suppose a standard basis has been obtained for C_p. Thus $C_p = Z_p \oplus Y_p$ as in the enunciation of the theorem and, moreover,

$$(v_p^1, \ldots, v_p^{\beta_p}, k_p^1 w_p^1, \ldots, k_p^{\gamma_p} w_p^{\gamma_p})$$

is a basis for $C_{p+1}\partial = B_p$, where each

$$k_p^i > 1 \quad \text{and} \quad k_p^i \mid k_p^{i+1} \quad (i = 1, \ldots, \gamma_p - 1).$$

Now, by Theorem 5.1.10, $C_{p+1} = Z_{p+1} \oplus B_p \lambda$, where $\lambda \partial = 1$. For ∂_{p+1} maps C_{p+1} onto B_p with kernel Z_{p+1}, and B_p, being a subgroup of C_p, is free abelian. Put $B_p \lambda = Y_{p+1}$, $v_p^i \lambda = d_{p+1}^i$, $(k_p^i w_p^i)\lambda = e_{p+1}^i$. We thus obtain a decomposition of Y_{p+1} as $D_{p+1} \oplus E_{p+1}$ and $d_{p+1}^i \partial = v_p^i$, $e_{p+1}^i \partial = k_p^i w_p^i$. We now treat Z_{p+1} and its subgroup $B_{p+1} = C_{p+2}\partial$ exactly as we treated Z_m and C_{m+1}, by appealing to Theorem 5.1.1, and thus obtain a standard basis for C_{p+1}. Since $C_{n+1} = 0$, it is clear that $V_n = 0$, $W_n = 0$ and the proof is complete. ∎

Let $C_p = C_p(K)$, where K is a finite oriented complex. We may read off the homology groups of $|K|$ from Theorem 5.1.15, namely,

5.1.16 Corollary. *The p-th Betti number of* $|K|$ *is* ρ_p *and the p-th torsion coefficients are* $k_p^1, \ldots, k_p^{\gamma_p}$.]

Thus we see that the homology groups of $|K|$ are *calculable in a finite number of steps*, since the construction of a standard basis is an effective procedure. Since the homology groups of $|K|$ are topological invariants, we thus have the possibility of demonstrating, in a finite number of steps, that two compact polyhedra are not homeomorphic.

Notice that the non-zero elements of W_p are the cycles we have described as being weakly homologous to zero. The group $\overline{W}_p$ of 2.5.1 is $V_p \oplus W_p$, and ${}_wH_p \cong A_p$.

Let $\phi : C \to \Gamma$ be a chain map. Then if standard bases are chosen for C and Γ (assumed to be geometric *fg*-complexes) we see that ϕ maps $Z_p(C)$ to $Z_p(\Gamma)$ and $B_p(C)$ to $B_p(\Gamma)$; moreover, ϕ maps $V_p(C) \oplus W_p(C)$ to $V_p(\Gamma) \oplus W_p(\Gamma)$, for if $z \in C$ is a cycle such that kz bounds then $z\phi$ is a cycle such that $k(z\phi)$ bounds. Thus ϕ induces $\phi_* : H_*(C) \to H_*(\Gamma)$ and ${}_w\phi_* : {}_wH_*(C) \to {}_wH_*(\Gamma)$. It is not necessary that ϕ maps $A_p(C)$ into $A_p(\Gamma)$. On the other hand, if $\iota : A_p \to Z_p$, $\kappa : Z_p \to A_p$ are the injection and projection, and if

$$w : Z_p \to Z_p/(V_p \oplus W_p) = {}_wH_p$$

is the projection, then $\iota w : A_p \cong {}_wH_p$ with inverse $w^{-1}\kappa$ and the diagram

5.1.17
$$\begin{array}{ccc} A_p(C) & \overset{\iota w}{\cong} & {}_wH_p(C) \\ {\scriptstyle \iota\phi\kappa}\downarrow & & \downarrow{\scriptstyle {}_w\phi_*} \\ A_p(\Gamma) & \overset{\iota w}{\cong} & {}_wH_p(\Gamma) \end{array}$$

commutes; here we write $\iota\phi\kappa$ for $\iota(C)\,\phi\kappa(\Gamma)$. The proof of commutativity is easy and is left to the reader. The distinction we draw between A_p and ${}_wH_p$ may strike the reader as artificial, but this is not so. For ${}_wH_p$ is *canonically* determined by C (without reference to basis), whereas A_p is defined by means of a standard basis. The link between them is given precisely by (5.1.17).

We may apply Theorem 5.1.15 and 5.1.17 to prove the *Hopf Trace Theorem*, applications of which will be found in Appendix 1 to this chapter. If F is a free *fg*-group and $\phi : F \to F$ is an endomorphism, then we may associate with ϕ an integer called its *trace*. It is defined as follows.

Let $(a_1, \ldots, a_n)$ be a basis for F. Then $a_i\phi = \sum_j \phi_{ij}a_j$, where the ϕ_{ij} are integers and $\operatorname{tr}(\phi)$, the *trace* of ϕ, is $\sum_i \phi_{ii}$. It is a familiar theorem in linear algebra that $\operatorname{tr}(\phi)$ depends only on ϕ and not on the choice of basis. Now ${}_wH_p$ is a free *fg*-group if C is a geometric *fg*-complex. Thus if $\phi : C \to C$ is an (endomorphic) chain map, then $\operatorname{tr}(\phi_p)$ and $\operatorname{tr}({}_w\phi_{*p})$ are defined.

We suppose now that C is a *finite* geometric chain complex. Thus C is a geometric *fg*-complex with bottom dimension m and top dimension $n < \infty$.

5.1.18 Theorem. (*The Hopf Trace Theorem*). *If $\phi : C \to C$ is a chain map of the finite geometric chain complex C, then*

$$\sum_m^n (-1)^p \operatorname{tr}(\phi_p) = \sum_m^n (-1)^p \operatorname{tr}({}_w\phi_{*p}).$$

Choose a standard basis for C. Let $\iota_{X,p} : X_p \to C_p$, $\kappa_{X,p} : C_p \to X_p$ be the injection and projection, $X = A, V, W, D$ or E, and let $\phi_{X,p} = \iota_{X,p}\phi_p\kappa_{X,p}$, so that $\phi_{X,p}$ is an endomorphism of X_p. Then

5.1.19 $\operatorname{tr}(\phi_p) = \operatorname{tr}(\phi_{A,p}) + \operatorname{tr}(\phi_{V,p}) + \operatorname{tr}(\phi_{W,p}) + \operatorname{tr}(\phi_{D,p}) + \operatorname{tr}(\phi_{E,p})$.

Consider the standard basis element $v_p^i \in V_p$. Then $d_{p+1}^i\partial = v_p^i$. Let $d_{p+1}^i\phi$, expressed in terms of the standard basis, assign to d_{p+1}^i the coefficient λ_i. Since $\partial\phi = \phi\partial$, it follows that $v_p^i\phi$ assigns to v_p^i the coefficient λ_i. From this we immediately deduce that

5.1.20 $$\operatorname{tr}(\phi_{V,p}) = \operatorname{tr}(\phi_{D,p+1}).$$

This holds for all p in $-\infty < p < \infty$. Similarly, consider the basis element w_p^i; $e_{p+1}^i\partial = k_p^i w_p^i$. Let $e_{p+1}^i\phi$ assign to e_{p+1}^i the coefficient μ_i. Then $(k_p^i w_p^i)\phi$ assigns to w_p^i the coefficient $k_p^i\mu_i$. We may divide by k_p^i since the chain groups are free abelian and we see that $w_p^i\phi$ assigns to w_p^i the coefficient μ_i. It follows that

5.1.21 $$\operatorname{tr}(\phi_{W,p}) = \operatorname{tr}(\phi_{E,p+1}),$$

for all p. From 5.1.19–5.1.21 we immediately deduce

5.1.22 $$\sum_m^n (-1)^p \operatorname{tr}(\phi_p) = \sum_m^n (-1)^p \operatorname{tr}(\phi_{A,p}).$$

Now $\phi_{A,p}$ is just the map $\iota\phi\kappa$ of 5.1.17. By the commutativity of that diagram we see that $\operatorname{tr}(\phi_{A,p}) = \operatorname{tr}({}_w\phi_{*p})$ and the theorem is proved. ∎

A special case of the Hopf Trace Theorem is obtained by taking $\phi = 1$. If C_p is of rank α_p, and A_p of rank ρ_p, we obtain

5.1.23 $$\sum_m^n (-1)^p \alpha_p = \sum_m^n (-1)^p \rho_p.$$

In particular, take $C = C(K)$, where K is a finite (unaugmented) simplicial complex of dimension n. Let K have α_p simplexes of dimension p and let the pth Betti number be ρ_p. The number $\sum_0^n (-1)^p \alpha_p$ is called the *Euler-Poincaré characteristic* of K, and written $\chi(K)$.

5.1.24 Corollary. *The Euler-Poincaré characteristic* $\chi(K)$ *is equal to* $\sum_0^n (-1)^p \rho_p$, *the alternating sum of the Betti numbers of* $|K|$. *In particular, it is a topological invariant.* ∎

We need not make explicit how the results of this and subsequent sections may be applied to the relative homology theory. In so far as our results in this chapter are purely algebraic, the process of relativization is formal.

5.2 Homology with general coefficients and contrahomology

The group of integers has been consistently picked out as playing a privileged role as coefficient group for homology. We will now justify this preference by showing that the homology and contrahomology groups of a geometric complex with arbitrary coefficient or value group are completely determined by its homology groups with integer coefficients. We prove this theorem for general (i.e. possibly infinite) complexes in the next section, using certain invariant operations on abelian groups. In this section we confine attention to geometric *fg*-complexes and argue from a standard basis for the complex.

First we wish to make quite precise the algebraic notion underlying that of forming linear combinations of simplexes with coefficients in an abelian group G (see 2.2). It turns out that the fundamental notion required here—and in many other places in algebraic topology—is that of the *tensor product* of two abelian groups.

5.2.1 Definition. The *tensor product*, $A \otimes B$, of the two abelian groups A, B is the abelian group generated by all pairs $a \otimes b$, $a \in A$, $b \in B$, subject to the relations

5.2.2 $(a_1 + a_2) \otimes b = a_1 \otimes b + a_2 \otimes b, \quad a \otimes (b_1 + b_2) = a \otimes b_1 + a \otimes b_2.$

Notice that $a\otimes 0 = 0\otimes b = 0$, and $(ka)\otimes b = a\otimes(kb) = k(a\otimes b)$ for any integer k. Notice also that in general not every element of $A\otimes B$ is expressible as $a\otimes b$.

We observe that $\otimes$ is in a sense commutative and associative; there are in fact canonical isomorphisms between $A\otimes B$ and $B\otimes A$ and between $A\otimes(B\otimes C)$ and $(A\otimes B)\otimes C$.

5.2.3 Proposition. *If A is generated by the elements a_α, then every element of $A\otimes B$ is expressible in the form $\sum_\alpha a_\alpha\otimes b_\alpha$, $b_\alpha \in B$, the summation being finite.*‡

Clearly such elements belong to $A\otimes B$ and it follows from 5.2.2 that they form a subgroup. Thus to prove the proposition it is sufficient to show that if $a \in A$, $b\in B$, then $a\otimes b$ is expressible in the given form. But $a = \sum_\alpha \lambda_\alpha a_\alpha$, finite sum, so that, by 5.2.2, $a\otimes b = \sum_\alpha a_\alpha \otimes \lambda_\alpha b$. ▌

Let K be a finite oriented complex. Then 5.2.3 shows that the group§ $C_p(K; G)$ is just $C_p(K)\otimes G$. We now proceed to make precise the way in which we have been entitled to speak of 'extending' boundary homomorphisms and chain maps from $C(K)$ to $C(K; G)$.

Let $\phi : A \to A'$, $\psi : B \to B'$ be homomorphisms of abelian groups.

5.2.4 Proposition. *There is a homomorphism $\theta : A\otimes B \to A'\otimes B'$ determined by*

$$(a\otimes b)\,\theta = a\phi\otimes b\psi, \quad a\in A,\ b\in B.$$

Since θ is defined on the generators of $A\otimes B$, it is only necessary to verify that θ preserves the relations in $A\otimes B$; but this is quite obvious. ▌ We write $\theta = \phi\otimes\psi$.

5.2.5 Proposition. $(\phi_1\otimes\psi_1)\,(\phi_2\otimes\psi_2) = \phi_1\phi_2\otimes\psi_1\psi_2.$

The meaning and proof of this proposition are clear. ▌ The reader should now recognize that the extension of $\partial : C_p(K) \to C_{p-1}(K)$ to a boundary operator $C_p(K; G) \to C_{p-1}(K; G)$ is precisely $\partial\otimes 1$. Then $(\partial\otimes 1)\,(\partial\otimes 1) = \partial\partial\otimes 1 = 0\otimes 1 = 0$. Similarly, the extension of $\phi : C_p(K) \to C_p(L)$ to a chain map $C_p(K; G) \to C_p(L; G)$ is $\phi\otimes 1$ and $\phi\otimes 1$ is a chain map since $(\phi\otimes 1)\,(\partial\otimes 1) = \phi\partial\otimes 1 = \partial\phi\otimes 1 = (\partial\otimes 1)\,(\phi\otimes 1)$. The reader is advised to verify the generalizations that have been made in previous chapters from integer coefficients to general coefficients; for example, if ϕ is a chain equivalence, so is $\phi\otimes 1$.

‡ I.e. $b_\alpha = 0$ for all but a finite number of values of α.
§ This also holds if K is infinite.

Since we wish to compute from $C_p(K)\otimes G$ to obtain $H_p(K; G)$ we must make a further study of the properties of tensor products.

5.2.6 Lemma. *Let A be the direct sum of the groups A_α, and let B be the direct sum of the groups B_β. Then $A\otimes B$ is the direct sum of the groups $A_\alpha\otimes B_\beta$.*

Let $i_\alpha : A_\alpha\to A$, $j_\beta : B_\beta\to B$ be the injections and let $p_\alpha : A\to A_\alpha$, $q_\beta : B\to B_\beta$ be the projections. Then there are homomorphisms $i_\alpha\otimes j_\beta : A_\alpha\otimes B_\beta\to A\otimes B$, $p_\alpha\otimes q_\beta : A\otimes B\to A_\alpha\otimes B_\beta$. Moreover,

$$(i_\alpha\otimes j_\beta)(p_{\alpha'}\otimes q_{\beta'}) = i_\alpha p_{\alpha'}\otimes j_\beta q_{\beta'} = 0$$

unless $\alpha = \alpha'$, $\beta = \beta'$, and $(i_\alpha\otimes j_\beta)(p_\alpha\otimes q_\beta) = 1$.

Let $\Sigma = \sum\limits_{\alpha,\beta} A_\alpha\otimes B_\beta$. Then a homomorphism $\Phi : \Sigma\to A\otimes B$ is given by $\Phi \mid A_\alpha\otimes B_\beta = i_\alpha\otimes j_\beta$. The lemma asserts that Φ is an isomorphism. To prove that Φ is an epimorphism it is sufficient to show that if $a\in A$, $b\in B$, then $a\otimes b\in \Sigma\Phi$. Now $a = \sum\limits_\alpha a_\alpha i_\alpha$, $b = \sum\limits_\beta b_\beta j_\beta$. Thus $a\otimes b = \sum\limits_{\alpha,\beta}(a_\alpha\otimes b_\beta)(i_\alpha\otimes j_\beta) = (\sum\limits_{\alpha,\beta} a_\alpha\otimes b_\beta)\Phi$.

It remains to show that Φ is monomorphic. Let $\Psi : A\otimes B\to\Sigma$ be given by $(a\otimes b)\Psi = \sum\limits_{\alpha,\beta} ap_\alpha\otimes bq_\beta$. The sum is finite since $ap_\alpha = 0$ for almost all α and Ψ is plainly homomorphic. We show that $\Phi\Psi = 1$. It is sufficient to prove this on a generator of $A_{\alpha_0}\otimes B_{\beta_0}$. But

$$\begin{aligned}(a_{\alpha_0}\otimes b_{\beta_0})\Phi\Psi &= (a_{\alpha_0}\otimes b_{\beta_0})(i_{\alpha_0}\otimes j_{\beta_0})\Psi\\ &= \sum_{\alpha,\beta}(a_{\alpha_0}\otimes b_{\beta_0})(i_{\alpha_0}\otimes j_{\beta_0})(p_\alpha\otimes q_\beta),\end{aligned}$$

and we have observed that $(i_{\alpha_0}\otimes j_{\beta_0})(p_\alpha\otimes q_\beta) = 0$ unless $\alpha=\alpha_0$, $\beta=\beta_0$, and $(i_{\alpha_0}\otimes j_{\beta_0})(p_{\alpha_0}\otimes q_{\beta_0}) = 1$. Thus $(a_{\alpha_0}\otimes b_{\beta_0})\Phi\Psi = a_{\alpha_0}\otimes b_{\beta_0}$ and $\Phi\Psi = 1$. It follows immediately that Φ is a monomorphism; for if $\gamma\in\Sigma$ and $\gamma\Phi = 0$ then $\gamma = \gamma\Phi\Psi = 0$. Thus Φ is an isomorphism with inverse Ψ. ▮

Before applying this lemma, we wish to meet the possible objection that the proof was unnecessarily fussy and the lemma obvious. It is necessary to be cautious in handling tensor products, and it is certainly not true in general that if A' is a subgroup of A then $A'\otimes B$ is a subgroup of $A\otimes B$; we deal explicitly with this in the next section. On the other hand, 5.2.6 does entitle us to regard $A'\otimes B$ as a subgroup, indeed direct factor, of $A\otimes B$ when A' is a direct factor of A. (The definition of a direct factor may be found just after I.2.4).

5.2.7 Theorem. *Let A be a free abelian group of rank‡ ρ. Then $A \otimes B$ is isomorphic with the direct sum of ρ copies of B. If $\{a_\alpha\}$ is a basis for A, then an isomorphism is induced by $a_\alpha \otimes b \to b^\alpha$, where $b \to b^\alpha$ is the copying isomorphism of B onto the copy B^α.*

In the light of Lemma 5.2.6, it is sufficient to prove this when $\rho = 1$, i.e. when A is cyclic infinite. But the map $b \to 1 \otimes b$ is obviously an isomorphism of B onto $J \otimes B$. ▮ Theorem 5.2.7 makes it clear that we may regard $A \otimes B_0$ as a subgroup of $A \otimes B$ if $B_0 \subseteq B$ and A is *free*. It shows that, if K has α_p simplexes of dimension p, then $C_p(K; G)$ is isomorphic to the direct sum of α_p copies of G. It is important to notice that the definition of $C_p(K; G)$ does *not* depend on a choice of basis for $C_p(K)$, but this isomorphism does.

Let C be a geometric *fg*-complex. Then $C \otimes G$ may be given the structure of a chain complex by defining $(C \otimes G)_p = C_p \otimes G$ and using the boundary operator $\partial \otimes 1$. If confusion is not to be feared we write ∂ for $\partial \otimes 1$ and proceed to compute $H_*(C \otimes G)$ by means of the standard basis of Theorem 5.1.15 for C.

By 5.2.6, we have

$$C_p \otimes G = (A_p \otimes G) \oplus (V_p \otimes G) \oplus (W_p \otimes G) \oplus (D_p \otimes G) \oplus (E_p \otimes G).$$

We now investigate $Z_p(C \otimes G)$, the group of p-cycles of $C \otimes G$. Since Z_p is a direct factor of C, we may regard $Z_p \otimes G$ as a subgroup of $C \otimes G$ and obviously $Z_p \otimes G \subseteq Z_p(C \otimes G)$. Consider $\sum_i d_p^i \otimes g_i + \sum_j e_p^j \otimes g_j'$, the general element of $D_p \otimes G \oplus E_p \otimes G$. Its boundary is

$$\sum_i v_{p-1}^i \otimes g_i + \sum_j k_{p-1}^j w_{p-1}^j \otimes g_j'.$$

By Theorem 5.2.7, this is zero if and only if each $g_i = 0$ and each $k_{p-1}^j g_j' = 0$. Let us write ${}_kG$ for the subgroup of G consisting of elements g such that $kg = 0$. We have thus proved that

5.2.8 $Z_p(C \otimes G) = (A_p \otimes G) \oplus (V_p \otimes G) \oplus (W_p \otimes G) \oplus (E_p'(C \otimes G))$,

where $E_p'(C \otimes G)$ is a subgroup of $E_p \otimes G$ and

5.2.9 $$E_p'(C \otimes G) \cong \sum_{j=1}^{\gamma_{p-1}} {}_{k_{p-1}^j}G.$$

We next consider $B_p(C \otimes G)$. Now

$$B_p(C \otimes G) = (C_{p+1} \otimes G)\,\partial = (B_p \otimes G)\,\bar{\imath},$$

‡ ρ need not be finite.

where $\bar{\iota} = \iota \otimes 1 : B_p \otimes G \to C_p \otimes G$ and $\iota : B_p \to C_p$ is the embedding. Now $B_p = V_p \oplus (k_p^1 w_p^1, \ldots, k_p^{\gamma_p} w_p^{\gamma_p})$. Thus, by 5.2.6,

5.2.10 $$B_p(C \otimes G) = (V_p \otimes G) \oplus \sum_{i=1}^{\gamma_p} (w_p^i) \otimes k_p^i G,$$

where $W_p \otimes G = \sum_{i=1}^{\gamma_p} (w_p^i) \otimes G$, and $(w_p^i) \otimes k_p^i G$ is regarded as a subgroup of $(w_p^i) \otimes G$.

Now $H_p(C \otimes G)$ is the factor group of $Z_p(C \otimes G)$ by $B_p(C \otimes G)$. If w generates a cyclic infinite group and we map $(w) \otimes G$ isomorphically onto G by sending $w \otimes g$ to g, then the subgroup $(w) \otimes kG$, where k is an integer, is mapped onto kG. Thus there is an isomorphism of factor groups
$$(w) \otimes G / (w) \otimes kG \cong G/kG.$$

We shall write‡ G_k for G/kG; the definitions of ${}_kG$ and G_k may be remembered from the exact sequence

5.2.11 $$0 \to {}_kG \to G \xrightarrow{k} G \to G_k \to 0,$$

where the middle homomorphism is multiplication by the integer k.

We deduce the main theorem of this section from 5.2.8 and 5.2.10.

5.2.12 Theorem. $H_p(C \otimes G) \cong (A_p \otimes G) \oplus \sum_{i=1}^{\gamma_p} G_{k_p^i} \oplus \sum_{j=1}^{\gamma_{p-1}} {}_{k_{p-1}^j}G.$ ▮

The reader should verify that the first two terms on the right may be combined as $H_p(C) \otimes G$. This follows because they were obtained as the cokernel of the mapping $B_p \otimes G \to Z_p \otimes G$ induced by the injection $B_p \to Z_p$.

5.2.13 Corollary. *Let $|K|$ be a compact polyhedron. Then $H_p(|K|; G)$ is entirely determined by G and the p-th Betti number of $|K|$ and the p-th and $(p-1)$-th torsion coefficients.* ▮

We remark that $T_p(K)$ is definable solely in terms of $C_{p+1}(K)$ and $C_p(K)$ and $\partial : C_{p+1} \to C_p$; there would be some sense, in fact, in writing it as $T_{p+\frac{1}{2}}(K)$. It is a special feature of J (and, from the point of view of symmetry, an unsatisfactory one) that $H_p(|K|; J)$ depends on the torsion coefficients for dimension '$(p+\frac{1}{2})$' and not for '$(p-\frac{1}{2})$'. The formulation, however, for $H_p(|K|; G)$ does something to restore the symmetry in that the $(p-1)$th torsion coefficients are involved.

Let $G = J_m$, generated by 1_m, say. Then ${}_kG = J_{(k,m)}$; indeed, if $(k, m) = d$ and $m = dm_0$, then ${}_kG$ is generated by $m_0 1_m$ and $m_0 1_m$ is of order d. Also $G_k = J_{(k,m)}$, being generated by the image in G_k of 1_m. Also $A_p \otimes J_m \cong (A_p)_m$, the direct

‡ This is consistent with the notation J_k for the cyclic group of order k.

sum of ρ_p copies of J_m, where ρ_p is the rank of A_p. Thus we deduce from Theorem 5.2.12

5.2.14 Corollary. $H_p(C \otimes J_m) \cong (A_p)_m \oplus \sum_{i=1}^{\gamma_p} J_{(k_p^i, m)} \oplus \sum_{j=1}^{\gamma_{p-1}} J_{(k_{p-1}^j, m)}$. ∎

5.2.15 *Example.* Consider the real projective space P^n. If r is odd and $0 < r \leqslant n$ then the homology class of ζ_r generates $H_r(P^n)$, which is J_2 (or J if $r = n$), and $H_{r-1}(P^n)$ has no torsion. Thus $H_r(P^n; J_2) = J_2$, generated by the class of ζ_r regarded as a cycle mod 2. If r is even and $0 < r \leqslant n$, then $H_r(P^n) = 0$ and $H_{r-1}(P^n) = J_2$; it follows from 5.2.14 that $H_r(P^n; J_2) = J_2$. Moreover, the generator is the class of the generator of the E'_p term in 5.2.8, which is precisely ζ_r, regarded as a chain with coefficients in J_2. We conclude (as we might have done, of course, by direct reasoning) that $H_r(P^n; J_2) = J_2$, $0 \leqslant r \leqslant n$, the generator in dimension r being the class of ζ_r (see 3.9.6).

A second important example of a coefficient group is the group, R_1, of real numbers reduced mod 1. If $G = R_1$, then ${}_kG = J_k$. For the elements of ${}_kG$ are just the residue classes mod 1 of the rationals $0, \frac{1}{k}, \frac{2}{k}, \ldots, \frac{k-1}{k}$. Also $G_k = 0$; for if $g \in R_1$ and k is a non-zero integer, then $g = k\frac{g}{k}$ so that $R_1 = kR_1$.

5.2.16 Corollary. $H_p(C \otimes R_1) \cong A_p \otimes R_1 \oplus \sum_{j=1}^{\gamma_{p-1}} J_{k_{p-1}^j}$. ∎

Let K be a finite simplicial complex and let $H_p(K)$ be expressed as $A_p \oplus T_p$, where T_p is the torsion subgroup. Then we deduce from 5.1.16 and 5.2.16

5.2.17 Theorem. $H_p(K; R_1) \cong (A_p \otimes R_1) \oplus T_{p-1}$. ∎

It follows that the homology groups of a finite complex with arbitrary coefficients may be deduced from the homology groups with coefficients in R_1.

We turn now to contrahomology. Our object will be to obtain a result analogous to Theorem 5.2.12. The technique is similar to that above, the role of the tensor product construction $A \otimes B$ being taken by that of $A \pitchfork B$, the *group of homomorphisms* of A into B. Many of the remarks dual to those made for tensor products are already familiar; we begin with the duals of Proposition 5.2.5 and Lemma 5.2.6.

First we fix notation. A homomorphism $\phi : A' \to A$ induces a homomorphism $\phi^{\cdot} : A \pitchfork B \to A' \pitchfork B$, this will be written, as before, as a left homomorphism. A homomorphism $\psi : B \to B'$ induces a right homomorphism $\psi_{\cdot} : A \pitchfork B \to A \pitchfork B'$. Together ϕ and ψ induce a homomorphism $A \pitchfork B \to A' \pitchfork B'$ which we may write $\phi \pitchfork \psi$. Thus $\phi^{\cdot} = \phi \pitchfork 1$, $\psi_{\cdot} = 1 \pitchfork \psi$. We shall write $\phi \pitchfork \psi$ on the *left*, unless otherwise stated.

5.2.5c Proposition. $(\phi_1 \pitchfork \psi_1)(\phi_2 \pitchfork \psi_2) = \phi_1\phi_2 \pitchfork \psi_2\psi_1$. ∎

The reader will notice the orders in which homomorphisms are composed on the right-hand side; whereas $\otimes$ is a fully covariant operation, $\pitchfork$ is partly covariant and partly contravariant.

5.2.6c Lemma. *Let A be a finite direct sum of the groups A_α, and let B be a finite direct sum of the groups B_β. Then $A \pitchfork B$ is the direct sum of the groups*‡ $A_\alpha \pitchfork B_\beta$.

We use the notation of 5.2.6. There are homomorphisms

$$i_\alpha \pitchfork q_\beta : A \pitchfork B \to A_\alpha \pitchfork B_\beta, \quad p_\alpha \pitchfork j_\beta : A_\alpha \pitchfork B_\beta \to A \pitchfork B$$

and
$$(i_\alpha \pitchfork q_\beta)(p_{\alpha'} \pitchfork j_{\beta'}) = i_\alpha p_{\alpha'} \pitchfork j_{\beta'} q_\beta;$$

hence
$$\begin{aligned}(i_\alpha \pitchfork q_\beta)(p_{\alpha'} \pitchfork j_{\beta'}) &= 0 \quad \text{unless } \alpha = \alpha' \text{ and } \beta = \beta', \\ &= 1 \quad \text{if } \alpha = \alpha' \text{ and } \beta = \beta'.\end{aligned}$$

Let $\Sigma = \sum_{\alpha, \beta} A_\alpha \pitchfork B_\beta$. Define $\Phi^c : \Sigma \to A \pitchfork B$ by $\Phi^c \mid A_\alpha \pitchfork B_\beta = p_\alpha \pitchfork j_\beta$, and define $\Psi^c : A \pitchfork B \to \Sigma$ by $\Psi^c = \sum_{\alpha, \beta} i_\alpha \pitchfork q_\beta$. Φ^c and Ψ^c will be written on the left. We prove that

$$\Phi^c \Psi^c = 1 : A \pitchfork B \to A \pitchfork B, \quad \Psi^c \Phi^c = 1 : \Sigma \to \Sigma.$$

It is readily seen that the former is equivalent to $\sum_{\alpha, \beta} p_\alpha i_\alpha \pitchfork q_\beta j_\beta = 1$. But $\sum_\alpha p_\alpha i_\alpha = 1$, $\sum_\beta q_\beta j_\beta = 1$, so that the assertion $\Phi^c \Psi^c = 1$ follows. The proof that $\Psi^c \Phi^c = 1$ follows the lines of proof that $\Phi \Psi = 1$ in 5.2.6. Thus Φ^c is an isomorphism with inverse Ψ^c. ∎

5.2.7c Theorem. *Let A be a free abelian group of finite rank ρ. Then $A \pitchfork B$ is isomorphic with the direct sum*§ *of ρ copies of B. If $\{a_\alpha\}$ is a basis for A, then an isomorphism from $A \pitchfork B$ to $\sum_\alpha B^\alpha$ is given by $\phi \to \sum_\alpha (a_\alpha \phi)^\alpha$, where $b \to b^\alpha$ is the copying isomorphism of B onto the copy B^α.* ∎

Let C be a geometric fg-complex. We recall that $C \pitchfork G$ is the contrachain complex with $(C \pitchfork G)^p = C_p \pitchfork G$ and the contraboundary operator $\delta = \partial^* = \partial \pitchfork 1$. By 5.2.6c, we have

$$C_p \pitchfork G = (A_p \pitchfork G) \oplus (V_p \pitchfork G) \oplus (W_p \pitchfork G) \oplus (D_p \pitchfork G) \oplus (E_p \pitchfork G).$$

Notice that, for example, $A_p \pitchfork G$ is embedded in $C_p \pitchfork G$ by regarding a homomorphism $A_p \to G$ as a homomorphism $C_p \to G$ sending $V_p \oplus W_p \oplus D_p \oplus E_p$ to zero. Now z^p is a p-contracycle if and only if it vanishes on all p-boundaries (2.6.7). Thus the elements of

$$(A_p \pitchfork G) \oplus (D_p \pitchfork G) \oplus (E_p \pitchfork G)$$

‡ The restriction to finite direct sums is natural; the dual notion to direct sum is *direct product* and 5.2.6c is a special case of the following:

Let A be the direct sum of the groups A_α, and let B be the direct product of the groups B_β. Then $A \pitchfork B$ is the direct product of the groups $A_\alpha \pitchfork B_\beta$.

§ If ρ is not finite we replace 'direct sum' by 'direct product'.

are contracycles, only $0 \in V_p \pitchfork G$ is a contracycle, and $z^p \in W_p \pitchfork G$ is a contracycle if and only if $(w_p^i, z^p) \in k_p^i G$ for each i. Now under the isomorphism $J \pitchfork G \cong G$ given by $\phi \to 1\phi$, the subgroup of $J \pitchfork G$ consisting of homomorphisms mapping J into $G_0 \subseteq G$ is mapped onto G_0. Thus we have proved

5.2.8c $\quad Z^p(C \pitchfork G) = (A_p \pitchfork G) \oplus (W'_p(C \pitchfork G)) \oplus (D_p \pitchfork G) \oplus (E_p \pitchfork G),$

where $W'_p(C \pitchfork G)$ is a subgroup of $W_p \pitchfork G$ and

5.2.9c
$$W'_p(C \pitchfork G) \cong \sum_{i=1}^{\gamma_p} k_p^i G.$$

We now consider $B^p(C \pitchfork G)$. Clearly if $b^p \in B^p(C \pitchfork G)$, then b^p vanishes on $Z_p(C)$. Thus b^p induces $b' : B_{p-1}(C) \to G$ by the rule

5.2.18
$$(c_p \partial)b' = (c_p, b^p).$$

We prove a lemma which enables us to write down $B^p(C \pitchfork G)$:

5.2.19 Lemma. *Let C be a chain complex and let $b^p \in C_p \pitchfork G$ vanish on $Z_p(C)$. Then b^p is a contraboundary if and only if b', defined by* 5.2.18, *may be extended to C_{p-1}.*

Notice that we make no special assumptions about C. Now suppose there exists $c^{p-1} \in C_{p-1} \pitchfork G$ with $c^{p-1} \mid B_{p-1} = b'$. Then

$$(c_p, \delta c^{p-1}) = (c_p \partial, c^{p-1}) = c_p \partial b' = (c_p, b^p),$$

for all $c_p \in C_p$, so that $\delta c^{p-1} = b^p$ and b^p is a contraboundary. Conversely, if $b^p = \delta c^{p-1}$, then $c_p \partial b' = (c_p, b^p) = (c_p, \delta c^{p-1}) = (c_p \partial, c^{p-1})$, so that $b' = c^{p-1} \mid B_{p-1}$. ∎

Now if b^p vanishes on all p-cycles, then $b^p \in (D_p \pitchfork G) \oplus (E_p \pitchfork G)$ and the associated b' is given by

$$v_{p-1}^i b' = (d_p^i, b^p), \qquad i = 1, \ldots, \beta_{p-1};$$
$$(k_{p-1}^j w_{p-1}^j)\, b' = (e_p^j, b^p), \quad j = 1, \ldots, \gamma_{p-1}.$$

Clearly b' may be extended to C_{p-1} if and only if it may be extended to $V_{p-1} \oplus W_{p-1}$; and b' may be extended to $V_{p-1} \oplus W_{p-1}$ if and only if $(e_p^j, b^p) \in k_{p-1}^j G$ for each j. Thus, by 5.2.19, we have

5.2.10c
$$B^p(C \pitchfork G) = (D_p \pitchfork G) \oplus \sum_{j=1}^{\gamma_{p-1}} (e_p^j) \pitchfork k_{p-1}^j G,$$

where $E_p \pitchfork G = \sum_{j=1}^{\gamma_{p-1}} (e_p^j) \pitchfork G$, and $(e_p^j) \pitchfork k_{p-1}^j G$ is regarded as a subgroup of $(e_p^j) \pitchfork G$.

Since the isomorphism $(e) \pitchfork G \cong G$ maps $(e) \pitchfork kG$ to kG it induces an isomorphism of the factor groups

$$(e) \pitchfork G/(e) \pitchfork kG \cong G_k.$$

We deduce from 5.2.8c and 5.2.10c

5.2.12c Theorem. If C is a geometric fg-complex,

$$H^p(C \pitchfork G) \cong (A_p \pitchfork G) \oplus \sum_{i=1}^{\gamma_p} {}_{k_p^i}G \oplus \sum_{j=1}^{\gamma_{p-1}} G_{k_{p-1}^j}. \;\blacksquare$$

The reader should verify that the first two terms on the right may be combined as $H_p(C) \pitchfork G$. This follows because they were obtained as the subgroup of $Z_p(C) \pitchfork G$ consisting of homomorphisms vanishing on $B_p(C)$.

5.2.13c Corollary. *Let $|K|$ be a compact polyhedron. Then $H^p(|K|; G)$ is entirely determined by G and the p-th Betti number of $|K|$ and the p-th and $(p-1)$-th torsion coefficients.* ∎

If we put $G = J$ we deduce from Theorem 5.2.12c

5.2.20 Theorem. *If C is a geometric fg-complex,*

$$H^p(C \pitchfork J) \cong A_p \oplus T_{p-1}. \;\blacksquare$$

The notation is the usual one. Thus the $(p-1)$th torsion coefficients are, as it were, the pth torsion 'contra-efficients'.

If we put $G = J_m$ we get

5.2.14c Corollary. $H^p(C \pitchfork J_m) \cong (A_p)_m \oplus \sum_{i=1}^{\gamma_p} J_{(k_p^i,\, m)} \oplus \sum_{j=1}^{\gamma_{p-1}} J_{(k_{p-1}^j,\, m)}.$ ∎

It follows that if $|K|$ is a compact polyhedron then

5.2.21 $$H_p(|K|; J_m) \cong H^p(|K|; J_m);$$

the isomorphism, however, depends on choosing generators for the groups. If m is prime, then each side of 5.2.21 is a vector space of finite dimension over J_m and the isomorphism is that obtaining between a vector space and its dual.

We may generalize the last remark by considering an arbitrary *field* of coefficients. If $\mathscr{F}$ is a field, then ${}_k\mathscr{F} = \mathscr{F}_k = \mathscr{F}$ if the characteristic‡ of $\mathscr{F}$ divides k, and ${}_k\mathscr{F} = \mathscr{F}_k = 0$ otherwise. Thus, from 5.2.12 and 5.2.12c, we see that $H_p(C \otimes \mathscr{F})$ and $H^p(C \pitchfork \mathscr{F})$ are vector spaces over $\mathscr{F}$, whose dimension is the sum of ρ_p and the number of torsion coefficients in dimensions p and $(p-1)$ which are divisible by the characteristic of $\mathscr{F}$. Moreover, since a contrachain in $C_p \pitchfork \mathscr{F}$ can evidently be identified with a linear function on $C_p \otimes \mathscr{F}$, we see that the vector spaces $H_p(C \otimes \mathscr{F})$ and $H^p(C \pitchfork \mathscr{F})$ may also be regarded as the duals of each other.

‡ The *characteristic* of $\mathscr{F}$ is the smallest positive integer p such that $p\mathscr{F} = 0$, provided this exists; otherwise the characteristic is by definition zero. If the characteristic is $p > 0$, then p is prime.

We might define the *p-th Betti number* mod q of $|K|$, where q is any prime, as the dimension of the vector space $H_p(|K|; J_q)$. If $q=2$, this is called the p-th *connectivity number* of $|K|$.

We now return to the contrahomology analogues of our results in homology.

5.2.15*c* *Example.* Consider the real projective space P^n. We deduce from 5.2.12*c* that $H^p(P^n; J_2) = J_2$ for all p such that $0 \leqslant p \leqslant n$. Moreover, by studying the way the isomorphism 5.2.12*c* is obtained, it is clear that each contrahomology group is generated by the class of the homomorphism sending the appropriate ζ_r to the non-zero element of J_2.

Now put $G = R_1$ in 5.2.12*c*. We obtain

5.2.16*c* Corollary. $H^p(C \pitchfork R_1) \cong (A_p \pitchfork R_1) \oplus \sum_{i=1}^{\gamma_p} J_{k_p^i}$. **]**

5.2.17*c* Theorem. *If* $|K|$ *is a compact polyhedron, then*

$$H^p(|K|; R_1) \cong (A_p \pitchfork R_1) \oplus T_p. \;]$$

The groups J and R_1 are dual groups in the sense of Pontryagin; if J is given the discrete topology and R_1 the circle topology, then each is the group of continuous characters ‡ of the other; and $H_p(K)$, $H^p(K; R_1)$ are continuous characters of each other. We do not pursue this important question further.

Our closing concern in this section is with the contrahomology *ring*. Corollaries 5.2.13 and 5.2.13*c* show that the homology and contrahomology groups of a compact polyhedron $|K|$, with coefficients or values in an arbitrary abelian group, are computable. If the values are taken in a ring, then the contrahomology ring is defined. Although there is in general no rule for determining $R^*(K; R)$, where R is an arbitrary ring, in terms of $R^*(K; J)$, none the less the contrahomology ring for arbitrary R is computable. Let us write $C^\cdot(K)$, $R^*(K)$ for $C^\cdot(K; J)$, $R^*(K; J)$. We may take standard contrachain bases in $C^\cdot(K)$ and express any element of $C^\cdot(K; R)$ as a linear combination of basis elements of $C^\cdot(K)$ with coefficients in R. The cup products of basis elements can be calculated mechanically from an ordering of the vertices of K and from this the cup product of any two such linear combinations can be read off. We are, in fact, saying that, if $C^\cdot(K)$, $C^\cdot(K; R)$ are regarded as rings, then the cup products in $C^\cdot(K)$ are calculable and§ $C^\cdot(K; R) \cong C^\cdot(K) \otimes R$, where $C^\cdot(K) \otimes R$ is given the obvious contrachain ring structure||; and therefore $R^*(K; R)$ is calculable.

‡ The character group of G is the subgroup of $G \pitchfork R_1$ consisting of continuous homomorphisms; it is itself topologized. See Pontryagin, *Topological Groups*.

§ This isomorphism holds only in general for finite complexes.

|| See 5.2.23.

In order to establish 5.2.25 below we now make the isomorphism between $C^{\cdot}(K;R)$ and $C^{\cdot}(K)\otimes R$ explicit. Let F be a free fg-group and let $F^{\cdot} = F \pitchfork J$. For any group G, let

$$\kappa : F^{\cdot}\otimes G \to F \pitchfork G$$

be given by $(x, \kappa(y^{\cdot}\otimes g)) = (x, y^{\cdot})\, g$, $x \in F$, $y^{\cdot} \in F^{\cdot}$, $g \in G$.

5.2.22 Lemma. *κ is an isomorphism.*

Let $(f_1, \ldots, f_m)$ be a basis for F, let $G^{(i)}$, $i = 1, \ldots, m$, be a copy of G and let $g \to g^{(i)}$ be the copying isomorphism. Define $\lambda : F^{\cdot}\otimes G \to \sum_i G^{(i)}$ by $\lambda(y^{\cdot}\otimes g) = \sum_i (f_i, y^{\cdot})\, g^{(i)}$ and $\mu : F \pitchfork G \to \sum_i G^{(i)}$ by $\mu(\phi) = \sum_i (f_i, \phi)^{(i)}$. Then λ and μ are isomorphisms‡ (5.2.7 and 5.2.7 c) and $\mu\kappa = \lambda$. ∎

Thus if ω is an ordering of the finite simplicial complex K there is a group isomorphism

$$\kappa : C^{\cdot}_{\omega}(K)\otimes G \cong C^{\cdot}_{\omega}(K; G).$$

We give $C^{\cdot}_{\omega}(K)\otimes G$ the structure of a contrachain complex in the obvious way (writing δ for $\delta\otimes 1$). If G is a ring, then $C^{\cdot}_{\omega}(K)\otimes G$ is a contrachain ring (see 4.2.1.) if we define

$$(d\otimes g)\,(e\otimes g') = (d \cup e)\otimes gg', \quad d, e \in C^{\cdot}_{\omega}(K), g, g' \in G.$$

5.2.23 Proposition. (i) *κ is a contrachain map;* (ii) *if G is a ring then κ is a ring isomorphism.*

Let $c \in C^{\omega}(K)$, $d \in C^{\cdot}_{\omega}(K)$, $g \in G$; then

$$\begin{aligned}(c, \delta\kappa(d\otimes g)) &= (c\partial, \kappa(d\otimes g)) = (c\partial, d)\, g \\ &= (c, \delta d)\, g = (c, \kappa(\delta d\otimes g)) = (c, \kappa\delta(d\otimes g)).\end{aligned}$$

Thus $\delta\kappa = \kappa\delta$ and (i) is proved.

Let R be a ring, $d \in C^p_{\omega}(K)$, $e \in C^q_{\omega}(K)$, $r, s \in R$; then (ii) is proved if we can show that

5.2.24 $$\kappa((d \cup e)\otimes rs) = \kappa(d\otimes r) \cup \kappa(e\otimes s).$$

We compute the value of each side of 5.2.24 on the ordered $(p+q)$-simplex $a^0 \ldots a^{p+q}$

$$\begin{aligned}(a^0 \ldots a^{p+q}, \kappa((d \cup e)\otimes rs)) &= (a^0 \ldots a^{p+q}, d \cup e)\, rs \\ &= (a^0 \ldots a^p, d)\,(a^p \ldots a^{p+q}, e)\, rs \\ &= (a^0 \ldots a^p, \kappa(d\otimes r))\,(a^p \ldots a^{p+q}, \kappa(e\otimes s)) \\ &= (a^0 \ldots a^{p+q}, \kappa(d\otimes r) \cup \kappa(e\otimes s)). \ \blacksquare\end{aligned}$$

‡ Notice that κ, unlike λ and μ, is independent of choice of basis in F.

The explicit form of the isomorphism $\kappa : C^{\cdot}_{\omega}(K)\otimes R \cong C^{\cdot}_{\omega}(K; R)$ enables us to draw the following important conclusion, which should be contrasted with Example 4.3.4:

5.2.25 Theorem. *If K is a finite complex and $H_*(K)$ is free, then $R^*(K; R) \cong R^*(K)\otimes R$.*

Since $H_p(K)$ is free, $B^{\omega}_p(K)$ is a direct factor of $C^{\omega}_p(K)$. Thus we may apply 5.2.22 with $F = C^{\omega}_p(K)/B^{\omega}_p(K)$ and obtain

$$\kappa : Z^p_{\omega}(K)\otimes R \cong Z^p_{\omega}(K; R),$$

which may be thought of as being obtained by restricting the isomorphism $\kappa : C^p_{\omega}(K)\otimes R \cong C^p_{\omega}(K; R)$. Similarly by taking

$$F = C^{\omega}_p(K)/Z^{\omega}_p(K)$$

we obtain
$$\kappa : B^p_{\omega}(K)\otimes R \cong B^p_{\omega}(K; R);$$

for we remark that, by 5.2.19, if $H_{p-1}(K)$ is torsion-free then a p-contrachain is a p-contraboundary if and only if it vanishes on all p-cycles. The result now follows by passing to quotient rings. ∎

5.3 Free and divisible groups

Before we extend to infinite complexes the results of the last section, we interpolate some central properties of free abelian groups, together with dual properties of divisible groups. Free groups will be more important to us than divisible groups, but we make some use of the latter and a parallel exposition may help the reader to remember one set of properties from the other. We are far from suggesting that one set of proofs will be any guide to the other set. Corresponding theorems will be given the same number affixed by *f* for free or *d* for divisible.

5.3.1*f* Theorem. *A subgroup of a free abelian group is free abelian.*

The reader content to confine his attention to *fg*-groups will require no further proof of this proposition.

Let F be free abelian, let R be a subgroup of F, and let $\{f_\alpha\}$ be a basis for F; we consider families $\rho = \{r_\beta\}$ of elements of R with the property (a) that they are linearly independent and (b) that if S_ρ is the group they generate and G_ρ is the group generated by those elements f_α involved in the expressions for the elements $r_\beta \in \rho$, then

$$G_\rho \cap R = S_\rho$$

(notice that, for any family ρ, $G_\rho \cap R \supseteq S_\rho$). We order these families ρ by inclusion and it is not difficult to see that they form an inductive set of families. Moreover, the set is non-empty since it contains the empty family $\{\ \}$. We may thus invoke Zorn's Lemma (I.3.1). Let ρ be a maximal family; writing S, G for S_ρ, G_ρ, we prove that $S = R$ and thus establish the theorem.

Suppose $S \neq R$; among all the elements of $R - S$ choose the set of those involving fewest extra f_α's and, in this set, choose one element, say z, having the smallest coefficient k for such an f_α, say $z = kf_1 + \ldots$, where f_1 is 'extra'. Then certainly $\rho^+ = \{\rho, z\}$ is linearly independent. Let $S^+ = S_{\rho^+}$, $G^+ = G_{\rho^+}$; we shall have arrived at the desired contradiction when we have shown that $G^+ \cap R \subseteq S^+$. Suppose then that $y \in G^+ \cap R$; then $y = lf_1 + \ldots$ and the minimality of k shows that $k \mid l$, say $l = km$. Then the minimality of the set from which z was chosen shows that $y - mz \in G \cap R = S$ and $y \in S^+$. ▮

An abelian group D is said to be *divisible* if for any positive integer n and any element $d_0 \in D$ there is at least one element $d \in D$ such that $nd = d_0$. In fact each $d_0 \in D$ can be "divided" (although not in general uniquely) by any positive integer. The group of rationals and the group of real numbers are examples of divisible groups in which the division is unique; R_1, the group of reals mod 1, is an example in which division is not unique.

5.3.1*d* Theorem. *The quotient group of a divisible group is divisible.*

If G is the quotient group let $\phi : D \rightarrow G$ be the epimorphism of the divisible group to G. For any $g_0 \in G$ let d_0 be some element of D such that $d_0\phi = g_0$; then for any n there is an element $d \in D$ such that $nd = d_0$ and consequently $n(d\phi) = d_0\phi = g_0$. ▮

5.3.2*f* Theorem. *Any abelian group G can be presented as the quotient group of a free group.*

Construct the free group F on generators $\{f_g\}$, where the indexing set is G itself. The transformation $f_g \rightarrow g$ determines a homomorphism which is plainly epimorphic. ▮

This presentation of G is called the *standard* presentation. The kernel of the epimorphism is of course, in view of 5.3.1*f*, a free group.

5.3.2*d* Theorem. *Any abelian group G can be embedded as a subgroup in a divisible group.*

Let $0 \to R \xrightarrow{\lambda} F \xrightarrow{\mu} G \to 0$ be an exact sequence in which F is free, so that μ is a presentation of G. Then F is the direct sum of cyclic infinite groups; each such factor J may be embedded in Q, the group of the rationals. In this way F is embedded in the direct sum D of copies of Q, which is divisible. Then $F/R\lambda \subseteq D/R\lambda$ and $D/R\lambda$ is, by 5.3.1d, divisible. Since $G \cong F/R\lambda$, the theorem is proved. ∎

We may describe the result by saying that there is an exact sequence

$$0 \leftarrow E \xleftarrow{\lambda'} D \xleftarrow{\mu'} G \leftarrow 0,$$

in which D (and therefore also E) is divisible. We shall describe μ' as a *rationalization* of G. If in 5.3.2d we take the standard presentation of G, the embedding of G that ensues may be called the *standard rationalization* of G.

For the next theorem we need the definition of a *projective* abelian group; the definition and word are both applied to other algebraic systems than abelian groups. A group P is *projective* if, given any epimorphism $\phi : A \to B$ and any homomorphism $\beta : P \to B$, there is a homomorphism $\alpha : P \to A$ such that $\alpha\phi = \beta$.

5.3.3f Theorem. *An abelian group is projective if and only if it is free.*

Let F be free and consider the diagram‡:

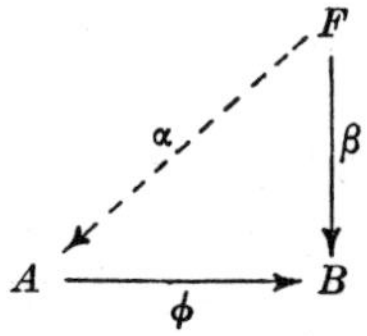

If $\{f_\lambda\}$ is a basis for F, to each f_λ we associate some $a_\lambda \in A$ such that $a_\lambda\phi = f_\lambda\beta$; this is possible because ϕ is epimorphic. The transformation $f_\lambda \to a_\lambda$ determines a homomorphism $\alpha : F \to A$, since F is free and $\{f_\lambda\}$ a basis, and plainly $\alpha\phi = \beta$.

Consider now a projective group P; let $\phi : F \to P$ be a presentation of P, and consider

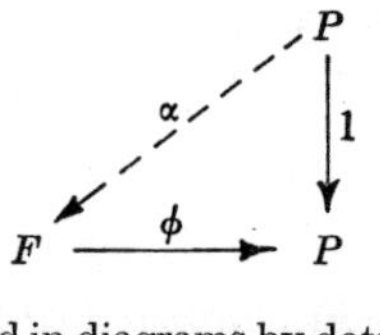

‡ Homomorphisms represented in diagrams by dotted arrows are not initially given but are constructed in the proof.

Since P is projective, there is a homomorphism $\alpha : P \to F$ such that $\alpha\phi = 1$. Then certainly α is monomorphic and embeds P as a subgroup of F, so that P is, by 5.3.1f, free. ∎

The definition dual to that of projective is that of injective. A group G is *injective* if, given any monomorphism $\phi : B \to A$ and any homomorphism $\beta : B \to G$, then is a homomorphism $\alpha : A \to G$ such that $\phi\alpha = \beta$.

5.3.3*d* Theorem. *An abelian group is injective if and only if it is divisible.*

Let D be a divisible group and suppose given $\phi : B \to A$ and $\beta : B \to D$ as above. Since ϕ is a monomorphism we lose no generality is supposing B to be a subgroup of A; we must prove that the homomorphism β defined on the subgroup can be extended to a homomorphism α of A. We consider the set of all pairs $(B_\lambda, \beta_\lambda)$, where $B \subseteq B_\lambda \subseteq A$, and $\beta_\lambda : B_\lambda \to D$ is an extension of β; this means that $\beta_\lambda | B = \beta$. We order‡ these pairs by the rule $(B_\lambda, \beta_\lambda) \prec (B_\mu, \beta_\mu)$ if $B_\lambda \subseteq B_\mu$ and β_μ is an extension of β_λ. This is an inductive ordering, so that, by Zorn's lemma, there is a maximal pair $(\bar{B}, \bar{\beta})$; we shall show that $\bar{B} = A$. If not, let $x \in A - \bar{B}$. If x is independent of $\bar{B}$ (i.e. if $nx \in \bar{B}$ implies $n = 0$), we may extend $\bar{\beta}$ to a homomorphism of $\bar{B} + (x)$, the subgroup generated by $\bar{B}$ and x, by defining the image of x to be zero. If, on the other hand, for some $n > 0$, $nx \in \bar{B}$, let n_0 be the least positive integer such that $n_0 x \in \bar{B}$. Choose $d \in D$ such that $n_0 d = (n_0 x)\bar{\beta}$; then $\bar{\beta}$ can be extended to $\bar{B} + (x)$ by defining the image of x to be d; it is easy to verify that $\bar{\beta}$ so defined is single-valued and homomorphic. In either case $\bar{\beta}$ has been extended beyond $\bar{B}$, contrary to the maximality of $(\bar{B}, \bar{\beta})$; wo may therefore conclude that $\bar{B} = A$ and that $\bar{\beta}$ may be taken for the extension α of β. So D is injective.

Conversely let I be injective, let $a \in I$ and let n be a positive integer. Then consider the diagram

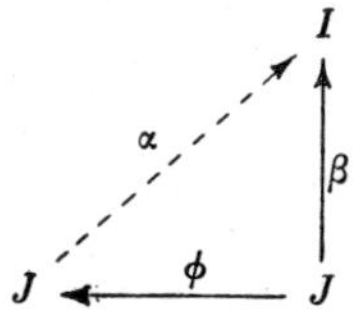

in which β maps 1 into a and ϕ maps 1 into n. Then there is a homomorphism $\alpha : J \to I$ such that $\phi\alpha = \beta$, since I is injective. So $a = (1)\beta = (1)\phi\alpha = (n)\alpha = n.(1)\alpha$. This proves I to be divisible. ∎

‡ This is, of course, a partial ordering.

Of the notions of this section it is those of 'projective' and 'injective' that generalize most usefully to other algebraic systems than abelian groups. The value of 5.3.3 (f and d) lies in the fact that it is the projective property of a free group and the injective property of a divisible group that are important in homological algebra (see, for example, 5.1.10).

Although the notion of 'free' can be generalized, it is not true that the analogue of 5.3.3f is in general true. In view of 5.3.3 (f and d), 5.3.1 and 5.3.2 may be restated for abelian groups in terms of 'projective' and 'injective'; in general contexts the analogue of 5.3.1, so restated, is false, whereas the analogue of 5.3.2 is true.

We offer the following result as an interesting application of the projective property of a free group; many of the ideas used in the proof occur again in the next section.

5.3.4 Theorem.‡ *Let C, C' be free chain complexes and let*

$$\sigma : H_*(C) \to H_*(C')$$

be a homomorphism. Then there is a chain map $\phi : C \to C'$ with $\phi_* = \sigma$.

We have a diagram

$$\begin{array}{ccccccccc} 0 & \longrightarrow & B_n & \xrightarrow{\lambda_n} & Z_n & \xrightarrow{\mu_n} & H_n & \longrightarrow & 0 \\ & & \downarrow{\scriptstyle \pi_n} & & \downarrow{\scriptstyle \rho_n} & & \downarrow{\scriptstyle \sigma_n} & & \\ 0 & \longrightarrow & B'_n & \xrightarrow{\lambda'_n} & Z'_n & \xrightarrow{\mu'_n} & H'_n & \longrightarrow & 0, \end{array}$$

where $H_n = H_n(C)$, $H'_n = H_n(C')$ and the horizontal sequences express the identities $Z_n/B_n = H_n$, $Z'_n/B'_n = H'_n$. Since C is free, Z_n is free (5.3.1f) and so (5.3.3f) we may define ρ_n such that $\rho_n \mu'_n = \mu_n \sigma_n$. Then ρ_n determines π_n such that $\pi_n \lambda'_n = \lambda_n \rho_n$. There is also an exact sequence $0 \to Z_n \xrightarrow{\xi_n} C_n \xrightarrow{\eta_n} B_{n-1} \to 0$, derived from the boundary operator $\partial_n : C_n \to C_{n-1}$, and (5.1.10) a homomorphism $\kappa_n : B_{n-1} \to C_n$ such that

5.3.5 $$C_n = Z_n \xi_n \oplus B_{n-1} \kappa_n;$$

moreover, $\kappa_n \eta_n = 1$ and $\partial_n = \eta_n \lambda_{n-1} \xi_{n-1}$. We have similar results for C', and we define $\phi_n : C_n \to C'_n$ by

$$z_n \xi_n \phi_n = z_n \rho_n \xi'_n, \quad b_{n-1} \kappa_n \phi_n = b_{n-1} \pi_{n-1} \kappa'_n.$$

‡ Our attention was drawn to this theorem by J. H. C. Whitehead.

Obviously $\phi_n \partial'_n = \partial_n \phi_{n-1} (=0)$ on $Z_n \xi_n$; on $B_{n-1}\kappa_n$ we have

$$\begin{aligned} b_{n-1}\kappa_n \phi_n \partial'_n &= b_{n-1}\pi_{n-1}\kappa'_n \partial'_n \\ &= b_{n-1}\pi_{n-1}\lambda'_{n-1}\xi'_{n-1} = b_{n-1}\lambda_{n-1}\rho_{n-1}\xi'_{n-1} \\ &= b_{n-1}\lambda_{n-1}\xi_{n-1}\phi_{n-1} = b_{n-1}\kappa_n \partial_n \phi_{n-1}. \end{aligned}$$

Thus ϕ is a chain map; plainly ϕ induces ρ and hence σ. ∎

5.4 Homology and contrahomology in infinite complexes

In this section we obtain expressions for $H_n(K; G)$, $H^n(K; G)$ in terms of the groups $H_p(K)$ and we make no finiteness assumption on K. In fact, our results are explicitly applicable to arbitrary free chain complexes and make no use of the theory of standard bases. The contrast between 5.2 and this section echoes the historical development of the subject. The pioneers tried to express all invariants in terms of a finite list of integers (for example, Betti numbers and torsion coefficients); later writers, however, found it more natural and elegant to regard whole groups as being the invariants.

The key to the method lies in a closer analysis of the tensor product. We thus begin with such an analysis; some further properties of the tensor product will be given in the next section.

Let $A' \xrightarrow{\lambda} A \xrightarrow{\mu} A'' \to 0$ be an exact sequence of abelian groups; we recall that this means that μ is epimorphic, and the kernel of μ is the image of λ. We may take tensor products with B and obtain a sequence $A' \otimes B \xrightarrow{\lambda\otimes 1} A \otimes B \xrightarrow{\mu\otimes 1} A'' \otimes B \to 0$. We shall write $\bar{\lambda}$, $\bar{\mu}$ for $\lambda\otimes 1$, $\mu\otimes 1$, and prove

5.4.1 Theorem. *The sequence* $A' \otimes B \xrightarrow{\bar{\lambda}} A \otimes B \xrightarrow{\bar{\mu}} A'' \otimes B \to 0$ *is exact.*

Notice that we do *not* assert that $\bar{\lambda}$ is monomorphic even where λ is monomorphic. There are three assertions to be proved:

(i) $\bar{\mu}$ *is epimorphic.* It is sufficient to show that $a''\otimes b$ is in the image of $\bar{\mu}$ for each $a'' \in A''$, $b \in B$; but μ is epimorphic so that $a'' = a\mu$ for some $a \in A$ and $a''\otimes b = (a\otimes b)\bar{\mu}$.

(ii) $\bar{\lambda}\bar{\mu} = 0$. Now $\bar{\lambda}\bar{\mu} = (\lambda\otimes 1)(\mu\otimes 1) = \lambda\mu\otimes 1 = 0\otimes 1 = 0$.

(iii) $\bar{\mu}^{-1}(0) \subseteq (A'\otimes B)\bar{\lambda}$. To prove this we first recall the notion of a bilinear pairing $\theta : A \times B \to G$ (4.1) and observe that any such bilinear pairing θ determines a homomorphism $\bar{\theta} : A \otimes B \to G$ such that $(a\otimes b)\bar{\theta} = (a,b)\theta$. We observe further that if $\kappa : A'' \to A$ is any map (not necessarily homomorphic) such that $\kappa\mu = 1$, if $\eta : A \to A/A'\lambda$ is the projection and if $\mu' : A/A'\lambda \to A''$ is the isomorphism induced by μ, then $\kappa\eta : A'' \to A/A'\lambda$ is the isomorphism inverse to μ'. The method of proof is to imitate this situation for the sequence

$$A' \otimes B \xrightarrow{\bar\lambda} A \otimes B \xrightarrow{\bar\mu} A'' \otimes B \to 0.$$

Let $\Gamma = (A' \otimes B)\bar\lambda$. Then, since $\bar\lambda\bar\mu = 0$, $\bar\mu$ induces

$$\bar\mu' : (A \otimes B)/\Gamma \to A'' \otimes B,$$

in the sense that $\bar\mu = \zeta\bar\mu'$, where $\zeta : A \otimes B \to (A \otimes B)/\Gamma$ is the projection, We must prove that $\bar\mu'$ is an isomorphism; we already know it is an epimorphism. Define $\theta : A'' \times B \to A \otimes B/\Gamma$ by $(a'', b)\theta = (a''\kappa \otimes b)\zeta$. Then θ *is bilinear*; it is obvious that θ is linear in the second variable. so we must verify that

$$(a_1'' + a_2'', b)\theta - (a_1'', b)\theta - (a_2'', b)\theta = 0.$$

Now $((a_1'' + a_2'')\kappa - a_1''\kappa - a_2''\kappa)\mu = a_1'' + a_2'' - a_1'' - a_2'' = 0$, so that

$$(a_1'' + a_2'')\kappa - a_1''\kappa - a_2''\kappa \in \mu^{-1}(0) = A'\lambda.$$

Thus

$$\begin{aligned} &(a_1'' + a_2'', b)\theta - (a_1'', b)\theta - (a_2'', b)\theta \\ &= ((a_1'' + a_2'')\kappa \otimes b)\zeta - (a_1''\kappa \otimes b)\zeta - (a_2''\kappa \otimes b)\zeta \\ &= (((a_1'' + a_2'')\kappa - a_1''\kappa - a_2''\kappa) \otimes b)\zeta \\ &= (a'\lambda \otimes b)\zeta, \quad \text{for some } a' \in A', \\ &= (a' \otimes b)\bar\lambda\zeta = 0, \quad \text{since } \Gamma\zeta = 0. \end{aligned}$$

It follows that θ determines a homomorphism‡ $\bar\theta : A'' \otimes B \to A \otimes B/\Gamma$ such that $(a'' \otimes b)\bar\theta = (a''\kappa \otimes b)\zeta$. We prove that $\bar\mu'\bar\theta = 1$; this will establish that $\bar\mu'$ is an isomorphism with inverse $\bar\theta$. We first remark that, for all $a \in A$, $(a\mu\kappa - a)\mu = a\mu - a\mu = 0$, so that $a\mu\kappa - a = a'\lambda$ for some $a' \in A'$. Thus

$$\begin{aligned} (a \otimes b)\zeta\bar\mu'\bar\theta &= (a \otimes b)\bar\mu\bar\theta = (a\mu \otimes b)\bar\theta = (a\mu\kappa \otimes b)\zeta \\ &= (a \otimes b)\zeta + (a'\lambda \otimes b)\zeta = (a \otimes b)\zeta, \end{aligned}$$

since $(a'\lambda \otimes b)\zeta = (a' \otimes b)\bar\lambda\zeta = 0$. Since $(A \otimes B)/\Gamma$ is generated by the elements $(a \otimes b)\zeta$, this proves that $\bar\mu'\bar\theta = 1$. We immediately infer that $\bar\mu'$ is monomorphic—and so isomorphic—so that the kernel of $\bar\mu$ is Γ. ∎

‡ $\bar\theta$ is the analogue of $\kappa\eta$ in the original sequence.

The reader may better appreciate the significance of this theorem if we point out the consequence:

5.4.2 $$(A \otimes B)/(A' \otimes B)\bar{\lambda} \cong (A/A'\lambda) \otimes B.$$

It is tempting, when λ is monomorphic, to forget the maps λ and $\bar{\lambda}$ and write 5.4.2 as $(A \otimes B)/(A' \otimes B) \cong (A/A') \otimes B$. Unfortunately, this way of writing is unacceptable because $\bar{\lambda}$ is not necessarily monomorphic. For example, if $A = J$, $A' = 2J$, $B = J_2$, then $A' \otimes B = J_2$ whereas $(A' \otimes B)\bar{\lambda} = 0$. For if J is the group of integers and b generates B then $A' \otimes B$ is generated by $2 \otimes b$, but, *in* $A \otimes B$, $2 \otimes b = 1 \otimes 2b = 1 \otimes 0 = 0$.

We now suppose λ monomorphic and examine the kernel of $\bar{\lambda}$, confining attention to the case in which A is free. Then $\mu : A \to A''$ is a presentation of A''. We adopt a more appropriate notation for this case and consider the exact sequence

5.4.3 $$0 \to R \overset{\lambda}{\to} F \overset{\mu}{\to} A \to 0,$$

where F is free. We shall refer to the *sequence* 5.4.3 as a presentation of A. By taking tensor products with B, we obtain an exact sequence

$$R \otimes B \overset{\bar{\lambda}}{\to} F \otimes B \overset{\bar{\mu}}{\to} A \otimes B \to 0,$$

and we write $A *_\mu B$ for the kernel of $\bar{\lambda}$. We are going to prove that $A *_\mu B$ is independent, up to isomorphism, of the choice of μ. We first obtain some properties of $A *_\mu B$ which are of a kind typical of homological algebra.

Let A', B' be groups and let $\phi : A \to A'$, $\psi : B \to B'$ be homomorphisms; let A be presented by 5.4.3 and A' by

$$0 \to R' \overset{\lambda'}{\to} F' \overset{\mu'}{\to} A' \to 0.$$

We prove

5.4.4 Theorem. *There are canonical homomorphisms*

$$\bar{\tau}(\phi; \mu, \mu') : A *_\mu B \to A' *_{\mu'} B, \quad \kappa(\psi; \mu) : A *_\mu B \to A *_\mu B'$$

such that

(i) $\bar{\tau}(1; \mu, \mu) = 1 : A *_\mu B \to A *_\mu B$;

(ii) $\bar{\tau}(\phi; \mu, \mu')\bar{\tau}(\phi'; \mu', \mu'') = \bar{\tau}(\phi\phi'; \mu, \mu'')$, *where* $\phi' : A' \to A''$ *and* μ'' *is a presentation of* A'';

(iii) $\kappa(1; \mu) = 1 : A *_\mu B \to A *_\mu B$;

(iv) $\kappa(\psi; \mu)\kappa(\psi'; \mu) = \kappa(\psi\psi'; \mu)$, where $\psi' : B' \to B''$;

(v) $\kappa(\psi; \mu)\bar{\tau}(\phi; \mu, \mu') = \bar{\tau}(\phi; \mu, \mu')\kappa(\psi; \mu') : A *_\mu B \to A' *_{\mu'} B'$.

We consider first the definition of $\bar{\tau} = \bar{\tau}(\phi; \mu, \mu')$. Given

$$\begin{array}{ccccccccc} 0 & \to & R & \xrightarrow{\lambda} & F & \xrightarrow{\mu} & A & \to & 0 \\ & & \downarrow{\scriptstyle\theta} & & \downarrow{\scriptstyle\chi} & & \downarrow{\scriptstyle\phi} & & \\ 0 & \to & R' & \xrightarrow{\lambda'} & F' & \xrightarrow{\mu'} & A' & \to & 0, \end{array}$$

we apply 5.3.3f to show that there exists a homomorphism $\chi : F \to F'$ with $\chi\mu' = \mu\phi$. Then $\lambda\chi\mu' = \lambda\mu\phi = 0$, so that $\lambda\chi$ maps R into $R'\lambda'$ and determines a homomorphism $\theta : R \to R'$ such that $\theta\lambda' = \lambda\chi$. We now take tensor products with B and obtain the diagram

5.4.5

$$\begin{array}{ccccccccccc} 0 & \to & A *_{\mu} B & \xrightarrow{\bar{\omega}} & R \otimes B & \xrightarrow{\bar{\lambda}} & F \otimes B & \xrightarrow{\bar{\mu}} & A \otimes B & \to & 0 \\ & & \downarrow{\scriptstyle\bar{\tau}} & & \downarrow{\scriptstyle\bar{\theta}} & & \downarrow{\scriptstyle\bar{\chi}} & & \downarrow{\scriptstyle\bar{\phi}} & & \\ 0 & \to & A' *_{\mu'} B & \xrightarrow{\bar{\omega}'} & R' \otimes B & \xrightarrow{\bar{\lambda}'} & F' \otimes B & \xrightarrow{\bar{\mu}'} & A' \otimes B & \to & 0, \end{array}$$

where $\bar{\omega}$, $\bar{\omega}'$ are injections and $\bar{\tau}$ is the restriction of $\bar{\theta}$ to $A *_{\mu} B$; it is clear that $\bar{\tau}$ maps $A *_{\mu} B$ into $A' *_{\mu'} B$ since $\bar{\omega}\bar{\theta}\bar{\lambda}' = \bar{\omega}\bar{\lambda}\bar{\chi} = 0$.

The homomorphism $\bar{\tau}$ has the properties required. The main step in showing this is to prove that $\bar{\tau}$ is canonical in that it does not depend on the choice of the map χ 'lifting' ϕ. Since χ must satisfy $\chi\mu' = \mu\phi$, we may alter χ by a homomorphism $\gamma\lambda'$, where γ is an arbitrary homomorphism of F into R'. Then θ is altered by $\lambda\gamma$, since

$$(\theta + \lambda\gamma)\,\lambda' = \lambda(\chi + \gamma\lambda').$$

Thus $\bar{\theta}$ is altered by $\bar{\lambda}\bar{\gamma}$; but $\bar{\omega}\bar{\lambda}\bar{\gamma} = 0$, so that $\bar{\tau}$ is unaffected and we may write $\bar{\tau} = \bar{\tau}(\phi; \mu, \mu')$.

Property (i) is obvious; for we may take each vertical map in 5.4.5 to be the identity. Now let $0 \to R'' \xrightarrow{\lambda''} F'' \xrightarrow{\mu''} A'' \to 0$ be a presentation of A'' and let $\phi' : A' \to A''$. If $\chi' : F' \to F''$ 'lifts' ϕ', i.e. if $\chi'\mu'' = \mu'\phi'$, then $\chi\chi' : F \to F''$ lifts $\phi\phi'$. Clearly, if we carry out the process of defining $\bar{\tau}(\phi\phi';\ \mu, \mu'')$ by means of this choice of lifting map, we obtain the map $\bar{\tau}(\phi; \mu, \mu')\,\bar{\tau}(\phi'; \mu', \mu'')$. Thus property (ii) is verified.

To obtain $\kappa = \kappa(\psi; \mu)$, we simply write down the diagram

5.4.6

$$\begin{array}{ccccccccccc} 0 & \to & A *_{\mu} B & \xrightarrow{\bar{\omega}} & R \otimes B & \xrightarrow{\bar{\lambda}} & F \otimes B & \xrightarrow{\bar{\mu}} & A \otimes B & \to & 0 \\ & & \downarrow{\scriptstyle\kappa} & & \downarrow{\scriptstyle 1\otimes\psi} & & \downarrow{\scriptstyle 1\otimes\psi} & & \downarrow{\scriptstyle 1\otimes\psi} & & \\ 0 & \to & A *_{\mu} B' & \xrightarrow{\bar{\omega}} & R \otimes B' & \xrightarrow{\bar{\lambda}} & F \otimes B' & \xrightarrow{\bar{\mu}} & A \otimes B' & \to & 0. \end{array}$$

Let us write $\bar{\psi}$ for $1 \otimes \psi$. Then the homomorphism κ is just the restriction of $\bar{\psi} : R \otimes B \to R \otimes B'$ to $A *_{\mu} B$. Since $\bar{\omega}\bar{\psi}\bar{\lambda} = \bar{\omega}\bar{\lambda}\bar{\psi} = 0$,

κ maps $A*_\mu B$ into $A*_\mu B'$. Property (iii) is obvious and property (iv) follows immediately from $\overline{\psi\psi'} = \overline{\psi}\,\overline{\psi}'$.

Finally, we wish to prove the commutativity of the diagram

$$\begin{array}{ccc} A*_\mu B & \xrightarrow{\kappa} & A*_\mu B' \\ \downarrow{\scriptstyle\overline{\tau}} & & \downarrow{\scriptstyle\overline{\tau}} \\ A'*_{\mu'} B & \xrightarrow{\kappa} & A'*_{\mu'} B'. \end{array}$$

But this diagram is obtained from the diagram

$$\begin{array}{ccc} R\otimes B & \xrightarrow{\overline{\psi}} & R\otimes B' \\ \downarrow{\scriptstyle\overline{\theta}} & & \downarrow{\scriptstyle\overline{\theta}} \\ R'\otimes B & \xrightarrow{\overline{\psi}} & R'\otimes B' \end{array}$$

by restriction, and this last diagram is obviously commutative. ∎

A further notation suggests itself by analogy with tensor products. Let us write $\phi *_{\mu\mu'} \psi : A*_\mu B \to A'*_{\mu'} B'$ for $\kappa(\psi;\mu)\,\overline{\tau}(\phi;\mu,\mu')$. In this notation

5.4.7
$$\begin{cases} \overline{\tau}(\phi;\mu,\mu') = \phi *_{\mu\mu'} 1, \\ \kappa(\psi;\mu) = 1 *_{\mu\mu} \psi, \\ \text{and}\quad (\phi *_{\mu\mu'} \psi)(\phi' *_{\mu'\mu''} \psi') = \phi\phi' *_{\mu\mu''} \psi\psi'. \end{cases}$$

5.4.8 Corollary. *Let* $0 \to R_i \xrightarrow{\lambda_i} F_i \xrightarrow{\mu_i} A \to 0$, $i = 0, 1$, *be two presentations of* A; *then there is an isomorphism* $\overline{\tau} : A*_{\mu_0} B \cong A*_{\mu_1} B$.

In fact let $\mu_i : F_i \to A$, $i = 0, 1, 2$ be three presentations. We write $\overline{\tau}(1;\mu_i,\mu_j)$ as $\overline{\tau}(\mu_i,\mu_j)$ and deduce from 5.4.4 (ii) that

5.4.9 $$\overline{\tau}(\mu_0,\mu_1)\,\tau(\mu_1,\mu_2) = \overline{\tau}(\mu_0,\mu_2) : A*_{\mu_0} B \to A*_{\mu_2} B.$$

Putting $\mu_2 = \mu_0$ and applying 5.4.4 (i), we have

$$\overline{\tau}(\mu_0,\mu_1)\,\overline{\tau}(\mu_1,\mu_0) = 1,$$

and similarly $$\overline{\tau}(\mu_1,\mu_0)\,\overline{\tau}(\mu_0,\mu_1) = 1.$$

It follows that $\overline{\tau} = \overline{\tau}(\mu_0,\mu_1) : A*_{\mu_0} B \simeq A*_{\mu_1} B$, with inverse $\overline{\tau}(\mu_1,\mu_0)$. ∎

If we do not wish to stress the presentation, we shall write $A*B$ for $A*_\mu B$ and call $A*B$ the *torsion product*‡ of A and B. Any two models of $A*B$ are related by the canonical isomorphism $\overline{\tau}$ of 5.4.8 and homomorphisms $\phi : A \to A'$, $\psi : B \to B'$ induce

$$\phi * \psi : A*B \to A'*B'.$$

‡ The notation Tor (A, B) for $A*B$ is also often used.

This remark is justified by the commutativity of the diagram

$$\begin{CD} A *_{\mu_0} B @>{\bar{\tau}(\mu_0,\, \mu_1)}>> A *_{\mu_1} B \\ @V{\phi *_{\mu_0 \mu_0'} \psi}VV @VV{\phi *_{\mu_1 \mu_1'} \psi}V \\ A' *_{\mu_0'} B' @>>{\bar{\tau}(\mu_0',\, \mu_1')}> A' *_{\mu_1'} B' \end{CD}$$

The notations 5.4.7 quickly show that each composite homomorphism from $A *_{\mu_0} B$ to $A' *_{\mu_1'} B'$ is equal to $\phi *_{\mu_0 \mu_1'} \psi$.

The group $A * B$ may also be given a more concrete definition for those who prefer. We may use, for each A, the standard (abelian) presentation of **5.3.2***f*. This is in most cases absurdly large and quite unsuitable for computation; but it does give a uniquely defined $A * B$ (and homomorphisms $\phi * \psi$) to which the groups $A *_\mu B$ are canonically isomorphic. An even more concrete definition is suggested in Exercises, Q. 10.

5.4.10 Proposition. *If A is the direct sum of groups A_α and B is the direct sum of groups B_β, then $A * B$ is the direct sum of the groups $A_\alpha * B_\beta$.*

In the light of 5.2.6, the proof can safely be left to the reader. ❚

5.4.11 Proposition. *If A or B is free, then $A * B = 0$.*

First suppose A free; then $0 \to 0 \xrightarrow{\lambda} A \xrightarrow{\mu} A \to 0$, where $\mu = 1$, is a presentation of A and obviously $A *_\mu B = 0$.

Now suppose B free and let $0 \to R \xrightarrow{\lambda} F \xrightarrow{\mu} A \to 0$ be a presentation of A. From the remark after 5.2.7 and the commutativity of $\otimes$ it follows directly that, if $\lambda : R \to F$ is monomorphic and B is free, then $\bar{\lambda} : R \otimes B \to F \otimes B$ is also monomorphic, whence $A *_\mu B = 0$. ❚

5.4.12 Proposition. $J_m * J_n \cong J_{(m,\, n)}$.

Present J_m as $0 \to mJ \xrightarrow{\lambda} J \xrightarrow{\mu} J_m \to 0$. Then $J_m *_\mu J_n$ is the kernel of $\bar{\lambda} : mJ \otimes J_n \to J \otimes J_n$. Any element of $mJ \otimes J_n$ is expressible (uniquely) as $mk \otimes b$, where b is a fixed generator of J_n and $0 \leqslant k < n$; and $mk \otimes b$ lies in the kernel of $\bar{\lambda}$ if and only if $n \mid mk$. The mapping $mk \otimes b \to kb$ is an isomorphism of $mJ \otimes J_n$ with J_n under which $J_m *_\mu J_n$ is mapped to the subgroup ${}_m(J_n)$. But we have seen that ${}_m(J_n) = J_{(m,\, n)}$ so the proposition is proved. ❚

Propositions 5.4.10–5.4.12, together with Theorem 5.1.4—or, preferably, 5.1.8—enable $A * B$ to be computed when A, B are *fg*. The reader will have remarked the symmetric role played by A and B in Propositions 5.4.10–5.4.12; in the next section we shall see that indeed $A * B \cong B * A$ for all A, B.

5.4.13 Theorem (*Universal coefficient theorem*). *Let C be a free chain complex; then $H_p(C\otimes G) \cong (H_p(C)\otimes G)\oplus(H_{p-1}(C) * G)$.*

In view of the pitfalls that surround a casual application of $\otimes G$ to a subgroup, we exercise pedantic care in this proof, distinguishing, for example, B_{p-1} as a group from its isomorphic image as a subgroup of C_{p-1}.

We observe that ∂ determines an epimorphism $\eta : C_p \to B_{p-1}$ such that $\eta\iota = \partial$, where $\iota : B_{p-1}\to C_{p-1}$ is the embedding; moreover, there is an exact sequence

$$0 \to Z_p \xrightarrow{\xi} C_p \xrightarrow{\eta} B_{p-1} \to 0$$

for each p, where ξ is the embedding. There is also an exact sequence

$$0 \to B_p \xrightarrow{\lambda} Z_p \xrightarrow{\mu} H_p \to 0$$

for each p, where λ is the embedding and $\lambda\xi = \iota$.

Now B_{p-1}, being a subgroup of C_{p-1}, is by 5.3.1f free. It thus follows from 5.1.10 that there exists $\kappa : B_{p-1}\to C_p$ with $\kappa\eta = 1$ and such that $C_p = Z_p\xi\oplus B_{p-1}\kappa$ (compare the proof of 5.3.4). Then

5.4.14 $\quad C_p\otimes G = (Z_p\xi\otimes G)\oplus(B_{p-1}\kappa\otimes G) = (Z_p\otimes G)\,\bar\xi\oplus(B_{p-1}\otimes G)\,\bar\kappa,$

where $\bar\xi$, $\bar\kappa$ are monomorphic.

We now calculate $Z_p(C\otimes G)$; certainly $(Z_p\otimes G)\,\bar\xi\bar\partial = 0$, where $\bar\partial = \partial\otimes 1$, so that $(Z_p\otimes G)\,\bar\xi \subseteq Z_p(C\otimes G)$. Consider the diagram of homomorphisms

$$C_p\otimes G \xleftarrow{\bar\kappa} B_{p-1}\otimes G \xrightarrow{\bar\lambda} Z_{p-1}\otimes G \xrightarrow{\bar\xi} C_{p-1}\otimes G.$$

Now if $x\in B_{p-1}\otimes G$, $x\bar\kappa\bar\partial = x\bar\kappa\bar\eta\bar\iota = x\bar\iota = x\bar\lambda\bar\xi$. Thus $x\bar\kappa\in Z_p(C\otimes G)$ if and only if $x\bar\lambda\bar\xi = 0$, but $\bar\xi$ is monomorphic, so that $x\bar\kappa\in Z_p(C\otimes G)$ if and only if $x\in\bar\lambda^{-1}(0) = H_{p-1} *_\mu G$. In the light of 5.4.14 we have proved that

5.4.15 $$Z_p(C\otimes G) = (Z_p\otimes G)\,\bar\xi\oplus(H_{p-1} *_\mu G)\,\bar\kappa.$$

Further,

$$B_p(C\otimes G) = (C_{p+1}\otimes G)\,\bar\partial = (C_{p+1}\otimes G)\,\bar\eta\bar\iota = (B_p\otimes G)\,\bar\iota = (B_p\otimes G)\,\bar\lambda\bar\xi.$$

Thus $H_p(C\otimes G) \cong (Z_p\otimes G/(B_p\otimes G)\,\bar\lambda)\oplus(H_{p-1}(C) * G)$. We invoke 5.4.2 to complete the proof. ∎

The reader is strongly advised to compare Theorem 5.4.13 with Theorem 5.2.12, and to note that the first two terms in the expression for $H_p(C\otimes G)$ in 5.2.12 correspond to $H_p(C)\otimes G$, the third term corresponding to $H_{p-1}(C) * G$. The advantage of 5.4.13 lies not only in its succinctness, but also in its invariance.

We can bring out the invariance (i.e. the independence of a choice of basis in C) more strongly as follows. First note that $H_p(C \otimes G), H_p(C) \otimes G, H_{p-1}(C) * G$ are all defined without reference to a basis in C. Any function of C so defined we may call *canonical*. Then

5.4.16 Theorem. *If C is free, there is an an exact sequence*

$$0 \to H_p(C) \otimes G \xrightarrow{\rho} H_p(C \otimes G) \xrightarrow{\rho'} H_{p-1}(C) * G \to 0,$$

where ρ, ρ' are canonical. Moreover, the image of ρ is a direct factor in $H_p(C \otimes G)$.

We use the notation of 5.4.13. Then $Z_p \otimes G$ is a subgroup of $Z_p(C \otimes G)$ and we obtain ρ by factoring out the image of $B_p \otimes G$. The boundary operator may be regarded as mapping $C_p \otimes G$ onto $B_{p-1} \otimes G$; the restriction to $Z_p(C \otimes G)$ maps the latter onto $H_{p-1}(C) * G$, and $B_p(C \otimes G)$ is mapped to zero. Thus ρ' is induced as the map of the factor group $Z_p(C \otimes G)/B_p(C \otimes G)$ onto $H_{p-1}(C) * G$. The remaining assertions of the theorem are proved in the course of demonstrating 5.4.13. ▌ A right inverse of ρ is obtained by means of a projection $C_p \to Z_p$, and a left inverse of ρ' is obtained by means of the map $\kappa : B_{p-1} \to C_p$.

We make some more remarks about tensor and torsion products as an aid to computation.

5.4.17 Proposition. *If A is a torsion group and B is divisible then $A \otimes B = 0$.* ▌

We remark next that the tensor product has a local character in the following sense:

5.4.18 Lemma. *Let $a_i \in A, b_i \in B, i = 1, \ldots, n$, and suppose $\sum_i a_i \otimes b_i = 0$; then there exist fg-groups $A_0 \subseteq A$, $B_0 \subseteq B$ such that $a_i \in A_0$, $b_i \in B_0$, $i = 1, \ldots, n$, and $\sum a_i \otimes b_i = 0$ in $A_0 \otimes B_0$.*

Let us write (A, B) for the free abelian group generated by pairs $(a, b), a \in A, b \in B$.

Examination of the relations in $A \otimes B$ (Definition 5.2.1) shows that $\sum_i a_i \otimes b_i = 0$ in $A \otimes B$ if and only if $\sum_i (a_i, b_i)$ is a *finite* sum of elements of (A, B) of the form $(a, b+b') - (a, b) - (a, b')$ and elements of the form $(a+a', b) - (a, b) - (a', b)$. We take such an expression for $\sum_i (a_i, b_i)$ and let A_0 be the subgroup of A generated by the first members of all pairs occurring in the expression and define B_0 similarly. Then evidently A_0 and B_0 have the desired property. ▌

We say that A is *locally free* if every *fg*-subgroup is free. This, in view of 5.1.3, is equivalent to the condition that the group A is torsion-free; or that the torsion subgroup is null. Thus, for example, the group of rationals is locally free (but not free).

5.4.19 Proposition. *Let* $\iota : A' \to A$ *be a monomorphism. Then* $\bar{\iota} : A' \otimes B \to A \otimes B$ *is a monomorphism if* (i) $A'\iota$ *is a direct factor in* A *or* (ii) B *is locally free.*

Case (i) is contained in 5.2.6. Case (ii) with B free is a direct consequence of 5.2.7.

Now suppose that B is locally free. Let $a'_i \in A'$, $b_i \in B$ and let $\sum_i a'_i \otimes b_i = 0$ in $A \otimes B$. Then there exists, by 5.4.18, an *fg*-subgroup $B_0 \subseteq B$ such that $b_i \in B_0$, all i, and $\sum_i a'_i \otimes b_i = 0$ in $A \otimes B_0$. But $\bar{\iota} : A' \otimes B_0 \to A \otimes B_0$ is monomorphic since B_0 is free. Thus $\sum_i a'_i \otimes b_i = 0$ in $A' \otimes B_0$. *A fortiori*, $\sum_i a'_i \otimes b_i = 0$ in $A' \otimes B$. ∎

5.4.20 Corollary. $A * B = 0$ *if* B *is locally free.* ∎

Notice that any field of characteristic 0 is locally free as an additive group. In particular, we may consider homology groups on an infinite abstract complex K. If $\mathscr{F}$ is a field of characteristic zero, we have ‡

5.4.21 $$H_p(K; \mathscr{F}) \cong H_p(K) \otimes \mathscr{F} \cong {}_wH_p(K) \otimes \mathscr{F}.$$

5.4.22 *Example.* If, for some integer m, $mA = {}_mB = 0$, then $A * B = 0$.

It follows from 5.4.18 that it is sufficient to prove this when A is finitely generated. Then A is the direct sum of cyclic groups whose orders divide m. It then follows from 5.4.10 that it is sufficient to take $A = J_k$, where $k \mid m$. In this case we present A as $0 \to J \xrightarrow{\lambda} J \xrightarrow{\mu} A \to 0$, where $1\lambda = k$. Identifying $J \otimes B$ with B, $\bar{\lambda} : B \to B$ is just multiplication by k. Since ${}_mB = 0$ and $k \mid m$, it is clear that $kb = 0$ implies $b = 0$. Thus $\bar{\lambda}$ is monomorphic so that $A *_\mu B = 0$.

Finally, before turning to contrahomology, we seek the generalized version of 5.2.16, taking R_1 as the coefficient group. We have

$$H_p(C \otimes R_1) \cong (H_p \otimes R_1) \oplus (H_{p-1} * R_1).$$

Now the exact sequence $0 \to T_p \to H_p \to {}_wH_p \to 0$ yields an exact sequence

$$T_p \otimes R_1 \to H_p \otimes R_1 \to {}_wH_p \otimes R_1 \to 0$$

by 5.4.1. Thus, by 5.4.17, $H_p \otimes R_1 \cong {}_wH_p \otimes R_1$. We shall prove in the next section that $H_{p-1} * R_1 = T_{p-1} * R_1 \cong T_{p-1}$, and so we shall have

5.4.23 Proposition. $H_p(C \otimes R_1) \cong ({}_wH_p \otimes R_1) \oplus T_{p-1}$. ∎

This is certainly as near as we can get to 5.2.16. If ${}_wH_p$ were free it would express the same result (it is, indeed, the invariant form of 5.2.16). However, it is not always true that the factor group of a group by its torsion subgroup is free—an easy example is provided by the group of rationals.

‡ Recall that ${}_wH_p = H_p / T_p$, where T_p is the torsion subgroup of H_p.

We now repeat the programme of this section for contrahomology. Given an exact sequence $A' \overset{\lambda}{\rightarrow} A \overset{\mu}{\rightarrow} A'' \rightarrow 0$ we obtain a sequence $A' \pitchfork B \overset{\lambda^{\cdot}}{\leftarrow} A \pitchfork B \overset{\mu^{\cdot}}{\leftarrow} A'' \pitchfork B \leftarrow 0$. We have ‡

5.4.1c Theorem. *The sequence* $A' \pitchfork B \overset{\lambda^{\cdot}}{\leftarrow} A \pitchfork B \overset{\mu^{\cdot}}{\leftarrow} A'' \pitchfork B \leftarrow 0$ *is exact.*

The proof, noticeably easier than that of 5.4.1 is left to the reader. ▌ Here $\lambda^{\cdot}$ is not necessarily epimorphic even when λ is monomorphic; not every homomorphism $A' \rightarrow B$ can be extended to A.

Let $0 \rightarrow R \overset{\lambda}{\rightarrow} F \overset{\mu}{\rightarrow} A \rightarrow 0$ be a presentation of A. We write $A \dagger_\mu B$ for the cokernel of $\lambda^{\cdot} : F \pitchfork B \rightarrow R \pitchfork B$.

As in the case of $*_\mu$, we shall prove $\dagger_\mu$ to be independent, up to isomorphism, of the choice of μ. Again we enunciate first some properties of $A \dagger_\mu B$.

Let A', B' be groups and let $\phi : A \rightarrow A'$, $\psi : B \rightarrow B'$ be homomorphisms and $0 \rightarrow R' \overset{\lambda'}{\rightarrow} F' \overset{\mu'}{\rightarrow} A' \rightarrow 0$ a presentation of A'; then

5.4.4c Theorem. *There is a canonical left homomorphism*

$$\tau^{\cdot}(\phi; \mu, \mu') : A' \dagger_{\mu'} B \rightarrow A \dagger_\mu B$$

and a right homomorphism $\kappa(\psi; \mu) : A \dagger_\mu B \rightarrow A \dagger_\mu B'$, *such that*

(i) $\tau^{\cdot}(1;\ \mu, \mu) = 1 : A \dagger_\mu B \rightarrow A \dagger_\mu B$;

(ii) $\tau^{\cdot}(\phi; \mu, \mu')\tau^{\cdot}(\phi'; \mu', \mu'') = \tau^{\cdot}(\phi\phi'; \mu, \mu'')$, *where* $\phi' : A' \rightarrow A''$ *and* μ'' *is a presentation of* A'';

(iii) $\kappa(1; \mu) = 1 : A \dagger_\mu B \rightarrow A \dagger_\mu B$;

(iv) $\kappa(\psi; \mu)\kappa(\psi'; \mu) = \kappa(\psi\psi'; \mu)$, *where* $\psi' : B' \rightarrow B''$;

(v) *for any* $\xi \in A' \dagger_{\mu'} B$,

$$(\tau^{\cdot}(\phi; \mu, \mu')\xi)\kappa(\psi; \mu) = \tau^{\cdot}(\phi; \mu, \mu')(\xi\kappa(\psi; \mu')) \in A \dagger_\mu B'.$$

The diagrams arising in the course of the proof may be taken to be

$$\begin{array}{ccccccccc}
0 & \longrightarrow & R & \overset{\lambda}{\longrightarrow} & F & \overset{\mu}{\longrightarrow} & A & \longrightarrow & 0 \\
 & & \downarrow{\scriptstyle\theta} & & \downarrow{\scriptstyle\chi} & & \downarrow{\scriptstyle\phi} & & \\
0 & \longrightarrow & R' & \overset{\lambda'}{\longrightarrow} & F' & \overset{\mu'}{\longrightarrow} & A' & \longrightarrow & 0
\end{array}$$

and

5.4.5c

$$\begin{array}{ccccccccccc}
0 & \longleftarrow & A \dagger_\mu B & \overset{\omega^{\cdot}}{\longleftarrow} & R \pitchfork B & \overset{\lambda^{\cdot}}{\longleftarrow} & F \pitchfork B & \overset{\mu^{\cdot}}{\longleftarrow} & A \pitchfork B & \longleftarrow & 0 \\
 & & \uparrow{\scriptstyle\tau^{\cdot}} & & \uparrow{\scriptstyle\theta^{\cdot}} & & \uparrow{\scriptstyle\chi^{\cdot}} & & \uparrow{\scriptstyle\phi^{\cdot}} & & \\
0 & \longleftarrow & A' \dagger_{\mu'} B & \overset{\omega'^{\cdot}}{\longleftarrow} & R' \pitchfork B & \overset{\lambda'^{\cdot}}{\longleftarrow} & F' \pitchfork B & \overset{\mu'^{\cdot}}{\longleftarrow} & A' \pitchfork B & \longleftarrow & 0
\end{array}$$

‡ In 5.4.1 we might have written $\lambda_{\cdot}$, $\mu_{\cdot}$ for $\bar{\lambda}$, $\bar{\mu}$ to bring out the duality.

and

5.4.6*c*

$$\begin{array}{ccccccccccc}
0 & \longleftarrow & A \dagger_\mu B & \xleftarrow{\omega^\cdot} & R \pitchfork B & \xleftarrow{\lambda^\cdot} & F \pitchfork B & \xleftarrow{\mu} & A \pitchfork B & \longleftarrow & 0 \\
 & & \Big\downarrow \kappa & & \Big\downarrow 1 \pitchfork \psi & & \Big\downarrow 1 \pitchfork \psi & & \Big\downarrow 1 \pitchfork \psi & & \\
0 & \longleftarrow & A \dagger_\mu B' & \xleftarrow{\omega^\cdot} & R \pitchfork B' & \xleftarrow{\lambda^\cdot} & F \pitchfork B' & \xleftarrow{\mu^\cdot} & A \pitchfork B' & \longleftarrow & 0
\end{array}$$

There are some small modifications to be made to the proof of 5.4.4, some of which arise from the fact that many of these homomorphisms are left homomorphisms. We leave them to the reader. ▌

A natural notation again suggests itself: let us write

$$\phi \dagger_{\mu\mu'} \psi : A' \dagger_{\mu'} B \rightarrow A \dagger_\mu B'$$

for the composition of $\tau^\cdot(\phi; \mu, \mu')$ with $\kappa(\psi; \mu)$. Since it is the composition of a left and a right homomorphism, there is no doctrinaire rule for deciding whether to regard it is a left or a right homomorphism; as in the case of $\phi \pitchfork \psi$ we shall normally regard it as a left homomorphism. We observe that

5.4.7*c*
$$\begin{cases} \tau^\cdot(\phi; \mu, \mu') = \phi \dagger_{\mu\mu'} 1, \\ \kappa(\psi; \mu) = 1 \dagger_{\mu\mu} \psi, \\ (\phi \dagger_{\mu\mu'} \psi')(\phi' \dagger_{\mu'\mu''} \psi) = (\phi\phi' \dagger_{\mu\mu''} \psi\psi') : A'' \dagger_{\mu''} B \rightarrow A \dagger_\mu B''. \end{cases}$$

5.4.8*c* Corollary. *Let* $0 \rightarrow R_i \xrightarrow{\lambda_i} F_i \xrightarrow{\mu_i} A \rightarrow 0$, $i = 0, 1$, *be two presentations of* A; *then there is a left isomorphism* $\tau^\cdot : A \dagger_{\mu_1} B \cong A \dagger_{\mu_0} B$.

We take $\tau^\cdot = \tau^\cdot(1; \mu_0, \mu_1)$. ▌

When we do not wish to stress the presentation we shall write $A \dagger B$ for $A \dagger_\mu B$, and we call $A \dagger B$ the *extension group*‡ of A by B. Any two models are related by a canonical isomorphism $\tau^\cdot$, as in 5.4.8*c*, and homomorphisms $\phi : A \rightarrow A'$, $\psi : B \rightarrow B'$ induce

$$\phi \dagger \psi : A' \dagger B \rightarrow A \dagger B'.$$

This remark is justified by the commutativity of the diagram

$$\begin{array}{ccc}
A \dagger_{\mu_0} B' & \xleftarrow{\tau^\cdot(1;\, \mu_0, \mu_1)} & A \dagger_{\mu_1} B' \\
\Big\uparrow \phi \dagger_{\mu_0 \mu_0'} \psi & & \Big\uparrow \phi \dagger_{\mu_1 \mu_1'} \psi \\
A' \dagger_{\mu_0'} B & \xleftarrow{\tau^\cdot(1;\, \mu_0', \mu_1')} & A' \dagger_{\mu_1'} B
\end{array}$$

The relations 5.4.7*c* show that the two composite homomorphisms from $A' \dagger_{\mu_1'} B$ to $A \dagger_{\mu_0} B'$ are both equal to $\phi \dagger_{\mu_0 \mu_1'} \psi$.

‡ The usual symbol for $A \dagger B$ is Ext (A, B). The name is justified in Appendix 2.

5.4.10c Proposition. *If A is the finite direct sum of groups A_α and B is the finite direct sum of groups B_β, then $A \dagger B$ is the direct sum of the groups $A_\alpha \dagger B_\beta$.* ❚

5.4.11c Proposition. *If A is free or B divisible then $A \dagger B = 0$.*

The assertion is trivial if A is free. It follows, for B divisible, immediately from Theorem 5.3.3d which asserts that every homomorphism into a divisible group may be extended. ❚

5.4.12c Proposition. $J_m \dagger J_n = J_{(m,n)}$, $J_m \dagger J = J_m$, $J \dagger J_n = 0$.

Present J_m as $0 \to mJ \overset{\lambda}{\to} J \overset{\mu}{\to} J_m \to 0$. Then $J_m \dagger_\mu J_n$ is the cokernel of $\lambda^{\bullet} : J \pitchfork J_n \to mJ \pitchfork J_n$. The map $\gamma \to (m)\gamma$, $\gamma \in mJ \pitchfork J_n$ maps $mJ \pitchfork J_n$ isomorphically onto J_n and the subgroup $\lambda^{\bullet}(J \pitchfork J_n)$ is mapped onto mJ_n. Thus $J_m \dagger J_n \cong (J_n)_m \cong J_{(m,n)}$. The second assertion is proved similarly; in fact, we observe that $J_m \dagger B \cong B_m$ for any B. The third assertion, included for completeness, is a special case of 5.4.11c. ❚

5.4.13c Theorem. *Let C be a free chain complex; then*

$$H^p(C \pitchfork G) \cong (H_p(C) \pitchfork G) \oplus (H_{p-1}(C) \dagger G).$$

We prove this with rather less pedantry than we displayed in our proof of 5.4.13. We have $C_p \pitchfork G = (Z_p \pitchfork G) \oplus (B_{p-1}\kappa \pitchfork G)$, and $\kappa^{\bullet} : B_{p-1}\kappa \pitchfork G \cong B_{p-1} \pitchfork G$. Now the elements of $Z^p(C \pitchfork G)$ are just those contrachains vanishing on B_p. Thus every element of $B_{p-1}\kappa \pitchfork G$ vanishes on Z_p and is therefore a contracycle, and an element of $Z_p \pitchfork G$ is a contracycle if and only if it vanishes on B_p. The subgroup of $Z_p \pitchfork G$ consisting of those homomorphisms which vanish on B_p is isomorphic to $H_p \pitchfork G$ (this is an application of 5.4.1c, and is anyway obvious). Indeed, writing $0 \to B_r \overset{\lambda}{\to} Z_r \overset{\mu}{\to} H_r \to 0$ as in 5.4.13, we have

5.4.15c $$Z^p(C \pitchfork G) = \mu^{\bullet}(H_p(C) \pitchfork G) \oplus (B_{p-1}\kappa \pitchfork G).$$

Since contraboundaries vanish on cycles, $B^p(C \pitchfork G) \subseteq B_{p-1}\kappa \pitchfork G$. Lemma 5.2.19 asserts that $\xi \in B_{p-1}\kappa \pitchfork G$ is a contraboundary if and only if $\kappa^{\bullet}\xi$ may be extended to C_{p-1}. Since Z_{p-1} is a direct factor in C_{p-1}, we have $\xi \in B^p(C \pitchfork G)$ if and only if $\kappa^{\bullet}\xi \in \lambda^{\bullet}(Z_{p-1} \pitchfork G)$. Thus

$$\kappa^{\bullet} B^p(C \pitchfork G) = \lambda^{\bullet}(Z_{p-1} \pitchfork G)$$

and

$$H^p(C \pitchfork G) \cong (H_p(C) \pitchfork G) \oplus (B_{p-1} \pitchfork G / \lambda^{\bullet}(Z_{p-1} \pitchfork G))$$

$$= (H_p(C) \pitchfork G) \oplus (H_{p-1}(C) \dagger G). \text{ ❚}$$

The reader is strongly advised to compare Theorem 5.4.13*c* with Theorem 5.2.12*c*, and to note that the first two terms in the expression for $H^p(C \pitchfork G)$ in 5.2.12*c* correspond to $H_p(C) \pitchfork G$, the third term corresponding to $H_{p-1}(C) \dagger G$.

5.4.16*c* Theorem. *Let C be as in 5.4.13c. Then there is an exact sequence*

$$0 \leftarrow H_p(C) \pitchfork G \overset{\rho}{\leftarrow} H^p(C \pitchfork G) \overset{\rho'}{\leftarrow} H_{p-1}(C) \dagger G \leftarrow 0,$$

where ρ, ρ' are canonical. Moreover, the image of ρ' is a direct factor in $H^p(C \pitchfork G)$.

The embedding $Z_p \subseteq C_p$ leads to a homomorphism $C_p \pitchfork G \to Z_p \pitchfork G$ which, restricted to $Z^p(C \pitchfork G)$, is an epimorphism to $\mu^{\cdot}(H_p(C) \pitchfork G)$, annihilating $B^p(C \pitchfork G)$. This defines ρ. The boundary $C_p \to B_{p-1}$ leads to a homomorphism $B_{p-1} \pitchfork G \to C_p \pitchfork G$ whose image is contained in $Z^p(C \pitchfork G)$ and which sends $\lambda^{\cdot}(Z_{p-1} \pitchfork G)$ to $B^p(C \pitchfork G)$. This defines ρ'. The remaining assertions of the theorem are proved in the course of demonstrating 5.4.13*c*. ▮

If $\mathscr{F}$ is a field of characteristic zero, then $\mathscr{F}$ is divisible as an additive group. Thus, by 5.4.11*c*, for contrahomology on an abstract complex,

5.4.21*c* $$H^p(K;\ \mathscr{F}) \cong H_p(K) \pitchfork \mathscr{F} \cong {}_wH_p \pitchfork \mathscr{F}.$$

Again, we may take a divisible group as value group and obtain an isomorphism similar to (5.4.21*c*). Thus, in particular, taking the reals mod 1, we get

5.4.23*c* Proposition $H^p(C \pitchfork R_1) \cong H_p(C) \pitchfork R_1$. ▮

Suppose that C is a free *fg*-complex and let $C^{\cdot} = C \pitchfork J$. Then $C^{\cdot}$ is a free contrachain complex and $\kappa : C^{\cdot} \otimes G \to C \pitchfork G$ is a contrachain isomorphism (5.2.22, 5.2.23). The proof of Theorem 5.4.13 may be applied to the contrachain complex $C^{\cdot} \otimes G$ and the conclusion transferred to the complex $C \pitchfork G$ by means of the contrahomology isomorphism induced by κ. The conclusion then is

5.4.24 Theorem. *If C is a free fg-complex and if $C^{\cdot} = C \pitchfork J$, then*

$$H^p(C \pitchfork G) \cong (H^p(C^{\cdot}) \otimes G) \oplus (H^{p+1}(C^{\cdot}) * G.) ▮$$

This result applies in particular if $C = C(K)$, where K is a finite complex.

5.4.25 Corollary. *If C is a geometric chain complex and $H_*(C)$ is fg in each dimension, then the conclusion of 5.4.24 holds.*

For we may plainly construct a geometric *fg*-complex D such that $H_*(D) \cong H_*(C)$; it then follows from 5.3.4 and 3.10.1 that D is chain equivalent to C. Since 5.4.24 holds for D it also holds for C. ▮

5.5 The products $\otimes$, $*$, $\pitchfork$, $\dagger$.

In this section we investigate the products‡ introduced in the previous section from a purely algebraic standpoint. That is to say, the results we obtain are not necessarily the algebraic parts of statements about the homology of polyhedra. On the other hand, these results are all related to and suggested by the notion of the homology theory of chain complexes.

Our first theorem will be applied to the homology theory of polyhedra in the next section. We prove it now because it yields as special cases two exact sequences, one involving $\otimes$ and $*$, the other involving $\pitchfork$ and $\dagger$.

In order that our theorem may be applicable to both chain and contrachain complexes, we adopt a suggestion made in chapter 3 and talk of a *differential group*. This is an abelian group A, together with a right endomorphism d satisfying $d^2 = 0$. A homomorphism, $\phi : A \to A'$, of differential groups is understood to commute with d.

If $Z(A) = d^{-1}(0)$, $B(A) = Ad$, then $B(A) \subseteq Z(A) \subseteq A$; we define $H(A) = Z(A)/B(A)$ as the *homology group* of A. A homomorphism $\phi : A \to A'$ induces $\phi_* : H(A) \to H(A')$.

5.5.1 Theorem. *Let* $0 \to A \overset{\lambda}{\to} B \overset{\mu}{\to} C \to 0$ *be an exact sequence of differential groups. Then there is an exact triangle*§

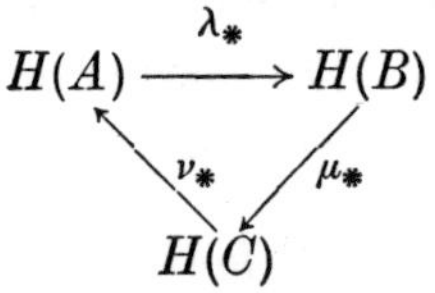

We first define ν_*. Let $c \in Z(C)$ and let $\{c\}$ be its homology class. Since μ is onto C, there exists $b \in B$ with $b\mu = c$. Then

$$bd\mu = b\mu d = cd = 0,$$

so that $bd = a\lambda$ for some $a \in A$. Moreover, $a \in Z(A)$ since

$$ad\lambda = a\lambda d = bd^2 = 0$$

and λ is monomorphic. We define $\{c\}\nu_* = \{a\}$. The definition of ν_* has involved two choices: first we selected a cycle from a homology

‡ We describe $\otimes$, $*$, $\pitchfork$, $\dagger$ as products since they associate a group with each ordered pair of groups. In the language of categories these are functors of two variables.

§ That is, the kernel of each homomorphism is the image of the preceding one.

class in $H(C)$, and, secondly, we selected an element b with $b\mu = c$. We now show that ν_* does not depend on these choices. Let us vary b. Thus if also $b'\mu = c$, then $b = b' + a''\lambda$ so that $bd = b'd + a''d\lambda$. Then if $b'd = a'\lambda$, we have $a = a' + a''d$ so that $\{a\} = \{a'\}$. Now vary c within $\{c\}$. Thus, if $c' \in \{c\}$, then $c = c' + c''d = c' + b''\mu d$. Thus $(b - b''d)\mu = c'$, and $(b - b''d)d = a\lambda$; c' therefore gives rise to the same $\{a\}$ as does c. This shows that ν_* is a mapping from $H(C)$ to $H(A)$ and it is obviously homomorphic.

We now prove that the triangle is exact.

(i) *Exactness at* $H(B)$. Since $\lambda\mu = 0$, we have $\lambda_*\mu_* = (\lambda\mu)_* = 0$. Conversely, let $\{b\} \in H(B)$ and $\{b\}\mu_* = 0$. Thus for some $c' \in C$, $b\mu = c'd = b'\mu d = b'd\mu$, for some $b' \in B$. Then $b - b'd = a\lambda$ and $a \in Z(A)$. Thus $\{b\} = \{b - b'd\} = \{a\lambda\} = \{a\}\lambda_*$, so that

$$\mu_*^{-1}(0) \subseteq H(A)\,\lambda_*.$$

(ii) *Exactness at* $H(C)$. Consider $\{b\}\mu_*\nu_*, = \{b\mu\}\nu_*$. Inspection of the definition of ν_* shows that this is the homology class of a where $a\lambda = bd$; but $bd = 0$, so that $\mu_*\nu_* = 0$. Conversely, suppose $\{c\}\nu_* = 0$. Then $c = b\mu$, $bd = a\lambda$, and $a = a'd$. But then $(b - a'\lambda)d = 0$ and $(b - a'\lambda)\mu = c$ so that $\{c\} = \{b - a'\lambda\}\mu_*$ and $\nu_*^{-1}(0) \subseteq H(B)\mu_*$.

(iii) *Exactness at* $H(A)$. Consider $\{c\}\nu_*\lambda_*$. Then $c = b\mu$, $bd = a\lambda$, $\{c\}\nu_* = \{a\}$, $\{c\}\nu_*\lambda_* = \{a\lambda\}$. But $a\lambda = bd$ so that $\{c\}\nu_*\lambda_* = 0$. Conversely, suppose $\{a\}\lambda_* = 0$. Then $a\lambda = bd$ and $b\mu d = bd\mu = a\lambda\mu = 0$. Thus $b\mu \in Z(C)$ and $\{b\mu\}\nu_* = \{a\}$, so that $\lambda_*^{-1}(0) \subseteq H(C)\nu_*$. ∎

Special cases of this theorem have been proved in 2.9 and used in chapters 2, 3 and 4. Those cases were distinguished from the general case in that A was a direct factor in B; this led to the definition of a homomorphism $\nu : C \to A$ inducing ν_*. Several further applications will be made in the next section. In algebraic topology we are particularly interested in the cases in which the differential groups are essentially chain or contrachain complexes. Thus A is a chain complex if $A = \sum_p A_p$ and $A_p d \subseteq A_{p-1}$. If A is a chain complex, then $H_*(A)$ is graded in the usual way as $\sum_p H_p(A)$. If λ, μ are chain maps of chain complexes, then ν_* lowers degree by 1 since its definition involves a single application of d. Thus we have

5.5.2 Corollary. *An exact sequence* $0 \to A \xrightarrow{\lambda} B \xrightarrow{\mu} C \to 0$ *of chain maps of chain complexes induces an exact sequence*

$$\dots \to H_p(A) \xrightarrow{\lambda_*} H_p(B) \xrightarrow{\mu_*} H_p(C) \xrightarrow{\nu_*} H_{p-1}(A) \to \dots. \; ∎$$

Similarly, we have

5.5.2*c* Corollary. *An exact sequence* $0 \to A \overset{\lambda}{\to} B \overset{\mu}{\to} C \to 0$ *of contrachain maps of contrachain complexes induces an exact sequence*

$$\ldots \to H^p(A) \overset{\lambda_*}{\to} H^p(B) \overset{\mu_*}{\to} H^p(C) \overset{\nu_*}{\to} H^{p+1}(A) \to \ldots . \;\rule{1ex}{1.5ex}$$

The application which we now make requires some preliminary remarks. Let

$$0 \to L \overset{\phi}{\to} M \overset{\psi}{\to} N \to 0$$

be an exact sequence of groups and let $0 \to R \overset{\lambda}{\to} F \overset{\mu}{\to} A \to 0$ be a presentation of A. We obtain a commutative diagram

5.5.3

$$\begin{array}{ccccccccc}
 & & 0 & & 0 & & 0 & & \\
 & & \downarrow & & \downarrow & & \downarrow & & \\
 & & A * L & \xrightarrow{1*\phi} & A * M & \xrightarrow{1*\psi} & A * N & & \\
 & & \downarrow{\scriptstyle \overline{\omega}} & & \downarrow{\scriptstyle \overline{\omega}} & & \downarrow{\scriptstyle \overline{\omega}} & & \\
0 & \longrightarrow & R \otimes L & \xrightarrow{\overline{\phi}} & R \otimes M & \xrightarrow{\overline{\psi}} & R \otimes N & \longrightarrow & 0 \\
 & & \downarrow{\scriptstyle \overline{\lambda}} & & \downarrow{\scriptstyle \overline{\lambda}} & & \downarrow{\scriptstyle \overline{\lambda}} & & \\
0 & \longrightarrow & F \otimes L & \xrightarrow{\overline{\phi}} & F \otimes M & \xrightarrow{\overline{\psi}} & F \otimes N & \longrightarrow & 0 \\
 & & \downarrow{\scriptstyle \overline{\mu}} & & \downarrow{\scriptstyle \overline{\mu}} & & \downarrow{\scriptstyle \overline{\mu}} & & \\
 & & A \otimes L & \xrightarrow{\overline{\phi}} & A \otimes M & \xrightarrow{\overline{\psi}} & A \otimes N & & \\
 & & \downarrow & & \downarrow & & \downarrow & & \\
 & & 0 & & 0 & & 0 & &
\end{array}$$

in which $\overline{\lambda} = \lambda \otimes 1$, $\overline{\mu} = \mu \otimes 1$, $\overline{\phi} = 1 \otimes \phi$, $\overline{\psi} = 1 \otimes \psi$, and $\overline{\omega}$ embeds the kernel of $\overline{\lambda}$.

5.5.4 Theorem. *If* $0 \to L \overset{\phi}{\to} M \overset{\psi}{\to} N \to 0$ *is an exact sequence of groups, there is an exact sequence*

$$0 \to A * L \overset{1*\phi}{\to} A * M \overset{1*\psi}{\to} A * N \overset{\nu}{\to} A \otimes L \overset{\overline{\phi}}{\to} A \otimes M \overset{\psi}{\to} A \otimes N \to 0.$$

If $0 \to R \overset{\lambda}{\to} F \overset{\mu}{\to} A \to 0$ is a presentation of A, let C be the chain complex in which $C_0 = F$, $C_1 = R$, $C_p = 0$ for $p \neq 0, 1$ and $\partial_1 = \lambda$. Then $\overline{\phi} = 1 \otimes \phi$ determines a chain map $\overline{\phi}. : C \otimes L \to C \otimes M$ and $\overline{\psi}$ a chain map $\overline{\psi}. : C \otimes M \to C \otimes N$. We assert that

$$0 \to C \otimes L \overset{\overline{\phi}.}{\to} C \otimes M \overset{\overline{\psi}.}{\to} C \otimes N \to 0$$

is exact. It is exact, except possibly at $C \otimes L$, from 5.4.1; since, however, F and hence also R are free (5.3.1*f*), the monomorphy of $\overline{\phi}.$ follows from 5.4.19 (ii) and the commutativity of $\otimes$.

Since C is a free chain complex we may apply 5.4.13 to deduce that $H_0(C\otimes L) = H_0(C)\otimes L = A\otimes L$ and that

$$H_1(C\otimes L) \cong H_0(C) * L \cong A * L$$

and that $H_p(C\otimes L)$ is null for $p \neq 0, 1$, with similar results for M and N. The result now follows from 5.5.2, since $\overline{\phi}_{*0} = 1\otimes\phi = \overline{\phi}$ and $\overline{\phi}_{*1} = 1 * \phi$ and similarly with ψ. ∎

The reader may verify that ν is the 'zigzag' homomorphism $\overline{\omega}\overline{\psi}^{-1}\overline{\lambda}\overline{\phi}^{-1}\overline{\mu}$. From this it is easy to see that the sequence of 5.5.4 is natural in the following sense: given two presentations μ, μ' of A, the diagram

5.5.5

$$\begin{array}{ccc} A *_{\mu} N & \xrightarrow{\nu} & A\otimes L \\ \Big\downarrow \bar{\tau} & \nearrow \nu' & \\ A *_{\mu'} N & & \end{array}$$

is commutative.

5.5.6 Theorem. *$A * B$ may be defined by means of a presentation of B; more precisely if, $0 \to S \xrightarrow{\phi} G \xrightarrow{\psi} B \to 0$ is a presentation of B, then $A * B \cong \overline{\phi}^{-1}(0)$, where $\overline{\phi} = 1\otimes\phi : A\otimes S \to A\otimes G$.*

We apply 5.5.4 with S, G, B for L, M, N. Since G is free $A * G = 0$ (5.4.11). Thus ν maps $A * B$ isomorphically onto the kernel of $\overline{\phi}$. ∎

5.5.7 Corollary. $A * B \cong B * A$.

This follows from 5.5.6 and the natural isomorphism

$$A\otimes B \cong B\otimes A. \;∎$$

Indeed, if $\eta = \eta(A, B)$ is the natural isomorphism $\eta : A\otimes B \cong B\otimes A$, then an isomorphism $\rho = \rho(\mu, \psi) : A *_{\mu} B \cong B *_{\psi} A$ is defined by the commutativity of the diagram

$$\begin{array}{ccc} A *_{\mu} B & \xrightarrow{\overline{\omega}} & R\otimes B \\ \Big\downarrow \rho & & \Big\downarrow \eta \\ B *_{\psi} A & \xrightarrow{\nu} & B\otimes R. \end{array}$$

Moreover, the isomorphism ρ is natural in the sense that, if $\alpha : A \to A'$, $\beta : B \to B'$ are maps, then the diagram

$$\begin{array}{ccc} A *_{\mu} B & \xrightarrow{\alpha *_{\mu\mu'} \beta} & A' *_{\mu'} B' \\ \Big\downarrow \rho(\mu, \psi) & & \Big\downarrow \rho(\mu', \psi') \\ B *_{\psi} A & \xrightarrow{\beta *_{\psi\psi'} \alpha} & B' *_{\psi'} A' \end{array}$$

is commutative. We leave the proof of this assertion as a straightforward exercise for the reader.

5.5.8 Corollary. *$A * B = 0$ if A is locally free.*

See 5.4.20. ▌

5.5.9 Proposition. *If T is a torsion group‡, then $T * R_1 \cong T$.*

Let R be the group of reals and J the subgroups of integers, so that $R_1 \cong R/J$. Then we apply 5.5.4 with T for A and J, R, R_1 for L, M, N. In the resulting exact sequence

$$0 \to T * J \to T * R \to T * R_1 \xrightarrow{\nu} T \otimes J \to T \otimes R \to T \otimes R_1 \to 0,$$

we know from 5.4.20 that $T * R$ is null, since R is locally free, and from 5.4.17 that $T \otimes R$ is null. Therefore ν is an isomorphism; and, since of course $T \otimes J \cong T$ (see 5.2.7), the result follows. ▌

5.5.10 Corollary (*of proof*). *If Q_1 is the group of rationals* mod 1, *then for any torsion group T, $T * Q_1 \simeq T$.* ▌

We can now give the promised proof of 5.4.23. If A is any abelian group and T its torsion subgroup and B any abelian group, we have from 5.5.4 the exact sequence

$$0 \to B * T \to B * A \to B * (A/T) \to \ldots;$$

but A/T is locally free, so that by 5.4.20 $B * (A/T)$ is null and therefore $B * T \cong B * A$. Reversing factors and applying 5.5.9,

$$A * R_1 \cong T * R_1 \cong T;$$

in particular $H_{p-1} * R_1 \cong T_{p-1}$. ▌

We now dualize diagram 5.5.3.

5.5.3 c

$$\begin{array}{ccccccccc}
 & & 0 & & 0 & & 0 & & \\
 & & \uparrow & & \uparrow & & \uparrow & & \\
 & & A \dagger L & \xrightarrow{1 \dagger \phi} & A \dagger M & \xrightarrow{1 \dagger \phi} & A \dagger N & & \\
 & & \uparrow{\scriptstyle \omega^{\cdot}} & & \uparrow{\scriptstyle \omega^{\cdot}} & & \uparrow{\scriptstyle \omega^{\cdot}} & & \\
0 & \longrightarrow & R \pitchfork L & \xrightarrow{\phi_{.}} & R \pitchfork M & \xrightarrow{\psi_{.}} & R \pitchfork N & \longrightarrow & 0 \\
 & & \uparrow{\scriptstyle \lambda^{\cdot}} & & \uparrow{\scriptstyle \lambda^{\cdot}} & & \uparrow{\scriptstyle \lambda^{\cdot}} & & \\
0 & \longrightarrow & F \pitchfork L & \xrightarrow{\phi_{.}} & F \pitchfork M & \xrightarrow{\psi_{.}} & F \pitchfork N & \longrightarrow & 0 \\
 & & \uparrow{\scriptstyle \mu^{\cdot}} & & \uparrow{\scriptstyle \mu^{\cdot}} & & \uparrow{\scriptstyle \mu^{\cdot}} & & \\
 & & A \pitchfork L & \xrightarrow{\phi_{.}} & A \pitchfork M & \xrightarrow{\psi_{.}} & A \pitchfork N & & \\
 & & \uparrow & & \uparrow & & \uparrow & & \\
 & & 0 & & 0 & & 0 & &
\end{array}$$

‡ Recall that R_1 is the group of reals mod 1.

5.5.4*c* Theorem. *There is an exact sequence*

$$0 \to A \pitchfork L \overset{\phi.}{\to} A \pitchfork M \overset{\psi.}{\to} A \pitchfork N \overset{\nu.}{\to} A \dagger L \overset{1 \dagger \phi}{\to} A \dagger M \overset{1 \dagger \psi}{\to} A \dagger N \to 0.$$

The proof is formally the same as that of 5.5.4, the rôle of 5.4.19 in that proof being played here by 5.3.3*f*.]

Now consider a rationalization of B

5.5.11 $$0 \to B \overset{\rho}{\to} I \overset{\sigma}{\to} D \to 0,$$

where I is divisible. We know from 5.3.2*d* that every group B admits such a rationalization. By 5.4.11*c*, $A \dagger I = 0$; so that, putting B, I, D for L, M, N in 5.5.4*c*, we get

5.5.6*c* Theorem. *$A \dagger B$ may be defined by means of a rationalization of B; more precisely, $A \dagger B \cong \operatorname{coker} \sigma.$, where $\sigma. : A \pitchfork I \to A \pitchfork D$ is induced by σ in* 5.5.11.]

This result replaces the symmetry of $A * B$; the latter may be obtained by means of a presentation of A or B, while $A \dagger B$ may be obtained by a presentation of A or a rationalization of B. We thus obtain a further diagram and exact sequence as follows.

Let $0 \to L \overset{\phi}{\to} M \overset{\psi}{\to} N \to 0$ be exact. We combine this sequence with 5.5.11 to obtain

5.5.3*c*′

$$\begin{array}{ccccccccc}
 & & 0 & & 0 & & 0 & & \\
 & & \downarrow & & \downarrow & & \downarrow & & \\
 & & L \pitchfork B & \overset{\phi^{\cdot}}{\longleftarrow} & M \pitchfork B & \overset{\psi^{\cdot}}{\longleftarrow} & N \pitchfork B & & \\
 & & \downarrow{\scriptstyle \rho.} & & \downarrow{\scriptstyle \rho.} & & \downarrow{\scriptstyle \rho.} & & \\
0 & \longleftarrow & L \pitchfork I & \overset{\phi^{\cdot}}{\longleftarrow} & M \pitchfork I & \overset{\psi^{\cdot}}{\longleftarrow} & N \pitchfork I & \longleftarrow & 0 \\
 & & \downarrow{\scriptstyle \sigma.} & & \downarrow{\scriptstyle \sigma.} & & \downarrow{\scriptstyle \sigma.} & & \\
0 & \longleftarrow & L \pitchfork D & \overset{\phi^{\cdot}}{\longleftarrow} & M \pitchfork D & \overset{\psi^{\cdot}}{\longleftarrow} & N \pitchfork D & \longleftarrow & 0 \\
 & & \downarrow{\scriptstyle \omega.} & & \downarrow{\scriptstyle \omega.} & & \downarrow{\scriptstyle \omega.} & & \\
 & & L \dagger B & \overset{\phi \dagger 1}{\longleftarrow} & M \dagger B & \overset{\psi \dagger 1}{\longleftarrow} & N \dagger B & & \\
 & & \downarrow & & \downarrow & & \downarrow & & \\
 & & 0 & & 0 & & 0 & &
\end{array}$$

Note that the middle rows are exact by 5.4.11*c* and 5.3.1*d*. We then obtain

5.5.4*c*′ Theorem. *There is an exact sequence*

$$0 \leftarrow L \dagger B \overset{\phi \dagger 1}{\leftarrow} M \dagger B \overset{\psi \dagger 1}{\leftarrow} N \dagger B \overset{\nu^{\cdot}}{\leftarrow} L \pitchfork B \overset{\phi^{\cdot}}{\leftarrow} M \pitchfork B \overset{\psi^{\cdot}}{\leftarrow} N \pitchfork B \leftarrow 0. \;]$$

This sequence of course differs from that of 5.5.4*c* but each is 'dual' to 5.5.4.

We now prove a result which brings out a relation between homology and contrahomology. Let Γ be any (abelian) group and let us write $A^\cdot$ for $A \pitchfork \Gamma$.

5.5.12 Theorem.‡ *There is a natural isomorphism* $\gamma : A \pitchfork B^\cdot \cong (A \otimes B)^\cdot$.

We define γ as a left isomorphism by $(a \otimes b)\gamma\phi = b(a\phi)$, for any $\phi \in A \pitchfork B^\cdot$. We leave to the reader the verification that γ is an isomorphism. The *naturality* of γ is expressed in the assertion that, if $\alpha : A \to A'$, $\beta : B \to B'$ are any homomorphisms, the diagram

$$\begin{array}{ccc}
(A \otimes B)^\cdot & \xleftarrow{\ \gamma\ } & A \pitchfork B^\cdot \\
\uparrow{\scriptstyle (\alpha \otimes \beta)^\cdot} & & \uparrow{\scriptstyle \alpha \pitchfork \beta^\cdot} \\
(A' \otimes B')^\cdot & \xleftarrow{\ \gamma\ } & A' \pitchfork B'^\cdot
\end{array}$$

is commutative. ∎

5.5.13 Theorem. *There is a natural homomorphism* $\chi : A \dagger B^\cdot \to (A * B)^\cdot$ *which is an isomorphism if* Γ *is divisible.*

Let $0 \to R \xrightarrow{\lambda} F \xrightarrow{\mu} A \to 0$ be a presentation of A, inducing exact sequences

$$0 \to A * B \xrightarrow{\bar\omega} R \otimes B \xrightarrow{\bar\lambda} F \otimes B \xrightarrow{\bar\mu} A \otimes B \to 0$$

and

$$0 \leftarrow A \dagger B^\cdot \xleftarrow{\omega^\cdot} R \pitchfork B^\cdot \xleftarrow{\lambda^\cdot} F \pitchfork B^\cdot \xleftarrow{\mu^\cdot} A \pitchfork B^\cdot \leftarrow 0.$$

We use γ to obtain the diagram

5.5.14

$$\begin{array}{ccccccccc}
 & & (A * B)^\cdot & \xleftarrow{\ \bar\omega^\cdot\ } & (R \otimes B)^\cdot & \xleftarrow{\ \bar\lambda^\cdot\ } & (F \otimes B)^\cdot & \xleftarrow{\ \bar\mu^\cdot\ } & (A \otimes B)^\cdot & & \\
 & & \uparrow{\scriptstyle \chi} & & \cong\uparrow{\scriptstyle \gamma} & & \cong\uparrow{\scriptstyle \gamma} & & \cong\uparrow{\scriptstyle \gamma} & & \\
0 & \longleftarrow & A \dagger B^\cdot & \xleftarrow{\ \omega^\cdot\ } & R \pitchfork B^\cdot & \xleftarrow{\ \lambda^\cdot\ } & F \pitchfork B^\cdot & \xleftarrow{\ \mu^\cdot\ } & A \pitchfork B^\cdot & \longleftarrow & 0
\end{array}$$

which is commutative because γ is natural; we do not assert that the upper row is exact in general. We shall define χ as the unique homomorphism preserving commutativity; this is a valid definition if $\bar\omega^\cdot \gamma \ker \omega^\cdot = 0$. Let $\omega^\cdot x = 0$; then, since the lower row is exact, there is $y \in F \pitchfork B^\cdot$ such that $\lambda^\cdot y = x$. Then $\bar\omega^\cdot \gamma x = \bar\omega^\cdot \gamma \lambda^\cdot y = \bar\omega^\cdot \bar\lambda^\cdot \gamma y$; since $\bar\omega\bar\lambda = 0$, $\bar\omega^\cdot \bar\lambda^\cdot = 0$ and $\bar\omega^\cdot \gamma x = 0$. We may therefore now define χ to be the (left) homomorphism such that $\chi\omega^\cdot = \bar\omega^\cdot\gamma$.

The sense in which χ is natural is obvious; the mechanical verification will be omitted.

We now turn to the case of Γ being divisible. We shall in fact prove that, if $L \xrightarrow{\phi} M \xrightarrow{\psi} N$ is exact, then if Γ is divisible $L^\cdot \xrightarrow{\phi^\cdot} M^\cdot \xrightarrow{\psi^\cdot} N^\cdot$ is also exact. It is obvious from 5.5.14 that this will suffice to prove χ isomorphic. We therefore prove

5.5.15 Lemma.§ *If* $L \xrightarrow{\phi} M \xrightarrow{\psi} N$ *is exact and* Γ *is divisible, then* $L^\cdot \xleftarrow{\phi^\cdot} M^\cdot \xleftarrow{\psi^\cdot} N^\cdot$ *is also exact.*

‡ This may also be written as $A \pitchfork (B \pitchfork \Gamma) \cong (A \otimes B) \pitchfork \Gamma$; in this form it has a family resemblance to 4.4.2, namely, $(c \frown d) \frown e = c \frown (d \cup e)$.

§ The dual of this (that, if F is a free group, operation by $\otimes F$ on a sequence preserves exactness) is a direct consequence of 5.2.7.

Certainly $\phi^{\cdot}\psi^{\cdot} = 0$, since $\phi\psi = 0$. We now prove that, if $\xi \in M^{\cdot}$ and $\phi^{\cdot}\xi = 0$, then there is $\eta \in N^{\cdot}$ such that $\psi^{\cdot}\eta = \xi$. Now ξ is a homomorphism $\xi : M \to \Gamma$ such that $\phi\xi = 0$.

$$\begin{array}{ccccc} L & \xrightarrow{\phi} & M & \xrightarrow{\psi} & N \\ & & \downarrow \xi & \swarrow \eta & \\ & & \Gamma & & \end{array}$$

So $\xi^{-1}(0) \supseteq L\phi = \psi^{-1}(0)$; therefore there is a homomorphism $\eta_0 : M\psi \to \Gamma$ such that $\psi\eta_0 = \xi$. But Γ is divisible and therefore injective, and the homomorphism η_0 of the subgroup $M\psi$ of N may be extended to a homomorphism $\eta : N \to \Gamma$, and certainly $\psi\eta = \xi$; this is to say that $\psi^{\cdot}\eta = \xi$.]]

5.5.16 Theorem. *If C is a free chain complex and Γ is divisible, then*

$$(H_p(C \otimes G))^{\cdot} \cong H^p(C \pitchfork G^{\cdot}).$$

This follows immediately from 5.4.13, 5.4.13*c*, 5.5.12 and 5.5.13.]

5.5.17 Corollary. *If $|K|$ is a polyhedron and Γ is divisible, then*

$$(H_p(|K|; G))^{\cdot} \cong H^p(|K|; G^{\cdot}). \;]$$

If we take G to be the integers, then $G^{\cdot} \cong \Gamma$.

5.6 Exact sequences

Theorem 5.5.1 is central to the application of exact sequences to homology theory. Applications were made in 5.5 and will be made in 5.7. Before giving two fundamental applications in this section, we wish once more to raise the question of naturality. We are concerned in particular here with the naturality of ν_* in 5.5.1, and we express it as follows. Let

$$\begin{array}{ccccccccc} 0 & \longrightarrow & A & \xrightarrow{\lambda} & B & \xrightarrow{\mu} & C & \longrightarrow & 0 \\ & & \downarrow \alpha & & \downarrow \beta & & \downarrow \gamma & & \\ 0 & \longrightarrow & A' & \xrightarrow{\lambda'} & B' & \xrightarrow{\mu'} & C' & \longrightarrow & 0 \end{array}$$

be a commutative diagram of homomorphisms of differential groups, in which each row is exact. Then

5.6.1 Theorem. *The diagram*

$$\begin{array}{ccccccccccc} \cdots & \longrightarrow & H(A) & \xrightarrow{\lambda_*} & H(B) & \xrightarrow{\mu_*} & H(C) & \xrightarrow{\nu_*} & H(A) & \longrightarrow & \cdots \\ & & \downarrow \alpha_* & & \downarrow \beta_* & & \downarrow \gamma_* & & \downarrow \alpha_* & & \\ \cdots & \longrightarrow & H(A') & \xrightarrow{\lambda'_*} & H(B') & \xrightarrow{\mu'_*} & H(C') & \xrightarrow{\nu'_*} & H(A') & \longrightarrow & \cdots \end{array}$$

is commutative.

This is evident for the first two squares. For the last we must recall the definition of ν_*. Let $\{z\} \in H(C)$ and let $z = b\mu$, $b \in B$. Then if $bd = a\lambda$, $\{z\}\nu_* = \{a\}$. Now $\{z\}\gamma_* = \{z\gamma\}$ and $z\gamma = b\mu\gamma = b\beta\mu'$. Also $b\beta d = bd\beta = a\lambda\beta = a\alpha\lambda'$. Thus

$$\{z\gamma\}\nu'_* = \{a\alpha\} \quad \text{or} \quad \{z\}\gamma_*\nu'_* = \{a\}\alpha_* = \{z\}\nu_*\alpha_*.$$

Thus commutativity holds round the third square. ∎

This means that wherever we use 5.5.1 to deduce an exact sequence, we can use 5.6.1 to deduce a homomorphism of exact sequences. This applies, for example, to Theorems 5.5.4, 5.5.4c and 5.5.4c'. We shall not explicitly append to those results which follow from 5.5.1 the consequences of 5.6.1, but recommend the reader to formulate them himself.

We now make further applications of Theorem 5.5.1; each application will, in fact, specialize 5.5.2 or 5.5.2c.

Let C be a chain complex and let C^0 be a subcomplex of C such that each C_p^0 is a direct factor in C_p. Let $\hat{C}$ be the factor complex C/C^0. We write

$$0 \to C^0 \xrightarrow{\lambda} C \xrightarrow{\mu} \hat{C} \to 0.$$

Let G be an abelian group. We obtain exact sequences

$$0 \to C^0 \otimes G \to C \otimes G \to \hat{C} \otimes G \to 0,$$

and

$$0 \leftarrow C^0 \pitchfork G \leftarrow C \pitchfork G \leftarrow \hat{C} \pitchfork G \leftarrow 0.$$

We apply 5.5.1 to obtain

5.6.2 Theorem. *There are exact sequences*

$$\dots \to H_p(C^0 \otimes G) \xrightarrow{\lambda_*} H_p(C \otimes G) \xrightarrow{\mu_*} H_p(\hat{C} \otimes G) \xrightarrow{\nu_*} H_{p-1}(C^0 \otimes G) \to \dots$$

and

$$\dots \leftarrow H^p(C^0 \pitchfork G) \xleftarrow{\lambda^*} H^p(C \pitchfork G) \xleftarrow{\mu^*} H^p(\hat{C} \pitchfork G) \xleftarrow{\nu^*} H^{p-1}(C^0 \pitchfork G) \leftarrow \dots. \ ∎$$

The exact sequences of 2.9 are just special cases of the sequences of this theorem; if K_0 is a subcomplex of K, then $C_p(K_0)$ is certainly a direct factor in $C_p(K)$ and $C(K)/C(K_0) = C(K, K_0)$.

5.6.3 Proposition. *If D, E, F are chain complexes such that $F \subseteq E \subseteq D$, then there is an exact sequence*

$$\dots \to H_p(E/F) \xrightarrow{\lambda_*} H_p(D/F) \xrightarrow{\mu_*} H_p(D/E) \xrightarrow{\nu_*} H_{p-1}(E/F) \to \dots,$$

where $\lambda : E/F \to D/F$ *and* $\mu : D/F \to D/E$ *are the chain maps induced by the injections of* E, F *in* D, F *and of* D, F *in* D, E.

The reader is referred to 3.1 for the definition of the factor complex. In view of 5.5.1 it is only necessary to check the exactness of the sequence $0 \to E/F \overset{\lambda}{\to} D/F \overset{\mu}{\to} D/E \to 0$. ▌

The exact sequence of 5.6.3 is called the *homology sequence of the triple D, E, F of chain complexes.*

5.6.4 Corollary. *If* K, L, M *are complexes such that* $M \subseteq L \subseteq K$, *there is an exact sequence*

$$\ldots \to H_p(L, M; G) \overset{\lambda_*}{\to} H_p(K, M; G) \overset{\mu_*}{\to} H_p(K, L; G) \overset{\nu_*}{\to} H_{p-1}(L, M; G) \to \ldots. \;▌$$

This is called the *homology sequence of the triple K, L, M of complexes.*

Analogous statements of course hold for contrachain complexes and contrahomology.

Now let C be a *free* chain complex and let $0 \to G' \overset{\lambda}{\to} G \overset{\mu}{\to} G'' \to 0$ be an exact sequence of coefficient groups. Then

$$0 \to C \otimes G' \to C \otimes G \to C \otimes G'' \to 0$$

is an exact sequence of chain complexes and

$$0 \to C \pitchfork G' \to C \pitchfork G \to C \pitchfork G'' \to 0$$

is an exact sequence of contrachain complexes. We deduce from 5.5.1

5.6.5 Theorem (*The coefficient sequences*). *If* $0 \to G' \overset{\lambda}{\to} G \overset{\mu}{\to} G'' \to 0$ *is exact and if* C *is free, then the sequences*

$$\ldots \to H_p(C \otimes G') \overset{\lambda_*}{\to} H_p(C \otimes G) \overset{\mu_*}{\to} H_p(C \otimes G'') \overset{\nu_*}{\to} H_{p-1}(C \otimes G') \to \ldots$$

$$\ldots \to H^p(C \pitchfork G') \overset{\lambda_*}{\to} H^p(C \pitchfork G) \overset{\mu_*}{\to} H^p(C \pitchfork G'') \overset{\nu_*}{\to} H^{p+1}(C \pitchfork G') \to \ldots$$

are exact. ▌

An interesting and important example of this is provided by the exact sequence

$$0 \to J_m \overset{\lambda}{\to} J_{m^2} \overset{\mu}{\to} J_m \to 0,$$

where, if J_m is generated by a and J_{m^2} by b, we take $a\lambda = mb$, $b\mu = a$. If we take $C = C(K)$ we have an exact sequence

$$\textbf{5.6.6} \quad \ldots \to H_p(K; J_m) \overset{\lambda_*}{\to} H_p(K; J_{m^2}) \overset{\mu_*}{\to} H_p(K; J_m) \overset{\nu_*}{\to} H_{p-1}(K; J_m) \to \ldots.$$

It is interesting to analyse ν_* in this case. Given an element of $H_p(K; J_m)$, we pick a representative cycle z_p and interpret‡ the coefficients of z_p as elements of J_{m^2}. We then take the boundary of z_p, as an element of $C_p(K; J_{m^2})$, and divide the resulting coefficients by m. We reinterpret the coefficients as elements of J_m and take the homology class. This gives us $\{z_p\}\nu_*$. Thus we may write the operation symbolically as $\frac{1}{m}\partial$. For example, if $|K| = P^n$, n-dimensional real projective space, then

$$\nu_* : H_p(K; J_2) \cong H_{p-1}(K; J_2), \quad p \text{ even}, 0 \leqslant p \leqslant n,$$
$$H_p(K; J_2)\,\nu_* = 0, \quad p \text{ odd}.$$

A second interesting and important example, related to the preceding one, is provided by the exact sequence

$$0 \to J \xrightarrow{\lambda'} J \xrightarrow{\mu'} J_m \to 0,$$

where λ' multiplies by m and $1\mu' = a$, generating J_m. We get an exact sequence

5.6.7 $$\ldots \to H_p(K; J) \xrightarrow{\lambda'_*} H_p(K; J) \xrightarrow{\mu'_*} H_p(K; J_m) \xrightarrow{\nu'_*} H_{p-1}(K; J) \to \ldots.$$

Again, ν'_* can be interpreted as $\frac{1}{m}\partial$; we leave to the reader a more accurate description of it. Moreover, ν_* in (5.6.6) and ν'_* in (5.6.7) are related precisely by

5.6.8 $$\nu'_*\mu'_* = \nu_*.$$

Of course, (5.6.6) and (5.6.7) have analogues in contrahomology§, namely,

5.6.6c $$\ldots \to H^p(K; J_m) \xrightarrow{\lambda_*} H^p(K; J_{m^2})$$
$$\xrightarrow{\mu_*} H^p(K; J_m) \xrightarrow{\nu_*} H^{p+1}(K; J_m) \to \ldots$$

and

5.6.7c $$\ldots \to H^p(K; J) \xrightarrow{\lambda'_*} H^p(K; J)$$
$$\xrightarrow{\mu'_*} H^p(K; J_m) \xrightarrow{\nu'_*} H^{p+1}(K; J) \to \ldots,$$

and (5.6.8) holds when interpreted for contrahomology.

The homomorphisms ν_* are called the *Bockstein boundary and contraboundary operators*. It follows immediately from 5.6.1 that they commute with homology homomorphisms induced by maps§.

‡ More precisely, we replace each coefficient of z_p by a counterimage under μ.

§ An application is to be found in Appendix 3 to this chapter.

We now gather together some important results on exact sequences. We start by proving

5.6.9 Lemma. *Let* ϕ *be a homomorphism of the exact sequence* $A_0 \xrightarrow{\alpha_0} A_1 \xrightarrow{\alpha_1} A_2 \xrightarrow{\alpha_2} A_3$ *into the exact sequence* $B_0 \xrightarrow{\beta_0} B_1 \xrightarrow{\beta_1} B_2 \xrightarrow{\beta_2} B_3$. *Then* (i) *if* $\phi_0 : A_0 \to B_0$ *is epimorphic, the kernel of* ϕ *is exact at* $\ker \phi_2$; (ii) *if* $\phi_3 : A_3 \to B_3$ *is monomorphic, the cokernel of* ϕ *is exact at* $\operatorname{coker} \phi_1$.

We prove (i). We have a diagram

$$\begin{array}{ccccccc}
M_0 & \xrightarrow{\mu_0} & M_1 & \xrightarrow{\mu_1} & M_2 & \xrightarrow{\mu_2} & M_3 \\
\downarrow{\scriptstyle \psi_0} & & \downarrow{\scriptstyle \psi_1} & & \downarrow{\scriptstyle \psi_2} & & \downarrow{\scriptstyle \psi_3} \\
A_0 & \xrightarrow{\alpha_0} & A_1 & \xrightarrow{\alpha_1} & A_2 & \xrightarrow{\alpha_2} & A_3 \\
\downarrow{\scriptstyle \phi_0} & & \downarrow{\scriptstyle \phi_1} & & \downarrow{\scriptstyle \phi_2} & & \downarrow{\scriptstyle \phi_3} \\
B_0 & \xrightarrow{\beta_0} & B_1 & \xrightarrow{\beta_1} & B_2 & \xrightarrow{\beta_2} & B_3
\end{array}$$

where ψ_i embeds $M_i = \ker \phi_i$ in A_i and μ_i is the restriction of α_i, all i. Obviously $\mu_1 \mu_2 = 0$. Conversely, let $x\mu_2 = 0, x \in M_2$. Then $x\psi_2\alpha_2 = x\mu_2\psi_3 = 0$, so that $x\psi_2 = y\alpha_1$ for some $y \in A_1$. Thus $y\phi_1\beta_1 = y\alpha_1\phi_2 = x\psi_2\phi_2 = 0$, so that $y\phi_1 = z\beta_0$ for some $z \in B_0$. But ϕ_0 is epimorphic, so we have $y\phi_1 = u\phi_0\beta_0 = u\alpha_0\phi_1$ for some $u \in A_0$; but then $y - u\alpha_0 = v\psi_1$ for some $v \in M_1$. It follows that

$$v\mu_1\psi_2 = v\psi_1\alpha_1 = y\alpha_1 = x\psi_2,$$

whence $x = v\mu_1$ and (i) is proved. (ii) may be deduced similarly. ∎

The conclusions of (i) and (ii) do not, however, follow without some restriction on ϕ. Consider, for example,

$$\begin{array}{ccccccccc}
0 & \longrightarrow & 0 & \longrightarrow & G & \xrightarrow{1} & G & \longrightarrow & 0 \\
 & & \downarrow & & \downarrow{\scriptstyle 1} & & \downarrow & & \\
0 & \longrightarrow & G & \xrightarrow{1} & G & \longrightarrow & 0 & \longrightarrow & 0;
\end{array}$$

if $G \neq 0$, then neither the kernel nor the cokernel sequence is exact.

5.6.10 Corollary. *Let* ϕ *be a homomorphism of the infinite exact sequence*

$$\ldots \to A_i \xrightarrow{\alpha_i} A_{i+1} \to \ldots$$

into the infinite exact sequence

$$\ldots \to B_i \xrightarrow{\beta_i} B_{i+1} \to \ldots.$$

Then (i) *if* ϕ *is epimorphic,* $\ker \phi$ *is exact;* (ii) *if* ϕ *is monomorphic,* $\operatorname{coker} \phi$ *is exact.*

This is an immediate consequence of 5.6.9. ∎ We remark that any finite exact sequence beginning and ending with the null group may, in an obvious way, be regarded as an infinite sequence.

The second consequence we draw from 5.6.9 is the celebrated 5-*lemma*.

5.6.11 Theorem. (5-*lemma*). *Suppose given the commutative diagram*

$$\begin{array}{ccccccccc} A_0 & \xrightarrow{\alpha_0} & A_1 & \xrightarrow{\alpha_1} & A_2 & \xrightarrow{\alpha_2} & A_3 & \xrightarrow{\alpha_3} & A_4 \\ \downarrow{\scriptstyle \phi_0} & & \downarrow{\scriptstyle \phi_1} & & \downarrow{\scriptstyle \phi_2} & & \downarrow{\scriptstyle \phi_3} & & \downarrow{\scriptstyle \phi_4} \\ B_0 & \xrightarrow{\beta_0} & B_1 & \xrightarrow{\beta_1} & B_2 & \xrightarrow{\beta_2} & B_3 & \xrightarrow{\beta_3} & B_4 \end{array}$$

where each row is exact. If ϕ_0, ϕ_1, ϕ_3, ϕ_4, are isomorphisms, so is ϕ_2.

Since ϕ_0 is epimorphic, $\ker \phi$ is exact at $\ker \phi_2$; but

$$\ker \phi_1 = \ker \phi_3 = 0,$$

so that, by exactness, $\ker \phi_2 = 0$. Similarly, since ϕ_4 is monomorphic, $\operatorname{coker} \phi$ is exact at $\operatorname{coker} \phi_2$; but $\operatorname{coker} \phi_1 = \operatorname{coker} \phi_3 = 0$ so that $\operatorname{coker} \phi_2 = 0$. ∎

Observe that the conclusion of the theorem follows under the weaker hypothesis that ϕ_1 and ϕ_3 are isomorphisms and ϕ_0 is an epimorphism, ϕ_4 a monomorphism.

The 5-lemma is usually applied to a situation in which the homomorphisms ϕ_0, ϕ_3, ϕ_6, ... are all of one kind (and known to be isomorphisms) and the homomorphisms ϕ_1, ϕ_4, ϕ_7, ... are also all of one kind (and known to be isomorphisms). Thus, let $f : |K|, |K_0| \to |L|\ |L_0|$, be a map such that $f_* : H_n(|K|) \cong H_n(|L|)$, $f_* : H_n(|K_0|) \cong H_n(|L_0|)$ for $n \leqslant m$.

5.6.12 Corollary. *Under these circumstances,*

$$f_* : H_n(|K|, |K_0|) \cong H_n(|L|, |L_0|), \quad n \leqslant m.$$

For we apply the 5-lemma to the commutative diagram

$$\begin{array}{ccccc} H_n(|K_0|) & \xrightarrow{\lambda_*} & H_n(|K|) & \xrightarrow{\mu_*} & H_n(|K|, |K_0|) \\ \downarrow{\scriptstyle f_*} & & \downarrow{\scriptstyle f_*} & & \downarrow{\scriptstyle f_*} \\ H_n(|L_0|) & \xrightarrow{\lambda_*} & H_n(|L|) & \xrightarrow{\mu_*} & H_n(|L|, |L_0|) \end{array}$$

$$\begin{array}{ccccc} \xrightarrow{\nu_*} & H_{n-1}(|K_0|) & \xrightarrow{\lambda_*} & H_{n-1}(|K|) \\ & \downarrow{\scriptstyle f_*} & & \downarrow{\scriptstyle f_*} \\ \xrightarrow{\nu_*} & H_{n-1}(|L_0|) & \xrightarrow{\lambda_*} & H_{n-1}(|L|). \end{array}$$ ∎

5.7 Tensor products of chain complexes

It will be shown in chapter 8 that, to compute the homology groups of $|K| \times |L|$, it is sufficient to consider a chain complex obtained from $C(K)$ and $C(L)$ by a process of taking tensor products. In this section we show how the homology groups of such a *tensor product of chain complexes* may be computed in terms of the homology groups of the original chain complexes.

5.7.1 Definition. Let C, C' be two chain complexes. We define their *tensor product* $C \otimes C'$ to be the chain complex D such that

$$D_p = \sum_{m+n=p} C_m \otimes C'_n,$$

$$(c_m \otimes c'_n)\partial = c_m \partial \otimes c'_n + (-1)^m c_m \otimes c'_n \partial, \quad c_m \in C_m, \; c'_n \in C'_n.$$

We verify that $\partial\partial = 0$ in $C \otimes C'$; for

$$\begin{aligned}(c_m \otimes c'_n)\partial\partial &= (c_m \partial \otimes c'_n)\partial + (-1)^m (c_m \otimes c'_n \partial)\partial \\ &= c_m \partial\partial \otimes c'_n + (-1)^{m-1} c_m \partial \otimes c'_n \partial \\ &\quad + (-1)^m c_m \partial \otimes c'_n \partial + c_m \otimes c'_n \partial\partial \\ &= 0.\end{aligned}$$

Since evidently‡ $D_p \partial \subseteq D_{p-1}$, we see that $D = C \otimes C'$ is a chain complex.

Where it seems advisable to distinguish notationally between the boundaries in C, C', $C \otimes C'$, they will be written as ∂, ∂', $\partial^{\otimes}$. As an example, let G be an abelian group and let G be regarded as a chain complex in which $G_0 = G$, $G_p = 0$, $p \neq 0$. Then $C \otimes G$ is a chain complex and the notation is consistent with our previous definition of $C \otimes G$.

We first record some basic facts about $C \otimes C'$; those proofs that are elementary are left to the reader.

5.7.2 Proposition. (i) *If C, C' are free, so is $C \otimes C'$;* (ii) *if C, C' are geometric, so is $C \otimes C'$;* (iii) *if C, C' are fg-complexes, so is $C \otimes C'$.* ∎

5.7.3 Proposition. (i) *The map $\mu = \mu(C, C') : C \otimes C' \to C' \otimes C$, given by*

$$(c_m \otimes c'_n)\mu = (-1)^{mn} c'_n \otimes c_m, \quad c_m \in C_m, \quad c'_n \in C'_n,$$

is a chain isomorphism. (ii) *The map $\alpha : (C \otimes C') \otimes C'' \to C \otimes (C' \otimes C'')$ given by $((c \otimes c') \otimes c'')\alpha = c \otimes (c' \otimes c'')$ is a chain isomorphism.*

‡ Notice that this inclusion holds but in general $(C_m \otimes C'_n)\partial$ is contained in neither $C_{m-1} \otimes C'_n$ nor $C_m \otimes C'_{n-1}$.

The reader should verify that μ and α are chain maps; he will then discover why the factor $(-1)^{mn}$ is inserted into the definition of μ. ∎

5.7.4 Proposition. *If $\phi : C \to E$, $\phi' : C' \to E'$ are chain maps, so is*

$$\phi \otimes \phi' : C \otimes C' \to E \otimes E'. ∎$$

5.7.5 Proposition. *If $\phi \simeq \psi : C \to E$, $\phi' \simeq \psi' : C' \to E'$, then*

$$\phi \otimes \phi' \simeq \psi \otimes \psi' : C \otimes C' \to E \otimes E'.$$

To prove this, we first show that if $\phi \simeq 0$, then $\phi \otimes \phi' \simeq 0$. We shall introduce the notation $\epsilon : C \to C$ (where C is any chain complex) for the group automorphism given by $c_m \epsilon = (-1)^m c_m$. Then

5.7.6 $$\epsilon^2 = 1, \quad \epsilon\partial = -\partial\epsilon, \quad \epsilon\phi = \phi\epsilon, \quad \epsilon\Delta = -\Delta\epsilon,$$

where $\Delta : C \to E$ is a chain homotopy.

Moreover, in $C \otimes C'$,

5.7.7 $$\partial^{\otimes} = \partial \otimes 1 + \epsilon \otimes \partial'.$$

Now suppose $\phi = \partial\Delta + \Delta\partial$. Then

$$\begin{aligned} \partial^{\otimes}(\Delta \otimes \phi') + (\Delta \otimes \phi')\,\partial^{\otimes} &= (\partial \otimes 1 + \epsilon \otimes \partial')(\Delta \otimes \phi') + (\Delta \otimes \phi')(\partial \otimes 1 + \epsilon \otimes \partial') \\ &= \partial\Delta \otimes \phi' + \epsilon\Delta \otimes \partial'\phi' + \Delta\partial \otimes \phi' + \Delta\epsilon \otimes \phi'\partial' \\ &= (\partial\Delta + \Delta\partial) \otimes \phi', \quad \text{by 5.7.6,} \\ &= \phi \otimes \phi'. \end{aligned}$$

Thus $\phi \otimes \phi' \overset{\Delta\otimes\phi'}{\simeq} 0$. Similarly if $\phi' \overset{\Delta'}{\simeq} 0$, then $\phi \otimes \phi' \overset{\epsilon\phi\otimes\Delta'}{\simeq} 0$. The proposition now follows by linearity. We conclude, in fact, that if $\phi \overset{\Delta}{\simeq} \psi$, $\phi' \overset{\Delta'}{\simeq} \psi'$, then $\phi \otimes \phi' \overset{\Delta\otimes\phi' + \epsilon\psi\otimes\Delta'}{\simeq} \psi \otimes \psi'$. ∎

5.7.8 Corollary. *If C is acyclic and geometric then $C \otimes C'$ is acyclic.*

For, by 3.4.3, $1_C \simeq 0_C : C \to C$. Thus

$$1_{C\otimes C'} = 1_C \otimes 1_{C'} \simeq 0_C \otimes 1_{C'} = 0_{C\otimes C'}.$$

5.7.9 Corollary. *If $\phi : C \simeq E$ with inverse ψ and $\phi' : C' \simeq E'$ with inverse ψ', then $\phi \otimes \phi' : C \otimes C' \simeq E \otimes E'$ with inverse $\psi \otimes \psi'$.* ∎

To state and prove the main theorem we introduce the notations

$$H_n = H_n(C), \quad H = H_*(C), \quad H'_n = H_n(C'), \quad H' = H_*(C'),$$

$$H^{\otimes}_n = H_n(C \otimes C'), \quad H^{\otimes} = H_*(C \otimes C'); \quad Z' = Z(C'), \quad B' = B(C').$$

Let us write
$$0 \to B \overset{\lambda}{\to} Z \overset{\mu}{\to} H \to 0,$$
$$0 \to B' \overset{\lambda'}{\to} Z' \overset{\mu'}{\to} H' \to 0,$$

for the homomorphisms implicit in $H = Z/B$, $H' = Z'/B'$.

5.7.10 Lemma. *The kernel of* $\mu \otimes \mu' : Z \otimes Z' \to H \otimes H'$ *is*
$$(B \otimes Z')\bar{\lambda} + (Z \otimes B')\bar{\lambda}'.$$

Obviously the kernel of $\mu \otimes \mu'$ contains the given subgroup of $Z \otimes Z'$. To prove the converse, we invoke 5.4.1 to obtain a commutative diagram of exact sequences

$$\begin{array}{ccccccc}
B \otimes B' & \xrightarrow{\bar{\lambda}} & Z \otimes B' & \xrightarrow{\bar{\mu}} & H \otimes B' & \longrightarrow & 0 \\
\downarrow{\scriptstyle \bar{\lambda}'} & & \downarrow{\scriptstyle \bar{\lambda}'} & & \downarrow{\scriptstyle \bar{\lambda}'} & & \\
B \otimes Z' & \xrightarrow{\bar{\lambda}} & Z \otimes Z' & \xrightarrow{\bar{\mu}} & H \otimes Z' & \longrightarrow & 0 \\
\downarrow{\scriptstyle \bar{\mu}'} & & \downarrow{\scriptstyle \bar{\mu}'} & & \downarrow{\scriptstyle \bar{\mu}'} & & \\
B \otimes H' & \xrightarrow{\bar{\lambda}} & Z \otimes H' & \xrightarrow{\bar{\mu}} & H \otimes H' & \longrightarrow & 0 \\
\downarrow & & \downarrow & & \downarrow & & \\
0 & & 0 & & 0 & &
\end{array}$$

Now $\mu \otimes \mu' = \bar{\mu}'\bar{\mu}$. Suppose $x\bar{\mu}'\bar{\mu} = 0$, $x \in Z \otimes Z'$. Then
$$\begin{aligned} x\bar{\mu}' &= y\bar{\lambda}, \qquad y \in B \otimes H', \\ &= u\bar{\mu}'\bar{\lambda}, \qquad u \in B \otimes Z', \\ &= u\bar{\lambda}\bar{\mu}'. \end{aligned}$$

Thus
$$x = u\bar{\lambda} + v\bar{\lambda}', \quad v \in Z \otimes B'. \;\rule{0.5em}{0.8em}$$

The boundary formula in $C \otimes C'$ shows that if $z \in Z$, $z' \in Z'$, then $z \otimes z' \in Z(C \otimes C')$ and that, if $b \in B$, $b' \in B'$, then $b \otimes z'$, $z \otimes b' \in B(C \otimes C')$. It follows from 5.7.10 that if $\iota : Z \to C$, $\iota' : Z' \to C'$ are the embeddings, then $\iota \otimes \iota'$ induces a homomorphism

5.7.11 $$\eta : H \otimes H' \to H^{\otimes}.$$

We graduate $H \otimes H'$ by $(H \otimes H')_p = \sum_{m+n=p} H_m \otimes H'_n$. Then η is degree-preserving. We write η_p for the component of η of degree p, and state the main theorem.

5.7.12 Theorem. *Let C or C' be free; then η_p is a monomorphism whose cokernel is isomorphic to $\sum_{m+n=p-1} H_m * H'_n$. If both C and C' are free, the image of η_p is a direct factor of $H_p^{\otimes}$.*

There is no real loss of generality in supposing that C is free; then Z is a direct factor in C so that, if $Z \otimes C'$ is the chain complex obtained by regarding Z as a chain complex with zero boundary and if $\iota : Z \to C$ is the embedding, then $\bar{\iota} : Z \otimes C' \to C \otimes C'$ is a chain monomorphism. Let $B \otimes C'$ be the chain complex with the usual graded structure of tensor products (Definition 5.7.1) but with boundary operator $-\epsilon \otimes \partial'$. Then the epimorphism $\bar{\partial} = \partial \otimes 1 : C \otimes C' \to B \otimes C'$ is a chain map since

$$\begin{aligned} \partial^{\otimes}\bar{\partial} = (\partial \otimes 1 + \epsilon \otimes \partial')(\partial \otimes 1) &= \epsilon\partial \otimes \partial' \\ &= -\partial\epsilon \otimes \partial' \quad (5.7.6) \\ &= (\partial \otimes 1)(-\epsilon \otimes \partial') \\ &= \bar{\partial}(-\epsilon \otimes \partial'). \end{aligned}$$

Clearly $\ker \bar{\partial} = (Z \otimes C')\bar{\iota}$, so that we have an exact sequence

$$0 \to Z \otimes C' \xrightarrow{\bar{\iota}} C \otimes C' \xrightarrow{\bar{\partial}} B \otimes C' \to 0.$$

We note that $\bar{\partial}$ lowers degree by 1; even so, however, we may apply 5.5.1 to obtain an exact homology sequence

5.7.13 $$\ldots \to H_p(Z \otimes C') \xrightarrow{\iota_*} H_p^{\otimes} \xrightarrow{\partial_*} H_{p-1}(B \otimes C') \xrightarrow{\nu_*} H_{p-1}(Z \otimes C') \to \ldots.$$

Consider $H_p(Z \otimes C')$. Since Z is free abelian, it follows readily from 5.4.1 and 5.4.19 that

$$Z_p(Z \otimes C') = \sum_{m+n=p} Z_m \otimes Z'_n, \quad B_p(Z \otimes C') = \sum_{m+n=p} Z_m \otimes B'_n,$$

and hence that $$H_p(Z \otimes C') = \sum_{m+n=p} Z_m \otimes H'_n.$$

Similarly $$H_p(B \otimes C') = \sum_{m+n=p} B_m \otimes H'_n.$$

We now examine $\nu_* : H_p(B \otimes C') \to H_p(Z \otimes C')$. A representative in $B \otimes C'$ of a typical element of $B_m \otimes H'_n$ may be written $\sum_i b_m^i \otimes z_n'^i$. Let $b_m^i = c_{m+1}^i \partial$; then $(\sum_i c_{m+1}^i \otimes z_n'^i)\bar{\partial} = \sum_i b_m^i \otimes z_n'^i$ and also

$$(\sum_i c_{m+1}^i \otimes z_n'^i)\,\partial^{\otimes} = \sum_i b_m^i \otimes z_n'^i,$$

now regarded as an element of $C \otimes C'$; but this is, of course, the $\bar{\iota}$-image of $\sum_i b'_m \otimes z'^i_n$ *regarded as an element of* $Z \otimes C'$. Thus ν_* is simply the homomorphism induced by the embedding $B \subseteq Z$. We revive the notation λ for this embedding and rewrite 5.7.13 as

5.7.14 $$\ldots \xrightarrow{\lambda_*} \sum_{m+n=p} Z_m \otimes H'_n \xrightarrow{\iota_*} H^{\otimes}_p \xrightarrow{\partial_*} \sum_{m+n=p-1} B_m \otimes H'_n \xrightarrow{\lambda_*} \sum_{m+n=p-1} Z_m \otimes H'_n \xrightarrow{\iota_*} \ldots.$$

It is clear that the cokernel of λ_{*p} is just $\sum_{m+n=p} H_m \otimes H'_n$ and that ι_{*p} induces precisely the map η_p of this cokernel into $H^{\otimes}_p$. Moreover, the kernel of $B_m \otimes H'_n \xrightarrow{\lambda_*} Z_m \otimes H_n$ is $H_m *_\mu H'_n$. Thus we obtain from 5.7.14 the exact sequence

$$0 \to \sum_{m+n=p} H_m \otimes H'_n \xrightarrow{\eta_p} H^{\otimes}_p \xrightarrow{\theta_p} \sum_{m+n=p-1} H_m * H'_n \to 0,$$

where θ_p is induced by ∂_{*p}.

It only remains to show that if C' is also free η_p has a right inverse; for it will then follow from 5.1.9′ that η_p maps onto a direct factor in $H^{\otimes}_p$. Let C' be free and let $\kappa : C \to Z$, $\kappa' : C' \to Z'$ be right inverses of ι, ι'. Consider $(c \otimes c')\,\partial^{\otimes}(\kappa \otimes \kappa') = c\partial \otimes c'\kappa' + c\epsilon\kappa \otimes c'\partial'$, where $c\partial$, $c'\partial'$ are to be regarded as elements of Z, Z'. It follows from 5.7.10 that $\kappa \otimes \kappa'$ maps $B(C \otimes C')$ into the kernel of $\mu \otimes \mu'$ and thus induces a map

$$\zeta : H^{\otimes} \to H \otimes H',$$

such that $\eta\zeta = 1$. ∎

The reader will have noticed the similarity of this result to that of Theorem 5.4.13; although G, regarded as a chain complex, is not, in general, free, the argument above holds for $C \otimes G$.

5.7.15 Corollary. *If H or H' is torsion-free, then*

$$\eta : H \otimes H' \cong H^{\otimes}. \;∎$$

5.7.16 Corollary. *Let $\mathscr{F}$ be a field of characteristic zero; then*

$$H(C \otimes C' \otimes \mathscr{F}) \cong H \otimes H' \otimes \mathscr{F}. \;∎$$

The formula

5.7.17 $$H^{\otimes}_p \cong \Big(\sum_{m+n=p} H_m \otimes H'_n\Big) \oplus \Big(\sum_{m+n=p-1} H_m * H'_n\Big)$$

for the homology of free chain complexes is called the *Künneth formula*. It leads to the relation of the homology groups of a topological product to those of the factors. Corollaries 5.7.15 and 5.7.16 show that the formula may simplify.

A further important case in which the formula simplifies is that in which we take coefficients J_q, where q is a prime. Since $J_q \cong J_q \otimes J_q$ we get a chain isomorphism $C \otimes C' \otimes J_q \cong (C \otimes J_q) \otimes (C' \otimes J_q)$. Arguing as before, we get an exact sequence

$$\ldots \xrightarrow{\lambda_*} \sum_{m+n=p} Z_m(C \otimes J_q) \otimes H_n(C' \otimes J_q) \xrightarrow{\iota_*} H_p(C \otimes C' \otimes J_q)$$
$$\xrightarrow{\partial_*} \sum_{m+n=p-1} B_m(C \otimes J_q) \otimes H_n(C' \otimes J_q) \xrightarrow{\lambda_*} \ldots.$$

However, λ_* is monomorphic, since $Z_m(C \otimes J_q)$ is a vector space over J_q and $B_m(C \otimes J_q)$ is a subspace and therefore a direct factor. We conclude

5.7.18 Theorem. *If q is a prime, then*

$$H_p(C \otimes C' \otimes J_q) \cong \sum_{m+n=p} H_m(C \otimes J_q) \otimes H_n(C' \otimes J_q). \rbrack$$

Assuming the basic theorem on topological products (see chapter 8), we conclude

5.7.19 Corollary. *Let $|K|$, $|L|$ be polyhedra. Then*‡

(i) $H_p(|K| \times |L|) \cong \sum_{m+n=p} H_m(|K|) \otimes H_n(|L|) \oplus \sum_{m+n=p-1} H_m(|K|) * H_n(|L|)$;

(ii) *if $\mathscr{F}$ is a field of characteristic zero,*

$$H_p(|K| \times |L|; \mathscr{F}) \cong \sum_{m+n=p} H_m(|K|) \otimes H_n(|L|) \otimes \mathscr{F};$$

(iii) *if q is a prime,* $H_p(|K| \times |L|; J_q) \cong \sum_{m+n=p} H_m(|K|; J_q) \otimes H_n(|L|; J_q)$.]

5.7.20 *Example.* Consider the torus $S^1 \times S^1$. We conclude from 5.7.19 that

$$H_0(S^1 \times S^1) = J \otimes J = J,$$
$$H_1(S^1 \times S^1) \cong H_1(S^1) \otimes H_0(S^1) \oplus H_0(S^1) \otimes H_1(S^1) = (J \otimes J) \oplus (J \otimes J) = J \oplus J,$$
$$H_2(S^1 \times S^1) \cong H_1(S^1) \otimes H_1(S^1) = J \otimes J = J,$$
$$H_p(S^1 \times S^1) = 0, \quad p > 2.$$

5.7.21 *Example.* Consider $P^2 \times P^2$, where P^2 is the projective plane. Then

$$H_0(P^2 \times P^2) = J \otimes J = J,$$
$$H_1(P^2 \times P^2) \cong H_1(P^2) \otimes H_0(P^2) \oplus H_0(P^2) \otimes H_1(P^2)$$
$$= (J_2 \otimes J) \oplus (J \otimes J_2) = J_2 \oplus J_2,$$
$$H_2(P^2 \times P^2) \cong H_1(P^2) \otimes H_1(P^2) = J_2 \otimes J_2 = J_2,$$
$$H_3(P^2 \times P^2) \cong H_1(P^2) * H_1(P^2) = J_2 * J_2 = J_2,$$
$$H_p(P^2 \times P^2) = 0, \quad p > 3.$$

‡ The homology groups may all be regarded as simplicial or singular; the groups on the right of the isomorphisms may be computed from K and L. In fact, 5.7.19 holds in singular homology for any spaces.

The particular interest in this example lies in the case of H_3. In the notation of 3.9, ζ_2 is a 2-chain on P^2 such that $\zeta_2\partial = 2\zeta_1$. Then $(\zeta_2 \otimes \zeta_1 + \zeta_1 \otimes \zeta_2)\partial = 2(\zeta_1 \otimes \zeta_1 - \zeta_1 \otimes \zeta_1) = 0$. On the other hand, $(\zeta_2 \otimes \zeta_2)\partial = 2(\zeta_1 \otimes \zeta_2 + \zeta_2 \otimes \zeta_1)$, yielding a generator, $\{\zeta_2 \otimes \zeta_1 + \zeta_1 \otimes \zeta_2\}$ of $H_3(P^2 \times P^2) = J_2$.

We may use 5.7.18 to compute $H_p(P^2 \times P^2; J_2)$ immediately; we obtain

$$H_0(P^2 \times P^2; J_2) = J_2, \quad H_1 = J_2 \oplus J_2,$$

$$H_2 = J_2 \oplus J_2 \oplus J_2, \quad H_3 = J_2 \oplus J_2, \quad H_4 = J_2, \quad H_p = 0, \quad p > 4.$$

The Künneth formula may, in fact, be generalized for arbitrary coefficients. We state the result in the form

5.7.22 Theorem. *If $|K|$, $|L|$ are polyhedra, then*

$$H_p(|K| \times |L|; G) \cong \sum_{m+n=p} H_m(|K|; H_n(|L|; G)).\ ]$$

(See Exercises, Q. 14.)

Similarly we may generalize 5.7.18 and 5.7.19 (iii). Let A, B be vector spaces over a field $\mathscr{F}$. Then we may define $A \otimes_{\mathscr{F}} B$ to be the group obtained from $A \otimes B$ by adding the relations

$$\alpha a \otimes b = a \otimes \alpha b, \quad a \in A, b \in B, \quad \alpha \in \mathscr{F}.$$

Then a simpler argument than that of 5.7.12 shows that

5.7.23 $$H_p(C \otimes_{\mathscr{F}} C') \cong \sum_{m+n=p} H_m(C) \otimes_{\mathscr{F}} H_n(C'),$$

where C, C' are vector-space chain complexes; that is, each C_n, C'_n is a vector space over $\mathscr{F}$ and the boundary operators are vector space homomorphisms.

Now let C, C' be free chain complexes. Then $C \otimes \mathscr{F}$, $C' \otimes \mathscr{F}$ are vector space chain complexes and clearly

$$(C \otimes \mathscr{F}) \otimes_{\mathscr{F}} (C' \otimes \mathscr{F}) \cong C \otimes C' \otimes \mathscr{F}.$$

Thus

5.7.24 Theorem. *If $\mathscr{F}$ is a field and C, C' are geometric chain complexes,*

$$H_p(C \otimes C' \otimes \mathscr{F}) \cong \sum_{m+n=p} H_m(C \otimes \mathscr{F}) \otimes_{\mathscr{F}} H_n(C' \otimes \mathscr{F}).\ ]$$

Notice that 5.7.18 is a special case of 5.7.24 since $\otimes_{J_q} = \otimes$.

5.7.25 Corollary. *For any field $\mathscr{F}$*

$$H_p(|K| \times |L|; \mathscr{F}) \cong \sum_{m+n=p} H_m(|K|; \mathscr{F}) \otimes_{\mathscr{F}} H_n(|L|; \mathscr{F}).\ ]$$

We may obtain the contrahomology groups of a topological product by using the Künneth formula and applying 5.4.13*c*. A direct Künneth formula is, however, available in contrahomology if C and C' are free

fg-complexes (compare 5.4.24). For let C, C' be free fg-complexes and let $C^{\cdot} = C \pitchfork J$, $C'^{\cdot} = C' \pitchfork J$. Then $C^{\cdot}$, $C'^{\cdot}$ are free contrachain complexes and $C^{\cdot} \otimes C'^{\cdot}$, defined entirely analogously to 5.7.1, is again a free contrachain complex. The argument proving 5.7.12 now establishes the isomorphism

5.7.26 $$H^p(C^{\cdot} \otimes C'^{\cdot}) \cong \sum_{m+n=p} H^m(C^{\cdot}) \otimes H^n(C'^{\cdot}) \oplus \sum_{m+n=p+1} H^m(C^{\cdot}) * H^n(C'^{\cdot}).$$

This formula acquires topological interest when we observe that $C^{\cdot} \otimes C'^{\cdot}$ and $(C \otimes C')^{\cdot}$ are contrachain isomorphic. In fact, if $C = C(K)$, $C' = C(L)$, it is proved in chapter 8 that $(C \otimes C') \pitchfork A$ is a suitable contrachain complex for computing $H^*(|K| \times |L|; A)$. The contrachain isomorphism $C^{\cdot} \otimes C'^{\cdot} \cong (C \otimes C')^{\cdot}$ is a special case of a contrachain map which will prove useful (in chapter 9) in a more general context and we now give the (general) definition.

Let C, C' be chain complexes, let G, G', A be abelian groups and let $\eta : G \otimes G' \to A$ be a homomorphism. Then

$$C \pitchfork G, \quad C' \pitchfork G', \quad (C \pitchfork G) \otimes (C' \pitchfork G'), \quad (C \otimes C') \pitchfork A$$

are contrachain complexes; we define a left homomorphism

$$\pi : (C \pitchfork G) \otimes (C' \pitchfork G') \to (C \otimes C') \pitchfork A$$

by

5.7.27 $$(c \otimes c', \pi(u \otimes u')) = ((c, u) \otimes (c', u'))\,\eta,$$
$$c \in C,\ c' \in C',\ u \in C \pitchfork G,\ u' \in C' \pitchfork G'.$$

5.7.28 Proposition. *π is a contrachain map.*

It is clearly enough to prove this with $A = G \otimes G'$, $\eta = 1$. Then

$$\begin{aligned}(c \otimes c', \delta\pi(u \otimes u')) &= ((c \otimes c')\,\partial, \pi(u \otimes u')) = (c\partial \otimes c' + c\epsilon \otimes c'\partial', \pi(u \otimes u')) \\ &= (c\partial, u) \otimes (c', u') + (c\epsilon, u) \otimes (c'\partial', u') \\ &= (c, \delta u) \otimes (c', u') + (c, \epsilon u) \otimes (c', \delta' u') \\ &= (c \otimes c', \pi(\delta u \otimes u' + \epsilon u \otimes \delta' u') = (c \otimes c', \pi\delta(u \otimes u')).\ \rrbracket\end{aligned}$$

5.7.29 Proposition. *If C, C' are free fg-complexes and $\eta : G \otimes G' \to A$ is an isomorphism, then $\pi : (C \pitchfork G) \otimes (C' \pitchfork G') \to (C \otimes C') \pitchfork A$ is an isomorphism of contrachain complexes.*

Let C_p have rank m_p and C'_q have rank m'_q. Then π maps

$$(C_p \pitchfork G) \otimes (C'_q \pitchfork G') \quad \text{to} \quad (C_p \otimes C'_q) \pitchfork A$$

and $C_p \otimes C'_q$ has rank $m_p m'_q$. If $\sum_{i=1}^{m_p} G_i$ is the sum of m_p copies of G and $\sum_{j=1}^{m'_q} G'_j$, $\sum_{k=1}^{m_p} \sum_{l=1}^{m'_q} A_{kl}$ are similarly defined, then π may be regarded as mapping $G_i \otimes G'_j$ to A_{ij} and, so regarded, is nothing other than the isomorphism η. ▌

If R is a ring and $\eta : R \otimes R \to R$ is the homomorphism (of additive groups)‡ given by

$$(r_1 \otimes r_2)\eta = r_1 r_2,$$

then π is a contrachain map

5.7.30 $$\pi : (C \pitchfork R) \otimes (C' \pitchfork R) \to (C \otimes C') \pitchfork R.$$

In particular, if $R = J$ then η is an isomorphism, so that, by 5.7.29, π is a contrachain isomorphism when C and C' are free *fg*-complexes; we have

5.7.31 $$\pi : C^{\cdot} \otimes C'^{\cdot} \cong (C \otimes C')^{\cdot}.$$

We combine this with 5.7.26 to obtain

5.7.12*c* Theorem. *If C, C' are free fg-complexes then*

$$H^p((C \otimes C')^{\cdot}) \cong \sum_{m+n=p} H^m(C^{\cdot}) \otimes H^n(C'^{\cdot}) \oplus \sum_{m+n=p+1} H^m(C^{\cdot}) * H^n(C'^{\cdot}). \;▌$$

Theorem 5.7.12*c* has a Corollary 5.7.19*c* for compact $|K|$, $|L|$ which is easily formulated by consulting 5.7.19. We deduce from 5.7.31

5.7.16*c* Corollary. *If C, C' are free fg-complexes and H or H' is torsion-free, then*

$$\pi^* : H^*(C^{\cdot}) \otimes H^*(C'^{\cdot}) \cong H^*((C \otimes C')^{\cdot}). \;▌$$

5.7.18*c* Theorem. *If C, C' are free fg-complexes, and q is a prime, then*

$$\pi^* : H^*(C \pitchfork J_q) \otimes H^*(C' \pitchfork J_q) \cong H^*((C \otimes C') \pitchfork J_q). \;▌$$

Finally, we record

5.7.22*c* Theorem. *If $|K|$, $|L|$ are polyhedra, then*

$$H^p(|K| \times |L|; G) \cong \sum_{m+n=p} H^m(|K|; H^n(|L|; G)). \;▌$$

5.7.25*c* Theorem. *If $\mathscr{F}$ is a field and $|K|$, $|L|$ are compact,*

$$H^p(|K| \times |L|; \mathscr{F}) \cong \sum_{m+n=p} H^m(|K|; \mathscr{F}) \otimes_{\mathscr{F}} H^n(|L|; \mathscr{F}). \;▌$$

‡ Actually η is a *ring* homomorphism if R is commutative; we exploit this fact in chapter 9.

5.8 Appendix 1: Applications of the Hopf Trace Theorem

We return to topology in this appendix to draw interesting consequences from the Hopf Trace Theorem (5.1.18).

Let $|K|$ be a compact polyhedron and let $f : |K| \to |K|$ be a map of $|K|$ into itself. Then $\sum_0^\infty (-1)^p \operatorname{tr}({}_w f_{*p})$ is called the *Lefschetz number* of f and written Λ_f. In this definition ${}_w f_{*p}$ operates on ${}_w H_p(K)$. The main consequence of Theorem 5.1.18 is

5.8.1 Theorem. *If f has no fixed points, then $\Lambda_f = 0$.*

Let ρ be a metric on $|K|$; since $|K|$ is compact and for all $x \in |K|$, $x \neq xf$, it follows that $\delta = \text{g.l.b. } \rho(x, xf) > 0$. Subdivide K to $K^{(r)}$ so that $\mu(K^{(r)}) < \frac{1}{3}\delta$, and let $v : K^{(q)} \to K^{(r)}$, $q \geqslant r$, be a simplicial approximation to f, inducing $\phi : C(K^{(q)}) \to C(K^{(r)})$. If

$$\psi = \phi\chi^{(q-r)} : C(K^{(q)}) \to C(K^{(q)}),$$

then $f_* \chi_*^{(q)} = \chi_*^{(q)} \psi_*$, by 3.6.2. Thus

$$\sum_0^\infty (-1)^p \operatorname{tr}({}_w\psi_{*p}) = \sum_0^\infty (-1)^p \operatorname{tr}({}_w f_{*p}) = \Lambda_f.$$

Since $\rho(x|v|, xf) < \frac{1}{3}\delta$, it follows that $\rho(x|v|, x) > \frac{2}{3}\delta$. Let $s \in K^{(q)}$ and let $t \in K^{(r)}$ be its carrier. Then for any $x \in \bar{s}$, $y \in \bar{t}$, $\rho(x, y) < \frac{1}{3}\delta$, so that $\rho(x|v|, y) > \frac{1}{3}\delta$. This implies that $sv \neq t$. From this it follows that, for each σ in $K^{(q)}$, $\sigma\psi$ is a chain in which σ appears with coefficient 0. Thus $\operatorname{tr} \psi_p = 0$ for each p, so that $\sum_0^\infty (-1)^p \operatorname{tr}(\psi_p) = 0$. By the Hopf Trace Theorem, $\sum_0^\infty (-1)^p \operatorname{tr}({}_w\psi_{*p}) = 0$, whence $\Lambda_f = 0$. ∎

5.8.2 Corollary. *If f is homotopic to a map without fixed points, then $\Lambda_f = 0$.*

For if $f \simeq g$ then $\Lambda_f = \Lambda_g$. ∎

The converse of 5.8.2 is false. For if $|K|$ is a figure 8 and f maps the top loop by a reflexion in a vertical line and winds the bottom loop on itself with degree 2, then $\Lambda_f = 0$, but every map homotopic to f has a fixed point in the top loop.

We now draw consequences from Theorem 5.8.1.

5.8.3 Theorem (*Brouwer Fixed Point Theorem*). *If E^n is a closed n-cell* (i.e. *a homeomorph of a closed n-simplex $\bar{s}_n$*), *then every map of E^n into itself has a fixed point.*

Now $H_*(E^n) \cong H_*(\bar{s}_n)$; but $H_0(\bar{s}_n) = J$, $H_p(\bar{s}_n) = 0$, $p > 0$. If $f: E^n \to E^n$ is a map, then any chain map inducing f_* is proper, so that f_{*0} is the identity automorphism. In this case ${}_wH_* = H_*$ so that $\Lambda_f = 1$, and f has a fixed point. ∎

Let $f: S^n \to S^n$ be a map. Then $H_n(S^n) = J$, generated by a, say, and $af_* = da$ for some integer d. We call d the *degree* of f. It depends, of course, on the homotopy class of f; and, conversely, one may prove (but not easily) that two maps $S^n \to S^n$ are homotopic if they have the same degree.

5.8.4 Theorem. *If $f: S^n \to S^n$ has no fixed points it is of degree $(-1)^{n+1}$.*

For it is easy to compute Λ_f if f has degree d, namely, $\Lambda_f = 1 + (-1)^n d$. Thus if f has no fixed points $\Lambda_f = 0$ and $d = (-1)^{n+1}$. ∎

5.8.5 Corollary. *If n is even, the identity map of S^n to itself cannot be deformed to a map without fixed points.*

For if $i: S^n \to S^n$ is the identity map and can be deformed to a map without fixed points, then $\Lambda_i = 0$ and i has degree $(-1)^{n+1}$; but this is false if n is even, since i has degree 1. ∎

Notice that the antipodal map $S^n \to S^n$ given by

$$(u_1, \ldots, u_{n+1}) \to (-u_1, \ldots, -u_{n+1})$$

is of degree $(-1)^{n+1}$ and is without any fixed points.

A *continuous vector field* over S^n is a set of vectors lying in R^{n+1}, tangential to S^n and one at each point of S^n and such that the length and direction of the vector at $x \in S^n$ vary continuously with x.

5.8.6 Theorem. *There exists a continuous field of unit vectors over S^n if and only if n is odd.*

First, let $n = 2m-1$. Then if $x = (u_1, \ldots, u_{2m}) \in S^{2m-1} \subseteq R^{2m}$, let $v(x)$ be the unit vector $(-u_{m+1}, \ldots, -u_{2m}, u_1, \ldots, u_m)$. Then $v(x)$ is orthogonal to the radius vector at x and varies continuously with x. Thus v is a continuous field of unit vectors over S^{2m-1}.

Now let n be even. We shall show that every continuous vector field over S^n possesses a zero vector. Let v be a vector field over S^n, let y be the end point of the vector $v(x)$ and let Oy, where O is the centre of S^n, meet S^n in xf. Then f is a continuous map $S^n \to S^n$. Let y_t divide xy in the ratio $1-t : t$ and let Oy_t cut S^n in xf_t. Then $f_t: S^n \to S^n$ is a homotopy between i and $f: S^n \to S^n$. Thus $f \simeq i$, whence, by Corollary 5.8.5, f has a fixed point. For such a point x, $v(x)$ clearly has zero length. ∎

5.9 Appendix 2: The group Ext (A, B)

Let $\ldots A_0 \xrightarrow{\alpha_0} A_1 \xrightarrow{\alpha_1} A_2 \xrightarrow{\alpha_2} A_3 \xrightarrow{\alpha_3} A_4 \rightarrow \ldots$ be an exact sequence of abelian groups. Then A_2 contains a subgroup $A_1\alpha_1 \cong A_1/A_0\alpha_0$ and $A_2/A_1\alpha_1 \cong \alpha_3^{-1}(0)$. Thus A_2 is in some sense an extension‡ of $\alpha_3^{-1}(0)$ by $A_1/A_0\alpha_0$, i.e. of the kernel of α_3 by the cokernel of α_0. Thus if we know α_0 and α_3 we know A_2 'up to a group extension'. This situation is often encountered in algebraic topology (for example, we may know $\lambda_* : H_*(K_0) \rightarrow H_*(K)$ and seek $H_*(K, K_0)$) and thus attention is directed to the problem of classifying extensions of a given factor group by a given subgroup.

Let A, B be given and let $0 \rightarrow B \xrightarrow{\lambda} E \xrightarrow{\mu} A \rightarrow 0$ be an exact sequence of abelian groups; we then say that E, and also the sequence itself, is an *extension*§ *of A by B*. If $0 \rightarrow B \xrightarrow{\lambda'} E' \xrightarrow{\mu'} A \rightarrow 0$ is another extension and $\phi : E \rightarrow E'$ is an isomorphism inducing the identity on A and B, i.e. such that

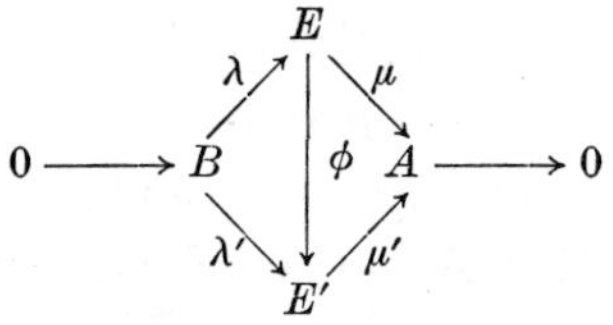

is commutative, we shall say that E and E' are *equivalent* extensions. We shall use the fact, deducible from the 5-lemma, that the commutativity alone implies that ϕ is an isomorphism. We write Ext (A, B) for the set of equivalence classes of extensions of A by B.

5.9.1 Theorem. *There exists a* $(1, 1)$ *correspondence between the members of* Ext (A, B) *and* $A \dagger B$.

Let $0 \rightarrow R \xrightarrow{\xi} F \xrightarrow{\eta} A \rightarrow 0$ be a presentation of A. Then $A \dagger_\eta B$ is the cokernel of $\xi^{\cdot} : F \pitchfork B \rightarrow R \pitchfork B$. We identify $A \dagger B$ with $A \dagger_\eta B$.

We first define a mapping $\Phi : A \dagger B \rightarrow \text{Ext}(A, B)$. Given $\gamma : R \rightarrow B$, let D be the subgroup of $F \oplus B$ consisting of elements $(r\xi, -r\gamma)$, $r \in R$, and let $E_\gamma = (F \oplus B)/D$. Define $\lambda : B \rightarrow E_\gamma$ by $b\lambda = \{0, b\}$ and $\mu : E_\gamma \rightarrow A$ by $\{f, b\}\mu = f\eta$, where $\{f, b\}$ is the coset of D containing (f, b), $f \in F$, $b \in B$. Then μ is single-valued since $\xi\eta = 0$ and the sequence

$$0 \rightarrow B \xrightarrow{\lambda} E_\gamma \xrightarrow{\mu} A \rightarrow 0 \tag{5.9.2}$$

is exact. To prove this, observe first that if $b\lambda = 0$, then $r\xi = 0$, $-r\gamma = b$ for some $r \in R$. But then $r = 0$, so $b = 0$ and the sequence is therefore exact at B. Next μ is onto A since η is onto A. Finally, let $\{f, b\}\mu = 0$. Then $f\eta = 0$, $f = r\xi$ and $\{f, b\} = \{0, b + r\gamma\} = (b + r\gamma)\lambda$ so that $\mu^{-1}(0) \subseteq B\lambda$. Since $\lambda\mu = 0$, the exactness of 5.9.2 is proved.

Now suppose $\gamma' : R \rightarrow B$ represents the same element of $A \dagger B$ as γ. Then $\gamma - \gamma' = \xi^{\cdot}\beta = \xi\beta$ for some $\beta : F \rightarrow B$. Construct $0 \rightarrow B \xrightarrow{\lambda'} E_{\gamma'} \xrightarrow{\mu'} A \rightarrow 0$ as in

‡ This use of the word extension (which is standard) may seem to some readers to be a solecism; for here the extension does not contain the group extended as a subgroup.

§ Or *abelian* extension, for emphasis.

5.9.2 and write $\{f, b\}'$ for a typical element of $E_{\gamma'}$. Define $\phi : E_\gamma \to E_{\gamma'}$ by $\{f, b\}\phi = \{f, b + f\beta\}'$. Then ϕ is single-valued since, for $r \in R$,

$$-r\gamma + r\xi\beta = -r\gamma + r(\gamma - \gamma') = -r\gamma'.$$

Moreover, $$b\lambda\phi = \{0, b\}\phi = \{0, b\}' = b\lambda',$$

and $$\{f, b\}\phi\mu' = \{f, b + f\beta\}'\mu' = f\eta = \{f, b\}\mu.$$

Thus E_γ and $E_{\gamma'}$ are equivalent, so that $\gamma \to E_\gamma$ induces a mapping

$$\Phi : A \dagger B \to \operatorname{Ext}(A, B).$$

We now define a mapping $\Psi : \operatorname{Ext}(A, B) \to A \dagger B$. Given $0 \to B \xrightarrow{\lambda} E \xrightarrow{\mu} A \to 0$, we may, using the fact that F is free, 'lift' the identity map of A and obtain a commutative diagram

5.9.3
$$\begin{array}{ccccccccc}
0 & \longrightarrow & R & \xrightarrow{\xi} & F & \xrightarrow{\eta} & A & \longrightarrow & 0 \\
 & & \downarrow{\scriptstyle\rho} & & \downarrow{\scriptstyle\sigma} & & \downarrow{\scriptstyle 1} & & \\
0 & \longrightarrow & B & \xrightarrow{\lambda} & E & \xrightarrow{\mu} & A & \longrightarrow & 0
\end{array}$$

We may alter σ by any $\kappa\lambda$, $\kappa : F \to B$; but then we alter ρ by $\xi\kappa$ so that the element of $A \dagger B$ represented by ρ is uniquely determined. If E' is equivalent to E and $\phi : E \cong E'$ is the equivalence, then we may lift $1 : A \to A$ to $\sigma\phi : F \to E'$ and $\rho\lambda' = \xi\sigma\phi$. Thus E' yields the same element of $A \dagger B$, so that $E \to \rho$ induces $\Psi : \operatorname{Ext}(A, B) \to A \dagger B$.

We show that Φ and Ψ are mutual inverses. Given $\gamma : R \to B$, define $\sigma : F \to E_\gamma$ by $f\sigma = \{f, 0\}$. Then $f\sigma\mu = f\eta$, so that σ 'lifts' $1 : A \to A$. Also $r\xi\sigma = \{r\xi, 0\}$, $r\gamma\lambda = \{0, r\gamma\}$, so that $r\xi\sigma = r\gamma\lambda$ and σ determines $\gamma : R \to B$. This shows that $\Phi\Psi = 1$.

Now consider 5.9.3 and form $0 \to B \xrightarrow{\lambda_\rho} E_\rho \xrightarrow{\mu_\rho} A \to 0$. Define $\phi : E_\rho \to E$ by $\{f, b\}\phi = f\sigma + b\lambda$. Then ϕ is single-valued since $\xi\sigma = \rho\lambda$; also $b\lambda_\rho\phi = b\lambda$ and $\{f, b\}\phi\mu = f\sigma\mu = f\eta = \{f, b\}\mu_\rho$. Thus E and E_ρ are equivalent so that $\Psi\Phi = 1$. ∎

By means of Ψ and Φ we may induce a group operation in $\operatorname{Ext}(A, B)$. Thus, for $\alpha, \beta \in \operatorname{Ext}(A, B)$, we define $\alpha + \beta = (\alpha\Psi + \beta\Psi)\Phi$. This definition ensures that Φ and Ψ are isomorphisms. On the other hand, the group $\operatorname{Ext}(A, B)$ has been discussed by algebraists independently of the isomorphism with $A \dagger B$. The standard law of addition is as follows: given $0 \to B \xrightarrow{\lambda_i} E_i \xrightarrow{\mu_i} A \to 0$, $i = 1, 2$, let D be the subgroup of $E_1 \oplus E_2$ consisting of pairs (e_1, e_2) with $e_1\mu_1 = e_2\mu_2$ and let D_0 be the subgroup of D consisting of pairs $(b\lambda_1, -b\lambda_2)$. Define $E = D/D_0$ and $\lambda : B \to E$, $\mu : E \to A$ by $b\lambda = \{b\lambda_1, 0\}$, $\{e_1, e_2\}\mu = e_1\mu_1$. Then

$$0 \to B \xrightarrow{\lambda} E \xrightarrow{\mu} A \to 0$$

is exact. For if $b\lambda = 0$, then there exists $b' \in B$ with $b\lambda_1 = b'\lambda_1$, $b'\lambda_2 = 0$; but then $b = b'$, $b' = 0$, so that $b = 0$. Next, μ is onto A since μ_1, μ_2 are onto A. Finally, $\lambda\mu = 0$ and, conversely, if $\{e_1, e_2\}\mu = 0$, then

$$e_1 = b_1\lambda_1, \quad e_2 = b_2\lambda_2, \quad \{e_1, e_2\} = \{b_1\lambda_1, b_2\lambda_2\} = \{b_1\lambda_1 + b_2\lambda_1, 0\} = (b_1 + b_2)\lambda.$$

Given any extension $0 \to A \xrightarrow{\lambda} E \xrightarrow{\mu} B \to 0$, we shall write $[E]$ for the equivalence class of the extension. We now show that, for the extension just defined, $[E]$ depends only on $[E_1]$ and $[E_2]$. Given

$$\begin{array}{ccccccc} & & & E_i & & & \\ & & \lambda_i \nearrow & & \searrow \mu_i & & \\ 0 \longrightarrow & B & & \downarrow \phi_i & & A & \longrightarrow 0, \\ & & \lambda_i' \searrow & & \nearrow \mu_i' & & \\ & & & E_i' & & & \end{array}$$

$i = 1, 2$, define $\phi : E \to E'$ by $\{e_1, e_2\}\phi = \{e_1\phi_1, e_2\phi_2\}'$—the notations should be obvious. Then ϕ is single-valued since $b\lambda_1\phi_1 = b\lambda_1'$, $b\lambda_2\phi_2 = b\lambda_2'$, and maps to E' since $e_1\phi_1\mu_1' = e_1\mu_1 = e_2\mu_2 = e_2\phi_2\mu_2'$. Moreover, $b\lambda\phi = \{b\lambda_1\phi_1, 0\}' = b\lambda'$ and $\{e_1, e_2\}\phi\mu' = e_1\phi_1\mu_1' = e_1\mu_1 = \{e_1, e_2\}\mu$. We define

5.9.4 $$[E_1] + [E_2] = [E]$$

and show that this definition agrees with that given in the previous paragraph by means of Φ and Ψ. To show the definitions agree it is clearly sufficient to demonstrate

5.9.5 Lemma. $[E_1]\Psi + [E_2]\Psi = [E]\Psi$.

Given

$$\begin{array}{ccccccccc} 0 & \longrightarrow & R & \xrightarrow{\xi} & F & \xrightarrow{\eta} & A & \longrightarrow & 0 \\ & & \downarrow \rho_i & & \downarrow \sigma_i & & \downarrow 1 & & \\ 0 & \longrightarrow & B & \xrightarrow{\lambda_i} & E_i & \xrightarrow{\mu_i} & A & \longrightarrow & 0 \end{array}$$

$i = 1, 2$, ρ_i represents $[E_i]\Psi$ and $\rho_1 + \rho_2$ represents $[E_1]\Psi + [E_2]\Psi$. Define $\sigma : F \to E$ by $f\sigma = \{f\sigma_1, f\sigma_2\}$. Then σ maps F into E since $f\sigma_1\mu_1 = f\sigma_2\mu_2 = f\eta$; for the same reason $f\sigma\mu = f\eta$, so that σ 'lifts' back $1 : A \to A$. Also

$$r\xi\sigma = \{r\xi\sigma_1, r\xi\sigma_2\} = \{r\rho_1\lambda_1, r\rho_2\lambda_2\} = \{r\rho_1\lambda_1 + r\rho_2\lambda_1, 0\} = r(\rho_1 + \rho_2)\lambda.$$

Thus σ determines $\rho_1 + \rho_2 : R \to B$, so that $[E]\Psi$ is represented by $\rho_1 + \rho_2$. ▌

We point out that there are technical advantages in inducing a group structure in $\operatorname{Ext}(A, B)$ by means of Φ and Ψ. For, if we had simply started from 5.9.4 we should have met considerable difficulty in establishing the associative law and the existence of inverses. Moreover, the group $A \dagger B$ is often easier to compute than $\operatorname{Ext}(A, B)$.

The zero element of $\operatorname{Ext}(A, B)$ is identified as the Φ-image of $0 \in A \dagger B$. Now we shall say that the extension $0 \to B \xrightarrow{\lambda} E \xrightarrow{\mu} A \to 0$ is *trivial* if there exists $\bar{\mu} : A \to E$ with $\bar{\mu}\mu = 1$. We shall show that the trivial extensions form a class, indeed the zero of $\operatorname{Ext}(A, B)$. We first recall from 5.1.9′ that E is a trivial extension if and only if $B\lambda$ is a direct factor in E.

5.9.6 Proposition. *If E is trivial, then E is equivalent to E' if and only if E' is trivial.*

If $E = B\lambda \oplus A\bar{\mu}$, $E' = B\lambda' \oplus A\bar{\mu}'$, define $\phi : E \to E'$ by $(b\lambda, a\bar{\mu})\phi = (b\lambda', a\bar{\mu}')$. Then ϕ is obviously an equivalence. Conversely, given $\bar{\mu} : A \to E$ with $\bar{\mu}\mu = 1$ and an equivalence $\phi : E \to E'$, define $\bar{\mu}' : A \to E'$ by $\bar{\mu}' = \bar{\mu}\phi$. Then

$$\bar{\mu}'\mu' = \bar{\mu}\phi\mu' = \bar{\mu}\mu = 1. \;▌$$

Thus we may speak of the equivalence class of trivial extensions.

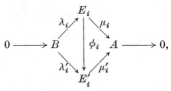

5.9.7 Proposition. *The class of trivial extensions is the zero of* $\operatorname{Ext}(A, B)$.

Given

$$\begin{array}{ccccccccc}
0 & \longrightarrow & R & \xrightarrow{\ \xi\ } & F & \xrightarrow{\ \eta\ } & A & \longrightarrow & 0 \\
 & & \Big\downarrow{\scriptstyle\rho} & & \Big\downarrow{\scriptstyle\sigma} & & \Big\downarrow{\scriptstyle 1} & & \\
0 & \longrightarrow & B & \xrightarrow{\ \lambda\ } & E & \underset{\bar{\mu}}{\overset{\mu}{\rightleftarrows}} & A & \longrightarrow & 0
\end{array}$$

where $\bar{\mu}\mu = 1$, let $\kappa : E \to B$ be the projection so that $\lambda\kappa = 1$; this exists since $B\lambda$ is a direct factor in E. Then $\xi\sigma\kappa = \rho\lambda\kappa = \rho$, so that $\sigma\kappa$ extends ρ and ρ represents $0 \in A \dagger B$. Thus the Ψ-image of the class of trivial extensions is zero, which establishes the proposition. ∎

The reader may find it instructive to interpret the homomorphisms of the Ext groups induced by homomorphisms $A \to A'$ or $B \to B'$.

5.10 Appendix 3: Lens spaces

For each pair of coprime integers, p, q $(p > 0)$ there is a 3-dimensional manifold $L(p, q)$; these manifolds are called Lens Spaces. The reader will find an attractive account of these spaces in Seifert and Threlfall [6], §§ 60, 61, 77. We may define $L(p, q)$ as follows. Take a closed region in R^2 bounded by a regular

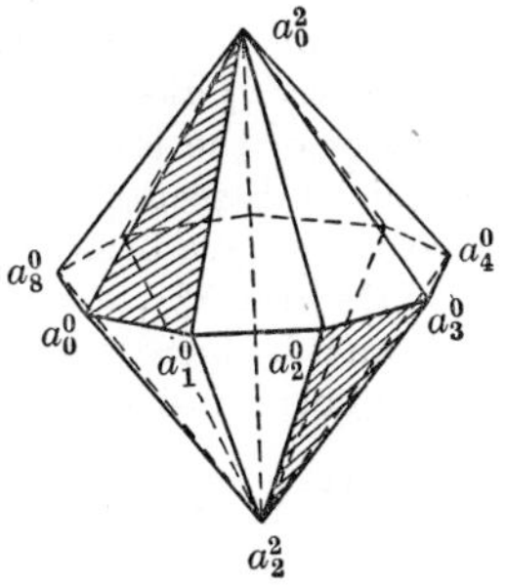

Fig. 5.1

p-sided polygon $a_0^0, a_1^0, \ldots, a_{p-1}^0$; we shall regard the subscript in a_i^0 as being an element of J_p, so that a_i^0 is defined for all i. Join this region to two points of R^3, a_0^2 and a_q^2, one on each side of R^2, to form a solid double pyramid P on the given polygonal base. The frontier of P is covered by triangles $a_i^0 a_{i+1}^0 a_j^2$, $j = 0$ or q; $L(p, q)$ is obtained from P by identifying certain points on its frontier. There is a unique linear order-preserving homeomorphism between $a_i^0 a_{i+1}^0 a_0^2$ and $a_{i+q}^0 a_{i+q+1}^0 a_q^2$; two points related under such a homeomorphism, $0 \leqslant i \leqslant p-1$, are to be identified. The sloping faces of one pyramid are matched with those of the other after a twist of $2\pi q/p$. The picture shows P for $p = 9$, $q = 2$, with a typical pair of identified faces shaded. The reader will be able to verify quickly that the identifications prescribed imply that the edges $a_i^0 a_{i+1}^0$ and $a_j^0 a_{j+1}^0$ are identified for all i, j; in fact any simplex is to be identified with any other simplex which can be obtained from itself by increasing all its subscripts (in J_p) by the same amount. Evidently if $(q - q')$ is divisible by p, $L(p, q) \cong L(p, q')$.

It is not immediately obvious that $L(p,q)$ is a manifold: clearly every point arising from an interior point of P has a Euclidean neighbourhood in $L(p,q)$, but some of the frontier points seem in danger of having more complicated neighbourhoods. There is, in fact, little difficulty in seeing that a frontier point of P not lying on the polygon has a Euclidean neighbourhood: this is composed of two 'hemispherical' neighbourhoods in P clapped together to form a Euclidean neighbourhood. Now P may be dissected like an orange into tetrahedra of the form $a_i^0 a_{i+1}^0 a_0^2 a_q^2$; using the identifications to be made on the faces of P these may be reassembled afresh to make a ring of tetrahedra round an axis formed from identifying all the edges $a_i^0 a_{i+1}^0$. This produces another double pyramid P', on whose faces identifications are to be made if the space $L(p,q)$ is to be recovered. It turns out, in fact, that the sloping faces of one pyramid are to be matched with those of the other after a twist of $2\pi q'/p$, where $qq' = 1$ (as elements of J_p). This process not only assures us that points on the polygon of P also have Euclidean neighbourhoods in L, but also that $L(p,q) \cong L(p,q')$, where $qq' = 1$. Reflexion (mental or in a mirror) shows that $L(p,q) \cong L(p,-q)$, so that $L(p,q) \cong L(p,q')$ if $q' = \pm q^{\pm 1}$ (as elements of J_p).

It has been shown that $q' = \pm q^{\pm 1}$ is not only sufficient but also necessary for $L(p,q)$ and $L(p,q')$ to be homeomorphic. Reidemeister‡ proved the condition necessary for the spaces to be combinatorially equivalent and a consequence of a result of Moise is that therefore the condition is necessary for homeomorphism. In what follows on Lens Spaces we shall be showing how contrahomology theory contributes to the problem of the *homotopy* classification of Lens Spaces, the classification by homotopy type.

We dissect $L(p,q)$ as a Ψ-complex K as follows. In P place a vertex a_i^1 at the midpoint of $a_i^0 a_{i+1}^0$ and a vertex a_0^3 at the centre of the regular polygon; P is then the union of tetrahedra of the forms $a_i^0 a_i^1 a_j^2 a_0^3$ and $a_{i+1}^0 a_i^1 a_j^2 a_0^3$, for $i \in J_p$ and $j=0$ or q. This simplicial dissection of P yields a pseudodissection K of $L(p,q)$ after the identifications have been made; the simplexes of K are the faces of the tetrahedra $\alpha_i^0 \alpha_i^1 \alpha_j^2 \alpha_0^3$ and $\alpha_{i+1}^0 \alpha_i^1 \alpha_j^2 \alpha_0^3$, where any face is identical with another face obtainable from it by increasing all its subscripts by the same amount. (We introduce the symbols α instead of a to distinguish between K and the dissection of P.)

The homology and contrahomology can be calculated from a block dissection, with a single block in each dimension. For $\bar{e}_3$ we take the whole of $|K|$; for $\bar{e}_2$ those closed simplexes not abutting on α_0^3; for $\bar{e}_1$ the pair of simplexes joining α^0 to α^1; and for e_0 the vertex α^0. We may so orient that $\epsilon_3 \partial = 0$, $\epsilon_2 \partial = p\epsilon_1$, $\epsilon_1 \partial = 0$. Therefore $H_r(L; J_p) \cong J_p$ for $0 \leqslant r \leqslant 3$, and $H^r(L; J_p) \cong J_p$ for $0 \leqslant r \leqslant 3$. Since also $H_1(L; J) \cong J_p$, the integer p is a homotopy invariant of $L(p,q)$; a necessary condition, therefore, that $L(p,q) \simeq L(p',q')$ is that $p = p'$.

5.10.1 Theorem. *A necessary condition that $L(p,q) \simeq L(p,q')$ is that qq' or $-qq'$ be a quadratic residue* mod p.

This means that, for some $m \in J_p$, $qq' = \pm m^2$; or, equivalently, for some m, $q' = \pm m^2 q$.

There is no real loss of generality in assuming that $0 < q < p$.

In the ordered pseudodissection K of $L(p,q)$ we define contrachains $c^1 \in C^1(K; J)$, $z^2 \in C^2(K; J)$ and $z^3 \in C^3(K; J)$ as follows:

(i) c^1 takes the value $v(i-j)$ on $\alpha_i^r \alpha_j^s$, where $v(k)$ is the non-negative integer $< p$ in the residue class mod p defined by k.

‡ *Abh. Hamb. Sem.* **11** (1935), 102.

(ii) z^2 takes the value zero on all simplexes except those of the form $\alpha_i^r\alpha_q^2\alpha_0^3$ $(r=0$ or 1; $0\leqslant i<q)$ or $\alpha_0^0\alpha_{p-1}^1\alpha_0^t$ $(t=2$ or $3)$; on these simplexes it takes the value 1. (Notice that the identity rules imply that z^2 also takes the value 1 on $\alpha_q^0\alpha_{q-1}^1\alpha_q^2$).

(iii) z^3 takes the value zero on all simplexes except $\alpha_1^0\alpha_0^1\alpha_q^2\alpha_0^3$, on which it takes the value 1.

We invite the reader to verify that $\delta c^1 = pz^2$ and $c^1 \cup z^2 \sim qz^3$. The former verification is somewhat long but automatic; the latter is short and amounts only to showing that $c^1 \cup z^2$ has zero value except on the q tetrahedra $\alpha_i^0\alpha_{i-1}^1\alpha_q^2\alpha_0^3$, $0\leqslant i<q$, on each of which it takes the value $+1$.

Let $c_{(p)}^1$, $z_{(p)}^2$, $z_{(p)}^3$ be the images of c^1, z^2, z^3 under the homomorphism $J \to J_p$; then, since $\delta c^1 = pz^2$, $\{z_{(p)}^2\}$ is the Bockstein contraboundary ν_* (see 5.6.8) of $\{c_{(p)}^1\}$. If $\{c_{(p)}^1\} = \xi^1 \in H^1(L(p,q); J_p)$ and $\xi^3 = \{z_{(p)}^3\} \in H^3(L; J_p)$, then we have shown that

$$\xi^1.(\xi^1\nu_*) = q\xi^3.$$

We observe now that ξ^1 generates $H^1(L; J_p) = J_p$. For

$$z_1 = \alpha_1^0\alpha_0^1 - \alpha_0^0\alpha_0^1 \in Z_1(L);$$

if $z_1^{(p)}$ is its image in $Z_1(L; J_p)$, $(z_1^{(p)}, c_{(p)}^1) = 1$, so that $(\{z_1^{(p)}\}, \xi^1) = 1$ and the order of ξ^1 is p; this proves that ξ^1 generates $H^1(L; J_p)$.

Next we prove that $\{z^3\}$ generates $H^3(L)$. For

$$z_3 = \sum_i (\alpha_{i+1}^0\alpha_i^1\alpha_q^2\alpha_0^3 - \alpha_i^0\alpha_i^1\alpha_q^2\alpha_0^3 - \alpha_{i+1}^0\alpha_i^1\alpha_0^2\alpha_0^3 + \alpha_i^0\alpha_i^1\alpha_0^2\alpha_0^3) \in Z_3(L);$$

but $(z_3, z^3) = 1$ so that $(\{z_3\}, \{z^3\}) = 1$ and $\{z^3\}$ generates $H^3(L) = J$.

Suppose now that $\tilde{L} = L(p,q')$ and that $f : L \to \tilde{L}$ is a homotopy equivalence; since f^* is isomorphic $f^*\{\tilde{z}^3\} = \pm\{z^3\}$ so that $f^*\tilde{\xi}^3 = \pm\xi^3$. Since ξ^1 generates $H^1(L; J_p)$, $f^*\tilde{\xi}^1 = m\xi^1$, for some $m \in J_p$. Then

$$f^*(q'\tilde{\xi}^3) = f^*(\tilde{\xi}^1.(\tilde{\xi}^1\nu_*)) = f^*\tilde{\xi}^1.f^*(\tilde{\xi}^1\nu_*) = f^*\tilde{\xi}^1.((f^*\tilde{\xi}^1)\nu_*)$$

(see the remark following 5.6.7c)

$$= m\xi^1.m\xi^1\nu_* = m^2q\xi^3 = \pm m^2qf^*\tilde{\xi}^3.$$

Therefore $q' = \pm qm^2$ for some $m \in J_p$. ∎

This theorem proves that, for instance, $L(5,1)$ and $L(5,2)$ are of different homotopy type.

J. H. C. Whitehead‡ has proved that this necessary condition is also sufficient for $L(p,q)$ and $L(p,q')$, to be of the same homotopy type; the homotopy classification of Lens spaces is therefore complete. It is perhaps interesting to observe that $L(7,1)$ and $L(7,2)$, for instance, are two non-homeomorphic 3-dimensional manifolds that are of the same homotopy type.

EXERCISES

1. Find the rank and torsion coefficients of the following abelian groups:

	generators	relations
(i)	a, b, c	$2a = 3b,\quad 2b = 3c,\quad 2c = 3a;$
(ii)	a, b	$12a = 0,\quad 6a = 15b;$
(iii)	a, b, c, d	$3a + 6b - 3c = 0,\quad 4a + 2b - 3d = 0.$

Find an appropriate set of generators in each case.

‡ *Ann. Math.* **42** (1941), 1197.

2. The chain complex (C, ∂) is defined as follows:

$C_0 = (c_0^1, c_0^2, c_0^3)$;

$C_1 = (c_1^1, c_1^2, c_1^3, c_1^4)$; $c_1^1\partial = c_0^2 - c_0^1$, $c_1^2\partial = 3c_0^3 - 3c_0^1$,
$c_1^3\partial = c_0^3 - c_0^1$, $c_1^4\partial = 3c_0^3 - 3c_0^1$;

$C_2 = (c_2^1, c_2^2, c_2^3, c_2^4, c_2^5)$; $c_2^1\partial = c_2^2\partial = 0$, $c_2^3\partial = 4c_1^2 - 36c_1^3 + 8c_1^4$,
$c_2^4\partial = 2c_1^2 - 18c_1^3 + 4c_1^4$, $c_2^5\partial = -9c_1^3 + 3c_1^4$;

$C_3 = (c_3^1, c_3^2, c_3^3)$; $c_3^1\partial = c_2^1 - 2c_2^2 + 2c_2^3 - 4c_2^4$,
$c_3^2 = c_2^1 - 7c_2^2 + 7c_2^3 - 14c_2^4$,
$c_3^3 = 5c_2^1 - 15c_2^2 + 15c_2^3 - 30c_2^4$;

$C_n = 0$, otherwise.
Calculate the homology groups of (C, ∂).

3. If the chain complex C is as in Q. 2, calculate the homology groups of $C \otimes G$ and the contrahomology groups of $C \pitchfork G$ when (i) $G = J_m$, (ii) $G = Q$, the group of rationals, (iii) $G = R_1$, the group of reals mod 1.

4. If $\overline{W}^p$ is the group of p-contracycles of $C^{\cdot}$ which are weakly contrahomologous to zero, prove that $\overline{W}^p$ is the annihilator of Z_p. [Here $C^{\cdot} = C \pitchfork J$.]

5. Prove that the tensor product $C = A \otimes B$ of two abelian groups A, B is characterized (up to isomorphism) by the following two properties:

(i) there is a bilinear map $\phi: A \times B \to C$ and C is generated by the ϕ-images of the elements of $A \times B$;

(ii) given any bilinear map $\theta: A \times B \to D$ there is a homomorphism $\psi: C \to D$ such that $\phi\psi = \theta$.

6. If a, b are elements of infinite order in A, B prove that $a \otimes b$ is of infinite order in $A \otimes B$.

7. Prove that the abelian group A is divisible if and only if every inclusion $A \subseteq B$ embeds A as a direct factor. What is the 'dual' property for free groups?

8. The subgroup A' of A is said to be *pure* if, whenever $a' \in A'$ and $a' = na_0$, $a_0 \in A$, then there exists $a_0' \in A'$ with $a' = na_0'$. Prove that the injection $A' \otimes B \to A \otimes B$ is a monomorphism for all B if and only if A' is a pure subgroup of A.

9. Prove that the injection $A' \otimes B \to A \otimes B$ is a monomorphism for all pairs (A, A'), where A' is a subgroup of A, if and only if B is locally free.

10. Let A, B be groups and let G be the group generated by symbols $(a, b)_m$ where $a \in A$, $b \in B$, $ma = 0$, $mb = 0$, subject to the two distributive relations and $(a, b)_{hk} = (ha, b)_k$, $hka = 0$, $kb = 0$; $(a, b)_{hk} = (a, hb)_k$, $ka = 0$, $hkb = 0$. Prove that $G \cong A * B$.

11. (i) Prove that $A \dagger B = 0$ for all A if and only if B is divisible.

(ii) Prove that $A \dagger B = 0$ for all B if and only if A is free.

(iii) Prove that if A is finitely generated then $A \dagger J = 0$ if and only if A is free.

12. (i) Show that $A * B \neq 0$ if A and B have J_p as subgroup.

(ii) Show that $A \dagger B \neq 0$ if A has J_p as subgroup and B has J_p as quotient group.

13. Establish the existence of a natural homomorphism

$$\kappa:(U\pitchfork V)\otimes W \to U\pitchfork(V\otimes W),$$

and prove that it is an isomorphism if U or W are free *fg* groups. Is it always an isomorphism?

14. Prove that $A*(B*C) \cong (A*B)*C$ and that

$$(A*B)\otimes C \oplus (A\otimes B)*C \cong A*(B\otimes C) \oplus A\otimes(B*C),$$

for any abelian groups A, B, C. Hence prove that

$$H_n(C\otimes C';G) \cong \sum_{p+q=n} H_p(C;H_q(C';G))$$

for any free chain complexes C, C'.

15. Prove that $(A*B)\dagger C \cong A\dagger(B\dagger C)$ and that

$$(A\otimes B)\dagger C \oplus (A*B)\pitchfork C \cong A\pitchfork(B\dagger C) \oplus A\dagger(B\pitchfork C),$$

for any abelian groups A, B, C. Hence prove that

$$H^n(C\otimes C';G) \cong \sum_{p+q=n} H^p(C;H^q(C';G))$$

for any free chain complexes C, C'.

16. Let

$$\begin{array}{ccccccccc}
 & & 0 & & 0 & & 0 & & \\
 & & \downarrow & & \downarrow & & \downarrow & & \\
0 & \to & A_1 & \to & A_2 & \to & A_3 & \to & 0 \\
 & & \downarrow & & \downarrow & & \downarrow & & \\
0 & \to & B_1 & \to & B_2 & \to & B_3 & \to & 0 \\
 & & \downarrow & & \downarrow & & \downarrow & & \\
0 & \to & C_1 & \to & C_2 & \to & C_3 & \to & 0 \\
 & & \downarrow & & \downarrow & & \downarrow & & \\
 & & 0 & & 0 & & 0 & &
\end{array}$$

be a commutative diagram in which each row and column is an exact sequence of differential groups. Then there are defined (see 5.5.1) homomorphism

$$\nu^C_*:H(C_3)\to H(C_1),\ \nu^1_*:H(C_1)\to H(A_1),\ \nu^3_*:H(C_3)\to H(A_3),\ \nu^A_*:H(A_3)\to H(A_1).$$

Prove that $\nu^C_*\nu^1_* = -\nu^3_*\nu^A_*$.

17. Let π be a group of *fixpoint-free* homeomorphisms of S^n; that is, if $\xi, \eta \in \pi$, and $x \in S^n$,

$$\xi\cdot x \in S^n,$$
$$\xi\cdot(\eta\cdot x) = (\xi\eta)\cdot x,$$
$$\xi\cdot x = x \quad \text{if and only if } \xi = 1.$$

Show that, if n is even, π contains at most 2 elements.

18. Deduce the exactness of the Mayer–Vietoris sequence of a simplicial triad $(K; K_1, K_2)$ from 5.5.1 (see chap. 2, Exercises, Q. 7).

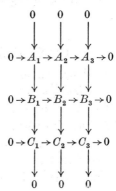

6

THE FUNDAMENTAL GROUP AND COVERING SPACES

In this chapter we define a *group* (not an *abelian group*) associated with any topological space X and any base point $x_0 \in X$; it is due to Poincaré and will be called the *fundamental group of* X *on* x_0. It differs in many ways from a homology group as so far defined: it is not in general commutative; it is defined for any space; it is from its definition evidently a topological invariant; and it arises from the ideas of homotopy. This group is usually written as $\pi_1(X, x_0)$ and is the special case, for $n=1$, of the homotopy group $\pi_n(X, x_0)$. It has, however, a distinctive character among homotopy groups which entitles it to consideration on its own.

6.1 Definitions of the fundamental group

We start with some definitions. Given $r \geqslant 0$ let I_r be the closed segment $[0, r]$, $0 \leqslant u \leqslant r$, and let $\dot{I}_r = 0 \cup r$. A *path*‡ *in* X is a pair§ (f, r), where f is a map $I_r \to X$. The points $(0)f$, $(r)f$ are called the *initial* and *final* points of f; r may be called the *length*‖ of the path and we often use f instead of (f, r) to stand for a path. The set of paths in X has a natural multiplicative structure; if the final point of f is the initial point of g, we formalize the notion of the path consisting of f followed by g as follows:

6.1.1 Definition. The *product* $(f * g, r+s)$ of (f, r) and (g, s) is defined if $(r)f = (0)g$ and is given by the rule

$$\begin{aligned}(u)f*g &= (u)f, &&\text{if } 0 \leqslant u \leqslant r,\\ &= (u-r)g, &&\text{if } r \leqslant u \leqslant r+s.\end{aligned}$$

(We invoke I. 1.6 to establish this as a path.)

‡ Some authors distinguish between a *path*, a continuous mapping of I_r, and an *arc*, a homeomorphism.

§ The authors owe to J. C. Moore their awareness of this approach to the fundamental group; he is, however, not to be held responsible for the details of the treatment. The rival and more usual approach deals only with paths of length 1.

‖ Some readers may prefer to think in terms of *duration* rather than length.

6.1.2 Proposition. *If $f*g$ and $g*h$ are defined, $(f*g)*h$ and $f*(g*h)$ are defined and are equal.* ∎

We shall in this case write $f*g*h$ for either.

6.1.3 Definition. The *reverse* $(\bar{f}, r)$ of (f, r) is given by the rule

$$(u)\bar{f} = (r-u)f, \quad 0 \leqslant u \leqslant r.$$

6.1.4 Proposition. $\overline{f*g} = \bar{g}*\bar{f}$ *and* $\bar{\bar{f}} = f$. ∎

Given a path $f = (f, r)$, let $f_{(s)} = (f_{(s)}, s)$, $s > 0$, be the path given by

6.1.5 $$uf_{(s)} = \frac{ru}{s}f.$$

Thus $f_{(s)}$ is a stretched or shrunken version of f. Notice that $(f_{(k)})_{(l)} = f_{(l)}$.

For each point $x \in X$ there is a path $e(x) = e$ of length 0 at x; then $e_{(s)}(x) = e_{(s)}$ is the constant path of length s at x. Clearly, if the products are defined, $e*f = f$ and $f*e = f$, and $e_{(r)}*e_{(s)} = e_{(r+s)}$.

We shall be mainly, but not exclusively, interested in *closed paths* or *loops*, that is those paths whose initial and final points coincide; if this common end point is x_0, we say that this is a closed path *on* x_0. Clearly for any two loops f, g on x_0, $f*g$ is defined; the set $\Omega(X, x_0)$ of those loops forms, indeed, a semi-group.

The fundamental group $\pi_1(X, x_0)$ arises from $\Omega(X, x_0)$ by dividing its elements into equivalence classes, using an equivalence relation corresponding roughly to the relation of being homotopic. Since, however, paths of different lengths are maps into X of different spaces, the standard definition of homotopy will not be satisfactory; we shall, for instance, want to consider as equivalent the paths f and $f_{(s)}$.

We list some elementary facts about homotopies rel $\dot{I}_r$; we shall often contract 'homotopic rel $\dot{I}_r$' to 'homotopic' and the symbol

$$\text{'}f \simeq g \operatorname{rel} \dot{I}_r\text{'} \quad \text{to} \quad \text{'}f \simeq g\text{'}.$$

6.1.6 Proposition. *If f, g are paths in X of length r and if $f \simeq g \operatorname{rel} \dot{I}_r$ then $f_{(s)} \simeq g_{(s)} \operatorname{rel} \dot{I}_s$.*

For if $f \overset{F}{\simeq} g$, then $f_{(s)} \overset{F_{(s)}}{\simeq} g_{(s)}$, where

$$(u, t)F_{(s)} = (ru/s, t)F, u \in I_s, t \in I. \ ∎$$

Since $(f_{(k)})_{(l)} = f_{(l)}$, it follows that $f_{(k)} \simeq g_{(k)}$ if and only if $f_{(l)} \simeq g_{(l)}$.

6.1.7 Proposition. *Let f, f' be homotopic paths in X of length r, and g, g' homotopic paths of length s. Then* (i) *if $f*g$ is defined, $f'*g'$ is defined and $f*g \simeq f'*g'$*; (ii) *the reverse paths $\bar{f}, \bar{f}'$ are homotopic.*

We have $f \overset{F}{\simeq} f'$ rel $\dot{I}_r$, $g \overset{G}{\simeq} g'$ rel $\dot{I}_s$. Then $0g' = 0g = rf = rf'$ so that $f' * g'$ is defined, and, in an obvious notation,

$$f*g \overset{F*G}{\simeq} f'*g' \operatorname{rel} \dot{I}_{r+s},$$
$$\bar{f} \overset{\bar{F}}{\simeq} \bar{f}' \operatorname{rel} \dot{I}_r. \;\blacksquare$$

6.1.8 Proposition. *If f is a path in X of length r, then*

$$e_{(s)} * f \simeq f_{(r+s)} \simeq f * e_{(s)}.$$

If $r = 0$, then $e_{(s)} * f = f_{(r+s)} = f * e_{(s)}$ $(= e_{(s)})$ so we may assume $r > 0$. Consider then the homotopy $\Lambda : I_{r+s} \times I \to I_{r+s}$ given by

$$\begin{aligned}(u,t)\Lambda &= (r+st)\,u/r, \quad 0 \leqslant u \leqslant r,\\ &= u(1-t)+(r+s)t, \quad r \leqslant u \leqslant r+s.\end{aligned}$$

The reader will easily verify that $f_{(r+s)} \overset{\Lambda f_{(r+s)}}{\simeq} f * e_{(s)}$, rel $\dot{I}_{r+s}$. The other assertion is proved similarly (or by reversing paths). $\blacksquare$

6.1.9 Definition. A loop (f, r) on x_0 is *null-homotopic* if it is homotopic rel $\dot{I}_r$ to $e_{(r)}$. The set of all null-homotopic loops on x_0 is referred to as $N(X, x_0) = N(x_0)$.

6.1.10 Proposition. *If f is a path with initial point x_0, then $f * \bar{f} \in N(x_0)$.*

Suppose f is of length r. The assertion is trivial if $r = 0$, so we assume $r > 0$. Given $t \in I$, let f_t be the path $f \mid I_{rt}$. Then $(f_t)_{(r)} * (\bar{f}_t)_{(r)}$ is obviously a homotopy of paths connecting $e_{(2r)}$ and $f * \bar{f}$. $\blacksquare$

Let $E(X, x_0) = E(x_0)$ be the set of paths in X with initial point x_0, so that $N(x_0) \subseteq \Omega(x_0) \subseteq E(x_0)$. If $f, g \in E(X, x_0)$, we say that f, g are *equivalent paths* and write $f \equiv g$, if $f * \bar{g}$ is defined and $\in N(x_0)$. Clearly if $f \equiv g$ then f, g have the same end point. Thus in particular if $f \in \Omega(x_0)$ then $g \in \Omega(x_0)$. We prove

6.1.11 Theorem. *The relation $f \equiv g$ is an equivalence relation.*

This is an immediate consequence of

6.1.12 Lemma. *$f \equiv g$ if and only if $f_{(1)} \simeq g_{(1)}$.*

If f, g are of length r, s, the assertion is trivial if $r = s = 0$, so we suppose $r + s > 0$. Let $f \equiv g$; then $f * \bar{g} \simeq e_{(r+s)}$. Thus

$$\begin{aligned}f_{(r+2s)} &\simeq f * e_{(2s)}, && \text{by 6.1.8,}\\ &\simeq f * \bar{g} * g, && \text{by 6.1.10 and 6.1.7 (i),}\\ &\simeq e_{(r+s)} * g, && \text{by 6.1.7 (i),}\\ &\simeq g_{(r+2s)}, && \text{by 6.1.8.}\end{aligned}$$

We deduce from 6.1.6 that $f_{(1)} \simeq g_{(1)}$. Conversely, suppose that $f_{(1)} \simeq g_{(1)}$; from 6.1.6 we infer that $f_{(r+s)} \simeq g_{(r+s)}$. Thus

$$
\begin{aligned}
(f * \bar{g})_{(r+2s)} &\simeq f * \bar{g} * e_{(s)}, && \text{by 6.1.8,} \\
&\simeq f * e_{(s)} * \bar{g}, && \text{by 6.1.8,} \\
&\simeq f_{(r+s)} * \bar{g}, && \text{by 6.1.8,} \\
&\simeq g_{(r+s)} * \bar{g}, && \text{by 6.1.7 (i),} \\
&\simeq e_{(r)} * g * \bar{g}, && \text{by 6.1.8,} \\
&\simeq e_{(r+2s)}, && \text{by 6.1.10 and 6.1.7 (i).}
\end{aligned}
$$

A final application of 6.1.6 yields $f * \bar{g} \simeq e_{(r+s)}$. ▮ We remark that this lemma enables us immediately to infer that $f \equiv f_{(1)}$, so that every equivalence class contains paths of length 1. We may also infer from 6.1.8 that $e_{(s)} * f \equiv f \equiv f * e_{(s)}$.

We shall write $\{f\}$ for the equivalence class of the path f. From 6.1.7 (ii) and 6.1.12 we infer that $\{\bar{f}\}$ depends only on $\{f\}$; we may write $\{f\}^{-1}$, the *reverse* of $\{f\}$, for $\{\bar{f}\}$.

6.1.13 Proposition. *If $f \in E(x_0)$ and $g \in E(x_1)$, where x_1 is the final point of f, and if $f \equiv f'$, $g \equiv g'$, then*

$$f * g \equiv f' * g'.$$

Let f, g, f', g' be of length r, s, r', s'. Then

$$f * g * \overline{f' * g'} = f * g * \bar{g}' * \bar{f}' \simeq f * e_{(s+s')} * \bar{f}' \simeq f * \bar{f}' * e_{(s+s')} \simeq e_{(r+r'+s+s')}. \text{ ▮}$$

It follows that we may define $\{f\}\{g\}$ unambiguously to be $\{f * g\}$.

6.1.14 Theorem. *The classes of loops on x_0 form a group under the product $\{f\}\{g\} = \{f * g\}$; the unity element is $N(x_0) = \{e\}$ and the inverse of $\{f\}$ is $\{\bar{f}\}$.*

Associativity follows from the associativity in $\Omega(x_0)$. Since $f * e = f$, $\{e\}$ is certainly the unity element; and 6.1.10 shows that $\{\bar{f}\}$ is the inverse of $\{f\}$. ▮

The group of 6.1.14 is called the *fundamental group* $\pi_1(X, x_0)$ *of* X *on* x_0; x_0 is called the *base point* of $\pi_1(X, x_0)$. In this chapter we shall normally write $\pi(X, x_0)$ for $\pi_1(X, x_0)$, even sometimes contracting further to π. Elements of the fundamental group will normally be represented by greek letters. It will be shown that the fundamental group is not in general commutative; the fundamental group, for instance, of a figure 8 with any base point is the free group on two generators.

6.1.15 *Example.* Let A be a convex set in R^n and x_0 any point of A; then $\pi(A, x_0) = 1$.

We use the convexity of A to prove that any loop on x_0 is null-homotopic. If (f, r) is such a loop, we define $F : I_r \times I \to A$ by the rule

$$(u, t)F = (1-t).(uf) + t.x_0.$$

Observe that it is from the convexity of A that we know that

$$(1-t).(uf) + t.x_0 \in A.$$

Since we have proved all loops on x_0 null-homotopic, $\Omega(A, x_0) = N(x_0)$ and there is but a single equivalence class, which gives the result. The essential fact, of course, that is used here is the contractibility of A to x_0.

We close this section by mentioning some alternative ways of defining the fundamental group.

(i) The most usual method confines attention to paths of length 1, that is, maps $f : I \to X$. For two such f, g the path $(f * g)_{(1)}$ is defined to be the product of f and g. This product is inconvenient in that it has no unity elements and fails to be associative; but it has the advantage that the equivalence relation between paths can be taken to be that determined by homotopy rel $\dot{I}$.

(ii) Some authors use only maps $f : I \to X$ but immediately identify f with hf when h is any homeomorphism: $I, 0, 1 \to I, 0, 1$. With this identification the product as defined in (i) is associative but lacks unity elements; there is also the minor embarrassment that a path is now a class of maps and the reader has frequently to verify that definitions are independent of the map chosen to represent a class.

(iii) Since the fundamental group is determined in terms of loops, it is possible to consider maps of an oriented circle S^1 with base-point b. There is a unique linear orientation-preserving map $k : I, \dot{I} \to S^1, b$ which is a homeomorphism from $I - \dot{I}$ to $S^1 - b$; it may in fact be regarded as the identification map of I, identifying $\dot{I}$ to b. For any $f : I, \dot{I} \to X, x_0$ there is determined (compare 1.5.6) a map $g : S^1, b \to X, x_0$ such that $f = kg$. Two loops f_0, f_1 are homotopic rel $\dot{I}$ if and only if the associated maps $g_0, g_1 : S^1, b \to X, x_0$ are homotopic rel b; further $g : S^1, b \to X, x_0$ is null-homotopic if and only if g can be extended to a map of the circular disc in R^2 bounded by S^1. We have preferred a definition of π in terms of loops rather than of maps of S^1 (a) because anyway we need to define the product of paths, e.g. in 6.2.1, and (b) because the definition of the product of two elements arises more naturally in terms of paths.

6.2 Role of the base-point

Since only points which may be joined to x_0 by paths are concerned in the construction of $\pi(X, x_0)$ it is clear that the fundamental group is a property of the path component of x_0 in X, that is, of the subspace of X consisting of points which may be joined to x_0 by paths. It is therefore sensible, in discussing the fundamental group, to restrict ourselves to path-connected spaces. We note that the polyhedron of a simplicial complex K is path-connected if and only if K is connected. For first, if K is connected, any two vertices of K can be joined by a path running along the 1-section of K; since any point of $|K|$ can be connected by a path to a vertex of K, $|K|$ is path-connected. Then conversely, if $|K|$ is path-connected, any two vertices a, b of K are connected by a path in $|K|$; we apply the simplicial approximation theorem to replace this map of I_r to $|K|$ by a simplicial map of some subdivision of I_r to K^1: this simplicial map provides in a natural way a 1-chain whose boundary is $b-a$. Thus K is connected.

Let us now assume that X is path-connected and let x_0, x_1 be points of X. We shall compare $\pi(X, x_0)$ and $\pi(X, x_1)$ and show, in fact, that they are isomorphic. To this end we choose a path w from x_0 to x_1 and for each loop f_0 on x_0 we define $f_0\phi = \overline{w}*f_0*w$, a loop on x_1, and for each loop f_1 on x_1 we define $f_1\overline{\phi} = w*f_1*\overline{w}$, a loop on x_0. Then 6.1.13 shows that, if $f_0 \equiv f_0'$, $f_0\phi \equiv f_0'\phi$, so that ϕ and $\overline{\phi}$ induce transformations $\phi_* : \pi(X, x_0) \to \pi(X, x_1)$ and $\overline{\phi}_* : \pi(X, x_1) \to \pi(X, x_0)$. We assert ($a$) that ϕ_*, $\overline{\phi}_*$ are homomorphisms and (b) that $\phi_*\overline{\phi}_*$ and $\overline{\phi}_*\phi_*$ are the identity automorphisms; it will then follow that ϕ_* is an isomorphism and $\overline{\phi}_*$ its inverse. To prove (a) we simply observe that $\overline{w}*f_0*g_0*w \equiv (\overline{w}*f_0*w)*(\overline{w}*g_0*w)$, by 6.1.10; thus ϕ_*, and so by symmetry $\overline{\phi}_*$, is homomorphic; (b) follows from symmetry and the fact that $w*(\overline{w}*f_0*w)*\overline{w} \equiv f_0$, again by 6.1.10.

We also note that, if $w \equiv v$ and v induces ψ_*, then $\phi_* = \psi_*$, since $\overline{w}*f_0*w \equiv \overline{v}*f_0*v$ by 6.1.13; and finally that the isomorphism induced by the product of two paths is the product of their isomorphisms.

We have now proved

6.2.1 Theorem. *Let X be a path-connected space and let x_0, x_1 be points of X; then a class ζ of paths from x_0 to x_1 induces a right isomorphism $\zeta_*: \pi(X, x_0) \to \pi(X, x_1)$. If η is a class of paths from x_1 to x_2 then $\zeta_*\eta_* = (\zeta\eta)_*$.* ∎

6.2.2. Corollary. *Let $\alpha, \beta \in \pi(X, x_0)$, then $(\alpha)\beta_* = \beta^{-1}\alpha\beta$.*

We simply take $x_1 = x_0$ in the definition of β_*. ∎

6.2.3 Corollary. *If ζ, η are two classes of paths from x_0 to x_1, $\zeta_* = \eta_*$ if and only if $\zeta^{-1}\eta$ is in the centre of $\pi(X, x_1)$.*

Let v, w be paths in the classes ζ, η and f_0 be a loop on x_0; then

$$\overline{v}*f_0*v \in \{f_0\}\zeta_* \quad \text{and} \quad \overline{w}*f_0*w \in \{f_0\}\eta_* \quad \text{and} \quad \overline{v}*w \in \zeta^{-1}\eta.$$

If now $\zeta^{-1}\eta$ is in the centre of $\pi(X, x_1)$,

$$\begin{aligned}\bar{v} * f_0 * v, &\equiv \overline{(\bar{v} * w)} * (\bar{v} * f_0 * v) * (\bar{v} * w) \\ &= (\bar{w} * v) * (\bar{v} * f_0 * v) * (\bar{v} * w) \\ &\equiv \bar{w} * f_0 * w.\end{aligned}$$

This proves that $\zeta_* = \eta_*$. Since any loop on x_1 is equivalent to one of the form $\bar{v} * f_0 * v$ for some loop f_0 on x_0, the converse also follows. ∎

Thus, if $\pi(X, x_1)$ is commutative, we always have $\zeta_* = \eta_*$; if on the other hand, $\pi(X, x_1)$ is not commutative, it can certainly happen that $\zeta_* \neq \eta_*$. Theorem 6.2.1 shows that $\pi(X, x_0)$ and $\pi(X, x_1)$ are isomorphic, but 6.2.3 shows that there is no canonical isomorphism between them. We speak of the *fundamental group* $\pi(X)$ of the path-connected space X, meaning a group of which $\pi(X, x_0)$, $\pi(X, x_1)$, ... are isomorphic copies; this is often a convenient form of words, but the interesting theorems normally refer to fundamental groups associated with base-points. We can, however, properly use this term in

6.2.4 Definition. A path-connected space is said to be *simply-connected* if its fundamental group consists of the unity element alone.

A space is therefore simply-connected if all its loops are null-homotopic. The statement of 6.1.15 could be replaced by 'a convex set in R^n is simply-connected'.

6.2.5 Theorem. *Any* $f : X, x_0 \to Y, y_0$ *induces a homomorphism* $f_* : \pi(X, x_0) \to \pi(Y, y_0)$, *and for any* $g : Y, y_0 \to Z, z_0$, $(fg)_* = f_* g_*$.

Now f induces a transformation from $\Omega(X, x_0)$ to $\Omega(Y, y_0)$ that preserves the product and the reversal; further, the image of $N(X, x_0)$ is contained in $N(Y, y_0)$, for the image of a null-homotopic loop is a null-homotopic loop; the transformation therefore preserves the relation of equivalence. Clearly then f induces $f_* : \pi(X, x_0) \to \pi(Y, y_0)$ which is a homomorphism. It is, finally, obvious that $(fg)_* = f_* g_*$. ∎

6.2.6 Theorem. *If* $f_0 \overset{F}{\simeq} f_1 : X \to Y$, *if* $x_0 f_i = y_i$, $i = 0, 1$, *and if* w *is the path given by* $(t)\, w = (x_0, t)\, F$, *then* $f_{1*} = f_{0*}\{w\}_* : \pi(X, x_0) \to \pi(Y, y_1)$.

If (h, r) is a loop in X on x_0, we must prove that $hf_1 \equiv \bar{w} * hf_0 * w$ or that $(hf_1) * \bar{w} * \overline{(hf_0)} * w \in N(y_1)$.

Consider the rectangle $I_r \times I$, which is mapped into Y by the map G given by $(u, t)\, G = (uh, t)\, F$. In $I_r \times I$ consider the paths (of lengths r, 1, r and 1) which traverse linearly the edges from A to B, from B

to C, from C to D and from D to A. Let their product be the loop v based on A; then $vG = hf_1 * \overline{w} * \overline{hf_0} * w$. But by 6.1.15, the loop v is null-homotopic in $I_r \times I$, so that its image is null-homotopic in Y. ❚

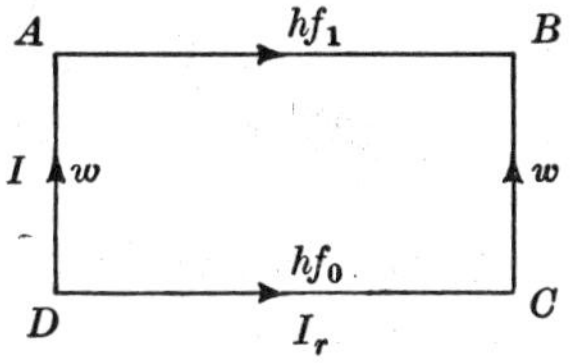

Fig. 6.1

6.2.7 Theorem. *The fundamental group of a path-connected space is an invariant of homotopy type.*

Let X, Y be path-connected spaces and let $f : X \to Y$ and $g : Y \to X$ be a homotopy equivalence and a homotopy inverse of it; suppose that $x_0 f = y_0$, $y_0 g = x_1$ and $x_1 f = y_1$. We consider $f_*^{(0)} : \pi(X, x_0) \to \pi(Y, y_0)$ and $f_*^{(1)} : \pi(X, x_1) \to \pi(Y, y_1)$ together with $g_* : \pi(Y, y_0) \to \pi(X, x_1)$ and we prove that g_* is an isomorphism. Now by 6.2.5

$$f_*^{(0)} g_* = (fg)_* : \pi(X, x_0) \to \pi(X, x_1).$$

But $fg \simeq 1$, since g is a homotopy inverse of f, so that, by 6.2.6, $f_*^{(0)} g_* = \{w\}_*$, where w is the track of x_0 in X under some homotopy deforming 1 into fg; by 6.2.1, $\{w\}_*$ is an isomorphism, so that g_* is epimorphic. Similarly, $g_* f_*^{(1)}$ is also an isomorphism, so that g_* is monomorphic; these results combine to imply that g_* is isomorphic. ❚ Of course $f_*^{(0)}, f_*^{(1)}$ are also isomorphic.

We observe that, if X is path-connected and $X \simeq Y$, then Y is also path-connected; we therefore never leave the class of path-connected spaces in passing from a path-connected space to a space of the same homotopy type.

6.3 Calculation of the fundamental group of a polyhedron

We have now defined the fundamental group of a space, shown it to be an invariant of homotopy type and discussed its behaviour under maps. We have, however, given no rules for calculating it. In this section we repair this omission in case the space is a polyhedron, by providing a technique for writing down the fundamental group of a polyhedron in terms of generators and relations. Our procedure is to define a group for a complex, to show that this group is, in fact, isomorphic with the fundamental group of its polyhedron and to describe how that group can be given in terms of generators and relations.

Let K be a complex with vertices $a^0, a^1, \ldots$ (not necessarily finite in number). We define an *edge-path* of K to be a sequence of vertices $w = a^{i_0}a^{i_1}\ldots a^{i_r}$ with $r \geqslant 0$, in which each successive pair (not necessarily distinct) span a simplex of K. If $w' = a^{i_r}\ldots a^{i_{r+s}}$ is another edge-path, whose initial vertex is the final vertex of w, we define $w * w'$ to be $a^{i_0}\ldots a^{i_r}a^{i_{r+1}}\ldots a^{i_{r+s}}$. We define $\overline{w}$, the reverse of w, to be $a^{i_r}\ldots a^{i_1}a^{i_0}$. This formal apparatus is somewhat similar to that of 6.1 (and not by accident), and clearly the product of edge-paths is associative (when defined) and an analogue of 6.1.4 holds.

We now define an *allowable operation* on edge-paths. If three consecutive vertices $a^{i_{k-1}}a^{i_k}a^{i_{k+1}}$ of w (not necessarily distinct) span a simplex of K, the middle vertex a^{i_k} can be removed and conversely a vertex may be inserted in an edge-path between any consecutive pair with which it spans a simplex of K; it is in addition allowable to replace the edge-path $a^{i_0}a^{i_0}$ by the edge-path a^{i_0} and vice versa. We plainly set up an equivalence relation, $\equiv$, between edge-paths by declaring two to be in the same class if one may be obtained from the other by a sequence of allowable operations. This classification has the property that, if $w_1 \equiv w_1'$, $\overline{w}_1 \equiv \overline{w}_1'$ and if further $w_2 \equiv w_2'$ and $w_1 * w_2$ is defined, then $w_1' * w_2'$ is defined and $w_1 * w_2 \equiv w_1' * w_2'$. We may therefore define a product between equivalence classes, written by juxtaposition, and define a reverse of an equivalence class, written as an inverse, by

6.3.1 $$\{w_1\}\{w_2\} = \{w_1 * w_2\} \quad \text{and} \quad \{w\}^{-1} = \{\overline{w}\}.$$

Notice that $w_1 * w_2 * \overline{w}_2 * w_3 \equiv w_1 * w_3$.

If a^0 is any vertex of K, we may consider *edge-loops on* a^0, namely, those edge-paths whose initial and final vertices are both a^0. We note that any edge-path equivalent to an edge-loop on a^0 is itself an edge-loop on a^0. It is easy to see that the equivalence classes of edge-loops on a^0 form a group under the product of 6.3.1 the unity element being $\{a^0\}$ and the inverse being that defined in 6.3.1. We call this group the *edge-group of K on* a^0 and write it as $\pi(K, a^0)$.

If $v: K \to L$ is a simplicial map and $w = a^{i_0}\ldots a^{i_r}$ an edge-path of K, we define $(w)v$ to be $a^{i_0}v\ldots a^{i_r}v$, which is an edge-path of L, and we observe that $(\overline{w})v = \overline{(w)v}$ and $(w * w')v = (w)v * (w')v$. Further, if $w \equiv w'$ then $(w)v \equiv (w')v$, since an allowable operation in K gives rise in the obvious way to an allowable operation in L. Thus, if $b^0 = a^0 v$, v induces a homomorphism $v_* : \pi(K, a^0) \to \pi(L, b^0)$ and, with the obvious meaning, $(v_1 v_2)_* = v_{1*} v_{2*}$.

We now define a transformation ρ from the edge-paths of K to the paths of $|K|$ which preserves the elements of structure. Let $M_0 = I_0$ and let M_r be the 1-dimensional complex which dissects I_r into r segments of unit length (r being an integer > 0). The edge-path $w = a^{i_0} \dots a^{i_r}$ determines a simplicial map $v_w : M_r \to K$ under which the vertex at the integer point j is mapped to a^{i_j}; ρ is then defined by

$$(w)\rho = |v_w|,$$

a path of length r in $|K|$ which visits in turn the vertices of w. It is again obvious that $(w * w')\rho = (w\rho) * (w'\rho)$ and that $(\overline{w})\rho = (\overline{w\rho})$; and since a single allowable operation on w leads to an edge-path w' such that $w\rho * (\overline{w'\rho}) = (w * \overline{w}')\rho$ is null-homotopic in $|K|$ it follows that, if $w \equiv w'$, $(w)\rho \equiv (w')\rho$. Notice that we have systematically used for edge-paths symbols already defined for paths and that the observations of the previous sentence imply that this should give rise to no misconceptions. They also imply that ρ induces a transformation from the equivalence classes of edge-paths of K to the equivalence classes of paths of $|K|$ which respects the product and the reversal, and so, on restricting attention to edge-loops on a^0, a homomorphism

$$\rho_* : \pi(K, a^0) \to \pi(|K|, a^0).$$

Moreover, ρ_* is natural in the sense that the diagram

$$\begin{array}{ccc} \pi(K, a^0) & \xrightarrow{\rho_*} & \pi(|K|, a^0) \\ \downarrow{\scriptstyle v_*} & & \downarrow{\scriptstyle |v|_*} \\ \pi(L, b^0) & \xrightarrow{\rho_*} & \pi(|L|, b^0) \end{array}$$

is commutative.

6.3.2 Theorem. *ρ_* is an isomorphism.*

The proof is just an exercise in the use of the simplicial approximation theorem (1.8.1). We start by proving ρ_* epimorphic. Let α be any element of $\pi(|K|, a^0)$, a class of loops on a^0; then by 1.8.1 α contains a loop which is a simplicial map v of some dissection L of I into K, the images of 0 and 1 both being a^0. If L has $(r+1)$ vertices, let $u : M_r \to L$ be the simplicial order-preserving homeomorphism. Then $uv : M_r \to K$ determines a loop $|uv|$ in the same class as $|v|$ and so belonging to α. But any simplicial map v' of M_r to K determines unmistakably an edge-path w of K such that $(w)\rho = |v'|$; $|uv|$ is therefore the image under ρ of an edge-loop on a^0 and α belongs to the image of ρ_*. This proves that ρ_* is epimorphic.

We now prove ρ_* to be monomorphic by proving that, if w is an edge-loop on a^0 and $(w)\rho$ is null-homotopic in $|K|$, then w is equivalent to the edge-loop a^0; this implies that the kernel of ρ_* consists only of the unity element of $\pi(K, a^0)$ and hence that ρ_* is monomorphic. Consider any edge-path $w = a^{i_0} \dots a^{i_r}$; $w\rho$ is a simplicial map of M_r into K. Suppose now that M_r is further subdivided in any way to give a complex M_r' and that $v : M_r' \to K$ is a simplicial approximation to $|w\rho|$; then the sequence of vertices of M_r' in increasing order is an

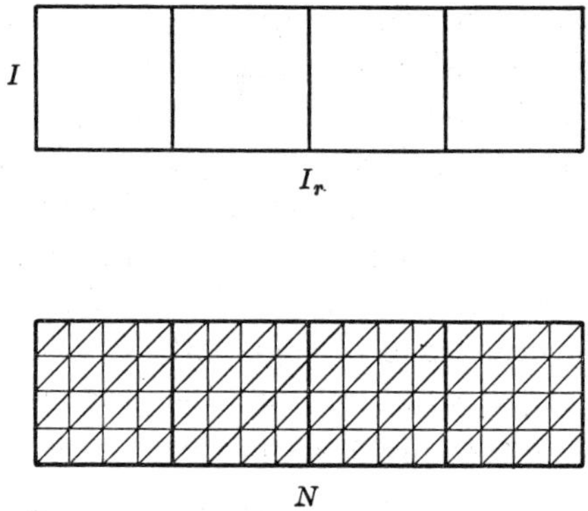

Fig. 6.2

edge-path of M_r' whose image under v is an edge-path w' of K and $w' \equiv w$. For w' is obtained from w by replacing each pair $a^{i_j}a^{i_{j+1}}$ by a string of vertices each of which is either a^{i_j} or $a^{i_{j+1}}$ and which starts with a^{i_j} and ends with $a^{i_{j+1}}$; such a string is evidently equivalent to $a^{i_j}a^{i_{j+1}}$. Consider now the case in which w is an edge-loop on a^0 and $|w\rho|$ is null-homotopic in $|K|$. Then there is a map $F : I_r \times I \to |K|$ such that

$$(t, 0)F = (t)|w\rho| \text{ for } 0 \leqslant t \leqslant r \text{ and } (t, 1)F = (0, u)F = (r, u)F = a^0.$$

By a trivial modification of 1.8.1 we may subdivide $I_r \times I$ in a manner sufficiently indicated in fig. 6.2 to give a complex N which admits a simplicial map that approximates F. Now in N the edge-path formed by the successive vertices in $I_r \times 0$ is equivalent to the edge-path consisting of the vertices (in order) of $0 \times I$ followed by those of $I_r \times 1$ followed (in reverse order) by those of $r \times I$; this equivalence is established by applying in some suitable order an admissible operation for each triangle of the complex N. Their images therefore under u are equivalent edge-loops in K. The former is an edge-loop w' already proved equivalent to w; the latter is an edge-loop each of whose vertices is a^0, which is plainly equivalent to a^0. This proves that, if wp is null-homotopic, $w \equiv a^0$. ∎

We note that only the structure of K^2 is involved in the definition of $\pi(K, a^0)$. In fact we have

6.3.3 Corollary. *Let ι_λ be the injection $\pi(|K^\lambda|, a^0) \to \pi(|K|, a^0)$; then ι_1 is epimorphic and ι_λ is isomorphic for $\lambda \geqslant 2$.* ∎

Theorem 6.3.2 reduces the problem of calculating the fundamental group of a polyhedron to a purely combinatorial one. We now give a comparatively practical procedure for writing down a system of generators and relations for such a group. We shall assume K to be connected, since otherwise we confine attention anyway to the component of the base-point, and we calculate $\pi(K, a^0)$, the edge group on a^0. We say that a space is 1-*connected* if it is connected and simply connected, and that a complex K is 1-*connected* if $|K|$ is 1-connected.

6.3.4 Proposition. *If K is a connected complex and K_0 a 1-connected subcomplex, there is a 1-connected subcomplex L such that $K_0 \subseteq L \subseteq K$ and that L contains all the vertices of K.*

No assumption is made about the finiteness of K and we use Zorn's Lemma for the proof. The 1-connected subcomplexes of K form a partially ordered set under inclusion; moreover, this set is inductive, in that any totally ordered subset has a least upper bound in the set, in this case the union of the elements of the set. For, if $\{K_\lambda\}$ is a totally ordered set of 1-connected subcomplexes and $\bigcup_\lambda K_\lambda = K_\mu$, we prove that K_μ is itself 1-connected; it is then certainly a least upper bound for $\{K_\lambda\}$. First we prove K_μ connected. Let a, a' be vertices of K_μ; then, for some λ, $a \in K_\lambda$, and for some λ', $a' \in K_{\lambda'}$; if (say) $K_\lambda \subseteq K_{\lambda'}$, $a \cup a' \subseteq K_{\lambda'}$ and $K_{\lambda'}$ is connected. There is therefore an edge-path connecting a and a' in $K_{\lambda'}$ and so, *a fortiori*, in K_μ; since any two vertices of K_μ are connected by an edge-path in K_μ, K_μ is connected. Next we prove K_μ simply connected by a similar argument. Let $w = a^{i_0}a^{i_1} \ldots a^{i_{r-1}}a^{i_0}$ be an edge-loop in K_μ; for some $\lambda_1, \ldots, \lambda_r$ the edges $a^{i_0}a^{i_1}, \ldots, a^{i_{r-1}}a^{i_0}$ belong to $K_{\lambda_1}, \ldots, K_{\lambda_r}$ respectively. If K_λ is the largest of the (totally ordered) set $K_{\lambda_1}, \ldots, K_{\lambda_r}$, w is an edge-loop in K_λ; but K_λ is simply connected so that w is equivalent to an edge-loop of length zero in K_λ and so, *a fortiori*, also in K_μ. Hence K_μ is simply-connected and so 1-connected.

Since the partially ordered set of 1-connected subcomplexes is inductive any element K_0 is contained in a maximal element L. Since L is maximal, the adjunction of a closed 1-simplex not already in L must give a subcomplex not 1-connected; the adjunction of a 1-simplex with one end-point in L and one in $K-L$ would, however, give a 1-connected subcomplex strictly containing L, so that there can be no such 1-simplex and L must contain all the vertices of the component of K in which it lies. But K is connected and L has therefore the required property. ∎

It is sometimes convenient to confine attention to 1-dimensional 1-connected complexes or *trees*; we can use the method of 6.3.4 to prove that any tree in K is contained in a tree T that contains all the vertices of K.

If K is a complex and L a subcomplex, let $G = G(K, L)$ be the group generated by elements g^{ij} one for each $[a^i a^j] \in K^\Omega$, subject to the relations

6.3.5
$$\begin{cases} \text{(i)} \quad g^{ij} = 1 \quad \text{if} \quad [a^i a^j] \in L^\Omega, \\ \text{and (ii)} \quad g^{ij} g^{jk} = g^{ik} \quad \text{if} \quad [a^i a^j a^k] \in K^\Omega. \end{cases}$$

We note that (ii) implies that $g^{ii} = 1$ by putting $j = i$, and then that $g^{ij} = (g^{ji})^{-1}$ by putting $k = i$.

6.3.6 Theorem. *If L is a 1-connected subcomplex of the connected complex K containing all the vertices of K, $\pi(K, a^0) \cong G(K, L)$.*

For each vertex $a^i \in K$ we select an edge-path x^i in L from a^0 to a^i, choosing for x^0 the edge-path a^0; such edge-paths exist since L is connected and contains all the vertices of K. For each $[a^i a^j] \in K^\Omega$, we define v^{ij} to be $x^i * a^i a^j * \bar{x}^j$, an edge-loop on a^0. We observe that, if $w = a^0 a^{i_1} a^{i_2} \dots a^{i_{r-1}} a^0$, then $w \equiv v^{0 i_1} * v^{i_1 i_2} * \dots * v^{i_{r-1} 0}$; if therefore $\xi^{ij} = \{v^{ij}\} \in \pi(K, a^0)$, $\{w\} = \xi^{0 i_1} \xi^{i_1 i_2} \dots \xi^{i_{r-1} 0}$. This shows that the elements ξ^{ij} generate $\pi(K, a^0)$.

We next verify that

6.3.7
$$\begin{cases} \text{(i)} \quad \xi^{ij} = 1 \quad \text{if} \quad [a^i a^j] \in L^\Omega, \\ \text{and (ii)} \quad \xi^{ij} \xi^{jk} = \xi^{ik} \quad \text{if} \quad [a^i a^j a^k] \in K^\Omega. \end{cases}$$

If $[a^i a^j] \in L^\Omega$, v^{ij} is an edge-loop in L, which is simply-connected; therefore $v^{ij} \equiv a^0$ in L and so, *a fortiori*, in K and consequently $\xi^{ij} = 1$. This establishes (i) of 6.3.7. If now $[a^i a^j a^k] \in K^\Omega$, $v^{ij} * v^{jk} \equiv v^{ik}$, as the reader may verify; hence $\xi^{ij} \xi^{jk} = \xi^{ik}$ and (ii) of 6.3.7 is established.

The transformation $g^{ij} \to \xi^{ij}$ therefore determines an epimorphism $\phi : G(K, L) \to \pi(K, a^0)$. We prove ϕ to be also monomorphic by constructing a homomorphism $\psi : \pi(K, a^0) \to G$ such that $\phi\psi = 1$. To that end we define a function θ on the edge-paths of K with values in G by the rule that, if $w = a^{i_0} a^{i_1} \ldots a^{i_r}$, $w\theta = g^{i_0 i_1} g^{i_1 i_2} \ldots g^{i_{r-1} i_r}$ (and that, if $w = a^i$, $w\theta = 1$). If $v \equiv w$, $v\theta = w\theta$; for an allowable operation on w leads to an edge-loop whose θ-image is equal to $w\theta$ in view of (ii) of 6.3.5. Clearly $(v * w)\,\theta = v\theta\,.\,w\theta$ and $\bar{v}\theta = (v\theta)^{-1}$; for $(g^{ij})^{-1} = g^{ji}$. Consequently θ, restricted to the edge-loops on a^0, induces a homomorphism $\psi : \pi(K, a^0) \to G$ by the rule that $\{w\}\,\psi = w\theta$.

Now $g^{ij}\phi\psi = \xi^{ij}\psi = v^{ij}\theta = x^i\theta\,.\,(a^i a^j)\,\theta\,.\,(\bar{x}^j)\,\theta$. But x^i, $\bar{x}^j$ are edge-paths in L, so that by (i) of 6.3.5, $x^i\theta = \bar{x}^j\theta = 1$; hence

$$g^{ij}\phi\psi = (a^i a^j)\,\theta = g^{ij}.$$

Since $\phi\psi$ is the identity on the generators of G and is homomorphic, $\phi\psi = 1 : G \to G$ and ϕ is monomorphic. This proves that

$$\phi : G \cong \pi(K, a^0). \;\rbrack$$

6.3.8 Proposition. *If the vertices of K are ordered and if L is a subcomplex of K, $G(K, L)$ is generated by the elements g^{ij} for which $i<j$ and $[a^i a^j] \in K^\Omega - L^\Omega$, subject to the relations $g^{ij} g^{jk} = g^{ik}$ if $i<j<k$ and $a^i a^j a^k \in K - L$, where g^{rs} is to be interpreted as 1 if $a^r a^s \in L$.*

First all the generators originally proposed for G are expressible in terms of the restricted set: if $[a^i a^j] \in L^\Omega$, $g^{ij} = 1$; if $i > j$ and $[a^i a^j] \in K^\Omega - L^\Omega$, $g^{ij} = (g^{ji})^{-1}$ and g^{ji} belongs to the restricted set. The restricted set therefore generates G.

Finally all the relations (ii) of 6.3.5 are consequences of the restricted set of relations. If $[a^i a^j a^k] \in L^\Omega$ the corresponding relation is $1\,.\,1 = 1$; if a^i, a^j, a^k span a simplex of $K - L$ of dimension < 2, the corresponding relation gives a relation between generators of the restricted set holding in any group. If a^i, a^j, a^k span a 2-simplex of $K - L$, but it is not true that $i<j<k$, the corresponding relation is equivalent to the relation arising from $a^i a^j a^k$ in their correct order.]

6.3.9 Corollary. *If K is a connected complex and L is a 1-connected subcomplex containing all the vertices of K, and if the vertices of K are ordered, $\pi(K, a^0)$ is isomorphic to the group generated by elements ξ^{ij}, one for each of the ordered 1-simplexes $a^i a^j$ of $K - L$, subject to the relations $\xi^{ij}\,\xi^{jk} = \xi^{ik}$, there being one such relation for each ordered 2-simplex $a^i a^j a^k$ of $K - L$, where ξ^{rs} is to be interpreted as 1 if $a^r a^s \in L$.*]

It is this corollary that provides usually the most practical way of calculating $\pi(|K|)$.

In all this we have nowhere assumed that the complex K is finite. For that case, however, our argument proves

6.3.10 Corollary. *The fundamental group of a compact polyhedron is finitely generated and finitely related.* ∎

6.3.11 Corollary. *If K is a 1-dimensional complex, $\pi(|K|)$ is free.*

For it is generated by the generators ξ^{ij}, one for each 1-simplex of $K-L$, and there are no relations, since there are no 2-simplexes. ∎

(A definition of a free group is given in [20] ch. 7).

6.3.12 *Example.* If $X = S^1 \vee S^1 \vee \ldots \vee S^1$, k circles with a single common point x_0, $\pi(X)$ is a free group on k generators, where k need not be finite.

We dissect each S^1 as a hollow triangle, x_0 being a vertex in each case, and we choose for L the union of the closed 1-simplexes having x_0 as a vertex. There are then k 1-simplexes of $K-L$ and the result follows.

6.3.13 *Example.* If Y is the space formed by removing from a disk the interiors of k disjoint circles, $\pi(Y)$ is a free group on k generators.

For Y is of the homotopy type of X of 6.3.12.

6.3.14 *Example.* If $n > 1$, $\pi(S^n) = 1$.

The reader is recommended to prove this by the above methods, dissecting S^n as $\dot{s}_{n+1}$. Another argument is, however, available. By 6.3.3 $\pi(\dot{s}_{n+1}) \cong \pi(\bar{s}_{n+1})$, since $n+1 > 2$; $\bar{s}_{n+1}$ is contractible and so simply connected.

6.4 Further theorems and calculations

In this section we prove two theorems that have some value in calculations.

6.4.1 Theorem. *If X, Y are path-connected spaces,*

$$\pi(X \times Y) \cong \pi(X) \times \pi(Y),$$

the direct product of $\pi(X)$ and $\pi(Y)$.

It is easy to show that $X \times Y$ is path-connected. We take base-points x_0, y_0, (x_0, y_0) for X, Y, $X \times Y$ and suppress them from the notation for simplicity. These are obvious projections $p : X \times Y \to X$,

$q : X \times Y \to Y$; p, for instance, is defined by the rule $(x, y)p = x$. Further, if W is any space and $f : W \to X$, $g : W \to Y$ are given, these determine a product map $(f, g) : W \to X \times Y$ such that $(f, g)p = f$ and $(f, g)q = g$; in fact, $(w)(f, g) = (wf, wg)$. Now p, q induce

$$p_* : \pi(X \times Y) \to \pi(X), \quad q_* : \pi(X \times Y) \to \pi(Y)$$

and these determine

$$(p_*, q_*) : \pi(X \times Y) \to \pi(X) \times \pi(Y),$$

given by $(\gamma)(p_*, q_*) = (\gamma p_*, \gamma q_*)$, for any $\gamma \in \pi(X \times Y)$. We shall prove that $r_* = (p_*, q_*)$ is an isomorphism. First we prove r_* epimorphic. If $(\alpha, \beta) \in \pi(X) \times \pi(Y)$, let $f : I, \dot{I} \to X, x_0$ and $g : I, \dot{I} \to Y, y_0$ be loops representing α, β; then (f, g) is a loop in $X \times Y$ such that $(f, g)p = f$ and $(f, g)q = g$, so that if $\gamma = \{(f, g)\} \in \pi(X \times Y)$, $\gamma p_* = \alpha$ and $\gamma q_* = \beta$. Hence $\gamma r_* = (\alpha, \beta)$ and r_* is epimorphic. Finally, we prove r_* monomorphic, by considering any $\gamma \in \pi(X \times Y)$ for which $\gamma r_* = 1$ and proving $\gamma = 1$. Let $h : I, \dot{I} \to X \times Y, (x_0, y_0)$ represent γ, then we know that hp, hq are null-homotopic rel $\dot{I}$. There exist, therefore, homotopies $F : I \times I \to X$, $G : I \times I \to Y$ which deform hp, hq to the constant paths; then $(F, G) : I \times I \to X \times Y$ deforms h to the constant path, so that $\gamma = 1$. ∎

6.4.2 *Example.* The fundamental group of the torus is free abelian on two generators.

For a torus is the topological product of two circles.

The second theorem, for which we shall find more applications than for 6.4.1, follows:

6.4.3 Theorem. (*Van Kampen's theorem for polyhedra*) *K is a connected complex and K_0, K_1, K_2 are connected subcomplexes, such that $K_1 \cup K_2 = K$ and $K_1 \cap K_2 = K_0$; a^0 is a vertex of K_0; the injection maps $i_r : K_0 \to K_r$, $r = 1, 2$, induce*

$$i_{r*} : \pi(K_0, a^0) \to \pi(K_r, a^0).$$

Then $\pi(K)$ is the free product of $\pi(K_1)$ and $\pi(K_2)$ with $\pi(K_0)\, i_{1*}$ amalgamated to $\pi(K_0)\, i_{2*}$ by means of the mapping $\alpha_0 i_{1*} \to \alpha_0 i_{2*}$ for all $\alpha_0 \in \pi(K_0)$.

In other words, $\pi(K)$ is obtained from the free product of $\pi(K_1)$ and $\pi(K_2)$ by adding the relations $\alpha_0 i_{1*} = \alpha_0 i_{2*}$ for all $\alpha_0 \in \pi(K_0)$.

We select first a tree T_0 in K_0, containing all the vertices of K_0; this may be extended to trees T_i maximal in K_i; then $T = T_1 \cup T_2$

is a tree containing all the vertices of K and $T_1 \cap T_2 = T_0$. The generators of $\pi(K)$ may be divided into three classes:

(i) ξ_0^{ij}, one for each $a^i a^j \in K_0 - T_0$,

(ii) ξ_1^{ij}, one for each $a^i a^j \in K_1 - (K_0 \cup T_1)$,

(iii) ξ_2^{ij}, one for each $a^i a^j \in K_2 - (K_0 \cup T_2)$.

The relations similarly fall into three classes:

(i) $R(\xi_0) = 1$, one for each 2-simplex of K_0,

(ii) $R(\xi_0, \xi_1) = 1$, one for each 2-simplex of $K_1 - K_0$,

(iii) $R(\xi_0, \xi_2) = 1$, one for each 2-simplex of $K_2 - K_0$.

Now the generators and relations of types (i) and (ii) define $\pi(K_1)$ and those of types (i) and (iii) define $\pi(K_2)$. To complete the proof we adopt a formal device: for each ξ_0^{ij} of type (i) we take two generators, ξ_{01}^{ij} and ξ_{02}^{ij}. Then $\pi(K)$ is the group generated by $\{\xi_{01}\}, \{\xi_{02}\}, \{\xi_1\}, \{\xi_2\}$, subject to the relations of five types:

$$\begin{aligned} &\text{(i)} & R(\xi_{01}) &= 1, \\ &\text{(ii)} & R(\xi_{01}, \xi_1) &= 1, \\ &\text{(iii)} & R(\xi_{02}) &= 1, \\ &\text{(iv)} & R(\xi_{02}, \xi_2) &= 1, \\ &\text{(v)} & \xi_{01}^{ij} &= \xi_{02}^{ij}. \end{aligned}$$

The group with these generators and the relations (i)–(iv) is by definition the *free product* of $\pi(K_1)$ and $\pi(K_2)$. Relations (v) are just the relations which identify two elements of $\pi(K_1)$ and $\pi(K_2)$ that arise from the same element of $\pi(K_0)$. ∎

6.4.4 Corollary. *If K_2 is simply connected, $\pi(K)$ is obtained from $\pi(K_1)$ by adjoining the relations $\xi_0^{ij} = 1$.*

For $\xi_{02}^{ij} = 1$ and $\xi_2^{ij} = 1$, since K_2 is simply-connected, and the relations (iii), (iv) may be replaced by these. We may then use only the generators and relations for $\pi_1(K_1)$ augmented by the relations (replacing (v)) $\xi_0^{ij} = 1$. ∎

6.4.5 Corollary. *If K_0 is simply-connected, $\pi(K)$ is the free product of $\pi(K_1)$ and $\pi(K_2)$.* ∎

This applies in particular when K_0 is a single point so that $|K| = |K_1| \vee |K_2|$.

Somewhat surprisingly it is not in general true for topological spaces; it may happen that $\pi(X_1 \vee X_2)$ is not isomorphic to the free product of $\pi(X_1)$ and $\pi(X_2)$. H. B. Griffiths (*Quart. J. Math.* **5** (1954), 175) has given a disturbing example of simply-connected spaces‡ X_1, X_2 for which $X_1 \vee X_2$ is not simply-connected. The spaces X_1, X_2 in this example enjoy bizarre local properties near their common point. A necessary and sufficient condition for the validity of van Kampen's theorem for general spaces has been given by P. Olum, *Ann. Math.* **68** (1958), 658.

We may apply 6.4.4 in a rather useful way. Suppose that K is a complex and $w = a^{i_0}a^{i_1}\dots a^{i_{r-1}}a^{i_0}$ is an edge-loop with no two consecutive vertices identical; we formalize the notion of attaching§ a 2-cell to K along w. Let $|L|$ be the convex subset of R^2 bounded by a regular polygon P of r sides; then, if $b^0, b^1, \dots, b^{r-1}$ are the vertices of P, the vertex transformation $b^k \to a^{i_k}$ defines a simplicial map $v: P \to K$. On the points of $|K| \cup |L|$ we set up the equivalence relation which identifies $y \in |P|$ with $y|v| \in |K|$ and we call the identification space X. If $|L|$ is dissected as the join of an interior point to P, this dissection together with the dissection K provides a pseudodissection M of X; then M' is a simplicial complex such that $|M'| = X$. In M' choose a 2-simplex s in L such that $\bar{s} \cap P = b^0$ and let $M' - s = M_1$, $\bar{s} = M_2$, $\dot{s} = M_0$. Then $|M_1|$ can be retracted by deformation onto $|K|$, the points of $|L| - s$ being projected radially from the barycentre of s into $|P|$ and the points of $|K|$ remaining fixed, so $\pi(M_1) \cong \pi(K)$. Certainly $\pi(M_2) = 1$ and we apply 6.4.4 to deduce that $\pi(X)$ is obtained from $\pi(M_1)$ by adjoining the relation $\alpha = 1$, where α is the generator of $\pi(\dot{s})$. Under the retraction $|M_1| \to |K|$ the image of α is the element of $\pi(K)$ determined by w; hence $\pi(X)$ is obtained from $\pi(K)$ by adjoining the relation $\{w\} = 1$. In short, the process of attaching a 2-cell along w has the effect on $\pi(|K|)$ of obliterating the element determined by w.

This enables us to give generators and relations for the compact 2-dimensional manifolds.‖ Any such manifold (other than S^2) is obtainable by attaching a 2-cell to $S^1_1 \vee S^1_2 \vee \dots \vee S^1_k$, where S^1_j is a circle, $1 \leqslant j \leqslant k$. If w_j is one of the two edge-loops on the common vertex a^0 which go once round S^1_j, then we may construct an orientable compact 2-dimensional manifold by taking $k = 2p$ and defining

$$w = (w_1 * w_2 * \overline{w}_1 * \overline{w}_2) * \dots * (w_{2p-1} * w_{2p} * \overline{w}_{2p-1} * \overline{w}_{2p})$$

and attaching a 2-cell along w. The fundamental group of $S^1_1 \vee \dots \vee S^1_{2p}$ is the free group generated by $\xi_1, \dots, \xi_{2p}$, so that the fundamental

‡ In fact X_1 and X_2 are both cones on the space of Example 6.6.10, and they are joined at the origin.

§ For a more general notion of attaching cells, see J. H. C. Whitehead, *Ann. Math.* **42** (1941), 409.

‖ See, for instance, Seifert and Threlfall [6].

group of this manifold is the group generated by $\xi_1, \ldots, \xi_{2p}$ with the single relation $(\xi_1\xi_2\xi_1^{-1}\xi_2^{-1})\ldots(\xi_{2p-1}\xi_{2p}\xi_{2p-1}^{-1}\xi_{2p}^{-1}) = 1$. A compact non-orientable 2-dimensional manifold can be constructed by attaching a 2-cell to $S_1^1 \vee \ldots \vee S_k^1$ along the edge-loop

$$w = (w_1 * w_1) * \ldots * (w_k * w_k);$$

its fundamental group is therefore that generated by $\xi_1, \ldots, \xi_k$ with the single relation $\xi_1^2\xi_2^2\ldots\xi_k^2 = 1$. In particular P^2 is the compact non-orientable manifold with $k = 1$; so $\pi(P^2) = J_2$.

6.4.6 Theorem. *If G is a group, there is a polyhedron whose fundamental group is isomorphic to G.*

Let G be given by generators $\xi_1, \ldots, \xi_m, \ldots$ and relations

$$R_1(\xi) = 1, \quad \ldots, \quad R_n(\xi) = 1, \quad \ldots;$$

we start from $S^1 \vee \ldots \vee S_m^1 \vee \ldots$ associating ξ_m with w_m, and we attach 2-cells along the edge-loops $R_1(w), \ldots, R_n(w), \ldots$; the 2-cells have of course mutually disjoint interiors. This space is given the fine topology; namely, a set is closed if and only if its intersection with the image of each 2-cell is closed in that image. ∎

In case the number of generators and relations is finite, the polyhedron we have constructed is, of course, compact.

We close this section by stating the relation between $\pi(K)$ and $H_1(K)$, for K connected. Any edge-loop w of K on a^0 determines in an obvious way a 1-cycle $z(w)$ of K^Ω. Clearly $z(w_1 * w_2) = z(w_1) + z(w_2)$, $z(\overline{w}) = -z(w)$ and, if $w_1 \equiv w_2$, then $z(w_1) \sim z(w_2)$ in K^Ω. This association therefore determines a homomorphism

$$\theta : \pi(K) \to H_1(K).$$

6.4.7 Theorem. *The homomorphism $\theta : \pi(K) \to H_1(K)$ is epimorphic, if K is connected, and the kernel is the commutator subgroup of $\pi(K)$.*

In other words, $H_1(K)$ is $\pi(K)$ made abelian. The reader will easily find a proof that θ is epimorphic and that the kernel of θ contains the commutator subgroup; it is a little harder to prove that the kernel

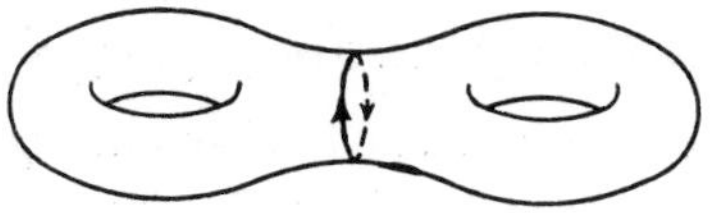

Fig. 6.3

contains no other elements. We give no proof here because chapter 8 contains a proof of the corresponding result for arbitrary spaces (see 8.8.3).

Fig. 6.3 gives an illustration of a loop on a 'double torus' (an orientable 2-manifold with $p = 2$) that is not null-homotopic but whose corresponding 1-cycle is homologous to zero. Theorem 6.4.7 assures us that the element of the fundamental group determined by this loop lies in the commutator subgroup.

6.5 Covering spaces

In this section we deal with general topological spaces‡; there would be no gain in simplicity of concept from restricting attention to polyhedra. Let $\tilde{X}$, X then be path-connected topological spaces; we say that a map $p : \tilde{X} \to X$ is a *covering map* if it is onto X and if each point $x \in X$ has a neighbourhood $U(x)$ such that the points $\tilde{x}_1, \tilde{x}_2, \ldots,$ of $p^{-1}(x)$ have neighbourhoods $\tilde{U}(\tilde{x}_1), \ldots,$ in $\tilde{X}$ with the properties that

(i) $p \,|\, \tilde{U}(\tilde{x}_i)$ is a homeomorphism of $\tilde{U}(\tilde{x}_i)$ onto $U(x)$, for each $\tilde{x}_i \in p^{-1}(x)$,

(ii) $\bigcup_i \tilde{U}(\tilde{x}_i) = p^{-1}\, U(x)$, and

(iii) if $\tilde{x}_i \neq \tilde{x}_j$, $\tilde{U}(\tilde{x}_i) \cap \tilde{U}(\tilde{x}_j) = \varnothing$.

We may call $U(x)$, $\tilde{U}(\tilde{x}_i)$ *canonical neighbourhoods*. If there is a covering map $p : \tilde{X} \to X$, we say that $\tilde{X}$ is a *covering space* of X with *covering map* (or projection) p.

We lose nothing in supposing all canonical neighbourhoods to be open sets; in that case we may restate (i), (ii) and (iii) by saying that $p^{-1}U(x)$ breaks up as the union of disjoint open sets $\tilde{U}(\tilde{x}_i)$ each of which p maps homeomorphically onto $U(x)$, so that a covering map is locally homeomorphic.

We shall see later that certain reasonable conditions ensure the existence of non-trivial covering spaces for a non-simply-connected space. We use the adjective 'non-trivial' in view of the obvious fact that every path-connected space is a covering space of itself with the identity map for projection.

6.5.1 *Example.* Consider the real line R^1, $-\infty < u < \infty$, and S^1, the circle $|z| = 1$. Then $(u)p = e^{2\pi i u}$ defines a covering map $p : R^1 \to S^1$.

6.5.2 *Example.* Consider S^2 and P^2; P^2 may be regarded as arising from S^2 by identifying all antipodal pairs and the identification map $k : S^2 \to P^2$ is a covering map. Similarly, S^n is a covering space of P^n.

‡ Notice, in particular, that we assume no separation axiom; for instance, the spaces need not be Hausdorff.

6.5.3 *Example.* The points of S^1, $|z| = 1$, are expressible as $e^{i\theta}$, $-\pi \leqslant \theta \leqslant \pi$; the mapping $p : S^1 \to S^1$ given by $e^{i\theta}p = e^{2i|\theta|}$ is *not* a covering map (see fig. 6.4). It is not possible to find $U(1)$ and $\tilde{U}(1)$, $\tilde{U}(-1)$ to satisfy (i), (ii) and (iii). If we take for $U(1)$ the set $|\theta| < \epsilon$, $p^{-1}(U)$ consists of two intervals $\tilde{U}(1)$, $\tilde{U}(-1)$, namely, the sets $|\theta| < \frac{1}{2}\epsilon$ and $|\theta \pm \pi| < \frac{1}{2}\epsilon$, each of which is indeed homeomorphic to $U(1)$: but p is not a homeomorphism from $\tilde{U}(1)$ to $U(1)$.

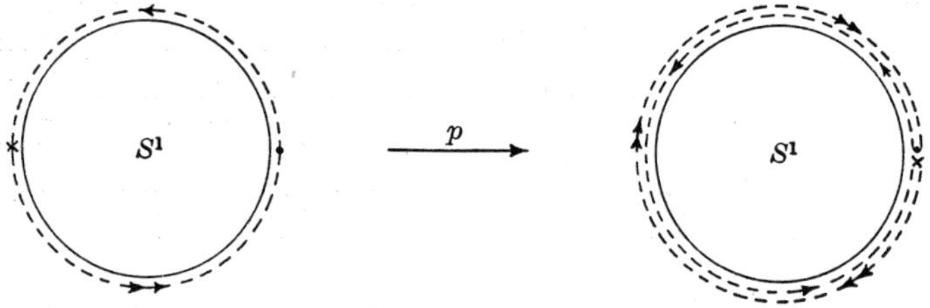

Fig. 6.4

We remark that a Riemann 'covering surface' is not in general, in this sense, a covering space for the Riemann sphere.

We now study some of the properties of covering spaces.

6.5.4 Proposition. *If $p : \tilde{X} \to X$ is a covering map, for each $x \in X$ $p^{-1}(x)$ is discrete.*

Since the canonical neighbourhoods $\tilde{U}(\tilde{x}_i)$ are disjoint by (iii) each $\tilde{x}_i$ is an open set of $p^{-1}(x)$. ∎

6.5.5 Theorem. *If $p : \tilde{X} \to X$ is a covering map and $\tilde{v}, \tilde{w} : I, 0 \to \tilde{X}, \tilde{x}_0$ have the property that $\tilde{v}p = \tilde{w}p$ then $\tilde{v} = \tilde{w}$.*

Let ρ be the least upper bound of the set of real numbers $r \leqslant 1$ such that $\tilde{v}\,|\,I_r = \tilde{w}\,|\,I_r$; the set is not empty since it contains 0 and ρ is therefore well defined. We observe first that $\rho\tilde{v} = \rho\tilde{w}$, so that $\tilde{v}\,|\,I_\rho = \tilde{w}\,|\,I_\rho$. (For, if not, $\rho > 0$ and $\rho\tilde{v}$, $\rho\tilde{w}$ have disjoint canonical neighbourhoods, by (iii); this conflicts with the fact that $r\tilde{v} = r\tilde{w}$ for $0 \leqslant r < \rho$.) Suppose now, for a contradiction, that $\rho < 1$ and that $\rho\tilde{v} = \rho\tilde{w} = \tilde{x}$. For some canonical neighbourhood $\tilde{U}$ of $\tilde{x}$, $p\,|\,\tilde{U} : \tilde{U} \to U$ is a homeomorphism; and for some $\delta > 0$ the segment $[\rho, \rho+\delta]$ is mapped both by $\tilde{v}$ and by $\tilde{w}$ into $\tilde{U}$. But, since $\tilde{v}p = \tilde{w}p$, for $\rho \leqslant t \leqslant \rho+\delta$, and p is (1, 1) on $\tilde{U}$, it follows that $\tilde{v}\,|\,I_{\rho+\delta} = \tilde{w}\,|\,I_{\rho+\delta}$, contradicting the definition of ρ. This contradiction disproves the supposition that $\rho < 1$ and proves, therefore, that $\tilde{v} = \tilde{w}$. ∎

If, for any space W, $\tilde{h} : W \to \tilde{X}$ and $h : W \to X$ are related by the rule that $\tilde{h}p = h$, we say that $\tilde{h}$ *lifts* h. Theorem 6.5.5 states that a path in X has only one path that lifts it with given initial point.

6.5.6 Corollary. *If W is path-connected and $\tilde{h}, \tilde{k} : W \to \tilde{X}$ both lift $h : W \to X$ and if, for some $w_0 \in W$, $w_0\tilde{h} = w_0\tilde{k}$, then $\tilde{h} = \tilde{k}$.* ∎

6.5.7 Theorem. *Let $p : \tilde{X} \to X$ be a covering map and let K be a simplicial complex (not necessarily finite) with subcomplex L. Let $\tilde{f}_0 : |K| \to \tilde{X}$ be a map and let $\tilde{g}_t : |L| \to \tilde{X}$, $f_t : |K| \to X$ be homotopies such that*

$$\tilde{g}_0 = \tilde{f}_0 \,|\, |L|, \quad \tilde{g}_t p = f_t \,|\, |L| \quad \text{and} \quad \tilde{f}_0 p = f_0.$$

Then there exists a homotopy $\tilde{f}_t : |K| \to \tilde{X}$ with

$$\tilde{g}_t = \tilde{f}_t \,|\, |L| \quad \text{and} \quad \tilde{f}_t p = f_t.$$

Moreover, $\tilde{f}_t$ is unique if K is connected.

This theorem (without the assertion of uniqueness) will be familiar to students of fibre spaces‡; it may be called the *lifting homotopy theorem* for covering spaces or the *covering homotopy theorem*, since it asserts that the homotopy f_t may be 'lifted' into the covering space $\tilde{X}$.

We may plainly re-express the assertion of the theorem as follows: we are given maps $\tilde{G} : (|K| \times 0) \cup (|L| \times I) \to \tilde{X}$ and $F : |K| \times I \to X$ such that F is an extension of $\tilde{G}p$. Then we may extend $\tilde{G}$ to $\tilde{F} : |K| \times I \to \tilde{X}$ such that $\tilde{F}p = F$, and $\tilde{F}$ is unique if K is connected. We shall give a proof adapted to this re-expression. It will proceed by first establishing the theorem in a very special case, then in a more general case and finally in complete generality. Of course, the uniqueness assertion presents no difficulty; if K is connected then $|K|$, and hence $|K| \times I$, is path-connected and we apply 6.5.6. Thus only the existence of $\tilde{F}$ will be in question in the subsequent argument.

We first take $K = \bar{s}$, $L = \dot{s}$, where s is a simplex of arbitrary dimension, and we suppose that I may be subdivided at

$$0 = t_0 < t_1 < \ldots < t_m = 1$$

in such a way that $(\bar{s} \times [t_{i-1}, t_i])\, F$ lies in some canonical neighbourhood $U(x_{i-1})$, for $i = 1, \ldots, m$. Suppose inductively that $\tilde{G}$ has been extended to $\tilde{F}_i : (\bar{s} \times [0, t_i]) \cup (\dot{s} \times I) \to \tilde{X}$ so that $\tilde{F}_i p = F \,|\, (\bar{s} \times [0, t_i]) \cup (\dot{s} \times I)$ and that $0 \leqslant i < m$. Now the canonical neighbourhoods $\tilde{U}(\tilde{x}_{i,1}), \ldots$ are disjoint and $(\bar{s} \times t_i) \cup (\dot{s} \times [t_i, t_{i+1}])$ is connected; there is therefore a single canonical neighbourhood $\tilde{U}(\tilde{x}_i)$ in which its image under $\tilde{F}_i$

‡ See P. J. Hilton [9], chapter 5.

lies. Let $q_i : U(x_i) \to \tilde{U}(\tilde{x}_i)$ be the inverse of $p \mid \tilde{U}(\tilde{x}_i)$; then we define $\tilde{F}_{i+1} : (\bar{s} \times [0, t_{i+1}]) \cup (\dot{s} \times I) \to \tilde{X}$ by

$$\tilde{F}_{i+1} \mid (\bar{s} \times [0, t_i]) \cup (\dot{s} \times I) = \tilde{F}_i,$$

and
$$\tilde{F}_{i+1} \mid \bar{s} \times [t_i, t_{i+1}] = Fq_i.$$

These partial maps agree where their closed domains overlap and so, by I. 1.6 they combine to define the map $\tilde{F}_{i+1}$. We observe that $\tilde{F}_{i+1}p = \tilde{F}_i p$ or $Fq_i p$, and that in either case $\tilde{F}_{i+1}p = F$. This establishes the induction and the special case of the theorem.

Secondly, we again take $K = \bar{s}$, $L = \dot{s}$, but make no further assumption. However, since $|K|$ is compact, we may subdivide K so finely (to $K^{(r)}$, say) that, for each simplex $t \in K^{(r)}$, $\bar{t}$ has the property, relative to the map F, of the first special case. Step-by-step application of the result in that case to the simplexes of $K^{(r)} - L^{(r)}$ (in non-decreasing order of dimension) leads to the required map $\tilde{F}$.

Finally, we consider the general case. Writing K_n for $K^n \cup L$ we may apply the result of the case above to extend $\tilde{G}$ first to

$$\tilde{G}_0 : (|K| \times 0) \cup (|K_0| \times I) \to \tilde{X}$$

and then, in succession, to maps

$$\tilde{G}_n : (|K| \times 0) \cup (|K_n| \times I) \to \tilde{X}$$

such that F extends $\tilde{G}_n p$. The maps $\tilde{G}_n$ combine to give the required map $\tilde{F}$; for $|K| \times I = \bigcup_n ((|K| \times 0) \cup (|K_n| \times I))$ and the topology of $|K| \times I$ ensures the continuity of $\tilde{F}$. ∎

We draw some immediate consequences from this important theorem.

6.5.8 Corollary. *Let K be a connected complex, L a subcomplex and y a point of $|L|$. Let $\tilde{f}_0, \tilde{f} : |K| \to \tilde{X}$ be maps such that $y\tilde{f}_0 = y\tilde{f}$ and $\tilde{f}_0 p \simeq \tilde{f}p \operatorname{rel} |L|$. Then $\tilde{f}_0 \simeq \tilde{f} \operatorname{rel} |L|$.*

Define $\tilde{g}_t : |L| \to \tilde{X}$ by $\tilde{g}_t = \tilde{f}_0 \mid |L|$. Then by 6.5.7 there is a homotopy $\tilde{f}_t : |K| \to \tilde{X} \operatorname{rel} |L|$ with $\tilde{f}_1 p = \tilde{f}p$. But $y\tilde{f}_1 = y\tilde{f}_0 = y\tilde{f}$ so that 6.5.6 ensures that $\tilde{f}_1 = \tilde{f}$. ∎

6.5.9 Proposition. *Any path in X starting at x_0 may be lifted (uniquely) to a path in $\tilde{X}$ starting at $\tilde{x}_0$. Moreover, if $v \simeq w : I, 0, 1 \to X, x_0, x_1 \operatorname{rel} \dot{I}$ and $\tilde{v}$, $\tilde{w}$ are the paths in $\tilde{X}$ starting at $\tilde{x}_0$ lifting v, w, then $\tilde{v}$, $\tilde{w}$ have the same final point and $\tilde{v} \simeq \tilde{w} \operatorname{rel} \dot{I}$.*

Let $v : I_r, 0 \to X, x_0$ be a path starting at x_0. Then plainly $e_{(r)} \simeq v \operatorname{rel} 0$. Let $\tilde{e}_{(r)}$ be the constant path of length r at $\tilde{x}_0$. Applying

6.5.7 (with $|K| = I_r$, $|L| = 0$), we infer a homotopy $\tilde{e}_{(r)} \simeq \tilde{v} \operatorname{rel} 0$ covering the homotopy $e_{(r)} \simeq v$. Thus in particular $0\tilde{v} = \tilde{x}_0$ and $\tilde{v}p = v$, so that $\tilde{v}$ is the required path lifting v.

Next we apply 6.5.8 with $|K| = I$, $|L| = \dot{I}$, $y = 0$. We immediately deduce $\tilde{v} \simeq \tilde{w} \operatorname{rel} \dot{I}$; this of itself implies that $\tilde{v}$, $\tilde{w}$ have the same final point. ▮

6.5.10 Theorem. *The covering map* $p : \tilde{X}, \tilde{x}_0 \to X, x_0$ *induces a monomorphism* $p_* : \pi(\tilde{X}, \tilde{x}_0) \to \pi(X, x_0)$.

The force of this theorem lies in the fact that p_* is monomorphic. This is an immediate consequence of 6.5.9 (together with the fact that every equivalence class of loops contains one of unit length). ▮

Any loop on x_0 lifts to a path of $\tilde{X}$ with initial point $\tilde{x}_0$; the loops on x_0 belonging to elements of $\pi(\tilde{X}, \tilde{x}_0)\, p_*$ are by 6.5.9 just those which lift to *closed* paths. It should be noticed that if $\tilde{x}_0$, $\tilde{x}_0'$ are distinct points of $p^{-1}(x_0)$, then a loop on x_0 may lift to a loop on $\tilde{x}_0$ but to a path on $\tilde{x}_0'$ which is not a loop. This observation is enriched by the following theorem.

6.5.11 Theorem. *Let* $\tilde{x}_0$, $\tilde{x}_0'$ *be points of* $p^{-1}(x_0)$; *then the groups* $\pi(\tilde{X}, \tilde{x}_0)\, p_*$, $\pi(\tilde{X}, \tilde{x}_0')\, p_*$ *are conjugate‡ subgroups of* $\pi(X, x_0)$. *Conversely, if* π_0 *is conjugate to* $\pi(\tilde{X}, \tilde{x}_0)\, p_*$, *there exists* $\tilde{x}_0' \in p^{-1}(x_0)$ *such that* $\pi(\tilde{X}, \tilde{x}_0')\, p_* = \pi_0$.

Let $\tilde{w}$ be any path in $\tilde{X}$ from $\tilde{x}_0$ to $\tilde{x}_0'$; note that $\tilde{X}$ is by definition path-connected. Then each element of $\pi(\tilde{X}, \tilde{x}_0')$ has a representative loop $\overline{\tilde{w}} * \tilde{v} * \tilde{w}$, where $\tilde{v}$ is a loop on $\tilde{x}_0$, and all such loops represent elements of $\pi(\tilde{X}, \tilde{x}_0')$. Now $\tilde{w}p$ is a loop (in X) on x_0; let ρ be its class in $\pi(X, x_0)$. Then $\{\overline{\tilde{w}} * \tilde{v} * \tilde{w}\}\, p_* = \rho^{-1}\{\tilde{v}p\}\rho$, so that

$$\pi(\tilde{X}, \tilde{x}_0')\, p_* = \rho^{-1}[\pi(\tilde{X}, \tilde{x}_0)\, p_*]\rho.$$

Conversely, let $\pi_0 = \rho^{-1}[\pi(\tilde{X}, \tilde{x}_0)\, p_*]\rho$, for $\rho \in \pi(X, x_0)$. Let v be any loop on x_0 of the class ρ; this lifts to a path $\tilde{v}$ with initial point $\tilde{x}_0$; let its final point be $\tilde{x}_0'$, which of course lies in $p^{-1}(x_0)$. Then

$$\pi_0 = \pi(\tilde{X}, \tilde{x}_0')\, p_*. \quad ▮$$

6.5.12 Theorem. *The cardinal of the set* $p^{-1}(x_0)$ *is the index of* $\pi(\tilde{X}, \tilde{x}_0)\, p_*$ *in* $\pi(X, x_0)$, *and is independent of* x_0.

Let $\tilde{x}_0' \in p^{-1}(x_0)$ and let $\tilde{w}$ be a path from $\tilde{x}_0$ to $\tilde{x}_0'$. Then $\tilde{w}p$ is a loop in X on x_0. If $\tilde{w}'$ is another path from $\tilde{x}_0$ to $\tilde{x}_0'$ then $\tilde{w} * \overline{\tilde{w}'}$ is a loop on

‡ For the group-theoretical concepts of this section see [20].

$\tilde{x}_0$ so that $\{\tilde{w}p * \bar{\tilde{w}}'p\} \in \pi(\tilde{X}, \tilde{x}_0)p_*$. This shows that each point of $p^{-1}(x_0)$ determines a right coset of $\pi(\tilde{X}, \tilde{x}_0)p_*$ in $\pi(X, x_0)$. It is clear that this association is in fact a (1, 1) correspondence. For, first, given any right coset, we pick a loop in a class in this coset and lift it (by 6.5.9) to a path in $\tilde{X}$ from $\tilde{x}_0$ to some $\tilde{x}_0' \in p^{-1}(x_0)$; then $\tilde{x}_0'$ determines the given coset. Secondly, if $\tilde{x}_0'$, $\tilde{x}_0''$ determine the same right coset and if $\tilde{w}'$, $\tilde{w}''$ are paths from $\tilde{x}_0$ to $\tilde{x}_0'$, $\tilde{x}_0''$, then

$$\{\tilde{w}'p * \bar{\tilde{w}}''p\} \in \pi(\tilde{X}, \tilde{x}_0)p_*.$$

Thus, if $\tilde{w}$ is the path ending in $\tilde{x}_0'$ which lifts $\tilde{w}''p$, $\tilde{w}' * \bar{\tilde{w}}$ is a loop on $\tilde{x}_0$ so that $\tilde{w}$ and $\tilde{w}'$ (and therefore also $\tilde{w}''$) have the same initial point $\tilde{x}_0$; it follows that $\tilde{w} = \tilde{w}''$ so that $\tilde{x}_0' = \tilde{x}_0''$.

It remains to show that the cardinal of the set $p^{-1}(x_0)$ is independent of x_0. Given $x_1 \in X$, let v be a path from x_0 to x_1 and let $\tilde{v}$ be a path in $\tilde{X}$ from $\tilde{x}_0$ to $\tilde{x}_1$ that lifts v. Then p induces

$$p_{i*} : \pi(\tilde{X}, \tilde{x}_i) \to \pi(X, x_i) \quad (i = 0, 1),$$

and there is a commutative diagram

$$\begin{array}{ccc} \pi(\tilde{X}, \tilde{x}_0) & \overset{\{\tilde{v}\}_*}{\cong} & \pi(\tilde{X}, \tilde{x}_1) \\ \big\downarrow{\scriptstyle p_{0*}} & & \big\downarrow{\scriptstyle p_{1*}} \\ \pi(X, x_0) & \overset{\{v\}_*}{\cong} & \pi(X, x_1) \end{array}$$

This means that the isomorphism $\{v\}_*$ maps $\pi(\tilde{X}, \tilde{x}_0)p_{0*}$ onto $\pi(\tilde{X}, \tilde{x}_1)p_{1*}$. Thus the index of $\pi(\tilde{X}, \tilde{x}_0)p_{0*}$ in $\pi(X, x_0)$ is equal to the index of $\pi(\tilde{X}, \tilde{x}_1)p_{1*}$ in $\pi(X, x_1)$. ∎

If the cardinal of $p^{-1}(x_0)$ is a finite number m, we say that the covering is *m-fold*.

6.5.13 *Example.* S^2 covers P^2; $\pi(P^2) = J_2$, $\pi(S^2) = 1$. Thus the index of $\pi(S^2)p_*$ in $\pi(P^2)$ is 2 and 2 points of S^2 lie over each point of P^2, or S^2 is a two-fold covering of P^2.

We define a covering as *regular* if $\pi(\tilde{X}, \tilde{x}_0)p_*$ is self-conjugate in $\pi(X, x_0)$. We leave it as an easy exercise to the reader to verify that this definition does not depend on the choice of base-point $x_0 \in X$. Clearly a regular covering is characterized by the property that a loop in X lifts to a loop for every choice of initial point if it lifts to a loop for some choice of initial point.

6.6 Existence and uniqueness theorems for covering spaces

The preceding section may be described as giving the postulational approach to the theory of covering spaces; we now adopt the constructive approach. We have seen how a covering space $\tilde{X}$ picks out a class of conjugate subgroups of $\pi = \pi(X, x_0)$; we shall now show how to construct a covering space corresponding to any given conjugacy class of subgroups of π, provided that X satisfies certain local conditions. Of course, if we have constructed a covering space $\tilde{X}$ such that, for some $\tilde{x}_0 \in \tilde{X}$, $\pi(\tilde{X}, \tilde{x}_0)p_*$ is a given subgroup π_0, then we get all subgroups conjugate to π_0 by choosing appropriately the base-point in $\tilde{X}$.

Consider a covering space $\tilde{X}$ and base-point $\tilde{x}_0$ such that

$$\pi(\tilde{X}, \tilde{x}_0)p_* = \pi_0 \subseteq \pi.$$

There is a $(1, 1)$ correspondence between the points $\tilde{x}$ of $\tilde{X}$ and the set of paths in X starting from x_0 which lift to paths in $\tilde{X}$ starting from $\tilde{x}_0$ and ending at $\tilde{x}$. Two paths in X corresponding in this way to the same point $\tilde{x}$ have the property that one followed by the reverse of the other is a loop determining an element of π_0. This observation motivates the construction.

Suppose now that X, x_0 and $\pi_0 \subseteq \pi$ are given and that X is path-connected; we proceed to define a space $\tilde{X}_{\pi_0}$ and a map $p: \tilde{X}_{\pi_0} \to X$. We consider the set of all paths in X starting at x_0 and we say that two such, v and w, are *π_0-equivalent* if $v * \overline{w}$ is defined and $\{v * \overline{w}\} \in \pi_0$; that this determines an equivalence relation follows from the group properties of π_0. The points of $\tilde{X}_{\pi_0}$ are defined to be the π_0-equivalence classes, and the transformation $p: \tilde{X}_{\pi_0} \to X$ associates with each equivalence class the common final point of its paths. Since X is path-connected, p is onto X. We observe that $p^{-1}(x_0)$ is the set of right cosets of π_0 in π.

We define next a basis for the open sets of $\tilde{X}_{\pi_0}$. For each pair consisting of an open set U of X and a point $\tilde{x} \in \tilde{X}_{\pi_0}$ such that $\tilde{x}p \in U$, we define the subset $(U, \tilde{x})$ of $\tilde{X}_{\pi_0}$ as follows: Let v be a path in X from x_0 representing $\tilde{x}$; then $(U, \tilde{x})$ consists of the π_0-equivalence classes of *continuations* $v * w$ of v in U; that is, paths $v * w$ such that w is a path in U. This set $(U, \tilde{x})$ is clearly independent of the choice v of representative path for $\tilde{x}$.

6.6.1 Proposition. *If* $\tilde{x}_2 \in (U, \tilde{x}_1)$, *then* $(U, \tilde{x}_2) = (U, \tilde{x}_1)$.

If v_1 represents $\tilde{x}_1$ and $v_1 * w_2 = v_2$ is a continuation in U representing $\tilde{x}_2$, then any continuation $v_2 * w$ of v_2 in U is also a continuation $v_1 * (w_2 * w)$ of v_1 in U; and any continuation $v_1 * w$ of v_1 in U is equivalent to a continuation $v_2 * (\overline{w}_2 * w)$ of v_2 in U. ▌ This proposition implies that if $\tilde{x}_1$, $\tilde{x}_2$ both belong to $p^{-1}(U)$, then $(U, \tilde{x}_1)$ and $(U, \tilde{x}_2)$ coincide or are disjoint.

The topology of $\tilde{X}_{\pi_0}$ is defined by saying that the sets $(U, \tilde{x})$ form a basis for its open sets. We verify that with these open sets $\tilde{X}_{\pi_0}$ is a topological space. First $\tilde{X}_{\pi_0} = (X, \tilde{x})$ for any $\tilde{x}$, so that the whole space is an open set. Then if $\tilde{U}$, $\tilde{V}$ are basic open sets and $\tilde{x} \in \tilde{U} \cap \tilde{V}$, by 6.6.1 there are open sets U, V of X such that $\tilde{U} = (U, \tilde{x})$ and $\tilde{V} = (V, \tilde{x})$; if $\tilde{W} = (U \cap V, \tilde{x})$, then $\tilde{W}$ is a basic set and

$$\tilde{x} \in \tilde{W} \subseteq \tilde{U} \cap \tilde{V}.$$

This completes the verification.

6.6.2 Proposition. *The transformation* $p : \tilde{X}_{\pi_0} \to X$ *is continuous.*

If U is an open set of X and $\tilde{x}p \in U$, then $(U, \tilde{x})$ is an open set of $\tilde{X}_{\pi_0}$ containing $\tilde{x}$ and lying in $p^{-1}(U)$; so $p^{-1}(U)$ is open and p is continuous. ▌

If $v : I_r, 0 \to X, x_0$ is any path we may define its lifted path $\tilde{v}$ in $\tilde{X}_{\pi_0}$ by the rule

$$t\tilde{v} = \{v \mid I_t\}_{\pi_0} \quad (0 \leqslant t \leqslant r),$$

where $\{\ \}_{\pi_0}$ means the π_0-equivalence class determined by the path.

6.6.3 Proposition. *The transformation* $\tilde{v} : I_r, 0 \to \tilde{X}_{\pi_0}, \tilde{x}_0$ *is continuous.*

Here $\tilde{x}_0$ is the π_0-equivalence class containing the path of length zero at x_0. Let $t_0 \in I_r$ and let U be an open set of X containing $t_0 v = t_0 \tilde{v} p$. Since v is continuous and U open there is a positive δ such that, for all $t \in (t_0 - \delta, t_0 + \delta) \cap [0, r]$, $tv \in U$; then for all such t $t\tilde{v} \in (U, t_0 \tilde{v})$. For, if $t > t_0$, $v \mid I_t$ is a continuation in U of $v \mid I_{t_0}$ and if $t < t_0$ $v \mid I_{t_0}$ is a continuation in U of $v \mid I_t$. In the former case the conclusion is immediate, in the latter it follows from 6.6.1. This shows t_0 to be interior to $\tilde{v}^{-1}(U, t_0 \tilde{v})$ and $\tilde{v}$ therefore to be continuous. ▌

6.6.4 Corollary. $\tilde{X}_{\pi_0}$ *is path-connected.*

If $\tilde{x} \in \tilde{X}_{\pi_0}$ and v is a path in X from x_0 representing $\tilde{x}$, then $\tilde{v}$ is a path in $\tilde{X}_{\pi_0}$ from $\tilde{x}_0$ to $\tilde{x}$. ▌

We cannot, unfortunately, conclude in general that this space $\tilde{X}_{\pi_0}$ is a covering space of X under p, for p may fail to be locally homeomorphic. It becomes necessary therefore to impose restrictions on X.

6.6.5 Definition. A space X is said to be *locally path-connected* if given any $x \in X$ and any open set U such that $x \in U$ there is an open set V such that $x \in V \subseteq U$ and such that any two points of V can be joined by a path in U.

We may abbreviate (locally) path-connected to (l.) p.c. We remark that the property of being l.p.c. is a local topological invariant: that is, if X and Y are locally homeomorphic and X is l.p.c., then so is Y.

6.6.6 Proposition. *X is l.p.c. if and only if it has a basis of p.c. open sets.*

The sufficiency of the condition is clear. Let X then be l.p.c., and let U be an open set of X and $x \in U$; we must prove that there is a p.c. open set V such that $x \in V \subseteq U$. We define the set V to be the set of points that can be joined to x by a path in U; clearly $x \in V \subseteq U$ and we must prove V to be p.c. and open. Now, if $y \in V$, there is a path joining x to y in U and any point on such a path is itself joined to x by a path in U; hence the whole path lies in V and V is p.c. Suppose, finally, that $y \in V \subseteq U$; since X is l.p.c., there is an open set W such that $y \in W \subseteq U$ and such that all the points of W can be joined to y in U, which implies that all the points of W can be joined to x in U and that $W \subseteq V$. So y is interior to V and V is open. ∎

6.6.7 *Example.* The cone on the subset of R^1 consisting of the points $0, 1, \frac{1}{2}, \frac{1}{3}, \ldots, 1/n, \ldots$ is p.c. but not l.p.c.; a small open set containing 0 will contain no p.c. open set containing 0. If we take this space to be X and construct $\tilde{X}_{\pi_0}$ as above, taking for π_0 the subgroup of one element (in this case the whole group), if $x_0 = 0$ and $\tilde{x}_0$ is the π_0-class containing $e(x_0)$ the constant path of length 0 at x_0, then small neighbourhoods of $\tilde{x}_0$ will contain no point whose projection is one of the points $1/n$; p is therefore not locally homeomorphic and so not a covering map. The effect here of passing from X to $\tilde{X}_{\pi_0}$ is to pull apart the pencil of lines so that they no longer bunch together.

6.6.8 Definition. A space X is said to be *locally simply connected* (*in the weak sense*)‡ if every point x lies in an open set U such that all loops in U are null-homotopic in X. We may describe a subset A as *weakly simply connected* or w.s.c. if every loop in A is null-homotopic

‡ X is l.s.c. in the strong sense if for each point x and open set $U \ni x$ there is an open set V such that $x \in V \subseteq U$ and any loop in V is null-homotopic in U. The cone on Y, where Y is the space of Fig. 6.5, is a space which is l.s.c in the weak but not in the strong sense.

in X; we may abbreviate locally simply connected (in the weak sense) to l.s.c. Thus X is l.s.c. if it can be covered by w.s.c. open sets. The property of being l.s.c. is plainly a local topological invariant.

6.6.9 Proposition. *X is l.p.c. and l.s.c. if and only if it has a basis of p.c. and w.s.c. open sets.*

The condition is again plainly sufficient. For necessity we must prove that for any $x \in X$ and any open set $U \ni x$ there is a p.c. and w.s.c. open set W such that $x \in W \subseteq U$. Now by Definition 6.6.8 there is an open set $V \ni x$ which is w.s.c. and by 6.6.6 there is a p.c. open set W such that $x \in W \subseteq U \cap V$. Since $W \subseteq V$ it is also w.s.c. ∎

6.6.10. *Example.* Let Y be the subset of R^2 which is the union of the circles with centre $(1/n, 0)$ and radius $1/n$, $n = 1, 2, \ldots$.

If y_0 is the origin, there is no open set of Y containing y_0 which is w.s.c. so that Y is not l.s.c.

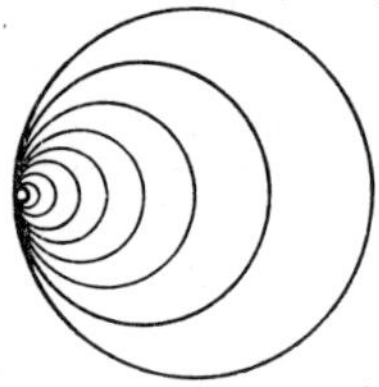

Fig. 6.5

If we construct $\tilde{Y}_{\pi_0}$ from Y taking for π_0 again the subgroup of just one element, and if $\tilde{y}_0$ is the class containing $e(y_0)$ then for any open set $V \ni y_0$, $(V, \tilde{y}_0)$ contains infinitely many points whose projections are y_0; for V contains whole circles and the transit of any such circle is a loop in V not determining an element of π_0. Here too, therefore, p fails to be a local homeomorphism and fails therefore also to be a covering map. The effect here of passing from Y to $\tilde{Y}_{\pi_0}$ is to unravel the local singularity at y_0; a local homeomorphism would preserve local features.

Examples 6.6.7 and 6.6.10 may reconcile the reader to the restriction imposed on X in the following theorem, the central theorem of this section:

6.6.11 Theorem. *If X is p.c., l.p.c., and l.s.c., and if $\tilde{X}_{\pi_0}$ is constructed as at the start of this section and $\tilde{x}_0$ is the π_0-equivalence class containing $e(x_0)$, then $p : \tilde{X}_{\pi_0} \to X$ is a covering map and*

$$\pi(\tilde{X}_{\pi_0}, \tilde{x}_0)\, p_* = \pi_0.$$

We know from 6.6.4 that $\tilde{X}_{\pi_0}$ is p.c. even in the general case. We define canonical neighbourhoods in X, $\tilde{X}_{\pi_0}$; those in X are the p.c. and w.s.c. open sets U; those in $\tilde{X}_{\pi_0}$ are the basic open sets of the form $(U, \tilde{x})$, where U is canonical in X. The canonical neighbourhoods in X form a basis by 6.6.9. Now let W be any open set of X, so that $(W, \tilde{x})$ is a basic open set of $\tilde{X}_{\pi_0}$; then if U is a p.c. and w.s.c. open set for which $\tilde{x}p \in U \subseteq W$, $(U, \tilde{x}) \subseteq (W, \tilde{x})$. It follows that the canonical neighbourhoods in $\tilde{X}_{\pi_0}$ also form a basis. We now proceed to verify properties (i), (ii) and (iii) of the definition of a covering map. Let

$$p_i = p \mid (U, \tilde{x}_i) : (U, \tilde{x}_i) \to U;$$

we prove p_i homeomorphic by proving that it is (1, 1), onto, open and continuous. It is onto U since U is p.c. Suppose $\tilde{y}, \tilde{y}' \in (U, \tilde{x}_i)$ such that $\tilde{y}p = \tilde{y}'p$. By 6.6.1, $(U, \tilde{x}_i) = (U, \tilde{y}) = (U, \tilde{y}')$; now $(U, \tilde{y})$ consists of π_0-classes of continuations in U of a representative path for $\tilde{y}$; since U is w.s.c. any continuation by a loop represents $\tilde{y}$ again, which implies that $\tilde{y}' = \tilde{y}$ and hence that p_i is (1, 1). Since p_i is a restriction of p, 6.6.2 implies that p_i is continuous. To prove that p_i is open we need only prove that, if $(V, \tilde{y})$ is a canonical neighbourhood in $(U, \tilde{x}_i)$, then $(V, \tilde{y})p$ is open in U. But, since V is p.c., $(V, \tilde{y})p_i = (V, \tilde{y})p = V$ which is open in X and therefore also open in U. This completes the verification of property (i). Let $y = \tilde{y}p \in U$, where U is a canonical neighbourhood of x and let v be a path representing $\tilde{y}$ and w a path in U from y to x. Then if $v * w$ represents $\tilde{x}$, $\tilde{y} \in (U, \tilde{x})$. This establishes property (ii). Property (iii) follows from 6.6.1 together with the fact already proved that p_i is (1, 1). The projection $p : \tilde{X}_{\pi_0} \to X$ is therefore a covering map.

The image $\pi(\tilde{X}_{\pi_0}, \tilde{x}_0)p_*$ is known to consist of those classes of loops on x_0 which lift to loops on $\tilde{x}_0$. Now if v is such a loop in X, $\tilde{v}$ as defined in 6.6.3 is the lifted path in $\tilde{X}_{\pi_0}$ starting at $\tilde{x}_0$; its final point is $\tilde{x}_0$ if and only if $\{v\}_{\pi_0} = \tilde{x}_0$, which is the same as to say that $\{v\} \in \pi_0$. ∎

Having proved that, under the restrictions imposed on X, there is a covering space of X corresponding to each subgroup of $\pi(X, x_0)$, we proceed to establish the uniqueness of the covering space; for this we shall be able to dispense with the requirement that X be l.s.c. The main lemma is

6.6.12 Lemma. *If* $q : \tilde{Y}, \tilde{y}_0 \to Y, y_0$ *is a covering map and*

$$f : X, x_0 \to Y, y_0$$

is a map of the p.c. and l.p.c. space X into Y such that

$$\pi(X,x_0)f_* \subseteq \pi(\tilde{Y},\tilde{y}_0)q_*,$$

then there exists a unique map $\tilde{f}: X, x_0 \to \tilde{Y}, \tilde{y}_0$ *such that* $\tilde{f}q = f$.

Let $x \in X$ and let v be path in X from x_0 to x. Then vf is a path in Y from y_0 to xf. By 6.5.9 we may lift this to a path in $\tilde{Y}$ from $\tilde{y}_0$ to some point $\tilde{y}$ such that $\tilde{y}q = xf$.

Now let v' be another path in X from x_0 to x. Then $v * \bar{v}'$ is a loop in X on x_0, so that $\{vf * \overline{v'f}\} \in \pi(X,x_0)f_*$. Since

$$\pi(X,x_0)f_* \subseteq \pi(\tilde{Y},\tilde{y}_0)q_*$$

it follows that $vf * \overline{v'f}$ lifts to a loop in $\tilde{Y}$ on $\tilde{y}_0$. This implies that, if we lift $v'f$ to a path in $\tilde{Y}$ starting at $\tilde{y}_0$, then the final point of this path is also $\tilde{y}$. Thus $\tilde{y}$ depends only on x (and not on the choice of v) and we define a transformation $\tilde{f}: X \to \tilde{Y}$ by $x\tilde{f} = \tilde{y}$. Then certainly $\tilde{f}q = f$ and, plainly, $x_0\tilde{f} = \tilde{y}_0$.

It remains to show that $\tilde{f}$ is continuous, since the uniqueness then follows from 6.5.6. If $x_1\tilde{f} = \tilde{y}_1$, let $\tilde{U}_1$ be any neighbourhood of $\tilde{y}_1$. If $\tilde{U}$ is a canonical neighbourhood of $\tilde{y}_1$, let $\tilde{W} = \tilde{U}_1 \cap \tilde{U}$ and let $W = \tilde{W}q$. Then $f^{-1}(W)$ is a neighbourhood of x_1 in X. Since X is l.p.c., we may find a p.c. neighbourhood V of x_1, with $V \subseteq f^{-1}(W)$. We assert that $V\tilde{f} \subseteq \tilde{W}$. For if $x \in V$ and v_1 is a path from x_0 to x_1, we may join x_0 to x by a continuation $v = v_1 * v_2$ of v_1 in V. Then $vf = v_1 f * v_2 f$ and $v_2 f$ is a path in W; it follows that $v_2 f$ lifts to a path in $\tilde{W}$ starting at $\tilde{y}_1$, so that vf lifts to a path in $\tilde{Y}$ starting at $\tilde{y}_0$ and ending in $\tilde{W}$. That is to say, $x\tilde{f} \in \tilde{W}$ and our assertion is proved. ∎

6.6.13 Corollary. *If X is l.p.c. and 1-connected then every map* $X, x_0 \to Y, y_0$ *can be lifted uniquely to a map* $X, x_0 \to \tilde{Y}, \tilde{y}_0$. ∎

Notice that the condition $\pi(X,x_0)f_* \subseteq \pi(\tilde{Y},\tilde{y}_0)q_*$ is necessary for the existence of $\tilde{f}$ in 6.6.12; for, if $\tilde{f}$ exists, then $\pi(X,x_0)f_* = \pi(X,x_0)\tilde{f}_* q_* \subseteq \pi(\tilde{Y},\tilde{y}_0)q_*$. Even more striking is the fact that we are not able simply to dispense with the requirement that X be l.p.c.

The following example is due to E. C. Zeeman

6.6.14 *Example.* The space $Y \subset R^2$ is the union of three pieces Y_0, Y_1, Y_2. Y_0 is the unit circle $u_1^2 + u_2^2 = 1$; Y_1, the fingers of Y, is the Euclidean join of $(1,0)$ to the set of points $(3, 1/n)$, $n = 1, 2, \ldots$; Y_2, the thumb of Y, is the arc $(u_1 - 2)^2 + u_2^2 = 1$, $u_2 \leqslant 0$, whose extremities are $y_0 = (1,0)$ and $y_1 = (3,0)$.

In fig. 6.6 the reader will find pictures of X, Y, $\tilde{Y}$. $\tilde{Y}$ is a two-handed version of Y as is X also; but in X the tip of each thumb (which is twice as long as Y_2) abuts on the finger-nails of the opposite hand. Under each of f and q the circle is wrapped twice round Y_0 and under each the sets of fingers are mapped isometrically to the fingers of Y; under q each thumb of $\tilde{Y}$ is also mapped isometrically to the thumb of Y, and under f each thumb is mapped by some homeomorphism to the thumb of Y. Certainly q is a (two-fold) covering and $\pi(X, x_0)f_* = \pi(\tilde{Y}, \tilde{y}_0)\,q_*$. The construction of 6.6.12 therefore determines a transformation $\tilde{f} : X, x_0 \to \tilde{Y}, \tilde{y}_0$ such that $\tilde{f}q = f$. Under $\tilde{f}$ the fingers at x_0 are mapped to the fingers at $\tilde{y}_0$ and the thumbnail x_1 to the thumbnail $\tilde{y}_1$. Now in X the sequence of right-hand finger nails converge to x_1, but their images under $\tilde{f}$ do not converge to $\tilde{y}_1$, the image of x_1. In this case therefore $\tilde{f}$ fails to be continuous. Note that X is not l.p.c. at x_1. It is useful to observe that in this example f is actually a covering map.

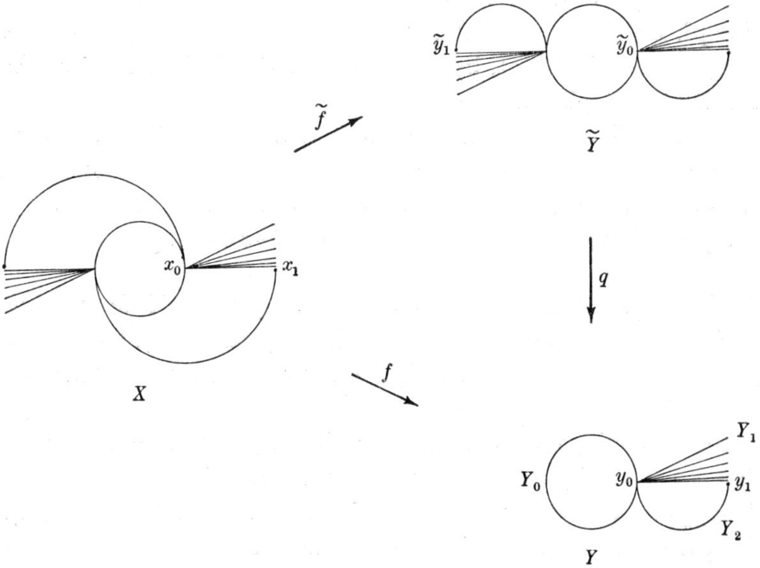

Fig. 6.6

6.6.15 *Example.* Suppose we have a covering map $q : \tilde{Y}, \tilde{y}_0 \to Y, y_0$; we consider maps $S^n, x_0 \to \tilde{Y}, y_0$ and $S^n, x_0 \to Y, y_0$ for $n \geq 2$. By 6.6.13 (since S^n is obviously l.p.c.) we know that every map $S^n, x_0 \to Y, y_0$ may be lifted to a map $S^n, x_0 \to \tilde{Y}, \tilde{y}_0$ and by 6.5.8 we know that if, for two maps $\tilde{f}_0, \tilde{f}_1 : S^n, x_0 \to \tilde{Y}, \tilde{y}_0$, $\tilde{f}_0 q \simeq \tilde{f}_1 q \operatorname{rel} x_0$, then $\tilde{f}_0 \simeq \tilde{f}_1 \operatorname{rel} x_0$. Thus the map q induces a (1, 1) correspondence q_* between homotopy classes of maps $S^n, x_0 \to \tilde{Y}, \tilde{y}_0$ and homotopy classes of maps $S^n, x_0 \to Y, y_0$. These classes are the elements of the n-th *homotopy groups*‡ $\pi_n(\tilde{Y}, \tilde{y}_0)$, $\pi_n(Y, y_0)$ and q_* is indeed an isomorphism between these groups.

A particularly interesting case is that in which $Y = S^1$. The real line R^1 is a covering space of S^1 and is, moreover, contractible rel u for any $u \in R^1$. Thus if $f : S^n, x_0 \to S^1, y_0$ is a map, f may be lifted into a contractible space and is thus nullhomotopic rel x_0. We have shown that every map $S^n \to S^1$, $n > 1$, is nullhomotopic. This fact is in striking contrast with the situation when we study maps $S^n \to S^m$, $n > m > 1$. The calculation of these—in general, highly non-trivial—homotopy groups of spheres has provided considerable employment for algebraic topologists.

‡ See Background to Part II.

We return now to the proof of uniqueness of the covering space.

6.6.16 Theorem‡ *If X is p.c. and l.p.c. and π_0 is a subgroup of $\pi(X, x_0)$ and if $p^i : \tilde{X}^i, \tilde{x}_0^i \to X, x_0$, $i = 1, 2$, are covering maps such that $\pi(\tilde{X}^1, \tilde{x}_0^1)\, p_*^1 = \pi(\tilde{X}^2, \tilde{x}_0^2)\, p_*^2 = \pi_0$, then there is a unique homeomorphism $h : \tilde{X}^1, \tilde{x}_0^1 \to \tilde{X}^2, \tilde{x}_0^2$ such that $hp^2 = p^1$.*

Since X is l.p.c., so is $\tilde{X}$. Thus we may apply 6.6.12 with $q = p^2$, $f = p^1$ to deduce the existence of a unique map $h : \tilde{X}^1, \tilde{x}_0^1 \to \tilde{X}^2, \tilde{x}_0^2$ with $hp^2 = p^1$. Similarly, there exists a map $k : \tilde{X}^2, \tilde{x}_0^2 \to \tilde{X}^1, \tilde{x}_0^1$ with $kp^1 = p^2$. Then $hkp^1 = p^1$. Thus hk and 1 are both maps $\tilde{X}^1, \tilde{x}_0^1 \to \tilde{X}^1, \tilde{x}_0^1$ lifting $1 : X \to X$ so that, by 6.5.6, $hk = 1$. Similarly, $kh = 1$ so that h is a homeomorphism. ❚

If $p : \tilde{X} \to X$ is a covering map, a *cover transformation* is an auto-homeomorphism $h : \tilde{X} \to \tilde{X}$ such that $hp = p$. The cover transformations obviously form a group H. We shall describe this situation by saying that H acts as a group of *right-operators* on $\tilde{X}$.

6.6.17 Proposition. *If X is l.p.c. and $p : \tilde{X} \to X$ is a regular covering map* (i.e. $\pi_0 = \pi(\tilde{X}, \tilde{x}_0)\, p_*$ *is self-conjugate in* π), *then for each point $\tilde{x}_0' \in p^{-1}(x_0)$ there is a unique cover transformation h such that $(\tilde{x}_0)\, h = \tilde{x}_0'$.*

The self-conjugacy of π_0 implies that $\pi(\tilde{X}, \tilde{x}_0')\, p_* = \pi_0$ and the existence and uniqueness of h follow from 6.6.16. ❚

The space X may then be regarded as the identification space $\tilde{X}/H$ arising from $\tilde{X}$ and the equivalence relation that relates $\tilde{x}$ with $\tilde{x}h$ for each $\tilde{x} \in \tilde{X}$ and each $h \in H$. We observe that each equivalence class is a discrete subspace of $\tilde{X}$ and that H operates without fixed points; this means that if, for some $\tilde{x} \in \tilde{X}$, $\tilde{x}h = \tilde{x}$, then h is the identity transformation. Since $hp = p$ this follows from 6.5.6.

We leave it as an exercise for the reader to state and prove a converse to 6.6.17.

A transformation γ from a group G to a group G' is said to be an *antihomomorphism* if, for all $g_1, g_2 \in G$, $(g_1 g_2)\gamma = (g_2\gamma)(g_1\gamma)$; if γ is (1, 1) on to G', we say that γ is an *anti-isomorphism* and write $\gamma : G \mathrel{\underset{\sim}{=}} G'$.

6.6.18 Theorem. *If X is l.p.c. and $p : \tilde{X} \to X$ is a regular covering map with $\pi_0 = \pi(\tilde{X}, \tilde{x}_0)\, p_*$ then the choice of $\tilde{x}_0$ and the correspondence $\tilde{x}_0' \to h$ of 6.6.17 induce an anti-isomorphism $\phi_* : \pi/\pi_0 \mathrel{\underset{\sim}{=}} H$.*

‡ A more general form (6.7.2) of this theorem is held over to the next section.

Each element $\alpha \in \pi$ determines a cover transformation as follows: if v is a loop on x_0 representing α and $\tilde{v}$ is the lifted path from $\tilde{x}_0$, its final point $\tilde{x}_0'$ lies in $p^{-1}(x_0)$ and so, by 6.6.17, determines a cover transformation under which $\tilde{x}_0$ is transformed into $\tilde{x}_0'$. This final point $\tilde{x}_0'$ is independent of the choice v of representative loop and we may call the cover transformation so defined h_α; note that h_α depends not only on α but also on the choice $\tilde{x}_0$ of base-point in $\tilde{X}$. If $\tilde{y}_0 \in p^{-1}(x_0)$, we consider $(\tilde{y}_0)\,h_\alpha$; we take any path $\tilde{w}$ from $\tilde{x}_0$ to $\tilde{y}_0$ and the path $\tilde{w}'$ which lifts $w = \tilde{w}p$ and starts from $\tilde{x}_0'$; it is the end-point $\tilde{y}_0'$ of $\tilde{w}'$ that is $(\tilde{y}_0)\,h_\alpha$. If now the loop w represents β in π, $\tilde{y}_0 = (\tilde{x}_0)\,h_\beta$ and we have shown that $\tilde{y}_0' = (\tilde{x}_0)\,h_\beta h_\alpha$. But we may consider the loop $v * w$, representing $\alpha\beta$; the lifted path starting from $\tilde{x}_0$ is $\tilde{v} * \tilde{w}'$ with final point $\tilde{y}_0'$ so that $(\tilde{x}_0)\,h_{\alpha\beta} = (\tilde{x}_0)\,h_\beta h_\alpha$. Therefore $h_{\alpha\beta} = h_\beta h_\alpha$ and we have determined an anti-homomorphism $\phi : \pi(X, x_0) \to H$ by the rule $\alpha\phi = h_\alpha$. Since $\tilde{X}$ is p.c., ϕ is epimorphic; its kernel is clearly π_0. Hence ϕ induces an anti-isomorphism $\phi_* : \pi/\pi_0 \approx H$. ❚

We observe that π itself operates on $\tilde{X}$ on the left through the anti-homomorphism ϕ. Notice that 6.6.18 asserts that $\phi_* : \pi/\pi_0 \approx H$, but that the particular anti-isomorphism exhibited was constructed with the help of the base-point $\tilde{x}_0$ in $p^{-1}(x_0)$. If, however, the point $\tilde{y}_0$ of the proof of 6.6.18 had been chosen for base-point and the associated anti-homomorphism written as $\psi : \pi \to H$, then since a path from $\tilde{y}_0$ to $\tilde{y}_0'$ is $\overline{\tilde{w}} * \tilde{v} * \tilde{w}'$ whose projection is $\overline{w} * v * w$, $\alpha\phi = h_\alpha = (\beta^{-1}\alpha\beta)\,\psi$. The anti-isomorphisms ϕ_*, ψ_* between π/π_0 and H are not in general the same; if, however, π_0 contains $[\pi, \pi]$, this construction provides a unique anti-isomorphism between π/π_0 and H, and π operates canonically on $\tilde{X}$. In this case of course π/π_0 is commutative, so that the anti-isomorphism is also an isomorphism and the distinction between left and right operation disappears.

6.7 The universal covering space

6.7.1 Definition. If $\tilde{X}$ is a covering space of X and is simply-connected it is said to be a *universal covering space* of X.

In this case $\pi(\tilde{X}, \tilde{x}_0)\,p_*$ consists of the unity element alone of π. We can therefore deduce from 6.6.16 that, if X is l.p.c., any two universal covering spaces are homeomorphic, which justifies the common practice of speaking about *the* universal covering space. If X is l.p.c. and l.s.c., we know from 6.6.11 that it has a universal covering space. The word 'universal' is explained by the Corollary to the next theorem.

6.7.2 Theorem. *If X is p.c. and l.p.c. and if $p^i : \tilde{X}^i, \tilde{x}_0^i \to X, x_0$, $i = 1, 2$, are covering maps such that $\pi^1 \subseteq \pi^2$, where $\pi^i = \pi(\tilde{X}^i, \tilde{x}_0^i)\, p_*^i$, then there is a covering map $h : \tilde{X}^1, \tilde{x}_0^1 \to \tilde{X}^2, \tilde{x}_0^2$ such that $hp^2 = p^1$.*

As in 6.6.16 (of which 6.7.2 is a generalization) we deduce the existence of a map $h : \tilde{X}^1, \tilde{x}_0^1 \to \tilde{X}^2, \tilde{x}_0^2$ such that $hp^2 = p^1$, and it remains to show that h is a covering map. If $U^i(x)$, $\tilde{U}^i(\tilde{x}^i)$ are canonical neighbourhoods for p^i, $i = 1, 2$, let $U(x)$ be a p.c. open set contained in $U^1(x) \cap U^2(x)$; if $\tilde{x}^i p^i = x$, let $\tilde{U}(\tilde{x}^i) = \tilde{U}^i(\tilde{x}^i) \cap (p^i)^{-1} U(x)$, $i = 1, 2$. Then $U(x)$, $\tilde{U}(\tilde{x}^i)$ are canonical neighbourhoods for p^i and the reader will readily verify that $\tilde{U}(\tilde{x}^2)$, $\tilde{U}(\tilde{x}^1)$ are canonical neighbourhoods for the covering map h. ∎

6.7.3 Corollary. *If X is p.c., l.p.c., and l.s.c. it has a universal covering space which is a covering space of all the covering spaces of X.* ∎

We note the following consequence of 6.6.18:

6.7.4 Corollary. *In the special case that $\tilde{X}$ is the universal covering space of the l.p.c. space X, then it is the fundamental group π of X that is mapped anti-isomorphically onto H.* ∎

If base-points $\tilde{x}_0$ and $x_0 = \tilde{x}_0 p$ are chosen, we may then regard π as acting as a group of left operators on $\tilde{X}$ without fixed points. However, π operates canonically on $\tilde{X}$ if and only if it is commutative.

The technique of passing from a space to its universal covering space has proved very useful in topology; the universal covering space is simply connected and is therefore easier to handle but it has higher homotopy groups isomorphic to those of the original space; compare 6.6.15. This process of 'killing' the first homotopy group (that is, the fundamental group) without changing the others has been generalized by fibre-space technique; this technique has the effect of 'killing' the lowest-dimensional non-vanishing homotopy group and leaving the rest unchanged.

6.7.5 *Example.* Let S^3 be the 3-sphere given by complex coordinates (z_0, z_1) subject to $z_0\bar{z}_0 + z_1\bar{z}_1 = 1$; let p, q be mutually prime integers and let $f : S^3 \to S^3$ be the rotation given by $(z_0, z_1)f = (z_0 e^{2\pi i/p}, z_1 e^{2q\pi i/p})$. Then f generates a group $\Gamma = J_p$ of rotations of S^3, operating on S^3 without fixed points. The factor space S^3/Γ is the lens space $L(p, q)$ (see 5.10) whose universal covering space is S^3. Since Γ is the cover transformation group, the fundamental group of $L(p, q)$ is J_p.

6.8 The covering space of a polyhedron

Our object is to triangulate the covering space of a connected polyhedron $|K|$ in such a way that the covering map is a simplicial map onto K.

We first prove that $|K|$ is l.p.c. and l.s.c. If K is locally finite, there is no difficulty in proving this; the complication of our proof arises because we make no assumption about K. If $x \in |K|$ and s is the carrier of x, then if $x \in U$, an open set of $|K|$, then $x \in U \cap |\text{st}(s)|$ which is also an open set of $|K|$. Now $|\text{st}(s)|$ is simply-connected so that any subset of it is w.s.c.; this proves that $|K|$ is l.s.c. To prove $|K|$ l.p.c., again let $x \in U$, an open set of $|K|$, and let $|K| - U = F$, a closed set and therefore having closed intersection with each closed simplex $\bar{t}$. We define another closed set $F' \supseteq F$ as follows: if $t \notin \text{st}(s)$, $\bar{t} \subseteq F'$; if $t \in \text{st}(s)$, $F' \cap \bar{t}$ is the shadow in $\bar{t}$ cast by $F \cap \bar{t}$ from a source of light at x. (More formally, $y \in F'$ if $y \notin |\text{st}(s)|$ or if $y \in |\text{st}(s)|$ and the closed segment xy intersects F.) Since for each $t \in K$, $F' \cap \bar{t}$ is a closed set of $\bar{t}$, F' is closed in $|K|$; evidently $F \subseteq F'$ and $x \notin F'$. Hence if $V = |K| - F'$, V is open and $x \in V \subseteq U$. But each point y of V is connected to x by a straight line path lying in V; so V is p.c. This proves that $|K|$ is l.p.c.

Suppose now that $p : \tilde{X}, \tilde{x}_0 \to |K|, a^0$ is a covering map and $\pi(\tilde{X}, \tilde{x}_0) p_* = \pi_0 \subseteq \pi(|K|, a^0)$. We refer to 6.3.2, which asserts that a canonical homomorphism $\rho_* : \pi(K, a^0) \to \pi(|K|, a^0)$ is an isomorphism and we write Q for the subgroup of $\pi(K, a^0)$ such that $Q\rho_* = \pi_0$. We shall triangulate $\tilde{X}$ by defining a complex $\tilde{K}_Q$ and simplicial map $q : \tilde{K}_Q \to K$ such that $|q|$ is a covering map and $\pi(\tilde{K}_Q, \tilde{a}^0) q_* = Q$, and then appeal to 6.6.16 and the naturality of ρ_*.

We start by giving the construction of $\tilde{K}_Q$ where Q is an arbitrary subgroup of $\pi(K, a^0)$, the construction being highly analogous to the construction of $\tilde{X}_{\pi_0}$. In the set of edge-paths of K starting from a^0, we set up an equivalence relation, saying that two such v, w are Q-equivalent if $\{v * \overline{w}\} \in Q$. We define the vertices $\tilde{a}$ of $\tilde{K}_Q$ to be the Q-equivalence classes; the transformation q from the vertices of $\tilde{K}_Q$ to those of K is defined by the rule that $\tilde{a}q$ is the final vertex of a representative edge-path of $\tilde{a}$.

We must now select the subsets of these vertices that are to span simplexes of $\tilde{K}_Q$. They are those subsets $\tilde{a}^{i_0}, \ldots, \tilde{a}^{i_n}$ of $\tilde{K}_Q$ such that (i) their images, $a^{i_0}, \ldots, a^{i_n}$, under q span a simplex in K, and (ii) if v_r is a representative edge-path for $\tilde{a}^{i_r}$, then $v_r * a^{i_r} a^{i_s}$ is a representative edge-path for $\tilde{a}^{i_s}$. Clearly any subset of a set of vertices spanning a simplex in $\tilde{K}_Q$ also satisfies these conditions and we have therefore defined an abstract complex $\tilde{K}_Q$ and the vertex transformation defines a simplicial map $q : \tilde{K}_Q \to K$.

6.8.1 Proposition. *If $a^{i_0}, \ldots, a^{i_n}$ span a simplex of K and if $\tilde{a}^{i_0} \in q^{-1}(a^{i_0})$, there is a unique set of vertices $\tilde{a}^{i_1}, \ldots, \tilde{a}^{i_n}$ such that $\tilde{a}^{i_r} q = a^{i_r}$, $r = 1, \ldots, n$, and that $\tilde{a}^{i_0}, \ldots, \tilde{a}^{i_n}$ span a simplex of $\tilde{K}_Q$.*

Let v be a representative edge-path for $\tilde{a}^{i_0}$; then the only vertex in $q^{-1}(a^{i_r})$ that spans a 1-simplex with $\tilde{a}^{i_0}$ is that represented by $v * a^{i_0} a^{i_r}$. This set $\tilde{a}^{i_0}, \ldots, \tilde{a}^{i_n}$ does span a simplex of $\tilde{K}_Q$; for we may represent $\tilde{a}^{i_r}$ by $v * a^{i_0} a^{i_r}$ and the continuation

$$v * a^{i_0} a^{i_r} * a^{i_r} a^{i_s} = v * a^{i_0} a^{i_r} a^{i_s} \equiv v * a^{i_0} a^{i_s}$$

which represents $\tilde{a}^{i_s}$; the second condition for a set of vertices to span in $\tilde{K}_Q$ is therefore satisfied as well as the first. ▮

This is a combinatorial analogue of 6.5.6.

6.8.2 Proposition. *If $\tilde{a}^0$ is the class of the edge-path a^0, then*

$$\pi(\tilde{K}_Q, \tilde{a}_0)\, q_* = Q.$$

We leave the proof as an exercise for the reader, referring him to the last part of the proof of 6.6.11. ▮

6.8.3 Proposition. *$|q|$ is a covering map.*

Since K is connected $|q|$ is onto $|K|$. We take for canonical neighbourhoods in $|K|$ and in $|\tilde{K}_Q|$ the stars of vertices. By 6.8.1 $|q|^{-1}\,|\mathrm{st}\,(a^i)|$ breaks up into a disjoint collection of stars in $|\tilde{K}_Q|$ of the various vertices in $q^{-1}(a^i)$, each such star in $\tilde{K}_Q$ being mapped homeomorphically onto $|\mathrm{st}\,(a^i)|$ by $|q|$. ▮

6.8.4 Theorem. *Any covering space $\tilde{X}$ of a polyhedron $X = |K|$ is the polyhedron of a complex of the same dimension as K.*

We adopt the notation of this section, choosing Q in $\pi(K, a^0)$ so that $Q\rho_* = \pi_0 = \pi(\tilde{X}, \tilde{x}_0)\, p_*$. It follows from the naturality of ρ_* that $\pi(|\tilde{K}_Q|, \tilde{a}^0)|q|_* = \pi_0$, and it then follows from 6.8.3 and 6.6.16 that there is a homeomorphism $h : |\tilde{K}_Q|, \tilde{a}_0 \to \tilde{X}, \tilde{x}_0$ such that $hp = |q|$; this homeomorphism imprints on $\tilde{X}$ a simplicial structure in terms of which p is a simplicial map. The assertion about dimension follows from the construction of $\tilde{K}_Q$. ▮

There is a somewhat unexpected deduction that can now be made.

6.8.5 Theorem. *Any subgroup of a free group is free.*

For any free group can be realized as the edge-group of a 1-dimensional complex; for each generator take a hollow triangle and let all these hollow triangles have a single common vertex. Any subgroup of this group is therefore by 6.8.4 the edge-group of a 1-dimensional complex, which is, by 6.3.11, free. ∎

In the proof of this theorem we reap an advantage from not having restricted ourselves to complexes that are finite, or even locally finite.

6.8.6 Proposition. *If $p: \tilde{X} \to X$ is a covering map of compact polyhedra and the index of $\pi(\tilde{X}, \tilde{x}_0)\, p_*$ in $\pi(X, x_0)$ is m, then*‡ $\chi(\tilde{X}) = m\chi(X)$.

In the light of 6.8.2–6.8.4 we may replace $p: \tilde{X} \to X$ by a simplicial map $q: \tilde{K}_Q \to K$, where $|K| = X$ and $Q\rho_* = \pi(\tilde{X}, \tilde{x}_0)\, p_*$. By 6.5.12 there are m vertices of $\tilde{K}_Q$ over each vertex of K and so by 6.8.1 m n-simplexes of $\tilde{K}_Q$ over each n-simplex of K. Thus, if K has α_n n-simplexes, $\tilde{K}_Q$ has $m\alpha_n$ and $\chi(\tilde{X}) = \chi(\tilde{K}_Q) = m\chi(K) = m\chi(X)$. ∎

6.8.7 Corollary. *A subgroup of finite index m of a free group on a finite number d of generators is free on $(m(d-1)+1)$ generators.*

The 1-dimensional complex for the given free group has characteristic $(1-d)$ and the appropriate covering complex has, therefore, characteristic $m(1-d)$. The number of free generators of its fundamental group is the number of its 1-simplexes not contained in a maximal tree T; now T contains all the vertices and one fewer 1-simplexes and the number of 1-simplexes not in T is therefore $m(d-1)+1$. ∎

6.9* Appendix: Fundamental group and covering groups of topological groups

In this appendix we consider briefly the important application of the two principal concepts of this section to topological groups.

6.9.1 Definition. A set G of elements $e, x, y, \ldots$ is called a *topological group* if (i) G is a group, (ii) G is a topological space, and (iii) the group operations in G are continuous. More precisely, (iii) asserts that the transformations $G \times G \to G$, $G \to G$, given by $(x, y) \to xy$, $x \to x^{-1}$, are continuous. A *homomorphism* of topological groups is a transformation which is homomorphic with respect to the group structures and continuous with respect to the topologies.

We now give some examples of topological groups:

(i) The set of real numbers, with the group structure of the additive group of real numbers and the topology of the real line R^1.

(ii) The set of complex numbers of unit modulus, with the group structure of the multiplicative group of complex numbers and the

‡ See 5.1.24 for the definition of $\chi(K)$, the Euler–Poincaré characteristic of K.

topology of the circle S^1. The transformation $p : R^1 \to S^1$, given by $up = e^{2\pi i u}$ (6.5.1) is a homomorphism of topological groups.

(iii) Consider the general linear group $GL(n, R)$ of non-singular linear transformations of R^n. If we fix a coordinate system in R^n, then $GL(n, R)$ is represented as the multiplicative group of $n \times n$ non-singular real matrices. Writing such a matrix as $M = I_n + (m_{ij})$, we may take the numbers m_{ij} as coordinates of M and thus topologize $GL(n, R)$ as a subspace of R^{n^2}. As topological subgroups we may take $SL(n, R)$, the group of matrices of determinant $+1$, $O(n)$ the orthogonal group, and $R(n) = SL(n, R) \cap O(n)$, the rotation group which acts transitively and effectively‡ on S^{n-1}.

(iv) Consider the general linear group $GL(n, C)$ in which we replace real numbers by complex numbers. As subgroups we have $SL(n, C)$, $U(n)$, the unitary group, and $SU(n) = SL(n, C) \cap U(n)$, the special unitary group.

(v) The centre of $GL(n, R)$ consists of matrices kI_n. Factoring by the centre we obtain $GP(n-1, R)$, the projective group operating on real projective $(n-1)$-space. We obtain $GP(n-1, C)$ in a similar fashion.

In all these examples the groups are also manifolds.

Now let G be a p.c. topological group with unity element e, which we take as base-point. In what follows all paths will have length 1 and we shall write $u *_1 v$ for $(u * v)_{(1)}$, where u, v are paths in G for which $u * v$ is defined. We shall also write x for the constant path at $x \in G$ if no confusion is to be feared.

Given paths u, v in G, let uv be the path given by $(t)\,uv = (tu)\,(tv), t \in I$ and let u^{-1} be the path given by $(t)\,u^{-1} = (tu)^{-1}$. Then the homotopy classes of uv, u^{-1} depend only on those of u, v. Clearly

6.9.2
$$(u *_1 v)\,(u' *_1 v') = uu' *_1 vv'.$$

6.9.3 Proposition. *If u, v are paths in G starting at e and ending at x, y respectively, then*

$$\{uv\} = \{u *_1 xv\} = \{v *_1 uy\}.$$

For

$$uv \equiv (u *_1 x)\,(e *_1 v) = u *_1 xv \quad \text{and} \quad uv \equiv (e *_1 u)\,(v *_1 y) = v *_1 uy. \;\blacksquare$$

‡ The operation is *transitive* in that, for any $x, y \in S^{n-1}$, there exists $\rho \in R(n)$ with $x\rho = y$; the operation is *effective* in that only the identity rotation leaves each point of S^{n-1} fixed.

6.9.4 Corollary. $\{\overline{x^{-1}u}\} = \{u^{-1}\}$.

For $\overline{x^{-1}\bar{u}} = x^{-1}u$ so that $u^{-1} *_1 \overline{x^{-1}\bar{u}} = u^{-1} *_1 x^{-1}u \equiv u^{-1}u = e$.]

6.9.5 Theorem. *In $\pi(G, e)$ the group operation is given by*

$$\{u\}\{v\} = \{uv\}$$

and the inverse by $\{u\}^{-1} = \{u^{-1}\}$.

This follows immediately from 6.9.3 and 6.9.4 with $x = y = e$.] Thus the group operation in $\pi(G, e)$ is directly related to the group operation in G.

6.9.6 Theorem. *$\pi(G, e)$ is commutative.*

This follows immediately from 6.9.3, with $x = y = e$.]

Now suppose that G is p.c., l.p.c. and l.s.c. (this certainly holds if G is a manifold); we construct, for $\pi_0 \subseteq \pi(G, e)$, a covering space $\tilde{G}_{\pi_0}$ as in 6.6. The points of $\tilde{G}_{\pi_0}$ are π_0-equivalence classes $\{u\}_{\pi_0}$ of paths in G starting at e. We define a product in $\tilde{G}_{\pi_0}$ by the rule

6.9.7 $$\{u\}_{\pi_0}\{v\}_{\pi_0} = \{uv\}_{\pi_0}.$$

This rule is unambiguous for if $u' \in \{u\}_{\pi_0}$, $v' \in \{v\}_{\pi_0}$, then $u' \equiv u_0 *_1 u$, $v' \equiv v_0 *_1 v$, where $\{u_0\} \in \pi_0$, $\{v_0\} \in \pi_0$, so that $u'v' \equiv u_0 v_0 *_1 uv$, whence $\{u'v'\}_{\pi_0} = \{uv\}_{\pi_0}$. Plainly $p : \tilde{G}_{\pi_0} \to G$ respects this product since $(1)\,uv = (1u)\,(1v)$.

6.9.8 Theorem. *$\tilde{G}_{\pi_0}$ is a topological group with respect to the multiplication* 6.9.7.

First we verify that $\tilde{G}_{\pi_0}$ is a group. Associativity is obvious, since $(uv)\,w = u(vw)$. The unity element is $\{e\}_{\pi_0}$ and $\{u\}_{\pi_0}^{-1} = \{u^{-1}\}_{\pi_0}$.

Next we verify that the group operations are continuous. Consider the open set $(W, \{uv\}_{\pi_0})$ of $\tilde{G}_{\pi_0}$, where u, v end in x, y and W is a neighbourhood of xy. Then there are neighbourhoods U, V of x, y such that $UV \subseteq W$; we show that $(U, \{u\}_{\pi_0})\,(V, \{v\}_{\pi_0}) \subseteq (W, \{uv\}_{\pi_0})$. For if $\{u *_1 u_0\} \in (U, \{u\}_{\pi_0})$, $\{v *_1 v_0\}_{\pi_0} \in (V, \{v\}_{\pi_0})$, then

$$\{u *_1 u_0\}_{\pi_0}\{v *_1 v_0\}_{\pi_0} = \{uv *_1 u_0 v_0\}_{\pi_0}$$

and $u_0 v_0$ is a path in W. This shows that the product 6.9.7 is continuous. That the inverse is continuous is immediate:

$$(U^{-1}, \{u^{-1}\}_{\pi_0})^{-1} = (U, \{u\}_{\pi_0}).\ ]$$

Let N be the kernel of $p : \tilde{G}_{\pi_0} \to G$. Then N is a discrete normal subgroup of $\tilde{G}_{\pi_0}$. Moreover, 6.9.7 shows that

6.9.9 $$N \cong \pi(G, e)/\pi_0.$$

Reference to 6.9.3 with $x = e$ shows that if u is a loop on e and v a path starting at e, then

$$\{uv\} = \{u *_1 v\} = \{vu\}.$$

A fortiori, $\{uv\}_{\pi_0} = \{vu\}_{\pi_0}$, so that N is central‡ in $\tilde{G}_{\pi_0}$. We sum up in

6.9.10 Theorem. *Let G be a p.c., l.p.c. and l.s.c. group and let*

$$\pi_0 \subseteq \pi(G, e).$$

Then there exists a topological group $\tilde{G}_{\pi_0}$ which is a covering space of G such that the covering map $p : \tilde{G}_{\pi_0} \to G$ is a homomorphism and $\pi(\tilde{G}_{\pi_0})p_* = \pi_0$. Moreover, the kernel of p is a discrete central subgroup N of $\tilde{G}_{\pi_0}$ isomorphic with $\pi(G, e)/\pi_0$. ▌

We add one remark: if K is a normal subgroup of the topological group H, we give H/K a topology, the factor-group topology, by declaring its open sets to be the images of the open sets of H under the natural mapping $H \to H/K$. Then if $\tilde{G}_{\pi_0}/N$ is given the factor-group topology, p induces a (topological) isomorphism of $\tilde{G}_{\pi_0}/N$ with G.

6.9.11 *Example.* Consider the rotation group $R(3)$ operating on S^2. One may prove§ that $R(3) \cong P^3$ (real projective 3-space) so that $\pi(R(3)) = J_2$ (6.4) and the universal covering space of $R(3)$ is S^3. We may take S^3 to be the space of quaternions with norm 1 and S^2 to be the subspace consisting of those quaternions whose real component is zero. Since $R(3)$ operates effectively on S^2, the covering map $p : S^3 \to R(3)$ is given when we describe how a quaternion q operates on S^2. If $q \in S^3$, $q_0 \in S^2$, it may be verified (by taking conjugates) that $q^{-1}q_0 q \in S^2$ and further that $q_0 \to q^{-1}q_0 q$ is a rotation; we thus let q operate on S^2 by the rule $(q_0)q = q^{-1}q_0 q$, and it is then clear that p is a homomorphism when S^3 is given the group structure of the multiplicative group of quaternions with norm 1. It is easy to see that $(q_0)q = q_0$ for all $q_0 \in S^2$ if and only if $q = \pm 1$ so that the kernel of p is J_2, as expected.

We may generalize this example by replacing $R(3)$ by $R(n)$, $n \geqslant 3$. We still have $\pi(R(n)) = J_2$, so that the universal covering group of $R(n)$ is again a 2-fold covering. This group is the so-called *spinor group*, Spin (n).

‡ This could also have been deduced on general grounds from the connectedness of $\tilde{G}_{\pi_0}$.

§ See N. E. Steenrod [18], p. 115.

EXERCISES

1. X is the space consisting of two points p, q, where p is open but q is not. Verify that X is path-connected.

2. Prove that if X is l.p.c., then its path-components are open and closed in X.

3. Prove that the subspace of the Euclidean plane consisting of the curve $y = \sin 1/x$, $0 < x < 1$, and the interval $[-1, 1]$ on the y-axis is not path-connected.

4. f and g are paths of unit length in X from x_0 to x_1 and $h: S^1 \to X$ is the map of the unit circle into X given by

$$\begin{aligned}(e^{\pi i t})\,h &= (1-t)f, \qquad 0 \leqslant t \leqslant 1,\\ &= (1+t)g, \qquad -1 \leqslant t \leqslant 0.\end{aligned}$$

Prove that $f \simeq g \operatorname{rel} \dot{I}$ if and only if h may be extended to the interior of S^1.

5. A circular hole is cut out of a torus. What is the fundamental group of the resulting space? Generalize to k holes.

6. Let f, g be unit loops in X on x_0 and let X_0 be a simply-connected subspace of X containing x_0. Let $F: I \times I \to X$ be a homotopy from f to g under which $\dot{I}$ stays in X_0. Prove that $\{f\} = \{g\} \in \pi(X, x_0)$.

7. Let K be a simplicial complex with k components and let SK, the suspension of K, be formed by joining K to two independent vertices. Show that $\pi(SK)$ is a free group on $(k-1)$ free generators.

8. Let $T = S_1^1 \vee S_2^1 \vee S_3^1 \vee S_4^1$, let w_i be a loop going once round S_i^1, $i = 1, 2, 3, 4$, and let X, Y be obtained from T by attaching a 2-cell to T along

$$w_1 * w_2 * \overline{w}_1 * \overline{w}_2 * w_3 * w_4 * \overline{w}_3 * \overline{w}_4,$$
$$w_1 * w_2 * w_3 * w_4 * \overline{w}_1 * \overline{w}_2 * \overline{w}_3 * \overline{w}_4,$$

respectively. Prove that $\pi(X) \cong \pi(Y)$. (For the topological significance of this result, see [6], p. 140 and p. 171.)

9. $\tilde{P}$ is a covering polyhedron of P and $\tilde{\tilde{P}}$ is a covering polyhedron of $\tilde{P}$. Prove that $\tilde{\tilde{P}}$ is a covering polyhedron of P.

10. Call $\tilde{X}$ a *weak covering space* of X if condition (iii) on a covering space is removed. Prove that Theorems 6.5.5 and 6.5.7 hold for weak covering spaces provided that it be assumed that the spaces under discussion are Hausdorff.

11. Prove that if $\tilde{X}$ is Hausdorff and a weak covering space of X such that $\pi(\tilde{X})$ is of finite index in $\pi(X)$, then $\tilde{X}$ is a covering space of X.

12. Prove that, with the obvious notation, $\tilde{X} \times \tilde{Y}$ is a covering space of $X \times Y$.

13. Let G_i be a subgroup of $\pi(S_i^1)$, $i = 1, 2$, and let G be the free product of G_1 and G_2. Describe the covering space of $S_1^1 \vee S_2^1$ corresponding to the subgroup G of $\pi(S_1^1 \vee S_2^1)$.

14. $\tilde{X}$ is a covering space of X with projection p, and base points $\tilde{x} \in \tilde{X}$, $x \in X$ are chosen with $\tilde{x}p = x$. X_0 is a subspace of X containing x and $\tilde{X}_0$ is the path-component of $p^{-1}(X_0)$ containing $\tilde{x}$. Prove that $\tilde{X}_0$ is a covering space of X_0. Describe $\pi(\tilde{X}_0, \tilde{x})$ in terms of p_* and the homomorphism $i_* : \pi(X_0, x) \to \pi(X, x)$. Hence obtain necessary and sufficient conditions on the pair (X, X_0) so that, for all $\tilde{X}$, $\tilde{X}$ is universal if and only if $\tilde{X}_0$ is universal. Give examples (i) where $\tilde{X}$ is universal but $\tilde{X}_0$ is not, (ii) where $\tilde{X}_0$ is universal but $\tilde{X}$ is not.

15. Let $\tilde{K}_Q$ be a regular covering complex of K and let $B = \pi(K)/Q$. Show how B acts as a group of left operators on $\tilde{K}_Q$ and hence as a group of left operators on $H_*(\tilde{K}_Q; G)$, $R^*(\tilde{K}_Q; G)$. Show that, by killing the operators on $C(\tilde{K}_Q)$, we recover $C(K)$.

Show by the example of the covering map $S^2 \to P^2$ that B may not act trivially on $H_*(\tilde{K}_Q)$.

[If B operates on the left on the abelian group A, we 'kill' the operators by introducing the relations $a = xa$ for all $a \in A$, $x \in B$. The group B operates *trivially* if the relations $a = xa$ are already satisfied in A.]

16. With the notation of Q. 15, let B be a finite group and let $\mathscr{F}$ be a field whose characteristic is prime to the order of B. Construct a chain map

$$k: C_n(K; \mathscr{F}) \to C_n(\tilde{K}_Q; \mathscr{F})$$

such that $kp = 1$. Hence prove that if B operates trivially on $H_*(\tilde{K}_Q; \mathscr{F})$, then

$$p_*: H_*(\tilde{K}_Q; \mathscr{F}) \cong H_*(K; \mathscr{F}),$$
$$p^*: R^*(K; \mathscr{F}) \cong R^*(\tilde{K}_Q; \mathscr{F}).$$

PART II

GENERAL HOMOLOGY THEORY

BACKGROUND TO PART II

In chapter 7 we shall make constant use of definitions and results from homotopy theory; they will also be used, but more sparingly, in chapters 8–10. This Background to Part II assembles the relevant facts about the theory. We reserve the title of 'Theorem' for those results which are substantial and difficult to prove; the proofs will be found in the references given. Propositions and corollaries should be deducible by the reader with the help of the sketched proofs which we append where necessary (in brackets).

1 Homotopy groups

Machinery. We recall that V^n stands for the *n-ball*, $u_1^2+\ldots+u_n^2 \leqslant 1$ in Euclidean space $R^n \subset H^\infty$ and that S^{n-1} stands for its frontier in R^n, the $(n-1)$-sphere $u_1^2+\ldots+u_n^2 = 1$. The point

$$x_* = (-1, 0, 0, \ldots)$$

lies in S^{n-1} for all $n \geqslant 1$ and is used as a *base-point*. S^{n-1} divides S^n into two closed hemispheres E_+^n, E_-^n; E_+^n is that for which $u_{n+1} \geqslant 0$ (see fig. II.1).

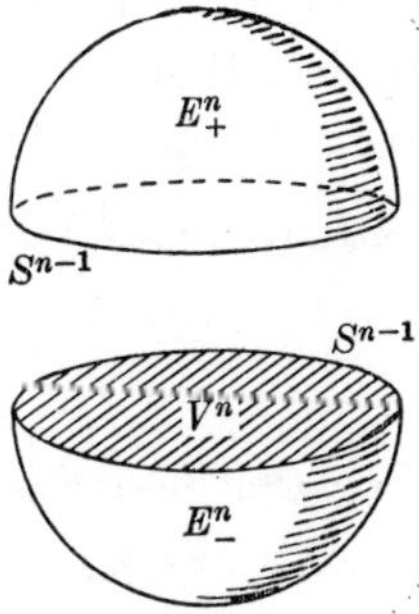

Fig. II.1

We use standard homeomorphisms $h_+ : V^n \to E_+^n$, $h_- : V^n \to E_-^n$ given by

II.1.1 $$(u_1, \ldots, u_n)h_\pm = (u_1, \ldots, u_n, \pm\sqrt{[1-u_1^2-\ldots-u_n^2]}).$$

Each of h_+, h_- is the identity on S^{n-1}. We shall often wish to regard $h_\pm$ as maps into S^n. The *reflexion* $\rho : R^{n+1}, S^n \to R^{n+1}, S^n$ is given by

II.1.2 $$(u_1, \ldots, u_{n+1})\rho = (u_1, \ldots, u_n, -u_{n+1}),$$

so that
$$h_+\rho = h_-.$$
The cylinder $S^{n-1}\times I_r$, $r>0$, is the subset of R^{n+1} given by
$$u_1^2+\ldots+u_n^2 = 1,\quad 0\leqslant u_{n+1}\leqslant r.$$
We shall later use a *standard map*
$$k_{(r)} = k : S^{n-1}\times I_r, (S^{n-1}\times 0)\cup(x_*\times I_r)\to V^n, x_*$$
given by

II. 1.3 $\quad (u_1,\ldots,u_n,u_{n+1})\,k_{(r)} = (-1+\lambda(1+u_1), \lambda u_2, \ldots, \lambda u_n),$

where $\lambda = u_{n+1}/r$. Plainly k does map $(S^{n-1}\times 0)\cup(x_*\times I_r)$ to x_* and is a homeomorphism between their complements; moreover, for any $x\in S^{n-1}$, $(x,r)\,k = x$ and $k^{-1}(S^{n-1}) = (S^{n-1}\times \dot{I}_r)\cup(x_*\times I_r)$ (see fig. II. 2).

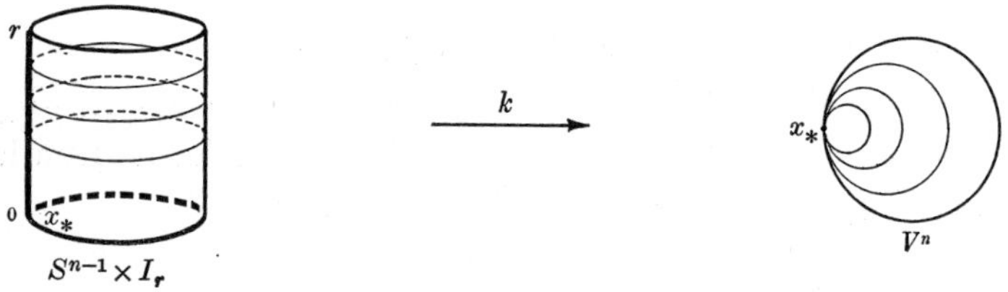

Fig. II. 2 ($n=2$).

The analytical details of the standard map k are given for completeness only; as always in homotopy theory, it is more important to understand what the map does than precisely how it does it.

We shall also use a *standard map*
$$j_{(r)} = j : S^{n-1}\times I_r, (S^{n-1}\times \dot{I}_r)\cup(x_*\times I_r)\to S^n, x_*$$
given by

II. 1.4
$$\begin{aligned}(u_1,\ldots,u_n,u_{n+1})\,j_{(r)} &= (u_1,\ldots,u_n,u_{n+1})\,k_{(\frac{1}{2}r)}h_+ &&\text{if}\quad 0\leqslant u_{n+1}\leqslant \tfrac{1}{2}r\\ &= (u_1,\ldots,u_n,r-u_{n+1})\,k_{(\frac{1}{2}r)}h_- &&\text{if}\quad \tfrac{1}{2}r\leqslant u_{n+1}\leqslant r.\end{aligned}$$

Again, j does map $(S^{n-1}\times \dot{I}_r)\cup(x_*\times I_r)$ to x_* and is a homeomorphism between their complements (see fig. II. 3). Moreover, j and k together determine a map

II. 1.5 $$q : V^n, S^{n-1}\to S^n, x_*$$

given by $kq = j$ and q maps $V^n - S^{n-1}$ homeomorphically onto $S^n - x_*$.

If Y, y_0 are a space and base-point, a *track* w (of length‡ r) of S^{n-1}, x_* in Y, y_0 is a map $w : S^{n-1}\times I_r, x_*\times I_r\to Y, y_0$. The maps

‡ We might allow $r=0$ here, but it simplifies subsequent statements not to do so.

$w_0, w_r : S^{n-1}, x_* \to Y, y_0$ given by $xw_\epsilon = (x, \epsilon)\, w$, $\epsilon = 0, r$, are the *initial* and *final* maps of w. The track w is *constant* if it is independent of $u \in I_r$. The *reverse track* $\overline{w}$ and the product $v * w$ of two tracks, defined if the final map of v is the initial map of w, are defined in an obvious way extending the definitions of 6.1. As there, we define an equivalence relation between tracks with initial map f by the rule that v (of length r) is equivalent to w (of length s) if and only if $v * \overline{w}$ is defined and homotopic, rel $(S^{n-1} \times \dot{I}_{(r+s)}) \cup (x_* \times I_{(r+s)})$, to the constant track with initial map f.

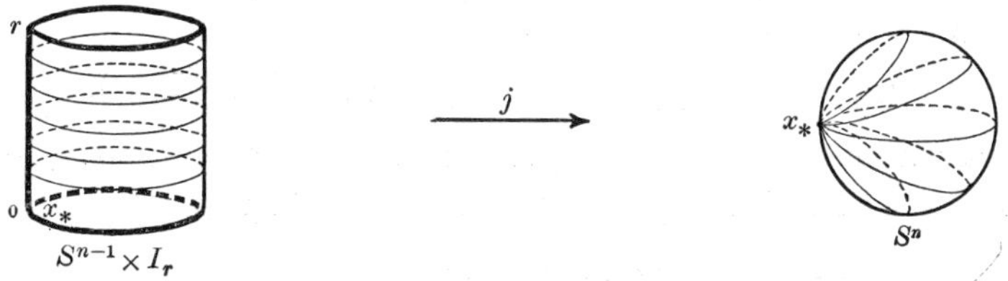

Fig. II. 3 ($n=2$).

We observe that k and j (II. 1.3, 4) are tracks of S^{n-1}, x_* in V^n, x_* and S^n, x_* respectively; plainly $\bar{j} = j\rho$ and II. 1.4 may be restated as

II. 1.6 $$j_{(r)} = k_{(\frac{1}{2}r)} h_+ * \overline{k_{(\frac{1}{2}r)} h_-}.$$

A *closed* track of S^{n-1}, x_* in Y *on* y_0 is a track whose initial and final maps are the constant map e sending S^{n-1} to y_0. As in 6.1.14 equivalence classes of closed tracks on y_0 form a group $\pi_n(Y, y_0)$, the *n-th homotopy group of* Y, y_0. If $n = 1$, this is the fundamental group. A map $g : Y, y_0 \to Z, z_0$ induces a homomorphism

$$g_* : \pi_n(Y, y_0) \to \pi_n(Z, z_0)$$

which depends only on the homotopy class of g. Moreover, $(fg)_* = f_* g_*$ for maps $f : X, x_0 \to Y, y_0$, $g : Y, y_0 \to Z, z_0$. We may deduce

II. 1.7 Proposition. *If $g : Y \to Z$ is a homotopy equivalence, g_* is an isomorphism.*

(As in 6.2.7, care must be taken with base-points. We require the proposition below, analogous to 6.2.1.)

II. 1.8 Proposition. *If v is a path in Y from y_0 to y_1, in the class ξ, v induces an isomorphism $\xi_* : \pi_n(Y, y_0) \to \pi_n(Y, y_1)$ depending only on ξ. If η is a class of paths from y_1 to y_2,*

$$(\xi\eta)_* = \xi_* \eta_* : \pi_n(Y, y_0) \cong \pi_n(Y, y_2).$$

(The path v determines a track of x_* in Y. Given a closed track of S^{n-1}, x_* in Y on y_0, the homotopy extension theorem enables us to deform it into a closed track of S^{n-1}, x_* in Y on y_1 by a deformation in which x_* describes the path v. The class of the final track is the ξ_*-image of the class of the original track.)

II. 1.9 Corollary. *If Y is path-connected and $y_0, y_1 \in Y$ then*

$$\pi_n(Y, y_0) \cong \pi_n(Y, y_1).$$

II. 1.10 Corollary. *$\pi_1(Y, y_0)$ acts as a group of right operators on $\pi_n(Y, y_0)$ and if $g : Y, y_0 \to Z, z_0$, then*

$$(\alpha\xi_*)\, g_* = (\alpha g_*)\, \xi g_*, \quad \alpha \in \pi_n(Y, y_0), \quad \xi \in \pi_1(Y, y_0).$$

(The second assertion, that g_* is an *operator* homomorphism, follows from the nature of the operation of ξ_*, described above.)

II. 1.11 Proposition. *If $n > 1$*, $\pi_n(Y, y_0)$ is commutative.

(The proof is sketched in II. 2, where the proposition is repeated as II. 2.8.)

We shall treat $\pi_n(Y, y_0)$, $n > 1$, as an abelian group, writing composition additively and the neutral element as 0. We shall even, by an abuse of notation, write $\pi_1(Y, y_0)$ as an additive group to avoid restating our results for $n = 1$.

Equivalent descriptions of elements of homotopy groups

Plainly, a track of S^{n-1}, x_* in Y, y_0 is closed (on y_0) if and only if it may be factored through j (II. 1.4). Indeed, any closed track w (of length r) on y_0 determines a *unique* map $w' : S^n, x_* \to Y, y_0$ such that $w = j_{(r)} w'$. This association induces a (1, 1) correspondence between elements of $\pi_n(Y, y_0)$ and homotopy classes of maps $S^n, x_* \to Y, y_0$. For any $f : S^n, x_* \to Y, y_0$, the symbol $\{f\}$ will be used either for its homotopy class or for the element of $\pi_n(Y, y_0)$ which it determines (namely, the class of jf). Then we have

II. 1.12 Proposition. *The map f can be extended over V^{n+1} if and only if $\{f\} = 0$.*

We may use the symbol $\pi_0(Y, y_0)$ for the set of homotopy classes of maps $S^0, x_* \to Y, y_0$; this is in (1, 1) correspondence with the path components of Y. The set $\pi_0(Y, y_0)$ has, in general, no group structure, but it has a distinguished element, the class of the constant map, which corresponds to the path component containing y_0. We say that $\pi_0(Y, y_0)$ is null if it consists of one element; this is to say, if Y is path-connected.

Plainly, there are maps $\phi_+, \phi_- : S^n, x_* \to S^n, x_*$ such that

$$E^n_- \phi_+ = E^n_+ \phi_- = x_* \quad \text{and} \quad \phi_+ \simeq 1,\ \phi_- \simeq 1.$$

It follows that in any homotopy class of maps $f : S^n, x_* \to Y, y_0$ there are maps f_+, f_-, such that $E^n_- f_+ = E^n_+ f_- = y_0$. Given two classes $\{f\}$, $\{g\}$, let $\{h\}$ be the class of the map $h : S^n, x_* \to Y, y_0$ given by $h \mid E^n_+ = f_+ \mid E^n_+$, $h \mid E^n_- = g_- \mid E^n_-$.

II. 1.13 Proposition. $\{h\} = \{f\} + \{g\}$.

(For $\{f\} + \{g\}$ is the class of

$$jf_+ * jg_- = kh_+ f_+ * \overline{kh_- f_+} * kh_+ g_- * \overline{kh_- g_-} \equiv kh_+ f_+ * \overline{kh_- g_-} = jh.)$$

This completes a description of the (familiar) definition in terms of maps of spheres.

An ordered pair of maps $g, g' : V^n, x_* \to Y, y_0$ that agree on S^{n-1} (that is, $g \mid S^{n-1} = g' \mid S^{n-1}$) determines an element $\alpha(g, g') \in \pi_n(Y, y_0)$ as follows: for any $r, r' > 0$, $k_{(r)} g * \overline{k_{(r')} g'}$ is a closed track of S^{n-1}, x_* on y_0; since its equivalence class does not depend on r, r' we may define $\alpha(g, g')$ to be $\{kg * \overline{kg'}\}$. For any $f : S^n, x_* \to Y, y_0$, clearly

$$\{f\} = \alpha(h_+ f, h_- f).$$

Also the maps g, g' above determine a map $\tilde{g} : S^n, x_* \to Y, y_0$ by the rule that $g = h_+ \tilde{g}$, $g' = h_- \tilde{g}$. Then

II. 1.14 $$\{\tilde{g}\} = \alpha(g, g').$$

II. 1.15 Proposition. *If $g, g', g'' : V^n, x_* \to Y, y_0$ agree on S^{n-1},*

$$\text{(i)}\ \alpha(g, g) = 0, \quad \text{(ii)}\ \alpha(g, g') = -\alpha(g', g),$$
$$\text{(iii)}\ \alpha(g, g') + \alpha(g', g'') = \alpha(g, g'').$$

(Clear from the definition of α.)

II. 1.16 Proposition. *Given $g : V^n, x_* \to Y, y_0$ and $\alpha_0 \in \pi_n(Y, y_0)$, there exists $g' : V^n, x_* \to Y, y_0$ with $\alpha(g, g') = \alpha_0$.*

(Let $w = kg$ and let v be a track representing $-\alpha_0$. Then $v * w$ is defined and $\{w * \overline{v * w}\} = \alpha_0$. Moreover, $v * w$ is a map

$$S^{n-1} \times I, (S^{n-1} \times 0) \cup (x_* \times I) \to Y, y_0$$

and so may be factored through k. Define g' by $kg' = v * w$. Then $g' \mid S^{n-1} = g \mid S^{n-1}$ and $\alpha(g, g') = \alpha_0$.)

These two propositions will be important in chapter 7. As a third description of an element of a homotopy group we observe that a map $g : V^n, S^{n-1} \to Y, y_0$ determines an element $\{g\}$ of $\pi_n(Y, y_0)$, namely, $\alpha(g, e)$, where e maps V^n to y_0.

For an arbitrary n-sphere and base-point we can use a homeomorphism with S^n, x_* to induce in the set of homotopy classes of its maps into Y, y_0 a group structure isomorphic with that of $\pi_n(Y, y_0)$. The isomorphism depends of course on the homeomorphism chosen; it can however be rendered unambiguous by introducing orientations. A similar remark applies to maps of n-cells.

Orientation

In 2.1 we defined the *orientation* of a simplex and drew attention to the resemblance to the choice of ordering for the coordinate axes in R^n. We can give substance to this resemblance by introducing a *standard orientation* into S^n. We recall from 3.9 the homeomorphism (by radial projection) $h : S^n \to \Sigma^n$, where Σ^n is the 'octahedron' covered by a complex there called $K^{(n)}$. We assign to the ordered n-simplex $a_+^1 a_+^2 \ldots a_+^{n+1}$ the orientation $+1$ and this orients each n-simplex of $K^{(n)}$ if oriented simplexes with a common $(n-1)$-face are to induce opposite orientations in that face. Thus Σ^n may be said to be 'oriented by the ordering of the axes in R^{n+1}'. It will be observed that to orient Σ^n is, essentially, to choose a generator of $H_n(\Sigma^n)$, for we have chosen the generating cycle in which $a_+^1 a_+^2 \ldots a_+^{n+1}$ appears with coefficient $+1$. Now we may extend h to a homeomorphism

$$h : V^{n+1}, S^n \to \Phi^{n+1}, \Sigma^n,$$

where Φ^{n+1} is the octahedral region bounded by Σ^n and we may triangulate Φ^{n+1} as the join of the origin to Σ^n. We *orient* Φ^{n+1} by assigning to the ordered $(n+1)$-simplex $Oa_+^1 a_+^2 \ldots a_+^{n+1}$ the orientation $+1$; as above, this is equivalent to choosing a generator of

$$H_{n+1}(\Phi^{n+1}, \Sigma^n).$$

II. 1.17 Definition.‡ An *orientation* in a (topological) n-sphere T^n is a choice of preferred generator of $H_n(T^n)$. An *orientation* in an n-cell E^n (see I. 1) is a choice of preferred generator of $H_n(E^n, \dot{E}^n)$, where $\dot{E}^n$ is the frontier of E^n. $\dot{E}^n$ is oriented *coherently* with E^n if

$$\nu_* : H_n(E^n, \dot{E}^n) \to H_{n-1}(\dot{E}^n)$$

‡ There are advantages in relating orientation directly to the *singular* homology groups. See the end of 8.6.

maps the preferred generator to the preferred generator. We may refer to the preferred generator as the *positive* generator.

II. 1.18 Definition. The *standard orientation* in S^n is the counterimage under h_* of the preferred generator of $H_n(\Sigma^n)$. The *standard orientation* in V^n is the counterimage under h_* of the preferred generator of $H_n(\Phi^n, \Sigma^{n-1})$. We now suppose that S^n and V^n have standard orientations.

II. 1.19 Proposition. *S^n is oriented coherently with V^{n+1}.*

(For $a_+^1 \dots a_+^{n+1}$ appears in the boundary of $Oa_+^1 \dots a_+^{n+1}$ with coefficient $+1$.)

If T_1^n, T_2^n are two oriented n-spheres, a homeomorphism $d: T_1^n \to T_2^n$ is said to be *orientation-preserving* or *orientation-reversing* according as d_* maps or does not map preferred generator to preferred generator; similarly for homeomorphisms of cells. Thus d is orientation-preserving (-reversing) if and only if it has degree $+1(-1)$.

To free ourselves from use of the particular model space S^n in defining $\pi_n(Y, y_0)$ we need the fundamental

II. 1.20 Theorem. (*Brouwer-Hopf*). *Two maps $f, g: S^n \to S^n$ are homotopic if and only if they have the same degree.*

('Only if' we know, since if $f \simeq g$ then $f_* = g_*$; for the harder 'if' part see H. Whitney, *Duke Math. J.* (1937), 46.)

II. 1.21 Corollary. *Two maps $f, g: S^n, x_* \to S^n, x_*$ are homotopic if and only if they have the same degree; in particular, any map of degree $+1$ and so any orientation-preserving homeomorphism $S^n, x_* \to S^n, x_*$ is homotopic to the identity map.*

II. 1.22 Corollary. *If T^n, w_* are an oriented n-sphere and base-point, a map $f: T^n, w_* \to Y, y_0$ determines an element, written as $\{f\}$, of $\pi_n(Y, y_0)$. If T^n is given the opposite orientation the same map determines $-\{f\}$.*

(If $d: S^n, x_* \to T^n, w_*$ is any orientation-preserving homeomorphism, $\{f\}$ is defined to be $\{df\} = \{jdf\} \in \pi_n(Y, y_0)$. That this is independent of the choice of d follows from II. 1.21. Indeed, we could choose for d any map of degree $+1$. Next we observe that $\rho: S^n \to S^n$ reverses orientation, for $a_+^1 a_+^2 \dots a_-^{n+1}$ receives the orientation -1 in the standard orientation of Σ^n. Thus, if T^n is given the opposite orientation, ρd is orientation-preserving and f now determines the element $\{j\rho df\} = \{\bar{j}df\} = \{\overline{jdf}\} = -\{jdf\} = -\{f\}$.)

If σ_{n+1} is an oriented $(n+1)$-simplex, its orientation picks out a generator of $H_{n+1}(\bar{s}_{n+1}, \dot{s}_{n+1})$ and we write $\dot{\sigma}_{n+1}$ for the n-sphere oriented coherently with σ_{n+1}. Thus if a^0 is a vertex of σ_{n+1} a map $f: \dot{\sigma}_{n+1}, a^0 \to Y, y_0$ determines an element $\{f\}$ of $\pi_n(Y, y_0)$ and $\{f\} = 0$ if and only if f is extendable over $\bar{\sigma}_{n+1}$.

As a further consequence of II. 1.20 we see that any map of degree $+1$ and so any orientation-preserving homeomorphism

$$V^n, S^{n-1} \to V^n, S^{n-1}$$

is homotopic to the identity map. Thus

II. 1.23 Corollary. *If E^n is an oriented n-cell, a map $f: E^n, \dot{E}^n \to Y, y_0$ determines an element $\{f\}$ of $\pi_n(Y, y_0)$. The same map of the same cell oppositely oriented determines $-\{f\}$.*

(Compare II. 1.22.) In particular, maps of $\bar{\sigma}_n, \dot{\sigma}_n$ or of‡ $I^n, \dot{I}^n$ into Y, y_0 determine elements of π_n (Y, y_0). This description is probably the one most commonly used.

In chapter 7 we shall be concerned with n-cells lying in an oriented n-sphere, T^n. We now explain what we mean by saying that such a cell E^n is *oriented by the orientation of T^n*.

Let $p \in T^n$; by triangulating T^n with p as a vertex we obtain an isomorphism $\mu_*: H_n(T^n) \simeq H_n^{T^n}(p)$ (see 3.7). We refer to the μ_*-image of the preferred generator of $H_n(T^n)$ as the *induced orientation of T^n at p*. It is clearly independent of the triangulation. Let E^n be an n-cell lying in T^n and containing p in its interior. There is then (3.7.6) an injection isomorphism $i_*: H_n^{E^n}(p) \simeq H_n^{T^n}(p)$. We refer to $\xi \in H_n^{E^n}(p)$ as *the induced orientation of E^n at p* if ξi_* is the orientation of T^n at p. By triangulating E^n as the join of p to a triangulation of $\dot{E}^n$ we may identify $H_n^{E^n}(p)$ with $H_n(E^n, \dot{E}^n)$. We thus pick out a generator of $H_n(E^n, \dot{E}^n)$ and hence an orientation of E^n. We call this the *orientation of E^n induced by T^n*. It is independent of the choice of triangulations in $T^n, \dot{E}^n$ and of the choice of point p. (For suppose q is interior to E^n and that q is also used to induce an orientation in E^n. There is then a homeomorphism $h: T^n, E^n \to T^n, E^n$ which is homotopic to the identity rel the complement of the interior of E^n and such that $ph = q$. We thus obtain a commutative diagram

$$\begin{array}{ccccccccc}
H_n(K) & \xrightarrow{\mu_*} & H_n^K(p) & \xleftarrow{i_*} & H_n^M(p) & = & H_n(M, M_0) & \xrightarrow{\nu_*} & H_{n-1}(M_0) \\
\downarrow j_* & & \downarrow h_* & & \downarrow h_* & & \downarrow h_* & & \downarrow k_* \\
H_n(L) & \xrightarrow{\mu_*} & H_n^L(q) & \xleftarrow{i_*} & H_n^N(q) & = & H_n(N, N_0) & \xrightarrow{\nu_*} & H_{n-1}(N_0)
\end{array},$$

‡ $I^n = I \times I \times \ldots \times I$ (n times); $\dot{I}^n$ is its frontier in R^n; we describe its orientation below.

where K, L are triangulations of T^n; M_0, N_0 are triangulations of $\dot{E}^n$; $M = pM_0$, $N = qN_0$ and j_*, k_* are induced by identity maps. Let $\bar{\imath}_* = i_*^{-1}$. Then if $\eta \in H_n(K)$ orients T^n, so does ηj_*. We have $\eta\mu_*\bar{\imath}_*h_* = \eta j_*\mu_*\bar{\imath}_*$, so that the two orientations induced in E^n correspond under h_*. But $h_*\nu_* = \nu_*k_*$, since

$$h_* : H_n(M, M_0) \rightarrow H_n(N, N_0)$$

is induced by the identity map so that the two induced orientations in E^n agree.)

We orient E^n_+, E^n_- by the orientations of S^n.

II. 1.24 Proposition. *E^n_+ is oriented coherently with S^{n-1} if and only if n is even, and h_+ is orientation-preserving if and only if n is even.*

(To orient E^n_+ in agreement with S^n is (identifying S^n with Σ^n) to assign to $a^1_+ \dots a^{n+1}_+$ the orientation $+1$; this assigns to $a^1_+ \dots a^n_+$ in S^{n-1} the orientation $(-1)^n$. This proves the first assertion, and the second is proved by remarking that h_+ is the identity on S^{n-1} and V^n is oriented coherently with S^{n-1}.)

The convention we have adopted for orienting E^n_+ and E^n_- is by no means claimed to be the most suitable in all branches of homotopy theory; we have chosen it because, as we have said, in chapter 7 we are often concerned with cells lying in an oriented sphere, and it then appears reasonable to orient the cells by the orientation of the sphere. Our convention will expose us to the conspicuous intrusion of the sign $(-1)^n$ in certain important formulae, as suggested by II. 1.24. This sign would be avoided if we oriented E^n_+ coherently with S^{n-1} or if we *defined* E^n_+ to be that hemisphere which received from S^n an orientation coherent with S^{n-1}. The reader is encouraged to trace the consequences in our subsequent work of such a redefinition of E^n_+.

Let E^m_1, E^n_2 be two oriented cells. Then $E^m_1 \times E^n_2$ is an $(m+n)$-cell and we now define the *product orientation* in $E^m_1 \times E^n_2$. Let $D^n \subset R^n$ be the cube given by $-1 \leqslant u_i \leqslant 1$, $i = 1, \dots, n$. We orient D^n by means of the 'radial' homeomorphism $V^n \rightarrow D^n$. We orient $D^m \times D^n$ by requiring the homeomorphism $D^m \times D^n \rightarrow D^{m+n}$ given by

$$((u_1, \dots, u_m), (v_1, \dots, v_n)) \rightarrow (u_1, \dots, u_m, v_1, \dots, v_n)$$

to be orientation-preserving. Let $h_1 : E^m_1 \rightarrow D^m$, $h_2 : E^n_2 \rightarrow D^n$ be homeomorphisms. Then $h_1 \times h_2 : E^m_1 \times E^n_2 \rightarrow D^m \times D^n$ is a homeomorphism and the *product orientation* in $E^m_1 \times E^n_2$ is that orientation such that $h_1 \times h_2$ is orientation-preserving if and only if h_1 and h_2 are both orientation-preserving or both orientation-reversing.

If I^n is the unit n-cube in R^n, given by $0 \leqslant u_i \leqslant 1$, $i = 1, \dots, n$, we orient I^n by requiring the 'stretching' homeomorphism $I^n \rightarrow D^n$ to be orientation-preserving.

II. 1.25 Proposition. *If $E^{m-1} \subseteq \dot{E}^m$ is oriented by $\dot{E}^m$ then $E^{m-1} \times E^n$ is oriented by $(E^m \times E^n)^{\cdot}$.*

II. 1.26 Proposition. *If $E^{n-1} \subseteq \dot{E}^n$ is oriented by $\dot{E}^n$ then $E^m \times E^{n-1}$ is oriented by $(E^m \times E^n)^{\cdot}$ if and only if m is even.*

(If $D^{p-1}_{\epsilon i}$ is the face of D^p given by $u_i = \epsilon$, $\epsilon = +1$ or -1, then $D^{p-1}_{\epsilon i}$ is oriented by $\dot{D}^p$ if and only if $\epsilon = +1$ and i is odd, or $\epsilon = -1$ and i is even. The two propositions may be proved by replacing $E^m \times E^n$ by D^{m+n} and invoking the quoted fact.) We use these propositions particularly when $E^n = I$ so that $E^{n-1} = 0$ or 1.

n-simple spaces

The path-connected space Y is said to be *n-simple* if, for some $y_0 \in Y$, $\pi_1(Y, y_0)$ operates trivially on $\pi_n(Y, y_0)$. This is equivalent to saying that, for any $y_1, y_2 \in Y$ the isomorphism $\pi_n(Y, y_1) \cong \pi_n(Y, y_2)$ induced by a path from y_1 to y_2 (see II. 1.8) is, in fact, independent of the choice of path; we may therefore abbreviate $\pi_n(Y, y_0)$ to $\pi_n(Y)$ if Y is n-simple (all groups $\pi_n(Y, y), y \in Y$ being *canonically* isomorphic). Notice that Y is n-simple for all n if Y is 1-connected.

II. 1.27 Proposition. *If Y is n-simple and T^n is an oriented n-sphere, any map $f : T^n \to Y$ determines an element $\{f\} \in \pi_n(Y)$; if E^n is an oriented n-cell, any map $g : E^n \to Y$ such that $\dot{E}^n g$ is a single point determines an element $\{g\} \in \pi_n(Y)$.*

Y will be n-simple throughout this subsection. We remark that in this case we need specify neither the base-point $y_0 \in Y$ nor the base-point $w_* \in T^n$. In particular, $(V^n \times I)^{\cdot}$, oriented coherently with $V^n \times I$, is an oriented n-sphere. Thus a map $(V^n \times I)^{\cdot} \to Y$ determines an element of $\pi_n(Y)$.

Let $g, g' : V^n \to Y$ and let $w : S^{n-1} \times I \to Y$ be a track in Y with initial map $g \mid S^{n-1}$ and final map $g' \mid S^{n-1}$. Then g, g', w combine to give a map $G_w : (V^n \times I)^{\cdot} \to Y$ and hence an element

$$\beta(g, g'; w) = \{G_w\} \in \pi_n(Y).$$

II. 1.28 Proposition. *If w is a constant track, then*

$$\beta(g, g'; w) = -\alpha(g, g').$$

(Notice that the hypothesis implies that $g \mid S^{n-1} = g' \mid S^{n-1}$. We define $p : (V^n \times I)^{\cdot} \to S^n$ by

$$(x, 0)\, p = xh_+, \quad x \in V^n,$$

$$(x, 1)\, p = xh_-, \quad x \in V^n,$$

$$(x, u)\, p = x, \qquad x \in S^{n-1}.$$

Now $V^n \times 0$ is oriented by $(V^n \times I)^{\cdot}$ if and only if n is odd and h_+ is orientation-preserving if and only if n is even (II. 1.26 and II. 1.24). Thus p has degree -1. Since $p\tilde{g} = G_w$ the assertion follows.)

II. 1.29 Proposition. *Let Y be n-simple, let $T^{n+1} = T$ be an oriented $(n+1)$-sphere and let $E_1, \ldots, E_s$ be $(n+1)$-cells lying in T^{n+1} and with disjoint interiors. Let each E_i be oriented by T and let $\dot{E}_i$ be oriented coherently with E_i. Let T_0 be the complement in T of the union of the interiors of the cells. Then, for any map $f : T_0 \to Y$, $\sum_{i=1}^{s} \{f \mid \dot{E}_i\} = 0$.*

(We first show that we may replace the cells E_i by disjoint closed simplexes of a triangulation of T. This will be proved when we have established a lemma. Let E^{n+1}, Q^{n+1} be cells in T, oriented by T, with Q in the interior of E. Using the linear structure of E we may retract $E - \operatorname{Int} Q$ onto $\dot{E}$. Let $g : \dot{Q} \to \dot{E}$ be obtained by restricting the retraction.

Lemma. *g has degree* $+1$.

Without real loss of generality, we may suppose that E is an $(n+1)$-ball whose centre a is interior to Q. Let $\tilde{E}^{n+1}$ be concentric with E and lying in Q and let T be triangulated with a a vertex and $|\mathrm{st}\,(a)| \subseteq \tilde{E}$. We obtain, by radial projection from a, a deformation $d_t : E, A \to E, A$, rel $\dot{E}$, where $A = E - \operatorname{Int} \tilde{E}$, such that $d_0 = 1$, $Ad_1 = \dot{E}$ and $d_1 \mid \dot{Q} = g$. Let $g_1 = d_1 \mid Q$, regarded as a map $Q, \dot{Q} \to E, \dot{E}$ and consider the diagram

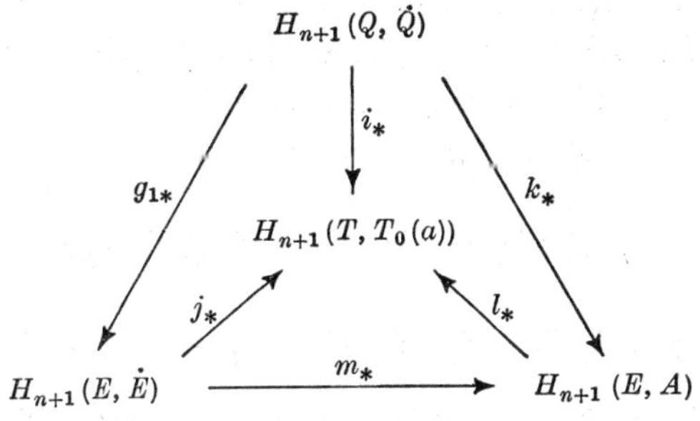

where, as in 3.7, $T_0(a) = T - \mathrm{st}\,(a)$ and all homomorphisms but g_{1*} are injections. We have to prove that $g_{1*}\, j_* = i_*$; but

$$g_{1*}\, j_* = g_{1*} m_* l_* = (g_1 m)_* l_* \quad \text{and} \quad g_1 m \simeq k.$$

Thus $g_{1*}\, j_* = k_* l_* = i_*$ and the lemma is proved.

We may thus make the stated simplifying assumption. It is then obvious that the theorem holds for $s = 1$. We next assume that it holds for $s = k-1$ and consider the case $s = k$ in its simplified form; $E_1, \ldots, E_k$ are simplexes of a given triangulation of T, all oriented by T.

Let E_1' be an $(n+1)$-simplex having a common n-face with E_1 and let $E_1' \neq E_2, \ldots, E_k$; then E_1' and $E_1 \cup E_1'$ are $(n+1)$-cells which we orient by T. Let $T_0'' = \overline{T_0 - E_1'}$ and $T_0' = T_0'' \cup E_1$; then $f \mid T_0''$ can be extended to $f' : T_0' \to Y$, since T_0'' is a retract of T_0'. Then

$$\{f \mid \dot{E}_1\} = \{f \mid (E_1 \cup E_1')^{\cdot}\} = \{f' \mid \dot{E}_1'\}$$

and, for $i \geqslant 2$, $f \mid \dot{E}_i = f' \mid \dot{E}_i$. We may repeat this proces of replacing a missing $(n+1)$-simplex by a neighbour and modifying the map, until E_1 has been replaced by $E_1^{\#}$, T_0 by $T_0^{\#} = \overline{T_0 - E_1^{\#}} \cup E_1$ and f by $f^{\#} : T_0^{\#} \to Y$ such that

(i) $f \mid \dot{E}_i = f^{\#} \mid \dot{E}_i$, for $i \geqslant 2$,

(ii) $\{f \mid \dot{E}_1\} = \{f^{\#} \mid \dot{E}_1^{\#}\}$, and

(iii) $E_1^{\#}$ has an n-face in common with some E_i, $i \geqslant 2$, say with E_2.

Then $E_1^{\#} \cup E_2$ is an $(n+1)$-cell E_{12} which we orient by T, and by II. 1.15 (iii)

$$\{f^{\#} \mid \dot{E}_1^{\#}\} + \{f^{\#} \mid \dot{E}_2\} = \{f^{\#} \mid \dot{E}_{12}\}.$$

Hence

$$\sum_1^k \{f \mid \dot{E}_i\} = \{f^{\#} \mid \dot{E}_{12}\} + \sum_3^k \{f^{\#} \mid \dot{E}_i\},$$

and the right-hand side is 0 by the inductive hypothesis.)

II. 1.30 Proposition. *Let Y be n-simple, let T^n be an oriented n-sphere, and let $E_1, \ldots, E_r$ be n-cells lying in T^n and with disjoint interiors, and let each E_i be oriented by T. Let T_0 be the complement in T of the union of the interiors of the cells. If $f, f' : T \to Y$ are maps and $G : T_0 \times I \to Y$ a homotopy between their restrictions to T_0, then*

$$(-1)^{n+1}(\{f\} - \{f'\}) = \sum_{i=1}^r \beta(g_i, g_i'; G_i),$$

where $\quad g_i = f \mid E_i, \quad g_i' = f' \mid E_i, \quad G_i = G \mid \dot{E}_i \times I.$

(We apply II. 1.29 to the oriented sphere $(E^{n+1} \times I)^{\cdot}$ and the cells $E^{n+1} \times 0$, $E^{n+1} \times 1$, $E_i \times I$, $i = 1, \ldots, r$, where $\dot{E}^{n+1} = T^n$. The maps f, f', G combine to give a map $H : (T \times \dot{I}) \cup (T_0 \times I) \to Y$ and

$$\{H \mid (E^{n+1} \times 0)^{\cdot}\} = \{f\}, \quad \{H \mid (E^{n+1} \times 1)^{\cdot}\} = \{f'\},$$

$$\{H \mid (E_i \times I)^{\cdot}\} = \beta(g_i, g_i'; G_i).$$

The signs in the final formula are consequences of II. 1.25 and II. 1.26.)

II. 1.31 Corollary. *Under the above hypotheses on Y and T, if f, f' agree on T_0,*

$$(-1)^n(\{f\}-\{f'\}) = \sum_{i=1}^{r} \alpha(g_i, g_i').$$

We remark that we can infer the *homotopy addition lemma* from II. 1.31; for if also $T_0 f = y_0$ then we may take $f' = e$ mapping T to y_0 and obtain

II. 1.32
$$(-1)^n\{f\} = \sum_{i=1}^{r} \{g_i\}.$$

The restriction that Y_0 be n-simple is superfluous here if $n > 1$. For the map f may be factored through the bunch, $S_1^n \vee \ldots \vee S_r^n$, of n-spheres with a single common point and the bunch is n-simple, being simply-connected; we may choose any point $w_* \in T_0$ as base-point.

Homotopy groups of spheres

II. 1.33 Proposition. $\pi_r(S^n) = 0$, $r < n$.

(Apply the simplicial approximation theorem.) A space Y is said to be *n-connected* if $\pi_r(Y) = 0$, $r = 0, 1, \ldots, n$. Thus we may say that S^n is $(n-1)$-connected.

II. 1.34 Proposition. $\pi_n(S^n) = J$ (see II. 1.20).

II. 1.35 Proposition. $\pi_n(S^1) = 0$, $n > 1$ (see 6.6.15).

II. 1.36 Theorem. $\pi_3(S^2) = J$; $\pi_{n+1}(S^n) = J_2$, $n \geqslant 3$.

(For an unoriginal proof of the first assertion of II. 1.36, see Hilton [9], p. 51. The classical (and original) proof of the second assertion may be found in H. Freudenthal, *Comp. Math.* **5** (1937), 299; it has also been deduced as a consequence of far-reaching algebraic methods, for example, in J.-P. Serre, *Ann. Math.* **58** (1953), 258. These methods are described in 10.6.)

2 Function spaces and loop-spaces

We noted in II. 1 the close analogy between the definition of the fundamental group (as a set of classes of closed *paths* with a certain composition law) and that of the nth homotopy group (as a set of classes of closed *tracks*). In fact $\pi_n(Y)$ may be regarded as the fundamental group of a certain *function space*. We define and discuss function spaces in this section and make applications in chapters 8 and 10.

Let Y^X be the set of maps of the space X into the space Y. Let $\mathscr{C}$ be the collection of compact subsets of X and let $\mathscr{U}$ be the collection of open subsets of Y. For each $C \in \mathscr{C}$, $U \in \mathscr{U}$ let $[C, U]$ be the subset of Y^X consisting of maps f such that $Cf \subseteq U$ and let $[\mathscr{C}, \mathscr{U}]$ be the collection of subsets $[C, U]$. We topologize Y^X by taking as a base for the open sets all finite intersections of sets of $[\mathscr{C}, \mathscr{U}]$; this is the *compact open topology.*

In chapter 1 we introduced a metric topology into the set Y^X if X is compact and Y metric by defining

$$\rho(f, f') = \underset{x \in X}{\text{l.u.b.}}\, \rho(xf, xf').$$

II.2.1 Proposition. *If X is compact and Y is metric the metric and compact-open topologies on Y^X coincide.*

We now use the symbol Y^X to mean the *function space* of maps $X \to Y$ with the compact-open topology.

Let A be a space and let $\bar{g} : A \to Y^X$, $g : X \times A \to Y$ be functions related by

II.2.2 $$x(a\bar{g}) = (x, a)\, g.$$

II.2.3 Proposition. *If g is continuous, $\bar{g}$ is continuous. Conversely, if $\bar{g}$ is continuous and X is Hausdorff and locally compact, then g is continuous.*

(Suppose g continuous; to prove $\bar{g}$ continuous it is sufficient to show that if $a \in \bar{g}^{-1}[C, U]$, then some neighbourhood of a is contained in $\bar{g}^{-1}[C, U]$. For each $x \in C$ there are open sets $U(x) \ni x$, $U_x(a) \ni a$ such that $(U(x) \times U_x(a))\, g \subseteq U$. From the open covering $\{U(x)\}$ of C select a finite covering $U(x_1), \ldots, U(x_k)$ and let $V(a) = \bigcap_{i=1}^{k} U_{x_i}(a)$. Then $(C \times V(a))\, g \subseteq U$ so that $V(a) \subseteq \bar{g}^{-1}[C, U]$ and $\bar{g}$ is continuous.

Conversely, suppose $\bar{g}$ continuous and let X be locally compact and Hausdorff. Let U be open in Y and let $(x, a)\, g \in U$. Since the elements of Y^X are continuous maps and X is locally compact and Hausdorff there is a compact neighbourhood $V(x) \ni x$ such that $(V(x) \times a)\, g \subseteq U$. Then $a\bar{g} \in [V(x), U]$ so that, by the continuity of $\bar{g}$, there is a neighbourhood $U(a) \in a$ such that $U(a)\, \bar{g} \subseteq [V(x), U]$; but then $(V(x) \times U(a))\, g \subseteq U$ and g is continuous.)

It follows that II. 2.2 establishes a canonical (1, 1) correspondence between maps $A \to Y^X$ and maps $X \times A \to Y$ if X is locally compact and Hausdorff. Thus, for example, a homotopy $H : I \times A \to Y$ corresponds to a map $\bar{H} : A \to Y^I$. The space Y^I is the *space of unit paths on* Y. For each t, $0 \leqslant t \leqslant 1$, there are maps $b_t : Y^I \to Y$, given by $vb_t = tv$, $v \in Y^I$. The maps b_t are continuous since $b^{-1}U = [t, U]$.

II. 2.4 Proposition. $b_0 \simeq b_1 : Y^I \to Y$.

(For the transformation $I \times Y^I \to Y$, given by $(t, v) \to tv$ is continuous, since it corresponds under II. 2.2 to the identity map $Y^I \to Y^I$; and it is plainly a homotopy from b_0 to b_1.)

Let $(Y, y_0)^{(S^{n-1}, x_*)}$ be the subspace of $Y^{S^{n-1}}$ consisting of those maps which send x_* to y_0. From II. 2.3 we deduce

II. 2.5 Proposition. $\pi_n(Y, y_0) \cong \pi_1((Y, y_0)^{(S^{n-1}, x_*)}, e)$.

Here e is the constant map to y_0. More generally we have

II. 2.6 $$\pi_n(Y, y_0) \cong \pi_p((Y, y_0)^{(S^q, x_*)}, e),$$

where $p+q = n$.

Let $P(Y)$ be the set of paths in Y. There is a (1, 1) function from $P(Y)$ into $Y^I \times R^+$, where R^+ is the interval $r \geqslant 0$ on the real line, given (see 6.1.5) by
$$(f, r)\,\phi = (f_{(1)}, r).$$

We topologize $P(Y)$ by requiring ϕ to be a homeomorphism onto its image and we topologize $E(Y, y_0)$, $\Omega(Y, y_0)$, $N(Y, y_0)$ as subspaces‡ of $P(Y)$. If E_1, Ω_1, N_1 refer to the analogous subspaces of *unit* paths there is plainly a homotopy equivalence
$$P(Y), E(Y), \Omega(Y), N(Y) \simeq Y^I, E_1(Y), \Omega_1(Y), N_1(Y).$$

The elements of $\pi_1(Y)$ are in (1, 1) correspondence with the path components of the *loop space* $\Omega(Y)$ (or $\Omega_1(Y)$). More generally we have (see II. 2.6)

II. 2.7 $$\pi_n(Y) \cong \pi_{n-1}(\Omega Y).$$

The space ΩY has most of the properties of a topological group (see 6.9), the exception being that there is no actual inverse (only a *homotopy* inverse). However, the arguments of 6.9 go through in so far as they establish that the law of composition in ΩY may be used to define the group operation in $\pi_1(\Omega Y)$—and so, indeed, in $\pi_r(\Omega Y)$—and that the fundamental group is commutative. We thus infer

II. 2.8 Proposition. *If* $n > 1$, $\pi_n(Y, y_0)$ *is commutative.*

‡ See 6.1.

3 Fibre spaces, relative homotopy groups, and exact homotopy sequences

Let X, B be spaces; a map $p: X \to B$ is a (*Serre*) *fibre map* if for each polyhedron $|K|$, map $f_0 : |K| \to X$ and homotopy $g_t : |K| \to B$ with $g_0 = f_0 p$, there is a homotopy $f_t : |K| \to X$ with $g_t = f_t p$. Thus p is a *fibre map* if and only if the '*lifting homotopy*' *property* holds for maps of polyhedra (compare 6.5; a covering map is a particular case of a fibre map). If we pick a base-point $b_0 \in B$ then $p^{-1}b_0$ is called the *fibre* (over b_0); X is called the *fibre space* or *total space* and B the *base-space* of the *fibration*. Applications of the statements in this section will be found in chapter 10.

As an example, consider the Hopf map $S^3 \to S^2$; represent points of S^3 by a pair of complex numbers (z_1, z_2) with $z_1\bar{z}_1 + z_2\bar{z}_2 = 1$, represent‡ S^2 as the complex projective line with homogeneous coordinates $[z_1, z_2]$ and the map is given by $(z_1, z_2) \to [z_1, z_2]$. This map is a fibre map and the fibres are great circles. For a further discussion see Hilton [9], p. 51 or Steenrod [18], p. 105.

II. 3.1 Proposition. *If B is path-connected and $p: X \to B$ is a fibre map then p maps onto B.*

II. 3.2 Proposition. *If $p: X \to B$ is a fibre map onto B and $|K|$ is contractible, then any map $|K| \to B$ may be 'lifted' into X.*

Fibre spaces are of particular importance in homotopy theory and arise, for example, in considering function spaces. Thus let $E(Y; A, B)$ be the space of paths in Y starting in $A \subseteq Y$ and ending in $B \subseteq Y$ and let $p: E(Y; A, B) \to A \times B$ associate with a path its initial and final points.

II. 3.3 Proposition. *p is a fibre map.*

(Serre's proof is reproduced on p. 65 of [9] using the equivalent space $E_1(Y; A, B)$ of unit paths. The proof is an easy application of II. 2.3.)

Let $A = y_0$. Then p may be regarded as a map $p: E(Y; y_0, B) \to B$ and, if $y_0 \in B$, the fibre over y_0 is ΩY. In particular, we may take $B = Y$ and conclude

II. 3.4 Proposition. *There is a fibration in which the total space is contractible, the base-space is Y and the fibre is ΩY.*

(For $E(Y)$ is certainly contractible, since each path may be withdrawn to its initial point y_0.)

‡ More precisely, replace S^2 by its homeomorph, the complex projective line.

We now define the *n-th relative homotopy group* of the pair Y, B, relative to the base-point $y_0 \in B$ by

II. 3.5 $$\pi_n(Y, B, y_0) = \pi_{n-1}(E(Y; y_0, B), e) \quad n \geqslant 2,$$

where e is the constant path at y_0; we shall generally suppress base-points in what follows. We shall not detail the properties of relative homotopy groups which are analogous to those for *absolute* homotopy groups given in II. 1. However, we shall need

II. 3.6 Proposition. *If $p : X \to B$ is a fibre map with fibre F (over the base-point $b_0 \in B$), then p induces an isomorphism*

$$\hat{p}_* : \pi_n(X, F) \cong \pi_n(B).$$

(This may be proved by reinterpreting the relative homotopy group as the set of homotopy classes of maps $V^n, S^{n-1}, x_* \to X, F, x_0$, where $x_0 \in F$ is a base-point. The group operation is such that the *relative* homotopy group $\pi_n(Y, y_0, y_0)$ in this interpretation coincides with the *absolute* homotopy group‡ $\pi_n(Y, y_0)$. Thus p is certainly homomorphic and it follows readily from the definition of a fibre map that $\hat{p}_*$ is isomorphic (see [9], p. 49).)

The map $p : E(Y; B) \to B$ evidently induces $p_* : \pi_n(Y, B) \to \pi_{n-1}(B)$. If we interpret $\pi_n(Y, B)$ as in the sketched proof above, in terms of maps $V^n, S^{n-1}, x_* \to Y, B, y_0$, then $\hat{p}_*$ is induced by restricting such maps to S^{n-1}; we shall use the symbol β_* for the homomorphism p_* so interpreted.

II. 3.7 Theorem. *Let (Y, B) be a pair. There is an exact sequence (the exact homotopy sequence of a pair)*

$$\ldots \to \pi_n(B) \xrightarrow{i_*} \pi_n(Y) \xrightarrow{j_*} \pi_n(Y, B) \xrightarrow{\beta_*} \pi_{n-1}(B) \to \ldots.$$

(Here i_* is induced by the injection $B, y_0 \to Y, y_0$ and j_* by the injection $Y, y_0 \to Y, B$. A proof is in Hilton [9], chapter 4.)

II. 3.8 Corollary. *Let $p : X \to B$ be a fibre map with fibre F. There is an exact sequence (the exact homotopy sequence of a fibration)*

$$\ldots \to \pi_n(F) \xrightarrow{i_*} \pi_n(X) \xrightarrow{p_*} \pi_n(B) \xrightarrow{\bar{\beta}_*} \pi_{n-1}(F) \to \ldots.$$

(Here $\bar{\beta}_*$ is defined by $\beta_* = \hat{p}_* \bar{\beta}_* : \pi_n(X, F) \to \pi_{n-1}(F)$ (see II. 3.6). Thus II. 3.8 is a consequence of II. 3.7. On the other hand, II. 3.7 may be deduced from II. 3.8 by considering $p : E(Y, B) \to B$ with fibre ΩY and invoking II. 2.7.)

‡ Using II. 3.5, the canonical isomorphism II. 2.7 intervenes.

7

CONTRAHOMOLOGY AND MAPS

7.1 Introduction

In this chapter we apply contrahomology theory to a particular case of a very general mathematical problem, indicating in 7.5 and 7.7 two directions in which further developments have been made. The general problem concerns a set X with subset A and another set Y; these sets have some structure, that of a group or a differentiable manifold or a topological space, for example. Given a transformation $g : A \to Y$, which preserves the structure, the problem is to determine whether g can be extended to a transformation $f : X \to Y$ which also preserves the structure.

The case of interest in topology is that in which X, Y are topological spaces and g, f are maps. We shall here be considering the special case in which X, A form a polyhedral pair. If L then is a subcomplex of the complex K (not necessarily finite), we suppose given a map $g : |L| \to Y$ and we look for an extension $f : |K| \to Y$. One method of trying to construct f springs early to mind; we can certainly extend g to a map $f^0 : |K^0 \cup L| \to Y$, selecting arbitrarily the images of the vertices of $K - L$, and we can then try to extend f^0 to a map $f^1 : |K^1 \cup L| \to Y$. If this proves to be impossible, we may profit by our experience of trying to extend f^0 and in the light of our difficulties reconsider our selection of f^0. In general, writing‡ P^n for $|K^n \cup L|$, we may suppose $f^n : P^n \to Y$ given and we may look for an extension to $f^{n+1} : P^{n+1} \to Y$; if there is no such extension, we may reconsider the extension f^n of f^{n-1} in the hope of finding another extension of f^{n-1} which admits an extension over P^{n+1}.

The method, which is due to Eilenberg, is to measure the obstruction to extending f^n over P^{n+1} as a contracycle of K, L with values in $\pi_n(Y)$ and for simplicity of treatment we assume Y to be n-simple; the case in which Y is not n-simple is discussed in 7.6. It will be proved

‡ The reader is warned that P^n is a polyhedron and that it is not (in general) n-dimensional; the superscript is intended to indicate that P^n is the polyhedron of the 'n-section mod L' of K. A similar remark applies to Q^n, R^n that appear later in the chapter. The reader is asked temporarily to forget the conventions that P^n stands for a real projective space and R^n for Euclidean space.

that f^{n-1} is extendable over P^{n+1} if and only if this *obstruction contracycle* is the contraboundary of a contrachain and that such a contrachain can be used to correct the errors (if any) in the extension f^n of f^{n-1}.

The extension problem has a close connexion with a homotopy problem. If $f_0, f_1 : |K| \to Y$ are two maps of $|K|$ and $G : |L| \times I \to Y$ is a partial homotopy, a homotopy connecting $g_0 = f_0 \mid |L|$ and $g_1 = f_1 \mid |L|$, we may want to know whether the partial homotopy G can be extended to a homotopy $F : |K| \to Y$ connecting f_0 and f_1. We may combine G, f_0, f_1 to give a map $H : (|K| \times \dot{I}) \cup (|L| \times I) \to Y$ and we ask whether H can be extended over $|K| \times I$. This is evidently a special case of the extension problem. In particular, if L is empty this question reduces to the question whether two given maps are homotopic. Another important application of these methods is that of measuring the obstruction to constructing a cross-section in a fibre space with polyhedral base-space; see Steenrod [18]. The last three sections of this chapter are devoted to informal prefaces to three important developments of obstruction theory.

7.2 The obstruction contracycle

We suppose now that K has been oriented by an ordering and that Y is n-simple and that $f = f^n : P^n = |K^n \cup L| \to Y$ is given. Let σ_{n+1} be an oriented $(n+1)$-simplex of K mod L; since $\dot{s}_{n+1} \subseteq P^n, f^n \mid \dot{\sigma}_{n+1}$ is defined and determines an element of $\pi_n(Y)$. Then f determines a contrachain‡ $c_f \in C^{n+1}(K, L; \pi_n(Y))$ by the rule

$$(\sigma_{n+1}, c_f) = \{f \mid \dot{\sigma}_{n+1}\}.$$

We observe from II. 1.12 that, for a simplex $\tau_{n+1} \in L$, $(\tau_{n+1}, c_f) = 0$ since $f \mid \dot{\tau}_{n+1}$ has an extension over τ_{n+1}, namely, $f \mid \tau_{n+1}$; c_f is therefore a contrachain of K mod L§. It is called the *obstruction contrachain*.

The reason for the name is given by

7.2.1 Proposition. *f^n may be extended over P^{n+1} if and only if $c_f = 0$.*

By II. 1.12, f^n can be extended over each s_{n+1} if and only if $c_f = 0$ and we invoke 1.10.4 to deduce that f^n can be extended over P^{n+1} if it can be extended severally over each s_{n+1}. ∎

‡ If $n = 1$, as Y is 1-simple $\pi_1(Y)$ is commutative and so, by abuse of language, may be regarded as an abelian group. The reader can get an adequate idea of the work of this section if he considers Y to be 1-connected and $n \geq 2$.

§ Here we are identifying a contrachain of K mod L with its image under $\mu^\cdot$ (2.8.8 c); we are in fact considering it to be a contrachain of K that takes the value 0 on L. This will be our point of view throughout this chapter.

7.2.2 Proposition. *If* $f_0 \simeq f_1 : P^n \to Y$, $c_{f_0} = c_{f_1}$.

For homotopic maps of $\dot{\sigma}_{n+1}$ determine the same element of $\pi_n(Y)$. ∎

7.2.3 Theorem. *The obstruction contrachain is a contracycle of*

$$C^{\cdot}(K, L; \pi_n(Y)).$$

We must prove that, for each $\sigma_{n+2} \in K - L$,

$$(\sigma_{n+2}\partial, c_f) = 0.$$

But this is a special case of II. 1.29, with $\dot{\sigma}_{n+2}$ for T^{n+1} and with $(-1)^i \sigma^i_{n+1}$ for E^{n+1}_i, where (as often) if $\sigma_{n+2} = a^0 a^1 \dots a^{n+2}$, we write σ^i_{n+1} for $a^0 \dots \hat{a}^i \dots a^{n+2}$. ∎

The problem of extending f^n over P^{n+1} is disposed of by 7.2.1 in the sense that 7.2.1 gives a necessary and sufficient condition for extendability in terms of the obstruction contracycle. Of far greater interest, however, is the problem of determining whether

$$f^{n-1} = f^n \mid P^{n-1}$$

can be extended over P^{n+1}. Here we know that f^{n-1} has an extension f^n over P^n, but we are prepared to take a different extension if this will lead to a further extension over P^{n+1}. The chief tool in this is the notion of the *difference contrachain* $d(f, f')$ of two extensions f, f' of the same $f^{n-1} : P^{n-1} \to Y$ over P^n, Y being still assumed to be n-simple.

Given the two maps $f, f' : P^n \to Y$ agreeing on P^{n-1}, for each $\sigma^i_n \in K$ we choose an orientation-preserving homeomorphism $h^i : V^n \to \sigma^i_n$. Then $h^i f, h^i f'$ agree on S^{n-1} and we define $\alpha^i(f, f')$ to be $\alpha(h^i f, h^i f')$, observing that the choice of h^i does not affect the value (see remark preceding II. 1.23). We define the *difference contrachain* $d(f, f')$ by the rule

7.2.4 $$(\sigma^i_n, d(f, f')) = \alpha^i(f, f').$$

For $\tau^j_n \in L$, f and f' agree on τ^j_n and (i) of II. 1.15 implies that $(\tau^j_n, d(f, f')) = 0$, so that $d(f, f')$ is indeed a contrachain of K mod L.

7.2.5 Proposition. *If* $f, f', f'' : P^n \to Y$ *agree on* P^{n-1}, *and if* Y *is* n-*simple*,

$$\text{(i)}\ d(f, f) = 0, \quad \text{(ii)}\ d(f, f') = -d(f', f),$$

$$\text{(iii)}\ d(f, f') + d(f', f'') = d(f, f'').$$

This follows at once from II. 1.15. ∎

7.2.6 Theorem. *If $f, f' : P^n \to Y$ agree on P^{n-1} and if Y is n-simple,*

$$c_f - c_{f'} = (-1)^n \delta d(f, f').$$

We consider an arbitrary $\sigma_{n+1} \in K$ and we prove that

$$(-1)^n (\sigma_{n+1}, c_f - c_{f'}) = (\sigma_{n+1}, \delta d(f, f'))$$

by proving that

$$(-1)^n \{f \mid \dot{\sigma}_{n+1}\} - (-1)^n \{f' \mid \dot{\sigma}_{n+1}\} = \sum_{i=0}^{n+1} (-1)^i \alpha(h^i f, h^i f'),$$

where $h^i : V^n \to \sigma_n^i$ is an orientation-preserving homeomorphism onto the face σ_n^i of σ_{n+1}. This, with slightly different terminology, is the statement of II. 1.31 with $\dot{\sigma}_{n+1}$ for T^n and $(-1)^i \sigma_n^i$ for E_i^n. ▮

As we remarked after II.1.24, the sign $(-1)^n$ can be avoided by different orientation conventions. It is not, however, a serious embarrassment.

This theorem shows that any two extensions over P^n of

$$f^{n-1} : P^{n-1} \to Y$$

have obstruction contracycles that are contrahomologous; we may therefore write $\{c_f\}$ as $\xi(f^{n-1})$, thus emphasizing that it does not depend on the choice f of extension over P^n. The next theorem gives a converse.

7.2.7 Theorem. *If Y is n-simple and $f : P^n \to Y$ has obstruction contracycle c_f and if $c_f \sim z$ in $C^{\cdot}(K, L; \pi_n(Y))$, then there is an extension f' of $f^{n-1} = f \mid P^{n-1}$ over P^n such that $c_{f'} = z$.*

Let $c_f - z = (-1)^n \delta d$ for $d \in C^n(K, L; \pi_n(Y))$. We shall interpret d as a difference contrachain; that is to say, we shall construct f' so that $d = d(f, f')$, from which the result follows. For each $\sigma_n^i \in K \bmod L$ we use II. 1.16 to define $f' \mid \overline{\sigma}_n^i$ so that $\alpha^i(f, f') = (\sigma_n^i, d)$; in P^{n-1} f' is defined to agree with f^{n-1} and thus the definition of f' is complete. ▮

From this follows the main theorem of the section.

7.2.8 Theorem. *If $f^{n-1} : P^{n-1} \to Y$, Y being n-simple, has an extension $f = f^n : P^n \to Y$, then f^{n-1} has an extension over P^{n+1} if and only if $c_f \sim 0$.*

If f^{n-1} is extendable over P^{n+1}, it has an extension f' over P^n that is extendable over P^{n+1}; by 7.2.1 $c_{f'} = 0$ and by 7.2.6 $c_f \sim c_{f'}$, so that $c_f \sim 0$. Conversely, if $c_f \sim 0$, by 7.2.7 there is an extension f' of f^{n-1} such that $c_{f'} = 0$; then by 7.2.1 f' is therefore extendable over P^{n+1}. ▮

If Y is $(n-1)$- and n-simple (for example, if Y is 1-connected), we can now answer the question whether $f^{n-1} : P^{n-1} \to Y$ is extendable over P^{n+1}. First $c_{f^{n-1}}$ must vanish; then for an arbitrary provisional extension $\tilde{f}$ of f^{n-1} over P^n, $c_{\tilde{f}}$ must be a contraboundary. In that case we use any contrachain whose contraboundary is $c_{\tilde{f}}$ as a guide to improve the extension $\tilde{f}$, replacing it (if necessary) by a different extension f that can itself be further extended over P^{n+1}.

The limitations of this must, however, be emphasized. Suppose, for instance, that $n = 1$ and that $P^0 = |K^0 \cup L|$ has been mapped by f^0 and that f^0 is extendable to $\tilde{f} = P^1 \to Y$ so that $c_{\tilde{f}} \sim 0$. We replace $\tilde{f}$ by f^1 and extend this to $f^2 : P^2 \to Y$. Our present methods can then only tell us whether $f^1 : P^1 \to Y$ has an extension over P^3. This may perhaps not be the case and yet there may at the same time be an extension of f^0 over P^3. We have in fact a method which allows us to revise the last step taken, but so far no method that takes advantage of the possibility of revising the penultimate step. We have therefore not established a general criterion for determining whether $g : |L| \to Y$ is extendable over $|K|$, only a partial result which may be of value in special cases.

7.3 The homotopy extension problem

As explained in the introductory section, the homotopy problem of extending to a homotopy of $|K|$ a partial homotopy of $|L|$ connecting two given maps of $|K|$ is an extension problem for the product $|K| \times I$. We suppose in fact that $H : (|K| \times \dot{I}) \cup (|L| \times I) \to Y$ is given and we look for an extension of H over $|K| \times I$. We shall write Q^n for $(|K| \times \dot{I}) \cup (P^{n-1} \times I)$ and we consider the case in which Y is n-simple and H has been extended to $F = F^n : Q^n \to Y$. As before we measure the obstruction to extending F^n over Q^{n+1}. There is a possible obstruction for each $\sigma_n \in K$ in that $F \mid (\bar{\sigma}_n \times I)^{\cdot}$ may not be extendable over $\bar{\sigma}_n \times I$. Now $F \mid (\bar{\sigma}_n \times I)^{\cdot}$ determines an element of $\pi_n(Y)$ and we define the *homotopy obstruction contrachain* $c_F \in C^n(K, L; \pi_n(Y))$ by

$$(\sigma_n, c_F) = \{F \mid (\bar{\sigma}_n \times I)^{\cdot}\}.$$

Since, for each $\tau_n \in L$, F is defined over $\bar{\tau}_n \times I$, $(\tau_n, c_F) = 0$, so that c_F is indeed a contrachain of $K \bmod L$. Notice that, whereas in the extension problem the obstruction lies in $C^{n+1}(K, L; \pi_n(Y))$ in the homotopy problem it lies in $C^n(K, L; \pi_n(Y))$. We could have located the obstruction in $C^{n+1}(K \times I, (K \times \dot{I}) \cup (L \times I); \pi_n(Y))$; there would, however, have been no advantage in doing this even if we had established a homology theory for $K \times I$.

7.3.1 Proposition. *$F: Q^n \to Y$, Y being n-simple, is extendable over Q^{n+1} if and only if $c_F = 0$.*]

7.3.2 Corollary. *If $\pi_n(Y) = 0$ for all $n \geqslant 0$, any two maps*

$$f_0, f_1 : |K| \to Y$$

are homotopic.

Certainly Y is n-simple for all n and a partial homotopy between $f_0 \big| |K^0|$ and $f_1 \big| |K^0|$ can be constructed, for Y is path-connected. If a partial homotopy between $f_0 \big| |K^{n-1}|$ and $f_1 \big| |K^{n-1}|$ has been constructed, the obstruction to extending this homotopy over $|K^n|$ lies in the null group $C^n(K; \pi_n(Y))$. The obstruction therefore vanishes and the homotopy can be extended. The union for all n of the partial homotopies so constructed gives a homotopy for $|K|$ even in the case when K is infinite-dimensional.]

7.3.3. Corollary. *A polyhedron $|K|$ is contractible if and only if $\pi_n(|K|) = 0$ for all $n \geqslant 0$.*

If $|K|$ is contractible, any map of S^n in $|K|$ is extendable over V^{n+1} so that $\pi_n(|K|)$ is null. If $\pi_n(|K|)$ is null for all $n \geqslant 0$, the identity map of $|K|$ in $|K|$ and the map of $|K|$ to any point of $|K|$ are homotopic by 7.3.2.]

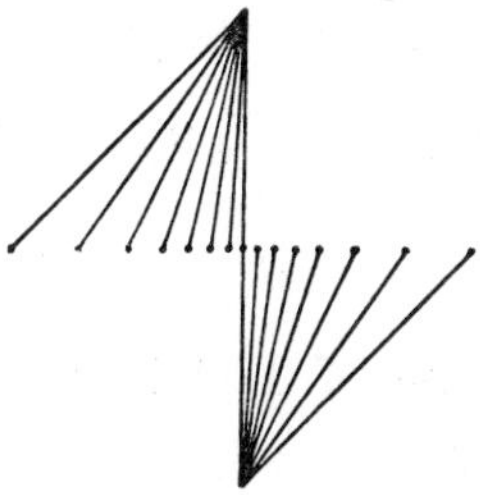

Fig. 7.1

There are spaces (necessarily not polyhedral) whose homotopy groups are all null and which are yet not contractible. Take for example the space of fig. 7.1, suggested to us by E. C. Zeeman. This is the join in R^2 of $(0, -1)$ to the set of points $0, 1, \frac{1}{2}, \frac{1}{3}, \ldots, 1/n, \ldots$ on the u_1-axis, together with the join of $(0, 1)$ to the set of points $0, -1, -\frac{1}{2}, -\frac{1}{3}, \ldots, -1/n, \ldots$ on the u_1-axis.

7.3.4 Theorem. *The homotopy obstruction contrachain is a contracycle of $C^{\cdot}(K, L; \pi_n(Y))$.*

The proof resembles that of 7.2.3. We prove that, for each $\sigma_{n+1} \in K$,

$$(\sigma_{n+1}, \delta c_F) = (\sigma_{n+1}\partial, c_F) = \Sigma(-1)^i \{F \mid (\bar{\sigma}_n^i \times I)^{\cdot}\} = 0.$$

In II. 1.29 we take $(\bar{\sigma}_{n+1}\times I)^{\cdot}$ for T^{n+1} and $(\bar{\sigma}_{n+1}\times 0)$, $(\bar{\sigma}_{n+1}\times 1)$ and $(\bar{\sigma}_n^i\times I)$ for $i=0,\ldots,n+1$ as the $(n+1)$-cells in T^{n+1}, with orientations $(-1)^n$, $(-1)^{n+1}$ and $(-1)^i$. Since $F\mid(\dot{\sigma}_{n+1}\times 0)$ and $F\mid(\dot{\sigma}_{n+1}\times 1)$ have extensions to $(\bar{\sigma}_{n+1}\times 0)$ and $(\bar{\sigma}_{n+1}\times 1)$,

$$\{F\mid(\dot{\sigma}_{n+1}\times 0)\}=\{F\mid(\dot{\sigma}_{n+1}\times 1)\}=0,$$

and the result follows. ▌ Notice the essential role played at the end of the proof by the fact that f_0, f_1 are defined over P^{n+1}.

As in the previous section we next consider two extensions F, F' of $F^{n-1}: Q^{n-1}\rightarrow Y$ over Q^n. On each $\bar{s}^i_{n-1}\times I$ they differ (if at all) only on the interior, agreeing on $(\bar{s}^i_{n-1}\times I)^{\cdot}\subseteq Q^{n-1}$. If $h^i: V^n\rightarrow\bar{\sigma}^i_{n-1}\times I$ is an orientation-preserving homeomorphism, h^iF, h^iF' agree on S^{n-1} and we write $\alpha^i(F,F')$ for $\alpha(h^iF,h^iF')$, the element being independent of the choice of h^i. The *homotopy difference contrachain* $d(F,F')\in C^{n-1}(K,L;\pi_n(Y))$ is defined by

$$(\sigma^i_{n-1}, d(F,F'))=\alpha^i(F,F').$$

7.3.5 Theorem. *If F, F' are two extensions over Q^n of $F^{n-1}: Q^{n-1}\rightarrow Y$, Y being n-simple, then*

$$c_F-c_{F'}=(-1)^n\,\delta d(F,F').$$

We take any $\sigma_n\in K$ and reduce the problem to that of proving that

$$(-1)^n\{F\mid(\bar{\sigma}_n\times I)^{\cdot}\}-(-1)^n\{F'\mid(\bar{\sigma}_n\times I)^{\cdot}\}=(\sigma_n\partial, d(F,F'))$$
$$=\sum_{i=0}^{n}(-1)^i(\sigma^i_{n-1}, d(F,F')).$$

But this is a special case of II. 1.31 with $(\bar{\sigma}_n\times I)^{\cdot}$ for T^n, and $(-1)^i(\bar{\sigma}^i_{n-1}\times I)$ for E^n_i. ▌

The analogue of 7.2.7, converse to 7.3.5, can be proved along the lines of the proof of 7.2.7 and hence

7.3.6 Theorem. *Let Y be n-simple, let L be a subcomplex of K and let $Q^n=(|K|\times\dot{I})\cup(|K^{n-1}\cup L|\times I)$; if maps $f_0, f_1: |K|\rightarrow Y$,*

$$F=F^n: Q^n\rightarrow Y$$

are given such that, for all $x\in|K|$, $(x,\epsilon)F=xf_\epsilon$, $\epsilon=0,1$, then

$$F^{n-1}=F\mid Q^{n-1}$$

is extendable over Q^{n+1} if and only if $c_F\sim 0$ in $C^{\cdot}(K,L;\pi_n(Y))$. ▌

This theorem thus deals with the case of two maps of $|K|$ and of a partial homotopy connecting them defined over $|K^{n-1}\cup L|$; it answers the question whether the restriction of that partial homotopy to $|K^{n-2} \cup L|$ is extendable to a partial homotopy of $|K^n \cup L|$.

There is a connexion between the homotopy obstruction contracycle and the difference contrachain of the previous section. If f_0, f_1 are extensions over P^n of $f^{n-1}: P^{n-1} \to Y$, we may ask whether the constant homotopy on P^{n-1} is extendable to a homotopy of P^n connecting f_0 and f_1; this question leads to the formulation of a homotopy obstruction element $c_F \in C^n(K, L; \pi_n(Y))$. For each $\sigma_n^i \in K$, let $h^i: V^n \to \sigma_n^i$ be an orientation-preserving homeomorphism; then in the general case of the homotopy obstruction contracycle,

$$(\sigma_n^i, c_F) = \beta(h^i f_0, h^i f_1; h^i F),$$

β being defined as before II. 1.28. In this special case, $h^i F$ is a constant track and we may apply II. 1.28 to deduce that

$$(\sigma_n^i, c_F) = \alpha(h^i f_1, h^i f_0) = (\sigma_n^i, d(f_1, f_0)),$$

so that

7.3.7 $$c_F = d(f_1, f_0).$$

A mild paradox here suggests itself; c_F is by 7.3.4 a contracycle, whereas $d(f_1, f_0)$ is in general not. This is easy to resolve. Here F refers to a (constant) partial homotopy connecting two maps of $P^n = |K^n \cup L|$, so that c_F is only known by 7.3.4 to be a contracycle of $C^\cdot(K^n \cup L, L; \pi_n(Y))$ and not a contracycle of $C^\cdot(K, L; \pi_n(Y))$. We can therefore only deduce about $d(f_1, f_0)$ that it is a contracycle of $C^\cdot(K^n \cup L, L; \pi_n(Y))$, and this is scarcely odd because it is an n-contrachain.

The condition that $d(f_1, f_0)$ be a contracycle of $C^\cdot(K, L; \pi_n(Y))$ comes at once from 7.2.6, namely, that $c_{f_0} = c_{f_1}$. This is always the case when f_0 and f_1 are both extendable over P^{n+1}.

7.3.8 Proposition. *If Y is n-simple and $f_0, f_1: P^n \to Y$ agree on P^{n-1}, then $d(f_1, f_0) \sim 0$ if and only if $f_0 \simeq f_1 \operatorname{rel} P^{n-2}$.*

By 7.3.7 and 7.3.6, $c_F \sim 0$ if and only if $F \mid (P^n \times \dot{I}) \cup (P^{n-2} \times I)$ is extendable over $P^n \times I$; since $F \mid P^{n-1} \times I$ is here the constant homotopy, this is equivalent to the condition that $f_0 \simeq f_1 \operatorname{rel} P^{n-2}$. ∎

In the next section we shall also need

7.3.9 Proposition. *If Y is n-simple and $F_0, F_1: Q^n \to Y$ agree on Q^{n-1}, then $d(F_1, F_0) \sim 0$ if and only if $F_0 \simeq F_1 \operatorname{rel} Q^{n-2}$.*

We sketch a proof analogous to that of 7.3.8. We may consider the obstruction to extending the constant homotopy Φ^n of Q^{n-1}

(connecting $F_0 \mid Q^{n-1}$ and $F_1 \mid Q^{n-1}$, which coincide) to a homotopy of Q^n connecting F_0 and F_1; this obstruction can be measured as an element $c_\Phi \in C^{n-1}(K, L; \pi_n(Y))$ in a way that should now be obvious. In a more general context we can prove that $c_\Phi \sim 0$ if and only if Φ restricted to a homotopy of Q^{n-2} is extendable to a homotopy of Q^n connecting F_0 and F_1; in this case such a homotopy is just a homotopy rel Q^{n-2}; we can also prove that, in this case of a constant homotopy Φ,

7.3.10 $$c_\Phi = d(F_1, F_0). \blacksquare$$

We shall make further use of the obstruction theory for homotopies of homotopies here adumbrated.

The ideas and language of obstruction theory are not limited to the present context. The reader may, for instance, be interested to refer to 3.4.3 and to satisfy himself that the main argument about acyclic carriers amounts to saying that there is no obstruction to constructing an appropriate chain homotopy.

7.4 Applications

In describing arguments involving obstruction theory we shall find it convenient to refer to the obstruction contracycle

$$c_f \in C^{n+1}(K, L; \pi_n(Y))$$

as the *primary obstruction* to extending $f = f^n : P^n \to Y$ over P^{n+1} and to refer to its contrahomology class

$$\xi(f^{n-1}) = \{c_f\} \in H^{n+1}(K, L; \pi_n(Y))$$

as the *secondary obstruction*‡ to extending $f^{n-1} = f \mid P^{n-1}$ over P^{n+1}. We shall use similar language for homotopy extension obstructions and the statement, for instance, 'there is no secondary obstruction' will mean that the secondary obstruction is the zero element.

We start with a situation at once too special and too general to arouse immediate interest. There are, however, two particular cases of the theorem, each of geometrical importance, and these will appear as corollaries of the theorem.

7.4.1 Theorem. *We suppose that Y is 1-connected and that there is an integer $m \geq 2$ (or that Y is 0-connected and n-simple for all n and that $m = 1$) such that*

(i) *for all $p = 2, \ldots, m-1$, (a) $H^{p-1}(K; \pi_p(Y))$ and (b) $H^p(K; \pi_p(Y))$ are null, and*

(ii) *for all $n > m$, (a) $H^n(K; \pi_n(Y))$ and (b) $H^{n+1}(K; \pi_n(Y))$ are null;*

‡ The reader is warned that some authors call the *secondary* obstruction the *first* obstruction.

then the homotopy classes of maps of $|K|$ *in* Y *stand in* $(1, 1)$ *correspondence with the elements of* $H^m(K; \pi_m(Y))$.

The idea of the proof is attractive and easy. Choose a base-point $y_0 \in Y$; for any $f: |K| \to Y$ there is no secondary obstruction to deforming f to a map which maps $|K^{m-1}|$ to y_0. There is, however, an obstruction to deforming this map to one mapping $|K^m|$ to y_0; this obstruction is a contracycle in $C^m(K; \pi_m(Y))$. A different deformation of f to a map which maps $|K^{m-1}|$ to y_0 determines an (in general) different but contrahomologous contracycle; the map f in this way determines an element of $H^m(K; \pi_m(Y))$. Clearly homotopic maps determine the same element; conversely, if two maps determine the same element, there is no secondary obstruction to deforming one to agree with the other on $|K^m|$ and then no secondary obstruction to deforming it to agree with the other totally; maps therefore giving the same element of $H^m(K; \pi_m(Y))$ are homotopic. Finally, for any element of $H^m(K; \pi_m(Y))$ a partial map $\tilde{f}^{m+1}: |K^{m+1}| \to Y$ can be constructed determining this element and there is no secondary obstruction to extending $f^m = \tilde{f}^{m+1} \big| |K^m|$ to a map of $|K|$ in Y also determining the given element. These assertions establish a $(1, 1)$ correspondence between the classes of maps of $|K|$ in Y and the elements of $H^m(K; \pi_m(Y))$.

For completeness we now give the technical details of this proof. We start by observing that, if $m \geqslant 2$, Y is simply connected and therefore n-simple for all n, so that whether $m > 1$ or $m = 1$, Y is n-simple. Now let e be the constant map of $|K|$ to y_0. We construct a provisional partial homotopy of $|K^0|$ connecting e and f; this is possible as Y is 0-connected. Calling the associated map $\tilde{F}^1: Q^1 \to Y$, we form $c_{\tilde{F}}$ a contracycle in $C^1(K; \pi_1(Y))$; here of course L is empty. If $m > 1$, by (i) (*b*) of 7.4.1 $H^1(K; \pi_1(Y))$ is null so that $c_{\tilde{F}}$ is a contraboundary and there is no secondary obstruction to extending $F^0: Q^0 \to Y$ over Q^2; here F^0 is the map which maps $|K| \times 0$ to y_0 and $|K| \times 1$ by the map f. Let $\tilde{F}^2: Q^2 \to Y$ be a provisional extension of F^0 and write $\tilde{F}^2 \mid Q^1$ as F^1; F^1 need not be regarded as provisional. In general if $\tilde{F}^p: Q^p \to Y$ has been constructed and $F^{p-1} = \tilde{F}^p \mid Q^{p-1}$, provided that $p < m$ there is no secondary obstruction to extending F^{p-1} over Q^{p+1}, since by (i) (*b*) $H^p(K; \pi_p(Y))$ is null. In this way we construct $\tilde{F}^m: Q^m \to Y$, giving a partial homotopy of $|K^{m-1}|$ connecting e and f. We shall make no further use of condition (i) (*b*).

The secondary obstruction to extending $F^{m-1} = \tilde{F}^m \mid Q^{m-1}$ over Q^{m+1} is an element of $H^m(K; \pi_m(Y))$; let us denote it as $\zeta(f; F^{m-1})$.

We now use (i) (a) to show that $\zeta(f; F^{m-1})$ does not depend on the choice of $F = F^{m-1} : Q^{m-1} \to Y$ extending F^0. Suppose in fact that $F' : Q^{m-1} \to Y$ is a different extension of F^0. Let R^p stand for

$$(Q^{m-1} \times \dot{I}) \cup (Q^{p-1} \times I)$$

and let $\Phi^1 : R^1 \to Y$ be the map associated with the partial constant homotopy on Q^0 connecting F and F'; thus $(x, 0)\,\Phi^1 = xF$ and $(x, 1)\,\Phi^1 = xF'$ for any $x \in Q^{m-1}$; and $(x, u)\,\Phi^1 = xF = xF'$ for any $x \in Q^0$. If $m > 1$, Φ^1 can be extended to a provisional map $\tilde{\Phi}^2 : R^2 \to Y$, since the primary obstruction lies in $C^0(K; \pi_1(Y))$ and $\pi_1(Y)$ is null. Suppose that $\tilde{\Phi}^p : R^p \to Y$, has been constructed extending Φ^0 and write $\tilde{\Phi}^p \mid R^{p-1}$ as Φ^{p-1}; then the secondary obstruction $\{c_\Phi\}$ to extending Φ^{p-1} over R^{p+1} lies in $H^{p-1}(K; \pi_p(Y))$; if $p < m$ this is null by (i) (a) and so by 7.3.10 and 7.3.9 a provisional extension $\tilde{\Phi}^{p+1}$ of Φ^{p-1} over R^{p+1} can be constructed. We thus construct

$$\tilde{\Phi}^m : R^m = Q^{m-1} \times I \to Y,$$

extending Φ^0; this is a homotopy rel Q^0 connecting F and F'. But clearly if $F \simeq F'$ rel Q^0, $\zeta(f; F) = \zeta(f\ ;\ F')$ and we have proved the independence of the secondary obstruction, now to be written as $\zeta(f)$, from the choice of partial homotopy of $|K^{m-2}|$.

Since evidently if $f \simeq g : |K| \to Y$, $\zeta(f) = \zeta(g)$, ζ induces a function ζ^* defined on the homotopy classes of maps of $|K|$ in Y with values in $H^m(K; \pi_m(Y))$. The arguments leading up to the definition of ζ^* have used conditions (i) of the theorem; we now use conditions (ii) to prove that ζ^* is (1, 1) onto $H^m(K; \pi_m(Y))$.

Suppose first that $\zeta^*\{f\} = \zeta^*\{g\}$; we shall use (ii) ($a$) to prove that $f \simeq g$, which will show that ζ^* is (1, 1) into $H^m(K; \pi_m(Y))$. Let

$$F = F^m : Q^m \to Y$$

be an extension of F^{m-1} where F^{m-1}, as before, gives a partial homotopy of $|K^{m-2}|$ connecting e and f; let $G^{m-1} : Q^{m-1} \to Y$ give a partial homotopy of $|K^{m-2}|$ connecting e and g and let $\tilde{G} = \tilde{G}^m : Q^m \to Y$ be a provisional extension of G^{m-1}. Since $\zeta(f) = \zeta(g)$, $c_F \sim c_{\tilde{G}}$ and by II. 1.16 there is an extension $G = G^m$ of G^{m-1} such that $c_F = c_G$. We now combine F and G to give $T = T^m : Q^m \to Y$, where T may be loosely described as $\bar{F} * G$ and precisely described by

$$(x, 0)\,T = xf \quad \text{and} \quad (x, 1)\,T = xg \quad \text{for} \quad x \in |K|,$$

$$(x, u)\,T = (x, 1-2u)\,F \quad \text{for} \quad x \in |K^{m-1}|, \quad 0 \leqslant u \leqslant \tfrac{1}{2}$$

$$= (x, 2u-1)\,G \quad \text{for} \quad x \in |K^{m-1}|, \quad \tfrac{1}{2} \leqslant u \leqslant 1.$$

Since $c_T = c_G - c_F = 0$, there is no primary obstruction to extending T^m provisionally to $\tilde{T}^{m+1} : Q^{m+1} \to Y$, giving a partial homotopy of $|K^m|$ connecting f and g. The secondary obstruction then to extending T^m over Q^{m+2} lies in $H^{m+1}(K; \pi_{m+1}(Y))$ which is null by (ii) (a); there is therefore a provisional extension $\tilde{T}^{m+2} : Q^{m+2} \to Y$ and we write $\tilde{T}^{m+2} \mid Q^{m+1}$ as T^{m+1}. Repeated application of this argument leads to a sequence $T^m, T^{m+1}, \ldots$ of maps of $Q^m, Q^{m+1}, \ldots$ in Y, each an extension of its predecessor. These combine to give a map $T : |K| \times I \to Y$ given by

$$xT = xT^r \quad \text{if} \quad x \in Q^r;$$

this is independent of the choice of r for a given point x, and the topology of $|K|$ ensures that T so defined is continuous. T is certainly an extension of $T^0 = T^m \mid Q^0$ and is therefore a homotopy connecting f and g. This finishes the demonstration that ζ^* is $(1, 1)$.

Finally, for the proof that ζ^* is onto $H^m(K; \pi_m(Y))$ we use (ii) (b), namely, that $H^{n+1}(K; \pi_n(Y))$ is null for $n > m$. Given an element $\{z\} \in H^m(K; \pi_m(Y))$ we shall construct $f : |K| \to Y$ such that $\zeta(f) = \{z\}$. We define $f^{m-1} : |K^{m-1}| \to Y$ to be the constant map to y_0 and define an extension $f^m : |K^m| \to Y$ such that $d(f^m, e) = z$, using II. 1.16. The primary obstruction to extending f^m over $|K^{m+1}|$ is

$$c_{f^m} = c_e + (-1)^m \delta d(f^m, e) = 0,$$

since $e \mid |K^m|$ is certainly extendable and z is a contracycle. Since there is no primary obstruction we may extend f^m to a provisional map $\tilde{f}^{m+1} : |K^{m+1}| \to Y$. The secondary obstruction to extending f^m over $|K^{m+2}|$ lies in $H^{m+2}(K; \pi_{m+1}(Y))$ which is null; there is therefore a provisional extension $\tilde{f}^{m+2} : |K^{m+2}| \to Y$ and we write $\tilde{f}^{m+2} \mid |K^{m+1}|$ as f^{m+1}. Proceeding as before, the secondary obstruction is always zero, and we construct a sequence of maps, $f^m, f^{m+1}, \ldots$ of $|K^m|$, $|K^{m+1}|, \ldots$, each an extension of its predecessor, which combine to give a map $f : |K| \to Y$ that extends f^m. But by the construction of f^m, the primary obstruction to extending the partial constant homotopy of $|K^{m-1}|$ to a partial homotopy of $|K^m|$ connecting e and f is $d(f, e) = z$ (see 7.3.7); so $\zeta(f) = \{z\}$ and $\zeta^*\{f\} = \{z\}$. This shows that an arbitrary element of $H^m(K; \pi_m(Y))$ is in the image of ζ^* and we have completed the proof that ζ^* is a $(1, 1)$ correspondence of the required kind. ∎

7.4.2 Corollary. (*Hopf's theorem*). *If K is m-dimensional* ($m \geq 1$), *the classes of maps of $|K|$ in S^m are in* $(1, 1)$ *correspondence with the elements of $H^m(K; J)$.*

If $m > 1$, S^m is 1-connected; S^1 is 0-connected and n-simple for all n; further, for all m, $\pi_m(S^m) = J$ (II. 1.34, II. 1.35). Conditions (i) of 7.4.1 are satisfied because $\pi_p(S^m)$ is null for $p < m$. To verify that conditions (ii) are also satisfied we observe that for $n > m$ plainly $C^n(K; \pi_n(Y)) = 0$. ∎

We remark that if we withdraw the dimension restriction on K we still get a map $\zeta^* : \pi(|K|; S^m) \to H^m(K)$, where $\pi(X; Y)$ is the set of homotopy classes of maps $X \to Y$ and the value group for contrahomology is J. Plainly a map $g : |K| \to |L|$ induces a map $g^\# : \pi(|L|; S^m) \to \pi(|K|; S^m)$, given by $g^\#\{f\} = \{gf\}$. Then ζ^* is *natural* in the sense that

7.4.3 $$g^*\zeta^* = \zeta^* g^\# : \pi(|L|; S^m) \to H^m(K).$$

For consider first a simplicial map $u : K \to L$. If $f : |L|, |L^{m-1}| \to S^m, y_0$ is a map and the obstruction to deforming it to a map sending $|L^m|$ to y_0 is $\zeta(f) \in Z^m(L)$, then clearly the obstruction to deforming

$$|u|f : |K|, |K^{m-1}| \to S^m, y_0$$

to a map sending $|K^m|$ to y_0 is $u^{\cdot}(\zeta(f))$; thus $u^{\cdot}(\zeta(f)) = \zeta(|u|f)$, whence

$$u^*\zeta^* = \zeta^* |u|^\#.$$

Now let $u : K^{(r)} \to L$ be a simplicial approximation to g, let $v : K^{(r)} \to K$ be a standard map and let $\chi : C(K) \to C(K^{(r)})$ be the subdivision chain map. Since $|u| \simeq g$, $|v| \simeq 1$, we have $|u|^\# = g^\#$, $|v|^\# = 1$ and a commutative diagram

$$\begin{array}{ccccc} H^m(L) & \xrightarrow{u^*} & H^m(K^{(r)}) & \xleftarrow{v^*} & H^m(K) \\ \uparrow{\scriptstyle \zeta^*} & & \uparrow{\scriptstyle \zeta^*_{(r)}} & & \uparrow{\scriptstyle \zeta^*} \\ \pi(|L|; S^m) & \xrightarrow{g^\#} & \pi(|K|; S^m) & \xleftarrow{1} & \pi(|K|; S^m) \end{array}$$

Now $g^* = \chi^* u^*$ and $\chi^* v^* = 1$. Thus

$$g^*\zeta^* = \chi^* u^* \zeta^* = \chi^* \zeta^*_{(r)} g^\# = \chi^* v^* \zeta^* g^\# = \zeta^* g^\#$$

and 7.4.3 is established. It follows that if K, L are m-dimensional and ζ^* is used to transfer the group structure from $H^m(K)$ to $\pi(|K|; S^m)$, then the induced contrahomology homomorphism g^* is transformed into the map $g^\#$, which will thus be homomorphic with respect to the induced group structure in $\pi(|K|; S^m)$.

The case of 7.4.2 for $m = 1$ was informally discussed in 2.7; in that case II. 1.35 allows us to deduce the consequence of 7.4.2 without any restriction on the dimension of K. This is also a special case of the next corollary.

7.4.4 Corollary. *If m is a positive integer and $\pi_n(Y)$ is null for all $n \neq m$ (and, if $m = 1$, Y is 1-simple), the classes of maps of $|K|$ in Y are in $(1, 1)$ correspondence with the elements of $H^m(K; \pi_m(Y))$.*

The conditions of the theorem follow directly from the conditions here imposed on Y. ∎

We observe that, for $|K|$ compact, a map $f : |K^m| \to S^m$ induces

$$f^* : H^m(S^m; J) \to H^m(K^m; J).$$

Defining η^m to be the positive generator of $H^m(S^m; J)$, we see that f determines an element $f^*\eta^m \in H^m(K^m; J)$. The reader is invited to prove that $\zeta^*\{f\} = f^*\eta^m$. The restriction on $|K|$ that it be compact can be removed by the use of singular contrahomology theory (see chapter 8).

A space having the property of Y in 7.4.4 is called an *Eilenberg–MacLane space* $K(\pi, m)$, where $\pi = \pi_m(Y)$. From II. 1.34, II. 1.35 we see that S^1 is a $K(J, 1)$; for $K(J_2, 1)$ we can take a suitably defined infinite-dimensional real projective space; for $K(J, 2)$ we can take infinite-dimensional complex projective space. Eilenberg–MacLane spaces for other values of m and other non-null groups π exist but are markedly less familiar. They have proved to be of very great importance in topology; since their homotopy features are analogous to the homology features of spheres, it is perhaps not surprising that they are useful spaces. We observe that they provide at once a geometrical picture for an element of $H^m(K; G)$, namely, that of a homotopy class of maps of $|K|$ into the space $K(G; m)$.

7.5* Maps of polyhedra into S^m

In this section we consider extensions to $|K|$ of maps of $|L|$ in S^m, where K has dimension $> m$. The earlier part is little more than a restatement of results already proved, but this leads to a discussion of the problem of extending the given map over $P^{m+2} = |K^{m+2} \cup L|$. We do not treat this problem fully but content ourselves with a few indications of Steenrod's method of attack.‡

We shall call a map of $|K|$ in S^m a *normal map* if the image of $|K^{m-1}|$ is x_*; any map is homotopic to a normal map and, if $L \subseteq K$ and $g : |L| \to S^m$ is normal, any extension of g over $|K|$ is homotopic rel $|L|$ to a normal map. A *normal homotopy* connecting two normal maps of $|K|$ is a homotopy rel $|K^{m-2}|$; a normal homotopy rel $|L|$ is a homotopy rel P^{m-2}. Any two homotopic normal maps of $|K|$ are connected by a normal homotopy and any two normal maps of $|K|$ homotopic rel $|L|$ are connected by a normal homotopy rel $|L|$. The various assertions of this paragraph are all provable by observing that the appropriate primary obstructions are all zero since $\pi_r(S^m)$ is null for $r < m$.

For any normal map $f : |K| \to S^m$, f agrees on $|K^{m-1}|$ with the constant map e to x_* and we define z_f to be $d(f, e)$, the difference contrachain in $C^m(K; \pi_m(S^m)) = C^m(K; J)$. When the value group is J we shall allow ourselves in this section to omit it; thus, for instance, we may write $C^m(K)$ for $C^m(K; J)$.

‡ See N. E. Steenrod, *Ann. Math.* **48** (1947), 290 and [19].

Since e and f are defined over $|K|$, $c_e = c_f = 0$ so that

$$\delta z_f = \delta d(f, e) = (-1)^n (c_f - c_e) = 0$$

and z_f is a contracycle of K. This is called the *characteristic contracycle* of f. If f' is a normal map homotopic to f, then $f' \simeq f \operatorname{rel} |K^{m-2}|$ and by 7.2.5 and 7.3.8. $z_{f'} - z_f = d(f', e) - d(f, e) = d(f', f) \sim 0$; homotopic normal maps have therefore contrahomologous characteristic contracycles. If therefore f is any map (not necessarily normal) we may associate with its homotopy class $\{f\}$ an element $\{z_f\} \in H^m(K)$. In fact if η^m is the positive generator of $H^m(S^m)$, then $\{z_f\} = f^*\eta^m$.

We now consider the problem of extending $g : |L| \to S^m$ over P^{m+1}.

7.5.1 Theorem. *If L is a subcomplex of K, a normal map $g : |L| \to S^m$ is extendable over P^{m+1} if and only if $z_g \in C^m(L)$, the characteristic contracycle of g, is extendable to a contracycle of K; this is equivalent to the condition that $\nu^*\{z_g\} = 0 \in H^{m+1}(K, L; J)$.*

We recall the exact sequence

$$0 \leftarrow C^m(L) \overset{\lambda^{\cdot}}{\leftarrow} C^m(K) \overset{\mu^{\cdot}}{\leftarrow} C^m(K, L) \leftarrow 0$$

and the contrahomology sequence

$$\ldots \overset{\mu^*}{\leftarrow} H^{m+1}(K, L) \overset{\nu^*}{\leftarrow} H^m(L) \overset{\lambda^*}{\leftarrow} H^m(K) \overset{\mu^*}{\leftarrow} H^m(K, L) \overset{\nu^*}{\leftarrow} \ldots;$$

we shall sometimes write $\hat{C}^m$ for $C^m(K, L)$ and similarly with $\hat{Z}$, $\hat{B}$, $\hat{H}$. The condition that z_g is extendable to a contracycle of K means that $z_g \in \lambda^{\cdot} Z(K)$; this may be seen to be equivalent to the condition that $\{z_g\} \in \lambda^* H(K)$, or, from the exactness of the contrahomology sequence, that $\nu^*\{z_g\} = 0$. We have therefore only to prove the first assertion of 7.5.1.

There is no obstruction to extending g to $f = f^m : P^m \to S^m$. Let z_f be the characteristic element of f; then z_f is an extension of z_g (i.e. $z_g = \lambda^{\cdot} z_f$) and is known to be a contracycle of $K^m \cup L$ but not in general a contracycle of K. On the other hand, if f is extendable to $f^{m+1} : P^{m+1} \to S^m$, we know that z_f is a contracycle of $K^{m+1} \cup L$ and therefore of K and conversely if z_f is a contracycle of K, f is extendable over P^{m+1}. We observe now that any contrachain in $C^m(K)$ that extends z_g can arise as the characteristic element of some extension of g over P^m; hence if z_g can be extended to a contracycle of K, there is an extension of g over P^m that is itself extendable over P^{m+1}. This proves that z_g is extendable to a contracycle if and only if g is extendable over P^{m+1}. ∎

The latter form of the statement of 7.5.1 has the advantage that it applies to maps of $|L|$ into S^m that are not normal, since it is stated in terms of contrahomology classes.

We now consider the *tertiary obstruction*, the obstruction to extending g over P^{m+2}, restricting attention to normal maps g that have (normal) extensions $\tilde{f} : P^{m+1} \to S^m$. If $f = \tilde{f} \mid P^m$, the secondary obstruction $\{c_{\tilde{f}}\}$ to extending f over P^{m+2} belongs to

$$H^{m+2}(K, L; \pi_{m+1}(S^m)) = H^{m+2}(K, L; J_2)$$

provided that $m \geqslant 3$; we may write this group as $\hat{H}^{m+2}_{(2)}$ and the obstruction as $\xi(f)$. In measuring the tertiary obstruction to extending g, the extension f of g over P^m is, of course, to be regarded as provisional; it turns out that, if $\tilde{f}' : P^{m+1} \to S^m$ is another extension of g and if $f' = \tilde{f}' \mid P^m$, then $\xi(f) - \xi(f')$ belongs to a subgroup of $\hat{H}^{m+2}_{(2)}$ written as $Sq^2 \hat{H}^m$, which can be calculated effectively. The coset, therefore, $[\xi(f)] \in \hat{H}^{m+2}_{(2)} / Sq^2 \hat{H}^m$ depends only on g and we shall write it as $\eta(g)$. It will appear that this is a satisfactory measure of the tertiary obstruction in that it is zero if and only if g can be extended over P^{m+2}.

Steenrod defines, by means of a variant of the cup-product, a homomorphism‡

$$Sq^i : \hat{H}^n \to \hat{H}^{n+i}_{(2)}$$

for any simplicial pair K, L and for each n, $i \geqslant 0$. It is this homomorphism for $n = m$ and $i = 2$ that determines the subgroup of $\hat{H}^{m+2}_{(2)}$ referred to above. If f, f' are as in the previous paragraph, then $\hat{d}(f, f')$ is defined because f and f' agree on P^{m-1}, and is a contracycle of $\hat{C}^m$ since f and f' are extendable over P^{m+1}; we define $\theta(f, f')$ to be $\{\hat{d}(f, f')\} \in \hat{H}^m$. Steenrod's principal result for the relative case is that

7.5.2 $$\xi(f) - \xi(f') = Sq^2 \theta(f, f').$$

This shows at once that $\xi(f) - \xi(f') \in Sq^2 \hat{H}^m$ and that

$$[\xi(f)] \in \hat{H}^{m+2}_{(2)} / Sq^2 \hat{H}^m$$

depends therefore only on g. We now deduce that $\eta(g) = 0$ if and only if g is extendable over P^{m+2}. First, if g is extendable over P^{m+2}, it has an extension f over P^m whose secondary obstruction $\xi(f)$ is zero; hence $\eta(g) = 0$. Conversely, suppose that $\eta(g) = 0$; let $\tilde{f} : P^{m+1} \to S^m$ be any normal extension of g and let $f = \tilde{f} \mid P^m$. Since $\eta(g) = 0$, $\xi(f) \in Sq^2(\hat{H}^m)$; let $\xi(f) = Sq^2 \theta$ and let $\hat{d}$ be a representative contra-

‡ In Steenrod's original notation this homomorphism would appear as Sq^{n-i}.

cycle for θ. Then by II. 1.16 we can find $f' : P^m \to S^m$ agreeing with f on P^{m-1} and such that $\hat{d}(f, f') = \hat{d}$; since $\hat{d}$ is a contracycle and f is extendable over P^{m+1}, so is f'. By 7.5.2, $\xi(f) - \xi(f') = Sq^2\theta$, so that $\xi(f') = 0$ and f', and hence g, is extendable over P^{m+2}.

In this way, for $m \geqslant 3$, a set of criteria is established for determining whether $g : |L| \to S^m$ is extendable over P^{m+2}:

$$\text{(i) } \nu^*\{z_g\} = 0 \quad \text{and} \quad \text{(ii) if so, } \eta(g) = 0.$$

The absolute form of 7.5.2 asserts that, if $f : |K^m| \to S^m$ is extendable over $|K^{m+1}|$ and if $\xi(f) \in H^{m+2}_{(2)}$ is the secondary obstruction,

7.5.3 $$\xi(f) = Sq^2\{z_f\}.$$

In this case therefore we have only to calculate $Sq^2\{z_f\}$ to determine whether or not f is extendable over $|K^{m+2}|$.

To establish 7.5.3 Steenrod uses the central properties of Sq^i and some model spaces M^{m+2} $(m \geqslant 2)$; M^4 is the complex projective plane and M^{m+2} is the suspension of M^{m+1} (its join to a pair of points). M^{m+2} may also be constructed as an identification space from V^{m+2} by identifying points on its boundary: let $h^3 : S^3 \to S^2$ be the Hopf map‡ and $h^{m+1} : S^{m+1} \to S^m$ be the suspension of $h^m : S^m \to S^{m-1}$ (so that for $m \geqslant 3$, $\{h^{m+1}\}$ is the non-zero element of $\pi_{m+1}(S^m)$); we define an equivalence relation on the points of V^{m+2} by declaring that $x \mathrm{R} x'$ if and only if $x = x'$ or $x, x' \in S^{m+1}$ and $xh^{m+1} = x'h^{m+1}$. We may take M^{m+2} to be the factor space with identification map $k^{m+2} : V^{m+2} \to M^{m+2}$ and we may regard $(S^{m+1})\, k^{m+2}$ to be S^m; so that $S^m \subset M^{m+2}$ and for $x \in S^{m+1}$, $xk^{m+2} = xh^{m+1}$. M^{m+2} plays a critical role in this argument because it may be regarded as the $(m+2)$-section of the Eilenberg-MacLane complex $K(J; m)$ referred to in 7.4, and we are here only concerned with $(m+2)$-sections.

If $i : S^m \to S^m$ is the identity map, we consider the secondary obstruction $\xi(i)$ to extending i to a map of M^{m+2} to S^m and verify that $\xi(i) = Sq^2\{z_i\}$, since each is the non-zero element of $H^{m+2}(M^{m+2}; J_2)$ §.

Now any map $g : S^{m+1} \to S^m \subset M^{m+2}$ can be extended to a map

$$g' : V^{m+2} \to M^{m+2} \quad (\text{for } m \geqslant 3);$$

if $\{g\} = 0 \in \pi_{m+1}(S^m)$, g can be extended indeed to a map $V^{m+2} \to S^m$; if $\{g\} \neq 0$, $g \simeq h^{m+1}$ and we can define an extension $g' \simeq k^{m+2} : V^{m+2} \to M^{m+2}$. Suppose now that $f : |K^m| \to S^m$ has no primary obstruction and let

$$f^{m+1} : |K^{m+1}| \to S^m \subset M^{m+2}$$

be an extension of f. Then f^{m+1} has an extension to a map $f' : |K^{m+2}| \to M^{m+2}$, since for each $s_{m+2} \in K$, $f^{m+1} \mid \dot{s}_{m+2}$ can be extended to a map of $\bar{s}_{m+2}$ in M^{m+2}.

‡ See II. 3.

§ Indeed, the critical property of Sq^2 in this and other applications is that $Sq^2\{z_i\}$ is the non-zero element of $H^{m+2}(M^{m+2}; J_2)$. This is brought out forcibly in the axiomatic treatment of Steenrod squares; see H. Cartan, *Comm. Math. Helv.* **29** (1955), 40).

Naturality properties of obstruction elements and of Sq^2 imply that

$$\begin{aligned}\xi(f) = f'^*\xi(i) &= f'^*Sq^2\{z_i\}\\ &= Sq^2 f'^*\{z_i\}\\ &= Sq^2\{z_f\}.\end{aligned}$$

This is the assertion of 7.5.3.

This is only an outline of Steenrod's work and is intended to arouse interest in the subject and in his work referred to at the start of the section.

7.6* Local systems of groups and obstruction theory in non-simple spaces

In the introductory section we stated that 'for simplicity of treatment' we would confine attention to spaces Y that are n-simple for the crucial values of n. Here we indicate the modifications needed in case Y fails to be n-simple.

7.6.1 Definition. If X is a path-connected space, a *local system of groups* $\{G_x\}_{x \in X}$ *on* X is a set of abelian groups G_x, one for each point $x \in X$, together with an isomorphism $w_G : G_{x_0} \to G_{x_1}$ for each path w from x_0 to x_1, subject to the conditions

(i) if $v \equiv w$, $v_G = w_G$, and

(ii) if $v * w$ is defined, $(v * w)_G = v_G w_G$.

An important example is provided by the homotopy groups $\pi_n(X, x)$ for $n \geqslant 2$ and the isomorphisms of II. 1.8.

A local system $\{G_x\}$ is said to be *trivial* if $v_G = w_G$ for any two paths v, w with the same initial and final points.

If $\{H\} = \{H_y\}$ is a local system of groups on Y, any map $f : X \to Y$ induces a local system $\{G\} = \{G_x\}$ on X by the rule that $G_x = H_{xf}$ and that $w_G = (wf)_H$. We may write $f^{-1}\{H\}$ for this induced system $\{G\}$.

If K is a complex there is a combinatorial form of 7.6.1.

7.6.2 Definition. If K is a connected complex, a *local system of groups* $\{G_a\}$ *on* K is a system of abelian groups G_a, one for each vertex $a \in K$, together with an isomorphism $w_G : G_{a^0} \to G_{a^1}$ for each edge-path w from a^0 to a^1, subject to the conditions

(i) if $v \equiv w$, $v_G = w_G$, and

(ii) if $v * w$ is defined, $(v * w)_G = v_G w_G$.

Clearly a local system on $|K|$ determines a local system on K, and it can be shown that the resulting local system on K can be used to determine topological invariants, that is, invariants of $|K|$ and the original system. A local system on K enables us to determine homology and contrahomology groups of K with this local system for coefficient groups, and these homology and contrahomology groups are topological invariants. We here give only the definition of the contrahomology group $H^*(K; \{G\})$.

Let the vertices of K be ordered and let $a(\sigma)$ mean the leading vertex of σ. Then $C^p(K; \{G\})$ is defined to be the set of functions c^p on the p-simplexes of K, the value on σ_p being drawn from $G_{a(\sigma_p)}$; this set $C^p(K; \{G\})$ is given the obvious abelian group structure. Next we define $\delta : C^p(K; \{G\}) \to C^{p+1}(K; \{G\})$ by the rule

$$(a^0 \dots a^{p+1}, \delta c^p) = (a^1 \dots a^{p+1}, c^p)(a^1 a^0)_G + \sum_{i=1}^{p+1} (-1)^i (a^0 \dots \widehat{a^i} \dots a^{p+1}, c^p).$$

Notice the isomorphism induced by the edge-path $(a^1 a^0)$ that appears in the first term in order that the value defined shall lie in G_{a^0}. In proving that $\delta\delta = 0$ we make use of the fact that

$$\begin{aligned}(a^2a^1)_G (a^1a^0)_G &= (a^2a^1a^0)_G \quad \text{by (ii) of 7.6.2,}\\ &= (a^2a^0)_G \quad \text{by (i) of 7.6.2.}\end{aligned}$$

The groups $C^p(K; \{G\})$ and homomorphisms δ form the contrachain complex $C^\cdot(K; \{G\})$. It is the contrahomology group of this complex that is defined to be $H^*(K; \{G\})$. If the local system is trivial then $H^*(K; \{G\}) = H^*(K; G)$, where G is the group at some arbitrary vertex.

If L is a subcomplex of K, we define $C^\cdot(K, L; \{G\})$ in the obvious way and hence $H^*(K, L; \{G\})$.

Suppose now that $n > 1$ and that a map $f : P^n \to Y$ is given. We measure the primary obstruction c_f to extending f over P^{n+1} as a contrachain with local coefficients $\{G\}$, where $G_a = \pi_n(Y, af)$. This is the system for $K^n \cup L$ derived from the system $f^{-1}\{\pi_n(Y, y)\}$ on P^n induced by f and the system $\{\pi_n(Y, y)\}$ on Y. Then if σ_{n+1} has leading vertex a we define c_f by

$$(\sigma_{n+1}, c_f) = \{f \mid (\dot{\sigma}_{n+1}, a)\} \in \pi_n(Y, af).$$

If $\tau_{n+1} \in L$, $(\tau_{n+1}, c_f) = 0$ since $f \mid \dot{\tau}_{n+1}$ has an extension over $\bar{\tau}_{n+1}$, namely, $f \mid \bar{\tau}_{n+1}$. The contrachain c_f is therefore an element of

$$C^{n+1}(K, L; \{G\}).$$

This is the form of the obstruction contrachain if Y is not n-simple and theorems analogous to those of the preceding sections of this chapter may be proved. The reader is referred to N. E. Steenrod, *Ann. Math.* **44** (1943), 610, for a general discussion of the relevant local homology and contrahomology theory, and to P. Olum, *Ann. Math.* **52** (1950), 1, for an account of obstruction theory in non-simple spaces.

7.7* Contrahomotopy and compression

Since by Hopf's theorem 7.4.2 the classes of maps of $|K^m|$ in S^m are in $(1, 1)$ correspondence with the elements of $H^m(K^m)$, this correspondence, which is canonical, allows us to introduce an abelian group structure into the set of classes of such maps. It is in fact possible to define a sensible group structure on the set of classes of maps of a space X in a space Y under far less special conditions. In particular, if X is a compactum of dimension $< 2m-1$ and Y is S^m, the classes form an abelian group first defined by Borsuk‡ and called by Spanier the mth (Borsuk) cohomotopy group of X (and by us the mth *contrahomotopy* group); it is written as $\pi^m(X)$ and the zero is the class of null-homotopic maps. These contrahomotopy groups have been proved§ by Spanier to have most of the familiar properties of contrahomology groups, the spectacular exception being that, if K is finite and n-dimensional, $\pi^m(|K|)$ is not in general defined if $m \leqslant \frac{1}{2}(n+1)$. If K is m-dimensional, then $\pi^m(|K|)$ is defined and is isomorphic to $H^m(K)$; this statement extends the result of 7.4.2.

We give no account of contrahomotopy theory, but in this section we outline an attractive application of it to a problem that is in a sense dual to the extension problem.

A generalized form of the extension problem may be stated as follows: if $j: X_0 \to X$ is an embedding and $f_0: X_0 \to Y$ is given, can we find $f: X \to Y$ such that $jf \simeq f_0$? (In our original presentation of extension problems the requirement is that $jf = f_0$; in view of the homotopy extension property of polyhedra, the two questions are equivalent if X, X_0 are a polyhedral pair and we prefer this restatement for the purpose of dualizing.) A dual problem may be stated as follows: if $j: Y_0 \to Y$ is an embedding and $f: X \to Y$ is given, can we find $f_0: X \to Y_0$ such that $f_0 j \simeq f$? In other words, can the map f be deformed (or compressed) to a map into the subspace Y_0? Contrahomotopy can be applied to suitable cases of compression problems.

Suppose in fact that X is a compact polyhedron of dimension $< 2n+1$ and that $f: X \to |L^{n+1}|$ is given, for some complex L; we consider whether f can be compressed into $|L^n|$, first allowing only deformations in $|L^{n+1}|$ and then allowing deformations in $|L^{n+2}|$. We suppose L to be oriented and for each $\sigma^i_{n+1} \in L$ we define $g^i: |L^{n+1}| \to S^{n+1}$ to be the map such that $(|L^{n+1}| - \sigma^i_{n+1})\, g^i$ is the base-point $z_0 \in S^{n+1}$ and that $g^i \mid \sigma^i_{n+1}$ is an orientation-preserving homeomorphism onto $S^{n+1} - z_0$; g^i is, in fact, a map that pinches $|L^{n+1}| - \sigma^i_{n+1}$ to a point and maps the resulting sphere onto S^{n+1}. Then $fg^i: X \to S^{n+1}$ determines an element $\{fg^i\} \in \pi^{n+1}(X)$. We now define the *compression obstruction chain* $c^f \in C_{n+1}(L; \pi^{n+1}(X))$ to be $\sum_i \{fg^i\}\, \sigma^i_{n+1}$. Note that X is compact so that, for all but a finite number of values of i, $(X)fg^i = z_0$ so that $\{fg^i\} = 0$; the sum is therefore a finite sum.

It turns out that c^f is a cycle, that it is zero if and only if f can be compressed into $|L^n|$ by a deformation in $|L^{n+1}|$, and that it is a boundary if and only if f can be compressed into $|L^n|$ by a deformation in $|L^{n+2}|$. These results‖ lend colour to our assertion that the compression problem is in a sense dual to the extension problem.

‡ See K. Borsuk, *C.R. Acad. Sci., Paris* 202 (1936), 1401.

§ See E. H. Spanier, *Ann. Math.* 50 (1949), 203.

‖ See E. H. Spanier and J. H. C. Whitehead, *Quart. J. Math.*, 6 (1955), 91.

EXERCISES

(Throughout K is a complex, L a subcomplex; and $P^n = |K^n \cup L|$, except in Q. 9.)

1. Suppose $\pi_i(Y) = 0$, $1 \leqslant i < n$. Show that any map $f: |L| \to Y$ is extendable to P^n. Show also that if $f_0, f_1: P^n \to Y$ then $c_{f_0} \sim c_{f_1}$ if and only if $f_0 \| |L| \simeq f_1 \| |L|$, and that if f_0 extends $f: |L| \to Y$, then f is extendable over P^{n+1} if and only if $\{c_{f_0}\} = 0$.

2. Let $h: X \to Y$ be a map of n-simple spaces. If $f: P^n \to X$ is a map, show that

$$c_{fh} = (c_f)\, h^{\cdot}_{*},$$

where $h^{\cdot}_{*}: C^{\cdot}(K, L; \pi_n(X)) \to C^{\cdot}(K, L; \pi_n(Y))$ is the value-group homomorphism induced by h. Similarly, show that if $f, f': P^n \to X$ agree on P^{n-1} then

$$d(fh, f'h) = d(f, f')\, h^{\cdot}_{*}.$$

3. Let the map $h: X \to Y$ of Q. 2 be a fibre map in which the fibre is connected and contractible in X. Deduce from II.3.8 that h_* maps $\pi_n(X)$ monomorphically on to a direct factor in $\pi_n(Y)$. Let $f: P^n \to X$ be such that $fh|P^{n-1}$ is extendable to P^{n+1}; show that $f|P^{n-1}$ is extendable to P^{n+1}. Similarly, show that if $f, f': |K| \to X$ agree on P^{n-1} and $fh|P^n \simeq f'h|P^n$, rel P^{n-2}, then

$$f|P^n \simeq f'|P^n, \text{rel}\, P^{n-2}.$$

Give an example of such a fibre map h in which neither X nor the fibre is contractible (over itself).

4. Let Y be n-simple and suppose $H^n(K, L; \pi_n(Y)) = 0$. Let $f_0, f_1: |K| \to Y$ be such that $f_0|P^{n-1} \simeq f_1|P^{n-1}$ rel A, where A is a subspace of $|L|$. Prove that $f_0|P^n \simeq f_1|P^n$ rel A.

5. Let n, q be integers and A, B abelian groups. A *contrahomology operation* $T = T(n, q, A, B)$ is a function assigning to each polyhedron X a transformation

$$T_X : H^n(X; A) \to H^q(X; B)$$

subject to the condition that, given a map $f: X \to Y$, the diagram

$$\begin{array}{ccc} H^n(X;A) & \xrightarrow{T_X} & H^q(X;B) \\ \uparrow f^* & & \uparrow f^* \\ H^n(Y;A) & \xrightarrow{T_Y} & H^q(Y;B) \end{array}$$

is commutative.

Let X_0 be an Eilenberg-MacLane polyhedron $K(A, n)$ (see 7.4). By the correspondence of 7.4.4 the class of the identity map $X_0 \to X_0$ corresponds to an element ϵ of $H^n(X_0; A)$. Show that the transformation $T \to T_{X_0}(\epsilon)$ sets up a (1, 1) correspondence between contrahomology operations $T(n, q, A, B)$ and elements of $H^q(X_0; B)$.

Use the Hurewicz theorem 8.8.5 to show that the operations of type (n, n, A, B) are just those induced by homomorphisms $A \to B$.

6. Let X be a connected polyhedron. Show that $\operatorname{cat} X \leqslant 2$ (see chap. 4, Exercises, Q. 6), if and only if the diagonal map $d: X \to X \times X$ (given by $xd = (x, x)$) is deformable into the bunch $X \vee X$. Deduce that if

$$q: X \times X \to X \# X = X \times X / X \vee X$$

is the identification map then $dq \simeq 0$ if $\operatorname{cat} X \leqslant 2$. Suppose that $\pi_1(X) \neq 1$, $H_n(X) = 0$, $n \geqslant 1$. Show that $dq \simeq 0$ but that $\operatorname{cat} X > 2$. [Hints: (*a*) use the result proved in chap. 8, Exercises, Q. 25; (*b*) show that if $\operatorname{cat} X \leqslant 2$, $\pi_1(X)$ is free.]

7. Let K, L be complexes with base vertices a, b and let G be a topological group. Show that there is no obstruction to extending a map $|K| \vee |L|$, $(a, b) \to G, e$ to $|K| \times |L|$. Let $\alpha \in \pi_p(Y, y_0)$, $\beta \in \pi_q(Y, y_0)$ be represented by maps $f: I^p, \dot{I}^p \to Y, y_0$, $g: I^q, \dot{I}^q \to Y, y_0$. Pick a base point w on $\dot{I}^p \times \dot{I}^q$ and define the map $h: (I^p \times I^q)^{\cdot}, w \to Y, y_0$ by

$$\begin{aligned} (x, x')\,h &= xf, \quad x \in I^p, \quad x' \in \dot{I}^q \\ &= x'q, \quad x \in \dot{I}^p, \quad x' \in Iq. \end{aligned}$$

Then h represents an element $\gamma \in \pi_{p+q-1}(Y, y_0)$ which depends only on α, β. This element is called the *Whitehead product* of α, β. Show that all Whitehead products vanish if Y is a topological group (or loop-space). By taking $p = 1$, or otherwise, show that a topological group is n-simple for all n. [Hint: represent γ as the image of a certain element of $\pi_{p+q-1}(S^p \vee S^q)$.]

8. Let K be ordered and let $y_0 \in Y$ be taken as base point. Indicate how, by confining attention to maps and homotopies which keep the vertices of K at y_0, one may build up an obstruction theory for maps into non-simple spaces without the introduction of local systems of coefficients. (We shall describe such maps and homotopies as *bound*.)

9. Classify the bound homotopy classes of maps of $|K|$ into P^n, where K is an n-complex and P^n is projective n-space. Show that P^n is n-simple if and only if n is odd.

10. The homotopy lifting property of fibre maps (see II.3) may be regarded as the vanishing of the obstruction to extending a homotopy. A map $p: X \to B$ is called a *locally trivial* fibre map if (i) the counterimages $p^{-1}(b)$, $b \in B$, are all homeomorphic to some space F, the *fibre*; and (ii) there exists an open covering $\{U\}$ of B, to each member of which corresponds a homeomorphism $\psi: U \times F \to p^{-1}(U)$ with $(b, y)\psi \in p^{-1}(b)$ if $b \in U$, $y \in F$. Adapt the argument of 6.5.7 to prove the existence of the extension $\tilde{f}_t$ of that theorem if $p: \tilde{X} \to X$ is a locally trivial fibre map. (Of course, a covering map is the special case of a locally trivial fibre map in which the fibre is discrete; and the conclusion of Q. 10 shows that a locally trivial fibre map *is* a fibre map in the Serre sense (II.3).)

11. The definability of the group structure in the contrahomotopy group $\pi^n(|K|)$ may be regarded as expressing the vanishing of the obstruction to a compression. It may be shown (by applying the relative Hurewicz theorem 8.8.8) that $\pi_r(S^p \times S^q, S^p \vee S^q) = 0$ if $r < p + q$. Show that if $\pi_r(Y, Y_0) = 0$, $r < m$, and if K is a complex with $\dim K \leqslant m - 1$, then every map $f: |K| \to Y$ may be compressed into a map $f': |K| \to Y_0$; and that if $\dim K \leqslant m-2$, then the homotopy class of f' is uniquely determined. Deduce that if $\alpha, \beta \in \pi^n(|K|)$, where $\dim K \leqslant 2n-2$, and if $f: |K| \to S^n$, $g: |K| \to S^n$ represent α, β, then there is a map $h': |K| \to S^n \vee S^n$ such that $h'j \simeq (f, g): |K| \to S^n \times S^n$, where $j: S^n \vee S^n \to S^n \times S^n$ is the inclusion map and $x(f,g) = (xf, xg)$, $x \in |K|$; moreover, the homotopy class of h' is determined by the classes α, β.

Let $k: S^n \vee S^n \to S^n$ be the 'folding' map given by $(x, x_0)k = (x_0, x)k = x$, where $x_0 \in S^n$ is the base point. Then $h = h'k$ determines an element of $\pi^n(|K|)$ which is, by definition, $\alpha + \beta$. Show that this is an abelian group structure in $\pi^n(|K|)$. [Hint: if f represents α and $\rho: S^n \to S^n$ is any map of degree -1, then $f\rho$ represents $-\alpha$.] Show that if $\dim K = n$ then the (1, 1) correspondence 7.4.2 may be strengthened to an isomorphism

$$H^n(K; J) = \pi^n(|K|).$$

Can the dimensionality restriction on K be replaced by a contrahomology condition in proving the definability of the group addition in $\pi^n(|K|)$?

8

SINGULAR HOMOLOGY THEORY

8.1 Description and scope of the theory

Up to now, we are only able to derive homology groups for a special kind of space, namely, a compact polyhedron. In chapter 2 we derived homology groups of a simplicial complex and in chapter 3 it was proved that, in a precise and natural sense, the groups are determined, in the case of a finite complex, by the underlying polyhedron. Thus the choice of triangulation of a given polyhedron was revealed as a mere administrative technique designed to associate with the polyhedron a suitable chain complex.

One of the great successes of Combinatorial Topology has been the extension of these techniques to produce homology theories for general topological spaces. A general homology theory may be regarded as satisfactory if it agrees with familiar homology theory on the category of polyhedra and if, moreover, certain central properties of simplicial homology theory hold good also for the general theory.

In this chapter one such homology theory is defined, the *singular theory*; it is defined by associating with each space a chain complex and then passing to the homology groups of that chain complex (and the contrahomology groups of the adjoint contrachain complex). A continuous map induces homology homomorphisms in an obvious way and we prove that homotopic maps induce the same homomorphisms. To prove this it is convenient to have the so-called *cubical* singular theory, as well as the *simplicial* singular theory, and the equivalence of these two theories is established. Singular theory is closely related to homotopy theory; there is a natural homomorphism from homotopy groups to singular homology groups. We shall also prove that singular homology theory is consistent with the simplicial homology theory for polyhedra. In this proof we are not confined to compact polyhedra so that we may deduce the topological invariance of the homology and contrahomology groups of a triangulation of a non-compact polyhedron.

There is at least one other valuable way of associating homology groups with general spaces, due to Čech. The Čech theory is also consistent with the familiar theory on polyhedra and differs from the singular theory on some spaces. It is briefly described in the Appendix to this chapter.

We now describe the (simplicial) singular theory‡. If X is a topological space, there is a reasonable intuitive notion of a 1-cycle on X in terms of closed paths; a closed path may be defined as a succession of maps of a closed 1-simplex into X, the last point of one image coinciding with the first point of its successor and the whole being closed by the condition that the first point of the first image coincide with the last point of the last. It is possible also to imagine a similar notion of a 2-cycle and a 1-boundary. These intuitive ideas get precise formulation in singular homology theory.

Let $v^0, v^1, \ldots, v^p, \ldots$ be independent points§ in Hilbert Space H^∞. The *standard ordered p-simplex*, t_p, is the closed ordered simplex with vertices $v^0, v^1, \ldots, v^p$. The *p-th singular chain group*, $\Delta_p(X)$, $p \geqslant 0$, is defined to be the free abelian group freely generated by the maps $f: t_p \to X$. When such a map f is regarded as a basis element of $\Delta_p(X)$, we shall normally write it f_p; then f_p is a *singular p-simplex* of X. We may *augment* $\Delta(X)$ to $\tilde{\Delta}(X)$ by putting $\tilde{\Delta}_{-1}(X) = J$; of course, $\Delta_{-1}(X) = 0$ and, in any case $\Delta_p(X) = \tilde{\Delta}_p(X) = 0$, $p < -1$.

We now define the boundary operator in $\Delta(X)$. Let

$$t^i_{p-1} = (v^0 v^1 \ldots \widehat{v^i} \ldots v^p), \quad i = 0, 1, \ldots, p, \quad p \geqslant 1.$$

Then for each i there is a unique linear (i.e. simplicial) map

$$V^i = V^i_{(p-1)} : t_{p-1} \to t_p,$$

given by

8.1.1
$$v^j V^i = v^j, \quad j < i$$
$$= v^{j+1}, \quad j \geqslant i,$$

which maps t_{p-1} onto t^i_{p-1}; and $V^i f$, being a map of t_{p-1} into X, is a singular $(p-1)$-simplex of X. We define $F^i_p : \Delta_p(X) \to \Delta_{p-1}(X)$ by linearity from

8.1.2
$$f_p F^i_p = (V^i f)_{p-1}.$$

We call F^i_p $(0 \leqslant i \leqslant p)$ the *i-th face operator* and may write f^i_{p-1} (or f^i) for $f_p F^i_p$. From 8.1.1 we readily deduce

8.1.3
$$V^i V^j = V^{j-1} V^i : t_{p-1} \to t_{p+1}, \quad \text{if} \quad i < j,$$

‡ We defer till chapter 9 the description of the singular contrahomology ring.

§ We may take v^p to be the point $(u_1, u_2, \ldots, u_n, \ldots)$ where $u_n = 0$, $n \neq p+1$, $u_{p+1} = 1$.

whence

8.1.4 $$F^jF^i = F^iF^{j-1} : \Delta_{p+1}(X) \to \Delta_{p-1}(X), \quad \text{if} \quad i<j.$$

We define $\partial_p : \Delta_p(X) \to \Delta_{p-1}(X)$, $p \geq 1$, by

8.1.5 $$\partial_p = \sum_{i=0}^{p} (-1)^i F_p^i;$$

the boundary operator $\tilde{\partial}$ in $\tilde{\Delta}(X)$ is the same except that

8.1.5a $$f_0 \tilde{\partial}_0 = 1.$$

From 8.1.4 it follows that $\partial_p \partial_{p-1} = 0$ if $p \geq 2$; it also follows immediately from 8.1.5 and 8.1.5a that $\tilde{\partial}_1 \tilde{\partial}_0 = 0$ in $\Delta(X)$. Thus we may define

8.1.6 Definition. The *singular (simplicial) chain complex*, $\Delta(X)$, of the space X is $\{\Delta_p(X); \partial_p\}$, where $\Delta_p(X) = 0$, $p<0$, and, if $p \geq 0$, $\Delta_p(X)$ is the free abelian group freely generated by the singular p-simplexes of X, and where ∂_p is given by 8.1.5.

The *augmented* singular (simplicial) chain complex $\tilde{\Delta}(X)$ differs only in that $\tilde{\Delta}_{-1}(X) = J$ and $\tilde{\partial}_0$ is given by 8.1.5a.

From $\Delta(X)$ we readily define the singular simplicial chain complex $\Delta(X; G)$ with coefficients in an arbitrary abelian group G and the singular simplicial contrachain complex $\Delta^{\cdot}(X; G)$. Thus, as for simplicial complexes, $\Delta(X; G) = \Delta(X) \otimes G$, $\Delta^{\cdot}(X; G) = \Delta(X) \pitchfork G$. Similarly we define $\tilde{\Delta}(X; G)$, $\tilde{\Delta}^{\cdot}(X; G)$.

8.1.7 Definition. The *singular (simplicial) homology group*, $H^s_*(X; G)$, of X with coefficients in G is $H_*(\Delta(X; G))$; the *reduced* singular (simplicial) homology group, $\tilde{H}^s_*(X; G)$ is $H_*(\tilde{\Delta}(X; G))$. The *singular (simplicial) contrahomology group*, $H^*_s(X; G)$, of X with values in G is $H^*(\Delta^{\cdot}(X; G))$; the *reduced* singular (simplicial) contrahomology group, $\tilde{H}^*_s(X; G)$, is $H^*(\tilde{\Delta}^{\cdot}(X; G))$.

We remark that the coefficient theorems (5.4.13 and 5.4.13c) hold and the relation between the groups‡ $H^s_*(X)$ and $\tilde{H}^s_*(X)$ is just that of 2.4.2 and 2.4.3.

Let $f : X \to Y$ be a map. Then f induces $f^{\Delta} : \Delta(X) \to \Delta(Y)$ by the rule

8.1.8 $$g_p f^{\Delta} = (gf)_p,$$

‡ We suppress J when it is the coefficient group for homology; this agrees with previous conventions.

where $g : t_p \to X$ is a singular p-simplex of X. Obviously f^Δ commutes with the face operators F^i and so is a chain map. Equally obviously, for $f_1 : X_1 \to X_2$ and $f_2 : X_2 \to X_3$,

8.1.9 $$(f_1 f_2)^\Delta = f_1^\Delta f_2^\Delta.$$

We write f_*, f^* for $f_*^\Delta, f^{\Delta *}$. Thus f induces

8.1.10 $$f_* : H_*^s(X; G) \to H_*^s(Y; G), \quad f^* : H_s^*(Y; G) \to H_s^*(X; G)$$

and

8.1.11 $$(f_1 f_2)_* = f_{1*} f_{2*}, \quad (f_1 f_2)^* = f_1^* f_2^*.$$

These statements hold for the reduced homology groups provided we put $f_{-1}^\Delta = 1$. The reader will recall that the uniqueness of f_* and the proof of 8.1.11 for *simplicial homology* in chapter 3 were far from trivial; for singular homology they are very elementary, and the topological invariance of the singular homology and contrahomology groups presents no difficulty.

We may relativize the definitions given above. If X_0 is a (not necessarily closed) subspace of X, the injection map $i : X_0 \to X$ induces $i^\Delta : \Delta(X_0) \to \Delta(X)$ and i^Δ embeds $\Delta(X_0)$ as a subcomplex of $\Delta(X)$. Moreover, each $\Delta_p(X_0)$ is a direct factor in $\Delta_p(X)$, so that the factor complex, $\Delta(X, X_0)$, is a free chain complex. We remark that $\Delta(X, X_0) = \Delta(X)/\Delta(X_0) = \tilde{\Delta}(X)/\tilde{\Delta}(X_0)$. Then the relative singular homology and contrahomology groups of the pair (X, X_0) are obtained from $\Delta(X, X_0)$; in symbols,

8.1.12 $$\begin{cases} H_*^s(X, X_0; G) = H_*(\Delta(X, X_0) \otimes G), \\ H_s^*(X, X_0; G) = H^*(\Delta(X, X_0) \pitchfork G). \end{cases}$$

There are then relativizations of 8.1.8–8.1.11 which the reader may supply. As in earlier chapters, relativizations will not always be made explicit though they will appear in final formulations.

From 5.6.1 and 5.6.2 we deduce

8.1.13 Theorem. *There are exact sequences*

$$\ldots \to H_p^s(X_0; G) \xrightarrow{\lambda_*} H_p^s(X; G) \xrightarrow{\mu_*} H_p^s(X, X_0; G) \xrightarrow{\nu_*} H_{p-1}^s(X_0; G) \to \ldots,$$

$$\ldots \leftarrow H_s^p(X_0; G) \xleftarrow{\lambda^*} H_s^p(X; G) \xleftarrow{\mu^*} H_s^p(X, X_0; G) \xleftarrow{\nu^*} H_s^{p-1}(X_0; G) \leftarrow \ldots.$$

Moreover, a map $f : X, X_0 \to Y, Y_0$ induces a homomorphism $(f_*^0, f_*, \hat{f}_*)$ of the singular homology sequence of (X, X_0) into that of (Y, Y_0) and a *homomorphism* $(f_0^*, f^*, \hat{f}^*)$ *of the singular contrahomology sequence of* (Y, Y_0) *into that of* (X, X_0). ∎

We close with a theorem which goes some way towards showing that homotopic maps induce the same homology homomorphisms. Given any space X, it will be convenient to take $I \times X$ rather than $X \times I$ for the cylinder on the base X and to regard CX, the cone on X, as obtained from $I \times X$ by pinching $1 \times X$ to a point. Let

$$k : I \times X \to CX$$

be the projection as in 1.5. Then X is *contractible* to $x_1 \in X$ if and only if there is a map $q : CX \to X$ with $(0, x)\,kq = x$, $(1, x)\,kq = x_1$, all $x \in X$. We recall‡ from 1.6.7 that a map $g : X \to Y$ determines a map $Cg : CX \to CY$ given by $k(Cg) = (1 \times g)\,k : I \times X \to CY$. We prove

8.1.14 Theorem. *The reduced singular homology and contrahomology groups of a contractible space* X *are null.*

Regard t_{p+1} as the Euclidean cone with vertex v^0 and base t_p^0. Then the map $h = h_{(p)} : Ct_p \to t_{p+1}$, given by

$$(u, y)\,kh = u\,.\,v^0 + (1-u)\,.\,yV^0, \quad u \in I,\ y \in t_p$$

is a homeomorphism (1.5.8).

Given $f : t_p \to X$, define $g = g(f) : t_{p+1} \to X$ by

$$hg = (Cf)\,q : Ct_p \to X.$$

The map g, though defined in a somewhat formal manner, is essentially just a null-homotopy of f. Indeed, throughout the proof of this theorem a simple picture will carry more conviction than the formal details we provide. Having defined the null-homotopy $g(f)$, we define as expected a chain homotopy $\Phi : \tilde{\Delta}(X) \to \tilde{\Delta}(X)$ by

$$f_p \Phi_p = g_{p+1}, \quad p \geqslant 0, \qquad 1\Phi_{-1} = g_0,$$

where g_0 maps $t_0\ (= v^0)$ to x_1, and we prove that

8.1.15 $$\Phi_p \tilde{\partial}_{p+1} + \tilde{\partial}_p \Phi_{p-1} = 1 : \tilde{\Delta}_p(X) \to \tilde{\Delta}_p(X).$$

An essential step in the proof is the observation that

8.1.16 $$h_{(p-1)} V^i = (CV^{i-1})\,h_{(p)} : Ct_{p-1} \to t_{p+1}, \quad i > 0.$$

This follows from the fact that the image under each (linear) map of the vertex of Ct_{p-1} is v^0 and that the image under each map of v^j is v^{j+1} if $j < i-1$ and is v^{j+2} otherwise.

‡ Notice that in the definition of Cg we have here $(1 \times g)$ rather than $(g \times 1)$ because of our choice of cylinder on X.

We now prove 8.1.15. If $p = -1$, then $\Phi_{-1}\tilde{\partial}_0 = 1$ and $\tilde{\partial}_{-1} = 0$, so that 8.1.15 is verified in this case.

Next we remark that, if $p \geqslant 0$,

8.1.17 $$\Phi_p F^0_{p+1} = 1.$$

For $f_p \Phi_p F^0_{p+1} = (V^0 g)_p$ and, for $y \in t_p$,

$$y V^0 g = (0, y)\, khg = (0, y)\, kC(f)\, q = (0, yf)\, kq = yf.$$

Since $v^0 V^1 g = v^0 g = x_1$, where $g = g(f)$, $f: t_0 \to x$, it follows that

$$\Phi_0 F^1_1 = \tilde{\partial}_0 \Phi_{-1},$$

so that 8.1.15 holds if $p = 0$.

Now take $p \geqslant 1$. Then, in the light of 8.1.17, 8.1.15 follows when we have proved

8.1.18 $$F^{i-1}_p \Phi_{p-1} = \Phi_p F^i_{p+1}, \quad i > 0.$$

Let $g' : t_p \to X$ be given by $hg' = (Cf^{i-1})\, q : Ct_{p-1} \to X$. Then $g'_p = f_p F^{i-1}_p \Phi_{p-1}$. Now

$$\begin{aligned} h_{(p-1)} g' &= (C(V^{i-1} f))\, q \\ &= (C V^{i-1})\, (Cf)\, q \\ &= (C V^{i-1})\, h_{(p)} g \\ &= h_{(p-1)}\, V^i g \quad \text{by 8.1.16.} \end{aligned}$$

Thus $g' = V^i g$ and hence $g'_p = g_{p+1} F^i_{p+1} = f_p \Phi_p F^i_{p+1}$. This proves 8.1.18 and completes the proof of 8.1.15. Thus Φ is a chain homotopy connecting the identity and zero chain maps and the theorem is proved. ∎ The chain homotopy Φ should be regarded as the algebraic analogue of the homotopy of maps $X \to X$ connecting the identity with the constant map $X \to x_1$. Indeed, a very slight modification of the proof establishes the following more general result.

8.1.19 Theorem. *If $f : X \to Y$ is homotopic to a constant map, then $f^\Delta \simeq 0 : \tilde{\Delta}(X) \to \tilde{\Delta}(Y)$.* ∎

8.2 The normalized singular chain complex

The singular chain complex $\Delta(X)$ contains a great many generators; for example, even if X consists of a single point, there is one generator in each dimension. It is possible to form an equivalent chain complex in which some of the more obviously degenerate singular simplexes are eliminated, and we now describe this.

For any $p \geqslant 1$ and any $i \leqslant p-1$ there is a unique linear (i.e. simplicial) map $T^i = T^i_{(p)} : t_p \to t_{p-1}$, given by

8.2.1 $$\begin{aligned} v^j T^i &= v^j, \quad j \leqslant i, \\ &= v^{j-1}, \quad j > i, \end{aligned}$$

which collapses to a point the edge $v^i v^{i+1}$ of t_p and all segments parallel to this edge. We define $D^i_{p-1} : \Delta_{p-1}(X) \to \Delta_p(X)$ by linearity from

8.2.2 $$f_{p-1} D^i_{p-1} = (T^i f)_p.$$

We call D^i_{p-1} $(0 \leqslant i \leqslant p-1)$ the *i-th degeneracy operator*. From 8.2.1 we readily deduce

8.2.3 $$T^j T^i = T^i T^{j-1} : t_{p+1} \to t_{p-1}, \quad \text{if} \quad i < j,$$

whence

8.2.4 $$D^i D^j = D^{j-1} D^i : \Delta_{p-1}(X) \to \Delta_{p+1}(X), \quad \text{if} \quad i < j.$$

For completeness we also record at this point the relations between the face and degeneracy operators; the proofs are automatic and will be omitted.

8.2.5 $$\begin{cases} D^j F^i = F^i D^{j-1}, & \text{if} \quad i < j; \\ D^i F^i = D^i F^{i+1} = 1; & \\ D^j F^i = F^{i-1} D^j, & \text{if} \quad i > j+1. \end{cases}$$

A free chain complex with a distinguished basis (consisting of 'simplexes') and provided with face and degeneracy operators F and D satisfying 8.1.4, 8.2.4 and 8.2.5 is called a *semi-simplicial complex* and this concept has proved very fruitful in modern research. Thus the set of singular simplexes of X together with the operators F and D constitutes a semi-simplicial complex called the singular complex of X. Notice that in passing from the singular complex to the chain complex $\Delta(X)$ considerable information is lost. In fact if X is a polyhedron the singular complex contains a complete set of invariants of homotopy type. Another example of a semi-simplicial complex is the total complex K^Ω, where F^i omits the ith vertex and D^i repeats it. See also the preamble to Exercise 11 at the end of the chapter.

Let $\Delta^\flat_p(X)$, $p \geqslant 1$, be the subgroup of $\Delta_p(X)$ which is the sum of the D^i-images of $\Delta_{p-1}(X)$, $i = 0, 1, \ldots, p-1$. Then, if $p \geqslant 2$,

$$\Delta^\flat_p(X)\,\partial \subseteq \Delta^\flat_{p-1}(X);$$

for if $f_p = g_{p-1} D^i = (T^i g)_p$, then obviously $f_p F^j \in \Delta^\flat_{p-1}(X)$, $j \neq i, i+1$, and $f_p F^i = f_p F^{i+1}$, whence $f_p \partial \in \Delta^\flat_{p-1}(X)$. A simpler argument shows that $\Delta^\flat_1(X)\,\partial = 0$. Thus if we define $\Delta^\flat_p(X) = 0$, $p \leqslant 0$, we obtain a chain subcomplex $\Delta^\flat(X)$ of $\Delta(X)$ (and, by abuse of language, of $\tilde{\Delta}(X)$). The singular simplexes belonging to $\Delta^\flat(X)$ are *degenerate* and we may call $\Delta^\flat(X)$ the 'flat' subcomplex of $\Delta(X)$. The factor complex

$\Delta^N(X) = \Delta(X)/\Delta^\flat(X)$ is called the *normalized* singular simplicial chain complex of X; it is generated by the classes of the non-degenerate singular simplexes.

Let $\mu : \Delta(X) \to \Delta^N(X)$ be the projection.

8.2.6 Theorem. *μ is a chain equivalence.*

This will be proved in 8.4. We make some remarks here which are vital to the proof of 8.2.6.

First we observe that μ is *natural* in the following sense. Let $f : X \to Y$ be a map; then f^Δ maps $\Delta^\flat(X)$ into $\Delta^\flat(Y)$ and so induces $f^N : \Delta^N(X) \to \Delta^N(Y)$ such that the diagram

8.2.7
$$\begin{array}{ccc} \Delta(X) & \xrightarrow{\mu} & \Delta^N(X) \\ \big\downarrow f^\Delta & & \big\downarrow f^N \\ \Delta(Y) & \xrightarrow{\mu} & \Delta^N(Y) \end{array}$$

is commutative.

Next, we observe that μ may be augmented to a chain map $\tilde{\Delta}(X) \to \tilde{\Delta}^N(X)$ by defining $1\mu = 1$.

The groups $H_*(\Delta^N(X))$, $H_*(\Delta^N(X) \otimes G)$, $H^*(\Delta^N(X) \pitchfork G)$ are called the *normalized* (singular homology or contrahomology) groups of X. We write them $H_*^N(X)$, $H_*^N(X; G)$, $H_N^*(X; G)$. If we replace $\Delta^N(X)$ by $\tilde{\Delta}^N(X)$ we get the *reduced* normalized groups.

Now suppose X contractible and consider the chain homotopy Φ defined in the proof of 8.1.14. Then

8.2.8 Lemma. $\Delta^\flat(X)\,\Phi \subseteq \Delta^\flat(X)$.

As the reader may readily prove (compare 8.1.16)

8.2.9
$$h_{(p)} T^{i+1} = (CT^i)\, h_{(p-1)} : Ct_p \to t_p.$$

Now let $f : t^p \to X$ and $f = T^i f'$. Then if $g = g(f)$, $g' = g(f')$, we have

$$h_{(p)} g = C(T^i)\, C(f')\, q = C(T^i)\, h_{(p-1)} g' = h_{(p)} T^{i+1} g',$$

so that
$$g = T^{i+1} g'.$$

This establishes 8.2.8. ∎ We conclude that Φ induces a chain homotopy $\Phi^N : \tilde{\Delta}^N(X) \to \tilde{\Delta}^N(X)$ connecting 1 and 0; thus we have proved

8.2.10 Theorem. *The reduced normalized singular homology and contrahomology groups of a contractible space are null.* ∎

Finally, we mention the *relative* normalized groups. If (X, X_0) is a pair, then
$$i^\Delta : \Delta(X_0), \Delta^\flat(X_0) \to \Delta(X), \Delta^\flat(X).$$

Moreover, $\Delta(X_0)\, i^{\Delta} \cap \Delta^{\flat}(X) = \Delta^{\flat}(X_0)\, i^{\Delta}$. Therefore $i^N: \Delta^N(X_0) \to \Delta^N(X)$ is a monomorphism and embeds $\Delta^N(X_0)$ as a direct factor in $\Delta^N(X)$. The factor complex, $\Delta^N(X, X_0) = \Delta^N(X)/\Delta^N(X_0)$, is thus a free abelian chain complex from which the relative normalized homology and contrahomology groups may be defined.

8.3 Cubical homology theory

For many purposes it is convenient to base the singular homology theory of a space X on maps into X not of simplexes but of cubes. It will be proved in the next section that the normalized chain complex arising from singular cubes is chain equivalent to $\Delta(X)$.

We proceed as in the definition of $\Delta(X)$. The *standard p-cube*, e_p, is the subset of H^∞ consisting of points $(u_1, u_2, \ldots, u_n, \ldots)$ such that $u_n = 0$, $n > p$, and $0 \leqslant u_n \leqslant 1$, $n \leqslant p$. For each $i \leqslant p$, e_p has two $(p-1)$-faces, e^{i0}_{p-1} and e^{i1}_{p-1}, where $e^{i\epsilon}_{p-1}$ ($\epsilon = 0$ or 1) is the subset of e_p given by $u_i = \epsilon$. We define a map $V^{i\epsilon} = V^{i\epsilon}_{(p-1)}: e_{p-1} \to e_p$ by

$$(u_1, u_2, \ldots, u_{p-1}, 0, \ldots)\, V^{i\epsilon}_{(p-1)} = (u_1, \ldots, u_{i-1}, \epsilon, u_i, \ldots, u_{p-1}, 0, \ldots);$$

thus $V^{i\epsilon}$ is clearly a homeomorphism onto $e^{i\epsilon}_{p-1}$.

Let $\square_p(X)$, $p \geqslant 0$, be the free abelian group generated by the set of all maps $f: e_p \to X$. Then f determines a *singular p-cube* f_p of X. We define $F^{i\epsilon} = F^{i\epsilon}_p: \square_p(X) \to \square_{p-1}(X)$ by $f_p F^{i\epsilon} = (V^{i\epsilon} f)_{p-1}$, and then we define

$$\partial: \square_p(X) \to \square_{p-1}(X), \quad p \geqslant 1,$$

by

8.3.1
$$\partial = \sum_{i=1}^{p} (-1)^i (F^{i0} - F^{i1}).$$

By inspection (or by writing down the commutation rules of the face operators $F^{i\epsilon}$, corresponding to 8.1.4) it may be verified that $\partial\partial = 0$.

The *cubical singular chain complex* $\square(X)$ of X is then $\{\square_p(X), \partial\}$. It may be augmented to $\tilde{\square}(X)$ by taking $\tilde{\square}_{-1}(X) = J$ and setting $f_0 \tilde{\partial}_0 = 1$ for all 0-cubes f_0.

However, $\square(X)$ has a curious feature which prevents us from calling its homology group a homology group of X. The feature is that it has too many non-bounding cycles. Thus

8.3.2 Theorem. *If X consists of a single point x_0, $H_p(\square(X)) = J$, $p \geqslant 0$.*

For each $p \geqslant 0$, there is a unique generator for $\square_p$, determined by the unique map $f: e_p \to x_0$. Since, for $p > 0$, $f_p F^{i0} = f_p F^{i1}$, it follows that each f_p is a cycle. Thus $Z_p(\square) = J$, $B_p(\square) = 0$, whence $H_p(\square) = J$, $p \geqslant 0$. ∎

This situation may be remedied by normalizing‡. For $p \geqslant 1$ and $i \leqslant p$, define $T^i = T^i_{(p)} : e_p \to e_{p-1}$ by§

$$(u_1, \ldots, u_p)\,T^i = (u_1, \ldots, u_{i-1}, u_{i+1}, \ldots, u_p)$$

and define $D^i = D^i_{p-1} : \square_{p-1}(X) \to \square_p(X)$ by

$$f_{p-1} D^i = (T^i f)_p.$$

There are then formulae analogous to 8.2.4 and 8.2.5. We define $\square^\flat_p(X)$ to be the subgroup of $\square_p(X)$ which is the sum of the D^i-images of $\square_{p-1}(X)$ and show that

$$\square^\flat_p(X)\,\partial \subseteq \square^\flat_{p-1}(X), \quad p \geqslant 2.$$

For if $f_p = g_{p-1} D^i$, then $f_p F^{j\epsilon} \in \square^\flat_{p-1}(X)$ unless $j = i$ and $f_p F^{i0} = f_p F^{i1}$. A simpler argument shows that $\square^\flat_1(X)\,\partial = 0$. We define $\square^\flat_0(X) = 0$ and so obtain a 'flat' subcomplex of $\square(X)$, generated by the *degenerate* singular cubes of X. The factor complex $\square^N(X) = \square(X)/\square^\flat(X)$ is called the *normalized* cubical singular chain complex of X; it is, of course, free. We write $H^c_*(X)$ for its homology group which we call the *cubical homology group* of X. Similarly, $\tilde{\square}^N(X) = \tilde{\square}(X)/\square^\flat(X)$ and $\tilde{H}^c_*(X) = H_*(\tilde{\square}^N(X))$ is the *reduced* cubical homology group of X. We do not stress these names (nor the obvious extensions to $H^c_*(X; G)$, $H^*_c(X; G)$, $\square^N(X, X_0)$, $H^c_*(X, X_0)$), since we shall prove in the next section that there exists a canonical isomorphism between $H^s_*(X)$ and $H^c_*(X)$. Thus cubical homology theory provides no new invariants of a space; however, we get certain clear advantages from using it. For example, in any question involving homotopy, cubes are more convenient than simplexes; $t_p \times I$ is not a simplex, but $e_p \times I$ is a cube. Again, cubes are convenient when topological products are in question; a singular p-cube of X and a singular q-cube of Y together determine a singular $(p+q)$-cube of $X \times Y$.

Note that the anomaly of Theorem 8.3.2 has been removed, as claimed, by normalizing. For if $X = x_0$, then $\square_p(X) = \square^\flat_p(X)$, $p > 0$, so that $\square^N_p(X) = 0$, $p > 0$. This establishes that a point has the 'right' cubical homology groups, namely,

8.3.3 Theorem. *The reduced cubical homology and contrahomology groups of a point are null.* ∎

‡ Recall that it has been claimed that the homology groups of $\Delta(X)$ are unaffected by normalizing.

§ We write $(u_0, \ldots, u_p)$ for $(u_0, \ldots, u_p, 0, \ldots)$.

We now establish some basic facts about cubical homology groups. Let $f : X \to Y$ be a map. Then f induces a chain map $f^{\Box} : \Box(X) \to \Box(Y)$ by the rule $g_p f^{\Box} = (gf)_p$, where g_p is a singular cube of X. Then obviously $\Box^{\flat}(X) f^{\Box} \subseteq \Box^{\flat}(Y)$ and, for $f_1 : X_1 \to X_2$, $f_2 : X_2 \to X_3$, $(f_1 f_2)^{\Box} = f_1^{\Box} f_2^{\Box}$. Thus $f^{\Box}$, and hence f, induces a chain map

8.3.4 $$f^N : \Box^N(X) \to \Box^N(Y),$$

such that ‡

8.3.5 $$(f_1 f_2)^N = f_1^N f_2^N.$$

We write f_* for f_*^N, so that f induces

8.3.6 $$f_* : H_*^c(X) \to H_*^c(Y),$$

such that

8.3.7 $$(f_1 f_2)_* = f_{1*} f_{2*}.$$

These statements may be extended to homology with arbitrary coefficients, to contrahomology, and to the relative case. If we are considering reduced homology groups we define $1 f_{-1}^{\Box} = 1$.

8.3.8 Theorem. *If $f^0 \simeq f^1 : X \to Y$, then $f_*^0 = f_*^1 : H_*^c(X) \to H_*^c(Y)$.*

Let $F : I \times X \to Y$ be a homotopy connecting f^0 and f^1; then

$$(\epsilon, x) F = x f^{\epsilon}, \quad \epsilon = 0 \text{ or } 1, \; x \in X.$$

Given $g : e_p \to X$, define $h = h(g) : e_{p+1} \to Y$ by

$$(u_1, \ldots, u_{p+1}) h = (u_1, (u_2, \ldots, u_{p+1}) g) F.$$

This defines a chain homotopy $\Lambda : \Box(X) \to \Box(Y)$, by the rule

$$g_p \Lambda = h_{p+1}.$$

Elementary computations show that $\Lambda F^{1\epsilon} = f^{\epsilon\Box}$ and $\Lambda F^{i+1,\epsilon} = F^{i\epsilon}\Lambda$, $i \geqslant 1$. Thus, by 8.3.1,

$$\begin{aligned}
\Lambda\partial &= \Lambda F^{11} - \Lambda F^{10} + \sum_{i=2}^{p+1} (-1)^i (\Lambda F^{i0} - \Lambda F^{i1}) \\
&= f^{1\Box} - f^{0\Box} - \sum_{i=1}^{p} (-1)^i (F^{i0}\Lambda - F^{i1}\Lambda) \\
&= f^{1\Box} - f^{0\Box} - \partial\Lambda.
\end{aligned}$$

It follows that Λ is a chain homotopy connecting $f^{0\Box}$ and $f^{1\Box}$. A further elementary computation shows that $D^i\Lambda = \Lambda D^{i+1}$. Thus $\Box^{\flat}(X)\Lambda \subseteq \Box^{\flat}(Y)$ and Λ induces a chain homotopy Λ^N connecting f^{0N} and f^{1N}. It follows that $f_*^0 = f_*^1$. ∎

‡ We may write $f^{\Box}$ for f^N if convenient.

8.3.9 Corollary. *If* $f^0 \simeq f^1 : X \to Y$, *then* $f^0_* = f^1_* : \tilde{H}^c_*(X) \to \tilde{H}^c_*(Y)$.

It is only necessary to define $1\tilde{\Lambda}_{-1} = 0$. ∎ The reader will observe that no attempt was made to prove the analogue of 8.3.8 for simplicial singular homology. As we have remarked, results involving homotopy are more conveniently handled using singular cubes and the analogue of 8.3.8 will be a consequence of the equivalence of the two theories, to be proved in the next section.

Applying 8.3.7 we deduce from 8.3.8 and 8.3.9

8.3.10 Corollary. *If* $f : X \to Y$ *is a homotopy equivalence, then*

$$f_* : H^c_*(X) \cong H^c_*(Y), \quad f_* : \tilde{H}^c_*(X) \cong \tilde{H}^c_*(Y). \;∎$$

From 8.3.3 and 8.3.10 we deduce, compare 8.1.14, 8.2.10

8.3.11 Corollary. *The reduced cubical homology and contrahomology groups of a contractible space are null.* ∎

8.4 Equivalence theorems

We have now suggested three candidates for the role of singular chain complex of a space X, namely, $\Delta(X)$, $\Delta^N(X)$ and $\square^N(X)$. In this section we prove that these three chain complexes are equivalent. To describe the concepts that play a basic role in the proof let us first consider two of the chain complexes, say $\Delta(X)$ and $\Delta^N(X)$.

Suppose that, for each space X, a chain map $\phi^X : \tilde{\Delta}(X) \to \tilde{\Delta}^N(X)$ is given. We say that ϕ^X is *proper* if $1\phi^X = 1$ and that the association $X \to \phi^X$ is *natural* if, for any map $f : X \to Y$,

8.4.1 $$f^\Delta \phi^Y = \phi^X f^N.$$

We say that the set of chain maps $\{\phi^X\}$ is *allowable* if each ϕ^X is proper and 8.4.1 holds. Notice that the set $\{\mu^X\}$ introduced in 8.2 has been shown to be allowable.

Now let $\{\psi^X\}$ be another allowable set of chain maps,

$$\psi^X : \tilde{\Delta}(X) \to \tilde{\Delta}^N(X),$$

and suppose that, for each X, Λ^X is a chain homotopy between ϕ^X and ψ^X. Then we say that Λ^X is *proper* if $1\Lambda^X = 0$ and that the association $X \to \Lambda^X$ is *natural* if, for any map $f : X \to Y$,

8.4.2 $$f^\Delta \Lambda^Y = \Lambda^X f^N.$$

We say that the set of chain homotopies $\{\Lambda^X\}$ is *allowable* if each Λ^X is proper and 8.4.2 holds.

Clearly, by suppressing $\tilde{\Delta}_{-1}$ and $\tilde{\Delta}^N_{-1}$, we obtain a chain homotopy‡ Λ^X between ϕ^X and ψ^X regarded as chain maps‡ $\Delta(X) \to \Delta^N(X)$; this follows from the fact that ϕ^X, ψ^X and Λ^X are proper.

Now let $C(X)$ be any one of $\Delta(X)$, $\Delta^N(X)$, $\square^N(X)$ and let also $D(X)$ be any one of $\Delta(X)$, $\Delta^N(X)$, $\square^N(X)$, not necessarily distinct from $C(X)$. It is then clear what is meant by saying that $\{\phi^X\}$, $\{\psi^X\}$ are allowable sets of chain maps $\tilde{C}(X) \to \tilde{D}(X)$ and that $\{\Lambda^X\}$ is an allowable set of chain homotopies between ϕ^X and ψ^X.

The equivalence theorem is a ready consequence of

8.4.3 Theorem. *There exists an allowable set of chain maps*

$$\phi^X : \tilde{C}(X) \to \tilde{D}(X)$$

and any two allowable sets of chain maps are connected by an allowable set of chain homotopies.

The method of proof is due to Eilenberg and MacLane§. There will be several further applications of this method in this and the next chapter. We describe the method as 'proof by acyclic models'; it is essentially a generalization of the 'proof by acyclic carriers' of chapter 3.

For definiteness we take the case in which $C(X) = \Delta(X)$, $D(X) = \square^N(X)$ and we indicate, after completing the proof in this case, the changes necessary if different choices are made for $C(X)$ and $D(X)$.

We first establish the existence of ϕ^X. We argue by induction, starting with $1\phi^X_{-1} = 1$, for all X, and suppose that a homomorphism $\phi^X_p : \tilde{\Delta}_p(X) \to \tilde{\square}^N_p(X)$ has been defined for all X and all $p < q$, where $q \geqslant 0$, satisfying (i) $\phi^X_p \partial = \partial \phi^X_{p-1}$; (ii) for any $f : X \to Y$, $f^\Delta_p \phi^Y_p = \phi^X_p f^N_p$. Notice that (ii) is satisfied if $p = -1$.

Now t_q is itself a topological space, indeed a contractible space. Thus by 8.3.11, $\tilde{\square}^N(t_q)$ is acyclic (the *acyclic model*). Let $i : t_q \to t_q$ be the identity map. Then i_q is a singular q-simplex of t_q and

$$i_q \partial \phi^{t_q}_{q-1} \in \square^N_{q-1}(t_q).$$

Moreover, $i_q \partial \phi^{t_q}_{q-1} \partial = i_q \partial\partial \phi^{t_q}_{q-2} = 0$. Thus $i_q \partial \phi^{t_q}_{q-1}$ is a $(q-1)$-cycle of $\tilde{\square}^N(t_q)$ and hence a boundary, say $i_q \partial \phi^{t_q}_{q-1} = d_q \partial$, $d_q \in \square^N_q(t_q)$. We define $i_q \phi^{t_q}_q = d_q$. The map ϕ^X_q is now determined by naturality. For let

‡ In this section we shall consistently use the same symbol to denote a proper transformation for unaugmented complexes and the associated proper transformation for the associated augmented complexes.

§ 'Acyclic models', *Amer. J. Math.* **75** (1953), 189.

$g : t_q \to X$ be a singular q-simplex of X. Then $g_q = i_q g^\Delta$, so that, if 8.4.1 is to hold,

8.4.4 $$g_q \phi_q^X = i_q g_q^\Delta \phi_q^X = i_q \phi_q^{t_q} g_q^N = d_q g_q^N.$$

We now prove that conditions (i) and (ii) hold if $p = q$. To prove (i) we observe that

$$\begin{aligned} g_q \phi_q^X \partial &= d_q g_q^N \partial = d_q \partial g_{q-1}^N \\ &= i_q \partial \phi_{q-1}^{t_q} g_{q-1}^N = i_q \partial g_{q-1}^\Delta \phi_{q-1}^X \quad \text{(from condition (ii))} \\ &= i_q g_q^\Delta \partial \phi_{q-1}^X = g_q \partial \phi_{q-1}^X. \end{aligned}$$

To prove (ii) we observe that

$$\begin{aligned} g_q f_q^\Delta \phi_q^Y &= (gf)_q \phi_q^Y \\ &= d_q (gf)_q^N, \quad \text{by definition 8.4.4,} \\ &= d_q g_q^N f_q^N = g_q \phi_q^X f_q^N. \end{aligned}$$

This completes the induction and shows that homomorphisms $\phi_p^X : \tilde{\Delta}_p(X) \to \tilde{\square}_p^N(X)$ may be defined for all X and all p satisfying (i) and (ii) and such that $1\phi_{-1}^X = 1$. Thus $\{\phi^X\}$ is an allowable set of chain maps $\tilde{\Delta}(X) \to \tilde{\square}^N(X)$.

Now let $\{\phi^X\}$, $\{\psi^X\}$ be two allowable sets of chain maps

$$\tilde{\Delta}(X) \to \tilde{\square}^N(X);$$

we establish the existence of an allowable set of chain homotopies $\{\Lambda^X\}$, connecting ϕ^X and ψ^X, by induction. We define $1\Lambda_{-1}^X = 0$, all X, and we suppose that a homomorphism $\Lambda_p^X : \tilde{\Delta}_p(X) \to \square_{p+1}^N(X)$ has been defined for all X and all $p < q$, where $q \geqslant 0$, satisfying

(iii) $\Lambda_p^X \partial + \partial \Lambda_{p-1}^X = \psi_p^X - \phi_p^X$;

(iv) for any $f : X \to Y$, $f_p^\Delta \Lambda_p^Y = \Lambda_p^X f_{p+1}^N$.

Then (iii) and (iv) are satisfied if $p = -1$.

Consider $i_q(\psi_q^{t_q} - \phi_q^{t_q} - \partial \Lambda_{q-1}^{t_q}) \in \square_q^N(t_q)$. Invoking (iii) with $p = q-1$, we readily prove that this q-chain is a cycle. Since $\tilde{\square}^N(t_q)$ is acyclic, there exists a $(q+1)$-chain $k_{q+1} \in \square_{q+1}^N(t_q)$ with

$$i_q(\psi_q^{t_q} - \phi_q^{t_q} - \partial \Lambda_{q-1}^{t_q}) = k_{q+1} \partial.$$

We define $i_q \Lambda_q^{t_q} = k_{q+1}$. The map Λ_q^X is now determined by naturality; as before, we see that if 8.4.2 is to hold, then

8.4.5 $$g_q \Lambda_q^X = k_{q+1} g_{q+1}^N.$$

Finally, we verify that conditions (iii) and (iv) hold if $p = q$. To prove (iii) we observe that

$$\begin{aligned} g_q(\Lambda_q^X \partial + \partial \Lambda_{q-1}^X) &= k_{q+1} g_{q+1}^N \partial + i_q g_q^\Delta \partial \Lambda_{q-1}^X \\ &= i_q(\psi_q^{t_q} - \phi_q^{t_q} - \partial \Lambda_{q-1}^{t_q}) g_q^N + i_q \partial g_{q-1}^\Delta \Lambda_{q-1}^X \\ &= i_q g_q^\Delta(\psi_q^X - \phi_q^X), \quad \text{invoking (ii) and (iv)} \\ &= g_q(\psi_q^X - \phi_q^X). \end{aligned}$$

The proof of (iv) with $p = q$ follows the same lines as the earlier proof of (ii). Thus the induction is complete and we have arrived at an allowable set of chain homotopies $\{\Lambda^X\}$ connecting ϕ^X and ψ^X.

The modifications required if $C(X)$ and $D(X)$ are given other values should be quite clear. If $C(X) = \square^N(X)$, we replace t_q by e_q, which is, of course, also a contractible topological space. If $D(X) = \Delta(X)$, we apply 8.1.14 in place of 8.3.11. If $D(X) = \Delta^N(X)$, we apply 8.2.10 in place of 8.3.11. ▮

8.4.6 Corollary. *Each chain map* $\phi^X : \tilde{C}(X) \to \tilde{D}(X)$ *of any allowable set is a chain equivalence.*

For let $\{\psi^X\}$ be an allowable set of chain maps $\psi^X : \tilde{D}(X) \to \tilde{C}(X)$. Then $\{\phi^X \psi^X\}$ is an allowable set of chain maps $\tilde{C}(X) \to \tilde{C}(X)$. But $\{1^X\}$ is obviously an allowable set of chain maps $\tilde{C}(X) \to \tilde{C}(X)$. Thus, by 8.4.3

$$\phi^X \psi^X \simeq 1^X.$$

Similarly

$$\psi^X \phi^X \simeq 1^X. \text{ ▮}$$

Let us call a set of chain maps $\phi^X : C(X) \to D(X)$ *allowable*‡ if the maps ϕ^X, when augmented to maps $\tilde{C}(X) \to \tilde{D}(X)$ by defining $1\phi^X = 1$, become an allowable set of chain maps $\tilde{C}(X) \to \tilde{D}(X)$. We make a similar definition for chain homotopies. We then infer from 8.4.3 and 8.4.6

8.4.7 Theorem. *Let* $C(X)$ *and* $D(X)$ *be any of* $\Delta(X)$, $\Delta^N(X)$, $\square^N(X)$. *Then*

(i) *there exists an allowable set of chain maps* $C(X) \to D(X)$;

(ii) *any allowable set of chain maps* $C(X) \to D(X)$ *consists of chain equivalences;*

(iii) *any two allowable sets of chain maps* $C(X) \to D(X)$ *are connected by an allowable set of chain homotopies.* ▮

‡ This is equivalent to asking that the maps ϕ^X preserve the Kronecker index (or commute with ϵ, see 4.4).

This is the equivalence theorem. It shows that the groups $H^s_*(X)$, $H_*(\Delta^N(X))$ and $H^c_*(X)$ are *canonically* isomorphic. Thus, for example, let $\{\phi^X\}$ be an allowable set of chain equivalences $\Delta(X) \simeq \square^N(X)$. Then ϕ^X_* is an isomorphism of $H^s_*(X)$ with $H^c_*(X)$ which does *not* depend on the choice of $\{\phi^X\}$ by 8.4.7 (iii). Moreover, if $f : X \to Y$, then

8.4.8 $$f_* \phi^Y_* = \phi^X_* f_*.$$

Thus the canonical isomorphisms ϕ^X_* may be used to identify $H^s_*(X)$ with $H^c_*(X)$ and to identify induced homomorphisms

$$f_* : H^s_*(X) \to H^s_*(Y)$$

with induced homomorphisms

$$f_* : H^c_*(X) \to H^c_*(Y).$$

We shall in fact so identify $H^s_*(X)$, $H^c_*(X)$ and $H^N_*(X)$ regarding them as realizations of the singular homology group of a space X. Henceforth $H^s_*(X)$ will often be used for any of these groups.

Theorem 8.2.6 follows immediately from 8.4.7 and the observation that $\mu^X : \Delta(X) \to \Delta^N(X)$ gives rise to an allowable set $\{\mu^X\}$ of chain maps (see 8.2.7). From the exact homology sequence associated with

$$0 \to \Delta^\flat(X) \to \tilde{\Delta}(X) \to \tilde{\Delta}^N(X) \to 0$$

we infer

8.4.9 Corollary. *$\Delta^\flat(X)$ is acyclic.* ∎

Theorem 8.4.7 is easily relativized. For if $X_0 \subseteq X$, it is apparent from the construction of $\{\phi^X\}$ that ϕ^X maps $C(X_0)$, regarded as a subcomplex of $C(X)$, into $D(X_0)$, regarded as a subcomplex of $D(X)$. Moreover, if $\{\phi^X\}$, $\{\psi^X\}$ are two set of chain maps,

$$\phi^X, \psi^X : C(X), C(X_0) \to D(X), D(X_0),$$

then Λ^X maps $C(X_0)$ into $D(X_0)$. Thus Theorem 8.4.7 relativizes immediately as

8.4.10 Theorem. *Let $C(X, X_0)$ and $D(X, X_0)$ be any of $\Delta(X, X_0)$, $\Delta^N(X, X_0)$, $\square^N(X, X_0)$. Then*

(i) *there exists an allowable set of chain maps $C(X, X_0) \to D(X, X_0)$;*

(ii) *any allowable set of chain maps $C(X, X_0) \to D(X, X_0)$ consists of chain equivalences;*

(iii) *any two allowable sets of chain maps $C(X, X_0) \to D(X, X_0)$ are connected by an allowable set of chain homotopies.* ∎

As in the absolute case, we identify $H^s_*(X, X_0; G)$, $H^N_*(X, X_0; G)$, $H^c_*(X, X_0; G)$, retaining the notation $H^s_*(X, X_0; G)$, and we identify $H^*_s(X, X_0; G)$, $H^*_N(X, X_0; G)$, $H^*_c(X, X_0; G)$, retaining the notation $H^*_s(X, X_0; G)$. We shall usually write $f_* : H^s_*(X, X_0) \to H^s_*(Y, Y_0)$ for the homomorphism induced by $f : X, X_0 \to Y, Y_0$ but we may write $\hat{f}_*$ to emphasize that relative homology groups are involved.

In particular, we wish to record the following consequence of our equivalence theorem, which exhibits the connection between the topological notion of homotopy and the combinatorial notion of chain homotopy.

8.4.11 Theorem. *If $f \simeq g : X, X_0 \to Y, Y_0$, then*

(i) $f^\Delta \simeq g^\Delta : \Delta(X, X_0) \to \Delta(Y, Y_0)$;

(ii) $f^N \simeq g^N : \Delta^N(X, X_0) \to \Delta^N(Y, Y_0)$;

(iii) $f^\square \simeq g^\square : \square^N(X, X_0) \to \square^N(Y, Y_0)$.

Assertion (iii) in the absolute case was proved in the course of establishing 8.3.8. Now let $\{\phi^{X, X_0}\}$ be an allowable set of chain equivalences $\phi^{X, X_0} : \square^N(X, X_0) \to \Delta(X, X_0)$. Then

$$\phi^{X, X_0} f^\Delta = f^\square \phi^{Y, Y_0} \simeq g^\square \phi^{Y, Y_0} = \phi^{X, X_0} g^\Delta,$$

and we deduce (i) from the fact that ϕ^{X, X_0} is a chain equivalence. Similarly we deduce (ii). ∎

In fact, as the reader will observe, assertions (i), (ii) and (iii) above are all equivalent.

8.5 The properties of singular homology

In this section we list the axioms of Eilenberg and Steenrod for homology theory; these axioms appear here as properties of singular homology. Similar properties hold for singular contrahomology and may be supplied by the reader.

The properties refer to the groups $H^s_*(X, X_0)$, the induced homomorphisms

$$f_* : H^s_*(X, X_0) \to H^s_*(Y, Y_0)$$

and to‡

$$\partial_* : H^s_*(X, X_0) \to H^s_*(X_0),$$

the homomorphism from the exact homology sequence of the pair (X, X_0) which lowers dimension by 1. We remark that it is an immediate consequence of 5.6.1 that the identification of $H^s_*(X, X_0)$, $H^N_*(X, X_0)$ and $H^c_*(X, X_0)$ leads to an identification of the exact homology sequences of the pair (X, X_0) obtained from the simplicial,

‡ The homomorphism ∂_* has previously been denoted more often by ν_*.

normalized simplicial and normalized cubical singular chain complexes. Thus we may refer unambiguously to the homomorphism ∂_* of the exact homology sequence.

The properties to be verified are

1. If $i : X, X_0 \to X, X_0$ is the identity map, then i_* is the identity automorphism.
2. If $f : X, X_0 \to Y, Y_0$, $g : Y, Y_0 \to Z, Z_0$, then $(fg)_* = f_* g_*$.
3. $f_* \partial_* = \partial_*(f \mid X_0)_*$.
4. The homology sequence is exact.
5. If $f^0 \simeq f^1 : X, X_0 \to Y, Y_0$, then $f^0_* = f^1_*$.
6. If U is an open subset of X and $\bar{U}$ is contained in the interior of X_0, and if $i : X - U, X_0 - U \to X, X_0$ is the inclusion map, then i_* is an isomorphism.
7. The homology groups of a point are zero in positive dimension.

Property 1 is obvious, for if $i = 1$, $i^\Delta = 1$; property 2 is 8.1.11 relativized; properties 3 and 4 are contained in 8.1.13; property 5 is 8.3.8 relativized; property 7 is contained in 8.1.14. It remains to establish property 6, the Excision Axiom. We shall base our proof that singular homology satisfies the Excision Axiom on a theorem which we shall also use to prove the consistency of singular homology and simplicial homology for polyhedra; this latter result will appear in the next section.

8.5.1 Theorem. *Let (X, X_0) be a pair and let $\{V_\lambda\}$ be a family of subsets of X whose interiors cover X. Define $\Delta^V(X)$ to be the subcomplex of $\Delta(X)$ generated by those singular simplexes f_p such that, for some λ, $t_p f \subseteq V_\lambda$ and let $\Delta^V(X_0) = \Delta(X_0) \cap \Delta^V(X)$. Then the injection*

$$j : \Delta^V(X)/\Delta^V(X_0) \to \Delta(X, X_0)$$

is a chain equivalence.

We remark first that it is clearly sufficient to consider the absolute case. For once we have proved that $j : \Delta^V(X) \to \Delta(X)$ is a chain equivalence, it will follow also that $j : \Delta^V(X_0) \to \Delta(X_0)$ is a chain equivalence, since, if $U_\lambda = V_\lambda \cap X_0$, then $\{U_\lambda\}$ is a family of subsets of X_0 whose interiors cover X_0 and $\Delta^U(X_0) = \Delta^V(X_0)$. The theorem then follows from the 5-lemma (5.6.11) and 3.10.1.

Thus we must prove that $j : \Delta^V(X) \simeq \Delta(X)$. A subset of X lying in some V_λ will be called *V-small*; we must show that the singular homology group of X can be calculated by confining attention to V-small singular simplexes. The idea of the proof is not difficult. If G_λ is the interior of V_λ then $\{G_\lambda\}$ is an open covering of X. Given $f: t_p \to X$, the sets $\{f^{-1}(G_\lambda)\}$ form an open covering of the compact metric space t_p. If δ is a Lebesgue number of the covering, we may subdivide t_p to $t_p^{(r)}$, say, so that $\mu(t_p^{(r)}) < \delta$. Then if s_p is any p-simplex of $t_p^{(r)}$, $f \mid s_p$ determines a V-small simplex and f itself determines a chain of $\Delta^V(X)$. This idea, however, for constructing a map $\Delta(X) \to \Delta^V(X)$, chain inverse to j, is too crude as it stands. In the first place we must be more precise as to how f determines a chain of $\Delta^V(X)$ and, secondly, we find that the suggested chain inverse is not a chain map. However, the necessary correction to this crude approximation will become clear when we have systematized the process of subdivision in $\Delta(X)$. This systematization is lengthy but straightforward, and the reader may be fortified in studying it by the knowledge that it is motivated here by the desire to make the simple idea which we have described work.

The procedure for introducing a subdivision chain map

$$\chi : \Delta(X) \to \Delta(X)$$

and a chain homotopy $\Lambda : \Delta(X) \to \Delta(X)$ such that $\Lambda\partial + \partial\Lambda = \chi - 1$ has affinities with the techniques of the previous section‡. We first consider $L(t_p)$, the subcomplex of $\Delta(t_p)$ generated by all *linear* maps. If $l: t_q \to t_p$ is a linear map it is determined by its *vertices* $a^i = v^i l \in t_p$, so that a linear map l_q may be given as an ordered set of points of t_p. Notice that if $l_q = a^0 \ldots a^q$, then $l_q \partial = \sum_{i=0}^{q} (-1)^i a^0 \ldots \hat{a^i} \ldots a^q$. If $b \in t_p$, we denote by bl_q the $(q+1)$-simplex $ba^0 \ldots a^q$ of $L(t_p)$ and extend this notation to $bc_q \in L_{q+1}(t_p)$, where c_q is a q-chain of $L(t_p)$. Let b^l be the barycentre of the vertices of l.

We define the subdivision chain map $\chi : L(t_p) \to L(t_p)$ inductively by

$$\text{(i)}\ \chi_0 = 1; \quad \text{(ii)}\ l_q \chi_q = b^l(l_q \partial \chi_{q-1}),\ q \geq 1.$$

It is easy to verify inductively that χ is a chain map. Let us write $|l_q|$ for $t_q l \subseteq t_p$. Then $|l_q|$ is the convex subset of t_p spanned by the vertices of l, so that $b^l \in |l_q|$. If $c_q \in L_q(t_p)$, let $|c_q|$ be the union of the sets $|l_q|$ such that l_q appears in c_q with non-zero coefficient, and let the *mesh* of c_q, $\mu(c_q)$, be the maximum of the diameters of these sets $|l_q|$. Then evidently $|c_q \chi| \subseteq |c_q|$, and by the argument of 1.4.5, $\mu(c_q \chi) \leq \dfrac{q}{q+1} \mu(c_q)$.

Clearly χ can be iterated; its rth iterate χ^r is again a chain map and

$$|c_q \chi^r| \subseteq |c_q|,\ \mu(c_q \chi^r) \leq \left(\frac{q}{q+1}\right)^r \mu(c_q).$$

‡ We shall, in fact, define maps χ^X, Λ^X which are natural in an obvious sense.

A chain homotopy connecting 1 and χ is given by the inductive rule

$$\text{(iii) } \Lambda_0 = 0; \quad \text{(iv) } l_q \Lambda_q = b^l(l_q(\chi - 1 - \partial\Lambda_{q-1})), \quad q \geqslant 1.$$

It is easy to verify inductively that $\Lambda\partial + \partial\Lambda = \chi - 1$. It follows that $(1+\chi+\ldots+\chi^{r-1})\,\Lambda$ is a chain homotopy connecting 1 and χ^r. We also observe, from (iv), that $|c_q \Lambda_q| \subseteq |c_q|$.

Let $l : t_p \to t_n$ be a linear map. Then l^Δ maps $L(t_p)$ into $L(t_n)$. Let us, temporarily, write ${}^p\chi, {}^p\Lambda$ for the maps χ, Λ defined above to stress their dependence on the dimension p of t_p. Then an easy induction argument shows that

8.5.2 $$({}^p\chi)\, l^\Delta = l^\Delta({}^n\chi), \quad ({}^p\Lambda)\, l^\Delta = l^\Delta({}^n\Lambda),$$

as maps $L(t_p) \to L(t_n)$. 8.5.2 applies, in particular, when we take for l a map V^j as defined by 8.1.1.

We are now ready to consider subdivision in $\Delta(X)$; as in 8.4, let i_p be the identity map of t_p, and define $\chi^X : \Delta(X) \to \Delta(X)$ by

8.5.3 $$f_p \chi^X = i_p \chi f^\Delta,$$

where, of course, $\chi = {}^p\chi$. We prove that χ^X is a chain map by applying 8.5.2.

For
$$\begin{aligned} f_p \partial \chi^X &= \sum_{j=0}^{p} (-1)^j (V^j f)_{p-1} \chi^X \\ &= \sum_{j=0}^{p} (-1)^j i_{p-1} ({}^{p-1}\chi)(V^j f)^\Delta \\ &= \sum_{j=0}^{p} (-1)^j i_{p-1} ({}^{p-1}\chi)(V^j)^\Delta f^\Delta \\ &= \sum_{j=0}^{p} (-1)^j i_{p-1} (V^j)^\Delta \chi f^\Delta, \quad \text{by 8.5.2,} \\ &= \left(\sum_{j=0}^{p} (-1)^j V^j\right) ({}^p\chi) f^\Delta \\ &= i_p \partial \chi f^\Delta \\ &= i_p \chi f^\Delta \partial \\ &= f_p \chi^X \partial. \end{aligned}$$

Let $|f_p| = t_p f \subseteq X$; and, if $c_p \in \Delta_p(X)$, let $|c_p|$ be the union of the sets $|f_p|$ such that f_p appears in c_p with non-zero coefficient. Then 8.5.3 ensures that $|f_p \chi^X| \subseteq |f_p|$. Thus, in particular, if f is V-small, $f\chi^X \in \Delta^V(X)$.

Let l_p be a linear map, $l_p \in L(t_p)$. Then $l_p = i_p l^\Delta$, so that

$$l_p \chi = i_p l^\Delta \chi = i_p \chi l^\Delta,$$

by 8.5.2. Thus

$$l_p \chi f^\Delta = i_p \chi l^\Delta f^\Delta = i_p \chi (lf)^\Delta = (lf)_p \chi^X = l_p f^\Delta \chi^X,$$

whence
$$\chi f^\Delta = f^\Delta \chi^X : L_p(t_p) \to \Delta_p(X).$$

It is then clear that if $(\chi^X)^s$ is the sth iterate of χ^X,

8.5.3$_s$
$$f_p(\chi^X)^s = i_p \chi^s f^\Delta.$$

Let δ be a Lebesgue number of the covering of t_p by the open sets $\{f^{-1}(G_\lambda)\}$. We may choose s so that $\mu(i_p \chi^s) < \delta$, since

$$\mu(i_p \chi^s) \leqslant \left(\frac{q}{q+1}\right)^s \mu(i_p).$$

Then 8.5.3$_s$ shows that, for such an $s, f_p(\chi^X)^s \in \Delta^V(X)$. We define $s(f)$, the *size* of f, to be the smallest integer s such that $f_p(\chi^X)^s \in \Delta^V(X)$.

By analogy with 8.5.3 we define $\Lambda^X : \Delta(X) \to \Delta(X)$ by

8.5.4
$$f_p \Lambda^X = i_p \Lambda f^\Delta,$$

where, of course,
$$\Lambda = {}^p\Lambda.$$

A similar computation to that given for χ^X shows that

$$\Lambda^X \partial + \partial \Lambda^X = \chi^X - 1.$$

Moreover, it is clear that $|f_p \Lambda^X| \subseteq |f_p|$, so that Λ^X maps $\Delta^V(X)$ to $\Delta^V(X)$.

We shall henceforth write χ, Λ for χ^X, Λ^X. To sum up, we have a chain map $\chi : \Delta(X), \Delta^V(X) \to \Delta(X), \Delta^V(X)$ and a chain homotopy $\Lambda : \Delta(X), \Delta^V(X) \to \Delta(X), \Delta^V(X)$ with $\Lambda\partial + \partial\Lambda = \chi - 1$; and with each $f_p \in \Delta(X)$ we have associated an integer $s(f)$ which is the least integer s such that $f_p \chi^s \in \Delta^V(X)$. Our crude plan was to define a chain inverse to $j : \Delta^V(X) \to \Delta(X)$ by mapping f_p to $f_p \chi^{s(f)}$. If this were a chain map we should try to construct a chain homotopy connecting f_p and $f_p \chi^{s(f)}$. Now $(1 + \chi + \ldots + \chi^{s-1})\Lambda$ connects 1 and χ^s. We consider the chain homotopy
$$\Phi : \Delta(X), \Delta^V(X) \to \Delta(X), \Delta^V(X),$$

given by
$$f_p \Phi = f_p(1 + \chi + \ldots + \chi^{s(f)-1})\Lambda.$$

Writing s for $s(f)$ and $s^{(j)}$ for $s(f^j_{p-1})$, we verify by straightforward computation that

8.5.5
$$f_p(\Phi\partial + \partial\Phi + 1) = f_p \chi^s - \sum_{j=0}^{p} (-1)^j \sum_{i=s^{(j)}}^{s-1} f^j_{p-1} \chi^i \Lambda.$$

Now $f_p\chi^s \in \Delta^V(X)$ and $f^j_{p-1}\chi^{s(j)} \in \Delta^V(X)$. Since $\Delta^V(X)\chi \subseteq \Delta^V(X)$ and $\Delta^V(X)\Lambda \subseteq \Delta^V(X)$, it follows that the right-hand side of 8.5.5 is a chain of $\Delta^V(X)$. We may therefore define a chain map

$$k : \Delta(X) \to \Delta^V(X)$$

by

8.5.6
$$f_p k = f_p\chi^s - \sum_{j=0}^{p} (-1)^j \sum_{i=s(j)}^{s-1} f^j_{p-1}\chi^i\Lambda.$$

By construction, $kj - 1 = \Phi\partial + \partial\Phi$; thus kj is a chain map since

$$\partial(\Phi\partial + \partial\Phi + 1) = (\Phi\partial + \partial\Phi + 1)\partial.$$

But j is a monomorphic chain map, so that k is itself a chain map. On the other hand, if $f_p \in \Delta^V(X)$ then $s = 0$, so that $jk = 1$ and the theorem is proved. ∎

8.5.7 Theorem. (*Excision theorem for singular homology.*) *If U is a subset of X whose closure is contained in the interior of X_0, then the injection $i : X - U, X_0 - U \to X, X_0$ induces an isomorphism*

$$i_* : H_*(X - U, X_0 - U) \cong H_*(X, X_0).$$

Let W be the interior of X_0. Then $\overline{U} \subseteq W$ so that $X - \overline{U}$ and W are open sets whose union is X. It follows that $X - U$ and X_0 are sets whose interiors together cover X. We apply 8.5.1 to the family $V = \{X - U, X_0\}$. Then $\Delta^V(X) = \Delta(X - U) + \Delta(X_0)$, $\Delta^V(X_0) = \Delta(X_0)$. Thus

$$j : (\Delta(X - U) + \Delta(X_0))/\Delta(X_0) \to \Delta(X, X_0)$$

is a chain equivalence. Let

$$h : \Delta(X - U, X_0 - U) \to (\Delta(X - U) + \Delta(X_0))/\Delta(X_0)$$

be the chain map induced by the injection

$$\Delta(X - U), \Delta(X_0 - U) \to \Delta(X - U) + \Delta(X_0), \Delta(X_0).$$

Since $\Delta(X - U) \cap \Delta(X_0) = \Delta(X_0 - U)$, h is an isomorphism (I.2.2); but $hj = i^\Delta$, so that i^Δ is a chain equivalence. ∎ The reader will remark that it is not necessary to suppose that U is open in X.

The hypotheses of Theorem 8.5.7 cannot be weakened by only requiring that U itself be contained in the interior of X_0, as the following example shows.

8.5.8 *Example.* Take $X = t_1$, the standard 1-simplex, so that a typical point of X is $\lambda_0 v^0 + \lambda_1 v^1$, $\lambda_0 \geqslant 0$, $\lambda_1 \geqslant 0$, $\lambda_0 + \lambda_1 = 1$. Let X_0 be the union of the subsets of X for which $2^{-2n} \leqslant \lambda_1 \leqslant 2^{-2n+1}$, $n \geqslant 1$, together with v^0, v^1. Let U be the interior of X_0, that is, the union of the subsets $2^{-2n} < \lambda_1 < 2^{-2n+1}$. Then i_1 is a singular 1-cycle of X mod X_0, since v^0 and v^1 belong to X_0, and determines an element $\gamma \in H_1^s(X, X_0)$. Now $X - U$ consists of the union of subsets $2^{-2n-1} \leqslant \lambda_1 \leqslant 2^{-2n}$, $n \geqslant 0$, together with v^0, and $X_0 - U$ is the set of end points of these intervals together with v^0. Then it may be seen that γ is not in the image under injection of $H_1^s(X-U, X_0-U)$.

This example shows that $H_*^s(X, X_0)$ is not determined by the topological space $X - X_0$ nor even by $\overline{X - X_0}$. A simpler example of the former fact is furnished by taking $X = t_1$, $X_0 = v^0 \cup v^1$ and obtaining Y, Y_0 by excising the whole of X_0. Then $H_1^s(X, X_0) = J$, but $H_1^s(Y, Y_0) = 0$, since $Y_0 = \phi$ and Y is contractible.

We now give an important application of the Excision theorem. Let $g : A \to Y_0$ be a map and let CA be a cone on A. Let Y be obtained from the (disjoint) union of CA and Y_0 by identifying $a \in A$ with $ag \in Y_0$ and let $f : CA, A \to Y, Y_0$ be the map which is the identity on $CA - A$ and agrees with g on A; thus f is the restriction to CA of the projection $CA \cup Y_0 \to Y$. We say that Y is obtained from Y_0 by *attaching the cone* CA by means of the *attaching map* g and f is called the *characteristic* map of the attachment; of particular importance is the case in which A is an $(n-1)$-sphere, when we speak‡ of attaching an n-cell to Y_0. We prove

8.5.9 Proposition. *The characteristic map $f : CA, A \to Y, Y_0$ induces homology and contrahomology isomorphisms.*

Let $k : A \times I \to CA$ identify the cylinder $A \times I$ to the cone CA so that $(a, 0)k = a$, $a \in A$. If $[u, v]$, $[u, v)$ are the subintervals $u \leqslant t \leqslant v$, $u \leqslant t < v$ of I, we shall write $A[u, v]$ for $(A \times [u, v])k$, $Y[u, v]$ for $(A \times [u, v])kf$ and similarly $A[u, v)$, $Y[u, v)$; then $CA = A[0, 1]$, $A = A[0, 0]$.

Consider the commutative diagram of maps

$$\begin{array}{ccccc}
A[\tfrac{1}{3}, 1], A[\tfrac{1}{3}, \tfrac{2}{3}] & \xrightarrow{i_1} & CA, A[0, \tfrac{2}{3}] & \xleftarrow{i_2} & CA, A \\
\downarrow{\scriptstyle f_1} & & \downarrow{\scriptstyle f_2} & & \downarrow{\scriptstyle f} \\
Y[\tfrac{1}{3}, 1], Y[\tfrac{1}{3}, \tfrac{2}{3}] & \xrightarrow{j_1} & Y, Y[0, \tfrac{2}{3}] \cup Y_0 & \xleftarrow{j_2} & Y, Y_0
\end{array}$$

where i_1, i_2, j_1, j_2 are embeddings and f_1, f_2 are restrictions of f. Passing to homology homomorphisms we see that i_{1*} is an isomorphism

‡ Following J. H. C. Whitehead.

since $A[\frac{1}{3}, 1], A[\frac{1}{3}, \frac{2}{3}]$ is obtained from $CA, A[0, \frac{2}{3}]$ by excising $A[0, \frac{1}{3})$. Similarly, j_{1*} is an isomorphism since we are here excising $Y[0, \frac{1}{3}) \cup Y_0$; we note that the closure of this set is $Y[0, \frac{1}{3}] \cup Y_0$ and the interior of $Y[0, \frac{2}{3}] \cup Y_0$ is $Y[0, \frac{2}{3}) \cup Y_0$. Further, i_{2*} is an isomorphism since we may deform the identity map on CA to a map retracting $A[0, \frac{2}{3}]$ onto A, rel A; similarly, j_{2*} is an isomorphism. Finally, f_{1*} is an isomorphism since f_1 is a homeomorphism. From these facts we deduce that f_{2*} and hence f_* are isomorphisms. Similarly f induces a contrahomology isomorphism f^*. ∎

We close with an application of 8.5.9. Given any space X, let X_+, X_- be two cones on X such that $X_+ \cap X_- = X$ and let

$$S(X) = X_+ \cup X_-.$$

Then $S(X)$ is called the *suspension* of X. Given $f: X \to Y$, let $f_+ : X_+ \to Y_+$, $f_- : X_- \to Y_-$ be extensions of f (see the diagram after 1.6.7). Then f_+, f_- combine to give a map $S(f): S(X) \to S(Y)$ called the *suspension* of f. Clearly if $g: Y \to Z$, then $S(fg) = S(f)\,S(g)$.

8.5.10 Theorem. *There is an isomorphism* $E_* : \tilde{H}^s_{p-1}(X) \cong \tilde{H}^s_p(SX)$ *such that, if* $f: X \to Y, E_*(S(f))_* = f_* E_*$.

From the homology sequence of (X_-, X) and the contractibility of X_-, we deduce

$$\nu_* : H^s_p(X_-, X) \cong \tilde{H}^s_{p-1}(X).$$

From the homology sequence of $(S(X), X_+)$ and the contractibility of X_+, we deduce

$$\mu_* : \tilde{H}^s_p(SX) \cong H^s_p(SX, X_+).$$

Let i_* be the injection $i_* : H^s_p(X_-, X) \to H^s_p(SX, X_+)$. We define E_* by $\nu_* E_* = i_* \mu_*$. The naturality property $E_*(Sf)_* = f_* E_*$ follows from the corresponding naturality properties of ν_*, i_*, μ_*.

It remains to prove that i_* is an isomorphism. Now we may regard SX as being obtained from X_+ by attaching the cone X_- by means of the injection map $X \to X_+$. The corresponding characteristic map is just the injection map $i: X_-, X \to SX, X_+$, so that 8.5.9 ensures that i_* is an isomorphism. ∎

8.6 The singular homology theory of a polyhedron

This section is devoted to the proof that the singular homology theory and the simplicial homology theory (cf. chapters 1–3) are equivalent on the category of polyhedra (or polyhedral pairs). Eilenberg and Steenrod, in [14], deduce this from the axioms 1–7 listed in the previous section, but we shall give a direct proof based on 8.5.1.

Let $|K|$ be a polyhedron, not necessarily compact. We shall define an obvious chain map $\alpha : C^{\Omega}(K) \to \Delta(|K|)$ and our object is to establish that α is a chain equivalence. The essential step is to define for $|K|$ a suitable family $\{V\}$ of subsets. We shall then define a chain map $\beta : C^{\Omega}(K) \to \Delta^{V}(|K|)$ and construct a chain inverse to β. It will then follow that, if $j : \Delta^{V}(|K|) \to \Delta(|K|)$ is the injection, then $\alpha = \beta j$ and so is a chain equivalence.

We proceed first to define the family $\{V\}$. As in 1.10, we describe points of $|K|$ by sets of barycentric coordinates $\{\lambda_j\}$ such that each $\lambda_j \geqslant 0$, $\Sigma \lambda_j = 1$, and the non-zero λ's correspond to vertices of a simplex of K. To each $s = (a^{i_0} \ldots a^{i_p}) \in K$ we associate a set $V_s \subseteq |K|$, namely, the set of points $\{\lambda_j\}$ such that

8.6.1 $$\max_{0 \leqslant r \leqslant p} (\lambda_{i_r}) > \max_{j \neq i_0, \ldots, i_p} (\lambda_j).$$

We now prove certain important properties of the family $\{V_s\}$.

8.6.2 Lemma. *The family $\{V_s\}$ is an open covering of $|K|$.*

Obviously $\bar{s} \subseteq V_s$, so that the family is a covering. We now prove that V_s is open. We recall that a subset of $|K|$ is closed if and only if its intersection with $\bar{t}$, for each $t \in K$, is closed. Equivalently, a subset of $|K|$ is open if and only if its intersection with each $\bar{t}$ is open in $\bar{t}$. We use this criterion to prove that V_s is open in $|K|$. Let

$$t = (a^{i_0} \ldots a^{i_q} a^{j_0} \ldots a^{j_r}),$$

where no j_k is one of $i_0, \ldots, i_p$. Then $\{\lambda_j\} \in \bar{t} \cap V_s$ if and only if

$$\lambda_j = 0, \quad j \neq i_0, \ldots, i_q, j_0, \ldots, j_r$$

and $$\max_{0 \leqslant h \leqslant q} \lambda_{i_h} > \max_{0 \leqslant k \leqslant r} \lambda_{j_k}.$$

Let $G_{h,k}$ be the open subset of $\bar{t}$ given by $\lambda_{i_h} > \lambda_{j_k}$. Then we have shown that $\bar{t} \cap V_s = \bigcap_{k=0,\ldots,r} \bigcup_{h=0,\ldots,q} G_{h,k}$, and is thus an open subset of $\bar{t}$. ∎

It follows from 8.5.1 that $j : \Delta^{V}(|K|) \simeq \Delta(|K|)$.

8.6.3 Lemma. *Each V_s is contractible.*

Since $\bar{s} \subseteq V_s$ and $\bar{s}$ is contractible, it is sufficient to demonstrate a deformation retraction of V_s onto $\bar{s}$.

Let $\{\lambda_j\} \in V_s$ and let $\sum_{r=0}^{p} \lambda_{i_r} = \mu$. Then $\mu > 0$ and we define $\{\bar{\lambda}_j\} \in \bar{s}$, the target point of $\{\lambda_j\}$, by

$$\bar{\lambda}_{i_r} = \frac{1}{\mu}\lambda_{i_r}, \quad r = 0, \ldots, p, \qquad \bar{\lambda}_j = 0, \quad j \neq i_0, \ldots, i_p.$$

Then the segment joining $\{\lambda_j\}$ to $\{\bar{\lambda}_j\}$ obviously lies in V_s, so that V_s may be retracted by deformation along these segments down onto $\bar{s}$. ▮

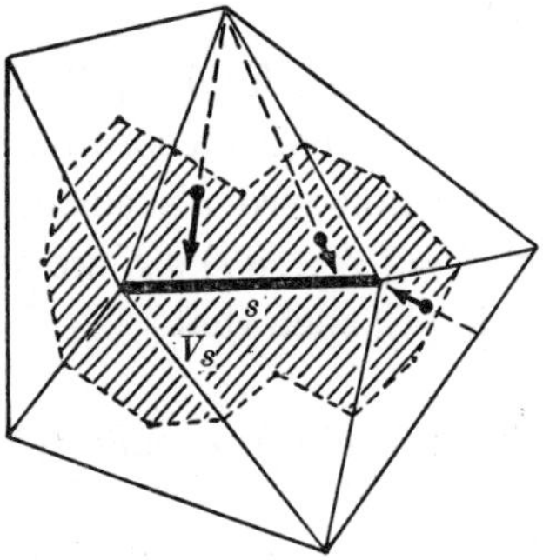

Fig. 8.1

8.6.4 Lemma. *If $A \subseteq |K|$ is V-small, then there exists a simplex $s(A)$ of lowest dimension with $V_{s(A)} \supseteq A$; moreover if $B \subseteq A$ then $s(B) \prec s(A)$.*

Let Σ be a collection of simplexes of K. The reader can verify that if $\bar{s} = \bigcap_{t \in \Sigma} \bar{t}$ then $V_s = \bigcap_{t \in \Sigma} V_t$ (where $V_{s_{-1}} = \varnothing$); for the largest λ's of a point in $\bigcap V_t$ must be associated with vertices belonging to each $t \in \Sigma$.

Let Σ be the collection of simplexes t of K such that $V_t \supseteq A$. Then if $\bar{s} = \bigcap_{t \in \Sigma} \bar{t}$, $A \subseteq \bigcap_{t \in \Sigma} V_t = V_s$, and, if $A \subseteq V_t$, $s \prec t$. Thus $s = s(A)$. If $B \subseteq A$, then $V_{s(A)} \supseteq A \supseteq B$, so that $s(B) \prec s(A)$. ▮

We now define a chain map $\beta : C^{\Omega}(K) \to \Delta^V(|K|)$ and a chain map $\gamma : \Delta^V(|K|) \to C^{\Omega}(K)$. A 'simplex' of K^{Ω} is an array $g = [a^{i_0} \ldots a^{i_p}]$, where the vertices, not necessarily distinct, span a simplex, s, of K. We may associate with g the linear map $l_g : t_p \to \bar{s}$ given by $v^r l_g = a^{i_r}$. Obviously l_g is V-small and the association $g \to l_g$ induces proper‡ chain maps $\alpha : C^{\Omega}(K) \to \Delta(|K|)$ and $\beta : C^{\Omega}(K) \to \Delta^V(|K|)$, given by

8.6.5 $$g\alpha = l_g, \quad g\beta = l_g;$$

plainly $\alpha = \beta j$.

To define γ, we first associate with each $x \in |K|$ a vertex $a(x)$ of $s(x)$, the smallest simplex in whose V-set x lies. Now let $f : t_p \to |K|$

‡ Recall the definition that precedes 3.4.11.

be V-small (that is, $t_p f$ is V-small). We define $\gamma : \Delta^V(|K|) \to C^\Omega(K)$ by linearity from

$$f_p\gamma = [a(v^0 f), \ldots, a(v^p f)].$$

Since $v^i f \in t_p f$, it follows from 8.6.4 that $s(v^i f) \prec s(t_p f) = s(f)$, say. Thus each $a(v^i f)$ is a vertex of $s(f)$ so that the right-hand side is an admissible array. It is clear that γ is a proper chain map.

8.6.6 Proposition. *$\beta : C^\Omega(K) \to \Delta^V(|K|)$ is a chain equivalence with chain inverse γ.*

If x is a vertex of K, then clearly $s(x) = x$, so that $\beta\gamma = 1$. On the other hand, if f_p is V-small, then $f_p\gamma\beta$ is the linear map l given by $v^r l = a(v^r f)$. Thus $f_p\gamma\beta$ is a linear map onto a face of $s(f)$. It follows that 1 and $\gamma\beta$ are both carried by the algebraic carrier function E, given by

$$E(f_p) = \Delta(V_{s(f)}).$$

Obviously E is a carrier function; for $t_{p-1} f^i_{p-1} \subseteq t_p f$ so that $V_{s(f^i_{p-1})} \subseteq V_{s(f)}$, and $E(f^i_{p-1}) \subseteq E(f_p)$. Moreover, by 8.6.3, E is acyclic. Since 1 and $\gamma\beta$ are proper‡, 3.4.5 yields $\gamma\beta \simeq 1$. ❚

8.6.7 Theorem. *$\alpha : C^\Omega(K) \to \Delta(|K|)$ is a chain equivalence.*

This follows immediately from 8.5.1 and 8.6.6. ❚

There is no difficulty in relativizing the proof (or the conclusion). We state without further comment:

8.6.8 Theorem. *There is a chain equivalence*

$$\alpha : C^\Omega(K), C^\Omega(K_0) \to \Delta(|K|), \Delta(|K_0|)$$

which induces an isomorphism of the homology sequence of the pair K, K_0 onto the singular homology sequence of the pair $|K|, |K_0|$ and an isomorphism of the singular contrahomology sequence of the pair $|K|, |K_0|$ onto the contrahomology sequence of the pair K, K_0. ❚

We note that α is a monomorphism of $C^\Omega(K)$ onto a direct factor in $\Delta(|K|)$. It is, moreover, clear that α is natural in the sense that if $v : K \to L$ is a simplicial map, then the diagram

$$\begin{array}{ccc} C^\Omega(K) & \xrightarrow{\alpha} & \Delta(|K|) \\ \downarrow{\scriptstyle v\Omega} & & \downarrow{\scriptstyle |v|\Delta} \\ C^\Omega(L) & \xrightarrow{\alpha} & \Delta(|L|) \end{array}$$

is commutative.

‡ Recall the definition that precedes 3.4.11.

Now let K be a *finite* complex. Then a map $f : |K| \to |L|$ induces

$$f_*^\Omega : H_*^\Omega(K) \to H_*^\Omega(L), \quad f_\Omega^* : H_\Omega^*(L; G) \to H_\Omega^*(K; G)$$

We prove

8.6.9 Theorem. $f_*^\Omega \alpha_* = \alpha_* f_*$, $f_\Omega^* \alpha^* = \alpha^* f^*$.

Let $K^{(r)}$ be the rth derived of K and let $\phi^{(r)} : C^\Omega(K^{(r)}) \to C^\Omega(K)$ be induced by a standard map $v^{(r)} : K^{(r)} \to K$. Then $|v^{(r)}| \simeq 1 : |K| \to |K|$ and $v^{(r)\Omega}\alpha = \alpha|v^{(r)}|^\Delta$. Thus $\phi_*^{(r)}\alpha_* = \alpha_*$. Now let $\chi^{(r)} : C^\Omega(K) \to C^\Omega(K^{(r)})$ be a subdivision chain map. Then $\chi_*^{(r)}\phi_*^{(r)} = 1$, so that $\alpha_* = \chi_*^{(r)}\alpha_*$. Let $u : K^{(r)} \to L$ be a simplicial approximation to f. Then by definition $f_*^\Omega = \chi_*^{(r)} u_*^\Omega$ and $u^\Omega\alpha = \alpha|u|^\Delta$. Moreover, $|u| \simeq f : |K| \to |L|$ so that $|u|_* = f_* : H_*^s(|K|) \to H_*^s(|L|)$. Putting these facts together

$$\begin{aligned} f_*^\Omega \alpha_* &= \chi_*^{(r)} u_*^\Omega \alpha_* \\ &= \chi_*^{(r)} \alpha_* |u|_* \\ &= \alpha_* f_*. \end{aligned}$$

The second assertion follows similarly. ▮

We may obviously interpret 8.6.9 for relative groups. We thus see that α_* is an isomorphism of the singular and simplicial homology theories for polyhedra, where the simplicial theory for compact polyhedra includes the homomorphism induced by continuous maps. In particular, Theorem 8.6.8 establishes the topological invariance of the homology and contrahomology sequences of simplicial pairs K, K_0. This had already been proved for finite complexes in chapter 3 by different methods. The method of this chapter is logically the more satisfactory, not only because it gives the result for infinite complexes, but also because it gives a single homology sequence for the polyhedral pair P, P_0 to which all homology sequences are canonically isomorphic. Thus we could have deferred all proofs of topological invariance to this chapter; we have, however, given the earlier proofs because they are conceptually simpler and provided, in their context, the answer to an immediate and unavoidable question.

We may, in the same way, establish the topological invariance of the local homology groups at a point of a polyhedron (see 3.7) by proving them isomorphic to certain singular homology groups. We define $H_*^X(p)$, the *singular homology group of* X *at a point* $p \in X$ by the formula

8.6.10 $$H_*^X(p) = H_*(X, X-p).$$

Now suppose K is a complex and a is a vertex of K. Then

$$\alpha_* : H_*^K(a) \cong H_*^s(|K|, |K_0(a)|),$$

and we may compose α_* with the injection

$$i_* : H_*^s(|K|, |K_0(a)|) \to H_*^{|K|}(a)$$

to produce a homomorphism $\tilde{\alpha}_* = \alpha_* i_* : H_*^K(a) \to H_*^{|K|}(a)$.

8.6.11 Proposition. *$\tilde{\alpha}_*$ is an isomorphism.*

We have only to prove that i_* is an isomorphism. Now $|K| - a$ may be retracted by deformation onto $|K_0(a)|$ by projecting from a. Thus the injection $H_*^s(|K_0(a)|) \to H_*^s(|K| - a)$ is an isomorphism and we deduce that i_* is an isomorphism by applying the 5-lemma to the homology sequences of the pairs $|K|, |K_0(a)|$ and $|K|, |K| - a$. ∎

8.6.12 Corollary. *The local homology groups at a point of an (arbitrary) polyhedron are topological invariants.* ∎

We note finally that, as for the global theory, $\tilde{\alpha}_*$ is natural with respect to homomorphisms induced by maps (compare 8.6.9). For let $f : |L|, b \to |K|, a$ be an l-map (see 3.7) and let f' be an l-approximation to f. Since $f \underset{l}{\simeq} f'$, we have $f \simeq f' : |L|, |L| - b \to |K|, |K| - a$ so that $f_* = f'_* : H_*^{|L|}(b) \to H_*^{|K|}(a)$. Thus there is no real loss of generality in supposing that f itself maps $|L_0(b)|$ into $|K_0(a)|$. Then f induces $f_*^\Omega : H_*^L(b) \to H_*^K(a)$, $f_* : H_*^s(|L|, |L_0(b)|) \to H_*^s(|K|, |K_0(a)|)$, and $\tilde{f}_* : H_*^{|L|}(b) \to H_*^{|K|}(a)$; and we have (8.6.9) $f_*^\Omega \alpha_* = \alpha_* f_*$ and, obviously, $i_* \tilde{f}_* = f_* i_*$. Thus $f_*^\Omega \tilde{\alpha}_* = \tilde{\alpha}_* \tilde{f}_*$. We have proved

8.6.13 Proposition. *Let $f : |L|, b \to |K|, a$ be an l-map. Then f induces $f_*^\Omega : H_*^L(b) \to H_*^K(a)$ and $\tilde{f}_* : H_*^{|L|}(b) \to H_*^{|K|}(a)$, and $f_*^\Omega \alpha_* = \tilde{\alpha}_* \tilde{f}_*$.* ∎

The reader is recommended to re-examine the discussion of orientation in II.1 using the notion of the singular homology group at a point; this notion leads to some simplification in the details of the discussion; for example, in the lemma contributing to the proof of II. 1.29.

8.7 Homology groups of topological products

In 5.7 we quoted the fact that, to compute the homology groups of $|K| \times |L|$ we may use the tensor product of the chain complexes $C(K)$, $C(L)$. We prove the statement here in the following general form. Let $C(X)$, $D(X)$ each stand for one of $\Delta(X)$, $\Delta^N(X)$, $\square^N(X)$ as in 8.4. Identifying the 0-simplex t_0 with the 0-cube e_0, we may identify

$\Delta_0(X)$, $\Delta_0^N(X)$ and $\Box_0^N(X)$ and thus speak unambiguously of $C_0(X)$, $D_0(X)$.

Let $\rho : C(X)\otimes C(Y)\to D(X\times Y)$ be a chain map. We shall say that ρ is *faithful* if $(f_0\otimes g_0)\rho = (f\times g)_0$, where $t_0(f\times g) = (t_0 f, t_0 g)\in X\times Y$. Let $k : X\to X_1$, $l : Y\to Y_1$ be arbitrary maps. Then there are chain maps, which we shall write $k\otimes l$, $k\times l$, where

$$k\otimes l : C(X)\otimes C(Y)\to C(X_1)\otimes C(Y_1),\quad k\times l : D(X\times Y)\to D(X_1\times Y_1).$$

A collection $\{\rho^{X,Y}\}$ of chain maps $\rho^{X,Y} : C(X)\otimes C(Y)\to D(X\times Y)$ is *allowable* if each $\rho^{X,Y}$ is faithful and for all maps k, l the *naturality* condition $\rho^{X,Y}(k\times l) = (k\otimes l)\rho^{X_1,Y_1}$ holds.

8.7.1 Theorem. (i) *There exists an allowable set of chain maps* $\rho^{X,Y}$;

(ii) *any allowable set consists of chain equivalences;*

(iii) *any two allowable sets are connected by an allowable set of chain homotopies.*

In (iii) a set of chain homotopies is *allowable* if it is natural in the obvious sense.

As the reader will expect, this is proved by an 'acyclic model' argument. For simplicity of notation we take $C(X) = D(X) = \Delta(X)$. Then define $\Delta(X\otimes Y)$ to be $\Delta(X)\otimes\Delta(Y)$ so that

$$\Delta_n(X\otimes Y) = \sum_{p+q=n}\Delta_p(X)\otimes\Delta_q(Y).$$

We define $\rho^{X,Y}$ inductively with respect to n, using the acyclicity of $\tilde{\Delta}(t_p\times t_q)$. We remark that $\Delta(X\otimes Y)$ may be augmented by taking $\tilde{\Delta}_{-1}(X\otimes Y) = J$ and $(f_0\otimes g_0)\tilde{\partial} = 1$. Then if ρ is faithful it is certainly proper. The details of the argument may be provided by any reader familiar with the proof of 8.4.3. Thus (i) is proved.

If $\{\rho\}$ and $\{\sigma\}$ are any two allowable sets of chain maps

$$\Delta(X\otimes Y)\to\Delta(X\times Y)$$

then we define inductively an allowable set of chain homotopies $\{\Lambda\}$ connecting them, starting with $\Lambda_0 = 0$. Again the details may be suppressed in the light of 8.4.3, and (iii) is proved.

Now we consider sets of chain maps $\{\bar{\rho}^{X,Y}\}$, where

$$\bar{\rho}^{X,Y} : \Delta(X\times Y)\to\Delta(X\otimes Y).$$

We say that $\bar{\rho}$ is *faithful* if $(f\times g)_0\bar{\rho} = f_0\otimes g_0$, where $f : t_0\to X$, $g : t_0\to Y$, and that the set $\{\bar{\rho}^{X,Y}\}$ is *allowable* if each $\bar{\rho}$ is faithful and the evident naturality condition is satisfied. Then the existence of an allowable set $\{\bar{\rho}^{X,Y}\}$ is established by induction with respect to dimension; in defining $\bar{\rho}$ on $\Delta_n(X \times Y)$, we consider the element

$$i_n \times i_n \in \Delta_n(t_n \times t_n)$$

and use the acyclicity of $\tilde{\Delta}(t_n \otimes t_n)$ (an immediate consequence of 5.7.5). In addition, we may prove any two allowable sets $\{\bar{\rho}\}$ and $\{\bar{\sigma}\}$ to be connected by an allowable set of chain homotopies $\{\bar{\Lambda}\}$; the set $\{\bar{\Lambda}\}$ is defined inductively starting with $\bar{\Lambda}_0 = 0$.

The reader will now be in no doubt as to how the argument is completed: $\{\rho^{X,Y}\bar{\rho}^{X,Y}\}$ and $\{1^{X,Y}\}$ are two allowable sets of chain maps $\Delta(X \otimes Y) \to \Delta(X \otimes Y)$, where the meaning of 'allowable' is clear, and they are evidently connected by an allowable set of chain homotopies. Similarly $\{\bar{\rho}^{X,Y}\rho^{X,Y}\}$ and $\{1^{X,Y}\}$ are connected by an allowable set of chain homotopies and (ii) is proved. The reader may further supply the modifications in the argument if C and D take different values among Δ, Δ^N, $\square^N$. ▮

8.7.2 Corollary. *There are canonical isomorphisms*

$$\rho_* : H_*(C(X \otimes Y)) \cong H_*(D(X \times Y)). \;▮$$

8.7.3 Proposition. *Let $C = \Delta$, Δ^N or $\square^N$; then*

$$(1 \otimes \rho^{Y,Z})\,\rho^{X,Y\times Z} \simeq (\rho^{X,Y} \otimes 1)\,\rho^{X\times Y,Z} : C(X \otimes Y \otimes Z) \to C(X \times Y \times Z).$$

Obviously Theorem 8.7.1 can be extended to any (finite) number of spaces $X, Y, Z, \ldots$, and allowable sets $\{\rho^{X,Y,Z,\ldots}\}$ of chain equivalences obtained, satisfying the analogue of (iii). In particular $(1 \otimes \rho^{Y,Z})\,\rho^{X,Y\times Z}$ and $(\rho^{X,Y} \otimes 1)\,\rho^{X\times Y,Z}$ are allowable sets of chain equivalences $C(X \otimes Y \otimes Z) \to C(X \times Y \times Z)$ and the proposition follows from (iii) ▮

8.7.4 Proposition. *Let $C = \Delta$, Δ^N or $\square^N$ and let*

$$\mu : C(X \otimes Y) \to C(Y \otimes X)$$

be the chain isomorphism of 5.7.3. *Then if $m : X \times Y \to Y \times X$ is given by $(x, y)\,m = (y, x)$, the diagram*

$$\begin{array}{ccc} C(X \otimes Y) & \xrightarrow{\mu} & C(Y \otimes X) \\ \downarrow{\scriptstyle \rho} & & \downarrow{\scriptstyle \rho} \\ C(X \times Y) & \xrightarrow{m} & C(Y \times X) \end{array}$$

is homotopy-commutative.

Here we use m for the chain map induced by m. We observe that $\{\rho^{X,Y}\}$ and $\{\mu^{X,Y}\rho^{Y,X}m^{Y,X}\}$ are allowable sets of chain maps

$C(X \otimes Y) \to C(X \times Y)$, so that, by 8.7.1, $\rho^{X,Y} \simeq \mu^{X,Y}\rho^{Y,X}m^{Y,X}$. Thus $\rho^{X,Y}m^{X,Y} \simeq \mu^{X,Y}\rho^{Y,X}$. ∎

A very convenient set of chain equivalences

$$\rho^{X,Y} : \square^N(X \otimes Y) \to \square^N(X \times Y)$$

may be given.‡ Let f_p be a singular p-cube of X, g_q a singular q-cube of Y. We define a singular $(p+q)$-cube, $f_p \times g_q$, of $X \times Y$ by

8.7.5 $(u_1, \ldots, u_{p+q})(f_p \times g_q) = ((u_1, \ldots, u_p)f_p, (u_{p+1}, \ldots, u_{p+q})g_q)$.

It is immediately clear that $f_p \times g_q$ is degenerate if f_p or g_q is degenerate, so that 8.7.5 induces a homomorphism

$$\rho^{X,Y} : \square^N(X \otimes Y) \to \square^N(X \times Y).$$

A straightforward computation shows that $\rho^{X,Y}$ is a chain map; moreover, it is evident that $\{\rho^{X,Y}\}$ is allowable.

We close by pointing out that 8.7.1, together with 8.6.7 and 5.7.9, enables us to compute the homology groups of topological products of compact polyhedra in a finite number of steps. For if $|K|$, $|L|$ are polyhedra and κ, λ are orientations of K, L, we have a 'chain' of chain equivalences

$$C^\kappa(K) \otimes C^\lambda(L) \xleftarrow{\theta^\kappa \otimes \theta^\lambda} C^\Omega(K) \otimes C^\Omega(L) \xrightarrow{\alpha^K \otimes \alpha^L}$$
$$\Delta(|K| \otimes |L|) \xrightarrow{\rho^{|K|,|L|}} \Delta(|K| \times |L|).$$

In fact $|K| \times |L|$ is again a polyhedron, but we do not use this fact, since we should calculate the homology group of $|K| \times |L|$ by calculating that of $C^\kappa(K) \otimes C^\lambda(L)$, using the Künneth formula 5.7.17.

8.8 The singular theory of n-connected spaces

By normalization we achieve a considerable reduction in the number of singular simplexes. In principle it is possible to reduce the number of singular simplexes to the minimum consistent with retaining the homology groups unaffected. The structure of such a *minimal complex* depends in an explicit way on the homotopy groups of the space. In this section we do not describe the minimal complex, but content ourselves with pointing out how the singular complex may be further reduced (i.e. beyond throwing out the degenerate simplexes) on the assumption that the space is n-connected §.

‡ In this respect singular cubes have a distinct advantage over singular simplexes.
§ The definition of n-connectedness is given after II. 1.33.

The singular theory we shall describe is really a relative theory for pairs (X, x_*) consisting of a space and base-point; in such a theory maps should be understood to map base-point to base-point.

We shall discuss in detail the singular theory of n-connected spaces based on normalized singular simplexes; the reader will easily provide the necessary modifications if $\Delta(X)$ or $\square^N(X)$ is in question. We say that the singular p-simplex f_p of X, x_* is *m-trivial* if $t'_m f = x_*$ for every m-dimensional face, t'_m, of t_p. Obviously $V^i f$ is m-trivial if f is m-trivial and thus the m-trivial simplexes generate a subcomplex,

$$^m\Delta(X, x_*) = {^m\Delta(X)},$$

of $\Delta(X)$. We define

8.8.1 $$^m\Delta^N(X) = {^m\Delta(X)}/{^m\Delta(X)} \cap \Delta^\flat(X).$$

Then we may regard $^m\Delta^N(X)$ as a subcomplex of $\Delta^N(X)$; let i be the injection.

8.8.2 Theorem. *If X is n-connected, then $i : {^n\Delta^N(X)} \to \Delta^N(X)$ is a chain equivalence.*

For the proof we give a procedure for deforming each singular simplex of X into an n-trivial singular simplex. Now a deformation of an image of t_p in X is a map $H : I \times t_p \to X$, and with this is associated in $(1, 1)$ correspondence a map $\bar{H} : t_p \to X^I$, given by the rule that $u(z\bar{H}) = (u, z)H$ for each $u \in I$ and $z \in t_p$ (see II. 2.2). For any $u \in I$ there is a map $b^u : X^I \to X$ given by the rule that, for each $w \in X^I$, $(w)b^u = (u)w$; we shall use b^0 and b^1 in particular and the fact that $b^0 \simeq b^1$ (see II. 2.4). The significance for us of the map b^u is clear from the fact that if $f_0 \overset{H}{\simeq} f_1$ then $\bar{H}b^u = f_u$; so in particular if H is a constant homotopy, $\bar{H}b^u$ is independent of u.

Instead of associating with each $f : t_p \to X$ a homotopy

$$H(f) : I \times t_p \to X$$

deforming f to an n-trivial map, we may associate with it the corresponding $\bar{H}(f) : t_p \to X^I$. This will prove convenient for two reasons: first the conditions we shall need to impose on the homotopies are more conveniently stated in terms of $\bar{H}$, and secondly we shall then pass to a chain map from $\Delta(X)$ to $\Delta(X^I)$ that is immediately describable in terms of $\bar{H}$. For this latter reason $\bar{H}$ must commute with the process of taking the ith face and because we are dealing with the normalized case it must also commute with the degeneracy operations.

We shall give an inductive definition of $\bar{H}$ subject to conditions (i)–(v) that follow, the motivation for which should now be clear.

We make now the inductive hypothesis that, for each $p < q$ and for each $f : t_p \to X$, $\bar{H}(f) : t_p \to X^I$ has been defined so that

(i) $\bar{H}(f)\, b^0 = f$;

(ii) $\bar{H}(f)\, b^1$ is n-trivial;

(iii) $V^i \bar{H}(f) = \bar{H}(V^i f)$, $0 \leqslant i \leqslant p$;

(iv) if $f = T^j g$, $0 \leqslant j \leqslant p-1$, $\bar{H}(f) = T^j \bar{H}(g)$;

(v) if f is n-trivial, $\bar{H}(f)\, b^u$ is independent of u.

Conditions (i) and (ii) say that $H(f)$ is a homotopy from f to an n-trivial map; conditions (iii) and (iv) ensure commutation with face and degeneracy operations; condition (v) is included to ensure that, if f is already n-trivial, the homotopy does not disturb it.

If $q = 0$, these conditions are vacuously satisfied. We proceed now to define $\bar{H}$ for each $f : t_q \to X$ in such a way that conditions (i)–(v) are satisfied with $p = q$.

In view of (iv) we have no choice as to the value of $\bar{H}(f)$ if $f = T^j g$ for $g : t_{q-1} \to X$. Perhaps also for some $k \neq j$ and some $h : t_{q-1} \to X$, $f = T^k h$; we shall leave it to the reader to verify that in that case $T^j \bar{H}(g) = T^k \bar{H}(h)$, so that $\bar{H}(f)$ is unambiguously defined. We observe also that this definition of $\bar{H}(f)$ when $f = T^j g$ satisfies (i), (ii), (iii) and (v); it satisfies (i) and (ii) fairly obviously, (iii) in view of the commutation rules 8.2.5 and (v) because, if $T^j g$ is n-trivial, so is g.

In view of (v) and (i) we have again no choice if f is n-trivial; $H(f)$ must be a constant homotopy in X and its initial map must be f. $\bar{H}(f)$ so defined plainly satisfies (ii); it satisfies (iii) since a face of an n-trivial simplex is itself n-trivial, and we have already observed that it satisfies (iv).

If now $f : t_q \to X$ is neither degenerate nor n-trivial we have a choice to make in defining $\bar{H}(f)$. We have, in fact, to choose a homotopy $H(f) : I \times t_q \to X$ such that

(i)′ the initial map is f;

(ii)′ the final map maps t_q^n, the n-section of t_q, to x_*;

(iii)′ the homotopy restricted to $I \times t_{q-1}^i$, $0 \leqslant i \leqslant q$, is given by $H(V^i f)$; more precisely $(1 \times V^i)\, H(f) = H(V^i f) : I \times t_{q-1} \to X$.

These restrictions on H provide partial maps of closed pieces of $(I \times t_q)^{\boldsymbol{\cdot}}$ and they agree on the intersections of these pieces. The restrictions (i)′ and (iii)′ agree on $(0 \times t_{q-1}^i)$, since by hypothesis $\bar{H}(V^i f)\, b^0 = V^i f$; the restrictions (ii)′ and (iii)′ agree on $(1 \times (t_{q-1}^i)^n)$, since by hypothesis $\bar{H}(V^i f)b^1$ is n-trivial; finally, the various restrictions included in (iii)′ agree in view of the commutativity laws 8.1.4. They combine therefore to give a map of $(0 \times t_q) \cup (I \times \dot{t}_q) \cup (1 \times t_q^n)$.

If $q \leqslant n$, $t_q^n = t_q$ and $H \mid (I \times t_q)^{\cdot}$ is completely prescribed; since X is n-connected this map of the frontier can be extended to the interior and we select one such extension as $H(f)$. If on the other hand $q > n$, $t_q^n \subseteq \dot{t}_q$ and H is prescribed on $(0 \times t_q) \cup (I \times \dot{t}_q)$; since this subset of $I \times t_q$ is a retract of $I \times t_q$ (compare 1.6.10), the given partial map can be extended to a map of $I \times t_q$ and we choose one such extension as $H(f)$. We have now defined $H(f)$ for each $f : t_q \to X$ in such a way that $\bar{H}(f)$ satisfies conditions (i)–(v). This shows that $\bar{H}(f)$ may be defined for all p and all maps $f : t_p \to X$ to satisfy (i)–(v). This completes the first and geometrical stage of the proof.

We now use $\bar{H}$ to define a chain map $\phi^\Delta : \Delta(X) \to \Delta(X^I)$. For each p we define $\phi_p^\Delta : \Delta_p(X) \to \Delta_p(X^I)$ by the rule that $f_p \phi_p^\Delta = (\bar{H}(f))_p$. These homomorphisms satisfy

(i)″ $\phi^\Delta b^{0\Delta} = 1$;
(ii)″ $\Delta_p(X)\phi^\Delta b^{1\Delta} \subseteq {}^n\Delta_p(X)$;
(iii)″ $\phi_p^\Delta F^i = F^i \phi_{p-1}^\Delta$, $0 \leqslant i \leqslant p$;
(iv)″ $\phi_{p-1}^\Delta D^j = D^j \phi_p^\Delta$, $0 \leqslant j \leqslant p-1$;
(v)″ if f is n-trivial, $f_p \phi^\Delta b^{1\Delta} = f_p \phi^\Delta b^{0\Delta} = f_p$.

We deduce from (iii)″ that ϕ^Δ is a chain map and from (iv)″ that it is a chain map $\phi^\Delta : \Delta(X), \Delta^\flat(X) \to \Delta(X^I), \Delta^\flat(X^I)$, and therefore induces a chain map $\phi^N : \Delta^N(X) \to \Delta^N(X^I)$. We now define

$$\gamma : \Delta^N(X) \to {}^n\Delta^N(X)$$

by the rule that, for each $c \in \Delta^N(X)$, $c\gamma = c\phi^N b^{1N}$. The distinction between γ and $\phi^N b^{1N}$ lies only in the chain complex that is the range of the map; in fact $\gamma i = \phi^N b^{1N} : \Delta^N(X) \to \Delta^N(X)$; since $\phi^N b^{1N}$ is a chain map so is γ.

We shall complete the proof now by showing that γ is a chain inverse to i, so that i will then have been proved a chain equivalence. First we have

$$\gamma i = \phi^N b^{1N} \simeq \phi^N b^{0N}, \quad \text{since} \quad b^1 \simeq b^0, \text{ (8.4.11 (i))}$$

$$= 1, \quad \text{since} \quad \phi^\Delta b^{0\Delta} = 1.$$

Then, if f is n-trivial, $f_p i\gamma = f_p \phi^N b^{1N} = f_p$, by (v)″. Hence $i\gamma = 1$. ∎

We shall feel free to use this theorem with $\Delta(X)$ or $\square^N(X)$ replacing $\Delta^N(X)$; the proofs for these cases are less troublesome than the proof we have given. In fact, we make an application now to path-connected spaces. If X is 0-connected, we may confine attention to 0-trivial cubes of X. Indeed, a map $f : e_1, \dot{e}_1 \to X, x_*$ represents a 1-cycle of ${}^0\square^N(X)$ and $Z_1({}^0\square^N(X))$ is thus generated by the 0-trivial 1-cubes. We will write I for e_1. Then $f : I, \dot{I} \to X, x_*$ represents an element of $\pi_1(X, x_*)$ and an element of $Z_1({}^0\square^N(X))$. Suppose $f \simeq f' \operatorname{rel} \dot{I}$; then there is a map $F : I \times I, \dot{I} \times I \to X, x_*$ with

$$(u_1, 0)F = u_1 f, \quad (u_1, 1)F = u_1 f', \quad (0, u_2)F = x_*, \quad (1, u_2)F = x_*.$$

We may regard F as a 0-trivial 2-cube; then $F\partial = f - f'$. It follows that $[f] \to \{f\}$, where [ ] refers to homotopy class and { } to homology class, is a map

$$\theta : \pi_1(X, x_*) \to H_1^s(X).$$

8.8.3 Theorem. *If X is 0-connected, θ is a homomorphism of $\pi_1(X, x_*)$ onto $H_1^s(X)$ whose kernel is the commutator subgroup of $\pi_1(X, x_*)$.*

Let $f^0, f^1 : I, \dot{I} \to X, x_*$ and in the notation of 6.1 let $f = (f^0 * f^1)_1$. Map $(I \times I)^{\cdot}$ to X by the rule $(0, u_2) \to u_2 f$, $(1, u_2) \to u_2 f^1$, $(u_1, 0) \to u_1 f^0$, $(u_1, 1) \to x_*$. Clearly this map is null-homotopic and so may be extended to $F : I \times I \to X$. Computing $F\partial$ in $\square^N(X)$, we find $F\partial = f^1 - f + f^0$. Thus $\{f^0 * f^1\} = \{f^0\} + \{f^1\}$ and θ is a homomorphism. Evidently θ is onto $H_1^s(X)$, which is generated by the classes $\{f\}$, and θ annihilates the commutator subgroup of $\pi_1(X, x_*)$, since $H_1^s(X)$ is abelian.

Finally, suppose that $f : I, \dot{I} \to X, x_*$ is null-homologous, say $f = \left(\sum_{i=1}^{k} n_i g_i\right)\partial$ in $\square^N(X)$, where g_i is a 0-trivial 2-cube of X. Put $V^{j\epsilon} g_i = g_i^{j\epsilon}$, $j = 1, 2$; $\epsilon = 0, 1$. Then $g_i \partial = -g_i^{10} + g_i^{11} + g_i^{20} - g_i^{21}$ and $(g_i^{10})^{-1} * g_i^{20} * g_i^{11} * (g_i^{21})^{-1}$ is null-homotopic. We write h_i for this loop. Then $f' = f * h_1^{-n_1} * \ldots * h_k^{-n_k}$ is a loop in the same class as f; either $f' = e$, a constant loop at x_*, or it may be written in the form $f_{i_1}^{w_1} * \ldots * f_{i_t}^{w_t}$, where each f_{i_j} is non-degenerate and each $w_j = \pm 1$ and $\sum_{j=1}^{t} w_j f_{i_j} = 0$ in $\square^N(X)$. Since $\square_1^N(X)$ is free abelian, it follows that each f_i occurs in $f_{i_1}^{w_1} * \ldots * f_{i_t}^{w_t}$ with total exponent zero. Thus in any case $\{f'\}$ is in the commutator subgroup of $\pi_1(X, x_*)$ and the theorem is proved. ∎

There is a generalization of 8.8.3 which is of fundamental importance in linking homology and homotopy theory; in particular, it provides one of the main steps in the modern techniques for computing homotopy groups.

Let X be a path-connected space and let $e_n = I^n$ be oriented as in II. 1; consider a map $f : e_n, \dot{e}_n \to X, x_*$. Then f determines an element α of $\pi_n(X, x_*)$ and an element z of $Z_n({}^0\square^N(X))$. As in the case of $n = 1$ it is easy to see that the homology class of z depends only on α so that there is defined in this way a map

8.8.4. $$\theta : \pi_n(X, x_*) \to H_n(X).$$

We must next prove that θ is a homomorphism. Let $f, g, h : e_n, \dot{e}_n \to X, x_*$ be maps such that $[f]+[g]+[h] = 0 \in \pi_n(X, x_*)$; we shall prove that $f_n + g_n + h_n \sim 0$ in ${}^0\square^N(X)$. In $\dot{e}_{n+1}$ choose three faces coherently oriented with e_{n+1}, say the images of $V^{2,0}$, $V^{1,1}$, $V^{3,1}$; map these to X by f', g', h' such that $f = V^{2,0}f'$, $g = V^{1,1}g'$, $h = V^{3,1}h'$, and map the remaining faces of $\dot{e}_{n+1}$ to x_*. These maps combine to give a map $k : \dot{e}_{n+1} \to X$ under which the $(n-1)$-section is mapped to x_*. The homotopy addition property II. 1.32 ensures that

$$[k] = [f'] + [g'] + [h'] = [f] + [g] + [h] = 0;$$

so that k can be extended to $\bar{k} : e_{n+1} \to X$. Certainly $\bar{k}_{n+1} \in {}^0\square(X)$ and

$$\bar{k}_{n+1}\partial - f_n - g_n - h_n \in {}^0\square^{\flat}(X).$$

Consequently $[f]\theta + [g]\theta + [h]\theta = 0$ and θ is therefore homomorphic. The homomorphism θ is called the *Hurewicz homomorphism*. A cycle whose homology class is in the image of θ is called a *spherical cycle*, and its class a *spherical homology class*.

8.8.5 Theorem. (*Hurewicz isomorphism theorem*). *If X is $(n-1)$-connected, $n > 1$, then θ is an isomorphism in dimension n.*

If X is $(n-1)$-connected we may confine attention to $(n-1)$-trivial cubes of X. Then plainly an $(n-1)$-trivial n-cube of X is just a map $e_n, \dot{e}_n \to X, x_*$ so that θ is epimorphic. To prove θ monomorphic we modify the argument of 8.8.3. If f is an $(n-1)$-trivial n-cube of X and $f = (\Sigma n_i g_i)\partial$, where g_i is an $(n-1)$-trivial $(n+1)$-cube, then for each i we consider the homotopy elements represented by the faces of g_i and use II. 1.32 to show that their sum (with suitable signs) is zero. It then follows that the homotopy element represented by f is zero. ∎ We emphasize that in this book we have only proved the Hurewicz isomorphism theorem modulo the Brouwer-Hopf theorem (II. 1.20).

8.8.6 Corollary. *If X is 1-connected and $H_r(X) = 0$ for $0 < r < n$, then $\pi_n(X) \cong H_n(X)$.*

Here the conclusion is that of 8.8.5, but the assumptions are formally different. The equivalence of the two sets of assumptions follows from 8.8.5. ∎

8.8.7 Corollary. *If K is a 1-connected acyclic complex, $|K|$ is contractible.*

This is a consequence of 8.8.6 and 7.3.3. ∎

For future reference we state without proof

8.8.8 Theorem. (*Relative Hurewicz isomorphism theorem*). *If $X_0 \subseteq X$ and $x_* \in X_0$, there is a canonical homomorphism $\hat{\theta} : \pi_n(X, X_0, x_*) \to H_n(X, X_0)$. If X and X_0 are both 1-connected and $\pi_r(X, X_0, x_*)$ is null for $r < n$, then $\hat{\theta}$ is isomorphic in dimension n.* ∎

8.9* Singular homology with local coefficients

In 7.6 we defined a local system of groups on a path-connected space and indicated how such a system on a polyhedron determines homology and contrahomology groups of an underlying complex with the local system as coefficients. Here we describe the corresponding singular homology groups for an arbitrary path-connected space and discuss an important application. We shall content ourselves with defining homology groups, confidently leaving the definition of the contrahomology groups to the reader.

Let $\{G_x\}$, $x \in X$, be a local system of groups on the path-connected space X. For any singular p-simplex $f: t_p \to X$, let us call $v^0 f \in X$ the leading vertex of f and let us write w^f for the f-image of the edge $v^0 v^1$ of t_p. By $\Delta_p(X; \{G_x\})$ we understand the set of formal finite sums $\sum_i g_i f_p^i$, where $g_i \in G_{v^0 f^i}$, with the obvious additive group structure. A boundary operator $\partial: \Delta_p(X; \{G_x\}) \to \Delta_{p-1}(X; \{G_x\})$ is defined by linearity from the rule

8.9.1 $$(gf_p)\partial = (g w_G^f)(V^0 f)_{p-1} + \sum_{i=1}^{p} (-1)^i g (V^i f)_{p-1}.$$

A straightforward computation (compare 7.6) shows that $\partial\partial = 0$ if $p > 1$. We thus obtain a chain complex $\Delta(X; \{G_x\})$ and the homology groups of this complex are called *the singular homology groups of X with local coefficients* $\{G\}$. We shall not detail the development of the local coefficient theory, since it would consist largely of automatic generalizations of results proved in this chapter. There are, however, points which we wish to bring out.

We first remark that we could use the normalized chain groups Δ_p^N instead of Δ_p and obtain isomorphic homology groups. This follows from the fact that our 'acyclic models' t_p are in fact simply connected, so that the local coefficients on t_p are trivial and the (reduced) homology groups of t_p, with local coefficient groups, vanish.

If we use cubes instead of simplexes then, of course, we must modify 8.9.1. We give details here since singular cubical homology with local coefficients is used in chapter 10. For any singular p-cube $f: e_p \to X$ let ${}^i w^f$ be the f-image of the edge of e_p from the origin O along the u_i-axis. By $\square_p(X; \{G_x\})$ we mean the set of formal finite sums $\sum g_i f_p^i$, where $g_i \in G_{Of^i}$, with the obvious additive abelian group structure. A boundary operator $\partial: \square_p(X; \{G_x\}) \to \square_{p-1}(X; \{G_x\})$ is defined by linearity from the rule

8.9.2 $$(gf_p)\partial = \sum_{i=1}^{p} (-1)^i g (V^{i0} f)_{p-1} - \sum_{i=1}^{p} (-1)^i (g\, {}^i w_G^f)(V^{i1} f)_{p-1}.$$

Again a straightforward computation (which we suppress) shows that $\partial\partial = 0$ if $p > 1$. Now suppose that $f = T^j h$ for some $h: e_{p-1} \to X$; then clearly $V^{i\epsilon} f$ is degenerate if $i \neq j$; moreover, ${}^j w^f$ is the point path at Of, so that $g\, {}^j w_G^f = g$. This shows that $\square^\flat(X; \{G_x\})$ is stable under ∂ so that, by passing to quotients, we obtain a chain complex $\square^N(X; \{G_x\})$ whose homology groups are naturally isomorphic with those of $\Delta(X; \{G_x\})$; we shall write $H_*(X; \{G\})$ for these homology groups.

Before describing the application to which we have referred we make a remark about local systems of groups. It is clear from 7.6.1 that a local system $\{G_x\}$ determines $\pi_1(X, x_*)$ as a group of right operators‡ on G_{x_*} for each x_*. Suppose, conversely, that, for a fixed x_* and abelian group G, we are given a rule whereby $\pi_1(X, x_*)$ acts as a group of right operators on G. For each $x \in X$ we choose a path $w(x)$ from x_* to x and to each path v from x to x' we associate the automorphism of G induced by $\{w(x) * v * \overline{w(x')}\} \in \pi_1(X, x_*)$. If we call this automorphism v_G, it is plain that conditions (i) and (ii) for a local system of groups are satisfied if we define $G_x = G$ for all $x \in X$. Clearly the choice of paths $w(x)$ does not affect the resulting homology groups (up to isomorphism); indeed, this choice is analogous to the choice of particular isomorphisms $G_{x_*} \cong G_x$ whereby to identify each G_x with G_{x_*}. We call $H_*(X; \{G\})$ *the homology groups with local coefficients in the* $\pi_1(X, x_*)$*-module* G. This notion turns up in chapter 10, where we also consider the case in which $\pi_1(X, x_*)$ operates on the *left* on G. For such an operation on the left we there define an operation on the right by the rule

$$g\xi = \xi^{-1}g, \quad g \in G, \ \xi \in \pi_1(X, x_*).$$

We now come to our application. We shall consider a fibre map $p: X \to B$ with fibre $F = p^{-1}(b_*)$, where B is path-connected and show how $\pi_1(B, b_*)$ operates on $H_*(F; G)$, where G is an *ordinary* coefficient group. As remarked above, this enables us to define a chain complex $\square^N(B; \{H_*(F; G)\})$ and homology groups $H_*(B; \{H_*(F; G)\})$. These homology groups play a central role in the theory described in 10.4–10.6.

Let $l: I \to B$ be a loop in B on b_*; we proceed to define a function $P_l: \square(F) \to \square(X)$, mapping p-cubes to $(p+1)$-cubes, with the properties

(i) $(y, 0)(fP_l) = yf$, $y \in e_p$;

(ii) $(y, t)(fP_l)p = tl$, $y \in e_p, t \in I$;

(iii) $V^{i\epsilon}(fP_l) = (V^{i\epsilon}f)P_l: e_p \to X$, $1 \leqslant i \leqslant p$;

(iv) $T^i(gP_l) = (T^i g)P_l: e_{p+1} \to X$, $1 \leqslant i \leqslant p$.

Here f is an arbitrary p-cube and g an arbitrary $(p-1)$-cube. We proceed by induction on p, supposing P_l defined on cubes of dimension $< p$,

‡ The precise definition of a group of operators may be found, if required, at the start of 10.7.

where $p \geqslant 0$. Let f be a p-cube; if it is degenerate, then fP_l is determined by condition (iv). If f is non-degenerate, we define $fP_l : e_{p+1} \to X$ on $(e_p \times 0) \cup (\dot{e}_p \times I)$ by means of (i) and (iii); call this partial map $h' : (e_p \times 0) \cup (\dot{e}_p \times I) \to X$. The map $l' : e_p \times I \to B$ given by $(y, t)\, l' = tl$, $y \in e_p$, is then an extension of $h'p$. The fundamental lifting property (see II. 3) for fibre maps ensures the existence of a map $fP_l : e_p \times I \to X$ extending h' and covering l' (in the sense that $(fP_l)\, p = l'$). Leaving the reader to fill in the details, we assert that this establishes the induction so that the function P_l may be defined. We may regard P_l as providing a systematic procedure for transporting cubes of F round the loop l of B. Given P_l we define $Q_l : \square(F) \to \square(F)$ by

8.9.3 $$y(fQ_l) = (y, 1)\,(fP_l).$$

Thus fQ_l is 'the image of f after transport round the loop l'. We may refer to P_l as an *l-transporter* and to Q_l as an *l-transport*.

8.9.4 Proposition. *Q_l is a chain map and $\square^\flat(F)\, Q_l \subseteq \square^\flat(F)$.*

This follows immediately from properties (iii) and (iv) of P_l. ∎

We deduce that Q_l induces a chain map

$$Q_l^N : \square^N(F; G) \to \square^N(F; G).$$

8.9.5 Proposition. *If $l \simeq m$, then $Q_l^N \simeq Q_m^N$.*

The reader should notice that we have not asserted the uniqueness of P_l. Thus 8.9.5 includes the substantial assertion that Q_{l*} depends only on l.

By hypothesis there is a map $T : I \times I \to B$ with $(t, 0)\, T = tl$, $(t, 1)\, T = tm$, $(0, u)\, T = (1, u)\, T = b_*$. We proceed by induction on p to construct a function H, mapping p-cubes of F to $(p+2)$-cubes of X with the properties

(i) $(y, 0, u)\,(fH) = yf$; $(y, t, 0)\,(fH) = (y, t)\,(fP_l)$;
$(y, t, 1)\,(fH) = (y, t)\,(fP_m)$;

(ii) $(y, t, u)\,(fH)\, p = (t, u)\, T$;

(iii) $V^{i\epsilon}(fH) = (V^{i\epsilon} f)\, H$;

(iv) $T^i(gH) = (T^i g)\, H$.

We suppose H constructed on cubes of dimension $< p$ and consider a p-cube f. If f is degenerate we use (iv) to define fH; if f is non-degenerate we use (i) and (iii) to define fH on

$$(e_p \times ((0 \times I) \cup (I \times \dot{I}))) \cup (\dot{e}_p \times I \times I).$$

Arguing as in the construction of P_l we extend this partial map to a map $fH : e_p \times I \times I \to X$ covering the map $T' : e_p \times I \times I \to B$ given by $(y, t, u)\, T' = (t, u)\, T$. Thus the induction is established.

We use H to define a chain homotopy $K : \square(F) \to \square(F)$ by the rule

$$fK = (-1)^p J,$$

where $(y, t)\, J = (y, 1, t)\, (fH)$. A simple computation shows that

$$K\partial + \partial K = Q_m - Q_l,$$

so that $Q_l \simeq Q_m$. Moreover, property (iv) of H ensures that

$$\square^\flat(F)\, K \subseteq \square^\flat(F).$$

Thus we may pass to the normalized complexes and deduce that $Q_l^N \simeq Q_m^N$. ∎

8.9.6 Proposition. *If l, m are loops in B on b_* of length 1 then $Q_l Q_m$ is‡ an $(l *_1 m)$-transport.*

Let $g = fQ_l$ and define $k : e_p \times I \to X$ by

$$\begin{aligned}(y, t)\, k &= (y, 2t)\, (fP_l), & 0 \leqslant t \leqslant \tfrac{1}{2},\\ &= (y, 2t - 1)\, (gP_m), & \tfrac{1}{2} \leqslant t \leqslant 1.\end{aligned}$$

Clearly k is a well-defined map and it is easy to verify that k satisfies conditions (i)–(iv) for $P_{l *_1 m}$. Since

$$(y, 1)\, k = (y, 1)\, (gP_m) = y(gQ_m) = y(fQ_l Q_m),$$

the proposition follows. ∎

8.9.7 Theorem. *$\pi_1(B, b_*)$ acts as a group of right operators on $H_*(F; G)$.*

This follows at once from 8.9.5 and 8.9.6. ∎

We remark that we have established that $\{H_*(F; G)\}$ constitutes a local system of groups in the form most suitable for the application in 10.4. We could however, just as easily have shown that the groups $\{H_*(p^{-1}b; G)\}$, $b \in B$, form a local system of groups in the sense of 7.6.

8.10 Appendix: Cech contrahomology theory

The singular homology theory is not the only homology theory for general spaces that has been devised; the most important other theory is the Čech theory. The Čech homology groups do not have, in this most general situation, the property (4) of 8.5; the homology sequence fails to be exact. The Čech contrahomology theory, on the other hand, has all the corresponding properties for arbitrary pairs of spaces and differs in general from the singular contrahomology theory (see Exercise 31). We here offer a highly descriptive account of this contrahomology theory; for a full treatment the reader is advised to consult Eilenberg and Steenrod [14], *Cartan Seminar Notes* [12], and C. H. Dowker, *Ann. Math.* **51** (1950), 278.

‡ We can, of course, define l-transporters for loops of any length.

If X is a space, a covering λ by open sets is a collection of open sets whose union is X; we shall be considering only coverings by open sets. With λ we associate N_λ, the *nerve of* λ, a (possibly infinite) simplicial complex; its vertices are the non-empty sets of λ and its simplexes are the finite collections of these sets which have non-empty intersection. We say that λ refines a covering μ ($\mu \prec \lambda$) if, to each set $U \in \lambda$, there is at least one set $V_U \in \mu$ such that $U \subseteq V_U$. When, for each U, V_U has been selected, we can define a simplicial map $\phi^{\lambda\mu} : N_\lambda \to N_\mu$ by the rule $(U)\phi^{\lambda\mu} = V_U$. It is easy to see that this vertex transformation is admissible and that a different selection $\{V'_U\}$ determines a simplicial map which is contiguous to $\phi^{\lambda\mu}$. If therefore λ refines μ, this relation determines canonical homomorphisms $\phi^{\lambda\mu}_* : H_*(N_\lambda) \to H_*(N_\mu)$ and

$$\phi^*_{\lambda\mu} : H^*(N_\mu) \to H^*(N_\lambda).$$

The relation $\prec$ is a partial ordering of the coverings; for, if $\lambda \prec \mu$ and $\mu \prec \nu$, then $\lambda \prec \nu$. Further, under these circumstances, $\phi^*_{\lambda\mu}\phi^*_{\mu\nu} = \phi^*_{\lambda\nu}$. It is also true that, for any pair λ, μ there is a covering that refines both λ and μ; we may take for ν the collection $\{U \cap V\}$ for all $U \in \lambda$ and $V \in \mu$. These two facts about the set of coverings are described by the statement that they form a *directed set* under $\prec$, and the established facts about the contrahomology groups by the statement that $\{H^*(N_\lambda), \phi^*_{\lambda\mu}\}$ form a *direct system of groups.*

We may form the *limit group* of any direct system of groups. In the present case we define an equivalence relation on the elements of the groups $\{H^*(N_\lambda)\}$ by the rule that $\alpha_\lambda \in H^*(N_\lambda)$ and $\beta_\mu \in H^*(N_\mu)$ are equivalent if and only if, for some ν, $\phi^*_{\nu\lambda}(\alpha_\lambda) = \phi^*_{\nu\mu}(\beta_\mu)$. It is easy to verify that this is, indeed, an equivalence relation. The equivalence classes are the elements of the limit group, and we define the sum of two classes to be the class containing the sum of a pair of elements, one from each class, belonging to the same contrahomology group; this determines the sum of two classes uniquely and, with this law of addition, the equivalence classes form an abelian group. This limit group is, by definition, the *Čech contrahomology group* $\check{H}^*(X)$. If we consider not $H^*(N_\lambda)$ but the contrahomology ring $R^*(N_\lambda)$ and proceed in this way, we can define a product between elements of $\check{H}^*(X)$ in terms of which the group becomes the *Čech contrahomology ring* $\check{R}^*(X)$.

We may fairly easily prove the consistency property for this ring, namely, that $\check{R}^*(|K|) \cong R^*(K)$ for a finite complex K. We remark that one covering for $|K|$ is provided by the stars of vertices of K and that the nerve of this star covering may be taken to be K itself. Further the star covering provided by K' is a refinement of this covering and any associated simplicial map $\phi : K' \to K$ is a standard map (see 1.7) which is known to induce a contrahomology ring isomorphism. Now the star coverings associated with $K^{(r)}$, $r = 0, 1, \ldots$, form a *cofinal* subset of the directed set of coverings of $|K|$; this means that, for any covering, there is an integer r such that λ is refined by the star covering of $K^{(r)}$. Any cofinal subset has itself a limit ring which is naturally isomorphic to the ring $\check{R}^*(|K|)$. This particular cofinal subset is totally ordered and each ϕ^* associated with it is an isomorphism, whence the limit group is isomorphic to $R^*(K)$, where we identify K with the nerve of its own star covering.

The definition we have given of $\check{R}^*(X)$ is virtually that given by Čech; however, there is another definition of the Čech contrahomology ring due to Alexander, the link between the two definitions being most elegantly forged by C. H. Dowker, *Ann. Math.* **56** (1952), 84. He observes that, given X and a covering λ by open sets $\{U\}$, N_λ has vertices $\{U\}$ and its simplexes are those finite subsets of $\{U\}$ having a common point $x \in X$. We may, however, consider a dually defined complex M_λ, whose vertices are the points $\{x\}$ of X and whose simplexes are those finite subsets of $\{x\}$ lying in a common $U \in \lambda$. He exploits the duality to prove that there is an isomorphism $i^*_\lambda : R^*(N_\lambda) \to R^*(M_\lambda)$. When $\mu \prec \lambda$, clearly M_λ is a subcomplex of M_μ and we may define the injection simplicial map $\theta^{\lambda\mu} : M_\lambda \to M_\mu$. An important property of the isomorphism i^*_λ is that $i^*_\lambda \phi^*_{\lambda\mu} = \theta^*_{\lambda\mu} i^*_\mu : R^*(N_\mu) \to R^*(M_\lambda)$. Now $\{R^*(M_\lambda), \theta^*_{\lambda\mu}\}$ form a direct system of rings and the isomorphisms $\{i^*_\lambda\}$ generate an isomorphism i^* between $\check{R}^*(X)$ and the limit ring of the system $\{R^*(M_\lambda), \theta^*_{\lambda\mu}\}$. We observe further that $\theta^{\lambda\mu}$ is unambiguously defined; we may therefore form a direct system of contrachain-ring-complexes $\{C^{\cdot}(M_\lambda); \theta^{\cdot}_{\lambda\mu}\}$ together with its limit contrachain-ring-complex. It can be proved that the limit ring of $\{R^*(M_\lambda)\}$ is the contrahomology ring of this limit complex.

Now M_λ is a subcomplex of the abstract complex M, whose vertices are the points of X and whose simplexes are all the finite subsets of these points. The contrachain-ring-complex $C^{\cdot}(M_\lambda)$ is the quotient ring of $C^{\cdot}(M)$ by the ideal $I_\lambda = C^{\cdot}(M, M_\lambda)$ of contrachains taking the value zero on all simplexes of M_λ. If $\mu \prec \lambda$, there is an injection $\psi_{\lambda\mu} : I_\mu \to I_\lambda$ and $\theta^{\cdot}_{\lambda\mu} : C^{\cdot}(M_\mu) \to C^{\cdot}(M_\lambda)$ is the quotient map induced by $1, \psi_{\lambda\mu} : C^{\cdot}(M),\ I_\mu \to C^{\cdot}(M), I_\lambda$; further the limit contrachain-ring-complex of $\{C^{\cdot}(M_\lambda); \theta^{\cdot}_{\lambda\mu}\}$ is isomorphic to the quotient of $C^{\cdot}(M)$ by the limit ideal I of the direct system $\{I_\lambda; \psi_{\lambda\mu}\}$ of ideals of $C^{\cdot}(M)$. Here, however, the limiting process can be easily seen to lead to the ideal $I = \bigcup_\lambda I_\lambda$.

We may now, therefore, give an alternative definition of $\check{R}^*(X)$ as being the contrahomology ring of the quotient contrachain-ring-complex

$$C^{\cdot}(M)/I = \check{C}^{\cdot}(X).$$

The elements of I admit an interesting geometrical description. We say that $c \in C^{\cdot}(M)$ is null on $X_0 \subseteq X$ if it takes the value zero on all simplexes of M all of whose vertices are points of X_0. The *support* $|c|$ of c is defined by the rule that $x \notin |c|$ if and only if c is null on some neighbourhood of x. Clearly this defines the complement of $|c|$ as an open set; $|c|$ is therefore closed. We now assert that a contrachain of $C^{\cdot}(M)$ belongs to I if and only if it has empty support. For, if c has empty support, for each $x \in X$ there is an open set $U_x \ni x$ such that c is null on U_x; this collection $\{U_x\}$ is a covering λ of X such that $c \in I_\lambda \subset I$. Conversely, if $c \in I$, for some λ, $c \in I_\lambda$; for any x, there is some $U_x \in \lambda$ such that $x \in U_x$; since $c \in I_\lambda$, c is null on U_x and $x \notin |c|$. As this holds for any x, c has empty support.

This definition of I could form the starting point for a discussion of Čech contrahomology. Most of the properties of the Čech theory follow more easily from this Alexander definition than from Čech's; an exception to this is the proof of consistency for polyhedra. The reader will find that some of the properties form the subject of Exercises 27–30 below.

EXERCISES

Natural equivalences in singular homology theory

1. Let r be a fixed integer and let $\square^{b,r}_p(X)$ be the subgroup of $\square_p(X)$ generated by those singular p-cubes of X which are degenerate on at least one of the last r coordinates. Prove that $\square^{b,r}(X)$ is a chain subcomplex of $\square(X)$. If $\square^{N,r}(X)$ is the quotient complex, prove that $\square^{N,r}(X) \simeq \square^N(X)$.

Carry out an analogous procedure for $\Delta(X)$.

2. Let $\Delta_n^0(X)$ be the intersection of the kernels of the homomorphisms $F_n^i : \Delta_n(X) \to \Delta_{n-1}(X)$, $i = 1, \ldots, n$. Show that F_n^0 maps $\Delta_n^0(X)$ into $\Delta_{n-1}^0(X)$ and that $F_n^0 F_{n-1}^0 | \Delta_n^0(X) = 0$. Let $\{H_n^0(X)\}$ be the homology groups of the chain-complex $\{\Delta_n^0(X), F_n^0\}$. Show that the inclusion map $\Delta_n^0(X) \to \Delta_n(X)$ is a chain equivalence (and so induces an isomorphism $H_n^0(X) \cong H_n(X)$).

Attaching cones, suspension, etc.

3. Let $g : X_1 \to X_2$ be a map. Let CA be attached to X_1 by $f_1 : A \to X_1$ and to X_2 by $f_1 g : A \to X_2$. Extend g to $g' : X_1 \cup_{f_1} CA \to X_2 \cup_{f_2} CA$ by defining g' to be the identity on $CA - A$. Prove that

$$g'_* : H_*(X_1 \cup_{f_1} CA, X_1) \to H_*(X_2 \cup_{f_2} CA, X_2)$$

is an isomorphism. Deduce that $g' : H_*(X_1 \cup_{f_1} CA) \cong H_*(X_2 \cup_{f_2} CA)$ if $g_* : H_*(X_1) \cong H_*(X_2)$.

4. Let (X, X_0) be a pair and let $X \cup CX_0$ be obtained from X by erecting a cone on X_0 (it is thus obtained by identification from the disjoint union of X and CX_0). Prove that $H_*(X, X_0) \cong \tilde{H}_*(X \cup CX_0)$. Deduce that if the pair (X, X_0) has the homotopy extension property (see Chap. 1, Exercises, Q. 12) and if X/X_0 is the space obtained from X by identifying X_0 to a point, then the identification map p induces a homology isomorphism

$$p_* : H_*(X, X_0) \cong \tilde{H}_*(X/X_0).$$

5. The *join* of X and Y is defined to be the space obtained from the disjoint union $X \cup (X \times Y \times I) \cup Y$ by the identifications

$$x = (x, y, 0), \quad y = (x, y, 1), \quad \text{all} \quad x \in X, \quad y \in Y.$$

Write $X*Y$ for the join. Show that X and Y are embedded in $X*Y$ as closed subspaces. By considering the space obtained from $X*Y$ by erecting cones on X and Y, or otherwise, prove that, if $n \geqslant 1$,

$$H_n(X*Y) \cong H_{n-1}(X \times Y, X \vee Y).$$

Hence compute the homology groups of the join of P^4 with P^2.

6. Let $C^{(m)}$ be complex projective space of m complex dimensions. Represent a point of $C^{(m)}$ by $(m+1)$ homogeneous complex coordinates $[z_0, z_1, \ldots, z_m]$ and represent a point of S^{2m+1} by $(m+1)$ complex coordinates $(z_0, z_1, \ldots, z_m)$ satisfying $z_0 \bar{z}_0 + \ldots + z_m \bar{z}_m = 1$. Let $f_m : S^{2m+1} \to C^{(m)}$ be the map given by

$$(z_0, z_1, \ldots, z_m) f_m = [z_0, z_1, \ldots, z_m].$$

Show that $C^{(m)}$ may be regarded as being obtained from $C^{(m-1)}$ by attaching the ball V^{2m} by means of the map f_{m-1}. Hence compute the homology and contra-homology groups of $C^{(m)}$.

Mayer-Vietoris sequence

7. (Whitehead-Barratt). Let

$$\begin{array}{ccccccccccc}
G_0 & \xrightarrow{\alpha_0} & G_1 & \xrightarrow{\alpha_1} & G_2 & \to \ldots \to & G_n & \xrightarrow{\alpha_n} & G_{n+1} & \to \ldots \\
\downarrow{\scriptstyle \phi_0} & & \downarrow{\scriptstyle \phi_1} & & \downarrow{\scriptstyle \phi_2} & & \downarrow{\scriptstyle \phi_n} & & \downarrow{\scriptstyle \phi_{n+1}} & \\
H_0 & \xrightarrow{\beta_0} & H_1 & \xrightarrow{\beta_1} & H_2 & \to \ldots \to & H_n & \xrightarrow{\beta_n} & H_{n+1} & \to \ldots
\end{array}$$

be a commutative diagram where the rows are exact sequences of abelian groups, ϕ_0 is an epimorphism and each ϕ_{3n}, $n > 0$, is an isomorphism. Prove that the sequence

$$\tilde{G}_0 \xrightarrow{\tilde{\tau}} G_1 \xrightarrow{\rho} H_1 \oplus G_2 \to \ldots \to G_{3n-2} \xrightarrow{\rho} H_{3n-2} \oplus G_{3n-1} \xrightarrow{\sigma} H_{3n-1} \xrightarrow{\tau} G_{3n+1} \to \ldots$$

is exact where $g\rho = g\alpha_{3n-2} + g\phi_{3n-2}$, $g \in G_{3n-2}$; $(h+g)\sigma = h\beta_{3n-2} - g\phi_{3n-1}$, $h \in H_{3n-2}$, $g \in G_{3n-1}$; $h\tau = h\beta_{3n-1}\phi_{3n}^{-1}\alpha_{3n}$, $h \in H_{3n-1}$; and $\tilde{G}_0 = \ker \phi_0\beta_0$, $\tilde{\tau} = \alpha_0 | \tilde{G}_0$.

Apply this theorem to prove that if X_0, X_1, X_2 are subspaces of X such that $X_1 \cup X_2 = X$, $X_1 \cap X_2 = X_0$, then there is an exact sequence (the *Mayer-Vietoris sequence*)

$$\ldots \to H_n(X_0) \to H_n(X_1) \oplus H_n(X_2) \to H_n(X) \to H_{n-1}(X_0) \to \ldots,$$

provided that one of the excision homomorphisms $H_*(X_1, X_0) \to H_*(X, X_2)$, $H_*(X_2, X_0) \to H_*(X, X_1)$ is an isomorphism.

Hurewicz homomorphism

8. Let $\theta : \pi_n(X, X_0) \to H_n(X, X_0)$ be the (relative) Hurewicz homomorphism. Show that θ commutes with homomorphisms induced by maps: precisely, if $f: X, X_0, x_0 \to Y, Y_0, y_0$ is a map, then the diagram

$$\begin{array}{ccc} \pi_n(X, X_0) & \xrightarrow{\theta} & H_n(X, X_0) \\ \downarrow f_* & & \downarrow f_* \\ \pi_n(Y, Y_0) & \xrightarrow{\theta} & H_n(Y, Y_0) \end{array}$$

is commutative.

9. Let $\beta_* : \pi_n(X, X_0) \to \pi_{n-1}(X_0)$, $\nu_* : H_n(X, X_0) \to H_{n-1}(X_0)$ be the 'boundary' homomorphisms in the homotopy and homology sequences respectively. Show that in the diagram

$$\begin{array}{ccc} \pi_n(X, X_0) & \xrightarrow{\beta_*} & \pi_{n-1}(X_0) \\ \downarrow \theta & & \downarrow \theta \\ H_n(X, X_0) & \xrightarrow[\nu_*]{} & H_{n-1}(X_0) \end{array}$$

we have $\beta_*\theta = \pm\theta\nu_*$.

10. Let $|K|$ be a 1-connected polyhedron. Consider for each n the homotopy sequence (in which we write K^p for $|K^p|$)

$$\ldots \to \pi_n(K^p, K^{p-1}) \xrightarrow{\beta_*^p} \pi_{n-1}(K^{p-1}) \xrightarrow{i_*^p} \pi_{n-1}(K^p) \xrightarrow{j_*^p} \pi_{n-1}(K^p, K^{p-1}) \to \ldots.$$

Define a chain complex (C_p, ∂_p) as follows:

$$C_p = 0,\ p < 2;\ C_2 = \pi_2(K^2)j_*^2;\ C_p = \pi_p(K^p, K^{p-1}),\ p > 2;$$

$$\partial_p = 0,\ p \leqslant 2;\ \partial_3 = \beta_*^3 j_*^2 : \pi_3(K^3, K^2) \to \pi_2(K^2)j_*^2;$$

$$\partial_p = \beta_*^p j_*^{p-1} : \pi_p(K^p, K^{p-1}) \to \pi_{p-1}(K^{p-1}, K^{p-2}),\ p > 3.$$

Verify that (C_p, ∂_p) is a chain complex and use the relative Hurewicz isomorphism theorem to prove that its homology groups are the homology groups of K.

Semi-simplicial complexes

A *semi-simplicial complex* K is a union $K = \bigcup_{q \geqslant 0} K_q$, where the K_q are disjoint sets, together with functions

$$F_q^i : K_q \to K_{q-1}, i = 0, 1, \ldots, q, \text{ and } D_q^i : K_q \to K_{q+1}, i = 0, 1, \ldots, q,$$

subject to the relations

$$F^j F^i = F^i F^{j-1}, \quad i < j$$

$$D^j D^i = D^{i-1} D^j, \quad i > j$$

$$D^j F^i = F^i D^{j-1}, \quad i < j$$

$$D^j F^j = D^j F^{j+1} = 1,$$

$$D^j F^i = F^{i-1} D^j, \quad i > j+1.$$

An element of K_q is called a *q-simplex* of K, the functions F^i are called *face operators* and the functions D^i *degeneracy operators*. A simplex is called degenerate if it is the image of a simplex under some D^i.

If K, L are semi-simplicial complexes a *semi-simplicial map* $u : K \to L$ is a function assigning to each simplex of K a simplex of L of the same dimension and commuting with the face and degeneracy operators. A *subcomplex* of a semi-simplicial complex is a subset of its simplexes closed under the face and degeneracy operators.

From the semi-simplicial complex K we define chain groups $C_q(K)$ in the obvious way; then the boundary $\partial_q : C_q(K) \to C_{q-1}(K)$ is given by $\partial_q = \Sigma(-1)^i F_q^i$, and the homology groups of K are defined. Then $u : K \to L$ plainly induces $u_* : H_*(K) \to H_*(L)$.

11. Let $S_q(X)$ be the set of singular q-simplexes of X. Then $S(X) = \bigcup_{q \geqslant 0} S_q(X)$ is a semi-simplicial complex with the face and degeneracy operators described in 8.1, 8.2, and is called the *singular complex* of X. It is plain that $H_*(S(X)) = H_*^s(X)$. Show that a map $f : X \to Y$ induces a semi-simplicial map $S(f) : S(X) \to S(Y)$, and that $S(fg) = S(f)S(g)$. Show that if $X \subseteq Y$, then $S(X)$ is a subcomplex of $S(Y)$.

12. Let K be a simplicial complex. Give K^Ω the structure of a semi-simplicial complex. Now order K by ω and consider K_ω^Ω, the semi-simplicial subcomplex of K^Ω generated by the simplexes of K^ω. Prove that the embedding $K_\omega^\Omega \subseteq K$ induces homology isomorphisms.

13. Let t_q be the standard closed q-simplex in Hilbert space and let G be an abelian group. Put $K_q = Z^n(t_q; G)$ and define $F^i : K_q \to K_{q-1}$, $D^i : K_q \to K_{q+1}$ by

$$uF^i = (V^i)^{\cdot} u,$$

$$uD^i = (T^i)^{\cdot} u, \quad u \in Z^n(t_q; G),$$

(see 8.1.1, 8.2.1). Show that K is a semi-simplicial complex (it is the semi-simplicial form of the Eilenberg-MacLane complex $K(G, n)$).

14. In which of the examples considered in Q. 11–13 is (*a*) a simplex determined by its proper faces, (*b*) a q-simplex degenerate provided that some $(q-1)$-face is degenerate, $q \geqslant 2$?

15. The semi-simplicial complex K has the *Kan property* if, given $(q-1)$-simplexes $\sigma^0, \ldots, \sigma^{k-1}, \sigma^{k+1}, \ldots, \sigma_q$ with $\sigma^i F^{j-1} = \sigma^j F^i$, $i<j$, $i, j \neq k$, there exists a q-simplex σ with $\sigma F^i = \sigma^i$, $i \neq k$. Which of the complexes considered in Q. 11–13 have the Kan property?

16. Given two semi-simplicial complexes K, L, we define the product $K \times L$ to be the complex M given by

$$M_q = K_q \times L_q, \quad F^i(M) = F^i(K) \times F^i(L), \quad D^i(M) = D^i(K) \times D^i(L).$$

Show that

$$S(X \times Y) = SX \times SY.$$

Let I be the semi-simplicial complex generated by the simplexes of an ordered closed 1-simplex (see Q. 12). Let $j_\epsilon : K \to K \times I$ be given by

$$\sigma_q j_\epsilon = (\sigma_q, \epsilon D^0 D^1 \ldots D^{q-1}), \quad \epsilon = 0, 1.$$

Show that j_ϵ embeds K as a subcomplex of $K \times I$. We say that two maps $f_0, f_1 : K \to L$ are *homotopic* if there exists a map $F : K \times I \to L$ such that $j_\epsilon F = f_\epsilon$, $\epsilon = 0, 1$. Show that homotopy is an equivalence relation if L is a Kan complex.

17. Formulate the notion of a *semi-cubical complex* and give examples.

n-connectedness, etc.

18. Prove the equivalent of Theorem 8.8.2 for the cubical complex.

19. A subcomplex of the semi-simplicial complex $S(X)$ (see Q. 11) is said to be *minimal* if, whenever, f, g are singular q-simplexes of X such that $f \simeq g \text{ rel } \dot{t}_q$, then $f = g$. Use the method of proof of Theorem 8.8.2 to show that $S(X)$ possesses a minimal subcomplex $M(X)$ which is equivalent to $S(X)$. Show that any two such minimal subcomplexes are isomorphic.

Pontryagin ring

20. Let G be a topological group. We give $\square(G)$ the structure of a graded ring by defining $f_p g_q = h_{p+q}$, where

$$(u_1, \ldots, u_{p+q}) h = (u_1, \ldots, u_p) f \,.\, (u_{p+1}, \ldots, u_{p+q}) g.$$

Show that this definition induces a ring structure in $H_*(G)$; this is the *Pontryagin ring* of G.

Show that we may also define a Pontryagin ring for a loop-space.

21. Use the technique introduced in 8.7 to define the Pontryagin ring via the singular simplicial chain complex; and show that the definitions coincide.

van Kampen's Theorem (see P. Olum, *Ann. Math.* **68** (1958), 658)

Let X be a path-connected space, $x_0 \in X$ a base-point, and π a group. We write $C^0(X, x_0; \pi)$, $C^1(X, x_0; \pi)$ for the sets of 0-contrachains and 1-contrachains of $X \bmod x_0$ with values in π. An element $z^1 \in C^1(X, x_0; \pi)$ is a 1-*contracycle*, if for any singular 2-simplex f_2,

$$(f_1^0, z^1)(f_1^1, z^1)^{-1}(f_1^2, z^1) = 1.$$

Two 1-contracycles $z^1, \bar{z}^1$ are called *contrahomologous* if there exists

$$c^0 \in C^0(X, x_0; \pi)$$

with

$$(f_1, \bar{z}^1) = (f_0^1, c^0)^{-1}(f_1, z^1)(f_0^0, c^0)$$

for all 1-simplexes f_1. We write, as usual, $z^1 \sim \bar{z}^1$.

22. Show that $z^1 \sim \bar{z}^1$ is an equivalence relation. If the equivalence classes are taken as elements of $H^1(X, x_0; \pi)$, establish a (1, 1) correspondence

$$H^1(X, x_0; \pi) \cong \pi_1(X, x_0) \pitchfork \pi.$$

23. Let A, B be path-connected subsets of $X = A \cup B$ such that $A \cap B$ is path-connected and contains x_0. If the inclusion $k : S(A) \cup S(B) \to S(X)$ induces a (1, 1) correspondence‡

$$k^* : H^1(X, x_0; \pi) \cong H^1(S(A) \cup S(B), x_0; \pi)$$

for all π, show that in the diagram

$$\begin{array}{ccccc} & & H^1(A, x_0; \pi) & & \\ & {\scriptstyle i_1^*} \nearrow & & \searrow {\scriptstyle j_1^*} & \\ H^1(X, x_0; \pi) & & & & H^1(A \cap B, x_0; \pi) \\ & {\scriptstyle i_2^*} \searrow & & \nearrow {\scriptstyle j_2^*} & \\ & & H^1(B, x_0; \pi) & & \end{array}$$

(*a*) $\ker i_1^* \cap \ker i_2^* = 0$ (notice that the sets H^1 contain a zero element); (*b*) if $j_1^* a = j_2^* b$, there exists a unique $c \in H^1(X, x_0; \pi)$ with $i_1^* c = a$, $i_2^* c = b$.

24. Deduce that under the hypotheses of Q. 23 $\pi_1(X, x_0)$ is the free product of $\pi_1(A, x_0)$ and $\pi_1(B, x_0)$ with elements in the images of $\pi_1(A \cap B, x_0)$ amalgamated (*van Kampen's formula*; see 6.4.3). Show that the hypotheses of Q. 23 are verified if A, B are sets whose interiors cover X.

25. Under the full hypotheses of Q. 4, deduce that if X and X_0 are path-connected

$$\pi_1(X/X_0) = \pi_1(X)/\overline{\pi_1(X)\, i_*},$$

where $i : X_0 \to X$ is the inclusion, and the bar denotes normal closure.

Let $|K|$, $|L|$ be compact connected polyhedra and let $|K| \sharp |L|$ be the identification space $|K| \times |L| / |K| \vee |L|$. Show that $|K| \sharp |L|$ is simply connected.

Čech contrahomology theory

26. Prove that, if $a, b \in C^\cdot(M)$, then

(i) $|a - b| \subseteq |a| \cup |b|$, (ii) $|\delta a| \subseteq |a|$, (iii) $|a \cup b| \subseteq |a| \cap |b|$.

Deduce that the contrachains with empty support form an ideal in $C^\cdot(M)$.

27. Writing $C^\cdot(M(X))$ as $A(X)$, prove that a map $g : X \to Y$ induces $g^\cdot : A(Y), I(Y) \to A(X), I(X)$ and hence $\check{g} : \check{C}(Y) \to \check{C}(X)$, $\check{g}^* : \check{R}^*(Y) \to \check{R}^*(X)$.

28. If $j : X_0 \to X$ is the injection of a subspace X_0 in X, define $A(X, X_0)$ to be the kernel of $j^\cdot : A(X) \to A(X_0)$ and $I(X, X_0) = A(X, X_0) \cap I(X)$. Prove that $\check{C}(X, X_0) = A(X, X_0)/I(X, X_0)$ is the kernel of $\check{j} : \check{C}(X) \to \check{C}(X_0)$.

29. If $\check{p}$ is the injection of $\check{C}(X, X_0)$ in $\check{C}(X)$, establish the exact sequence

$$\ldots \leftarrow \check{H}^{r+1}(X, X_0) \xleftarrow{\delta^*} \check{H}^r(X_0) \xleftarrow{j^*} \check{H}^r(X) \xleftarrow{p^*} \check{H}^r(X, X_0) \leftarrow \ldots.$$

30. Prove that $\check{R}^*(X, X_0)$ depends only on $X - X_0$.

31. If X is the subset of the Cartesian plane consisting of the set of points $(x, \sin 1/x)$ for $0 < x \leqslant 1$ and the set of points $(0, y)$ for $-1 \leqslant y \leqslant 1$, show that $H_s^0(X) = J \oplus J$ and $\check{H}^0(X) = J$.

‡ $S(X)$ is the singular complex of X (see Q. 11).

9

THE SINGULAR CONTRAHOMOLOGY RING

9.1 Definitions and properties

In this chapter we carry the discussion of singular contrahomology theory further by introducing a cup product into $\Delta^{\cdot}(X)$ which induces a ring structure in $H_s^*(X)$. We establish the principal properties of the singular contrahomology ring, we investigate the problem of computing the ring for a topological product, and, as an application of the product structure, we devote the last section of the chapter to discussing the *Hopf invariant* of maps of S^{2n-1} into S^n.

The cup-product structure in $\Delta^{\cdot}(X)$ is highly analogous to that in $C^{\cdot}_{\Omega}(K)$ and $C^{\cdot}_{\omega}(K)$. It is most succinctly described by introducing linear maps $\theta = \theta_{(r)}, \psi = \psi_{(r)} : t_r \to t_s$ where $r \leqslant s$, defined by the rules

9.1.1 $$v^i\theta = v^i, \quad v^i\psi = v^{s-r+i}, \quad i = 0, \ldots, r.$$

Thus θ (ψ) is an order preserving linear homeomorphism of t_r onto the first (last) r-face of t_s in a lexical ordering of r-faces.

Now let R be a ring and write $\Delta^{\cdot}(X)$ for $\Delta^{\cdot}(X; R)$. Let $d^p \in \Delta^p(X)$, $e^q \in \Delta^q(X)$. We define $d^p \cup e^q \in \Delta^{p+q}(X)$ by

9.1.2 $$(f_{p+q}, d^p \cup e^q) = (\theta_{(p)} f, d^p) . (\psi_{(q)} f, e^q).$$

In words, the value of $d^p \cup e^q$ on the singular $(p+q)$-simplex f is the value of d on the first p-face of f multiplied by the value of e on the last q-face. In considering the contrahomology ring it is convenient to regard $\Delta^{\cdot}(X)$ as $\sum_p \Delta^p(X)$. Thus a general element of $\Delta^{\cdot}(X)$ is a *non-homogeneous* contrachain $d = \sum_p d^p$, and we derive from 9.1.2 and the distributivity law the product formula for general contra-chains

9.1.2′ $$(f_n, d \cup e) = \sum_{p+q=n} (\theta_{(p)} f, d) . (\psi_{(q)} f, e).$$

9.1.3 Proposition

(i) $d \cup (e_1 + e_2) = d \cup e_1 + d \cup e_2$, $(d_1 + d_2) \cup e = d_1 \cup e + d_2 \cup e$;
(ii) $d \cup (e \cup f) = (d \cup e) \cup f$;
(iii) $\delta(d^p \cup e^q) = \delta d^p \cup e^q + (-1)^p d^p \cup \delta e^q$.

The proofs of these assertions follow closely the lines of the proofs of corresponding results in chapter 4. ▌ However, the reader is advised to prove (iii) by examining how the θ and ψ maps commute with the V^i maps defined in 8.1.

It follows that $\Delta^{\cdot}(X)$, enriched with the cup-product structure, is a contrachain ring in the sense of 4.2.1. The cup product therefore induces a product, to be written by juxtaposition, in $H_s^*(X)$; the contrahomology group with this product forms a ring $R_s^*(X)$—more precisely, $R_s^*(X; R)$—the *singular contrahomology ring* of X with values in R.

We may clearly relativize this notion to obtain $R_s^*(X, X_0)$. Moreover, it is easy to see that if $d^p, e^q \in \Delta^{\cdot}(X)$ and if either d^p or e^q vanishes on the singular simplexes of X_0, then so does $d^p \cup e^q$. This implies that $\Delta^{\cdot}(X, X_0)$ is a two-sided ideal in $\Delta^{\cdot}(X)$; the quotient ring of $\Delta^{\cdot}(X)$ by $\Delta^{\cdot}(X, X_0)$ is isomorphic with $\Delta^{\cdot}(X_0)$ furnished with its cup-product structure. Thus

$$0 \leftarrow \Delta^{\cdot}(X_0) \overset{\lambda^{\cdot}}{\leftarrow} \Delta^{\cdot}(X) \overset{\mu^{\cdot}}{\leftarrow} \Delta^{\cdot}(X, X_0) \leftarrow 0$$

is an exact sequence of contrachain ring maps, so that, in the induced contrahomology sequence, λ^* and μ^* are ring homomorphisms. Further, we may copy Theorem 4.2.7 in this more general situation. Given $\eta \in H^q(X)$, $\xi_0 \in H^p(X_0)$, $\xi \in H^p(X)$, $\hat{\xi} \in H^p(X, X_0)$ we may define $\xi_0 \eta \in H^{p+q}(X_0)$, $\xi\eta \in H^{p+q}(X)$, $\hat{\xi}\eta \in H^{p+q}(X, X_0)$, namely, $\xi_0\eta = \xi_0(\lambda^*\eta)$, $\xi\eta$ is the product in the ring $R^*(X)$, and $\hat{\xi}\eta = \{\hat{x} \cup y\}$, where $\hat{\xi}$ is the class of $\hat{x}$ and η is the class of y and $\{\ \}$ refers to contrahomology in $\Delta^{\cdot}(X, X_0)$; recall that $\hat{x} \cup y \in \Delta^{\cdot}(X, X_0)$, since $\Delta^{\cdot}(X, X_0)$ is an ideal in $\Delta^{\cdot}(X)$. Clearly if $\eta = \mu^*\hat{\eta}$ then $\hat{\xi}\eta = \hat{\xi}\hat{\eta}$. The element η determines homomorphisms

$$\cup_\eta : H^*(X_0) \to H^*(X_0),$$

$$\cup_\eta : H^*(X) \to H^*(X),$$

$$\cup_\eta : H^*(X, X_0) \to H^*(X, X_0),$$

which raise dimension by q.

9.1.4 Theorem. *The diagram*

$$\begin{array}{ccccccc}
\cdots \xleftarrow{\nu^*} & H^r(X_0) & \xleftarrow{\lambda^*} & H^r(X) & \xleftarrow{\mu^*} & H^r(X, X_0) & \\
 & \downarrow \cup_\eta & & \downarrow \cup_\eta & & \downarrow \cup_\eta & \\
\cdots \longleftarrow & H^{q+r}(X_0) & \longleftarrow & H^{q+r}(X) & \longleftarrow & H^{q+r}(X, X_0) &
\end{array}$$

$$\begin{array}{ccc}
\xleftarrow{\nu^*} & H^{r-1}(X_0) & \longleftarrow \cdots \\
 & \downarrow \cup_\eta & \\
\longleftarrow & H^{q+r-1}(X_0) & \longleftarrow \cdots
\end{array}$$

is commutative. ▌

We refer to the diagram of 9.1.4 as the *singular* (contrahomology) *product structure* of the pair (X, X_0).

Suppose $f_{p+q} \in \Delta^\flat(X)$, say $f = T^i g$. Then if $i \leqslant p-1$, $\theta_{(p)} f \in \Delta^\flat(X)$ and, if $p \leqslant i \leqslant p+q-1$, $\psi_{(q)} f \in \Delta^\flat(X)$. It follows that 9.1.2 induces a product in $\Delta_N^{\cdot}(X)$, from which a contrahomology ring $R_N^*(X)$ may be formed. Again we may relativize to $R_N^*(X, X_0)$, and it is clear that the natural map $\Delta(X, X_0) \to \Delta^N(X, X_0)$ induces a ring isomorphism of $R_N^*(X, X_0)$ with $R_s^*(X, X_0)$. Thus we may identify these two rings—and, indeed, the two product structures.

Let $g: X, X_0 \to Y, Y_0$ be a map. Then g induces

$$g_\Delta^{\cdot} = g_\Delta : \Delta^{\cdot}(Y, Y_0) \to \Delta^{\cdot}(X, X_0),$$

given by
$$(f_p, g_\Delta d^p) = (f_p g^\Delta, d^p),$$

where f_p is a singular p-simplex of X mod X_0 and $d^p \in \Delta^p(Y, Y_0)$. Then

9.1.5
$$g_\Delta(d^p \cup e^q) = g_\Delta d^p \cup g_\Delta e^q.$$

For

$$\begin{aligned}
&(f_{p+q}, g_\Delta(d^p \cup e^q)) \\
&\quad = (f_{p+q} g^\Delta, d^p \cup e^q) = ((fg)_{p+q}, d^p \cup e^q) \\
&\quad = (\theta_{(p)} fg, d^p) \,.\, (\psi_{(q)} fg, e^q) = ((\theta_{(p)} f) g^\Delta, d^p) \,.\, ((\psi_{(q)} f) g^\Delta, e^q) \\
&\quad = (\theta_{(p)} f, g_\Delta d^p) \,.\, (\psi_{(q)} f, g_\Delta e^q) = (f_{p+q}, g_\Delta d^p \cup g_\Delta e^q).
\end{aligned}$$

We deduce

9.1.6 Theorem. *A map $g: X, X_0 \to Y, Y_0$ induces a homomorphism of the product structure‡ of (Y, Y_0) into that of (X, X_0).* ▌

‡ It is understood that the structural element $\cup_\eta$ is transformed into the structural element $\cup_{g^*\eta}$, $\eta \in H^*(Y)$.

9.1.7 Corollary. *Pairs of the same homotopy type have isomorphic singular contrahomology product structures.* ∎

Consider the chain equivalence $\alpha : C^{\Omega}(K), C^{\Omega}(K_0) \to \Delta(|K|), \Delta(|K_0|)$ of 8.6.8. It is evident from the definition of α (8.6.5) that $\alpha^{\cdot}$ preserves cup products. We deduce

9.1.8 Theorem. *The chain equivalence*

$$\alpha : C^{\Omega}(K), C^{\Omega}(K_0) \to \Delta(|K|), \Delta(|K_0|)$$

induces an isomorphism of the singular product structure of $(|K|, |K_0|)$ *onto the product structure of* (K, K_0). ∎

Finally, we remark that a cap product may be defined between elements of $\Delta_{p+q}(X, X_0)$ and $\Delta^p(X, X_0)$ with values in $\Delta_q(X, X_0)$. We leave to the reader the task of making the definition and verifying that it induces a product between elements of $H_{p+q}(X, X_0)$ and $H^p(X, X_0)$ with values in $H_q(X, X_0)$ which is an invariant of homotopy type. Similarly, we leave to the reader the formulation of results of this chapter when the value ring is replaced by a bilinear pairing of two value groups to a third (see the last paragraph of 4.1).

9.2 Skew-commutativity of $R^*(X)$

For further study of the contrahomology ring, it is useful to characterize the cup product as a contrachain map

9.2.1 $$\cup : \Delta^{\cdot}(X; R) \otimes \Delta^{\cdot}(X; R) \to \Delta^{\cdot}(X; R).$$

For we recall from 5.7 that the tensor product of contrachain complexes may be given the structure of a contrachain complex; and 9.1.3 (iii) asserts that $\cup$ is a contrachain map.

As a contrachain map, $\cup$ admits an important factorization in the following way. We recall the contrachain map π defined in 5.7.27. As a special case, we have

$$\pi = \pi_{X,Y} : \Delta^{\cdot}(X; R) \otimes \Delta^{\cdot}(Y; R) \to \Delta^{\cdot}(X \otimes Y; R),$$

where $\Delta^{\cdot}(X \otimes Y; R)$ is the contrachain complex $\Delta(X \otimes Y) \pitchfork R$; π is given in this case by

9.2.2 $$(f \otimes g, \pi(d \otimes e)) = (f, d) \, . \, (g, e),$$

where f is a singular simplex of X, g is a singular simplex of Y, $d \in \Delta^{\cdot}(X; R)$, $e \in \Delta^{\cdot}(Y; R)$.

9.2.3 Proposition. *Let $k : X \to X_1$, $l : Y \to Y_1$ be maps. Then*

$$\pi_{X,Y}(k^{\cdot}_{\Delta} \otimes l^{\cdot}_{\Delta}) = (k^{\Delta} \otimes l^{\Delta})^{\cdot}\, \pi_{X_1, Y_1} : \Delta^{\cdot}(X_1) \otimes \Delta^{\cdot}(Y_1) \to \Delta^{\cdot}(X \otimes Y).$$

The proof is straightforward. ∎

We now define a map $D : \Delta(X) \to \Delta(X \otimes X)$ by

9.2.4. $$f_n D = \sum_{p+q=n} \theta_{(p)} f \otimes \psi_{(q)} f.$$

9.2.5 Proposition. (i) *D is a chain map;* (ii) *if $g : X \to Y$, then $D(g^{\Delta} \otimes g^{\Delta}) = g^{\Delta} D$.*

The proof of (i) follows closely the lines of the proof of the formula for $\delta(d \cup e)$—as, indeed, one would expect. By the same token, the proof of (ii) closely resembles that of 9.1.5. ∎ The relation between $\cup$ and D is simply expressed by

9.2.6 Theorem. $\cup = D^{\cdot} \pi_{X,X}$.

We have
$$(f_n, d \cup e) = \sum_{p+q=n} (\theta_{(p)} f, d) \,.\, (\psi_{(q)} f, e)$$
$$= \sum_{p+q=n} (\theta_{(p)} f \otimes \psi_{(q)} f, \pi(d \otimes e))$$
$$= (f_n D, \pi(d \otimes e)) = (f_n, D^{\cdot}\pi(d \otimes e)). \;∎$$

The properties of $\cup$ may be deduced from the characterization 9.2.6 and the properties of D and π; we may clearly deduce 9.1.3 and 9.1.5 from 9.2.3 and 9.2.5, but our main objective is to use 9.2.6 to deduce the skew-commutativity of $R^*(X)$.

We recall that we established in 8.7.1 the existence of a canonical class of chain equivalences $\rho^{X,Y} : \Delta(X \otimes Y) \to \Delta(X \times Y)$. Let $d^X = d : X \to X \times X$ be the diagonal map $xd = (x, x)$, $x \in X$. We invest the chain map D with geometrical significance by proving

9.2.7 Theorem. $D^X \rho^{X,X} \simeq (d^X)^{\Delta} : \Delta(X) \to \Delta(X \times X)$.

The argument is by acyclic models. We note that $D\rho = d^{\Delta}$ in dimension 0 and that $D\rho$ and d^{Δ} both satisfy the evident naturality condition

9.2.8 $$g^{\Delta}\phi = \phi(g \times g)^{\Delta}, \quad \phi = D\rho \text{ or } d^{\Delta},$$

for any $g : X \to Y$. We now construct a chain homotopy Λ_r^X, inductively with respect to r, starting with $\Lambda_0 = 0$, using the acyclicity of $\tilde{\Delta}(t_r \times t_r)$. The reader will easily supply the details. We remark that the chain homotopy is constructed to have the naturality property

9.2.9 $$g^{\Delta}\Lambda^Y = \Lambda^X (g \times g)^{\Delta}$$

for $g : X \to Y$. ∎

We recall the chain isomorphism $\mu : \Delta(X \otimes Y) \rightarrow \Delta(Y \otimes X)$ of 8.7.4, given by $(f_p \otimes g_q)\mu = (-1)^{pq} g_q \otimes f_p$, $f_p \in \Delta_p(X)$, $g_q \in \Delta_q(Y)$. We prove

9.2.10 Theorem. $D \simeq D\mu : \Delta(X) \rightarrow \Delta(X \otimes X)$.

We have $\mu\rho \simeq \rho m^{\Delta} : \Delta(X \otimes X) \rightarrow \Delta(X \times X)$ (8.7.4). Thus

$$D\mu\rho \simeq D\rho m^{\Delta} \simeq d^{\Delta} m^{\Delta} = (dm)^{\Delta} = d^{\Delta} \simeq D\rho,$$

since $dm = d$. The result follows from the fact that ρ is a chain equivalence. ▌ In addition, however, we remark that all homotopies in this argument are natural, so that the homotopy, Λ, connecting D and $D\mu$ may be chosen to have the naturality property‡

9.2.11 $$\Lambda^X (g^{\Delta} \otimes g^{\Delta}) = g^{\Delta} \Lambda^Y$$

for $g : X \rightarrow Y$.

9.2.12 Theorem. *If the value ring R is commutative, then $R^*(X; R)$ is skew-commutative; that is, for*

$$\xi \in H^p(X; R), \quad \eta \in H^q(X; R),$$

we have $$\xi\eta = (-1)^{pq} \eta\xi.$$

We first observe that, if R is commutative, then for $d \in \Delta^p(X; R)$, $e \in \Delta^q(Y; R)$, we have

9.2.13 $$\mu^{\cdot}\pi(d \otimes e) = (-1)^{pq} \pi(e \otimes d).$$

For if $f \in \Delta_p(X)$, $g \in \Delta_q(Y)$, then

$$\begin{aligned}(g \otimes f, \mu^{\cdot}\pi(d \otimes e)) &= ((g \otimes f)\mu, \pi(d \otimes e)) \\ &= (-1)^{pq} (f \otimes g, \pi(d \otimes e)) = (-1)^{pq} (f, d) . (g, e) \\ &= (-1)^{pq} (g, e) . (f, d) = (-1)^{pq} (g \otimes f, \pi(e \otimes d)).\end{aligned}$$

Then $\pi_{X,Y}$ induces $\pi^*_{X,Y} : H^*(X; R) \otimes H^*(Y; R) \rightarrow H^*(X \otimes Y; R)$ and, if $\xi \in H^p(X; R)$, $\eta \in H^q(Y; R)$,

9.2.14 $$\mu^* \pi^*_{X,Y}(\xi \otimes \eta) = (-1)^{pq} \pi^*_{Y,X}(\eta \otimes \xi).$$

Now let $X = Y$; then by 9.2.6, 9.2.10 and 9.2.14,

$$\begin{aligned}\xi\eta = D^*\pi^*(\xi \otimes \eta) = D^*\mu^*\pi^*(\xi \otimes \eta) &= (-1)^{pq} D^*\pi^*(\eta \otimes \xi) \\ &= (-1)^{pq} \eta\xi. \text{ ▌}\end{aligned}$$

‡ Of course, 9.2.10 and 9.2.11 could be proved by an acyclic model argument.

The reader will notice that this proof of skew-commutativity could be adapted to provide a direct proof of the skew-commutativity of $R^*_\Omega(K)$. The proof given in chapter 4 is not, however, readily adaptable to the more general situation; it was preferred in chapter 4 on the grounds of conceptual simplicity.

Theorem 9.2.12 relativizes in the obvious way; we leave the details to the reader.

9.3* Cup products in cubical contrahomology

We have observed that 9.1.2 may be interpreted as a cup-product formula in $\Delta^{\cdot}_N(X; R)$ and that we obtain a natural isomorphism between the rings R^*_s and R^*_N. In this section we define a cup product in $\square^{\cdot}_N(X; R)$ and prove that the induced contrahomology ring is naturally isomorphic to $R^*(X; R)$.

As in 9.2, we proceed by defining a chain map

$$D^\square : \square^N(X) \to \square^N(X \otimes X),$$

where $\square^N(X \otimes X)$ stands for $\square^N(X) \otimes \square^N(X)$; we then define

9.3.1 $$\cup = D^{\cdot}_\square \pi : \square^{\cdot}_N(X; R) \otimes \square^{\cdot}_N(X; R) \to \square^{\cdot}_N(X; R).$$

The contrachain map π, of course, is defined for any two chain complexes and so no difficulty arises in passing to cubical complexes. The proof however that $D^\square$ is a chain map involves us in some rather heavy computation, and we think it fair to inform the reader that our only application of this section is in 10.5 and 10.6.

We shall induce $D^\square$ by a map $D : \square(X) \to \square(X \otimes X)$. Let e_n be the standard n-cube and H an ordered subset $h_1, \ldots, h_p$ of the integers $1, 2, \ldots, n$; we define $\lambda^\epsilon_H : e_p \to e_n$ ($\epsilon = 0$ or 1) by the rule

9.3.2 $$(u_1, \ldots, u_p) \lambda^\epsilon_H = (v_1, \ldots, v_n),$$

where $v_i = \epsilon$ if $i \notin H$ and $v_{h_r} = u_r$, $r = 1, \ldots, p$. In words, λ^0_H is an isometry of e_p onto the p-face which contains the origin and lies in the subspace spanned by the coordinates $u_{h_1}, \ldots, u_{h_p}$; λ^1_H is an isometry of e_p onto the p-face which contains the point $(1, 1, \ldots, 1)$ and is parallel to this subspace.

We now define $D : \square(X) \to \square(X \otimes X)$ by

9.3.3 $$f_n D = \sum_H \rho_{HK} \lambda^0_H f \otimes \lambda^1_K f,$$

where K is the complementary set to H, and ρ_{HK} is the signature of the permutation HK of the integers $1, 2, \ldots, n$, and the summation is over all ordered subsets H of the integers $1, 2, \ldots, n$.

To establish the properties of D we must first write down the commutation laws of λ_H^ϵ with respect to the face and degeneracy operators. Given $H = (h_1, \ldots, h_p)$ we define $\hat{H}_\mu$ to be

$$(h_1, \ldots, h_{\mu-1}, h_{\mu+1}, \ldots, h_p)$$

and $\bar{H}_\mu$ to be $$(h_1, \ldots, h_{\mu-1}, h_{\mu+1}-1, \ldots, h_p-1).$$

If $j \notin H$ and $h_r < j < h_{r+1}$, we define H_j to be

$$(h_1, \ldots, h_r, h_{r+1}-1, \ldots, h_p-1);$$

this definition is extended to the cases $j < h_1$ and $h_p < j$ in the obvious way.

9.3.4 Proposition‡

(i) $V_\mu^\epsilon \lambda_H^\eta = \lambda_{H_\mu}^\eta V_{h_\mu}^\epsilon$, $\epsilon = 0$ or 1, $\eta = 0$ or 1;

(ii) $V_\mu^\epsilon \lambda_H^\epsilon = \lambda_{\hat{H}_\mu}^\epsilon$;

(iii) $\lambda_H^\epsilon = \lambda_{H_j}^\epsilon V_j^\epsilon$;

(iv) $\lambda_H^\epsilon T_{h_\mu} = T_\mu \lambda_{\bar{H}_\mu}^\epsilon$.

Notice that in these formulae H_j and $\bar{H}_\mu$ are to be regarded as ordered subsets of the integers $1, 2, \ldots, n-1$. The formulae are proved by direct computation; we are content to prove (i) and leave the verification of the rest to the conscientious reader (with the remark that (i) and (ii) imply (iii)).

Now

$$(u_1, \ldots, u_{p-1})\, V_\mu^\epsilon \lambda_H^\eta = (u_1, \ldots, u_{\mu-1}, \epsilon, u_\mu, \ldots, u_{p-1})\, \lambda_H^\eta = (v_1, \ldots, v_n),$$

where $v_{h_\lambda} = u_\lambda$, $\lambda < \mu$; $v_{h_\mu} = \epsilon$; $v_{h_\lambda} = u_{\lambda-1}$, $\lambda > \mu$; and $v_i = \eta$ otherwise. On the other hand $(u_1, \ldots, u_{p-1})\, \lambda_{\bar{H}_\mu}^\eta = (w_1, \ldots, w_{n-1})$, where $w_{h_\lambda} = u_\lambda$, $\lambda < \mu$; $w_{h_\lambda - 1} = u_{\lambda-1}$, $\lambda > \mu$; and $w_i = \eta$ otherwise. Then

$$(u_1, \ldots, u_{p-1})\, \lambda_{\bar{H}_\mu}^\eta V_{h_\mu}^\epsilon = (v_1', \ldots, v_n'),$$

where $v_i' = w_i$, $i < h_\mu$; $v_{h_\mu}' = \epsilon$; $v_i' = w_{i-1}$, $i > h_\mu$. It follows that $v_{h_\lambda}' = u_\lambda$, $\lambda < \mu$; $v_{h_\mu}' = \epsilon$; $v_{h_\lambda}' = u_{\lambda-1}$, $\lambda > \mu$; and $v_i' = \eta$ otherwise. Thus

$$(v_1, \ldots, v_n) = (v_1', \ldots, v_n')$$

and (i) is proved. ∎

We return now to 9.3.3. Suppose f_n is degenerate, say $f = T_i g$. Then, for each H, $i \in H$ or $i \in K$, so that, by 9.3.4 (iv), either $\lambda_H^0 f$ or $\lambda_K^1 f$ is degenerate. It follows that D maps $\square^\flat(X)$ into

$$\square(X) \otimes \square^\flat(X) + \square^\flat(X) \otimes \square(X)$$

‡ We write V_μ^ϵ for $V^{\mu\epsilon}$ for notational convenience; in conformity we write T_μ for T^μ.

and thus induces a map

$$D^{\square} : \square^N(X) \to \square^N(X \otimes X).$$

9.3.5 Proposition. *$D^{\square}$ is a chain map.*

We prove more, namely that D is a chain map. Consider first $f_n D\partial$. Then

$$\begin{aligned} f_n D\partial &= (\sum_H \rho_{HK} \lambda_H^0 f \otimes \lambda_K^1 f)\partial \\ &= \sum_H \rho_{HK} \left\{ \sum_{\mu=1}^{p} (-1)^{\mu} (V_\mu^0 \lambda_H^0 f - V_\mu^1 \lambda_H^0 f) \otimes \lambda_K^1 f \right. \\ &\quad \left. + (-1)^p \sum_{\mu=1}^{q} (-1)^{\mu} \lambda_H^0 f \otimes (V_\mu^0 \lambda_K^1 f - V_\mu^1 \lambda_K^1 f) \right\}, \end{aligned}$$

where H has p elements and $p+q=n$.

Now

$$V_\mu^0 \lambda_H^0 f \otimes \lambda_K^1 f = \lambda_{\hat{H}_\mu}^{0} f \otimes \lambda_K^1 f \quad \text{and} \quad \lambda_H^0 f \otimes V_\mu^1 \lambda_K^1 f = \lambda_H^0 f \otimes \lambda_{\hat{K}_\mu}^{1} f,$$

by 9.3.4 (ii). Let $\check{K}_\mu$ be the complementary set to $\hat{H}_\mu$. It follows from the above that $\lambda_{\hat{H}_\mu}^{0} f \otimes \lambda_K^1 f$ arises twice in the expression for $f_n D\partial$, once by deleting h_μ from H in the (H, K) term and once by deleting h_μ from $\check{K}_\mu$ in the $(\hat{H}_\mu, \check{K}_\mu)$ term. On its first occurrence it receives the sign $\rho_{HK}(-1)^{\mu}$ and on its second occurrence

$$\rho_{\hat{H}_\mu \check{K}_\mu} \times (-1)^{p-1} (-1)^{\alpha+1} (-1),$$

where $k_\alpha < h_\mu < k_{\alpha+1}$. By comparing the relevant permutations we see that

$$\rho_{\hat{H}_\mu \check{K}_\mu} = (-1)^{p-\mu+\alpha} \rho_{HK}.$$

It follows that $\lambda_{\hat{H}_\mu}^{0} f \otimes \lambda_K^1 f$ receives opposite signs on its two occurrences. Thus

9.3.6 $$\begin{aligned} f_n D\partial = \sum_H \rho_{HK} &\left\{ \sum_{\mu=1}^{p} (-1)^{\mu+1} V_\mu^1 \lambda_H^0 f \otimes \lambda_K^1 f \right. \\ &\left. + (-1)^p \sum_{\mu=1}^{q} (-1)^{\mu} \lambda_H^0 f \otimes V_\mu^0 \lambda_K^1 f \right\}. \end{aligned}$$

On the other hand,

9.3.7 $$\begin{aligned} f_n \partial D &= \sum_{j=1}^{n} (-1)^j (V_j^0 f - V_j^1 f) D \\ &= \sum_{j=1}^{n} (-1)^j \sum_{\tilde{H}} \rho_{\tilde{H}\tilde{K}} (\lambda_{\tilde{H}}^{0} V_j^0 f \otimes \lambda_{\tilde{K}}^{1} V_j^0 f - \lambda_{\tilde{H}}^{0} V_j^1 f \otimes \lambda_{\tilde{K}}^{1} V_j^1 f), \end{aligned}$$

where $\tilde{H}$ is an ordered subset of $1, 2, \ldots, n-1$ and $\tilde{K}$ the complementary set.

We now point out that with each triple $(\tilde{H}, \tilde{K}, j)$ we may associate a unique triple (H, K, μ) such that $\bar{H}_\mu = \tilde{H}$, $K_j = \tilde{K}$, and $j = h_\mu$. For if $\tilde{K} = \tilde{k}_1, \ldots, \tilde{k}_q$, we define $K = k_1, \ldots, k_q$ by $k_i = \tilde{k}_i$, $k_i < j$; $k_i = \tilde{k}_i + 1$, $k_i \geq j$. Then clearly K is the unique subset of $1, 2, \ldots, n$ such that $K_j = \tilde{K}$. Then H is the complement of K, $j \in H$, and $j = h_\mu$ for some (unique) μ. It is evident that $\bar{H}_\mu = \tilde{H}$.

Now by 9.3.4 (i) and (iii), $V^1_\mu \lambda^0_H f \otimes \lambda^1_K f = \lambda^0_{\bar{H}_\mu} V^1_{h_\mu} f \otimes \lambda^1_{K_j} V^1_j f$, and, similarly, $\lambda^0_H f \otimes V^0_\mu \lambda^1_K f = \lambda^0_{H_j} V^0_j f \otimes \lambda^1_{\bar{K}_\mu} V^0_{k_\mu} f$. It follows that there is a (1, 1) correspondence between the terms occurring on the right-hand sides of 9.3.6 and 9.3.7, corresponding terms differing at most in sign, so that all that remains is to show that corresponding terms occur in fact with the same sign. We are content to do this for the term $V^1_\mu \lambda^0_H f \otimes \lambda^1_K f$. We must show that

9.3.8 $$(-1)^\mu \rho_{HK} = (-1)^{h_\mu} \rho_{\bar{H}_\mu K_{h_\mu}}.$$

Suppose $k_\alpha < h_\mu < k_{\alpha+1}$. Then we have to compare the permutation

$$h_1, \ldots, h_{\mu-1}, h_\mu, \ldots, h_p, k_1, \ldots, k_\alpha, k_{\alpha+1}, \ldots, k_q$$

with the permutation

$$h_1, \ldots, h_{\mu-1}, h_{\mu+1} - 1, \ldots, h_p - 1, k_1, \ldots, k_\alpha, k_{\alpha+1} - 1, \ldots, k_q - 1, n.$$

Now

$$h_\mu, \ldots, h_p, k_{\alpha+1}, \ldots, k_q \quad \text{and} \quad h_{\mu+1} - 1, \ldots, h_p - 1, k_{\alpha+1} - 1, \ldots, k_q - 1, n$$

are permutations of the integers from h_μ to n of the *same* signature; for we may pass from the second to the first by adding one to obtain $h_{\mu+1}, \ldots, h_p, k_{\alpha+1}, \ldots, k_q, h_\mu$ and then bringing h_μ to the front, and each of the two steps alters the signature by $(-1)^{n-h_\mu}$. Thus the signatures of the two given permutations differ by $(-1)^\alpha$. But α is the number of k's less than h_μ so that $\alpha = h_\mu - \mu$ and 9.3.8 is proved. Similarly, we may prove that $\lambda^0_H f \otimes V^0_\mu \lambda^1_K f$ occurs in 9.3.6 and 9.3.7 with the same sign and 9.3.5 is proved. ∎

9.3.9 Proposition. *$D^\square$ is natural in the sense that the diagram*

$$\begin{array}{ccc}
\square^N(X) & \xrightarrow{D^\square} & \square^N(X \otimes X) \\
\downarrow{\scriptstyle g^\square} & & \downarrow{\scriptstyle g^\square \otimes g^\square} \\
\square^N(Y) & \xrightarrow{D^\square} & \square^N(Y \otimes Y)
\end{array}$$

commutes, where $g : X \to Y$. ∎

From 9.3.1 and 9.3.5 it follows that

$$\cup : \square^{\cdot}_N(X; R) \otimes \square^{\cdot}_N(X; R) \to \square^{\cdot}_N(X; R)$$

is a contrachain map which thus induces a multiplication in $H^*_c(X; R)$.

9.3.10 Theorem. *The multiplication in* $H_c^*(X; R)$ *converts it into a ring* $R_c^*(X; R)$. *Moreover, if* $\phi : \Delta(X) \to \square^N(X)$ *is a canonical chain equivalence* (see 8.4), *then* $\phi^* : R_c^*(X; R) \cong R_s^*(X; R)$.

Clearly all is proved if we show that ϕ^* is product-preserving. Consider the diagram

$$\begin{array}{ccc}
\square_N^{\cdot}(X; R) \otimes \square_N^{\cdot}(X; R) & \xrightarrow{\phi^{\cdot} \otimes \phi^{\cdot}} & \Delta^{\cdot}(X; R) \otimes \Delta^{\cdot}(X; R) \\
\downarrow{\scriptstyle \pi} & & \downarrow{\scriptstyle \pi} \\
\square_N^{\cdot}(X \otimes X; R) & \xrightarrow{(\phi \otimes \phi)^{\cdot}} & \Delta^{\cdot}(X \otimes X; R) \\
\downarrow{\scriptstyle D_{\square}^{\cdot}} & & \downarrow{\scriptstyle D_{\Delta}^{\cdot}} \\
\square_N^{\cdot}(X; R) & \xrightarrow{\phi^{\cdot}} & \Delta^{\cdot}(X; R)
\end{array}$$

We wish to prove that $\phi^{\cdot} D_{\square}^{\cdot} \pi \simeq D_{\Delta}^{\cdot} \pi(\phi^{\cdot} \otimes \phi^{\cdot})$. Certainly

$$\pi(\phi^{\cdot} \otimes \phi^{\cdot}) = (\phi \otimes \phi)^{\cdot} \pi;$$

this is a special case of the naturality of π. Thus it will be sufficient to prove

9.3.11 $$\phi D^{\square} \simeq D^{\Delta}(\phi \otimes \phi) : \Delta(X) \to \square^N(X \otimes X).$$

We prove 9.3.11 by an acyclic model argument. We remark that $\phi D^{\square}$ and $D^{\Delta}(\phi \otimes \phi)$ agree in dimension zero, and that each has the naturality property expressed by the commutativity of the diagram

$$\begin{array}{ccc}
\Delta(X) & \xrightarrow{k} & \square^N(X \otimes X) \\
\downarrow{\scriptstyle g^{\Delta}} & & \downarrow{\scriptstyle g^{\square} \otimes g^{\square}} \\
\Delta(Y) & \xrightarrow{k} & \square^N(Y \otimes Y)
\end{array}$$

where $g : X \to Y$ and $k = \phi D^{\square}$ or $D^{\Delta}(\phi \otimes \phi)$. We thus define a chain homotopy Λ_r^X inductively starting with $\Lambda_0 = 0$, using the acyclicity of $\tilde{\square}^N(t_r \otimes t_r)$. The details may be omitted. ∎ The reader will notice that, in a sense which we may make precise, we have proved that any natural equivalence between $\Delta(X)$ and $\square^N(X)$ induces an isomorphism between any natural product structure in $H^*(X)$ and any natural product structure in $H_s^*(X)$.

Theorem 9.3.10 may, of course, be relativized. The upshot of this section is that we may identify the cubical and simplicial singular contrahomology product structures and, where convenient, employ formula 9.3.3 to give an explicit cup product in $\square_N^{\cdot}(X; R)$; of course, 9.1.3 (iii) is the cup-product contraboundary formula in $\square_N^{\cdot}(X; R)$ as in $\Delta^{\cdot}(X; R)$; this follows from 9.3.1 and 9.3.5.

9.4 The contrahomology ring of a topological product

We recall the chain equivalence $\rho : \Delta(X \otimes Y) \simeq \Delta(X \times Y)$ of 8.7. Our object in this section is to invest $\Delta^{\cdot}(X \otimes Y; R)$ with a cup-product structure in such a way that ρ^* is a ring isomorphism. We suppress the symbol R for the value ring.

We will define the cup product as a contrachain map

$$\cup : \Delta^{\cdot}(X \otimes Y) \otimes \Delta^{\cdot}(X \otimes Y) \to \Delta^{\cdot}(X \otimes Y).$$

Indeed, as in 9.2 and 9.3, we shall define $\cup$ as the composition $D^{\cdot}_{X \otimes Y}\pi$, where

$$\pi : \Delta^{\cdot}(X \otimes Y) \otimes \Delta^{\cdot}(X \otimes Y) \to \Delta^{\cdot}(X \otimes Y \otimes X \otimes Y),$$

and it remains to define

9.4.1 $$D^{X \otimes Y} : \Delta(X \otimes Y) \to \Delta(X \otimes Y \otimes X \otimes Y).$$

We write μ for $1 \otimes \mu \otimes 1 : \Delta(X \otimes X \otimes Y \otimes Y) \to \Delta(X \otimes Y \otimes X \otimes Y)$ and define

9.4.2 $$D^{X \otimes Y} = (D^X \otimes D^Y)\mu.$$

This definition ensures that $D^{X \otimes Y}$ is a chain map, so that the definition

9.4.3 $$\cup = D^{\cdot}_{X \otimes Y}\pi$$

ensures that $\cup$ is a contrachain map.

We remark that $\Delta^{\cdot}(X \otimes Y)$ is *bigraded* in the sense that it is generated by homogeneous elements $d^{p,q}$ which take the value zero on $f_{p'} \otimes g_{q'}$ unless $p = p'$, $q = q'$.

9.4.4 Proposition. *If $d^{p,q}$, $e^{r,s}$ are homogeneous contrachains of $\Delta^{\cdot}(X \otimes Y)$, then $d^{p,q} \cup e^{r,s}$ is a homogeneous contrachain of degree $(p+r, q+s)$ and*

$$(f_{p+r} \otimes g_{q+s}, d \cup e) = (-1)^{qr}(\theta_{(p)}f \otimes \theta_{(q)}g, d)(\psi_{(r)}f \otimes \psi_{(s)}g, e).$$

Suppose f is a singular m-simplex of X and g a singular n-simplex of Y. Then

$$\begin{aligned}(f \otimes g, d \cup e) &= (f \otimes g, D^{\cdot}_{X \otimes Y}\pi(d \otimes e)) \\ &= ((f \otimes g)D^{X \otimes Y}, \pi(d \otimes e)) \\ &= ((fD^X \otimes gD^Y)\mu, \pi(d \otimes e)) \\ &= \Big(\sum_{\substack{t+u=m \\ v+w=n}}\sum (\theta_{(t)}f \otimes \psi_{(u)}f \otimes \theta_{(v)}g \otimes \psi_{(w)}g)\mu, \pi(d \otimes e)\Big) \\ &= \Big(\sum_{\substack{t+u=m \\ v+w=n}}\sum (-1)^{uv}\theta_{(t)}f \otimes \theta_{(v)}g \otimes \psi_{(u)}f \otimes \psi_{(w)}g, \pi(d \otimes e)\Big).\end{aligned}$$

Now $(\theta_{(t)} f \otimes \theta_{(v)} g \otimes \psi_{(u)} f \otimes \psi_{(w)} g,\ \pi(d \otimes e)) = 0$ unless $t = p$, $v = q$, $u = r$, $w = s$. Thus $d \cup e$ is homogeneous of degree $(p+r, q+s)$ and

$$(f_{p+r} \otimes g_{q+s}, d \cup e) = (-1)^{qr} (\theta_{(p)} f \otimes \theta_{(q)} g, d)\,(\psi_{(r)} f \otimes \psi_{(s)} g, e). ▮$$

Of course, this formula could be taken as a definition of $\cup$ in $\Delta^{\cdot}(X \otimes Y)$.

9.4.5 Proposition. *The diagram*

$$\begin{array}{ccc} \Delta(X \otimes Y) & \xrightarrow{D} & \Delta(X \otimes Y \otimes X \otimes Y) \\ \Big\downarrow{\scriptstyle f^{\Delta} \otimes g^{\Delta}} & & \Big\downarrow{\scriptstyle f^{\Delta} \otimes g^{\Delta} \otimes f^{\Delta} \otimes g^{\Delta}} \\ \Delta(X_1 \otimes Y_1) & \xrightarrow{D} & \Delta(X_1 \otimes Y_1 \otimes X_1 \otimes Y_1) \end{array}$$

is commutative, where $f : X \to X_1$, $g : Y \to Y_1$.

This is an easy consequence of 9.2.5 (ii). ▮

The next proposition is the key to this section. Readers who have ploughed through the previous section will be rewarded by learning that the analogous diagram for the cubical case is actually commutative if the obvious choice (8.7.5) of ρ is made. Although Proposition 9.4.6 follows directly from this, we give an independent proof.

9.4.6 Proposition. *The diagram*

$$\begin{array}{ccc} \Delta(X \otimes Y) & \xrightarrow{D^{X \otimes Y}} & \Delta(X \otimes Y \otimes X \otimes Y) \\ \Big\downarrow{\scriptstyle \rho^{X,Y}} & & \Big\downarrow{\scriptstyle \rho^{X,Y} \otimes \rho^{X,Y}} \\ \Delta(X \times Y) & \xrightarrow{D^{X \times Y}} & \Delta(X \times Y) \otimes (X \times Y) \end{array}$$

is homotopy-commutative: $\rho D^{X \times Y} \simeq D^{X \otimes Y}(\rho \otimes \rho)$.

We shall give a proof‡ based on proved results and using the popular procedure of argument by diagram.

Consider the diagram

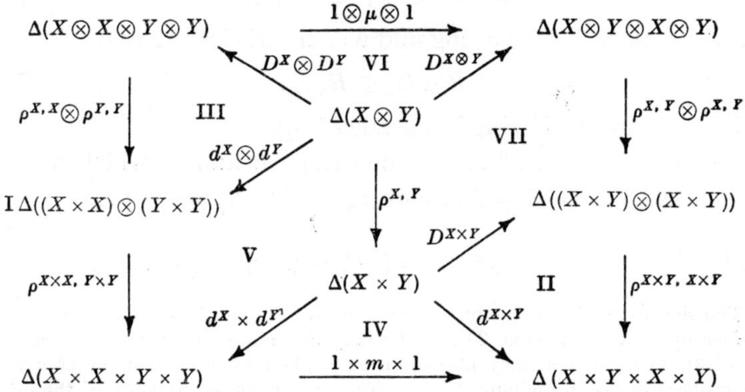

‡ As the reader would expect, an acyclic model argument is also available.

We wish to prove region VII homotopy-commutative. We assert that the remaining regions are homotopy-commutative:

I: (the exterior region): an easy application of 8.7.3 and 8.7.4.

II: 9.2.7.

III: 9.2.7 (5.7.5).

IV: this is manifestly commutative.

V: this is commutative by the naturality of ρ.

VI: this is commutative by definition of $D^{X\otimes Y}$.

From these facts we quickly infer that

$$\rho^{X,Y}D^{X\times Y}\rho^{X\times Y,\,X\times Y} \simeq D^{X\otimes Y}(\rho^{X,Y}\otimes\rho^{X,Y})\,\rho^{X\times Y,\,X\times Y}.$$

But ρ is a chain equivalence so that $\rho^{X,Y}D^{X\times Y} \simeq D^{X\otimes Y}(\rho^{X,Y}\otimes\rho^{X,Y})$. ∎

9.4.7 Theorem. *The multiplication in $H_s^*(X\otimes Y)$ induced by the cup product (in $\Delta^{\cdot}(X\otimes Y)$) turns it into a ring $R^*(X\otimes Y)$ and*

$$\rho^* : R^*(X\times Y) \cong R^*(X\otimes Y).$$

This follows immediately from 9.4.6 and the naturality of π (compare 9.3.10). ∎

We shall apply 9.4.7 in particular when X and Y are compact polyhedra. To prepare for this application we make some remarks about tensor products of rings and the map π.

The *tensor product* of two rings R and S, which we write $R\otimes S$, is the tensor product of their additive groups, together with the ring structure given by the multiplication rule

$$(r\otimes s)(r'\otimes s') = rr'\otimes ss'$$

on the generators and extended by distributivity to arbitrary elements of $R\otimes S$. This definition was implicit at the end of 5.2.

A *graded ring* R is a graded group $\sum_n R_n$ together with a multiplication which turns it into a ring and which satisfies the rule

$$R_m.R_n \subseteq R_{m+n}.$$

Thus a contrachain ring is a graded ring.

The *tensor product* of two graded rings R and S, which we write $R\otimes S$, is the graded ring T, where $T_n = \sum_{p+q=n} R_p\otimes S_q$ and‡

9.4.8 $$(r_p\otimes s_q)(r_{p'}\otimes s_{q'}) = (-1)^{qp'}(r_pr_{p'}\otimes s_qs_{q'}).$$

‡ The sign $(-)^{qp'}$ is an example of a standard 'gimmick'; we pass s_q past $r_{p'}$ and so suffer qp' changes of sign. The reader may find many other examples (e.g. 4.1.8 and 5.7.3). The gimmick may also often be applied when a chain or contrachain and an operator are interchanged; the operator is given as 'dimension' the amount by which it increases dimension; see 4.1.3 and 5.7.1.

If $C^{\cdot}$ and $D^{\cdot}$ are contrachain rings, their tensor product qua graded rings becomes a contrachain ring if we define (as usual)

$$\delta(c^p \otimes d^q) = \delta c^p \otimes d^q + (-1)^p c^p \otimes \delta d^q.$$

We have to show that

$$\delta((c^p \otimes d^q) \cup (c^r \otimes d^s)) = \delta(c^p \otimes d^q) \cup (c^r \otimes d^s) + (-1)^{p+q} (c^p \otimes d^q) \cup \delta(c^r \otimes d^s).$$

Now

$$\begin{aligned}
\delta((c^p \otimes d^q) \cup (c^r \otimes d^s)) &= (-1)^{qr} \delta((c^p \cup c^r) \otimes (d^q \cup d^s)) \\
&= (-1)^{qr} \{(\delta c^p \cup c^r + (-1)^p c^p \cup \delta c^r) \otimes (d^q \cup d^s) \\
&\quad + (-1)^{p+r} (c^p \cup c^r) \otimes (\delta d^q \cup d^s + (-1)^q d^q \cup \delta d^s)\} \\
&= (\delta c^p \otimes d^q) \cup (c^r \otimes d^s) \\
&\quad + (-1)^{p+q} (c^p \otimes d^q) \cup (\delta c^r \otimes d^s) \\
&\quad + (-1)^p (c^p \otimes \delta d^q) \cup (c^r \otimes d^s) \\
&\quad + (-1)^{p+q+r} (c^p \otimes d^q) \cup (c^r \otimes \delta d^s) \\
&= \delta(c^p \otimes d^q) \cup (c^r \otimes d^s) \\
&\quad + (-1)^{p+q} (c^p \otimes d^q) \cup \delta(c^r \otimes d^s).
\end{aligned}$$

After these definitions we return to the map

$$\pi : (C \pitchfork G) \otimes (D \pitchfork H) \to (C \otimes D) \pitchfork A,$$

associated with a homomorphism $\eta : G \otimes H \to A$; it is given by the formula

$$(c \otimes d, \pi(e \otimes f)) = ((c, e) \otimes (d, f))\, \eta, \quad c \in C,\ d \in D,\ e \in C \pitchfork G,\ f \in D \pitchfork H.$$

We shall be particularly concerned with the case in which G, H, A are rings and $\eta : G \otimes H \to A$ is a ring homomorphism; that is,

$$(g \otimes h)\, \eta\, (g' \otimes h')\, \eta = (gg' \otimes hh')\, \eta.$$

If A is a ring then in certain important cases the contrachain complex $(C \otimes D) \pitchfork A$ may be given a ring structure by a formula analogous to 9.4.4; we shall show that in those cases π is a contrachain-ring map if η is a ring-homomorphism. The cases in which we are particularly interested are that in which $C = C^\omega(K)$, $D = C^\omega(L)$ and that in which $C = C^\Omega(K)$, $D = C^\Omega(L)$. In both cases we may define a cup-product structure in $(C \otimes D) \pitchfork A$ as in 9.4.4. Moreover with these definitions

$$(\bar{\theta}^\omega_K \otimes \bar{\theta}^\omega_L)^{\cdot} : (C^\Omega(K) \otimes C^\Omega(L))^{\cdot} \to (C^\omega(K) \otimes C^\omega(L))^{\cdot},$$

and $\quad (\alpha_K \otimes \alpha_L)^{\cdot} : \Delta^{\cdot}(|K| \otimes |L|) \to (C^\Omega(K) \otimes C^\Omega(L))^{\cdot} \quad$ (see 8.6.7) are contrachain-*ring* equivalences; the reader is invited to verify these statements.

Now let the chain complexes C, D be singular chain complexes, total chain complexes or ordered chain complexes.‡ Then if R is a ring, the contrachain complex $C \pitchfork R$ has the cup-product ring structure. We prove

9.4.9 Proposition. *If G, H, A are rings and η is a ring homomorphism, then*

$$\pi : (C \pitchfork G) \otimes (D \pitchfork H) \to (C \otimes D) \pitchfork A$$

is a contrachain-ring map.

We have to show that π is product-preserving. Let $c^p, c^r \in C \pitchfork G$, $d^q, d^s \in D \pitchfork H$. We wish to show that

$$\pi((c^p \otimes d^q) \cup (c^r \otimes d^s)) = \pi(c^p \otimes d^q) \cup \pi(c^r \otimes d^s).$$

Now each side of the equation is homogeneous of degree $(p+r, q+s)$ so that it is enough to verify it on $f_{p+r} \otimes g_{q+s}$, where f, g are simplexes. Then

$$(c^p \otimes d^q) \cup (c^r \otimes d^s) = (-1)^{qr} (c^p \cup c^r) \otimes (d^q \cup d^s),$$

so that

$$\begin{aligned}(f \otimes g, \pi((c^p \otimes d^q) \cup (c^r \otimes d^s))) &= (-1)^{qr} (f \otimes g, \pi((c^p \cup c^r) \otimes (d^q \cup d^s))) \\ &= (-1)^{qr} \{(f, c^p \cup c^r) \otimes (g, d^q \cup d^s)\} \eta.\end{aligned}$$

On the other hand,

$$\begin{aligned}(f \otimes g, \pi(c^p \otimes d^q) \cup \pi(c^r \otimes d^s)) &= (-1)^{qr} (\theta_{(p)} f \otimes \theta_{(q)} g, \pi(c^p \otimes d^q)) \\ &\quad \times (\psi_{(r)} f \otimes \psi_{(s)} g, \pi(c^r \otimes d^s)) \quad (9.4.4) \\ &= (-1)^{qr} \{(\theta_{(p)} f, c^p) \otimes (\theta_{(q)} g, d^q)\} \eta \\ &\quad \times \{(\psi_{(r)} f, c^r) \otimes (\psi_{(s)} g, d^s)\} \eta \\ &= (-1)^{qr} \{(\theta_{(p)} f, c^p)(\psi_{(r)} f, c^r) \\ &\quad \otimes (\theta_{(q)} g, d^q)(\psi_{(s)} g, d^s)\} \eta \quad \text{since } \eta \text{ is a ring-homomorphism}, \\ &= (-1)^{qr} \{(f, c^p \cup c^r) \otimes (g, d^q \cup d^s)\} \eta. \ ∎\end{aligned}$$

9.4.10 Corollary. $\pi : (C \pitchfork R) \otimes (D \pitchfork R) \to (C \otimes D) \pitchfork R$ *is a contrachain-ring map if R is commutative.*

For in this case $\eta : R \otimes R \to R$ is defined by $(r_1 \otimes r_2)\eta = r_1 r_2$, and this is a ring homomorphism if R is commutative. ∎

‡ Thus $C = \Delta(X)$, $C\Omega(K)$ or $C^{\omega}(K)$; we could also of course have $C = \square^N(X)$ or $\Delta^N(X)$.

9.4.11 Theorem. *Let K, L be finite ordered simplicial complexes and let G, H, A be rings such that $G \otimes H \cong A$. Then there is a contrachain-ring equivalence*

9.4.12 $$C^{\cdot}_{\omega}(K; G) \otimes C^{\cdot}_{\omega}(L; H) \simeq \Delta^{\cdot}(|K| \times |L|; A).$$

For we have the sequence of ring equivalences

$$\Delta^{\cdot}(|K| \times |L|; A) \overset{\rho^{\cdot}}{\simeq} \Delta^{\cdot}(|K| \otimes |L|; A) \overset{(\alpha_K \otimes \alpha_L)^{\cdot}}{\simeq} (C^{\Omega}(K) \otimes C^{\Omega}(L)) \pitchfork A$$

$$\overset{(\bar{\theta}^{\omega}_K \otimes \bar{\theta}^{\omega}_L)^{\cdot}}{\simeq} (C^{\omega}(K) \otimes C^{\omega}(L)) \pitchfork A \overset{\pi}{\cong} C^{\cdot}_{\omega}(K; G) \otimes C^{\cdot}_{\omega}(L; H).$$

That π is a contrachain-ring isomorphism is guaranteed by 9.4.9 and 5.7.29. ▌ Clearly the contrachain ring $C^{\cdot}_{\omega}(K; G) \otimes C^{\cdot}_{\omega}(L; H)$ is comparatively convenient for calculating $R^*(|K| \times |L|; A)$.

Notice that 9.4.11 applies if $G = J$ and $H = A$. Moreover, we have

9.4.13 Theorem. *Let K, L be finite simplicial complexes, let K be without torsion and let R be a ring. Then*

$$R^*(|K| \times |L|; R) \cong R^*(K; J) \otimes R^*(L; R).$$

Clearly $J \otimes R \cong R$. Since K is without torsion, the contracycle group of $C^{\cdot}_{\omega}(K; J) \otimes C^{\cdot}_{\omega}(L; R)$ is $Z^{\cdot}_{\omega}(K; J) \otimes Z^{\cdot}_{\omega}(L; R)$. Thus the homomorphism $R^*(K; J) \otimes R^*(L; R) \to R^*(|K| \times |L|; R)$ induced by the contrachain-ring equivalence 9.4.12 is in this case an isomorphism. ▌ This theorem shows that, if K is without torsion, then the contrahomology ring $R^*(|K| \times |L|; R)$ is calculable from $R^*(K; J)$ and $R^*(L; R)$. In general, this is not true if K has torsion. We show this by an example which we owe to J. F. Adams.

9.4.14 *Example.* Let

$$|K_1| = P^2 \vee S^3, \quad |L_1| = P^2, \quad |K_2| = P^3, \quad |L_2| = P^2.$$

Then $\quad R^*(|K_1|; J) \cong R^*(|K_2|; J), \quad R^*(|L_1|; J) \cong R^*(|L_2|; J),$

but $$R^*(|K_1| \times |L_1|; J) \not\cong R^*(|K_2| \times |L_2|; J).$$

The complexes K_1 and K_2 figured in Example 4.3.4, where the isomorphism $R^*(|K_1|; J) \cong R^*(|K_2|; J)$ was established. A suitable contrachain ring for P^2 is provided by c^0, c^1, c^2, where $(c^1)^2 = c^2$, $\delta c^1 = 2c^2$ and $\{c^2\}$ generates $H^2(P^2; J) = J_2$; we obtain a contrachain ring for $P^2 \vee S^3$ by adjoining z^3, generating H^3, and in this ring $c^1 \cup c^2 = 0$. On the other hand, we obtain a contrachain ring for P^3 by adjoining c^3 with $c^1 \cup c^2 = c^3$.

In‡ dimension 2, $C^{\cdot}_{(1)} = C^{\cdot}(K_1) \otimes C^{\cdot}(L_1)$ is generated by $c^0 \otimes c^2$, $c^1 \otimes c^1$, and $c^2 \otimes c^0$; thus $Z^2_{(1)}$ is generated by $c^0 \otimes c^2$ and $c^2 \otimes c^0$ and $H^2_{(1)} = J_2 \oplus J_2$, generated by the classes of these contracycles. Now $C^3_{(1)}$ is generated by $c^1 \otimes c^2$, $c^2 \otimes c^1$ and $z^3 \otimes c^0$; thus $Z^3_{(1)}$ is generated by $c^1 \otimes c^2 - c^2 \otimes c^1$ and $z^3 \otimes c^0$ and $T^3_{(1)}$, the torsion subgroup of $H^3_{(1)}$, is J_2, generated by the class of $c^1 \otimes c^2 - c^2 \otimes c^1$. Then clearly $H^2_{(1)} . T^3_{(1)} = 0$.

On the other hand, if $C^{\cdot}_{(2)} = C^{\cdot}(K_2) \otimes C^{\cdot}(L_2)$, then $H^2_{(2)} = H^2_{(1)}$, and $C^3_{(2)}$ is generated by $c^1 \otimes c^2$, $c^2 \otimes c^1$ and $c^3 \otimes c^0$. Thus $Z^3_{(2)}$ is generated by $c^1 \otimes c^2 - c^2 \otimes c^1$ and $c^3 \otimes c^0$ and $T^3_{(2)}$, the torsion subgroup of $H^3_{(2)}$, is J_2, generated by the class of $c^1 \otimes c^2 - c^2 \otimes c^1$. Thus $H^2_{(2)} . T^3_{(2)}$ contains the class of the element $(c^0 \otimes c^2) \cup (c^1 \otimes c^2 - c^2 \otimes c^1) = -c^2 \otimes c^3$. This class is non-zero so that $H^2_{(2)} . T^3_{(2)} \neq 0$ and the statement 9.4.14 is verified.

We close this section with some remarks. First, if, under the hypotheses of 9.4.13, L is also without torsion, then, by 5.2.25,

$$R^*(|K| \times |L|; R) \cong R^*(K; J) \otimes R^*(L; J) \otimes R.$$

Secondly, if R is a commutative ring and $\eta : R \otimes R \to R$, given by $(r_1 \otimes r_2)\eta = r_1 r_2$, is a ring isomorphism, then 9.4.12 holds with G, H, A replaced by R. This would be the case if $R = J, J_m$ or Q, the field of rational numbers. In the last case we could infer from 9.4.12 that

$$R^*(|K| \times |L|; Q) = R^*(K; Q) \otimes R^*(L; Q).$$

In general if R is an arbitrary field $\mathscr{F}$, we may, by an argument closely parallel to that of 9.4.12, infer that

9.4.15 $$R^*(|K| \times |L|; \mathscr{F}) \cong R^*(K; \mathscr{F}) \otimes_{\mathscr{F}} R^*(L; \mathscr{F})$$

(compare 5.7.25*c*).

Thirdly, we remark that we may infer from 9.4.7 a ring isomorphism

9.4.16 $$\rho^* : R^*(X \times Y; \mathscr{F}) \cong R^*(X; R^*(Y; \mathscr{F})).$$

It must, however, be understood that $R^*(X; R^*(Y))$ is a contrahomology ring with values in a *graded* ring. If $\xi \in R^p(X; R^q(Y))$, $\xi' \in R^{p'}(X; R^{q'}(Y))$, then their product $\xi * \xi'$ in $R^*(X; R^*(Y))$ is related to their product $\xi\xi'$ as elements of the contrahomology ring of X with values in the *ungraded* ring $R^*(Y)$ by the rule

9.4.17 $$\xi * \xi' = (-1)^{qp'} \xi\xi'.$$

‡ We now suppress the symbol J for the value ring.

With this understanding, ρ^* is product-preserving with respect to the product $*$ and preserves degree in the sense that

$$\rho^* : H^n(X \times Y) \cong \sum_{p+q=n} H^p(X; H^q(Y)).$$

The sign change in 9.4.17 is just what should be expected in passing from rings to graded rings (see 9.4.8 and the accompanying footnote). We revert to 9.4.17 in 10.5.

9.5 The Hopf invariant

In 1931, H. Hopf opened up a new field of research in algebraic topology, the calculation of the homotopy groups of spheres, by proving that there exist infinitely many homotopy classes of maps from S^3 to S^2; in 1935 he extended his method to cover the general case of maps of S^{2n-1} to S^n. The method consists in attaching to each homotopy class of maps an integer now called the Hopf invariant.‡

Hopf introduced the integer invariant in the following terms. Let $f: K \to L$ be a simplicial map such that $|K| = S^3$, $|L| = S^2$. Let p, p' be interior to simplexes s_2, s_2' of L; then $|f|^{-1}(p)$ is a 1-cycle on a subdivision of K and consists of disjoint polygonal loops, each edge of which penetrates a tetrahedron of K. Moreover, the 1-cycles $|f|^{-1}(p)$, $|f|^{-1}(p')$ may be oriented in a way determined by given orientations of K, L, and they thus have a 'linking number' which describes the algebraic number of times either intersects a 2-chain bounded by the other. This linking number is the Hopf invariant which Hopf proved to depend only on the homotopy class of $|f| : S^3 \to S^2$.

The above definition is dependent, in the first instance, on the choice of triangulations K, L; we may immediately circumvent this dependence by using singular homology theory. Our formulation will be largely based on Steenrod's treatment (*Ann. Math.* **50** (1949), 954), first translating Hopf's definition (for maps $S^{2n-1} \to S^n$, $n > 1$) into singular contrahomology. The value ring is J throughout this section and will be omitted. We shall suppose that S^{2n-1} and S^n have been oriented; that is, we choose fixed generators ξ, η of $H^{2n-1}(S^{2n-1})$, $H^n(S^n)$.

Let $f: S^{2n-1} \to S^n$ be a map. Then f induces

$$f_\Delta = f_\Delta^{\boldsymbol{\cdot}} : \Delta^{\boldsymbol{\cdot}}(S^n) \to \Delta^{\boldsymbol{\cdot}}(S^{2n-1}).$$

Let u be a contracycle in the class η. Then $f_\Delta(u)$ is an n-contracycle of $\Delta^{\boldsymbol{\cdot}}(S^{2n-1})$, so that $f_\Delta(u) = \delta v$, $v \in \Delta^{n-1}(S^{2n-1})$. Also u^2, the cup square of u, is a $2n$-contracycle§ of $\Delta^{\boldsymbol{\cdot}}(S^n)$, so that $u^2 = \delta a$, $a \in \Delta^{2n-1}(S^n)$.

‡ *Math. Ann.* **104** (1931), 637; *Fund. Math.* **25** (1935), 427. We shall not be concerned in this section with any generalizations of the Hopf invariant.

§ In simplicial contrahomology of course we should have $u^2 = 0$, $a = 0$ and $v \cup f_\Delta u$ already a contracycle.

Then $v \cup f_\Delta u - f_\Delta a$ is a $(2n-1)$-contracycle of $\Delta^\cdot(S^{2n-1})$. For, by 9.1.3 and 9.1.5, $\delta(v \cup f_\Delta u - f_\Delta a) = f_\Delta u \cup f_\Delta u - f_\Delta u^2 = 0$. Thus

$$\{v \cup f_\Delta u - f_\Delta a\} = \gamma_0 \xi,$$

for some integer γ_0.

9.5.1 Proposition. *The integer γ_0 depends only on the map f.*

We give no proof here since the proposition is an obvious consequence of a redefinition of γ_0 that will be given later.

The integer γ_0 as determined above by f is the invariant as defined by Hopf (although he used simplicial homology). The definition given is inconvenient for proving general theorems about the invariant because it depends on choices (of u, v, a). We now give a definition of an invariant nature, that is, one which does not involve choices. We must first introduce the notion of *mapping cylinder*, due to J. H. C. Whitehead.

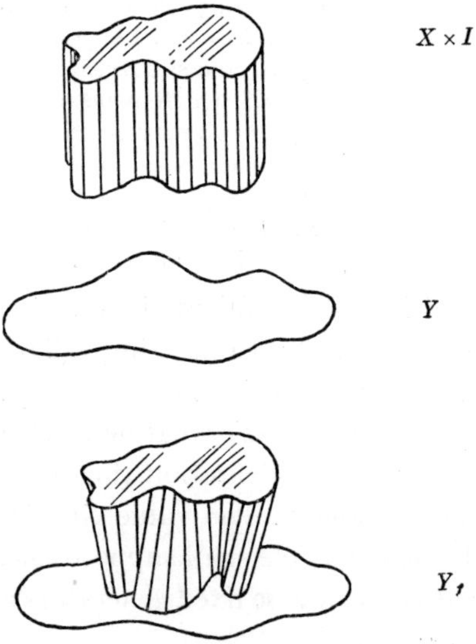

Fig. 9.1

Let $f: X \to Y$ be a map. Consider the space T, the disjoint union of $X \times I$ and Y. We introduce into T an equivalence relation generated by the rule that $(x, 0)\, R\, (xf)$ for all $x \in X$. Let Y_f be the identification space of R, let $k: T \to Y_f$ be the projection, let $i: X \to Y_f$ be given by $xi = (x, 1)\,k$ and let $p: Y_f \to Y$ be given by $(x, t)\,kp = xf$, $ykp = y$. Then it is not difficult to see that $ip = f$, that i embeds X homeomorphically in Y_f as a closed subset and that p is a homotopy equivalence (with inverse $k \mid Y$). We call Y_f the *mapping cylinder* of f. We may think of Y_f as a 'cylinder' with X as its top and with its base embedded in Y, the 'generators' being segments connecting x with xf (see fig. 9.1).

Now let

$$\begin{array}{ccc} X^0 & \xrightarrow{f^0} & Y^0 \\ \downarrow{\scriptstyle g} & & \downarrow{\scriptstyle h} \\ X^1 & \xrightarrow{f^1} & Y^1 \end{array}$$

be a commutative diagram of maps, $f^0h = gf^1$. Then there is an obvious map $j : Y^0_{f^0} \to Y^1_{f^1}$ such that the diagram

9.5.2
$$\begin{array}{ccccc} X^0 & \xrightarrow{i^0} & Y^0_{f^0} & \xrightarrow{p^0} & Y^0 \\ \downarrow{\scriptstyle g} & & \downarrow{\scriptstyle j} & & \downarrow{\scriptstyle h} \\ X^1 & \xrightarrow{i^1} & Y^1_{f^1} & \xrightarrow{p^1} & Y^1 \end{array}$$

is commutative; precisely, $(x^0, t)\, k^0 j = (x^0 g, t)\, k^1$ and $y^0 k^0 j = y^0 h k^1$, $x^0 \in X^0$, $y^0 \in Y^0$. We shall say that j *extends* (g, h) to $Y^0_{f^0}$.

9.5.3 Proposition. *Let $f^0 \simeq f^1 : X \to Y$. Then*

$$(Y_{f^0}, Xi^0) \simeq (Y_{f^1}, Xi^1).$$

Let $F : X \times I \to Y$ be a homotopy between f^0 and f^1. Let $g^\epsilon : X \to X \times I$ be given by $xg^\epsilon = (x, \epsilon)$, $\epsilon = 0, 1$, and let $j^\epsilon : Y_{f^\epsilon} \to Y_F$ extend $(g^\epsilon, 1)$ in the sense of 9.5.2, where $1 : Y \to Y$. It is easy to see that j^ϵ embeds Y_{f^ϵ} topologically in Y_F; we shall show that $(Y_F, (X \times I)\, i)$ can be retracted by deformation onto $(Y_{f^\epsilon}, Xi^\epsilon)$. The proposition will follow immediately; indeed we shall have the more precise result

9.5.4
$$j^\epsilon : (Y_{f^\epsilon}, Xi^\epsilon) \simeq (Y_F, (X \times I)\, i).$$

Now Y_F is obtained from $(X \times I \times I) \cup Y$ by identifying $(x, u, 0)$ with $(x, u)\, F$, and $(X \times I)\, i$ is the subspace $(X \times I \times 1)\, k$. Then j^ϵ embeds Y_{f^ϵ} as $((X \times \epsilon \times I) \cup Y)\, k$ and Xi^ϵ as $(X \times \epsilon \times 1)\, k$. Let $\rho^\epsilon_t : I \times I \to I \times I$ be a deformation retraction of $I \times I$ onto $(I \times 0) \cup (\epsilon \times I)$ under which $I \times 1$ is retracted over itself to $\epsilon \times 1$. We define $\sigma^\epsilon_t : Y_F \to Y_F$ by

$$(x, u, v)\, k\sigma^\epsilon_t = (x, (u, v)\, \rho^\epsilon_t)\, k, \quad yk\sigma^\epsilon_t = yk.$$

The diligent reader may verify that σ^ϵ_t is the required deformation retraction onto $(Y_{f^\epsilon}, Xi^\epsilon)$. ▮

We now come to the definition of the Hopf invariant. Let $f : S^{2n-1} \to S^n$ be a map, let S_f be the mapping cylinder of f, and let S^{2n-1} be identified with $S^{2n-1} i \subseteq S_f$. Let $\eta_f = p^* \eta$, so that η_f generates

$H^n(S_f)$. In the contrahomology sequence of the pair (S_f, S^{2n-1}) we have $\mu^* : H^n(S_f, S^{2n-1}) \cong H^n(S_f)$. Let $\mu^* \hat{\eta}_f = \eta_f$. Then the square $\hat{\eta}_f \hat{\eta}_f = \hat{\eta}_f^2$ is an element of $H^{2n}(S_f, S^{2n-1})$. Reference to the contrahomology sequence shows that $\nu^* : H^{2n-1}(S^{2n-1}) \cong H^{2n}(S_f, S^{2n-1})$. Thus $\hat{\eta}_f^2 = \gamma \nu^*(\xi)$, for some unique integer γ. The procedure for defining γ may be understood from the diagram

9.5.5 $$H^n(S^n) \underset{\cong}{\overset{p^*}{\to}} H^n(S_f) \underset{\cong}{\overset{\mu^*}{\leftarrow}} H^n(S_f, S^{2n-1}) \overset{Sq}{\to} H^{2n}(S_f, S^{2n-1}) \underset{\cong}{\overset{\nu^*}{\leftarrow}} H^{2n-1}(S^{2n-1}),$$

where $Sq\hat{\zeta} = \hat{\zeta}^2$; Sq is not in general a homomorphism. Since it is the only link in the diagram that is not isomorphic, it is sensible to regard γ as being determined by Sq. We prove

9.5.6 Theorem. $\gamma = -\gamma_0$.

We retain the notation used in defining γ_0. Let $u_f = p_\Delta(u)$, so that u_f represents η_f and let $a_f = p_\Delta(a)$. Then $f_\Delta(u)$ is the restriction of u_f to the singular simplexes of S^{2n-1} and similarly for $f_\Delta(a)$,

$$f_\Delta(u) = \lambda^\cdot(u_f), \quad f_\Delta(a) = \lambda^\cdot(a_f);$$

and $\lambda^\cdot(u_f) = \delta v$, where $\lambda : \Delta(S^{2n-1}) \to \Delta(S_f)$ is the injection. Let w be an extension of v to S_f, that is, $w \in \Delta^\cdot(S_f)$ and $\lambda^\cdot w = v$. Consider $u_f - \delta w$; then $u_f - \delta w$ is a contracycle of $\Delta^\cdot(S_f)$ vanishing on $\Delta^\cdot(S^{2n-1})$. We may thus regard it as a contracycle of $\Delta^\cdot(S_f, S^{2n-1})$ and it clearly represents $\hat{\eta}_f$. Thus $(u_f - \delta w) \cup u_f$ represents $\hat{\eta}_f \eta_f = \hat{\eta}_f^2$.

Now in $\Delta^\cdot(S^{2n-1})$

$$f_\Delta a - v \cup f_\Delta u = \lambda^\cdot(a_f) - \lambda^\cdot w \cup \lambda^\cdot(u_f) = \lambda^\cdot(a_f - w \cup u_f).$$

Thus $\nu^*\{f_\Delta a - v \cup f_\Delta u\}$ is the relative contrahomology class of

$$\delta(a_f - w \cup u_f) = u_f^2 - (\delta w \cup u_f) = (u_f - \delta w) \cup u_f.$$

It follows that $\nu^*(-\gamma_0 \xi) = \gamma \nu^*(\xi)$, whence $\gamma = -\gamma_0$. ∎

We observe that of course 9.5.6 establishes 9.5.1.

9.5.7 Definition.‡ The *Hopf invariant* of the map $f : S^{2n-1} \to S^n$, relative to given orientations ξ, η of S^{2n-1} and S^n, is that integer γ such that

$$\hat{\eta}_f^2 = \gamma \nu^*(\xi),$$

where $p^*\eta = \eta_f$, $\mu^*\hat{\eta}_f = \eta_f$ (see 9.5.5).

‡ In view of 9.5.6 our definition differs from Hopf's in sign.

We can evidently define the Hopf invariant of a map of any oriented homeomorph of S^{2n-1} to any oriented homeomorph of S^n.

We use this definition to prove the most immediately accessible facts about the Hopf invariant.

9.5.8 Theorem. *If n is odd, γ is zero.*

This follows from 9.5.7 and the skew-commutativity of the relative contrahomology product. ∎

9.5.9 Theorem. (i) *Let $g : S^n \to S^n$ be a map of degree d; then*

$$\gamma(fg) = d^2\gamma(f).$$

(ii) *Let $h : S^{2n-1} \to S^{2n-1}$ be a map of degree d; then*

$$\gamma(hf) = d\gamma(f).$$

We prove (i): let $m : S_f \to S_{fg}$ extend $(1, g)$ in the sense of 9.5.2 where $1 : S^{2n-1} \to S^{2n-1}$. We then get a commutative diagram

$$\begin{array}{ccccc}
H^n(S^n) & \xrightarrow{p^*} & H^n(S_f) & \xleftarrow{\mu^*} & H^n(S_f, S^{2n-1}) \\
\uparrow{\scriptstyle g^*} & & \uparrow{\scriptstyle m^*} & & \uparrow{\scriptstyle \hat{m}^*} \\
H^n(S^n) & \xrightarrow{p^*} & H^n(S_{fg}) & \xleftarrow{\mu^*} & H^n(S_{fg}, S^{2n-1})
\end{array}$$

$$\begin{array}{ccccc}
\xrightarrow{Sq} & H^{2n}(S_f, S^{2n-1}) & \xleftarrow{\nu^*} & H^{2n-1}(S^{2n-1}) \\
& \uparrow{\scriptstyle \hat{m}^*} & & \uparrow{\scriptstyle 1} \\
\xrightarrow{Sq} & H^{2n}(S_{fg}, S^{2n-1}) & \xleftarrow{\nu^*} & H^{2n-1}(S^{2n-1})
\end{array}$$

Then $g^*(\eta) = d\eta$, so that

$$m^*(\eta_{fg}) = d\eta_f, \quad \hat{m}^*(\hat{\eta}_{fg}) = d\hat{\eta}_f \quad \text{and} \quad \hat{m}^*(\hat{\eta}_{fg}^2) = d^2\hat{\eta}_f^2.$$

Assertion (i) follows immediately. The proof of (ii) uses the right-hand rectangle of a similar diagram and is simpler; we omit it. ∎

9.5.10 Corollary. *If the orientation of S^n is changed, the Hopf invariant is unaffected. If the orientation of S^{2n-1} is changed, the Hopf invariant is multiplied by -1.* ∎

9.5.11 Theorem. *$\gamma(f)$ depends only on the homotopy class of f.*

Let F be a homotopy between f^0 and $f^1 : S^{2n-1} \to S^n$. We write S_0 for S_{f^0} (and η_0, $\hat{\eta}_0$ similarly) and consider the map $j^0 = j : S_0 \to S_F$ as in 9.5.4.

We then have the commutative diagram

$$\begin{array}{ccccc}
H^n(S^n) & \xrightarrow{p^*} & H^n(S_0) & \xleftarrow{\mu^*} & H^n(S_0, S^{2n-1}) \\
\uparrow 1 & & \uparrow j^* & & \uparrow \hat{\jmath}^* \\
H^n(S^n) & \xrightarrow{p^*} & H^n(S_F) & \xleftarrow{\mu^*} & H^n(S_F, S^{2n-1}\times I)
\end{array}$$

$$\begin{array}{ccccc}
& \xrightarrow{Sq} & H^{2n}(S_0, S^{2n-1}) & \xleftarrow{\nu^*} & H^{2n-1}(S^{2n-1}) \\
& & \uparrow \hat{\jmath}^* & & \uparrow g^{0*} \\
& \xrightarrow{Sq} & H^{2n}(S_F, S^{2n-1}\times I) & \xleftarrow{\nu^*} & H^{2n-1}(S^{2n-1}\times I)
\end{array}$$

Then $j^*(\eta_F) = \eta_0$, $\hat{\jmath}^*(\hat{\eta}_F) = \hat{\eta}_0$, $\hat{\jmath}^*(\hat{\eta}_F^2) = \hat{\eta}_0^2$. Since, by 9.5.4, j is a homotopy equivalence (of pairs), it follows that, if $\bar{\xi}$ is the generator of $H^{2n-1}(S^{2n-1}\times I)$ such that $g^{0*}(\bar{\xi}) = \xi$, then $\hat{\eta}_F^2 = \gamma(f^0)\,\nu^*\bar{\xi}$.

We may use the same argument with g^1 replacing g^0. But $g^0 \simeq g^1$ so that $g^{1*}\bar{\xi} = \xi$, and as above $\hat{\eta}_F^2 = \gamma(f^1)\,\nu^*\bar{\xi}$. Thus $\gamma(f^0) = \gamma(f^1)$. ∎

9.5.12 Theorem. *The association $f \to \gamma(f)$ yields a homomorphism*

$$\gamma : \pi_{2n-1}(S^n) \to J.$$

Let $f_i : S^{2n-1} \to S^n$ be maps (sending base-point to base-point) representing $\alpha_i \in \pi_{2n-1}(S^n)$, $i = 1, 2$. Let $T = S_1^{2n-1} \vee S_2^{2n-1}$, let $q : S^{2n-1} \to T$ be the map pinching S^{2n-2} to the base-point and let $h : T \to S^n$ be given by $h \mid S_i^{2n-1} = f_i$. Then it follows from II. 1.13 that qh represents $\alpha_1 + \alpha_2$. Let j_1 embed S^{2n-1} in T as S_1^{2n-1}. Then $f_1 = j_1 h$. Let $m_1 : S_{f_1} \to S_h$ extend $(j_1, 1)$ where $1 : S^n \to S^n$. Then we have the commutative diagram

$$\begin{array}{ccccc}
H^n(S^n) & \xrightarrow{p^*} & H^n(S_{f_1}) & \xleftarrow{\mu^*} & H^n(S_{f_1}, S^{2n-1}) \\
\uparrow 1 & & \uparrow m_1^* & & \uparrow \hat{m}_1^* \\
H^n(S^n) & \xrightarrow{p^*} & H^n(S_h) & \xleftarrow{\mu^*} & H^n(S_h, T)
\end{array}$$

$$\begin{array}{ccccc}
& \xrightarrow{Sq} & H^{2n}(S_{f_1}, S^{2n-1}) & \xleftarrow{\nu^*} & H^{2n-1}(S^{2n-1}) \\
& & \uparrow \hat{m}_1^* & & \uparrow j_1^* \\
& \xrightarrow{Sq} & H^{2n}(S_h, T) & \xleftarrow{\nu^*} & H^{2n-1}(T)
\end{array}$$

Let $\gamma_i = \gamma(f_i)$ and let ξ_1, ξ_2 be the generators of $H^{2n-1}(T)$ corresponding to S_1^{2n-1}, S_2^{2n-1}. Clearly $\hat{m}_1^*(\hat{\eta}_h^2) = \hat{\eta}_{f_1}^2 = \nu^*(\gamma_1\xi)$ and $j_1^*\xi_1 = \xi$, $j_1^*\xi_2 = 0$. Now for some integers β_1, β_2, $\hat{\eta}_h^2 = \nu^*(\beta_1\xi_1 + \beta_2\xi_2)$; thus

$$\begin{aligned}
\nu^*(\gamma_1\xi) &= \hat{m}_1^*\nu^*(\beta_1\xi_1 + \beta_2\xi_2) \\
&= \nu^* j_1^*(\beta_1\xi_1 + \beta_2\xi_2) \\
&= \nu^*(\beta_1\xi).
\end{aligned}$$

Hence $\beta_1 = \gamma_1$.

Similarly we prove that $\beta_2 = \gamma_2$, so that

$$\hat{\eta}_h^2 = \nu^*(\gamma_1\xi_1 + \gamma_2\xi_2).$$

Put $f = qh$ and let $m : S_f \to S_h$ extend $(q, 1)$, where $1 : S^n \to S^n$. We have the commutative diagram

$$\begin{array}{ccccc}
H^n(S^n) & \xrightarrow{p^*} & H^n(S_f) & \xleftarrow{\mu^*} & H^n(S_f, S^{2n-1}) \\
\uparrow 1 & & \uparrow m^* & & \uparrow \hat{m}^* \\
H^n(S^n) & \xrightarrow{p^*} & H^n(S_h) & \xleftarrow{\mu^*} & H^n(S_h, T)
\end{array}$$

$$\begin{array}{ccccc}
\xrightarrow{Sq} & H^{2n}(S_f, S^{2n-1}) & \xleftarrow{\nu^*} & H^{2n-1}(S^{2n-1}) \\
& \uparrow \hat{m}^* & & \uparrow q^* \\
\xrightarrow{Sq} & H^{2n}(S_h, T) & \xleftarrow{\nu^*} & H^{2n-1}(T)
\end{array}$$

Clearly $\hat{m}^*(\hat{\eta}_h^2) = \hat{\eta}_f^2$. Moreover, q maps S^{2n-1} with degree 1 on each S_i^{2n-1} so that $q^*\xi_1 = q^*\xi_2 = \xi$. Thus

$$\hat{\eta}_f^2 = \hat{m}^*\nu^*(\gamma_1\xi_1 + \gamma_2\xi_2) = \nu^*q^*(\gamma_1\xi_1 + \gamma_2\xi_2) = (\gamma_1 + \gamma_2)\,\nu^*\xi$$

so that $\gamma(f) = \gamma_1 + \gamma_2$. ▌

We shall show that, if n is even, there exists a map $S^{2n-1} \to S^n$ of Hopf invariant 2. It will then follow from 9.5.9 (ii) or 9.5.12 that there exist maps of Hopf invariant $2d$ for any integer d. To achieve this object it is convenient to modify the characterization of the Hopf invariant.

Consider CS^{2n-1}, the cone on S^{2n-1}; let $g : S^{2n-1} \to S^n$ be a map and let Y be obtained by attaching CS^{2n-1} to S^n by means of g (see 8.5). We write f for the characteristic map and a for the image under f of the vertex of the cone; then by 8.5.9

$$f^* : H^*(Y, S^n) \to H^*(CS^{2n-1}, S^{2n-1})$$

is an isomorphism, so that $H^{2n}(Y, S^n) = J$, $H^p(Y, S^n) = 0$, $p \neq 2n$. It follows (since $n > 1$) from the exact sequence for Y, S^n that $H^p(Y) = 0$ unless $p = 0, n, 2n$ and that

9.5.13 $$H^p(Y) = J, \quad p = 0, n, 2n.$$

9.5.14 Proposition. *We may choose generators σ, τ for $H^n(Y)$, $H^{2n}(Y)$ so that $\sigma^2 = \gamma\tau$, where γ is the Hopf invariant of g.*

Consider the mapping cylinder S_g; reference to the definitions of S_g, CS^{2n-1} and Y show at once that there is a map $k : S_g, S^{2n-1} \to Y, a$ that is homeomorphic from $S_g - S^{2n-1}$ to $Y - a$. Simple excision and retraction arguments like those of 8.5.9 show that

$$k^* : H^*(Y, a) \to H^*(S_g, S^{2n-1})$$

is an isomorphism and so

$$k^* : R^*(Y, a) \cong R^*(S_g, S^{2n-1}).$$

Identifying $H^r(Y,a)$ with $H^r(Y)$, as we may, for $r \geqslant 1$, and using the fact that $\hat{\eta}_g$, $\nu^*\xi$ are generators of $H^n(S_g, S^{2n-1})$, $H^{2n}(S_g, S^{2n-1})$, we deduce that there are generators σ, τ of $H^n(Y)$, $H^{2n}(Y)$ such that $k^*\sigma = \hat{\eta}_g$, $k^*\tau = \nu^*\xi$. Since $\hat{\eta}_g^2 = \gamma\nu^*\xi$, $\sigma^2 = \gamma\tau$. ∎

We use the characterization of the Hopf invariant given by 9.5.14 to prove that the map we are about to define has Hopf invariant ± 2.

Consider the map $q : V^n, S^{n-1} \to S^n, x_*$ given by II. 1.5 which is a homeomorphism of the interior of the n-ball V^n onto $S^n - x_*$, and sends the frontier S^{n-1} to the base-point x_*. Now $V^n \times V^n$ is homeomorphic to a $2n$-ball; its frontier, $(V^n \times V^n)^{\cdot}$, is the union of $S^{n-1} \times V^n$ and $V^n \times S^{n-1}$ and is homeomorphic to S^{2n-1}. Consider the map

$$g : (V^n \times V^n)^{\cdot} \to S^n,$$

given by

$$\begin{aligned} (x,y)\,g &= xq, \quad (x,y) \in V^n \times S^{n-1} \\ &= yq, \quad (x,y) \in S^{n-1} \times V^n. \end{aligned}$$

Then g has a Hopf invariant $\gamma(g)$, defined since $(V^n \times V^n)^{\cdot}$ is oriented.

9.5.15 Theorem. $\gamma(g) = \pm 2$, *if n is even.*

We construct Y by attaching $V^n \times V^n$ to S^n by means of g. Consider now $S^n \times S^n$ and identify $S^n \vee S^n$ with the subset $(S^n \times x_*) \cup (x_* \times S^n)$ of $S^n \times S^n$. There is then a commutative diagram of maps

$$\begin{array}{ccc} V^n \times V^n, (V^n \times V^n)^{\cdot} & \xrightarrow{q \times q} & S^n \times S^n, S^n \vee S^n \\ & {\scriptstyle k} \searrow \quad \swarrow {\scriptstyle h} & \\ & Y, S^n & \end{array}$$

where k is the identification map (so that $k \,|\, (V^n \times V^n)^{\cdot} = g$), and h maps $S^n \times S^n - (S^n \vee S^n)$ homeomorphically onto $Y - S^n$ and $h \,|\, S^n \vee S^n$ is the 'folding' map

9.5.16 $$(x, x_*)\,h = (x_*, x)\,h = x, \quad x \in S^n.$$

Let σ, τ be as in 9.5.14. Then $H^n(S^n \times S^n) = J \oplus J$ with generators σ_1, σ_2 such that $h^*\sigma = \sigma_1 + \sigma_2$. It follows easily from 9.4.13 that $H^{2n}(S^n \times S^n) = J$, and is generated by $\sigma_1\sigma_2$ and that $\sigma_1^2 = \sigma_2^2 = 0$.

Moreover, it follows from 8.5.9 that $q \times q$ and k both induce (relative) contrahomology isomorphisms, so that

$$\hat{h}^* : R^*(Y, S^n) \cong R^*(S^n \times S^n, S^n \vee S^n),$$

and we immediately conclude from the exact sequences that

$$h^* : H^{2n}(Y) \cong H^{2n}(S^n \times S^n).$$

Thus we have $h^*(\tau) = \pm\sigma_1\sigma_2$. Now

$$h^*(\sigma^2) = (\sigma_1+\sigma_2)^2 = \sigma_1\sigma_2+\sigma_2\sigma_1 = 2\sigma_1\sigma_2,$$

since n is even. Thus $h^*(\sigma^2) = \pm 2h^*(\tau)$, so that $\sigma^2 = \pm 2\tau$, and the theorem follows from 9.5.14. ∎

This result leaves open the question of the existence of elements of odd Hopf invariant; by 9.5.12 and 9.5.15 this is equivalent to the existence of elements of Hopf invariant 1. Hopf (in the papers cited) showed that such elements exist‡ if $n = 2$, 4 or 8; Adem showed that no such elements exist if $n \neq 2^k$. J. F. Adams has recently completed the investigation by proving that Hopf's examples are the only cases where maps of Hopf invariant 1 exist. Adams' result imply (i) that S^1, S^3, S^7 are the only parallelizable spheres, and (ii) that there is no division algebra over the reals of order > 8.

We close with an informal observation. Let $g: S^{m-1} \to S^{n-1}$ and let $f = S(g)$, the *suspension* of g (see 8.5), so that $f: S^m \to S^n$. Then it is easy to see that, if $q: S^n \to S^n$ is a map of degree d, and if f represents $\alpha \in \pi_m(S^n)$, then fq represents $d\alpha$. From this, 9.5.9 (i) and 9.5.12, we see that $d^2\gamma(\alpha) = d\gamma(\alpha)$, all d, so that $\gamma(\alpha) = 0$. Thus the *Hopf invariant of a suspension element is zero*. The intimate relation between the Hopf invariant and the suspension homomorphism has proved of great consequence in homotopy theory.

9.6* Appendix: Naturality

At various places in this book, and particularly in chapter 8 and the present chapter, we have had occasion to remark that a certain transformation is 'natural'; each time we have stated explicitly in what property this naturality consists. Moreover, we have seen that, in the acyclic model arguments, the naturality property has been decisive in securing the truth of the assertion under consideration.

Reference to Eilenberg and Steenrod [14], chapter 4, will provide the reader with a precise statement of the naturality property. However, since many of our applications are not to be found in that reference, we repeat in this appendix the essential definitions and point out their relevance to the operations $(\pi, \rho, D, \ldots)$ which have been discussed in this and the previous chapter.

An *abstract category* is an associative groupoid with identities. That is, it is a collection, $\mathscr{C}$, of elements $\gamma_1, \gamma_2, \ldots$, such that a product $\gamma_1\gamma_2$ is defined for some pairs $\gamma_1, \gamma_2 \in \mathscr{C}$; an *identity* of $\mathscr{C}$ is an element ϵ such that $\epsilon\gamma_2 = \gamma_2$, $\gamma_1\epsilon = \gamma_1$, whenever $\epsilon\gamma_2$, $\gamma_1\epsilon$ are defined. Then $\mathscr{C}$ is to satisfy the following axioms:

($\mathscr{C}$ 1) (associative law) $\gamma_1(\gamma_2\gamma_3)$ is defined if and only if $(\gamma_1\gamma_2)\gamma_3$ is defined and then $\gamma_1(\gamma_2\gamma_3) = (\gamma_1\gamma_2)\gamma_3$. (We write either as $\gamma_1\gamma_2\gamma_3$.)

‡ The Hopf map $S^3 \to S^2$, mentioned as an example of a fibre map in II.3, has Hopf invariant 1. The Hopf maps $S^7 \to S^4$, $S^{15} \to S^8$, defined as in II.3 but with quaternions, Cayley numbers replacing complex numbers, also have Hopf invariant 1.

($\mathscr{C}$ 2) $\gamma_1\gamma_2\gamma_3$ is defined if $\gamma_1\gamma_2$ and $\gamma_2\gamma_3$ are defined.

($\mathscr{C}$ 3) For each γ, there exist identities ϵ, ϵ' such that $\epsilon\gamma$ and $\gamma\epsilon'$ are defined.

For example, the collection of continuous maps of topological spaces forms an abstract category. Indeed, we will refer to the elements of $\mathscr{C}$ as *maps*.

9.6.1 Proposition. *The identities ϵ, ϵ' given by $\mathscr{C}$ 3, are uniquely determined by γ.*

Let $\epsilon_1\gamma$, $\epsilon_2\gamma$ be defined, where ϵ_1, ϵ_2 are identities. Then $\epsilon_1(\epsilon_2\gamma)$ is defined, so that, by $\mathscr{C}$ 1, $\epsilon_1\epsilon_2$ is defined. Then $\epsilon_1 = \epsilon_1\epsilon_2 = \epsilon_2$. Thus ϵ is uniquely determined, and so also ϵ'. ∎

Let $\mathscr{I}$ be a collection of objects in a fixed (1, 1) correspondence with the identities of $\mathscr{C}$. If $\gamma \in \mathscr{C}$, let X, X' be the objects of $\mathscr{I}$ corresponding to the identities ϵ, ϵ' given by $\mathscr{C}$ 3. Then we may write‡

$$\gamma : X \to X',$$

and we may write 1_X, $1_{X'}$ for ϵ, ϵ'.

The abstract category $\mathscr{C}$, together with the collection $\mathscr{I}$ and the fixed (1, 1) correspondence, is called a *category*. Notice that the structure of a category is determined by that of the underlying abstract category—the objects are only introduced for convenience.

A map $\gamma : X \to X'$ is called an *equivalence* (in the category) if there exists a map $\gamma' : X' \to X$ such that $\gamma'\gamma$ and $\gamma\gamma'$ are identities.

We now give examples of categories; we write first the objects and then the maps:

(i) the category of topological spaces and continuous maps;

(ii) the category of simplicial complexes and simplicial maps;

(iii) the category of finite simplicial complexes and continuous maps; here the objects are simplicial complexes, and a 'map' from K to L is a continuous map from $|K|$ to $|L|$, or more precisely a triple (K, L, f), where $f : |K| \to |L|$;

(iv) the category of chain complexes and chain maps;

(v) the category of abelian groups and homomorphisms;

(vi) the category of graded abelian groups and graded homomorphisms;

(vii) the category of contrachain rings and contrachain-ring maps.

An equivalence in (i) is a homeomorphism, an equivalence in (iii)

‡ Notice the discrepancy with the Eilenberg–Steenrod notation, caused by our practice of writing operators on the *right*.

is a triple (K, L, h) where $h : |K| \cong |L|$; in the other examples, we call it an isomorphism (of the given sort).

Let $\mathscr{C}$ and $\mathscr{D}$ be categories and let T be a function which maps the objects of $\mathscr{C}$ to objects of $\mathscr{D}$ and the maps of $\mathscr{C}$ to maps of $\mathscr{D}$. Then T is called a *covariant functor* if the following conditions are satisfied:

(F 1) If $\gamma : X \to X'$, then $\gamma T : XT \to X'T$.

(F 2) $1_X T = 1_{XT}$.

(F 3) If $\gamma_1 \gamma_2$ is defined, then $(\gamma_1 \gamma_2) T = (\gamma_1 T)(\gamma_2 T)$.

We give examples of covariant functors:

(*a*) Δ, the functor‡ from category (i) to category (iv) which associates with a space X the singular chain complex $\Delta(X)$ and, with a map $f : X \to Y$, the chain map $f^\Delta : \Delta(X) \to \Delta(Y)$.

(*b*) H_*^Ω, the functor from category (iii) to category (vi) which associates with a simplicial complex K the graded abelian group $H_*^\Omega(K)$ and, with a map $f : |K| \to |L|$, the homomorphism

$$f_*^\Omega : H_*^\Omega(K) \to H_*^\Omega(L).$$

(*c*) $\otimes$, the functor which associates with a pair of abelian groups A, B their tensor product $A \otimes B$ and, with a pair of homomorphisms $\phi : A \to A'$, $\psi : B \to B'$, their tensor product $\phi \otimes \psi : A \otimes B \to A' \otimes B'$.

We are also concerned with *contravariant functors*; when such a function T from $\mathscr{C}$ to $\mathscr{D}$ is in question we have found it convenient (to our purposes) to write the maps of $\mathscr{D}$ as *left* operators. Then the only difference in the definition of a contravariant functor is that F 1 is replaced by

(F 1′) If $\gamma : X \to X'$, then $\gamma T : X'T \to XT$.

We give examples of contravariant functors:

(*d*) H_s^*, the functor from category (i) to category (vi) which associates with a space X the graded abelian group $H_s^*(X)$ and with $f : X \to Y$ the homomorphism $f^* : H_s^*(Y) \to H_s^*(X)$;

(*e*) $\pitchfork G$, the functor which associates with a chain complex C the contrachain complex $C \pitchfork G$ and with a chain map $\phi : C \to D$ the (adjoint) contrachain map $\phi^\cdot : D \pitchfork G \to C \pitchfork G$.

There are also functors of several variables§ and these may be mixed, in the same sense as tensors are mixed. Thus, for example, we may consider $\pitchfork$ as a functor of two variables, each taken from

‡ Had the ideas of category and functor been introduced early in this book, it would have been logical for us, for instance, to have written the singular chain complex of X as $(X)\Delta$ and the homology group of K as $(K)H_*$.

§ $\otimes$ may, of course, be interpreted as a functor of two variables; we interpreted it above as a functor of a single variable, taken from the category of ordered pairs of abelian groups.

the category of abelian groups, with values in the category of abelian groups. It is then contravariant in the first variable and covariant in the second.

9.6.2 Proposition. *If $T : \mathscr{C} \to \mathscr{D}$, $U : \mathscr{D} \to \mathscr{E}$ are functors, then TU is a functor. It is covariant if T and U are both covariant or both contravariant and it is contravariant otherwise.* ∎

Let S and T be two covariant functors from the category $\mathscr{C}$ to the category $\mathscr{D}$. Then a *natural transformation* from S to T is a function Γ, defined on the objects of $\mathscr{C}$ and taking as values maps of $\mathscr{D}$ such that

(N 1) If X is an object of $\mathscr{C}$, then $X\Gamma : XS \to XT$.

(N 2) If $\gamma : X \to X'$, then the diagram

$$\begin{array}{ccc} XS & \xrightarrow{X\Gamma} & XT \\ \downarrow{\scriptstyle \gamma S} & & \downarrow{\scriptstyle \gamma T} \\ X'S & \xrightarrow{X'\Gamma} & X'T \end{array}$$

is commutative.

If S and T are *contravariant* functors, (N 2) should be replaced by

(N 2′) If $\gamma : X \to X'$, then the diagram

$$\begin{array}{ccc} XS & \xrightarrow{X\Gamma} & XT \\ \uparrow{\scriptstyle \gamma S} & & \uparrow{\scriptstyle \gamma T} \\ X'S & \xrightarrow{X'\Gamma} & X'T \end{array}$$

is commutative.

Finally, we say that Γ is a *natural equivalence* (between the functors S and T) if $X\Gamma$ is an equivalence (in $\mathscr{D}$) for each $X \in \mathscr{C}$.

We give examples:

(α) Consider the functors Δ and $\square^N$ from the category, $\mathscr{C}$, of topological spaces to the category, $\mathscr{D}$, of chain complexes. The maps $\{\phi^X\}$ of 8.4 provide a natural transformation from Δ to $\square^N$. Let H be the homology functor from the category $\mathscr{D}$ to the category $\mathscr{E}$ of graded abelian groups. Then the maps $\{\phi_*^X\}$ provide a natural equivalence between the functors ΔH and $\square^N H$. This gives precision to the statement that the singular simplicial homology theory is equivalent to the singular cubical homology theory.

(β) Consider the functors Δ and $\Delta \otimes \Delta$ from the category $\mathscr{C}$ to the category $\mathscr{D}$ ($\mathscr{C}$, $\mathscr{D}$ as above and by definition $\Delta \otimes \Delta$ maps X to $\Delta X \otimes \Delta X$). Then D (9.2.4) is a natural transformation from Δ to $\Delta \otimes \Delta$. For 9.2.5 (i) asserts that $XD : \Delta X \to \Delta X \otimes \Delta X$ is in $\mathscr{D}$ and 9.2.5 (ii) asserts that N 2 is satisfied.

(γ) Let $\mathscr{D} \times \mathscr{D}$ be the category of ordered pairs of chain complexes and let $\mathscr{D}^{\cdot}$ be the category of contrachain complexes. If R is a ring we may consider the contravariant functors $S = (\pitchfork R) \otimes (\pitchfork R)$ and $T = \otimes . \pitchfork R$ from $\mathscr{D} \times \mathscr{D}$ to $\mathscr{D}^{\cdot}$, where $(C, D)\, S = (C \pitchfork R) \otimes (D \pitchfork R)$ and $(C, D)\, T = (C \otimes D) \pitchfork R$. Then π (see 5.7.27) is a natural transformation from S to T.

(δ) Let $\mathscr{P}$ be the category (iii) and let J be the functor from $\mathscr{P}$ to $\mathscr{C}$ which associates with each complex its polyhedron. Then (8.6.7 and 8.6.9) α_* is a natural equivalence from the functor H^{Ω} to the functor $J \Delta H$.

EXERCISES

On the definition of the cup product

1. Let κ be the Kronecker product

$$\kappa : H_p(X) \otimes H^p(X; A) \to A,$$

$$\kappa : H_q(Y) \otimes H^q(Y; B) \to B,$$

$$\kappa : H_r(X \otimes Y) \otimes H^r(X \otimes Y; G) \to G.$$

Let $\theta : A \otimes B \to G$ be a pairing inducing

$$\pi : H^p(X; A) \otimes H^q(Y; B) \to H^{p+q}(X \otimes Y; G).$$

Let

$$\eta : H_p(X) \otimes H_q(Y) \to H_{p+q}(X \otimes Y)$$

be as in 5.7.11. Now κ induces a product, which we again call κ,

$$\kappa : \{H_p(X) \otimes H_q(Y)\} \otimes \{H^p(X; A) \otimes H^q(Y; B)\} \to A \otimes B.$$

Prove that, writing maps on the left, $\kappa(\eta \otimes \pi) = \theta\kappa$. Deduce that the homology diagram corresponding to the cup product diagram

$$H^p(X) \otimes H^q(X) \xrightarrow{\pi} H^{p+q}(X \otimes X) \xrightarrow{D^*} H^{p+q}(X)$$

is

$$H_p(X) \otimes H_q(X) \xrightarrow{\eta} H_{p+q}(X \otimes X) \xleftarrow{D_*} H_{p+q}(X).$$

(We owe essentially to N. E. Steenrod this explanation of the absence of a product in homology corresponding to the cup product in contrahomology.)

2. Let K be a simplicial complex and let $K \times K$ be a cellular dissection of $|K| \times |K|$ by the cells $s \times t$, where s, t are simplexes of K. Let $d_0 : K \to K \times K$ be a cellular approximation to the diagonal map $d : |K| \to |K| \times |K|$, and let $m : |K| \times |K| \to |K| \times |K|$ be given by $(x, y)\, m = (y, x)$. Then m induces a cellular map $m : K \times K \to K \times K$. Show that we may find a sequence of homomorphisms $d_n : C(K) \to C(K \times K) = C(K \otimes K)$, $n = 0, 1, \ldots$, where d_n raises dimension by n and

$$d_n - (-1)^n d_n m = d_{n+1} \partial + (-1)^n \partial d_{n+1}.$$

Interpret the case $n = 0$.

3. Prove the skew-commutativity of $R^*(K)$ as in 9.2.

4. Let X, Y be spaces whose homology groups are finitely generated in each dimension, and let $R = J$ or J_m, qua ring. Prove that

$$\pi : \Delta^{\cdot}(X;R) \otimes \Delta^{\cdot}(Y;R) \to \Delta^{\cdot}(X \otimes Y;R)$$

is a contrachain ring equivalence. [Hint: use the proof of 5.4.25]

Computations

5. Given any m complex numbers (∞ being admitted), $z_1, \ldots, z_m$, let

$$a_0 z^m + a_1 z^{m-1} + \ldots + a_m = 0$$

be the equation whose roots are $z_1, \ldots, z_m$. Show that $(z_1, \ldots, z_m) \to [a_0, a_1, \ldots, a_m]$ determines a map $f : S^2 \times S^2 \times \ldots \times S^2 \to C^{(m)}$. Let S_m, the symmetric group on m objects, act in the obvious way on $S^2 \times S^2 \times \ldots \times S^2$ and let $SP^m(S^2)$ be the space obtained by identifying points of $S^2 \times S^2 \times \ldots \times S^2$ which correspond under the action of S_m; we call $SP^m(S^2)$ the *m-fold symmetric product* of S^2. Show that f induces a homeomorphism $SP^m(S^2) \cong C^{(m)}$. By considering the identification map $h : S^2 \times S^2 \times \ldots \times S^2 \to SP^m(S^2)$, compute the integral contrahomology ring of $C^{(m)}$. (The reader is intended to assume without proof the fact that h has degree $m!$) Hence prove that $C^{(m-1)}$ is not a retract of $C^{(m)}$.

6. If P^n is real projective n-space, compute the contrahomology rings over J and J_2 of $C^{(m)} \times P^n$.

7. Prove that if X is the suspension of some space Y and if $\alpha \in H^p(X;R)$, $\beta \in H^q(X;R)$, $p, q > 0$, then $\alpha\beta = 0$.

Functional cup product

8. Let

$$(\mathrm{D}) \qquad \begin{array}{ccccccccccccc}
\cdots \to & G_{n-1} & \xrightarrow{\gamma_{n-1}} & G_n & \xrightarrow{\gamma_n} & G_{n+1} & \xrightarrow{\gamma_{n+1}} & G_{n+2} & \xrightarrow{\gamma_{n+2}} & G_{n+3} & \to \cdots \\
& \downarrow{\scriptstyle \mu_{n-1}} & & \downarrow{\scriptstyle \mu_n} & & \downarrow{\scriptstyle \mu_{n+1}} & & \downarrow{\scriptstyle \mu_{n+2}} & & \downarrow{\scriptstyle \mu_{n+3}} & \\
\cdots \to & H_{n-1} & \xrightarrow{\eta_{n-1}} & H_n & \xrightarrow{\eta_n} & H_{n+1} & \xrightarrow{\eta_{n+1}} & H_{n+2} & \xrightarrow{\eta_{n+2}} & H_{n+3} & \to \cdots
\end{array}$$

represent a homomorphism of exact sequences. Let $g \in \ker \gamma_{n+2} \cap \ker \mu_{n+2}$. Then $g = g'\gamma_{n+1}$, $g'\mu_{n+1}\eta_{n+1} = g\mu_{n+2} = 0$, so $g'\mu_{n+1} = h\eta_n$, $h \in H_n$. Show that the passage from g to h determines a homomorphism

$$\rho : \ker \gamma_{n+2} \cap \ker \mu_{n+2} \to H_n/(G_n\mu_n + H_{n-1}\eta_{n-1}).$$

By interpreting (D) as the diagram of 9.1.4 obtain a homomorphism

$$\rho_\eta : \bar{H}^p(X) \to H^{p+q-1}(X_0)/(\cup_\eta H^{p-1}(X_0) + \lambda^* H^{p+q-1}(X)),$$

where $\bar{H}^p(X)$ is the subgroup of $H^p(X)$ consisting of elements ξ satisfying $\xi\eta = 0$, $\lambda^*\xi = 0$.

9. Let $f : X \to Y$ be a map and let $i : X \to Y_f$ embed X in the mapping cylinder. Let $\xi, \eta \in R^*(Y)$ be such that $\xi\eta = 0, f^*\xi = 0$ and let $p : Y_f \to Y$ be the retraction of 9.5. Then $(p^*\xi)(p^*\eta) = 0$ and $\lambda^*(p^*\xi) = 0$. Define

$$\langle f, \xi, \eta \rangle = \rho_{p^*\eta}(p^*\xi).$$

Prove that

$$\langle f, \xi, \eta \rangle \in H^{p+q-1}(X)/(\cup_{f^*\eta} H^{p-1}(X) + f^* H^{p+q-1}(Y))$$

if $\xi \in H^p(Y)$, $\eta \in H^q(Y)$. We call $\langle f, \xi, \eta \rangle$ the *functional cup product* of ξ and η, with respect to f.

Interpret the Hopf invariant as a functional cup product.

10. (i) Let $\xi_1\eta = \xi_2\eta = 0$ and $f^*\xi_1 = f^*\xi_2 = 0$ (notation of Q. 9). Show that

$$\langle f, \xi_1 + \xi_2, \eta \rangle = \langle f, \xi_1, \eta \rangle + \langle f, \xi_2, \eta \rangle.$$

What can be said about linearity with respect to η?

(ii) Suppose also that $g: W \to X$ and put $h = gf: W \to Y$. Show that g induces

$$g^{\#}: H^{p+q-1}(X)/(\cup_{f^*\eta} H^{p-1}(X) + f^* H^{p+q-1}(Y))$$
$$\to H^{p+q-1}(W)/(\cup_{h^*\eta} H^{p-1}(W) + h^* H^{p+q-1}(Y))$$

and that
$$\langle h, \xi, \eta \rangle = g^{\#}\langle f, \xi, \eta \rangle.$$

11. Deduce from Q. 10 (ii) that if a map $h: S^{2n-1} \to S^n$ can be factored through S^m, where $2n-1 > m > n$, then its Hopf invariant is zero.

Categories and functors

Express in the terminology of categories and functors the notions of Q. 12–18

12. The subdivision operation $K, K_0 \to K', K_0'$ and the chain map χ.

13. The chain equivalence $\bar{\theta}^\omega: C^\omega(K) \to C^\Omega(K)$ and the induced contrahomology ring isomorphism.

14. The exact contrahomology sequence of a pair (K, K_0).

15. The torsion product $A * B$ (i) as a function of A, (ii) as a function of A and B.

16. The isomorphism $\operatorname{Ext}(A, B) \cong A \dagger B$ of 5.9 (as a natural equivalence of two functors of two variables).

17. The exact triangle theorem (see 5.5.1)

18. The chain equivalence $\rho: \Delta(X \otimes Y) \to \Delta(X \times Y)$.

19 Show that the composition of two natural transformations (equivalences) is a natural transformation (equivalence). Give examples.

20. Prove that a partially ordered set $\{\Lambda, \prec\}$ is a category, whose objects are the elements $\lambda \in \Lambda$ and whose 'maps' are the ordered pairs $\lambda \prec \lambda'$. Prove that a covariant functor from a directed set to the category of groups is a direct system of groups.

10*

SPECTRAL HOMOLOGY THEORY AND HOMOLOGY THEORY OF GROUPS

We shall be mainly concerned in this chapter with two algebraic ideas; these ideas, however, have many geometrical applications, some of which we shall describe. Spectral theory, for instance, is applied to provide a connexion between the homology groups of fibre space, fibre and base, and a section on the homology theory of a space with operators gives an application both of spectral theory and of the homology theory of groups.

10.1 Filtration

An abelian group G may possess any or all of three additional elements of structure; two of these are familiar, namely, graduation and derivation, the third, filtration, is soon to be defined. It is the interplay of these structures that is studied in this chapter. We first give definitions of graduation and derivation appropriate to this context.

10.1.1 Definition. A *positive graduation* $\{G_n\}$ on an abelian group is a set of subgroups $G_0, G_1, \ldots, G_n, \ldots$ such that $G = G_0 \oplus G_1 \oplus \ldots = \sum_n G_n$. A group with a positive graduation is called a *graded group* or *g.-group* $(G; G_n)$. If $(G; G_n)$, $(G'; G'_n)$ are g.-groups a *homomorphism*

$$\phi : (G; G_n) \to (G'; G'_n)$$

is a homomorphism $\phi : G \to G'$ such that $G_n \phi \subseteq G'_n$ for all n; we may call such a homomorphism a *g.-homomorphism from G to G'*.

In this chapter we shall deal mainly with positive graduations and the adjective will normally be omitted. A positive graduation determines a graduation, namely that for which $G_n = 0$ for $n < 0$; in any formula related to a positive graduation, G_n is to be taken to be null for negative n. A g.-homomorphism often arises in practice from a set $\{\phi_n\}$ of homomorphisms $\phi_n : G_n \to G'_n$; any such set determines a g.-homomorphism and conversely. It is occasionally convenient to extend the definition of a g.-homomorphism by defining a homomorphism $\phi : G \to G'$ as a *g.-homomorphism of index s* if $G_n\phi \subseteq G'_{n+s}$ for all n.

10.1.2 Definition. A *differential group* or *d.-group* $(G; d)$ is an abelian group G and an endomorphism $d : G \to G$ such that $dd = 0$. If $(G; d)$, $(G'; d')$ are d.-groups a *homomorphism* $\phi : (G; d) \to (G'; d')$ is a homomorphism $\phi : G \to G'$ such that $\phi d' = d\phi$; we may call such a homomorphism a *d.-homomorphism from G to G'*.

A group may be a differential group and at the same time have a graduation; if these structural elements are suitably related it is then called a differential graded group.

10.1.3 Definition. A *differential graded group* or *d.g.-group of index s* $(G; G_n, d)$ is a graded group $(G; G_n)$ together with a g.-endomorphism d of index s such that $dd = 0$. If $(G'; G'_n, d')$ is another d.g.-group of the same index s, a *homomorphism* $\phi : (G; G_n, d) \to (G'; G'_n, d')$ is a g.-homomorphism $\phi : (G; G_n) \to (G'; G'_n)$ such that $\phi_n d' = d\phi_{n+s}$; we may call such a homomorphism a *d.g.-homomorphism* from G to G'.

We shall of course be almost exclusively concerned with the two cases $s = +1$ and $s = -1$, those of contrachain complexes and maps and of chain complexes and maps.

It is sometimes useful to extend the definition of a d.g.-homomorphism by defining a *d.g.-homomorphism of index t* to be a g.-homomorphism ϕ of index t such that $\phi_n d' = (-1)^{st} d\phi_{n+s} : G_n \to G'_{n+s+t}$. If t is odd such a d.g.-homomorphism is not a d.-homomorphism; this anomaly should cause no inconvenience as we shall have little interest in the case $t \neq 0$; we remark however that the map $\nu : C(K, K_0) \to C(K_0)$ of 2.8.15 is a d.g.-homomorphism of index -1.

We now introduce the notion of a filtration. It has a rather wide application; a construct may be said to be filtered if an increasing sequence of sub-constructs is selected which exhaust the whole construct.

10.1.4 Definition. An *increasing*‡ *filtration* $[A_p]$ on an abelian group G is a set§ of subgroups $(G)A_p, p = 0, 1, 2, \ldots$, such that $GA_0 \subseteq GA_1 \subseteq \ldots$, and that $\bigcup_p GA_p = G$. An abelian group with such a filtration is called a *filtered group* or *f.-group* $(G; A_p)$. If $(G'; A'_p)$ is also a filtered group a *homomorphism* $\phi : (G; A_p) \to (G'; A'_p)$ is a homomorphism $\phi : G \to G'$

‡ A decreasing filtration arises in studying contrahomology; see 10.5.

§ In this definition we regard A_p, for each p, as a function associating with the group G a subgroup $(G)A_p$. When later $(G)A_p$ is abbreviated to A_p, it is natural (and not misleading) to think of A_p as being the subgroup itself. The point of view of the definition, however, is useful in defining a filtration on a graded group in 10.3.

such that, for all p, $GA_p\phi \subseteq G'A'_p$. We may call such a homomorphism an *f.-homomorphism from G to G'*.

For convenience we also define GA_p for $p<0$ to be 0. When no confusion can arise we shall abbreviate GA_p to A_p and increasing filtration to filtration.

10.1.5 Definition. A filtration $[A_p]$ is called a *finite filtration of length* n if $A_n = G$.

It is important to distinguish at the outset between the notion of filtration and graduation; the former is an increasing nest of subgroups exhausting the group, the latter a sequence of subgroups with null mutual intersections whose sum is the whole group. Although these ideas are strongly contrasted there are useful ways of passing from one to the other.

10.1.6 Definition. If $(G; A_p)$ is an f.-group, the *associated graded group* $\mathrm{Gr}\,(G; A_p) = \mathrm{Gr}\,[A_p]$ is the direct sum $\sum\limits_0^\infty A_p/A_{p-1}$ with $(\mathrm{Gr}\,[A_p])_n = A_n/A_{n-1}$.

With any graded group $(G; G_n)$ we may also associate the filtered group $(G; \sum\limits_{n \leqslant p} G_n)$. There is plainly a canonical g.-isomorphism between $(G; G_n)$ and $\mathrm{Gr}\,[\sum\limits_{n\leqslant p} G_n]$, determined by the canonical isomorphism $G_n \cong \sum\limits_{m\leqslant n} G_m / \sum\limits_{m<n} G_m$; it will sometimes in fact be convenient to identify these groups. On the other hand, if we start with an f.-group $(G; A_p)$, pass to the associated graded group and from that to its associated filtered group, we may reach an f.-group not f.-isomorphic with $(G; A_p)$; indeed $\mathrm{Gr}\,[A_p]$ may not even be isomorphic as a group with G. This is shown by

10.1.7 *Example.* Let $G = J_4$, $A_0 = 2J_4$, $A_1 = A_2 = \ldots = G$; then $A_0 \cong J_2$, $A_1/A_0 \cong J_2$, so that $\mathrm{Gr}\,[A_p] \cong J_2 \oplus J_2 \ncong J_4$.

We shall return later to g.-groups and consider filtrations of g.-groups of much greater interest and generality than the filtration associated with the graduation. The important properties of a filtration, however, arise in the case of a differential group, whether graded or not.

10.1.8 Definition. A *differential filtered group* or *d.f.-group* $(G; A_p, \partial)$ is an f.-group $(G; A_p)$ with an f.-endomorphism‡ $\partial : G \to G$ such that

‡ Because increasing filtration is associated with homology we use ∂ rather than d for the differential.

$\partial\partial = 0$. A *d.f.-homomorphism* is an f.-homomorphism that is also a d.-homomorphism.

The condition that ∂ be an f.-endomorphism implies that $A_p \partial \subseteq A_p$ for each p. It is of course possible to consider a group on which a differential and a filtration are defined that are not related by this condition; much of what follows can be extended to cover such cases, but we shall reserve the term d.f.-group for the cases in which ∂ is an f.-endomorphism and shall not deal with the more general situation.

Various examples of d.f.-groups can be given. If C is a chain complex it is a g.-group and with the associated filtration it is a d.f.-group; here $A_p = \sum_{n \leqslant p} C_n$, so that $A_p \partial \subseteq A_{p-1}$. Since in general we only require that $A_p \partial \subseteq A_p$, this example may prove misleading. We may also consider (compare 10.9) a filtration $[K_p]$ on a complex K, a set of subcomplexes $K_0 \subseteq K_1 \subseteq \ldots$ such that $\bigcup K_p = K$; then $C = C(K)$ is filtered by the subgroups $A_p = C(K_p)$; since K_p is a subcomplex, $A_p \partial \subseteq A_p$.

We shall make great use of a further example:

10.1.9 *Example.* Let $C^{\otimes} = C \otimes C'$ be the tensor product of the chain complexes C, C'; here $C_n^{\otimes} = \sum_{p+q=n} C_p \otimes C'_q$, and (compare 5.7.7)

$$\partial^{\otimes} = \partial \otimes 1 + \epsilon \otimes \partial'$$

is the differential in $C^{\otimes}$. We take the filtration of $C^{\otimes}$ given by

$$A_p^{\otimes} = \sum_{k \leqslant p} C_k \otimes C'.$$

Then $(C^{\otimes};\, A_p^{\otimes}, \partial^{\otimes})$ is a d.f.-group. Here $A_p^{\otimes} \partial^{\otimes} \subseteq A_p^{\otimes}$, but in general $A_p^{\otimes} \partial^{\otimes} \nsubseteq A_{p-1}^{\otimes}$. This is a good example to bear in mind in studying d.f.-groups and will prove important in the topological applications.

10.2 The spectral sequence of a differential filtered group

In $(G;\, A_p, \partial)$ the interplay of the differential or boundary operator ∂ and the filtration gives rise in a natural way to a host of subgroups of G. As usual we write the kernel of ∂ as Z and call the elements of Z cycles and we write $G\partial$ as B, the subgroup of boundaries. We also consider for each q the subgroup $\partial^{-1}(A_q)$, the subgroup of 'cycles mod A_q', providing a second filtration of G, and for each p, r the

10.2.1

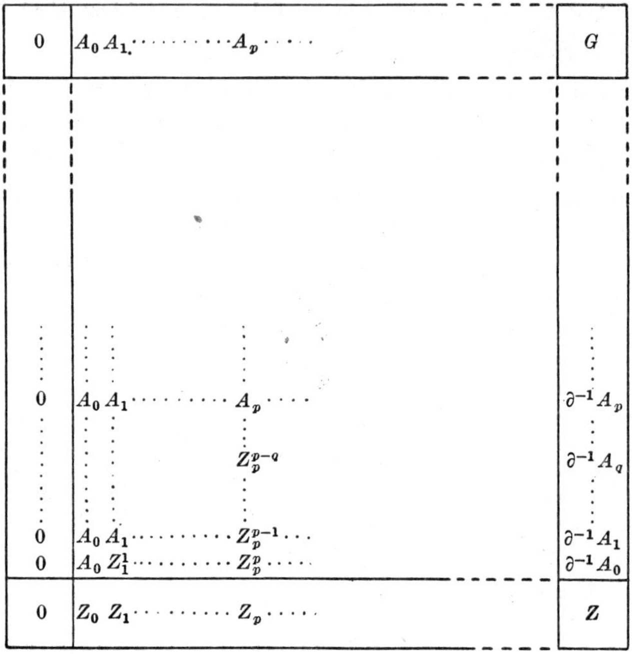
0
$A_0\, A_1 \cdots A_p \cdots$
G
0
$A_0\, A_1 \cdots A_p \cdots$
$\partial^{-1} A_p$
Z_p^{p-q}
$\partial^{-1} A_q$
0
$A_0\, A_1 \cdots Z_p^{p-1} \cdots$
$\partial^{-1} A_1$
0
$A_0\, Z_1^1 \cdots Z_p^p \cdots$
$\partial^{-1} A_0$
0
$Z_0\, Z_1 \cdots Z_p \cdots$
Z

10.2.2

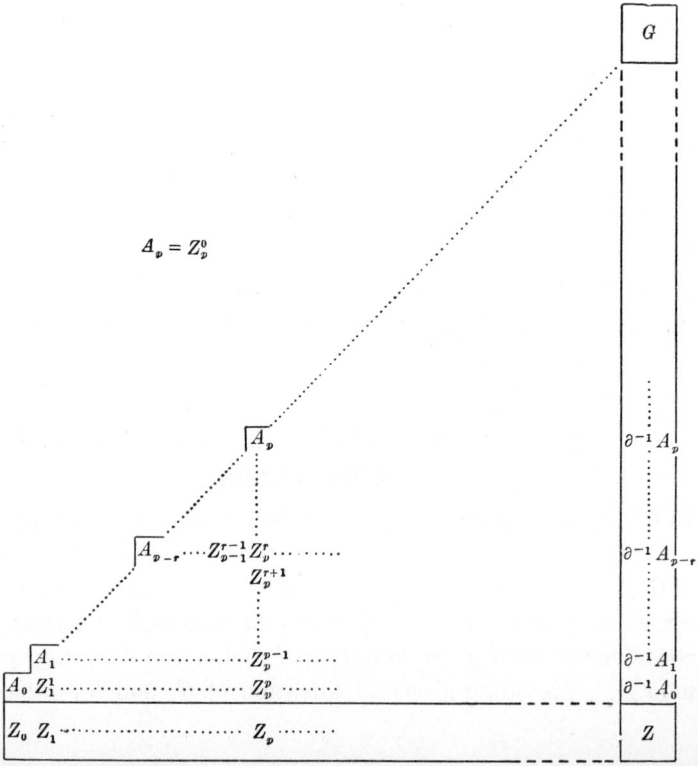
G
$A_p = Z_p^0$
A_p
$\partial^{-1} A_p$
$A_{p-r} \cdots Z_{p-1}^{r-1}\, Z_p^r \cdots$
Z_p^{r+1}
$\partial^{-1} A_{p-r}$
$A_1 \cdots Z_p^{p-1} \cdots$
$\partial^{-1} A_1$
$A_0\, Z_1^1 \cdots Z_p^p \cdots$
$\partial^{-1} A_0$
$Z_0\, Z_1 \cdots Z_p \cdots$
Z

subgroup $\partial^{-1}(A_{p-r}) \cap A_p$ of elements of A_p with boundaries in A_{p-r}; this is written‡ as Z_p^r. Since $\partial^{-1}A_p \supseteq A_p$, $Z_p^0 = A_p$, and since $A_{-q} = 0$ for $q > 0$, $Z_p^{p+q} = Z \cap A_p$ for $q > 0$; we write Z_p (or sometimes Z_p^∞) for $Z \cap A_p$, the group of cycles lying in A_p. The inclusion relation of these Z-subgroups of G may be shown in diagram 10.2.1.

10.2.3

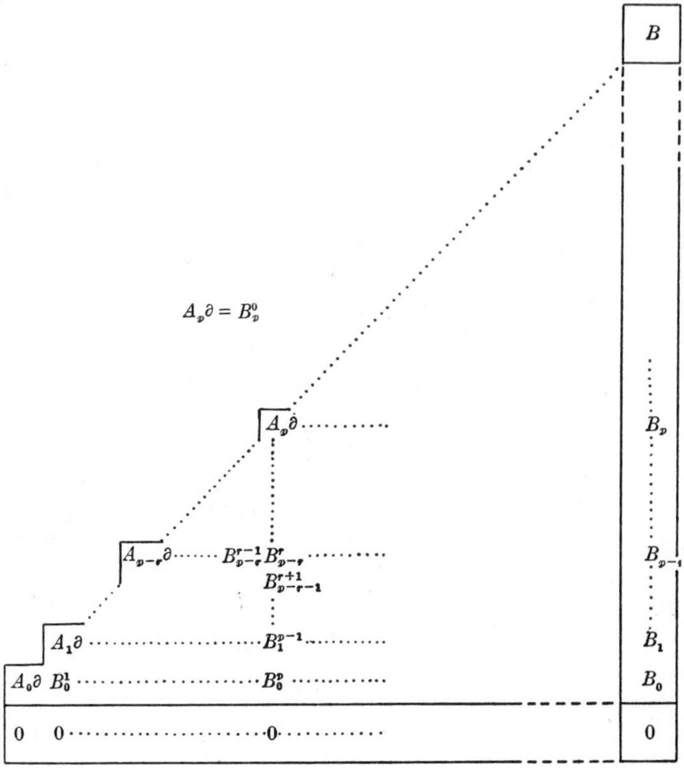

The two filtrations of G appear in a row at the top and in a column to the right of the main part of the diagram, and these filtering subgroups can be used to index the rows and columns. Each group in the diagram is the intersection of the groups indexing its row and its column. The groups in each row filter the indexing group of that row; the groups in the A_p column give a finite filtration for A_p of length p. A group in the diagram is a subgroup of any group appearing above it or to the right of it (or both).

‡ A common notation for this subgroup is C_p^r; we follow Cartan and Eilenberg [13] in writing Z_p^r, since the elements have some resemblance to cycles and because we want to be able to use C for a chain complex in contexts where a confusion with C_p^r might arise.

We shall in future omit the left-hand column and the elements above the diagonal, which give no interesting information, and shall therefore use the diagram 10.2.2.

We now operate on 10.2.2 with ∂ to form a diagram of groups $A_p \partial \cap A_q$; this group of elements of A_q that bound in A_p is written as B_q^{p-q}, so

10.2.4

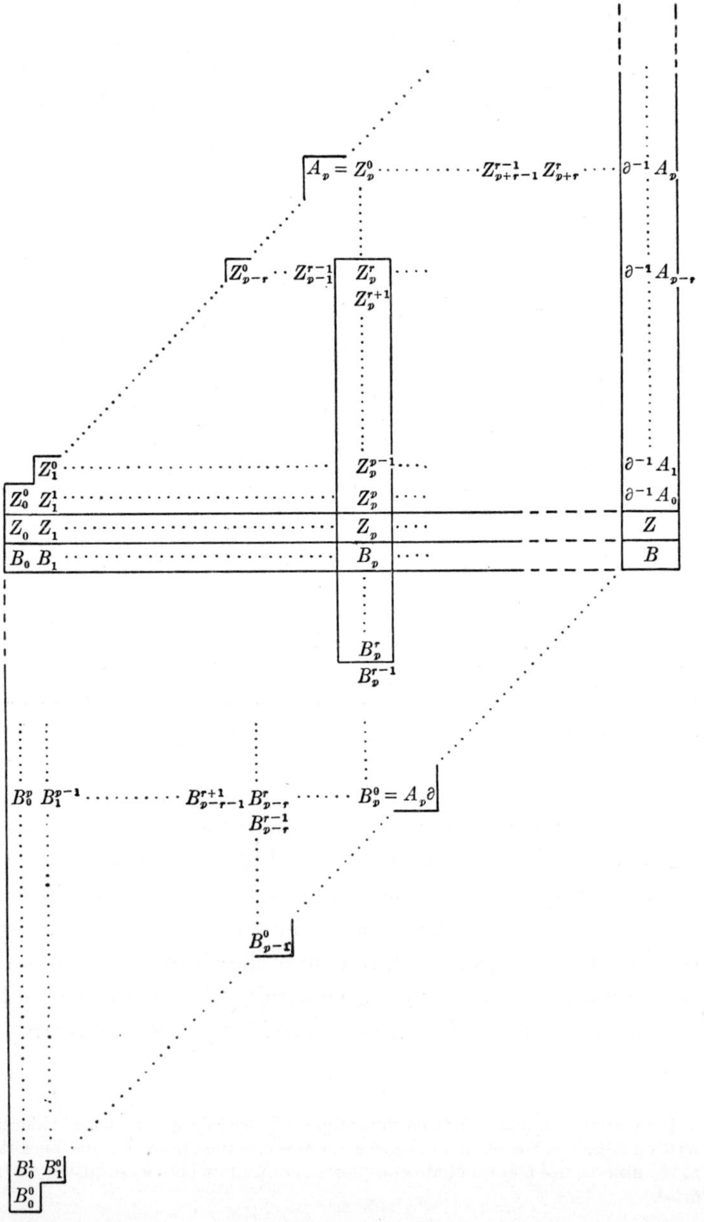

that $Z_p^{p-q}\partial = B_q^{p-q}$, or $Z_p^r\partial = B_{p-r}^r$. We write $B \cap A_q$ as B_q, the group of boundaries lying in A_q; since, for all r, $B_q^r \subseteq B_q$, the symbol B_q^∞ may be used for B_q. The B-diagram is diagram 10.2.3.

The notation is so chosen that the subscript shows the filtering sub-group in which all the elements lie. Inclusion relations between B-groups correspond in description to the relations for Z-groups, and the same description relates B-groups and Z-groups in the diagram due to Zeeman‡ obtained by omitting the bottom row of diagram 10.2.3, reflecting the diagram in its diagonal and placing it below diagram 10.2.2.

Diagram 10.2.4 can be used not only to describe inclusion relations; we may set up a useful correspondence between subgroups represented in the diagram and regions of the diagram. With each subgroup we associate the region to the left of and below its symbol (and including the symbol itself); in this way an inclusion between subgroups is pictured by an inclusion between their corresponding regions and the region for the intersection of two subgroups is the intersection of their regions. Further, we may picture the sum of two subgroups by the union of their regions and the quotient of a subgroup by an included subgroup by the difference of their regions. We here only suggest this as a pictorial aid although Zeeman has provided a precise justification for arguments based on it.

We shall say that an element of A_p has filtration p and that an element of A_p not in A_{p-1} has *exact filtration* p. For each p, r we now define a quotient group E_p^r which somewhat resembles a homology group for elements of exact filtraction p. The subgroup Z_p^r consists of 'cycles mod A_{p-r}' of filtration p, so that Z_p^r/Z_{p-1}^{r-1} may be thought of (loosely and temporarily) as a quotient group of cycles mod A_{p-r} of exact filtration p. We are in a sense ignoring elements of A_{p-r} in this choice of 'cycles'; to define the corresponding 'boundaries' we ignore A_{p+r}, in that we confine attention to those elements of A_p that bound in A_{p+r-1}, namely, to B_p^{r-1}. The quotient group that results from classifying cycles mod A_{p-r} of exact filtration p by boundaries in A_{p+r-1} is

$$E_p^r = Z_p^r/(Z_{p-1}^{r-1} + B_p^{r-1}).$$

The region of the Zeeman diagram representing E_p^r is the column outlined in 10.2.4, with Z_p^r at its head and B_p^r at its foot.

Now the differential ∂ induces a homomorphism of pairs

$$Z_p^r,\ (Z_{p-1}^{r-1} + B_p^{r-1}) \to Z_{p-r}^r,\ (Z_{p-r-1}^{r-1} + B_{p-r}^{r-1});$$

‡ E. C. Zeeman, 'On the filtered differential group', *Ann. Math.* **66** (1957), 557.

for $\quad Z_p^r\partial = B_{p-r}^r \subseteq Z_{p-r}^r, \quad Z_{p-1}^{r-1}\partial = B_{p-r}^{r-1} \quad$ and $\quad B_p^{r-1}\partial = 0;$

∂ therefore induces $\qquad \partial_p^r : E_p^r \to E_{p-r}^r.$

Since $\partial\partial = 0$, $\partial_p^r\partial_{p-r}^r = 0$. We may therefore define E^r to be $\sum_p E_p^r$, graduate it by the subgroups E_p^r, and combine the homomorphisms ∂_p^r into a g.-endomorphism ∂^r of index $(-r)$ with $\partial^r\partial^r = 0$; in this way we construct the d.g.-group $(E^r; E_p^r, \partial^r)$ of index $(-r)$, which will normally be written simply as E^r. An element of E_p^r is said to have *filtering degree* p, so that ∂^r lowers filtering degree by r. The d.g.-group E^r has a graded homology group $H(E^r)$; the graduation is given by

$$H_p(E^r) = \operatorname{Ker}\partial_p^r / E_{p+r}^r\partial^r.$$

10.2.5 Theorem. *The g.-groups $H(E^r)$ and E^{r+1} are isomorphic.*

We have in fact to prove that $H_p(E^r) \cong E_p^{r+1}$.

In the Zeeman diagram 10.2.6 we show E_{p+r}^r, E_p^r, E_{p-r}^r and $E_{p+r}^r\partial$, $E_p^r\partial$; here by $E_p^r\partial$ we mean the set of groups of the diagram that are ∂-images of the groups in the E_p^r region. The part of the E_p^r column from Z_p down is mapped to 0 by ∂ and all of the column except the square Z_p^r is mapped to the left of B_{p-r}^r; this suggests correctly that the kernel of ∂_p^r is represented by the beheaded column from Z_p^{r+1} down to and including B_p^r. Again $E_{p+r}^r\partial^r$ is represented by that portion of the region for $E_{p+r}^r\partial$ lying inside the region for E_p^r, namely, the single square B_p^r. The quotient group Ker $(\partial_p^r)/E_{p+r}^r\partial^r$ should therefore be represented by the column for E_p^r topped and tailed; but this column from Z_p^{r+1} to B_p^{r+1} (inclusive) is the column representing E_p^{r+1}.

We now give the formal argument. The ∂^r-cycles in E_p^r form the subgroup whose representative elements lie in

$$Z_p^r \cap \partial^{-1}(Z_{p-r-1}^{r-1} + B_{p-r}^{r-1}).$$

Now

$$\begin{aligned}\partial^{-1}(Z_{p-r-1}^{r-1} + B_{p-r}^{r-1}) &= \partial^{-1}Z_{p-r-1}^{r-1} + \partial^{-1}B_{p-r}^{r-1}, \quad \text{since} \quad B_{p-r}^{r-1} \subseteq G\partial \\ &= \partial^{-1}A_{p-r-1} + (Z + Z_{p-1}^{r-1}) \\ &= \partial^{-1}A_{p-r-1} + Z_{p-1}^{r-1}, \qquad \text{since} \quad Z \subseteq \partial^{-1}A_{p-r-1}.\end{aligned}$$

Hence

$$\begin{aligned}Z_p^r \cap \partial^{-1}(Z_{p-r-1}^{r-1} + B_{p-r}^{r-1}) &= Z_p^r \cap (\partial^{-1}A_{p-r-1} + Z_{p-1}^{r-1}) \\ &= Z_p^{r+1} + Z_{p-1}^{r-1}.\end{aligned}$$

In the final equality we have used the fact that $\cap$ and $+$ here obey a distributive law‡ since $Z_p^r \supseteq Z_{p-1}^{r-1}$. The ∂^r-cycle subgroup of E_p is therefore $(Z_p^{r+1}+Z_{p-1}^{r-1})/(Z_{p-1}^{r-1}+B_p^{r-1})$. Further, $Z_{p+r}^r\partial = B_p^r$, so that $E_{p+r}^r\partial^r$ is the set of elements of E_p^r represented by elements of B_p^r, namely, $(B_p^r+Z_{p-1}^{r-1})/(Z_{p-1}^{r-1}+B_p^{r-1})$. Hence, invoking I. 2.1,

$$H_p(E^r) \cong (Z_p^{r+1}+Z_{p-1}^{r-1})/(B_p^r+Z_{p-1}^{r-1}).$$

10.2.6

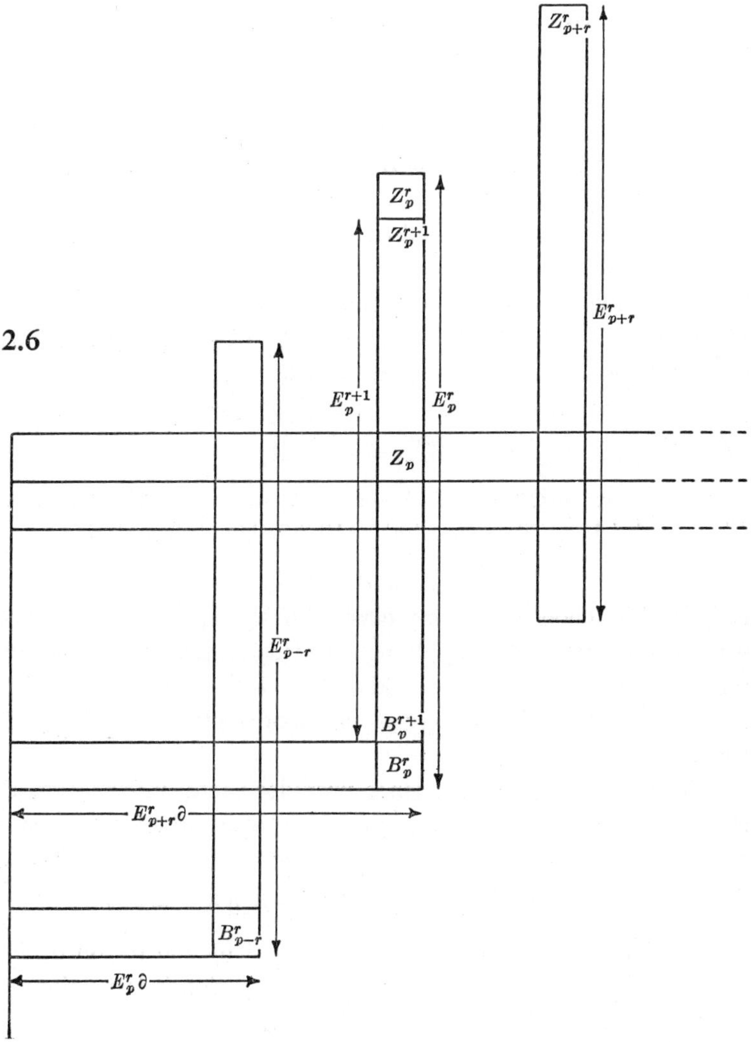

‡ In the terminology of lattice theory subgroups of abelian groups form a *modular lattice.*

Now $B_p^r + Z_{p-1}^r = Z_p^{r+1} \cap (B_p^r + Z_{p-1}^{r-1})$, the distributivity law holding again because $Z_p^{r+1} \supseteq B_p^r$; also

$$Z_p^{r+1} + (B_p^r + Z_{p-1}^{r-1}) = Z_p^{r+1} + Z_{p-1}^{r-1}.$$

The isomorphism theorem I. 2.2 shows that

$$(Z_p^{r+1} + Z_{p-1}^{r-1})/(B_p^r + Z_{p-1}^{r-1}) \cong Z_p^{r+1}/(Z_{p-1}^r + B_p^r) = E_p^{r+1}. \blacksquare$$

We shall henceforth feel free to identify groups connected by either of the fundamental isomorphisms I. 2.1 and I. 2.2 of group theory. We may thus restate 10.2.5 as

$$H(E^r) = E^{r+1}.$$

We define a *spectral sequence* to be a sequence of d.-groups (E^0, ∂^0), (E^1, ∂^1), ... such that $E^{n+1} = H(E^n, \partial^n)$ for all $n \geqslant 0$. The sequence $\{(E^r; E_n^r, \partial^r)\}$, $r = 0, 1, 2, \ldots$ of d.g.-groups is therefore by 10.2.5 a spectral sequence and is called the *spectral* (*homology*) *sqeuence* of the d.f.-group $(G; A_p, \partial)$.

We consider in detail the terms E^0, E^1 of this sequence. First $E_p^0 = Z_p^0/(Z_{p-1}^{-1} + B_p^{-1}) = A_p/A_{p-1}$ and $\partial_p^0 : E_p^0 \to E_p^0$ is just the endomorphism of A_p/A_{p-1} induced by ∂; so $E^0 = \mathrm{Gr}\,[A_p]$ and ∂^0 preserves graduation. Next $E_p^1 = Z_p^1/(Z_{p-1}^0 + B_p^0)$; $Z_{p-1}^0 = A_{p-1}$ so that

$$Z_p^1/Z_{p-1}^0 = Z(A_p/A_{p-1})$$

and B_p^0 is the subgroup of A_p of elements bounding in A_p; hence $E_p^1 = H(A_p/A_{p-1}, \partial^0)$ as asserted in 10.2.5. The differential

$$\partial_p^1 : H(A_p/A_{p-1}) \to H(A_{p-1}/A_{p-2})$$

is defined by choosing for any class of $H(A_p/A_{p-1})$ an element of A_p with boundary in A_{p-1}, applying ∂ to this element and treating the resulting element as a cycle mod A_{p-2}; it is thus precisely the homomorphism ν_* in the homology sequence of the triple (A_p, A_{p-1}, A_{p-2}) (see 5.6.3).

Whereas the spectral sequence provides useful information about the fine structure of a d.f.-group, it is also valuable to consider groups of a somewhat grosser nature. By analogy with E_p^r, we define

$$E_p^\infty = Z_p^\infty/(Z_{p-1}^\infty + B_p^\infty) = Z_p/(Z_{p-1} + B_p).$$

In the Zeeman diagram E_p^∞ is represented by the region of the Z_p square alone; for the remainder of the Z_p region is the union of the region for Z_{p-1} and the region for B_p.

We define the graded group E^∞ to be $\sum_p E_p^\infty$ and we establish a connexion between E^∞ and $H(G, \partial)$, which is represented by the Z row of the Zeeman diagram. The injection of A_p in G induces a homomorphism of $H(A_p)$ into $H(G)$, which is of course not in general monomorphic. We designate the image of this homomorphism by D_p, consisting of those homology classes of (G, ∂) that contain cycles in A_p; in the diagram $D_p \cong Z_p/B_p$ is represented by the Z strip from Z_0 to Z_p. The subgroups $[D_p]$ form a filtration of $H(G)$.

10.2.7 Theorem. *The g.-groups* E^∞ *and* $\operatorname{Gr}[D_p]$ *are isomorphic.*

We have only to prove that $E_p^\infty \cong D_p/D_{p-1}$; the diagram leaves no doubt as to the truth of this, each being represented by the single square of Z_p. Formally, $D_p \cong (Z_p + B)/B$, so that

$$D_p/D_{p-1} \cong (Z_p+B)/(Z_{p-1}+B) \cong Z_p/(Z_p \cap (Z_{p-1}+B))$$
$$= Z_p/(Z_{p-1}+B_p) = E_p^\infty. \,\rbrack$$

This shows that a full knowledge of $H(G, \partial)$ and its filtering subgroups D_p gives full knowledge of the g.-group E^∞. As observed in 10.1 the converse does not in general hold, but, as will be seen in special cases, information about E^∞ gives useful information about $H(G)$.

10.2.8 *Example.* Consider a chain complex $C = (C; C_n, \partial)$ and the associated filtration in which $A_p = \sum_{n \leqslant p} C_n$. Then $E_p^0 = A_p/A_{p-1} = C_p$; moreover, $\partial^0 : A_p/A_{p-1} \to A_p/A_{p-1}$ is trivial since $A_p\partial \subseteq A_{p-1}$; hence $(E^0; E_p^0, \partial^0)$ is isomorphic with $(C; C_p, 0)$. Since $\partial^0 = 0$, $E_p^1 = E_p^0 = C_p$: now $\partial_p^1 : A_p/A_{p-1} \to A_{p-1}/A_{p-2}$ is just $\partial_p : C_p \to C_{p-1}$, so that $E^1 \cong C$ as a d.g.-group. Since $E^2 = H(E^1, \partial^1)$, $E_p^2 = H_p(C)$. An element of E_p^2 is a class of cycles of C_p and ∂_p^2 is induced by ∂ and so is trivial; therefore $(E^2; E_p^2, \partial_p^2) \cong (H(C); H_p(C), 0)$. Similarly, ∂^3, ∂^4, ... are trivial, so that $E_p^2 = E_p^3 = \ldots = H_p(C)$. Now $D_p = \sum_{n \leqslant p} H_n(C)$, so that $E_p^\infty = D_p/D_{p-1} = H_p(C)$. To sum up,

$$E_p^0 = E_p^1 = C_p; \quad E_p^2 = E_p^3 = \ldots = E_p^\infty = H_p(C).$$

It is of course to be expected that a d.f.-homomorphism between two d.f.-groups will induce suitable homomorphisms of E^r and E^∞. This is made precise in

10.2.9 Theorem. *If G, G' are d.f.-groups, a d.f.-homomorphism $\phi : G \to G'$ induces for each $r \leqslant \infty$ a d.g.-homomorphism*

$$\phi^r : E^r(G) \to E^r(G')$$

and an f.-homomorphism

$$\hat{\phi} : (H(G); D_p) \to (H(G'); D'_p)$$

compatible with ϕ^∞.]

This proposition states that the rule associating with a d.f.-group its spectral sequence of d.g.-groups E^r is functorial in the sense of 9.6 and so also is the rule associating with it its homology group filtered by subgroups D_p.

10.3 Spectral theory for a differential filtered graded group

We start by defining the notion of a filtered graded group by imposing a filtration on a graded group $(G; G_n)$; from such a point of view it is natural to begin by filtering the separate subgroups G_n. If $(G_n) A_p$, $p=0, 1, \ldots$, is the filtration of G_n, we write $(G_n) A_p$ as ${}_nA_p$.

A filtration on each G_n provides a filtration of G by the obvious rule

$$A_p = \sum_n {}_nA_p.$$

We shall only be concerned with filtrations of graded groups that can be defined in this way from filtrations of the graduating subgroups. A necessary and sufficient condition on a filtration $[A_p]$ of G that it be so definable is that $A_p = \sum_n (G_n \cap A_p)$.

There is a further restriction on the filtration of graded groups that we shall consider, namely, that the filtration of each G_n be of length n; that is to say, that G_n is filtered by $0 \subseteq {}_nA_0 \subseteq {}_nA_1 \subseteq \ldots \subseteq {}_nA_n = G_n$. This condition is equivalent to the condition $G_n \subseteq A_n$. In 10.9 we give an application in which the filtrations do not satisfy this restriction; we shall thus need definitions both of restricted and of unrestricted filtered graded groups.

10.3.1 Definition. A *(restricted) filtered graded group or f.g.-group* $(G; G_n, A_p)$ is a graded group $(G; G_n)$ filtered by subgroups $[A_p]$ such that (i) $A_p = \sum_n (G_n \cap A_p)$ and (ii) $G_n \subseteq A_n$. An *f.g.-homomorphism* between two f.g.-groups G, G' is a homomorphism preserving graduation and filtration.

10.3.2 Definition. An *unrestricted filtered graded group* or *u.f.g.-group* $(G; G_n, A_p)$ is a graded group $(G; G_n)$ filtered by subgroups $[A_p]$ such that $A_p = \sum_n (G_n \cap A_p)$; a u.f.g.-homomorphism is a homomorphism between u.f.g.-groups that preserves graduation and filtration.

Notice that every f.g.-group is a u.f.g.-group and every f.g.-homomorphism a u.f.g.-homomorphism. Our reason for giving f.g.-groups priority over u.f.g.-groups is that in all our main applications the condition (ii) of 10.3.1 is present and decisive.

The reader should beware of supposing that the condition that $A_p = \Sigma(G_n \cap A_p)$ is automatically satisfied. If $G = G_0 \oplus G_1$, where $G_i = J$ generated by g_i, $i = 1, 2$, then we might take A_0 to be the subgroup generated by $g_0 + g_1$; in that case $\sum_n (G_n \cap A_0) = 0 \neq A_0$.

10.3.3 Definition. A *differential (restricted) filtered graded group* or *d.f.g.-group* $(G\,;\, G_n, A_p, \partial)$ is an f.g.-group $(G\,;\, G_n, A_p)$ with an f.g.-endomorphism ∂ of index -1 such that $\partial\partial = 0$. A *d.f.g.-homomorphism* is a homomorphism between d.f.g.-groups that preserves differential, filtration and graduation. Similarly, we define *d.u.f.g.-groups* and *d.u.f.g.-homomorphisms.*

It is occasionally convenient also to consider a *d.f.g.-homomorphism of index t*, which is an f.-homomorphism that is a d.g.-homomorphism of index t.

In a d.u.f.g.-group we write

$$ {}_nZ^r_p = G_n \cap Z^r_p, \quad {}_nB^r_p = G_n \cap B^r_p, $$

where Z^r_p, B^r_p are defined as in the previous section in terms of the differential and the filtration. Notice that, although $[{}_nA_p]$ is a filtration of G_n, we cannot define ${}_nZ^r_p$ in terms of this filtration alone as we have no differential on G_n (unless ∂ has index 0). This situation is familiar; in a chain complex C, Z_n is a subgroup of C_n defined not in terms of C_n alone but from $\partial_n : C_n \to C_{n-1}$.

10.3.4 Proposition. *In a d.u.f.g.-group* $Z^r_p = \sum_n {}_nZ^r_p$, $B^r_p = \sum_n {}_nB^r_p$. ∎

We next define $\quad {}_nE^r_p = {}_nZ^r_p/({}_nZ^{r-1}_{p-1} + {}_nB^{r-1}_p)$.

We may also write‡ ${}_nE^r_p$ as $E^r_{p,q}$, where $p+q = n$. In contexts where both notations are used, we may pass from one to the other without explicitly stating that $p+q = n$. An element of ${}_nE^r_p = E^r_{p,q}$ is said to have *total* degree n, *filtering* degree p and *complementary* degree q.

We define

10.3.5 $$ E^r = \sum_{p,q} E^r_{p,q}. $$

The differential ∂ induces

$$ {}_n\partial^r_p : {}_nE^r_p \to {}_{n-1}E^r_{p-r}, $$

or

$$ \partial^r_{p,q} : E^r_{p,q} \to E^r_{p-r,q+r-1}. $$

‡ Some authors use the notation $Z^r_{p,q}$ for ${}_nZ^r_p$ and $B^r_{p,q}$ for ${}_nB^r_p$; we prefer to keep the 'p, q' suffix notation for the case of a bigraduation and avoid the notation $B^r_{p,q}$, since $B^r_{p,q} \subseteq B^r_{p+1,q-1}$.

The homomorphisms $\partial^r_{p,q}$ combine to give an endomorphism

$$\partial^r : E^r \to E^r$$

such that $\partial^r \partial^r = 0$.

The group E^r is *bigraded* as $\sum_{p,q} E^r_{p,q}$ and ∂^r has index $(-r, r-1)$. Its homology group $H(E^r, \partial^r)$ is then also bigraded; we write ${}_nH_p(E^r)$ or $H_{p,q}(E^r)$ for the subgroup of homology classes containing ∂^r-cycles in ${}_nE^r_p = E^r_{p,q}$. The analogue of 10.2.5 now reads

10.3.6 Theorem. *For a d.u.f.g.-group the bigraded groups $H(E^r)$ and E^{r+1} are equal.*

The proof that ${}_nH_p(E^r) = {}_nE^{r+1}_p$ is an automatic modification of the proof of 10.2.5. ∎

We define

$$E^\infty_{p,q} = {}_nE^\infty_p = {}_nZ_p/({}_nZ_{p-1} + {}_nB_p),$$

and

$$E^\infty_p = \sum_n {}_nE^\infty_p,$$

so that E^∞ is bigraded by the subgroups ${}_nE^\infty_p$. The group $H(G)$ is graded by the subgroups $H_n(G)$ and filtered by the subgroups $[D_p]$ as in the previous section. Now $D_p = \sum_n (H_n \cap D_p)$ and, if $(G; G_n, A_p)$ is a restricted f.g. group, $H_n \subseteq D_n$; for D_p is the set of homology classes of $(G; \partial)$ containing elements in A_p, and H_n is the set of homology classes containing elements in G_n. This shows that $(H(G); H_n, D_p)$ is a u.f.g.-group, and is a restricted f.g. group if G is restricted. Just as an f.-group gives rise to a graded group (10.1.6), so a u.f.g.-group gives rise to a bigraded group and we can state and prove

10.3.7 Proposition. *The bigraded groups E^∞ and* $\mathrm{Gr}(H(G))$ *are isomorphic.*

It is only necessary to prove that ${}_nE^\infty_p \cong {}_nD_p/{}_nD_{p-1}$; the proof is essentially the same as that of 10.2.7. ∎

We now collect together some useful properties of d.f.g.-groups which do not hold for general d.u.f.g.-groups.

10.3.8 Proposition. *For a d.f.g.-group*

(i) ${}_nA_{n+m} = G_n$ *for all* $m \geqslant 0$;
(ii) ${}_nZ^r_p = {}_nZ_p$ *if* $r > p$; ${}_nB^r_p = {}_nB_p$ *if* $r \geqslant n-p+1$;
(iii) $E^r_{p,q} = 0$ *if* $p < 0$ *or* $q < 0$.

Since $G_n = {}_nA_n \subseteq {}_nA_{n+m} \subseteq G_n$, for $m \geqslant 0$, (i) is immediate. Next

$${}_nZ^r_p = G_n \cap A_p \cap \partial^{-1}A_{p-r} = G_n \cap A_p \cap \partial^{-1}0,$$

if $r > p$; hence ${}_nZ_p^r = {}_nZ_p$. Similarly, ${}_nB_p^r = G_n \cap A_p \cap A_{p+r}\partial$; now $G_n \cap A_{p+r}\partial = (G_{n+1} \cap A_{p+r})\partial = G_{n+1}\partial$ if $p+r \geqslant n+1$; hence

$${}_nB_p^r = G_{n+1}\partial \cap A_p = G_n \cap A_p \cap B = {}_nB_p,$$

and (ii) is proved. To prove (iii) we first show that $E_{p,q}^0 = 0$ if $p < 0$ or $q < 0$. If $p < 0$, ${}_nA_p = 0$ and ${}_nE_p^0$ is a quotient of ${}_nA_p$, so that $E_{p,q}^0 = 0$; if $q < 0$, $n < p$ and ${}_nA_p = {}_nA_{p-1}$, so that ${}_nE_p^0 = {}_nA_p/{}_nA_{p-1} = 0$. The result now follows from the fact that $E_{p,q}^r$ is a *subquotient group* (i.e. quotient of a subgroup) of $E_{p,q}^{r-1}$. ▮

We now give a theorem of great importance in the applications of spectral theory, where a connexion between the terms $E_{p,q}^r$ and the term $E_{p,q}^\infty$ is of interest.

10.3.9 Theorem. *For a d.f.g.-group*

$$E_{p,q}^r = E_{p,q}^\infty \quad \textit{if} \quad r > \max(p, q+1).$$

For $$E_{p,q}^r = {}_nZ_p^r/({}_nZ_{p-1}^{r-1} + {}_nB_p^{r-1});$$

if $r > p$, $${}_nZ_p^r = {}_nZ_p$$

and $${}_nZ_{p-1}^{r-1} = {}_nZ_{p-1},$$

and if $r > q+1$, $${}_nB_p^{r-1} = {}_nB_p,$$

all by 10.3.8 (ii). ▮

Observe that if $r > \max(p, q+1)$ then since ${}_nZ_p^r = {}_nZ_p$, each element of ${}_nZ_p^r$ is a cycle of $(G; \partial)$, so that each element of $E_{p,q}^r$ is a cycle of (E^r, ∂^r). Further, ${}_{n+1}E_{p+r}^r = 0$, since ${}_{n+1}Z_{p+r}^r = {}_{n+1}Z_{p+r-1}^r$ as $n+1 < p+r$; so that only the zero element of $E_{p,q}^r$ is a boundary in (E^r, ∂^r). This means that the passage from $E_{p,q}^r$ to $E_{p,q}^{r+1}$ is the identity.

Theorem 10.3.9 establishes the important fact that E^∞ is in a sense the limit of the bigraded groups $E^0, E^1, \ldots$; for the sequence $E_{p,q}^0$, $E_{p,q}^1, \ldots$ is constant from $r = \max(p+1, q+2)$ onwards, its value then being $E_{p,q}^\infty$. For each p, q, therefore, $\{E_{p,q}^r\}$ converges to $E_{p,q}^\infty$; but this convergence is not uniform, for the value of r after which the sequence becomes constant will in general tend to infinity as p, q tends to infinity.

10.3.10 Corollary. *If the bigraded groups E^r associated with a d.f.g.-group are all isomorphic for $r \geqslant r_0$, then their common value for $r \geqslant r_0$ is E^∞.* ▮

We remark that there is no proposition corresponding to 10.3.9 about the g.-groups E^r associated with a d.f.-group; 10.3.9 depends essentially on the finiteness of the filtration of each G_n. In a d.f.-group it is true that $Z_p^r = Z_p$

for $r > p$, because $A_{-m} = 0$ for $m > 0$; on the other hand, there is in general, given p, no value of r for which $B^r_p = B_p$. We can still draw the consequence that for $r > p$ every element of E^r_p is a cycle of $(E^r; \partial^r)$, so that the passage from E^r_p to E^{r+1}_p is that of taking a quotient group of the whole group. This is clear from the Zeeman diagram 10.2.4; increase of r shrinks the column representing E^r_p; it never drops the head of the column below $Z_p = Z^{p+1}_p = Z^{p+2}_p = \ldots$, so that for $r > p$ the passage from E^r_p to E^{r+1}_p is one only of removing the tail from the column. This process however in general never terminates. It may however be readily verified that the direct system of groups

$$E^{p+1}_p \rightarrow E^{p+2}_p \rightarrow \ldots,$$

in which the homomorphisms are the projections just described, has as direct limit a group canonically isomorphic to E^∞_p. If $\partial^r = 0$ for all $r \geqslant r_0$,

10.3.11 $$E^{r_0} = E^{r_0+1} = \ldots = E^\infty.$$

We apply this in 10.9.

10.3.12 *Example.* The chain complex $C^\otimes = C \otimes C'$ of 10.1.9 is a d.f.g.-group. As in 10.1.9 we define $A^\otimes_p$ to be $\sum\limits_{m \leqslant p} C_m \otimes C'$ and we define $C^\otimes_n$ to be $\sum\limits_{p+q=n} C_p \otimes C'_q$. Then ${}_nA^\otimes_p = \sum\limits_{m \leqslant p} C_m \otimes C'_{n-m}$; $[{}_nA^\otimes_p]$ provides a finite filtration of length n for $C^\otimes_n$, $A^\otimes_p \partial^\otimes \subseteq A^\otimes_p$ and $C^\otimes_n \partial^\otimes \subseteq C^\otimes_{n-1}$.

The reader is invited to verify that, if C, C' are geometric chain complexes, for $(C^\otimes; C^\otimes_n, A^\otimes_p, \partial^\otimes)$

(i) $E^0_{p,q} = C_p \otimes C'_q$;
(ii) $E^1_{p,q} \cong C_p \otimes H_q(C')$;
(iii) $E^2_{p,q} \cong H_p(C \otimes H_q(C'))$;
(iv) $E^2_{p,q} = E^3_{p,q} = \ldots = E^\infty_{p,q}$.

(The verification of (iv) is not easy.)

10.3.13 Theorem. *Let* $(G; G_n, A_p, \partial)$ *and* $(G'; G'_n, A'_p, \partial')$ *be d.f.g.-groups of the same index s and let* $\phi : G \rightarrow G'$ *be a d.f.g.-homomorphism; then* ϕ *induces* $\phi^r_{p,q} : E^r_{p,q} \rightarrow E'^r_{p,q}$, ${}_n\hat{\phi}_p : {}_nD_p \rightarrow {}_nD'_p$. *If for some r and all p, q each* $\phi^r_{p,q}$ *is an isomorphism, then each* ${}_n\hat{\phi}_p$ *is an isomorphism and in particular* ϕ *induces an isomorphism* $\phi_* : H(G) \cong H(G')$.

Since $\phi\partial' = \partial\phi$, $\phi^r\partial'^r = \partial^r\phi^r : E^r \rightarrow E'^r$, so that $\phi^r : E^r \rightarrow E'^r$ is a d.g.-homomorphism. Since ϕ is a d.f.g.-homomorphism it is plain that $\hat{\phi} : D \rightarrow D'$ is a g.-homomorphism. Now suppose ϕ^r isomorphic; the induced homology homomorphism $\phi^{r+1} : E^{r+1} \rightarrow E'^{r+1}$ is then also isomorphic and it follows that $\phi^k : E^k \rightarrow E'^k$ is isomorphic for all $k \geqslant r$. Choosing for each p, q a large enough k, it follows from 10.3.9 that

$$\phi^\infty_{p,q} : E^\infty_{p,q} \cong E'^\infty_{p,q}.$$

Now by 10.3.7, ${}_nE_p^\infty \cong {}_nD_p/{}_nD_{p-1}$, so that ${}_nE_0^\infty = {}_nD_0$ and similarly for E'^∞ and D'; hence

$$ {}_n\hat{\phi}_0 : {}_nD_0 \cong {}_nD'_0. $$

Fix n and suppose that, for some p, ${}_n\hat{\phi}_p : {}_nD_p \to {}_nD'_p$ is isomorphic, and consider the diagram

$$
\begin{array}{ccccccccc}
0 & \longrightarrow & {}_nD_p & \longrightarrow & {}_nD_{p+1} & \longrightarrow & {}_nE^\infty_{p+1} & \longrightarrow & 0 \\
 & & \Big\downarrow {\scriptstyle {}_n\hat{\phi}_p} & & \Big\downarrow {\scriptstyle {}_n\hat{\phi}_{p+1}} & & \Big\downarrow {\scriptstyle {}_n\phi^\infty_{p+1}} & & \\
0 & \longrightarrow & {}_nD'_p & \longrightarrow & {}_nD'_{p+1} & \longrightarrow & {}_nE'^\infty_{p+1} & \longrightarrow & 0.
\end{array}
$$

The horizontal homomorphisms form the exact sequences related to the statement that ${}_nE^\infty_{p+1} \cong {}_nD_{p+1}/{}_nD_p$ and the corresponding statement for the bottom row. The diagram is commutative and the 5-lemma shows that ${}_n\hat{\phi}_{p+1}$ is isomorphic. Since ${}_n\hat{\phi}_0$ is isomorphic it follows by induction that ${}_n\hat{\phi}_p$ is isomorphic for all n, p. In particular, ${}_nD_n \cong {}_nD'_n$ or $H_n(G) \cong H_n(G')$, the isomorphism being that induced by ϕ. ∎

The reader should not suppose that spectral sequences arise only from differential filtered groups. W. S. Massey‡ has described a situation that arises frequently in algebraic topology and which leads to a spectral sequence.

An *exact couple*

$$
\begin{array}{ccc}
A & \xrightarrow{\;f\;} & A \\
{\scriptstyle h}\nwarrow & & \swarrow{\scriptstyle g} \\
 & C &
\end{array}
$$

is an exact triangle of groups and homomorphisms in which two groups coincide. Given such an exact couple we may define a differential d on C by $d = hg$; notice that $dd = hghg = 0$ since $gh = 0$; thus (C, d) is a differential group.

We may now define the *derived couple*

$$
\begin{array}{ccc}
A' & \xrightarrow{\;f'\;} & A' \\
{\scriptstyle h'}\nwarrow & & \swarrow{\scriptstyle g'} \\
 & C' &
\end{array}
$$

of the original couple. Here $A' = Af$, $C' = H(C, d)$; $f' : A' \to A'$ is $f \mid Af$, $g' : A' \to C'$ is given by $(af)g' = \{ag\}$, and $h' : C' \to A'$ is given by $\{z\}h' = zh$, for $z \in Z(C)$. To verify that g' is single-valued, observe that if $af = 0$ then, for some $c \in C$, $a = ch$, so that $\{ag\} = \{chg\} = \{cd\} = 0$; to verify that $(af)g'$ is a homology class of cycles, notice that $agd = aghg = 0$, since $gh = 0$. Finally, we see that h' is single-valued since $cdh = chgh = 0$, and that $C'h' \subseteq A'$ because $zhg = zd = 0$ so that $zh \in Af$.

The derived couple may be informally represented as

$$
\begin{array}{ccc}
Af & \xrightarrow{\;f\;} & Af \\
{\scriptstyle h}\nwarrow & & \swarrow{\scriptstyle f^{-1}g} \\
 & H(C, hg) &
\end{array}
.
$$

‡ *Ann. Math.* **56** (1952), 363; ibid. **57** (1953), 248. See also H. Federer, *Trans. Amer. Math. Soc.* **82** (1956), 340.

If A, C are graded or bigraded groups and f, g, h are g.-homomorphisms, then d is a differential on the graded group C and the derived couple is again a couple of g.-groups and g.-homomorphisms.

10.3.14 Proposition. *The derived couple is exact.*

First $\{z\}h'f' = zhf = 0$; conversely, if $a'f' = 0$ and $a' = af$, then $(af)f = 0$ so that, for some $c \in C$, $af = ch$; then $cd = chg = afg = 0$ and c is therefore a cycle and $a' = \{c\}h'$.

Next $a'f'g' = a'fg' = \{a'g\} = 0$ since $a' \in Af$; conversely, if $a'g' = 0$ and $a' = af$, then $ag = cd$ for some $c \in C$; hence $ag = chg$ and $(a - ch) = a_1 f$ for some $a_1 \in A$, so that $a' = af = (a_1 f)f \in A'f'$.

Finally, if $a' = af$, $a'g'h' = afg'h' = \{ag\}h' = agh = 0$; conversely if $\{z\}h' = 0$, $zh = 0$ and so, for some $a \in A$, $z = ag$; hence $\{z\} = (af)g'$. ∎

We may thus repeat the process of derivation and obtain the second, ..., mth derived couple

$$\begin{array}{ccc} A^{(m)} & \xrightarrow{f^{(m)}} & A^{(m)} \\ & {\scriptstyle h^{(m)}}\nwarrow \quad \swarrow{\scriptstyle g^{(m)}} & \\ & C^{(m)} & \end{array};$$

putting $d^{(m)} = h^{(m)}g^{(m)}$ we obtain a sequence of differential groups $(C^{(m)}, d^{(m)})$ such that $H(C^{(m)}, d^{(m)}) = C^{(m+1)}$; this is a spectral sequence.

In the case that A, C are graded or bigraded, we have a spectral sequence of d.g. groups. If f, g, h are g.-homomorphisms of index p, q, r respectively, then it is easily verified that $f^{(m)}, g^{(m)}, h^{(m)}$ are g.-homomorphisms of index $p, q - mp, r$ respectively and $d^{(m)}$ is a g.-homomorphism of index $q + r - mp$.

It is not difficult to show that a d.f.-group gives rise to a graded exact couple whose spectral sequence is that of the d.f.-group; we take $A = \sum_p H(A_p)$, $C = \sum_p H(A_p, A_{p-1})$ and the homomorphisms f, g, h are essentially the homomorphisms of the exact homology sequence of the pair. Conversely, it may be shown that, given any exact couple, it is possible to construct a differential *doubly-infinite* filtered group

$$\ldots \subseteq (G)A_{p-1} \subseteq (G)A_p \subseteq (G)A_{p+1} \subseteq \ldots, \quad -\infty < p < \infty,$$

whose spectral sequence contains all the information obtainable from the spectral sequence of the exact couple.

An interesting example of an exact couple is the *homotopy exact couple of a simplicial complex*. Let K be a connected simplicial complex; let a_* be a base vertex in K for the homotopy groups and let it be suppressed from the notation. Consider for each p the homotopy exact sequence (see Q. 10, p. 357)

$$\ldots \to \pi_n(K^{p-1}) \xrightarrow{i_*} \pi_n(K^p) \xrightarrow{j_*} \pi_n(K^p, K^{p-1}) \xrightarrow{\beta_*} \pi_{n-1}(K^{p-1}) \to \ldots.$$

We define

$$\begin{aligned} {}_nA_p &= \pi_n(K^p) \quad \text{for } n \geq 2, p \geq 1 \quad (\text{and } {}_nA_p = 0 \text{ otherwise}), \\ {}_nC_p &= \pi_n(K^p, K^{p-1}) \quad \text{for } n \geq 3, p \geq 1 \\ &= \pi_2(K^p)j_* \quad \text{for } n = 2, p \geq 1 \quad (\text{and } {}_nC_p = 0 \text{ otherwise}). \end{aligned}$$

We then define $A = \sum_{n,p} {}_nA_p$, $C = \sum_{n,p} {}_nC_p$. In fact ${}_nA_1 = 0$, ${}_nC_1 = 0$, since the higher homotopy groups of 1-dimensional complexes are null. We define further

$$\begin{array}{llll} {}_nf_p : {}_nA_p \to {}_nA_{p+1} & \text{to be} & i_* : \pi_n(K^p) \to \pi_n(K^{p+1}) & \text{if } n \geqslant 2, p \geqslant 1; \\ {}_ng_p : {}_nA_p \to {}_nC_p & \text{to be} & j_* : \pi_n(K^p) \to \pi_n(K^p, K^{p-1}) & \text{if } n \geqslant 3, p \geqslant 1; \\ {}_2g_p : {}_2A_p \to {}_2C_p & \text{to be} & j_* : \pi_2(K^p) \to \pi_2(K^p) j_* & \text{if } p \geqslant 1; \\ {}_nh_p : {}_nC_p \to {}_{n-1}A_{p-1} & \text{to be} & \beta_* ; \pi_n(K^p, K^{p-1}) \to \pi_{n-1}(K^{p-1}) & \text{if } n \geqslant 3, p \geqslant 2. \end{array}$$

Other homomorphisms are necessarily zero and f, g, h are the homomorphisms determined by $\{{}_nf_p\}$, $\{{}_ng_p\}$, $\{{}_nh_p\}$. The exactness of the couple follows at once from the exactness of the homotopy sequence. Evidently f is a g.-homomorphism of index (0, 1), g is one of index (0, 0) and h is one of index $(-1, -1)$; consequently d has index $(-1, -1)$ and in the mth derived couple $f^{(m)}$ has index (0, 1), $g^{(m)}$ has index $(0, -m)$, $h^{(m)}$ has index $(-1, -1)$ and $d^{(m)}$ has index $(-1, -m-1)$.

Now ${}_nC_p$ (and hence ${}_nC_p^{(m)}$) is zero if $n<2$ or $p<2$ or $n<p$; we infer that ${}_nC_p^{(m)} = {}_nC_p^{(m+1)}$ if $m > \max(p-3, n-p)$. Then ${}_nC_p^{(\infty)}$, the limit‡ of ${}_nC_p^{(m)}$, is just the common value of the groups ${}_nC_p^{(m)}$ for $m > \max(p-3, n-p)$.

Let ${}_nD_m$ be the image of $\pi_n(K^m)$ in $\pi_n(K)$, $n > 1$; then

$$0 = {}_nD_0 = {}_nD_1 \subseteq \dots \subseteq {}_nD_n = \pi_n(K).$$

It is easy to see that ${}_nA_p^{(m)}$ is the image of $\pi_n(K^{p-m})$ in $\pi_n(K^p)$. It then follows readily from examination of the $(n-1)$th derived couple that

$${}_nC_p^{(\infty)} \cong {}_nD_p / {}_nD_{p-1};$$

in other words, the limit group of the spectral sequence is isomorphic to the graded group of the homotopy group of K suitably filtered.

10.4 Spectral theory of a map; fibre spaces

We shall now be concerned with path-connected spaces in each of which a base-point is selected and shall consider a map

$$f: X, x_* \to Y, y_*$$

between two such, x_*, y_* being the base-points. The subspace ('fibre') $F \subseteq X$ is defined to be $f^{-1}(y_*)$ and x_* is taken for base-point for F as well as for X. We use f to define a second graduation and hence a filtration in the d.g.-group ${}^0\square^N(X)$, where this, as in 8.8, means the normalized chain group generated by singular cubes all of whose vertices are mapped to x_*; it was there shown that ${}^0\square^N(X) \simeq \square^N(X)$. In fact we start by defining a second graduation on $\square(X)$ and prove that it induces a graduation and hence the associated filtration in ${}^0\square^N(X)$; we shall abbreviate ${}^0\square^N$ to C.

‡ Any spectral sequence has a limit; see Massey (loc. cit.).

We define‡ $\Box_{p,q}(X)$ to be the subgroup of $\Box(X)$ generated by singular cubes g_n where $n = p+q$ and $g : e_n \to X$ has the property that p is the least integer such that $(u_1, \ldots, u_n)gf$ is independent of all u_j with $j > p$. Clearly for each $g : e_n \to X$ this criterion determines an integer p such that $0 \leqslant p \leqslant n$, so that each generator of $\Box$ belongs to $\Box_{p,q}$ for some p, q. As in Example 10.3.12 we define a filtration $[T_p]$ in $\Box_n$ by defining $(\Box_n)T_p = {}_nT_p$ to be the subgroup of $\Box_n$ generated by elements of $\Box_{k,n-k}$ with $k \leqslant p$. Then ${}_nT_n = \Box_n$ and, if we define $T_p = \sum_n {}_nT_p$, $(\Box; \Box_n, T_p)$ is an f.g.-group.

10.4.1 Proposition. *If $g_n \in T_p$, $(V^{i,\epsilon}g)_{n-1} \in T_p$ for $1 \leqslant i \leqslant n$; indeed, for $i \leqslant p$, $(V^{i,\epsilon}g)_{n-1} \in T_{p-1}$.* ▮

This show that, if $g_n \in T_p$, $g_n\partial \in T_p$, and hence that $(\Box; \Box_n, T_p, \partial)$ is a d.f.g.-group.

If we confine attention to ${}^0\Box$ and define ${}^0T_p = {}^0\Box \cap T_p$ we see at once that $({}^0\Box; {}^0\Box_n, {}^0T_p, \partial)$ is also a d.f.g.-group. We now pass to the factor group ${}^0\Box^N = C$ and define a filtration on C by $A_p = {}^0T_p/({}^0T_p \cap {}^0\Box^\flat)$, and with these definitions $(C; C_n, A_p, \partial)$ is also a d.f.g.-group. C_n is generated by non-degenerate singular n-cubes, A_p by non-degenerate singular cubes of any dimension whose f-images are maps not depending on $u_{p+1}, \ldots$, and in each case the generators are cubes with all vertices mapped to x_*. We observe that ${}_nA_p$ is a direct factor in C; this has the important consequence that the groups ${}_nA_p \otimes G$ are naturally embedded as subgroups of $C \otimes G$ and so provide a d.f.g.-group $(C \otimes G; C_n \otimes G, A_p \otimes G, \partial \otimes 1)$. This enables us to extend results to the case of arbitrary coefficients.

10.4.2 Definition. The spectral (homology) sequence of $(C; C_n, A_p, \partial)$ is called the *spectral (homology) sequence of the map* $f : X, x_* \to Y, y_*$.

The most interesting case is that of a fibre map; we introduce the concept in full generality however, in order not to worry the reader with assumptions superfluous at this stage. We observe that the spectral sequence of a map is *not* an invariant of the homotopy class of the map.

Let us now suppose that $F = f^{-1}(y_*)$ is also path-connected; we proceed in this case to forge a connexion between the spectral sequence

‡ Compare J.-P. Serre, 'Homologie Singulière des Espaces Fibrés', *Ann. Math.* **54** (1951), 425. This is a four-star paper. His definition of degeneracy is slightly different, but the reader should have no difficulty in modifying his arguments appropriately. The modifications can in fact be found in R. Bott and H. Samelson, *Comm. Math. Helv.* **27** (1953), 320.

of the map f and the spectral sequence of $C(Y)\otimes C(F)$ as outlined in Example 10.3.12. We define ${}^0\square_{p,q}$ to be ${}^0\square \cap \square_{p,q}$ and proceed to define a set of homomorphisms

$$ {}^0\phi_{p,q} : {}^0\square_{p,q} \to {}^0\square_p(Y)\otimes {}^0\square_q(F). $$

Let g_n be a generator of ${}^0\square_{p,q}$; then we define $g_p^Y \in {}^0\square_p(Y)$ from the map

$$ (u_1, \dots, u_p)\, g^Y = (u_1, \dots, u_p, 0, \dots, 0)\, gf. $$

Since gf is independent of $u_{p+1}, \dots,$ it is in fact immaterial what values are inserted after u_p on the right-hand side; evidently g^Y maps the vertices of e_p to y_*.

Next we define $g_q^F \in {}^0\square_q(F, x_*)$ from the map

$$ (u_1, \dots, u_q)\, g^F = (0, \dots, 0, u_1, \dots, u_q)\, g. $$

Since gf is independent of the last q variables, $(u_1, \dots, u_q)\, g^F f = y_*$ so that we may regard g^F as a map of e_q in F; evidently it maps the vertices of e_q to x_*.

The definition of ${}^0\phi_{p,q}$ is now easy; if g_n is a generator of ${}^0\square_{p,q}$,

$$ g_n\, {}^0\phi_{p,q} = g_p^Y \otimes g_q^F. $$

Since ${}^0\square = \sum_{p,q} {}^0\square_{p,q}$, the homomorphisms ${}^0\phi_{p,q}$ combine to give a g.-homomorphism ${}^0\phi : {}^0\square \to {}^0\square(Y)\otimes {}^0\square(F)$.

If $g_n \in {}^0\square_{p,q}^\flat$, either $g^Y \in {}^0\square^\flat(Y)$ or $g^F \in {}^0\square^\flat(F)$, the former if the degeneracy is with respect to one of the variables $u_1, \dots, u_p$, and the latter if with respect to one of $u_{p+1}, \dots, u_n$. (In fact g_n cannot be degenerate with respect to u_p by the definition of $\square_{p,\,q}$.) We may therefore pass to ${}^0\square^N = C$ and ${}^0\phi$ induces a g.-homomorphism

$$ \phi : C(X) \to C(Y)\otimes C(F). $$

We graduate $C^\otimes = C(Y)\otimes C(F)$ by total degree and filter, as in 10.3.12, by defining $A_p^\otimes$ to be $\sum_{m\leqslant p} C_m(Y)\otimes C(F)$; then ϕ is an f.g.-homomorphism from $(C; C_n, A_p)$ to $(C^\otimes; C_n^\otimes, A_p^\otimes)$. It is not, however, in general true that ϕ commutes with the differentials‡, and it is therefore not in general a d.f.g.-homomorphism and does not therefore induce a homomorphism of the spectral sequence $\{E^r\}$ of C into the spectral sequence $\{E^{\otimes r}\}$ of $C^\otimes$. Partial results, however, in this direction are available.

‡ Indeed, the differential in $C(X)$ reflects the very twisting of the 'fibres' that distinguishes X from $Y \times F$. An important problem in the homology theory of fibre spaces is the construction of a suitable differential in $C(Y)\otimes C(F)$ with respect to which ϕ is a chain map.

10.4.3 Theorem. *The homomorphism ϕ induces a d.g.-homomorphism*

$$\phi^0 : (E^0; E^0_{p,q}, \partial^0) \to (E^{\otimes 0}; E^{\otimes 0}_{p,q}, \partial^{\otimes 0}).$$

Since ${}_nE^0_p = {}_nA_p / {}_nA_{p-1}$ and ϕ is an f.g.-homomorphism, ϕ certainly induces homomorphisms $\phi^0_{p,q} : E^0_{p,q} \to E^{\otimes 0}_{p,q}$ which combine to give a g.-homomorphism $\phi^0 : E^0 \to E^{\otimes 0}$; it remains therefore to prove that

$$\phi^0_{p,q} \partial^{\otimes 0} = \partial^0 \phi^0_{p,q-1}.$$

Now $E^{\otimes 0}_{p,q}$ is generated by the classes of elements $g_p \otimes h_q$, $g_p \in C_p(Y)$, $h_q \in C_q(F)$; since

$$(g_p \otimes h_q) \partial^{\otimes} = g_p \partial \otimes h_q + (-1)^p g_p \otimes h_q \partial \quad \text{and} \quad g_p \partial \otimes h_q \in {}_{n-1}A^{\otimes}_{p-1},$$

we may write $$(g_p \otimes h_q) \partial^{\otimes 0} = (-1)^p g_p \otimes h_q \partial,$$

where (as elsewhere) we use a symbol for a generator also for the class in the appropriate quotient group. Accordingly, if g_n is a generator of ${}_nA_p$,

10.4.4 $$g_n \phi^0_{p,q} \partial^{\otimes 0} = (-1)^p g^Y_p \otimes g^F_q \partial.$$

On the other hand,

$$g_n \partial = \sum_{\substack{i \leqslant p \\ \epsilon = 0,1}} (-1)^{i+\epsilon} (V^{i,\epsilon} g)_{n-1} + \sum_{\substack{i > p \\ \epsilon = 0,1}} (-1)^{i+\epsilon} (V^{i,\epsilon} g)_{n-1};$$

the first term on the right-hand side lies in A_{p-1} by 10.4.1, so that $g_n \partial^0 = \sum_{\substack{i>p \\ \epsilon=0,1}} (-1)^{i+\epsilon} (V^{i,\epsilon} g)_{n-1}$. Now it is easy to verify that, if $i > p$, $(V^{i,\epsilon} g)^Y = g^Y$ and $(V^{i,\epsilon} g)^F = V^{i-p,\epsilon}(g^F)$. Hence

$$g_n \partial^0 \phi^0_{p,q-1} = \sum_{\substack{i>p \\ \epsilon=0,1}} (-1)^{i+\epsilon} g^Y \otimes V^{i-p,\epsilon}(g^F) = (-1)^p g^Y \otimes g^F \partial.$$

This, together with 10.4.4, proves the result. ▌

10.4.5 Corollary. *ϕ induces a bigraded homomorphism*

$$\phi^1 : E^1 \to E^{\otimes 1} = C(Y) \otimes H(F). \; ▌$$

Notice that we do not assert that ϕ^1 is a d.-homomorphism and, indeed, this is in general not the case. On the other hand, Serre (loc. cit.) has proved that, if $f : X \to Y$ is a fibre map and Y is 1-connected then ϕ^1 is a d.g.-homomorphism, and that in that case ϕ^1 is in fact isomorphic. A consequence of this is that ϕ induces a g.-isomorphism $\phi^2 : E^2 \to H(Y; H(F))$. We offer first a discussion of the greatly simplified case in which $X = Y \times F$ and f projects X onto the first factor space Y; it provides a pattern for the argument in the case of a general fibre space.

10.4.6 Theorem. *If* $X = Y \times F$ *and* $f : X \to Y$ *is the projection, the homomorphism* ϕ^0 *of* 10.4.3, *is a chain equivalence.*

We define a chain map $\psi : C(Y) \otimes C(F) \to C(X)$. The formula 8.7.5 gives a chain map $\rho^{Y,F} : \square^N(Y) \otimes \square^N(F) \to \square^N(Y \times F)$, and, on restriction to ${}^0\square$, ρ induces a chain map which we shall designate by ψ. If g_p, h_q are generators of $C_p(Y), C_q(F)$, $(g_p \otimes h_q)\,\psi$ is the generator $(g \times h)_n$ of $C_n(X)$; clearly $(g \times h)_n \in A_p$ and ψ is accordingly a d.f.g.-homomorphism. We shall show that the induced d.g.-homomorphism

$$\psi^0 : (E^{\otimes 0}; E^{\otimes 0}_{p,q}, \partial^{\otimes 0}) \to (E^0; E^0_{p,q}, \partial^0)$$

is a chain inverse to ϕ^0, which will establish 10.4.6.

Let g_p, h_q be generators of $C_p(Y)$, $C_q(F)$, so that neither can be degenerate. Then

$$(g_p \otimes h_q)\,\psi\phi = (g \times h)_n\,\phi = (g \times h)^Y_{p'} \otimes (g \times h)^F_{q'},$$

where p' is the exact filtration of $(g \times h)$. Certainly $p' \leqslant p$; since g is not degenerate it follows that $p' \nless p$, so that $p' = p$ and

$$(g_p \otimes h_q)\,\psi\phi = g_p \otimes h_q.$$

This proves that $\psi\phi$ is the identity on the generators of $C(Y) \otimes C(F)$ and it is therefore the identity automorphism. It follows at once that $\psi^0\phi^0 = (\psi\phi)^0 = 1$. In the present case of a topological product we know in fact that ψ is a chain equivalence and so also ψ^0; what we have now proved shows that ϕ^0 is its chain inverse and so itself a chain equivalence. This argument, however, is not available in a general fibre space, and we proceed with a different argument that can be adapted to the general case.

We prove in fact that $\phi^0\psi^0 \simeq 1$ by considering $g_n\phi\psi$ for any generator of $\square_{p,q}$; such a g_n is a singular cube of exact filtration p. Then $g_n\phi\psi = (g^Y_p \otimes g^F_q)\,\psi = (g^Y \times g^F)_n$. We shall compare the maps g and $g^Y \times g^F : e^n \to X$ and shall construct a homotopy $G : e^n \times I \to X$ connecting them. A point of $e^n = e^p \times e^q$ will be written as (v, w), $v \in e^p$, $w \in e^q$. We then define G by

$$(v, w, t)\,G = ((v, w)\,gf, ((1-t)\,v, w)\,g\bar{f}),$$

where $\bar{f}$ is the projection $X \to F$ onto the second factor space.

Clearly $(v,w,0)G = (v,w)g$ and $(v,w,1)G = (v,w)(g^Y \times g^F)$. Since $e^n \times I = e^{n+1}$, G defines a singular $(n+1)$-cube of X and this cube has exact filtration p; for $(v,w,t)Gf = (v,w)gf$, which does not depend on w but does depend on u_p. The correspondence $g_n \to (-1)^n G_{n+1}$ provides therefore a set of homomorphisms $\Lambda_{p,q} : \square_{p,q} \to \square_{p,q+1}$ which combine to a g.-homomorphism of index $(0,1)$, $\Lambda : \square \to \square$. Since Λ preserves filtration, ${}_nT_p\Lambda \subseteq {}_{n+1}T_p$; evidently ${}^0\square\Lambda \subseteq {}^0\square$ and $\square^\flat\Lambda \subseteq \square^\flat$; we may therefore restrict to ${}^0\square$, normalize and pass from the bigraded to the f.g.-group, after which Λ induces an f.g.-homomorphism $\Gamma = \Gamma(\Lambda)$ of index 1,

$$\Gamma : (C; C_n, A_p) \to (C; C_{n+1}, A_p).$$

We consider g_n, a generator of ${}_nA_p$ and we observe that

(i) if $i \leqslant p$, $g_n F^{i,\epsilon}\Lambda$ and $g_n \Lambda F^{i,\epsilon} \in A_{p-1}$ (see 10.4.1);

(ii) if $p+1 \leqslant i \leqslant n$, $g_n F^{i,\epsilon}\Lambda = -g_n \Lambda F^{i,\epsilon}$ (the negative sign arising from the $(-1)^n$ in the definition of Λ);

(iii) $g_n \Lambda F^{n+1,1} = (-1)^n (g^Y \times g^F)_n$ and $g_n \Lambda F^{n+1,0} = (-1)^n g_n$.

Now the f.g.-endomorphism Γ induces a g.-endomorphism $\Gamma^0 : E^0 \to E^0$ of index $(1,0)$, since $E^0_{p,q} = {}_nA_p/{}_nA_{p-1}$, and the observations (i), (ii), (iii) imply that

$$(g_n)(\Gamma^0\partial^0 + \partial^0\Gamma^0) = g_n - (g^Y \times g^F)_n = g_n(1 - \phi^0\psi^0).$$

Hence $\phi^0\psi^0 \simeq 1$. ▮

Theorems 10.4.3 and 10.4.6 imply at once

10.4.7 Corollary. *If $X = Y \times F$, the homomorphism ϕ induces a g.-isomorphism $\phi^1 : E^1 \cong E^{\otimes 1} = C(Y) \otimes H(F)$.* ▮

The fact, however, that ψ is a d.f.g.-homomorphism implies

10.4.8 Theorem. *If $X = Y \times F$, the d.f.g.-homomorphism*

$$\psi : C(Y) \otimes C(F) \to C(X)$$

of the proof of 10.4.6 *induces an isomorphism*

$$\psi_* : H_*(C(Y) \otimes C(F)) \to H_*(X).$$

We apply 10.3.13; ψ is a d.f.g.-homomorphism and ψ^0 has been proved to be a chain equivalence, so that ψ^1 is isomorphic and the result is immediate. ▮

This gives an alternative proof, at any rate in case Y and F are path-connected, that $H_*(Y \times F)$ can be calculated from $C(Y) \otimes C(F)$. We remark also that 10.4.8 and 3.10.1 imply that ψ itself is a chain equivalence.

The projection $Y \times F \to Y$ is of course a fibre map; in the case of a general fibre map $p: X, x_* \to B, b_*$, the construction of a suitable $\psi: C(B) \otimes C(F) \to C(X)$, where $F = p^{-1}b_*$, is rather more awkward. As in the special case of 10.4.6 ψ is defined in terms of a rule which associates with a pair (g, h), where g is a p-cube of B and h is a q-cube of F, a $(p+q)$-cube, $(g \otimes h)\,\psi$, of X; all cubes, of course, are 0-trivial. The definition is made inductively with respect to q, using the fundamental 'lifting' property (II. 3) of a fibre map. It is important to remark that in defining ψ for $q = 0$ we use the path-connectedness of F to lift g into a 0-*trivial* cube of X. In this general case ψ is not a d.-homomorphism (only an f.g.-homomorphism), but ψ^0 is a d.g.-homomorphism. Similarly, a suitable homotopy G between g and $(g^B \otimes g^F)\,\psi$ is defined by an inductive procedure, again using II. 3. The detailed proofs may be found in Serre (loc. cit.) and the conclusion is

10.4.9 Theorem. *If $p: X, x_* \to B, b_*$ is a fibre map with path-connected fibre F, the homomorphism ϕ induces a d.g.-homomorphism $\phi^0: E^0 \to E^{\otimes 0}$ that is a chain equivalence; here $E^{\otimes 0}$ is the leading term in the spectral sequence of $C(B) \otimes C(F)$.* ∎

Hence we deduce

10.4.10 Corollary. *For a fibre map p as above, ϕ induces a g.-isomorphism $\phi^1: E^1 \to E^{\otimes 1} = C(B) \otimes H(F)$.* ∎

The map $\psi^1: C(B) \otimes H(F) \to E^1$ is the inverse of ϕ^1. It follows that if $C(B) \otimes H(F)$ is endowed with the differential $\psi^1 \partial^1 \phi^1$ then ϕ^1 becomes a chain isomorphism. To examine this differential, let g be a 0-trivial p-cube of B and let h^j be 0-trivial q-cubes of F such that $\sum_j y_j h^j$ is a q-cycle of F (with coefficients y_j in some coefficient group G). By direct computation, using the fact that $\sum_j y_j h^j$ is a cycle, it may be shown that

$$(g \otimes \Sigma y_j h^j)\,\psi \partial \phi = \sum_j \sum_{i=1}^{p} (-1)^i \left((V^{i0} g \otimes y_j h^j) - (V^{i1} g \otimes y_j (h^j Q_i)) \right);$$

here Q_i is the map from q-cubes of F to q-cubes of F given by

$$z(hQ_i) = (0, \ldots, 0, 1, 0, \ldots, 0, z)\,(g \otimes h)\,\psi, \quad z \in e_q,$$

where 1 is in the ith place. We now observe that, in the terminology of 8.9, Q_i is an ${}^iw^g$-transport. (Recall that ${}^iw^g$ is the loop of B which is the g-image of the edge of e_p from the origin along the ith axis.) It thus follows from 8.9.2 and 8.9.5–8.9.7 that $\psi^1 \partial^1 \phi^1$ is just the boundary in the chain complex $C(B; \{H(F)\})$ with local coefficients. Thus Serre proves the fundamental theorem

10.4.11 Theorem. *If $p : X, x_* \to B, b_*$ is a fibre map with path-connected fibre F, the homomorphism ϕ induces a g.-isomorphism*

$$\phi^2 : E^2_{p,q} \cong H_p(B; \{H_q(F)\}).$$

If B is 1-connected, then $\phi^2 : E^2_{p,q} \cong H_p(B; H_q(F))$. ∎

It is the special case of 10.4.11 when B is 1-connected that will concern us when we come to applications in 10.6. The form of the theorem suggests that the filtering degree p is associated with the base B and the complementary degree q with the fibre F. This suggestion can be strongly reinforced.

Returning to the general case of a d.f.g.-group, let $R = A_0$ and $R_n = {}_nA_0$. Then $R_n\partial = (G_n \cap A_0)\partial \subseteq G_{n-1} \cap A_0 = R_{n-1}$, so that $(R; R_n, \partial)$ is a d.g.-group, indeed a chain complex. Reference to the definition shows that ${}_nE^0_0 = R_n$ and ${}_nE^1_0 = H_n(R)$. For $r \geqslant 1$, each element of ${}_nE^r_0$ is a cycle, since ∂^r decreases the filtering degree; hence there is a sequence of epimorphisms

10.4.12 $$H_n(R) = {}_nE^1_0 \to {}_nE^2_0 \to {}_nE^3_0 \to \ldots.$$

For $r > n+1$, the epimorphism ${}_nE^r_0 \to {}_nE^{r+1}_0$ is an isomorphism (by 10.3.9) and each is equal to ${}_nE^\infty_0 = {}_nD_0 \subseteq H_n(G)$. There is therefore a sequence of homomorphisms

10.4.13 $$H_n(R) \to {}_nE^\infty_0 \to H_n(G),$$

of which the first is epimorphic and the second monomorphic. The composite homomorphism is that induced by the injection of $(R; R_n, \partial)$ as a subcomplex of $(G; G_n, \partial)$.

Now in the case of a fibre map $p : X \to B$ and its spectral sequence, $A_0 = C(F)$, being generated by singular cubes whose projections map to b_*; here then the chain complex R is the chain complex $C(F)$ and 10.4.13 may be identified with the sequence

10.4.14 $$H_n(F) \to E^\infty_{0,n} \to H_n(X),$$

the composite homomorphism being induced by the injection of F in X.

We now consider the term $E^r_{n,0}$ and write S_n for $E^1_{n,0}$ and S for $\sum_n S_n$. Since $E^1_{n,0}\partial^1 \subseteq E^1_{n-1,0}$, $(S; S_n, \partial^1)$ is a chain complex and $E^2_{n,0} \cong H_n(S)$. Now for $r > 1$, ∂^r increases the complementary degree, so that no element of $E^r_{n,0}$ is a boundary; there is therefore a sequence of monomorphisms

10.4.15 $$\ldots \to E^r_{n,} \to \ldots \to E^3_{n,0} \to E^2_{n,0} = H_n(S).$$

But $E^{n+1}_{n,0} = E^{\infty}_{n,0}$ by 10.3.9 and $E^{\infty}_{n,0} = {}_nD_n/{}_nD_{n-1} = H_n(G)/{}_nD_{n-1}$. There is accordingly a sequence of homomorphisms

10.4.16 $$H_n(G) \to E^{\infty}_{n,0} \to H_n(S)$$

of which the first is epimorphic and the second monomorphic. To characterize the composite of these two homomorphisms we remark that $E^1_{n,0}$ is a quotient group of ${}_nZ^1_n = G_n$; passage to the quotient for each n induces a d.g.-epimorphism

$$\pi : G \to S$$

and hence a homomorphism $\pi_* : H_n(G) \to H_n(S)$. This is the composite homomorphism for which we are looking.

In Serre's definition of degeneracy the singular n-cube g of X is degenerate if and only if $g_n \in \square_{n-1}(X)\,D^n$. If we write $\tilde{C}(X)$ for the resulting chain group of normalized 0-trivial chains, then a map $f : X, x_* \to Y, y_*$ induces a chain map $\tilde{f} : \tilde{C}(X) \to \tilde{C}(Y)$. In the case of a fibre map $p : X, x_* \to B, b_*$ it is evident from II. 3 that $\tilde{p}$ is epimorphic and we readily deduce that $\tilde{C}(B) \cong S$. Moreover, we know (Bott and Samelson, loc. cit.; see also Exercises Q.1 of chapter 8) that the natural epimorphism $\tilde{C} \to C$ is a chain equivalence. We may therefore rewrite 10.4.16 as

10.4.17 $$H_n(X) \to E^{\infty}_{n,0} \to H_n(B),$$

the composite homomorphism being induced by the projection $p : X \to B$.

The arguments leading up to 10.4.13 and 10.4.16 can be illuminated by reference to a Zeeman diagram. In the case of a d.f.g.-group there is *for each* n a diagram that is finite; we give the diagram for $n = 4$.

10.4.18
$$\begin{array}{ccccc}
 & & & {}_4Z^0_3 & {}_4Z^1_4 \\
 & & {}_4Z^0_2 & {}_4Z^1_3 & {}_4Z^2_4 \\
 & {}_4Z^0_1 & {}_4Z^1_2 & {}_4Z^2_3 & {}_4Z^3_4 \\
{}_4Z^0_0 & {}_4Z^1_1 & {}_4Z^2_2 & {}_4Z^3_3 & {}_4Z^4_4 \\
\hline
{}_4Z_0 & {}_4Z_1 & {}_4Z_2 & {}_4Z_3 & {}_4Z_4 \\
\hline
{}_4B^5_0 & {}_4B^4_1 & {}_4B^3_2 & {}_4B^2_3 & {}_4B^1_4 \\
{}_4B^4_0 & {}_4B^3_1 & {}_4B^2_2 & {}_4B^1_3 & {}_4B^0_4 \\
{}_4B^3_0 & {}_4B^2_1 & {}_4B^1_2 & {}_4B^0_3 & \\
{}_4B^2_0 & {}_4B^1_1 & {}_4B^0_2 & & \\
{}_4B^1_0 & {}_4B^0_1 & & & \\
{}_4B^0_0 & & & &
\end{array}$$

The lower part of the diagram is the image under ∂ (after reflexion in the diagonal) of the upper part of the diagram for $n = 5$. The group ${}_4Z_4^0$ is left out since it is equal to ${}_4Z_4^1$.

We now reduce this diagram to a skeleton, specifying certain regions of interest.

10.4.19

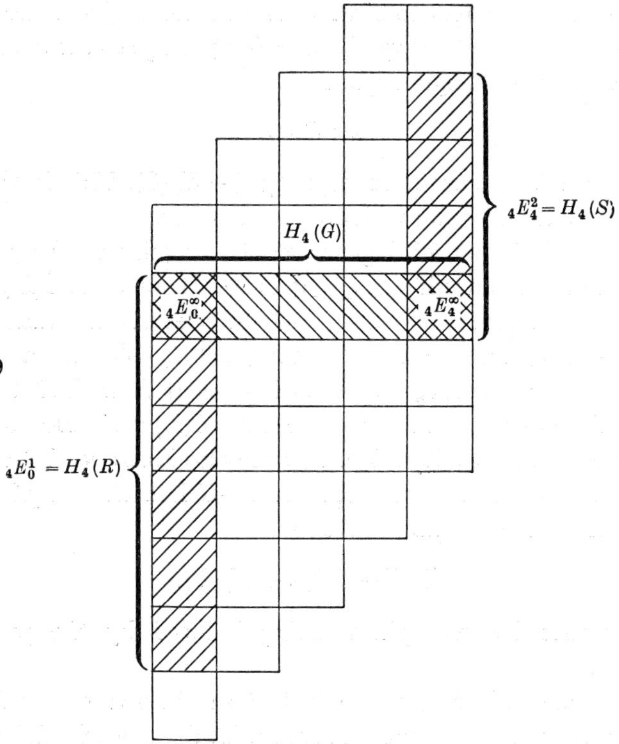

The epimorphism $H_n(R) \to {}_nE_0^\infty$ corresponds to the cutting down of the column for ${}_nE_0^1$ to its top square, which represents ${}_nE_0^\infty$; the monomorphism ${}_nE_0^\infty \to H(G)$ corresponds to the injection of the square representing ${}_nE_0^\infty$ into the row representing $H(G)$. Similar interpretations are available for the epimorphism and monomorphism of 10.4.16, as indicated in diagram 10.4.19.

10.5 Spectral contrahomology theory

The *increasing* filtration of 10.1 is well adapted to the study of chain groups and homology; for contrachain groups and contrahomology however, a decreasing filtration is appropriate.

10.5.1 Definition. A *decreasing filtration* $[A^p]$ on an abelian group G is a set of subgroups of G, $(G)\,A^p = A^p$, $p = 0, 1, \ldots$, such that

$$G = A^0 \supseteq A^1 \supseteq \ldots \supseteq A^p \supseteq \ldots$$

and that $\bigcap_p A^p = 0$. A *filtered group* or *f.-group with decreasing filtration* $(G; A^p)$ is a group G and a decreasing filtration $[A^p]$. An *f.-homomorphism* between such f.-groups is one preserving filtration.

It is convenient also to define A^p to be G for $p < 0$.

If $(G; G^n)$ is a g.-group, we may define A^p to be $\sum_{n \geqslant p} G^n$; this is the decreasing filtration associated with the g.-group G.

If $[A^p]$ is a decreasing filtration of G, the *associated graded group* $\mathrm{Gr}\,[A^p]$ is $\sum_{p=0}^{\infty} A^p/A^{p+1}$.

10.5.2 Definition. A *differential filtered group* or *d.f.-group with decreasing filtration* $(G; A^p, \delta)$ is a group G with decreasing filtration A^p and an *f.-endomorphism* δ such that $\delta\delta = 0$.

The following example gives the significant connexion between d.f.-groups with increasing and decreasing filtration.

10.5.3 *Example.* Let $(G; A_p, \partial)$ be a d.f.-group with increasing filtration and Q be any abelian group; then we may write $G^{\cdot}$ for $G \pitchfork Q$ and $\delta : G^{\cdot} \to G^{\cdot}$ for the endomorphism adjoint to ∂. The *annihilator* of A_{p-1} is the subgroup of elements of $G^{\cdot}$ which transform A_{p-1} to 0; we designate this group as A^p. Then $(G^{\cdot}; A^p, \delta)$ is a d.f.-group with decreasing filtration.

For first A^0 is the annihilator of $A_{-1} = 0$, so that $A^0 = G^{\cdot}$; then if an element of $G^{\cdot}$ annihilates A_p *a fortiori* it annihilates A_{p-1}, so that $A^p \supseteq A^{p+1}$. If $g^{\cdot} \in \bigcap_p A^p$ and $g \in G$, then, for some q, $g \in A_q$; since $g^{\cdot} \in A^{q+1}$, $(g, g^{\cdot}) = 0$; this shows that $g^{\cdot}$ annihilates the whole of G and so $g^{\cdot} = 0$; thus $\bigcap_p A^p = 0$. Certainly $\delta\delta = 0$ since $\partial\partial = 0$; and finally if $g \in A^p$, then $(A_{p-1}, \delta g^{\cdot}) = (A_{p-1}\partial, g^{\cdot}) \subseteq (A_{p-1}, g^{\cdot}) = 0$, so that $\delta g^{\cdot} \in A^p$.

When we come to define a decreasing filtration on a g.-group $(G; G^n)$ we proceed, as in the case of an increasing filtration, by defining a decreasing filtration $[{}^nA^p]$ on each G^n, of length n in the main case. The subgroups $A^p = \sum_p {}^nA^p$ then provide a decreasing filtration for G. If we start from a group G with a graduation $\{G^n\}$ and a decreasing filtration $[A^p]$, the conditions that the set of subgroups ${}^nA^p = G^n \cap A^p$ satisfy these restrictions are that (i) $A^p = \sum_n {}^nA^p$ and (ii) $G^n \cap A^{n+1} = 0$. Accordingly we have

10.5.4 Definition. A *(restricted) filtered graded group* or *f.g.-group with decreasing filtration*, $(G; G^n, A^p)$ is a graded group $(G; G^n)$ with a decreasing filtration $[A^p]$ such that (i) $A^p = \sum_n (G^n \cap A^p)$ and (ii) $G^n \cap A^{n+1} = 0$. An *f.g.-homomorphism* of G to another f.g.-group G' with decreasing filtration is a g.-homomorphism $\phi : G \to G'$ such that, for all p, $\phi A^p \subseteq A'^p$.

We may sometimes also speak of an *f.g.-homomorphism of index t* if ϕ is of index t. We may also define u.f.g.-groups with decreasing filtration (and u.f.g.-homomorphisms) by omitting condition (ii) of 10.5.4.

10.5.5 Definition. A *differential filtered graded group* or *d.f.g.-group with decreasing filtration* $(G; G^n, A^p, \delta)$ is an f.g.-group $(G; G^n, A^p)$ with decreasing filtration and a (left) f.g.-endomorphism δ of index $+1$ such that $\delta\delta = 0$.

In future we shall feel free to omit the statement that the filtration is decreasing when the index p of filtration is written as a superscript.

10.5.6 *Example.* Let $(G; G_n, A_p, \partial)$ be a d.f.g.-group (with increasing filtration) and let Q be any abelian group. As in 3.1 we form the graded group $G^{\cdot}$ by defining G^n to be $G_n \pitchfork Q$ and $G^{\cdot}$ to be $\sum_n G^n$. We then define ${}^nA^p \subseteq G^n$ to be the annihilator of ${}_nA_{p-1}$, whence ${}^nA^0 = G^n$ and ${}^nA^{n+1}$ is the annihilator of ${}_nA_n = G_n$, so that $G^n \cap A^{n+1} = {}^nA^{n+1} = 0$. Then $(G^{\cdot}; G^n, A^p, \delta)$ is a d.f.g.-group. If G is only given to be a d.u.f.g.-group, then $G^{\cdot}$ is a d.u.f.g.-group with decreasing filtration.

Spectral theory may be developed for a d.f.g.-group with decreasing filtration in a way quite analogous with the development of 10.3. We define

$$\begin{aligned} {}^nZ^p_r &= {}^nA^p \cap (A^{p+r})\,\delta^{-1}; \\ {}^nB^p_r &= {}^nA^p \cap \delta A^{p-r} = \delta({}^{n-1}Z^{p-r}_r); \\ E^{p,q}_r &= {}^nE^p_r = {}^nZ^p_r/({}^nZ^{p+1}_{r-1} + {}^nB^p_{r-1}). \end{aligned}$$

The differential δ induces $\delta^{p,q}_r : E^{p,q}_r \to E^{p+r,q-r+1}_r$, which combine to give a differential δ_r of index $(r, 1-r)$ on the bigraded group $(E_r; E^{p,q}_r)$.

10.5.7 Theorem. *For a d.u.f.g.-group of decreasing filtration the bigraded groups $H(E_r; \delta_r)$ and E_{r+1} are equal.* ∎

We define ${}^nZ^p$ as the subgroup of 'contracycles' lying in ${}^nA^p$ and ${}^nB^p$ as the subgroup of 'contraboundaries' lying in ${}^nA^p$, and we define

$$E_\infty^{p,q} = {}^nE_\infty^p = {}^nZ^p/({}^nZ^{p+1} + {}^nB^p).$$

Further we define ${}^nD^p$ to be the image of $H^n(A^p)$ in $H^n(G)$, and we can prove that

10.5.8 Theorem. ${}^nE_\infty^p = {}^nD^p/{}^nD^{p+1}$. ▮

We observe that, if $D^p = \sum_n {}^nD^p$, $(H(G); H^n(G), D^p)$ is a u.f.g.-group, and is restricted if G is restricted.

In analogy with 10.3.9 we find that

10.5.9 Theorem. *For a d.f.g.-group of index* $+1$ *of decreasing filtration,* $E_r^{p,q} = E_\infty^{p,q}$ *if* $r > \max(p, q+1)$.

We use the fact that $G^n \cap A^p = 0$ for $p > n$ to deduce that

$$G^n \cap (A^t)\,\delta^{-1} = G^n \cap Z,$$

if $t > n+1$. Thus if $p+r > n+1$,

$${}^nZ_r^p = G^n \cap A^p \cap (A^{p+r})\,\delta^{-1} = G^n \cap A^p \cap Z = {}^nZ^p,$$

and of course ${}^nZ_{r-1}^{p-1} = {}^nZ^{p-1}$; but $p+r > n+1$ if and only if $r > q+1$. Next

$$\begin{aligned} {}^nB_{r-1}^p = G^n \cap A^p \cap \delta A^{p-r+1} &= G^n \cap A^p \cap \delta G \quad \text{if} \quad r > p \\ &= {}^nB^p. \end{aligned}$$

Hence if $r > \max(p, q+1)$, $E_r^{p,q} = E_\infty^{p,q}$. ▮

We sum up the conclusions by saying that a d.f.g.-group with decreasing filtration leads to a bigraded spectral sequence (E_r, δ_r) converging to E_∞ the bigraded group associated with the f.g.-group $(H(G); H^n(G), D^p)$.

In view of its application to contrahomology and the importance of the cup product of contrachains, it is natural to inquire under what conditions a ring structure in G induces a ring structure in the groups of the spectral sequence. To this end we give

10.5.10 Definition. A *d.f.g.-ring* is a ring R with graduation $\{R^n\}$ and decreasing filtration $[A^p]$ and differential δ of index $+1$ such that $(R; R^n, A^p, \delta)$ is a d.f.g.-group and that

(i) if $x \in R^n$, $y \in R$, $\delta(xy) = (\delta x)y + (-1)^n x(\delta y)$;

(ii) $R^n . R^{n'} \subseteq R^{n+n'}$; and

(iii) $A^p . A^{p'} \subseteq A^{p+p'}$.

The conditions (i) and (ii) have appeared in chapter 4 in the definition of a contrachain ring. A d.f.g.-ring is a contrachain ring suitably filtered. A differential satisfying (i) and (ii) is called an *antiderivative*.

A d.u.f.g.-ring is similarly defined.

10.5.11 Theorem. *Each of the groups E_r of a d.f.g.-ring R may be given a ring structure such that*

(*a*) $E_r^{p,q}.E_r^{p',q'} \subseteq E_r^{p+p',q+q'}$;

(*b*) *δ_r is an antiderivative with respect to total degree; and*

(*c*) *E_{r+1} is the bigraded contrahomology ring of (E_r, δ_r).*

We show first that

10.5.12 $\quad Z_r^p.Z_{r'}^{p'} \subseteq Z_{r''}^{p''}$, where $p'' = p+p'$ and $r'' = \min(r,r')$.

It is plainly enough to prove this for homogeneous elements. If $x \in {}^nZ_r^p$, $x' \in Z_{r'}^{p'}$, then $xx' \in A^{p''}$ and

$$\delta(xx') = (\delta x)x' + (-1)^n x(\delta x') \in A^{p+r}.A^{p'} + A^p.A^{p'+r'} \subseteq A^{p''+r''};$$

hence $Z_r^p.Z_{r'}^{p'} \subseteq Z_{r''}^{p''}$.

Next we show that

10.5.13 $\quad B_{r-1}^p.Z_r^{p'} + Z_r^p.B_{r-1}^{p'} \subseteq Z_{r-1}^{p''+1} + B_{r-1}^{p''}$.

$$\begin{aligned} B_{r-1}^p.Z_r^{p'} &= \delta(Z_{r-1}^{p-r+1}).Z_r^{p'} \subseteq \delta(Z_{r-1}^{p-r+1}.Z_r^{p'}) + Z_{r-1}^{p-r+1}.\delta Z_r^{p'} \\ &\qquad\qquad \text{by (i) of 10.5.10,} \\ &\subseteq \delta(Z_{r-1}^{p''-r+1}) + Z_{r-1}^{p-r+1}.B_r^{p'+r} \text{ by 10.5.12,} \\ &\subseteq B_{r-1}^{p''} + Z_{r-1}^{p-r+1}.Z_\infty^{p'+r} \\ &\subseteq B_{r-1}^{p''} + Z_{r-1}^{p''+1} \qquad \text{by 10.5.12.} \end{aligned}$$

Similarly $Z_r^p.B_{r-1}^{p'} \subseteq Z_{r-1}^{p''+1} + B_{r-1}^{p''}$.

We now deduce that the product in R induces a product in E_r, with respect to which E_r of course has a ring structure (with unity element if R has one). In view of 10.5.12 it is plainly enough to prove that

10.5.14 $$({}^nZ_{r-1}^{p+1} + {}^nB_{r-1}^p).{}^{n'}Z_r^{p'} + {}^nZ_r^p.({}^{n'}Z_{r-1}^{p'+1} + {}^{n'}B_{r-1}^{p'}) \subseteq {}^{n''}Z_{r-1}^{p''+1} + {}^{n''}B_{r-1}^{p''},$$

where $n'' = n+n'$. But this follows at once from 10.5.12 and 10.5.13 and condition (ii) of 10.5.10. This proves (*a*) of the theorem; (*b*) follows from the fact that δ_r is induced by δ; and (*c*) follows from the fact that the group isomorphisms involved in proving that $H(E_r, \delta_r) \cong E_{r+1}$ are plainly ring isomorphisms with respect to the products in E_r and E_{r+1}. ∎

It follows from 10.5.10 (i) that $H^*(R)$ acquires a ring structure from R; it then follows from 10.5.10 (ii), (iii) that

$$^{n}D^{p} \,.\, ^{n'}D^{p'} \subseteq {}^{n+n'}D^{p+p'}.$$

Thus the associated bigraded group $\mathrm{Gr}\,[^{n}D^{p}]$ acquires a bigraded ring structure. It is, moreover, clear that 10.5.11 (*a*) applies with $r = \infty$ and we have

10.5.15 Proposition. *E_∞ is the limit ring of the bigraded rings E_r and the g.-isomorphism* 10.5.8 *is a ring isomorphism.* ∎

We now point to a situation to which this theory of d.f.g.-rings is relevant.

10.5.16 Theorem. *If $f: X, x_* \to Y, y_*$ is a map of path-connected spaces and if $(C^\cdot; C^n, A^p, \delta)$ is the d.f.g. group associated with $(C; C_n, A_p, \partial)$ of* 10.4 *as in Example* 10.5.6, *with Q a ring, then $C^\cdot$ is a d.f.g.-ring with respect to the cup product of* 9.3.

Here $C^n = {}^0\square^n_N(X)$ and A^p is the annihilator of A_{p-1}, the subgroup of ${}^0\square^N$ generated by cubes of filtration $(p-1)$. Clearly

$${}^0\square_N \cup {}^0\square_N \subseteq {}^0\square_N,$$

so that the product is defined in $C^\cdot$ and we can easily see that $C^\cdot$ is a d.f.g.-ring; for conditions (i) and (ii) of 10.5.10 are certainly satisfied and we have only to verify condition (iii) that $A^p \cup A^{p'} \subseteq A^{p''}$. Let $c \in A^p$, $c' \in A^{p'}$ and let g be a generator in $A_{p''-1}$; we shall prove that $(g, c \cup c') = 0$ and so establish that $c \cup c'$ annihilates $A_{p''-1}$ and lies in $A^{p''}$.

Let g be any singular cube of ${}^0\square(X)$ and λ^0_H, λ^1_K be as in the notation of 9.3, so that H, K are complementary subsets. Let the exact filtrations of $\lambda^0_H g$, $\lambda^1_K g$ be p, p' respectively, so that $\lambda^0_H gf$, $\lambda^1_K gf$ are not independent of u_p, $u_{p'}$. Then gf is not independent of u_{h_p}, $u_{k_{p'}}$; since $\max(h_p, k_{p'}) \geqslant p + p' = p''$, the exact filtration of g is not less than p''. If therefore $g \in A_{p''-1}$, either $\lambda^0_H g \in A_{p-1}$ or $\lambda^1_K g \in A_{p'-1}$; hence $(g, c \cup c') = \sum_H \rho_{HK}(\lambda^0_H g, c) \,.\, (\lambda^1_K g, c') = 0$, since for each H either $(\lambda^0_H g, c)$ or $(\lambda^1_K g, c')$ is zero. ∎

Serre's results about the spectral sequence of a fibre map (10.4.9–10.4.11) have their obvious analogues in contrahomology. However, 10.4.11*c* is enriched by showing that $\phi^\cdot_2$ preserves multiplicative structure.

Let R be a ring and let $(C(B)\otimes C(F)) \pitchfork R$ be given the ring structure defined by 9.4.3. A direct (but tedious!) computation shows that

$$\phi^{\cdot} : (C(B)\otimes C(F)) \pitchfork R \to C(X) \pitchfork R$$

is product-preserving. We conclude

10.5.17 Theorem. *If B is 1-connected and $p : X, x_* \to B, b_*$ is a fibre map with path-connected fibre F, the homomorphism*

$$\phi : C(X) \to C(B)\otimes C(F)$$

induces, for values in a field $\mathscr{F}$, a bigraded ring isomorphism

$$\phi_2^{\cdot} : R^*(B; R^*(F; \mathscr{F})) \cong E_2,$$

where $R^*(B; R^*(F; \mathscr{F}))$ has the structure of a contrahomology ring with values in a graded *ring* (see 9.4.17). ▌

From 10.5.17 and 9.2.12 we may readily deduce that the product in E_2 is skew-commutative, in that, if $c \in {}^nE_2$ and $c' \in {}^{n'}E_2$, then $cc' = (-1)^{nn'} c'c$. From this it follows that E_r is skew-commutative for all $r \geqslant 2$.

10.6 Spectral sequence of a fibre map: applications

We suppose throughout this section that we have a fibre map $p : X \to B$ with fibre F, such that X, F are 0-connected and B is 1-connected, and that we have associated with p a spectral homology sequence (E^r, ∂^r) and a spectral contrahomology sequence of rings (E_r, δ_r) such that

10.6.1 $$E^2_{p,q} \cong H_p(B; H_q(F)),$$

10.6.1 c $$E_2^{p,q} \cong H^p(B; H^q(F));$$

compare 10.4.11 and 10.5.17. The coefficient or value group will be J throughout this section and will be suppressed from the notation; our main results hold in fact also for any field of coefficients.

Let $i : F \to X$ be the injection map: we prove

10.6.2 Theorem. *If $H_p(B) = 0$ for $0 < p < p_0$, $H_q(F) = 0$ for $0 < q < q_0$, then there is an exact sequence*

$$H_{n_0-1}(F) \xrightarrow{i_*} H_{n_0-1}(X) \xrightarrow{p_*} H_{n_0-1}(B) \xrightarrow{t} H_{n_0-2}(F) \to \dots$$

$$\dots \to H_k(F) \xrightarrow{i_*} H_k(X) \xrightarrow{p_*} H_k(B) \xrightarrow{t} H_{k-1}(F) \to \dots \to H_1(X) \to 0,$$

where $n_0 = p_0 + q_0$.

From 10.6.1 and the conditions on $H_p(B)$, $H_q(F)$, if $0<p<n<n_0$, then ${}_nE^2_p = 0$.

Since ${}_nE^{r+1}_p$ is a subquotient group of ${}_nE^r_p$, this together with 10.3.9 implies that

10.6.3 $\quad {}_nE^r_p = 0 \quad \text{if} \quad 0<p<n<n_0 \quad \text{for} \quad r = 2, 3, \ldots, \infty.$

In the filtration $0 \subseteq {}_kD_0 \subseteq {}_kD_1 \subseteq \ldots \subseteq {}_kD_{k-1} \subseteq {}_kD_k = H_k(X)$, we know from 10.3.7 that ${}_kD_p/{}_kD_{p-1} \cong {}_kE^\infty_p$, and by 10.6.3 ${}_kE^\infty_p = 0$ for $0<p<k<n_0$. Hence

$$ {}_kE^\infty_0 \cong {}_kD_0 = {}_kD_1 = \ldots = {}_kD_{k-1} $$

and $$ H_k(X)/{}_kD_{k-1} = {}_kE^\infty_k \quad \text{for} \quad 0<k \leqslant n_0 - 1. $$

These results combine to give an exact sequence

10.6.4 $\quad 0 \to E^\infty_{0,k} \xrightarrow{\iota} H_k(X) \xrightarrow{\kappa} E^\infty_{k,0} \to 0 \quad \text{for} \quad 0<k\leqslant n_0 - 1.$

Now by 10.4.12 there is an epimorphism $E^r_{0,q} \to E^{r+1}_{0,q}$ whose kernel is ${}_{q+1}E^r_r\partial^r$; by 10.6.3, if $2 \leqslant r \leqslant q \leqslant n_0 - 2$ this is an isomorphism, and this sequence of isomorphisms establishes an isomorphism

$$ E^2_{0,q} \cong E^{q+1}_{0,q} \quad \text{for} \quad 1 \leqslant q \leqslant n_0 - 2. $$

Moreover the kernel of the epimorphism $E^{q+1}_{0,q} \to E^{q+2}_{0,q}$ is $E^{q+1}_{q+1,0}\partial^{q+1}$ for all q, and, by 10.3.9, $E^{q+2}_{0,q} \cong E^\infty_{0,q}$. These results may be summarized in the diagram

10.6.5
$$ \begin{array}{ccccccc} E^{q+1}_{q+1,0} & \xrightarrow{\partial^{q+1}} & E^{q+1}_{0,q} & \longrightarrow & E^{q+2}_{0,q} & \longrightarrow & 0 \\ & & \uparrow \cong & & \downarrow \cong & & \\ & & E^2_{0,q} & & E^\infty_{0,q} & & \end{array} $$

in which the row is exact, and the isomorphisms hold for $1 \leqslant q \leqslant n_0 - 2$.

Again by 10.4.15 there is a monomorphism $E^{r+1}_{p,0} \to E^r_{p,0}$ with cokernel the subgroup $E^r_{p,0}\partial^r$ of ${}_{p-1}E^r_{p-r}$. By 10.6.3 this cokernel is null for $2 \leqslant r \leqslant p-1 \leqslant n_0 - 1$, so that we have a sequence of isomorphisms combining to an isomorphism

$$ E^p_{p,0} \cong E^2_{p,0} \quad \text{for} \quad 2 \leqslant p \leqslant n_0. $$

Moreover, the cokernel of $E^{p+1}_{p,0} \to E^p_{p,0}$ is $E^p_{p,0}\partial^p$, and, by 10.3.9, $E^\infty_{p,0} \cong E^{p+1}_{p,0}$. Hence in the diagram

10.6.6
$$ \begin{array}{ccccccc} 0 & \longrightarrow & E^{p+1}_{p,0} & \longrightarrow & E^p_{p,0} & \xrightarrow{\partial^p} & E^p_{0,p-1} \\ & & \uparrow \cong & & \downarrow \cong & & \\ & & E^\infty_{p,0} & & E^2_{p,0} & & \end{array} $$

the row is exact, and the isomorphisms hold for $2 \leqslant p \leqslant n_0$.

Combining the diagrams 10.6.5 and 10.6.6 and completing to a commutative diagram with homomorphisms χ, $\bar{\partial}$, ψ we form the diagram

10.6.7

$$\begin{array}{ccccccccc} 0 & \longrightarrow & E^{k+1}_{k,0} & \longrightarrow & E^{k}_{k\,0} & \xrightarrow{\partial^k} & E^{k}_{0,k-1} & \longrightarrow & E^{k+1}_{0,k-1} & \longrightarrow & 0 \\ & & \uparrow\cong & & \downarrow\cong & & \uparrow\cong & & \downarrow\cong & & \\ 0 & \longrightarrow & E^{\infty}_{k,0} & \xrightarrow{\chi} & E^{2}_{k,0} & \xrightarrow{\bar{\partial}} & E^{2}_{0,k-1} & \xrightarrow{\psi} & E^{\infty}_{0,k-1} & \longrightarrow & 0 \end{array}$$

defined for $1 < k < n_0$, in which each row is exact.

The lower sequence of 10.6.7 and the sequence 10.6.4 combine to give the exact sequence

10.6.8 $$E^2_{0,n_0-1} \xrightarrow{\psi\iota} H_{n_0-1}(X) \xrightarrow{\kappa\chi} E^2_{n_0-1,0} \xrightarrow{\bar{\partial}} E^2_{0,n_0-2} \xrightarrow{\psi\iota} \ldots;$$

the sequence may start at E^2_{0,n_0-1} since the right-hand half of 10.6.7 comes from 10.6.5 and holds for $k \leqslant n_0-1$; it terminates at $H_1(X) \to 0$. The sequence of the theorem is obtained from 10.6.8 using 10.6.1 and the standard interpretations of $\psi\iota$ and $\kappa\chi$. ∎

Arguments of this sort are in practice generally carried out with the aid of illustrative diagrams, from which conclusions may be drawn by informal reasoning. We introduce the reader to this technique on pp. 435–443, where the diagram appropriate to 10.6.2 is given.

We see that $t : H_k(B) \to H_{k-1}(F)$ corresponds essentially to

$$\partial^k : E^k_{k,0} \to E^k_{0,k-1};$$

t is called the (*homology*) *transgression* in the fibration. If we make no special assumptions on B and F then $E^k_{k,0}$ $(k > 0)$ corresponds to the subgroup of $H_k(B)$ that is the image of $H_k(X, F)$ under the projection p; we call this homomorphism, induced by p, $\hat{p}_* : H_k(X,F) \to H_k(B)$. Further, $E^k_{0,k-1}$ corresponds to the quotient group $H_{k-1}(F)/(\ker \hat{p}_*)\,\nu_*$, where $\nu_* : H_k(X,F) \to H_{k-1}(F)$ is the homomorphism from the homology sequence for the pair X, F. Then

$$t : H_k(X,F)\,\hat{p}_* \to H_{k-1}(F)/(\ker \hat{p}_*)\,\nu_*$$

is given by

10.6.9 $$\xi\hat{p}_* t = [\xi\nu_*], \quad \xi \in H_k(X,F).$$

Conceptually t performs the following function; we consider a cycle z of B that is the projection of a relative cycle $\hat{y}$ of $X \bmod F$; with such a z we associate the boundary of $\hat{y}$ and regard this as a cycle of F. In this general case an element in $H_k(X,F)\,\hat{p}_*$ is called *transgressive.* In the k-diagram analogous to 10.4.19 the transgressive elements occupy the box immediately above that labelled ${}_kE_k^{\infty}$; they are mapped by transgression into the box immediately below that labelled ${}_{k-1}E_0^{\infty}$ in the $(k-1)$-diagram. Part of the assertion of 10.6.2 is that under the conditions of the theorem every element of $H_k(B)$ is transgressive for $2 \leqslant k \leqslant n_0 - 1$.

Notice that if we make no special assumptions on the homology of B and F then we do not get a sequence (exact or not) like 10.6.2 relating the homology groups of X, B and F; for, in general, no connecting homomorphism $H_k(B) \to H_{k-1}(F)$ is defined. There is always a sequence

$$\ldots \to H_k(X) \to E_{k,0}^k \to E_{0,k-1}^k \to H_{k-1}(X) \to \ldots$$

which is exact at $E_{k,0}^k$ and $E_{0,k-1}^k$; but it fails in general to be exact at $H_k(X)$ owing to the presence of non-zero groups $E_{p,q}^{\infty}$ with p, q both positive. Thus, for example, if we take $X = S^2 \times S^2$, $B = S^2$, $F = S^2$, then $H_4(X) = J(=E_{2,2}^{\infty})$, but $E_{0,4}^5 = E_{4,0}^4 = 0$.

We infer immediately from 10.6.2, from the exact homology sequence of the pair X, F and from the 5-lemma

10.6.10 Theorem. *If* $H_p(B) = 0$ *for* $0 < p < p_0$ *and* $H_q(F) = 0$ *for* $0 < q < q_0$, *then*

$$\hat{p}_* : H_k(X, F) \cong H_k(B), \quad 1 \leqslant k < n_0 = p_0 + q_0. \;\rule{0.5em}{0.8em}$$

Passing to contrahomology we have the corresponding results:

10.6.2 c Theorem. *If* $H^p(B) = 0$ *for* $0 < p < p_0$, $H^q(F) = 0$ *for* $0 < q < q_0$ *then there is an exact sequence*

$$H^{n_0-1}(F) \xleftarrow{i^*} H^{n_0-1}(X) \xleftarrow{p^*} H^{n_0-1}(B) \xleftarrow{t^\cdot} \ldots \leftarrow H^k(F) \xleftarrow{i^*}$$

$$H^k(X) \xleftarrow{p^*} H^k(B) \xleftarrow{t^\cdot} H^{k-1}(F) \leftarrow \ldots \leftarrow H^1(X) \leftarrow 0. \;\rule{0.5em}{0.8em}$$

In general, $t^\cdot$ is the map $\delta_k^\cdot$ from the subgroup

$$E_k^{0,k-1} = \{\hat{p}^* H^k(B)\}\, \nu^{*-1}$$

of $H^{k-1}(F)$ to the quotient group $E_k^{k,0} = H^k(B)/\ker \hat{p}^*$ of $H^k(B)$; it is called the (*contrahomology*) *transgression* and is given by $t^\cdot \eta = [\zeta]$, where $\hat{p}^*\zeta = \nu^*\eta$, $\eta \in H^{k-1}(F)$, $\zeta \in H^k(B)$. An element in

$$\{\hat{p}^* H^k(B)\}\, \nu^{*-1} = E_k^{0,k-1}$$

is called *transgressive.*

10.6.10 c Theorem. *If* $H^p(B) = 0$ *for* $0 < p < p_0$, $H^q(F) = 0$ *for* $0 < q < q_0$, *then*

$$\hat{p}^* : H^k(B) \cong H^k(X, F), \quad 1 \leqslant k < n_0. \;\rule{0.5em}{0.8em}$$

In our second application of 10.6.1, 10.6.1c we suppose that F is an m-sphere (it is sufficient, in fact, that F has the homology groups of S^m). We prove

10.6.11 Theorem. *If F is an m-sphere, there is an exact sequence*

$$\ldots \to H_{k+1}(X) \xrightarrow{p_*} H_{k+1}(B) \to H_{k-m}(B) \to H_k(X) \to \ldots.$$

The argument is similar to that of the proof of 10.6.2; an appropriate diagram is to be found on p. 439. From 10.6.1 we deduce that $E^2_{p,q} = 0$ for $q \neq m, 0$, and hence that $E^r_{p,q} = 0$ for $q \neq m, 0$ and $r = 2, 3, \ldots, \infty$. Hence if $k > m$, in the filtration

$$0 \subseteq {}_kD_0 \subseteq {}_kD_1 \subseteq \ldots \subseteq {}_kD_{k-m-1} \subseteq {}_kD_{k-m} \subseteq \ldots$$
$$\subseteq {}_kD_{k-1} \subseteq {}_kD_k = H_k(X)$$

all quotient groups except ${}_kD_{k-m}/{}_kD_{k-m-1}$ and ${}_kD_k/{}_kD_{k-1}$ are null, including ${}_kD_0/0$. Hence we deduce

$$E^\infty_{k-m,m} \cong {}_kD_{k-m} \cong \ldots \cong {}_kD_{k-1}$$

and $$H_k(X)/{}_kD_{k-1} \cong E^\infty_{k,0}.$$

This gives the exact sequence

10.6.12 $$0 \to E^\infty_{k-m,m} \to H_k(X) \to E^\infty_{k,0} \to 0.$$

For $r \geqslant 2$, $E^r_{k-m,m}\partial^r \subseteq E^r_{k-m-r,m+r-1} = 0$; so that every element of $E^r_{k-m,m}$ is a ∂^r-cycle. For $2 \leqslant r \leqslant m$, $E^r_{k-m+r,m-r+1} = 0$, so that no non-zero element of $E^r_{k-m,m}$ is a ∂^r-boundary. Hence

$$E^2_{k-m,m} \cong E^3_{k-m,m} \cong \ldots \cong E^{m+1}_{k-m,m}.$$

The ∂^{m+1}-boundaries in $E^{m+1}_{k-m,m}$ belong to $E^{m+1}_{k+1,0}\partial^{m+1}$ and

$$E^{m+2}_{k-m,m} \cong E^\infty_{k-m,m}.$$

There is therefore a diagram

10.6.13
$$\begin{array}{ccccccc}
E^{m+1}_{k+1,0} & \xrightarrow{\partial^{m+1}} & E^{m+1}_{k-m,m} & \longrightarrow & E^{m+2}_{k-m,m} & \longrightarrow & 0 \\
 & & \uparrow \cong & & \downarrow \cong & & \\
 & & E^2_{k-m,m} & & E^\infty_{k-m,m} & &
\end{array}$$

By similar arguments $E^\infty_{k+1,0}$ is mapped isomorphically to $E^{m+2}_{k+1,0}$ and $E^{m+2}_{k+1,0}$ is mapped monomorphically into $E^{m+1}_{k+1,0}$ with cokernel $E^{m+1}_{k+1,0}\partial^{m+1}$, and $E^{m+1}_{k+1,0} \cong E^2_{k+1,0}$; hence in the diagram

10.6.14
$$\begin{array}{ccccccc}
0 & \longrightarrow & E^{m+2}_{k+1,0} & \longrightarrow & E^{m+1}_{k+1,0} & \xrightarrow{\partial^{m+1}} & E^{m+1}_{k-m,m} \\
 & & \uparrow \cong & & \downarrow \cong & & \\
 & & E^\infty_{k+1,0} & & E^2_{k+1,0} & &
\end{array}$$

the row is exact.

We combine 10.6.13 and 10.6.14 and complete to the commutative diagram

10.6.15

$$\begin{array}{ccccccccc} 0 & \longrightarrow & E^{m+2}_{k+1,0} & \longrightarrow & E^{m+1}_{k+1,0} & \xrightarrow{\partial^{m+1}} & E^{m+1}_{k-m,m} & \longrightarrow & E^{m+2}_{k-m,m} & \longrightarrow & 0 \\ & & \uparrow \cong & & \downarrow \cong & & \uparrow \cong & & \downarrow \cong & & \\ 0 & \longrightarrow & E^{\infty}_{k+1,0} & \longrightarrow & E^{2}_{k+1,0} & \longrightarrow & E^{2}_{k-m,m} & \longrightarrow & E^{\infty}_{k-m,m} & \longrightarrow & 0 \end{array}$$

in which each row is exact. From 10.6.1 we see that $E^2_{k-m,m} \cong H_{k-m}(B)$ and we combine the lower row of 10.6.15 with 10.6.12 to deduce the exact sequence of the theorem. ∎

The proof provides interpretations of the homomorphisms of the sequence of 10.6.11. This sequence and its contrahomology form below are due to W. Gysin, *Comm. Math. Helv.* **14** (1941), 61; they are known as the Gysin sequences.

10.6.11 *c* Theorem. *If F is an m-sphere, there is an exact sequence*

$$\dots \leftarrow H^{k+1}(X) \overset{p^*}{\leftarrow} H^{k+1}(B) \overset{\phi}{\leftarrow} H^{k-m}(B) \leftarrow H^k(X) \leftarrow \dots . ∎$$

10.6.16 Corollary. *There exists no fibre map $p: X \rightarrow B$ where X is contractible, F is an m-sphere, $m \geqslant 1$, and B is the polyhedron of a finite-dimensional* 1*-connected complex.*

For we see from 10.6.11 that $H_k(B) = J$ for $k = (m+1)$, $2(m+1)$, $3(m+1)$, …. ∎

In our third application we suppose that B is an m-sphere, $m > 1$, obtaining homology and contrahomology sequences originally due to H. C. Wang, *Duke Math. J.* **16** (1949), 33. Confidently leaving the proof‡ to the reader, we state

10.6.17 Theorem. *If B is an m-sphere, $m > 1$, there is an exact Wang sequence*

$$\dots \rightarrow H_k(F) \overset{i_*}{\rightarrow} H_k(X) \rightarrow H_{k-m}(F) \overset{\theta}{\rightarrow} H_{k-1}(F) \rightarrow \dots . ∎$$

We record that θ is defined by means of the differential

$$\partial^m : E^m_{m,k-m} \rightarrow E^m_{0,k-1}.$$

‡ Serre (loc. cit.) deduces 10.6.2, 10.6.11, 10.6.17 from a single theorem giving conditions under which exact sequences are obtainable from spectral sequences with special properties.

Indeed, identifying $E^m_{m,k-m}$ with $H_m(B)\otimes H_{k-m}(F)$ and $E^m_{0,k-1}$ with $H_0(B)\otimes H_{k-1}(F)$ we may define θ by

10.6.18 $$(\sigma\otimes\xi)\,\partial^m = \iota\otimes\xi\theta,$$

where σ is a chosen generator of $H_m(B)$, ι is a chosen generator of $H_0(B)$ and $\xi\in H_{k-m}(F)$.

10.6.17 *c* Theorem. *If B is an m-sphere, $m>1$, there is an exact Wang sequence*

$$\ldots \leftarrow H^k(F) \overset{i^*}{\leftarrow} H^k(X) \leftarrow H^{k-m}(F) \overset{\theta^\cdot}{\leftarrow} H^{k-1}(F) \leftarrow \ldots. \;\rule{0.5em}{1em}$$

Moreover $\theta^\cdot$ is given by

10.6.18 *c* $$\delta_m(\iota\otimes\xi) = \sigma\otimes\theta^\cdot\xi,$$

where we apply 5.4.25 to 10.6.1*c* as $H_*(B)$ is finitely generated in each dimension, and where σ is a chosen generator of $H^m(B)$, ι is a chosen generator of $H^0(B)$ and $\xi\in H^{k-1}(F)$.

Now by 10.5.11 δ_m satisfies the usual product rule for a contrahomology differential. Thus

$$\delta_m((\iota\otimes\xi)(\iota\otimes\xi')) = \delta_m(\iota\otimes\xi).(\iota\otimes\xi')+(-1)^q(\iota\otimes\xi).\delta_m(\iota\otimes\xi'),$$

where $\dim\xi = q$. Hence

$$\begin{aligned}\sigma\otimes\theta^\cdot(\xi\xi') &= (\sigma\otimes\theta^\cdot\xi)(\iota\otimes\xi')+(-1)^q(\iota\otimes\xi)(\sigma\otimes\theta^\cdot\xi')\\ &= \sigma\otimes(\theta^\cdot\xi)\,\xi'+(-1)^q(-1)^{qm}\,\sigma\otimes\xi(\theta^\cdot\xi') \quad \text{(see 9.4.8)}.\end{aligned}$$

We have proved

10.6.19 Proposition. *If B is an m-sphere, $m>1$, and*

$$\theta^\cdot : H^{k-1}(F)\to H^{k-m}(F)$$

is given by 10.6.18*c*, *then*‡, *if m is odd,*

$$\theta^\cdot(\xi\xi') = (\theta^\cdot\xi)\,\xi'+\xi(\theta^\cdot\xi),$$

and, if m is even

$$\theta^\cdot(\xi\xi') = (\theta^\cdot\xi)\,\xi'+(-1)^q\xi(\theta^\cdot\xi'),$$

where $\dim\xi = q$. ▮

We apply 10.6.17 and 10.6.19 in two examples.

10.6.20 *Example.* It was pointed out in II. 3 that $p: E(B)\to B$ is a fibre map with contractible fibre space $E(B)$ and fibre $\Omega(B)$. Let

‡ Notice the 'gimmick' mentioned in the footnote on p. 374. The sign arises through permuting ξ of 'degree' q with $\theta^\cdot$ of 'degree' $(m-1)$.

$B = S^m$, $m > 1$, and let us compute the contrahomology ring of ΩS^m. From 10.6.17*c*, II.2.7 and the Hurewicz isomorphism theorem 8.8.5, it follows easily that $H^*(\Omega S^m) = 0$ except in dimensions $n = k(m-1)$, when $H^n = J$.

Now let m be odd; then we may prove by induction on $(k+l)$ that, if generators $\eta^{(k)}$ of $H^{k(m-1)}(\Omega S^m)$ are chosen, so that $\theta^{\cdot}\eta^{(k)} = \eta^{(k-1)}$, then

10.6.21 $$\eta^{(k)}\eta^{(l)} = (k, l)\, \eta^{(k+l)},$$

where (k, l) is the binomial coefficient $(k+l)!/k!\,l!$. For, using the inductive hypothesis and 10.6.19, we have

$$\begin{aligned}\theta^{\cdot}(\eta^{(k)}\eta^{(l)}) &= \eta^{(k-1)}\eta^{(l)} + \eta^{(k)}\eta^{(l-1)} \\ &= (k-1, l)\, \eta^{(k+l-1)} + (k, l-1)\, \eta^{(k+l-1)} \\ &= (k, l)\, \eta^{(k+l-1)} = \theta^{\cdot}((k, l)\, \eta^{(k+l)});\end{aligned}$$

but $\theta^{\cdot}$ is an isomorphism and 10.6.21 is verified. Thus $R^*(\Omega S^m)$ is what Cartan has described as an *algebra with divided powers*: if $\eta = \eta^{(1)}$, then we may 'divide' η^n by $n!$ and obtain $\eta^{(n)}$. As an algebra with divided powers, $R^*(\Omega S^m)$ may be said to be generated by $\eta^{(1)}$.

The case in which m is even is a little more complicated. The argument is entirely analogous to that given above; the difference in the conclusion can be traced to the fact that the form of 10.6.19 depends on the parity of m. We suppress the details, stating the result that $R^*(\Omega S^m)$ is then isomorphic to the tensor product of an exterior algebra generated by $\eta^{(1)}$ and an algebra with divided powers generated by $\eta^{(2)}$. In fact, $R^*(\Omega S^m)$ is generated qua ring, by the elements $\eta^{(r)}$ subject to the relations

$$(\eta^{(1)})^2 = 0, \quad \eta^{(1)}\eta^{(2k)} = \eta^{(2k)}\eta^{(1)} = \eta^{(2k+1)},$$

$$\eta^{(2k)}\eta^{(2l)} = (k, l)\, \eta^{(2k+2l)}.$$

Merely to demonstrate a useful diagrammatic approach to computations with spectral sequences, we shall recalculate the homology groups of ΩS^m without using 10.6.17 (but of course using 10.6.1). For this method we start by representing the groups $E^2_{p,q}$ by points (p, q) in the Cartesian plane. Then the horizontal axis ($q = 0$) represents the homology of the base and the vertical axis ($p = 0$) the homology of the fibre; moreover the differential ∂^2 is represented by a particular 'knight's move'‡ in the plane. The aim is to argue from partial knowledge of E^2 and E^∞ to more complete knowledge.

The passage from $E^2_{p,q}$ to $E^\infty_{p,q}$ is effected by repeatedly taking quotient groups of subgroups. The first subgroup is the kernel of

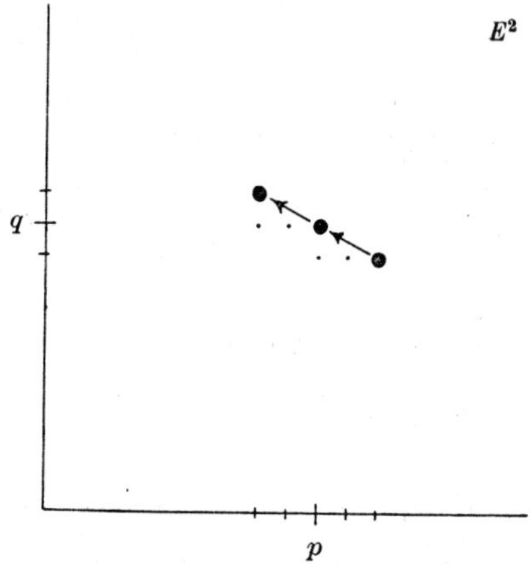

$\partial^2 : E^2_{p,q} \to E^2_{p-2,q+1}$ and the kernel of the projection onto $E^3_{p,q}$ is $E^2_{p+2,q-1}\partial^2$. The groups connected by ∂^2 may be joined in the diagram by arrows. The passage from $E^2_{p,q}$ to $E^3_{p,q}$ is an isomorphism if both of $E^2_{p-2,q+1}$ and $E^2_{p+2,q-1}$ are null or, of course, if $E^2_{p,q}$ is null. The passage,

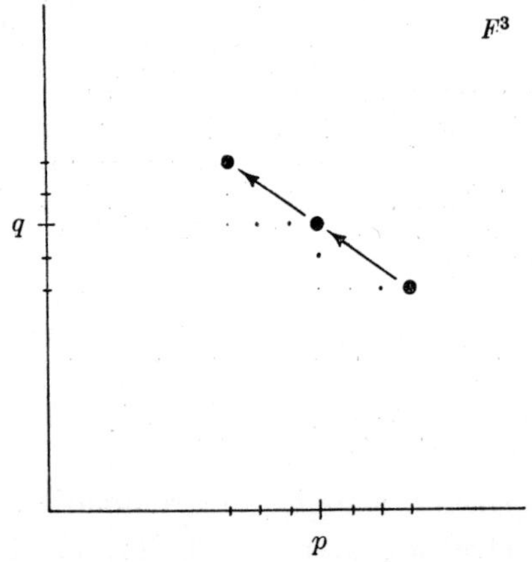

then, from E^2 to E^3 is isomorphic unless there is some pair of groups connected by ∂^2 neither of which is null. If, on the other hand, one of each such pair is null, we need make no distinction between E^2

‡ For definition, see for instance, H. Staunton, *Chess Praxis* (Bohn, 1860), p. 3.

and E^3. In the diagram, however, for E^3 the arrows must pass from (p, q) to $(p-3, q+2)$; in the diagram for E^r they pass from (p, q) to $(p-r, q+r-1)$—one may think of these informally as generalized knight's moves.

Suppose now that, for some p, q, we know that $E^{\infty}_{p,q}$ is null whereas $E^2_{p,q}$ is not. At some passage to quotient group of subgroup each element of $E^2_{p,q}$ must 'be killed'; death can be caused either by failure to be a ∂^r-cycle or by being a ∂^r-boundary, for some r; in either case the element killed is paired to a non-zero element by ∂^r. An element of $E^2_{p,q}$ that survives into $E^r_{p,q}$ is then in fact a representative of some

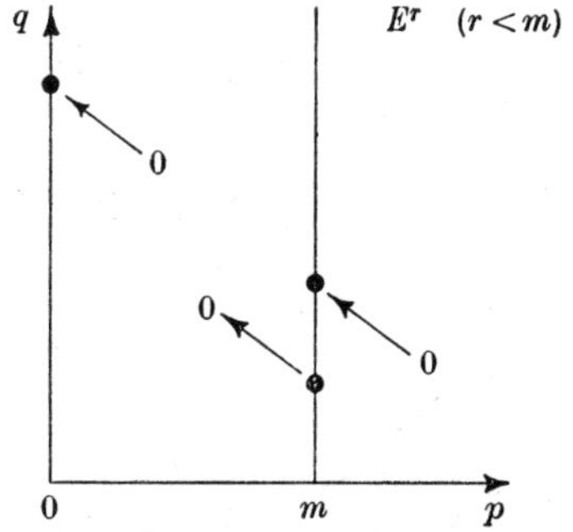

coset of elements of $E^2_{p,q}$, since the elements of $E^r_{p,q}$ are such cosets; if at the next stage this coset is killed, it is legitimate to speak of a representative element in $E^2_{p,q}$ as being killed.

Let us now apply this method to calculate $H_*(\Omega S^m)$, $m > 1$, considering as before the contractible fibre space $X = E(S^m)$, with base $B = S^m$ and fibre $F = \Omega(S^m)$. Here $H_n(X) = 0$ for $n > 0$ and so, by 10.3.7, ${}_nE^{\infty} = 0$ for $n > 0$. By 10.6.1 $E^2_{p,q} \simeq H_p(S^m; H_q(F))$; hence

$$E^2_{p,q} \cong H_q(F) \quad \text{if } p = 0 \text{ or } m$$
$$= 0 \qquad \text{otherwise.}$$

Hence $E^r_{p,q} = 0$ if $p \neq 0, m$, for $r = 2, 3, \ldots, \infty$.

Since ∂^r connects groups whose p-coordinates differ by r, the only value of $r \geqslant 2$ for which a pair of non-null groups can be connected by ∂^r is $r = m$. Therefore $E^2 \cong E^3 \cong \ldots \cong E^m$ and $E^{m+1} \cong \ldots \cong E^{\infty}$.

Now ${}_nE^{\infty} = 0$ for $n > 0$, so that $E^{m+1}_{m,q} = 0$ for $q \geqslant 0$ and $E^{m+1}_{0,m+q-1} = 0$ for $q > 1-m$. Any non-zero element in $E^2_{m,q}$ survives into $E^m_{m,q}$ and must then be killed, by being connected by ∂^m to a non-zero element of $E^m_{0,m+q-1}$; hence
$$\partial^m : E^m_{m,q} \to E^m_{0,m+q-1}$$

is monomorphic. Any non-zero element in $E^2_{0,q'}$ survives into $E^m_{0,q'}$, and must then be killed (if $q' > 0$), by being connected by ∂^m to an element of $E^m_{m,q'-m+1}$; hence

$$\partial^m : E^m_{m,q} \to E^m_{0,m+q-1}$$

is epimorphic for $q > 1-m$. Accordingly

10.6.22 $$\partial^m : E^m_{m,q} \cong E^m_{0,m+q-1} \quad \text{for} \quad q > 1-m.$$

We know that $E^m_{m,q} = 0$ for $q < 0$ and that $E^m_{m,0} = J$, since $H_0(F) = J$. Hence $E^m_{0,q'} = 0$ for $0 < q' < m-1$ and $E^m_{0,m-1} = J$; we also know that $E^m_{m,q} \cong E^m_{0,q} \cong H_q(F)$ for all q. It thus follows by an easy inductive argument that

$$H_q(F) = J \quad \text{if} \quad q = k(m-1) \text{ and } k \text{ is an integer} \geqslant 0,$$
$$= 0 \quad \text{otherwise.}$$

Indeed, the conclusion becomes quite obvious from inspection of the diagram.

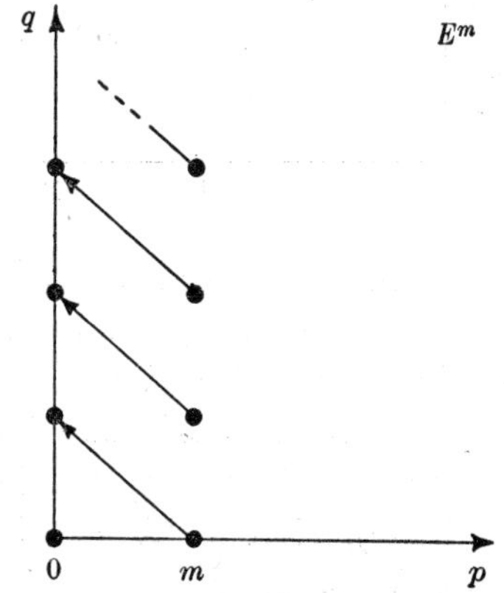

The diagrammatic approach to computations with spectral sequences—as, indeed, any approach—depends strongly on 10.6.1 in order that E^2 should be deducible from knowledge of $E^2_{p,0}$ (all p) and $E^2_{0,q}$ (all q). It is particularly, but certainly not exclusively, applicable when, as in our example, the E^∞ terms are trivial in the obvious sense. With practice the reader may well find himself able to handle a hybrid diagram incorporating, as it were, the E^2 terms and all the differentials ∂^r, $r \geqslant 2$. If we concentrate attention on a particular *isogonal* $p+q = n$ of the diagram, then the ${}_nE^\infty$ term consists of those parts of the $E^2_{p,q}$ terms on this isogonal which have survived the 'attacks' of the differentials ∂^r, $r \geqslant 2$.

The diagram appropriate to 10.6.2 is

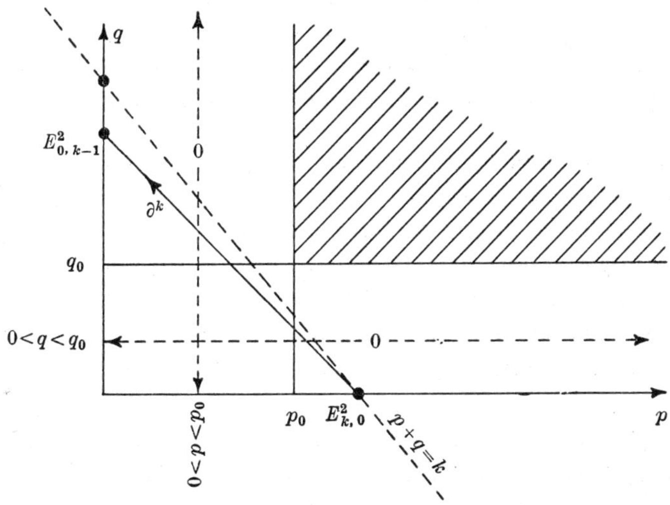

Here $k < p_0 + q_0$ so that the isogonal $p+q = k$ contains possibly non-zero terms only at $(k, 0)$ and $(0, k)$. All differentials ∂^r, $2 \leqslant r < k$, send $E^2_{k,0}$ to zero; and no non-zero differential ∂^r, $2 \leqslant r < k$, finishes up at $E^2_{0,k-1}$. Thus on the E^∞ diagram the isogonal $p+q = k$ contains just $\ker \partial^k$ at $(k, 0)$ and $\operatorname{coker} \partial^{k+1}$ at $(0, k)$.

The diagram appropriate to 10.6.11 is

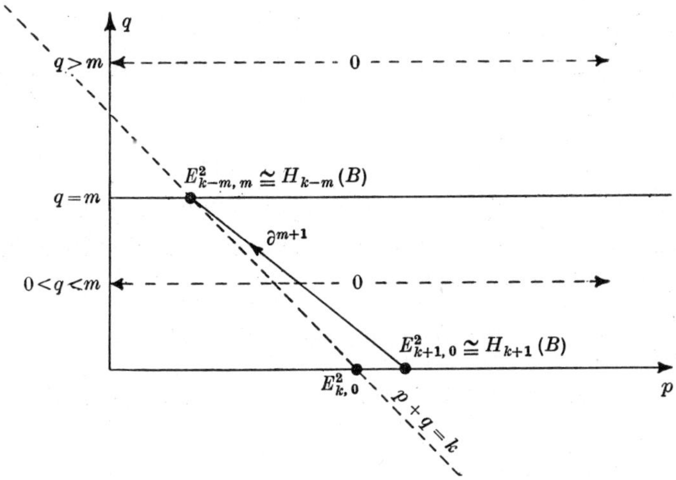

The isogonal $p+q = k$ contains possibly non-zero terms only at $(k, 0)$ and $(k-m, m)$. The differentials ∂^r, $2 \leqslant r \leqslant m$, connect $E^2_{k+1,0}$ to zero and no non-zero differential ∂^r, $2 \leqslant r \leqslant m$, finishes up at $E^2_{k-m,m}$. Thus on the E^∞ diagram the isogonal $p+q = k$ contains just $\ker \partial^{m+1}$ at $(k, 0)$ and $\operatorname{coker} \partial^{m+1}$ at $(k-m, m)$.

Similar diagrams are, of course, of service in contrahomology computations; we attach such a diagram to the proof of 10.6.24.

We now give a connected sequence of applications which show how spectral theory can be applied to the computation of the homotopy groups of spheres; in this we have the limited aim of whetting the reader's appetite and of preparing the ground for more systematic study. We make use in the next example of the fact not proved in this book that any two Eilenberg–MacLane spaces $K(\pi, n)$ for the same π and n have isomorphic singular homology groups.

10.6.23 *Example.* The contrahomology ring of $K(J, 2)$ is a polynomial ring generated by an element of $H^2(K(J, 2))$.

At the end of 7.4 we introduced $K(\pi, n)$ as a space with null homotopy groups except in dimension n and with nth homotopy group π. We observed there that S^1 was an example of a $K(J, 1)$. It follows from II. 2.7 that $\Omega K(\pi, n)$ is a $K(\pi, n-1)$, so that $\Omega K(J, 2)$ has the homotopy and homology groups of S^1. We apply II. 3.4, with $K(J, 2)$ for base and consider the contractible fibre space with base $K(J, 2)$ and fibre having the homology groups of S^1. To this situation the Gysin contrahomology sequence 10.6.11*c* applies.

The homomorphism ϕ of the Gysin sequence is defined by means of the differential δ_{m+1} (compare the definition of $\theta^\cdot$ in 10.6.18*c*). If we choose the sign‡ of ϕ in such a way that

$$\delta_{m+1}(\xi \otimes \sigma) = (-1)^{pm} \phi\xi \otimes \iota,$$

where $\xi \in H^p(B)$, σ is the chosen generator of $H^m(F)$ and ι is the chosen generator of $H^0(F)$, then an argument similar to but simpler than that establishing 10.6.19 shows that

$$\phi(\xi\eta) = (\phi\xi)\,\eta = (-1)^{(m+1)}\,\xi(\phi\eta).$$

In our particular case, $m = 1$ and X is contractible so that

$$\phi : H^{k-1}(B) \cong H^{k+1}(B).$$

‡ The homomorphism ϕ depends on the choice of isomorphisms $H^{k+1}(B) \cong E_2^{k+1,0}$, $H^{k-m}(B) \cong E_2^{k-m,m}$; in particular, the replacement of one of these isomorphisms by its negative changes the sign of ϕ.

Since $H^0(B) = J$ and (using the Hurewicz isomorphism theorem) $H^1(B) = 0$, it follows that

$$H^k(K(J,2)) = J \quad \text{if } k \text{ is even,}$$
$$= 0 \quad \text{if } k \text{ is odd.}$$

Let 1 be the unit element of $R^*(K(J,2))$; thus $\phi(1)$ generates H^2 and we write $\phi(1) = \xi$. Now $\phi(\xi^k) = \phi(1\xi^k) = (\phi 1)\,\xi^k = \xi^{k+1}$. It follows by induction on k that ξ^k generates H^{2k} and $R^*(K(J,2))$ is the polynomial ring generated by ξ. ▮

We have already remarked in 7.4 that an example of a $K(J,2)$ is provided by $C^{(\infty)}$, infinite dimensional complex projective space. To establish this we generalize the Hopf map (see II. 3) to a fibre map $S^{2n+1} \to C^{(n)}$, where $C^{(n)}$ is complex projective space of n complex dimensions. The fibres are great circles and we may apply II. 1.35 and II. 3.8 to prove that $\pi_2(C^{(n)}) = J$ and that $\pi_r(C^{(n)}) = 0$ if $r = 1$, or $2 < r < 2n+1$. Using an obvious injection of $C^{(n)}$ into $C^{(n+1)}$, passing to the limit and defining a suitable topology for $\bigcup_n C^{(n)}$ we obtain $C^{(\infty)}$; since we can prove that the injection of $C^{(n)}$ in $C^{(n+1)}$ induces homotopy isomorphisms in dimensions $< 2n$, it follows that $\pi_2(C^{(\infty)}) = J$ and that $\pi_r(C^{(\infty)}) = 0$ for $r \neq 2$.

10.6.24 *Example.* $\pi_4(S^3) = J_2$.

It was briefly mentioned in chapter 6 that we may generalize the technique in which the fundamental group is killed by passing to the universal covering space; namely, if B is $(n-1)$-connected there is a space X and fibre map $p : X \to B$ such that $p_* : \pi_r(X) \cong \pi_r(B)$ for $r > n$ and such that $\pi_r(X) = 0$, $r \leqslant n$. By II. 3.8 the fibre must then be a $K(\pi_n(B), n-1)$.

We consider the case in which $B = S^3$; the fibre is then a $K(J,2)$, whose contrahomology ring has been computed in 10.6.23. The Wang sequence 10.6.17*c*, together with the results of 10.6.19, can now be applied to compute $H^*(X)$. If ξ is the generator of $R^*(K(J,2))$ of 10.6.23, $\theta^{\cdot}(\xi^k) = k\xi^{k-1}$; hence $\theta^{\cdot} : H^{2k}(F) \to H^{2k-2}(F)$ is monomorphic with cokernel J_k. Moreover $H^{2k-1}(F) = H^{2k-1}(K(J,2)) = 0$. From the exactness of the Wang sequence

$$\ldots \leftarrow H^{2k+1}(F) \leftarrow H^{2k+1}(X) \leftarrow H^{2k-2}(F) \overset{\dot\theta}{\leftarrow} H^{2k}(F) \leftarrow H^{2k}(X) \leftarrow H^{2k-3}(F) \leftarrow \ldots,$$

we see that $H^{2k+1}(X) = J_k$ and $H^{2k}(X) = 0$ for $k > 0$.

The appropriate E_2 diagram is

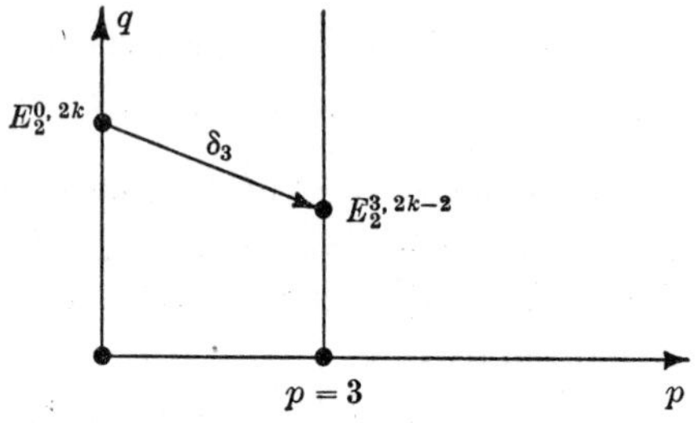

The groups concerned are all J and δ_3 is multiplication by k. Thus the E_∞ diagram is

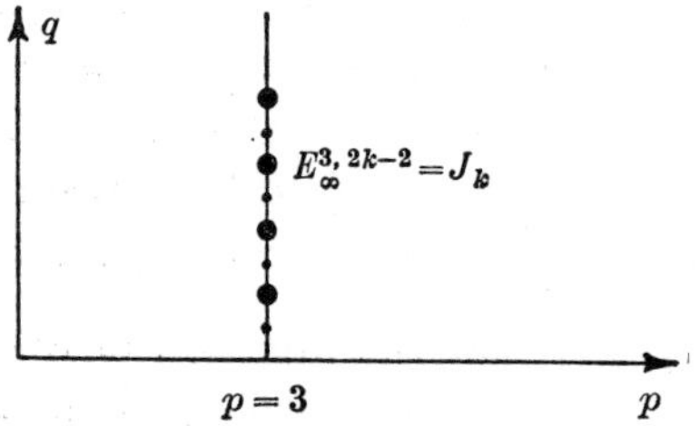

When the actual values of groups are to be inserted into diagrams it may be found more convenient to replace the 'points' (p, q) by boxes. Thus the E_∞ diagram may be represented as

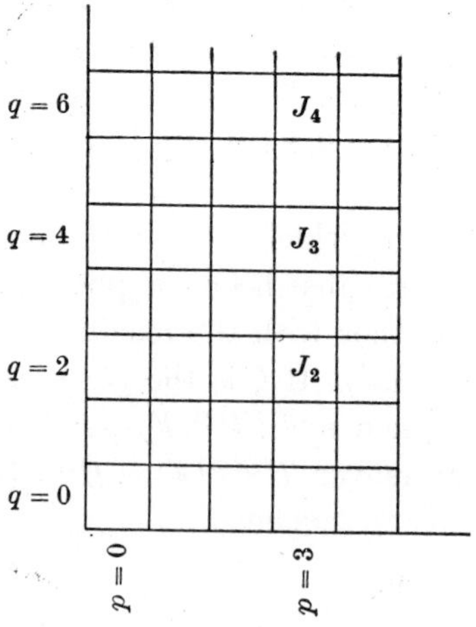

Returning to homology, these results imply that

10.6.25 $H_{2k}(X) = J_k$ and $H_{2k-1}(X) = 0$ for $k > 0$.

In particular $H_4(X) = J_2$. But, since $\pi_r(X) = 0$ for $r \leqslant 3$, by the Hurewicz isomorphism theorem $\pi_4(X) = H_4(X) = J_2$. Now

$$p_* : \pi_4(X) \cong \pi_4(B) = \pi_4(S^3),$$

hence $\pi_4(S^3) = J_2$. ∎

The next example makes use of the relative Hurewicz isomorphism theorem (8.8.8). We recall from 8.5 the suspension functor S which associates with a space X its suspension SX and with a map $f : X \to Y$ its suspension $Sf : SX \to SY$. Clearly $S(S^n) = S^{n+1}$ and it is not difficult to see that $f \to Sf$ induces a homomorphism

$$\sigma : \pi_n(S^m) \to \pi_{n+1}(S^{m+1}).$$

10.6.26 Theorem. *The Freudenthal suspension homomorphism*

$$\sigma : \pi_n(S^m) \to \pi_{n+1}(S^{m+1})$$

is isomorphic if $n < 2m-1$ and epimorphic if $n = 2m-1$.

We first describe a correspondence $\alpha \to \bar{\alpha}$ realizing the isomorphism $\kappa : \pi_{n+1}(S^{m+1}) \cong \pi_n(\Omega_1 S^{m+1})$ of II. 2.7. The closed track

$$j = j_{(1)} : S^n \times I \to S^{n+1}$$

of S^n, x_* in S^{n+1} on x_* given in II. 1.4 may be combined with any $f : S^{n+1}, x_* \to S^{m+1}, x_*$ to give

$$jf : S^n \times I \to S^{m+1},$$

and hence in the notation of II. 2

$$\overline{jf} : S^n \to (S^{m+1})^I.$$

For each $x \in S^n$, $x\overline{jf}$ is a closed loop (of length 1) based on x_*, so that the image of $\overline{jf}$ lies in $\Omega_1(S^{m+1})$, the space of loops of length 1. So with each $f : S^{n+1} \to S^{m+1}$ we associate in this way $\overline{jf} : S^n \to \Omega_1(S^{m+1})$ and so with each element $\alpha \in \pi_{n+1}(S^{m+1})$ an element $\bar{\alpha} \in \pi_n(\Omega_1 S^{m+1})$.

We may embed S^m in $\Omega_1 S^{m+1}$ by passing from $j : S^m \times I \to S^{m+1}$ to $\bar{j} : S^m \to \Omega_1(S^{m+1})$ and $\bar{j}$ induces $\bar{j}_* : \pi_n(S^m) \to \pi_n(\Omega_1 S^{m+1})$ such that $\bar{j}_* = \sigma\kappa$. Since certainly σ is an isomorphism on $\pi_m(S^m)$, so is $\bar{j}_*$. From the relative homotopy sequence of the pair $(\Omega_1 S^{m+1}, S^m)$ we infer that $\pi_m(\Omega_1 S^{m+1}, S^m) = 0$ and also, trivially, that

$$\pi_n(\Omega_1 S^{m+1}, S^m) = 0 \quad \text{for} \quad n < m.$$

The relative Hurewicz isomorphism theorem now implies that

$$H_n(\Omega_1 S^{m+1}, S^m) = 0 \quad \text{for} \quad n \leqslant m.$$

Now it was shown in 10.6.20 that $H_n(\Omega_1 S^{m+1}) = 0$ for $m < n < 2m$ and that $H_{2m}(\Omega_1 S^{m+1}) = J$; we can therefore deduce from the homology sequence of the pair $(\Omega_1 S^{m+1}, S^m)$ that $H_n(\Omega_1 S^{m+1}, S^m) = 0$ for $n < 2m$, and that $H_{2m}(\Omega_1 S^{m+1}, S^m) \cong H_{2m}(\Omega_1 S^{m+1}) = J$. A second application of the relative Hurewicz isomorphism theorem shows that $\pi_n(\Omega_1 S^{m+1}, S^m) = 0$ for $n < 2m$ and that $\pi_{2m}(\Omega_1 S^{m+1}, S^m) = 0$. The homotopy sequence for the pair now shows that $\bar{\jmath}_*$ is isomorphic on $\pi_n(S^m)$ for $n < 2m-1$ and epimorphic (with kernel a quotient of J) for $n = 2m-1$. Since $\bar{\jmath}_* = \sigma\kappa$ the conclusion follows. ∎

10.6.27 Corollary. *If* $m \geqslant 3$, $\pi_{m+1}(S^m) = J_2$.

For $\pi_4(S^3) = J_2$ and $4 < (2 \times 3) - 1$, so that $\pi_4(S^3) \cong \pi_5(S^4)$. In general if $m \geqslant 3$, $\pi_{m+1}(S^m) \cong \pi_{m+2}(S^{m+1})$. ∎

We make a final observation. Serre has shown that the Hurewicz isomorphism theorem may be applied simply to the p-primary components (see 5.1) of the groups concerned. Thus if the homology groups of the 1-connected space X are finite and have zero p-primary components in positive dimension less than n, then the homotopy groups are finite and have zero p-primary components in the same dimension, and the p-primary components of $H_n(X)$ and $\pi_n(X)$ are isomorphic. Thus, reverting to 10.6.25, we can conclude that the p-primary component of $\pi_n(S^3)$ is zero for $n < 2p$ and that that of $\pi_{2p}(S^3)$ is J_p. A crude consequence of this is that $\pi_n(S^3)$ is non-null for infinitely many n.

10.7 Homology and contrahomology of modules and groups

The notion of abstract homology and contrahomology groups plays a prominent role in various branches of mathematics. The original idea was suggested by Hurewicz's theorem that if $\pi_r(X) = 0$, $1 < r < n$, then $\pi_1(X)$ determines $H_r(X)$, $1 \leqslant r < n$, and the quotient group of $H_n(X)$ by the subgroup of spherical homology classes; the first result was Hopf's formulation of the case $n = 2$. Hopf then gave a general description of the homology groups of an abstract group π and proved that they are isomorphic with the homology groups of a space $X(\pi)$ with fundamental group π whose universal covering space is acyclic in all dimensions. The study was continued by Eckmann, Freudenthal, Eilenberg and MacLane, who defined contrahomology groups of an abstract group and then related them to the contrahomology groups of $X(\pi)$. (Of course, if X is a space such that $\pi_1(X) = \pi$ and $\pi_r(X) = 0$,

$1 < r < n$, then the homology groups of X in dimensions $< n$ are those of $X(\pi)$.) Eilenberg and MacLane made a penetrating investigation of both the topological significance and the algebraic nature of the contrahomology groups of groups and generalized the idea further in their study of the Eilenberg–MacLane complexes. The development we give here owes much to Eilenberg's lectures to the Paris seminar on algebraic topology in 1950–1.

Although we are mainly interested in the homology groups of groups, it seems natural to handle them as special cases of homology groups of modules. Our first objective in this section is to define the homology groups of *left* π-modules.

10.7.1 Definition. If R is a ring (not in general commutative) with unity element 1, a *left R-module A* is an abelian group together with a product $r.a$ defined for all $r \in R$, $a \in A$, with values in A, subject to

$$\text{(i)}\quad (r+r').a = r.a + r'.a;$$

$$\text{(ii)}\quad r.(a+a') = r.a + r.a';$$

$$\text{(iii)}\quad (rr').a = r.(r'.a);$$

$$\text{(iv)}\quad 1.a = a,$$

where $r, r' \in R$ and $a, a' \in A$.

When we need to refer to the abelian group underlying A stripped of its structure of operations from R, we may write it as $\breve{A}$. A *right R-module* is similarly defined, with products $a.r$ such that

$$a.(rr') = (a.r).r'.$$

Plainly a vector space over a field F is an F-module (left and right) under the usual product between a scalar and a vector, and the additive group of a ring R can be given the structure of an R-module (left or right), using the product in R.

We shall only consider the case in which the ring R is the *group ring* $J(\pi)$ of a group π. This is the *ring* whose elements are formal finite sums $\sum_i n_i g_i$, $n_i \in J$, $g_i \in \pi$, with the obvious addition and multiplication. We shall refer to a $J(\pi)$-module more briefly as a π-module. Clearly the module structure of A exhibits π as a group of left operators on $\breve{A}$ in that, for each $\xi \in \pi$, $a \in \breve{A}$, $\xi.a$ is defined as an element of $\breve{A}$ and

$$\text{(i)}\quad \xi.(a+a') = \xi.a + \xi.a';$$

$$\text{(ii)}\quad (\xi\xi')a = \xi.(\xi'.a);$$

$$\text{(iii)}\quad 1.a = a.$$

Conversely, if π acts as a group of left operators on an abelian group $\breve{A}$, we may define $r.a$ for any $r \in J(\pi)$, $a \in \breve{A}$, by the rule that

$$(\Sigma n_i \xi_i).a = \Sigma n_i(\xi_i.a).$$

With this definition $\breve{A}$ is turned into a π-module A. In case $\pi = 1$ it becomes a little delicate to distinguish between a π-module and a mere abelian group; we may therefore regard statements about π-modules as generalizations of statements about abelian groups.

If G is an abelian group, left operations from π can in general be defined in many different ways; in fact, any antihomomorphism of π into $Aut(G)$, the group of automorphisms of G, determines such an operation, an antihomomorphism being a transformation preserving the product with reversal of order. There is, in particular, always one such operation (corresponding to the antihomomorphism transforming π to the identity element), namely, that under which, for all $\xi \in \pi$ and $g \in G$, $\xi.g = g$. We may use the same symbol G to describe both the abelian group and this left π-module, which may be called the *trivial left π-module* on G, the operation being itself described as the *trivial operation*.

If A, B are left π-modules, a π-homomorphism $\phi: A \to B$ is a homomorphism $\breve{\phi}: \breve{A} \to \breve{B}$ that respects the operators, in that, for all $r \in J(\pi)$, $a \in \breve{A}$, $(r.a)\breve{\phi} = r.(a\breve{\phi})$. A left π-module P is said to be *π-projective* (compare 5.3) if, whenever a diagram of π-homomorphisms

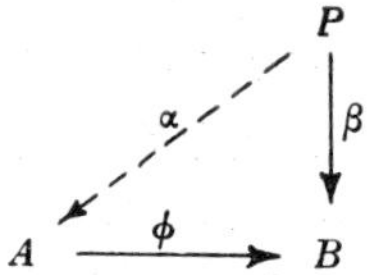

is given in which ϕ is epimorphic, there is a π-homomorphism $\alpha: P \to A$ such that $\alpha\phi = \beta$. A left π-module F is said to be *π-free* if it contains a set of elements $\{f_i\}$, called a π-basis for F, such that each element of F is uniquely expressible as a finite sum $\Sigma r_i f_i$, $r_i \in J(\pi)$, $f_i \in \{f_i\}$; this is the same as to say that the set of elements $\{\xi.f_i\}$, where ξ runs through the elements of π, constitutes a free basis for $\breve{F}$. As in 5.3 we may prove that any π-free module is π-projective.

10.7.2 Proposition. *A left π-module is π-projective if and only if it can be injected as a direct factor in a π-free module.*

First let F be π-free and let $F = F_1 \oplus F_2$; we must prove that F_1 is projective. Any π-homomorphism $\beta_1 : F_1 \to B$ can be extended to $\beta : F \to B$ by defining $(F_2)\beta$ to be 0. Now given

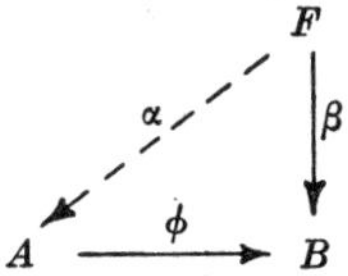

with ϕ epimorphic, since F is π-free and hence π-projective, it follows that there is a π-homomorphism $\alpha : F \to A$ such that $\alpha\phi = \beta$. Then if $\alpha_1 = \alpha \mid F_1$, $\alpha_1\phi = \beta_1$; this is enough to prove that F_1 is π-projective.

Next let P be π-projective and let F be the free π-module on generators $\{f_p\}_{p \in P}$, indexed by the elements of P. Then $f_p \to p$ determines a π-epimorphism $\phi : F \to P$. Consider the diagram

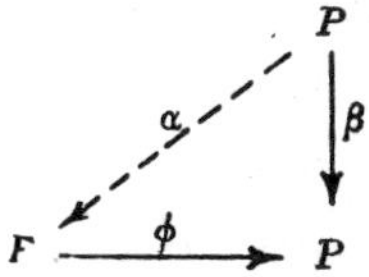

Since ϕ is epimorphic and P is π-projective, there is a π-homomorphism $\alpha : P \to F$ such that $\alpha\phi = 1 : P \to P$. As in 5.1.9, α injects P as a direct factor in F. ∎

Although at present the objects of interest are left π-modules, we shall make use also of right π-modules; we have spared the reader the corresponding statements of definitions and results.

Before we can define the homology group of a left π-module, we shall need some auxiliary notions. A *right π-complex* $(C; \partial, \epsilon)$ is a chain complex (C, ∂) in which $C_q = 0$ for $q < 0$, each C_q is a right π-module and ∂ is a π-endomorphism, together with an augmentation ϵ, which is a π-homomorphism $\epsilon : C_0 \to J$, such that $\partial_1\epsilon = 0 : C_1 \to J$; here, according to our convention, J stands for the trivial right π-module on J. This augmentation is a generalization of the augmentation ϵ introduced just before 3.4.14.

A π-chain map $\phi : (C; \partial, \epsilon) \to (C'; \partial', \epsilon')$ between two π-complexes is a chain map $\phi : (C; \partial) \to (C'; \partial')$ that is a π-homomorphism and has the additional property that $\epsilon = \phi_0\epsilon' : C_0 \to J$.

A π-complex C is said to be π-projective (π-free) if each C_q is π-projective (π-free).

We may define the homology group $H_*(C)$ for a π-complex C, as for ordinary chain complexes; these have the structure of π-modules, since ∂ is a π-endomorphism. A π-complex C is said to be *acyclic* if $H_q(C) = 0$ for $q \neq 0$ and ϵ induces a π-isomorphism of $H_0(C)$ onto J. An equivalent and more useful criterion for acyclicity is that the sequence

$$\to C_q \xrightarrow{\partial} C_{q-1} \xrightarrow{\partial} \dots \xrightarrow{\partial} C_1 \xrightarrow{\partial} C_0 \xrightarrow{\epsilon} J \to 0$$

should be exact.

We now pick out for a given group π a privileged right π-complex $C(\pi)$ which is π-free and acyclic. It may seem surprising that a π-complex with such innocuous properties should be singled out for privileged treatment; we use it however as an instrument for associating with an arbitrary left π-module A a chain complex (not a π-complex) whose homology groups are to be called the homology groups of A. Students of homological algebra may recognize $C(\pi)$ as a $J(\pi)$-projective resolution of J.

We start by defining $C_q(\pi)$ to be the free abelian group generated by the set of all ordered $(q+1)$-ples $(\xi_0, \xi_1, \dots, \xi_q)$, $\xi_i \in \pi$; the operation of π on $C_q(\pi)$ is determined by

$$(\xi_0, \dots, \xi_q) . \xi = (\xi_0 \xi, \dots, \xi_q \xi),$$

together with the linearity conditions. Thus $C_q(\pi)$ is a right π-module, with π-basis provided by the set of elements $(1, \xi_1, \dots, \xi_q)$, where 1 is the neutral element of π; $C_q(\pi)$ is therefore π-free.

The boundary operator ∂ is defined in the usual way by linear extension from the relations

$$(\xi_0, \dots, \xi_q) \partial = \sum_{i=0}^{q} (-1)^i (\xi_0, \dots, \widehat{\xi_i}, \dots, \xi_q).$$

By the usual argument $\partial\partial = 0$ and ∂ is plainly a π-homomorphism. The augmentation ϵ is given by

$$(\xi)\epsilon = 1$$

and extended by linearity. This is a π-homomorphism since π operates trivially on J, and

$$(\xi_0, \xi_1)\partial\epsilon = (\xi_1)\epsilon - (\xi_0)\epsilon = 0,$$

so that $\partial_1 \epsilon = 0$. This completes the verification that $(C(\pi); \partial, \epsilon)$ as defined is a π-free right π-complex.

Notice that $C_0 = J(\pi)$, where $J(\pi)$ is given the structure of a right π-module by the product in $J(\pi)$.

10.7.3 Theorem. *$C(\pi)$ is acyclic.*

The proof follows closely the proof that $K^{\Omega}(\bar{s})$ is acyclic (see the proof of 3.5.1). The π-structure is irrelevant and we shall regard $C(\pi)$

as a chain complex for this argument. We augment $C(\pi)$ to $\tilde{C}(\pi)$ by defining $\tilde{C}_{-1}(\pi)$ to be the cyclic group freely generated by ( ), the ordered set of no elements of π. We may identify $\tilde{C}_{-1}(\pi)$ with J by identifying ( ) with 1 and then $\tilde{\partial}_0$ is identified with ϵ. ∎

We next extend the notion of a tensor product of abelian groups and define $M \otimes_\pi N$ when M is a right π-module and N a left π-module, the group to be defined being an abelian group *without π-structure.* It is in fact the abelian group generated by pairs $m \otimes n$, $m \in M$, $n \in N$, subject to the relations

$$\text{(i)}\ (m+m')\otimes n = m\otimes n + m'\otimes n;$$

$$\text{(ii)}\ m\otimes(n+n') = m\otimes n + m\otimes n';$$

$$\text{(iii)}\ \text{for any } \xi \in \pi,\ (m.\xi)\otimes n = m\otimes(\xi.n).$$

Relations (i) and (ii) define $\breve{M} \otimes \breve{N}$, so that $M \otimes_\pi N$ is, in fact $\breve{M} \otimes \breve{N}$ modified by the additional relations (iii).

If C is a right π-complex and A a left π-module, then we define $C \otimes_\pi A$ to be a chain complex with an augmentation as follows:

$$(C\otimes_\pi A)_p = C_p \otimes_\pi A,$$

$$(c\otimes a)\bar{\partial} = (c\partial)\otimes a, \quad c \in C,\ a \in A,$$

$$(c\otimes a)\bar{\epsilon} = (c\epsilon)\otimes a \in J \otimes_\pi A.$$

It is easy to check that these definitions of $\bar{\partial}$, $\bar{\epsilon}$ are meaningful in that they respect the relations (i)–(iii) above. Hence the augmentation is a homomorphism $\bar{\epsilon} : (C\otimes_\pi A)_0 \to J \otimes_\pi A$. Now $J \otimes \breve{A}$ is canonically isomorphic with $\breve{A}$; $J \otimes_\pi A$ is therefore the group arising from $\breve{A}$ by imposing relation (iii). Since π operates trivially on J, these relations produce an abelian group, written as A_π, the quotient of $\breve{A}$ by the subgroup generated by the set of elements $a - (\xi.a)$, $\xi \in \pi$, $a \in A$. We shall not make any distinction between $J \otimes_\pi A$ and A_π.

For any left π-module A we now define $C(\pi; A)$ to be $C(\pi) \otimes_\pi A$ and

10.7.4 $$H_*(\pi; A) = H_*(C(\pi; A)).$$

Since $C_0(\pi) = J(\pi)$, $C_0(\pi; A) = J(\pi) \otimes_\pi A \cong \breve{A}$, the isomorphism being determined by the correspondence $\xi \otimes a \to \xi.a$; since $\bar{\partial}_0 = 0$, $Z_0(\pi; A) \cong \breve{A}$. Moreover, $B_0(\pi; A)$ is generated by elements

$$((\xi_0, \xi_1)\otimes a)\bar{\partial} = (\xi_1)\otimes a - (\xi_0)\otimes a$$

and is therefore canonically isomorphic to the kernel of the projection $\breve{A} \to A_\pi$; hence there is a canonical isomorphism $\tau_* : H_0(\pi; A) \to A_\pi$.

Any π-homomorphism $\phi : A \to B$ between left π-modules induces in an obvious way a chain map $\overline{\phi} : C(\pi; A) \to C(\pi; B)$ and hence a g.-homomorphism $\phi_* : H_*(\pi; A) \to H_*(\pi; B)$.

For any exact sequence $0 \to A' \overset{\lambda}{\to} A \overset{\mu}{\to} A'' \to 0$ of left π-modules and π-homomorphisms, since $C(\pi)$ is π-free, the sequence

$$0 \to C(\pi; A') \overset{\overline{\lambda}}{\to} C(\pi; A) \overset{\overline{\mu}}{\to} C(\pi; A'') \to 0$$

is also exact; compare 5.2.7 for the exactness at $C(\pi; A')$ and notice that the π-structure introduces no complications. Hence by 5.5.1, there is induced an exact sequence

$$\ldots \to H_q(\pi; A') \overset{\lambda_*}{\to} H_q(\pi; A) \overset{\mu_*}{\to} H_q(\pi; A'') \overset{\nu_*}{\to} H_{q-1}(\pi; A') \to \ldots.$$

We shall regard the groups $H_*(\pi; A)$ and the homomorphisms τ_*, ϕ_*, ν_* as a system, the homology system for left π-modules. We list properties of this system in

10.7.5 Theorem. *The homology system for left π-modules has the following properties:*

(i) *if $\phi : A \to A$ is the identity, ϕ_* is the identity;*

(ii) *if $\phi_1 : A_1 \to A_2$, $\phi_2 : A_2 \to A_3$ are π-homomorphisms,*

$$(\phi_1 \phi_2)_* = \phi_{1*} \phi_{2*};$$

(iii) *for any commutative diagram of left π-modules and π-homomorphisms,*

$$\begin{array}{ccccccccc} 0 & \longrightarrow & A' & \overset{\lambda}{\longrightarrow} & A & \overset{\mu}{\longrightarrow} & A'' & \longrightarrow & 0 \\ & & \downarrow{\scriptstyle \phi'} & & \downarrow{\scriptstyle \phi} & & \downarrow{\scriptstyle \phi''} & & \\ 0 & \longrightarrow & B' & \overset{\rho}{\longrightarrow} & B & \overset{\sigma}{\longrightarrow} & B'' & \longrightarrow & 0 \end{array}$$

in which each row is exact, the diagram

$$\begin{array}{ccc} H_q(\pi; A'') & \overset{\nu_*}{\longrightarrow} & H_{q-1}(\pi; A') \\ \downarrow{\scriptstyle \phi''_*} & & \downarrow{\scriptstyle \phi'_*} \\ H_q(\pi; B'') & \overset{\nu_*}{\longrightarrow} & H_{q-1}(\pi; B') \end{array}$$

is commutative;

(iv) *for any exact sequence $0 \to A' \overset{\lambda}{\to} A \overset{\mu}{\to} A'' \to 0$ of left π-modules and π-homomorphisms, the sequence*

$$\ldots \to H_q(\pi; A') \overset{\lambda_*}{\to} H_q(\pi; A) \overset{\mu_*}{\to} H_q(\pi; A'') \overset{\nu_*}{\to} H_{q-1}(\pi; A') \to \ldots$$

is exact;

(v) *for each left* π*-module* A, τ_* *is an isomorphism, and, if* $\phi_\pi : A_\pi \to B_\pi$ *is the homomorphism induced by the* π*-homomorphism* $\phi : A \to B$, *the diagram*

$$\begin{array}{ccc} H_0(\pi; A) & \xrightarrow{\tau_*} & A_\pi \\ \downarrow \phi_* & & \downarrow \phi_\pi \\ H_0(\pi; B) & \xrightarrow{\tau_*} & B_\pi \end{array}$$

is commutative;

(vi) *if* A *is* π*-free,* $H_q(\pi; A) = 0$ *for* $q \neq 0$.

The proof is much shorter than the statement. For (i) and (ii) are obvious; (iii) follows from 5.6.1; (iv) has already been stated to follow from modifications of 5.2.7 and 5.5.1; (v) has already been partly proved and the commutativity follows at once from the definitions. For (vi) we appeal again to a modification of 5.2.7; since

$$\ldots \to C_{q+1}(\pi) \xrightarrow{\partial} C_q(\pi) \xrightarrow{\partial} C_{q-1}(\pi) \to \ldots$$

is exact and A is π-free, the sequence

$$\ldots \to C_{q+1}(\pi; A) \xrightarrow{\bar\partial} C_q(\pi; A) \xrightarrow{\bar\partial} C_{q-1}(\pi; A) \to \ldots$$

is also exact and the result follows. ▮

It is profitable to notice an analogy between $H_*(\pi; A)$ and $H_*(X; G)$, and even more profitable if G is replaced by a local system of groups on X. The properties of 10.7.5 correspond to properties relating the various homology groups on a given space X with different local coefficient groups. It is natural then to consider whether a homomorphism $f : \pi \to \varpi$ between groups gives rise to homology homomorphisms; in the case of spaces and local systems of groups, a map $f : X \to Y$ induces homology homomorphisms only if the local system for X is that induced by f and the local system for Y. We may therefore expect to need the notion of a π-module associated with a homomorphism $f : \pi \to \varpi$ and a ϖ-module A.

Given a ϖ-module A and a homomorphism $f : \pi \to \varpi$ we may give $\breve{A}$ the structure of a π-module $f^{-1}A$ by defining

$$\xi . a = (\xi f) . a, \quad a \in \breve{A}, \ \xi \in \pi.$$

Given a ϖ-homomorphism $\phi : A \to B$ we may regard $\breve{\phi} : \breve{A} \to \breve{B}$ as a π-homomorphism $f^{-1}\phi : f^{-1}A \to f^{-1}B$. This notation, however, quickly palls and, whenever the context displays the group from which operations are drawn, we shall feel free to abbreviate $f^{-1}A$ to A and $f^{-1}\phi$

to ϕ; we may thus write $H_*(\pi; A)$ rather than $H_*(\pi; f^{-1}A)$. In case ϖ acts trivially this notation is of course the natural one as π also acts trivially; this case corresponds to the case of $H_*(X; G)$ where G is an honest coefficient group.

10.7.6 Theorem. *A homomorphism $f: \pi \to \varpi$ induces a set of homomorphisms $f_* = f_*^A : H_*(\pi; A) \to H_*(\varpi; A)$ for each left ϖ-module A, such that*

(vii) *the diagram*

$$\begin{array}{ccc} H_0(\pi; A) & \xrightarrow{\tau_*} & A_\pi \\ \downarrow{\scriptstyle f_*} & & \downarrow \\ H_0(\varpi; A) & \xrightarrow{\tau_*} & A_\varpi \end{array}$$

is commutative, where the unnamed homomorphism is that induced by the identity $\breve{A} \to \breve{A}$;

(viii) *for any ϖ-homomorphism $\phi: A \to B$, the diagram*

$$\begin{array}{ccc} H_*(\pi; A) & \xrightarrow{\phi_*} & H_*(\pi; B) \\ \downarrow{\scriptstyle f_*^A} & & \downarrow{\scriptstyle f_*^B} \\ H_*(\varpi; A) & \xrightarrow{\phi_*} & H_*(\varpi; B) \end{array}$$

is commutative, where the upper ϕ_* is induced by the π-homomorphism ϕ;

(ix) *for any exact sequence of ϖ-homomorphisms*

$$0 \to A' \to A \to A'' \to 0,$$

the diagram

$$\begin{array}{ccc} H_q(\pi; A'') & \xrightarrow{\nu_*} & H_{q-1}(\pi; A') \\ \downarrow{\scriptstyle f_*^{A''}} & & \downarrow{\scriptstyle f_*^{A'}} \\ H_q(\varpi; A'') & \xrightarrow{\nu_*} & H_{q-1}(\varpi; A') \end{array}$$

is commutative.

Before proving this theorem we observe that it asserts that f_* is a homomorphism of the system $(H(\pi; A); \tau_*, \phi_*, \nu_*)$ into the system $(H(\varpi; A); \tau_*, \phi_*, \nu_*)$.

The homomorphism $f: \pi \to \varpi$ induces, for a left ϖ-module A,

$$\bar{f} = \bar{f}^A : C(\pi; A) \to C(\varpi; A),$$

by the obvious rule that

$$((\xi_0, \ldots, \xi_p) \otimes_\pi a)\bar{f} = (\xi_0 f, \ldots, \xi_p f) \otimes_\varpi a.$$

We observe that

$$\begin{aligned}((\xi_0, \ldots, \xi_p)\,\xi \otimes_\pi a)\bar{f} &= ((\xi_0\xi)f, \ldots, (\xi_p\xi)f) \otimes_\varpi a \\ &= (\xi_0 f, \ldots, \xi_p f).\xi f \otimes_\varpi a \\ &= (\xi_0 f, \ldots, \xi_p f) \otimes_\varpi \xi f.a \\ &= ((\xi_0, \ldots, \xi_p) \otimes_\pi \xi.a)\bar{f};\end{aligned}$$

hence $\bar{f}$ respects the relations (iii) for $\otimes_\pi$. Then f^A_* is defined to be the homology homomorphism induced by $\bar{f}^A$.

Property (vii) follows from the commutativity of the diagram

$$\begin{array}{ccc} C_0(\pi) \otimes_\pi A & \xrightarrow{\bar{\epsilon}} & J \otimes_\pi A = A_\pi \\ \downarrow \bar{f} & & \downarrow \\ C_0(\varpi) \otimes_\varpi A & \xrightarrow{\bar{\epsilon}} & J \otimes_\varpi A = A_\varpi \end{array}$$

Property (viii) follows from the commutativity of

$$\begin{array}{ccc} C(\pi; A) & \xrightarrow{\bar{\phi}} & C(\pi; B) \\ \downarrow \bar{f}^A & & \downarrow \bar{f}^B \\ C(\varpi; A) & \xrightarrow{\bar{\phi}} & C(\varpi; B) \end{array}$$

Finally, to prove (ix) it is enough to observe that the given exact sequence gives rise to a commutative diagram

$$\begin{array}{ccccccccc} 0 & \longrightarrow & C(\pi; A') & \xrightarrow{\bar{\lambda}} & C(\pi; A) & \xrightarrow{\bar{\mu}} & C(\pi; A'') & \longrightarrow & 0 \\ & & \downarrow \bar{f} & & \downarrow \bar{f} & & \downarrow \bar{f} & & \\ 0 & \longrightarrow & C(\varpi; A') & \longrightarrow & C(\varpi; A) & \longrightarrow & C(\varpi; A'') & \longrightarrow & 0 \end{array}$$

in which the rows are exact because $C(\pi)$, $C(\varpi)$ are π-free, ϖ-free respectively. ∎

Although we have defined $H_*(\pi; A)$ by means of $C(\pi)$, in fact any other acyclic π-projective π-complex would have given rise to the same groups. This is the content of 10.7.9, which follows from

10.7.7 Theorem. *Let C be a π-projective right π-complex and D an acyclic right π-complex, then there is a π-map $\alpha : C \to D$ and any two such π-maps are chain-homotopic.*

We consider the diagram

$$\begin{array}{ccccccccc}
\cdots & \longrightarrow & C_{q+1} & \xrightarrow{\partial} & C_q & \xrightarrow{\partial} & C_{q-1} & \longrightarrow & \cdots \\
 & & \downarrow{\scriptstyle \alpha_{q+1}} & & \downarrow{\scriptstyle \alpha_q} & & \downarrow{\scriptstyle \alpha_{q-1}} & & \\
\cdots & \longrightarrow & D_{q+1} & \xrightarrow{\partial} & D_q & \xrightarrow{\partial} & D_{q-1} & \longrightarrow & \cdots
\end{array}$$

$$\begin{array}{ccccccccc}
\xrightarrow{\partial} & C_1 & \xrightarrow{\partial} & C_0 & \xrightarrow{\epsilon} & J & & \\
 & \downarrow{\scriptstyle \alpha_1} & & \downarrow{\scriptstyle \alpha_0} & & \downarrow{\scriptstyle 1} & & \\
\xrightarrow{\partial} & D_1 & \xrightarrow{\partial} & D_0 & \xrightarrow{\eta} & J & \longrightarrow & 0
\end{array}$$

in which the bottom row is exact and in which each C_q is π-projective; the π-homomorphisms α_q are to be constructed to give a commutative diagram.

Since η is epimorphic and C_0 is π-projective, there is a π-homomorphism α_0 such that $\alpha_0\eta = \epsilon$. Since $\partial\alpha_0\eta = \partial\epsilon = 0$,

$$C_1\partial\alpha_0 \subseteq \eta^{-1}(0) = D_1\partial$$

(exactness at D_0); as C_1 is π-projective, there is a π-homomorphism α_1 such that $\alpha_1\partial = \partial\alpha_0$. An inductive argument now readily establishes the existence of α_q for all $q \geqslant 0$ and so of $\alpha : C \to D$.

Suppose now that $\alpha, \beta : C \to D$ are two π-maps, we shall construct a chain homotopy $\Lambda : C \to D$ such that $\alpha - \beta = \partial\Lambda + \Lambda\partial$. First $(\alpha_0 - \beta_0)\eta = \epsilon - \epsilon = 0$; so $C_0(\alpha_0 - \beta_0) \subseteq D_1\partial$ and, as C_0 is π-projective, there is a π-homomorphism $\Lambda_0 : C_0 \to D_1$ such that $\Lambda_0\partial = \alpha_0 - \beta_0$. Suppose Λ_r defined for $r < q$; then $(\alpha_q - \beta_q - \partial\Lambda_{q-1})\partial = 0$ by the usual argument, so that $C_q(\alpha_q - \beta_q - \partial\Lambda_{q-1}) \subseteq D_{q+1}\partial$, from the exactness at D_q. Since C_q is π-projective there is a π-homomorphism $\Lambda_q : C_q \to D_{q+1}$ such that $\Lambda_q\partial = \alpha_q - \beta_q - \partial\Lambda_{q-1}$. Hence by induction $\alpha \simeq \beta$. ∎

The proof may be summarized by saying that the properties of C and D ensure that there is no obstruction to defining either α or Λ.

For any right π-complex C and left π-module A we define $H_*(C; A)$ to be $H_*(C \otimes_\pi A)$.

10.7.8 Corollary. *Under the conditions of the theorem there is a canonical homomorphism* $\alpha_*: H_*(C; A) \to H_*(D; A)$. ∎

Note that if

$$\alpha_* : H_*(C; A) \to H_*(D; A) \quad \text{and} \quad \beta_* : H_*(D; A) \to H_*(E; A)$$

are both canonical then $\alpha_*\beta_*$ is canonical and that $\alpha_* = 1$ if $C = D$.

We deduce

10.7.9 Corollary. *If C and D are both π-projective and both acyclic, there is a canonical isomorphism between $H_*(C; A)$ and $H_*(D; A)$.* ▌

The π-complex $C(\pi)$ is in fact simply one convenient π-projective acyclic π-complex. If D is any other such, we could define $H_*(D \otimes_\pi A)$ and homomorphisms analogous to τ_*, ϕ_*, ν_*, and the resulting system would be equivalent to the system we have defined.

We have now described the homology system of a *left* π-module. We may equally well define the homology system of a *right* π-module. For the chain complex $C(\pi)$ may clearly be given the structure of a left π-free acyclic complex by defining

10.7.10 $$\xi(\xi_0, \ldots, \xi_q) = (\xi\xi_0, \ldots, \xi\xi_q),$$

and we then define the homology groups of the right π-module A by way of the chain complex $A \otimes_\pi C(\pi)$; we write these groups as $H_*(A; \pi)$ to recall the order of the factors in the tensor product.

The distinction between left and right π-modules is, however, somewhat artificial‡. If A is a left π-module then, as mentioned briefly in 8.9, we may give $\check{A}$ the structure of a right π-module by defining

10.7.11 $$a\xi = \xi^{-1}a, \quad a \in \check{A},\ \xi \in \pi;$$

we call this module $\bar{A}$. Then $\bar{A}$ has the same homology groups as A; precisely,

10.7.12 Theorem. *There is a natural isomorphism Θ :*

$$\Theta : H_*(\bar{A}; \pi) \cong H_*(\pi; A).$$

We induce Θ by the map $\bar{A} \otimes_\pi C(\pi) \to C(\pi) \otimes_\pi A$ defined by linearity from the rule

$$a \otimes (\xi_0, \ldots, \xi_q) \to (\xi_0^{-1}, \ldots, \xi_q^{-1}) \otimes a.$$

It is plain that this defines, indeed, a chain isomorphism. ▌ Of course we may say more, namely, that the homology *systems* of A and $\bar{A}$ are naturally isomorphic. A consequence of 10.7.12 is that, from the point of view of homology, it is a matter of indifference whether we regard an abelian group as a trivial left π-module or a trivial right π-module.

If B is a right π-module we may define the left π-module $\bar{B}$ in the obvious way; we may extend the notation to π-complexes, noting that, if C is a projective acyclic π-complex, so is $\bar{C}$. If A is a left module

‡ In this respect they contrast with general R-modules, where the distinction is substantial.

and D a left complex we shall write $D\otimes_\pi A$ for $\bar{D}\otimes_\pi A$; we note that then

$$d\otimes a = \xi d\otimes \xi a, \quad d\in D,\ a\in A,\ \xi\in\pi.$$

If D is π-free and acyclic, then $H_*(D\otimes_\pi A) \cong H_*(\pi; A)$.

We now consider a geometrical situation that gives rise to such a left π-complex. Suppose that $X = X(\pi)$ is a p.c., l.p.c. and l.s.c. space and that $\pi_n(X) = 0$ if $n \neq 1$ and that $\pi_1(X) \cong \pi$. If $\tilde{X}$ is its universal covering space, $\pi_1(\tilde{X}) = 1$ and $\pi_n(\tilde{X}) \cong \pi_n(X) = 0$ for $n > 1$; hence by the Hurewicz isomorphism theorem $\tilde{H}_n(\tilde{X}) = 0$ for all n, so that the augmented singular complex $\tilde{\Delta}(\tilde{X})$ is acyclic.

Now $\Delta(\tilde{X})$ is a left π-complex. For by 6.7.4, π operates as a group of left operators on $\tilde{X}$; for any $f : t_n \to \tilde{X}$ and $\xi\in\pi$ we may therefore define $\xi.f : t_n \to \tilde{X}$ by the rule that, for all points $w\in t_n$

$$(w)(\xi.f) = \xi.(wf).$$

By defining $\xi.f_n$ to be $(\xi.f)_n$ and $f_0\epsilon$ to be 1 we give $\Delta(\tilde{X})$ the structure of a left π-complex; for $\Delta_n(\tilde{X})$ is now a π-module and ∂ is a π-homomorphism and $\partial_1\epsilon = 0$.

Moreover $\Delta_n(\tilde{X})$ is π-free. For each $x\in X$ we select some point $\tilde{x}\in p^{-1}(x)$, where $p : \tilde{X}\to X$ is the projection; a map $\tilde{f} : t_n \to \tilde{X}$ may be called a preferred map and $\tilde{f}_n$ a *preferred singular simplex* if $v^0\tilde{f}$ is the selected point for $v^0\tilde{f}p$. Any singular simplex of $\tilde{X}$ is uniquely expressible as $g.\tilde{f}_n$, where $g\in\pi$ and $\tilde{f}_n$ is preferred. The preferred singular n-simplexes of $\tilde{X}$ thus form a π-free basis for $\Delta_n(\tilde{X})$.

Suppose now that A is a left π-module, then

$$H_*(\pi; A) = H_*(\Delta(\tilde{X})\otimes_\pi A).$$

We can define an isomorphism (induced by $p : \tilde{X}\to X$)

$$\tilde{p}_n : \Delta_n(\tilde{X})\otimes_\pi A \to \Delta_n(X)\otimes \check{A}$$

by the rule that, if $\tilde{f}_n$ is preferred and $a\in A$

$$(\tilde{f}_n\otimes a)\,\tilde{p}_n = (\tilde{f}p)_n\otimes a.$$

This is well defined since it respects the relations in $\Delta_n(\tilde{X})\otimes_\pi A$. An inverse q_n can be defined by $(f_n\otimes a)\,q_n = \tilde{f}_n\otimes a$, where $\tilde{f}_n$ is the unique preferred singular n-simplex of $\tilde{X}$ such that $\tilde{f}p = f$. It is easy to verify that $\tilde{p}_n q_n$ and $q_n\tilde{p}_n$ are identity automorphisms and hence that $\tilde{p}_n$ is an isomorphism.

The isomorphisms $\tilde{p}_n$ do not in general combine to give a chain isomorphism between $\Delta(\tilde{X})\otimes_\pi A$ and $\Delta(X)\otimes\check{A}$; we use $\tilde{p}$ therefore to define a boundary operator from $\Delta_n(X)\otimes\breve{A}$ to $\Delta_{n-1}(X)\otimes\breve{A}$. Let $\tilde{f}_n$ be preferred and $a\in A$, then

$$(\tilde{f}_n\otimes a)\tilde{\partial} = \tilde{f}^0_{n-1}\otimes a+\sum_1^n(-1)^i(\tilde{f}^i_{n-1}\otimes a).$$

Now for $i\geqslant 1$, $\tilde{f}^i_{n-1}$ is preferred; in general, however, $\tilde{f}^0_{n-1}$ is not preferred. Let $\tilde{f}^0_{n-1}$ be $g(\tilde{f}).\tilde{f}_{n-1}$, where $\tilde{f}_{n-1}$ is preferred. Then if $\tilde{f}p = f$

10.7.13 $$(\tilde{f}_n\otimes a)\tilde{\partial}\tilde{p}_{n-1} = f^0_{n-1}\otimes(g(\tilde{f}))^{-1}.a+\sum_1^n(-1)^i f^i_{n-1}\otimes a.$$

The selection of a point $\tilde{x}\in p^{-1}(x)$ for each $x\in X$ is equivalent to a choice of a class of paths in X from x_* to x. As described in 8.9, this choice enables us to define, from the right module $\bar{A}$, a local system $\{\breve{A}_x\}$ of groups on X, and then 10.7.13 is just the boundary of $f_n\otimes a$ in $\Delta(X;\{\breve{A}_x\})$ (see 8.9.1).

We have now proved

10.7.14 Theorem. *If A is a left π-module and if $X = X(\pi)$ is a p.c., l.p.c., l.s.c. space such that $\pi_n(X) = 0$ for $n\neq 1$ and $\pi_1(X)\cong\pi$, then $H_*(\pi;A)\cong H_*(X;\{\breve{A}_x\})$ for some local system of coefficient groups $\{\breve{A}_x\}$.* ∎

If we replace A by the abelian group G we obtain

10.7.15 Corollary. *Under the same hypotheses on X,*

$$H_*(\pi;G)\cong H_*(X;G),$$

where G is any abelian group. ∎

We may generalize 10.7.14 and 10.7.15 in a way which makes them more useful.

10.7.16 Theorem. *If A is a left π-module and if X is a p.c., l.p.c. and l.s.c. space such that $\pi_n(X) = 0$ for $1<n<k$ and $\pi_1(X)\cong\pi$, then*

$$H_n(\pi;A)\cong H_n(X;\{\breve{A}\}),\quad n<k.$$

This may be deduced by topological arguments from 10.7.14. However, we prefer to derive it as an immediate consequence of an algebraic lemma, generalizing 10.7.9.

10.7.17 Lemma. *If C and D are both π-projective in dimensions $<k$ and both acyclic in dimensions $<k$, there is a canonical isomorphism between $H_n(C;A)$ and $H_n(D;A)$, $n<k$.*

In the light of 10.7.9, this will be proved when we have shown that we may construct a π-projective acyclic complex C' such that $C'_n = C_n$,

$n<k$, and $C'_k\partial' = C_k\partial$. Let us present $C_k\partial$ as the image of a π-free module C'_k and let us write $\partial'_k : C'_k \to C_{k-1}$ for the composition of the projection $C'_k \to C_k\partial$ with the injection $C_k\partial \subseteq C_{k-1}$. Let Z'_k be the kernel of ∂'_k; we present Z'_k as the image of a π-free module C'_{k+1} and write $\partial'_{k+1} : C'_{k+1} \to C'_k$ for the composite homomorphism. We proceed in this way to construct π-free modules C'_r and homomorphisms ∂'_r such that C' is a π-free acyclic complex with the required properties. This proves the lemma and hence 10.7.16. ▌▌

We remark that the technique of this proof is the technique whereby R-projective acyclic complexes can be constructed for a general ring R.

Although these theorems provide a geometrical setting for the homology groups of π-modules, we cannot make full use of them in developing the theory without proving that for each π such a space $X(\pi)$ exists. This is, indeed, the case (see J. H. C. Whitehead, *Ann. Math.* **50** (1949), 261). We do not, however, propose to reproduce a proof nor to use the theorem except in the case that π is a free group. If π is freely generated by the family $\{g_\lambda\}$ of elements, we form $\bigvee_\lambda S^1_\lambda$, a bunch of circles in (1, 1) correspondence with the generators. This is a polyhedron X and $\pi_1(X) \cong \pi$. Then $\tilde{X}$ is the polyhedron of a 1-connected 1-dimensional complex and hence by the Hurewicz isomorphism theorem $\pi_n(\tilde{X}) = 0$ for all n; this implies that $\pi_n(X) = 0$ for $n > 1$. Thus when π is free, $X(\pi)$ can be constructed as the polyhedron of a 1-dimensional complex.

10.7.18 Theorem. *If π is free and A is a left π-module, $H_n(\pi; A) = 0$ for $n > 1$.*

Since $X(\pi)$ is a polyhedron, its singular homology groups with any system of local coefficients are isomorphic with the corresponding groups of any underlying complex; such a complex is 1-dimensional and so has null homology groups in dimension > 1. ▌

In 10.7.9 we proved that $H_*(\pi; A)$ can be derived from any projective acyclic π-complex, but more than this is true. Suppose in fact that we have associated with each left π-module A a graded group $\bar{H}_*(\pi; A)$ and a homomorphism $\bar{\tau}_* : \bar{H}_0(\pi; A) \to A_\pi$; that with each π-homomorphism $\phi : A \to B$ we have associated a g.-homomorphism $\bar{\phi}_* : \bar{H}_*(\pi; A) \to \bar{H}_*(\pi; B)$ and with each exact π-sequence $0 \to A' \to A \to A'' \to 0$ a g.-homomorphism

$$\bar{\nu}_* : \bar{H}_*(\pi; A'') \to \bar{H}_*(\pi; A')$$

of index -1; suppose also that the system $\{\bar{H}_*; \bar{\tau}_*, \bar{\phi}_*, \bar{\nu}_*\}$ has the properties (i)–(vi) of 10.7.5. It can then be proved that this system is naturally equivalent to $\{H_*; \tau_*, \phi_*, \nu_*\}$ as already defined. One method of proof is to show that for any such systems $\bar{H}$ for π-modules and $\tilde{H}$ for ϖ-modules and for any homomorphism $f: \pi \to \varpi$, there is a unique set of homomorphisms

$$f_*(A) : \bar{H}_*(\pi; A) \to \tilde{H}_*(\varpi; A)$$

satisfying conditions similar to (vii), (viii) and (ix) of 10.7.6. From this, using the identity automorphism of π, the equivalence of any two systems for π-modules follows rather readily. In this proof it is natural to use the fact that any π-module can be presented as the epimorphic image of a π-free module.

We may carry out a programme dual to what we have done in this section to define the contrahomology groups of a left π-module A. We start from the *left* π-complex $C(\pi)$ (see 10.7.10). We then define the contrachain complex $C^{\cdot}(\pi; A) = C(\pi) \pitchfork_\pi A$. The nth component of this complex is the group $C^n(\pi; A) = C_n(\pi) \pitchfork_\pi A$, the symbol $\pitchfork_\pi$ meaning that we consider the abelian group of all π-homomorphisms; as in the case of $\otimes_\pi$ the resulting group has no π-structure; an element is determined by a function c of ordered $(n+1)$-ples from π with values in A, satisfying $(\xi(\xi_0, \ldots, \xi_n))c = \xi((\xi_0, \ldots, \xi_n)c)$. The contraboundary homomorphism $\delta: C_n(\pi) \pitchfork_\pi A \to C_{n+1}(\pi) \pitchfork_\pi A$ is adjoint to $\partial: C_{n+1}(\pi) \to C_n(\pi)$. Having thus defined $C(\pi) \pitchfork_\pi A$, we define the contrahomology group of the left π-module A to be

$$H^*(\pi; A) = H^*(C(\pi) \pitchfork_\pi A).$$

Since $C_1(\pi) \xrightarrow{\partial_1} C_0(\pi) \xrightarrow{\epsilon} J \to 0$ is exact so is

$$C_1(\pi) \pitchfork_\pi A \xleftarrow{\delta^0} C_0(\pi) \pitchfork_\pi A \xleftarrow{\epsilon^{\cdot}} J \pitchfork_\pi A \leftarrow 0.$$

Since there is evidently a canonical isomorphism $J \pitchfork_\pi A \cong A^\pi$, the subgroup of A consisting of all elements invariant under π, we infer that there is a canonical isomorphism $\tau_*: H^0(\pi; A) \cong A^\pi$.

Any π-homomorphism $\phi: A \to B$ between left π-modules induces in an obvious way a contrachain map $\bar{\phi}: C^{\cdot}(\pi; A) \to C^{\cdot}(\pi; B)$ and so a (right) homomorphism

$$\phi_*: H^*(\pi; A) \to H^*(\pi; B).$$

If $0 \to A' \xrightarrow{\lambda} A \xrightarrow{\mu} A'' \to 0$ is an exact sequence of left π-modules and π-homomorphisms the sequence

$$0 \to C^{\cdot}(\pi; A') \xrightarrow{\bar{\lambda}} C^{\cdot}(\pi; A) \xrightarrow{\bar{\mu}} C^{\cdot}(\pi; A'') \to 0$$

is also exact, since $C(\pi)$ is projective. Then by 5.5.2*c* we have an exact sequence of right homomorphisms

$$\ldots \to H^n(\pi; A') \xrightarrow{\lambda_*} H^n(\pi; A) \xrightarrow{\mu_*} H^n(\pi; A'') \xrightarrow{\nu_*} H^{n+1}(\pi; A') \to \ldots.$$

For a full description of the characteristic properties of the contrahomology system $\{H^*(\pi; A); \tau_*, \phi_*, \nu_*\}$ we need a further definition dual to that of a projective π-module. A left π-module I is said to be *π-injective* if, for each diagram of π-modules and π-homomorphisms

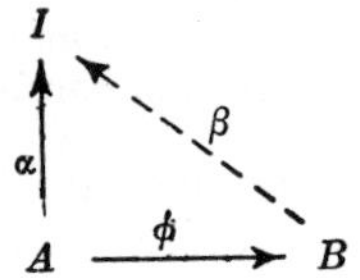

in which ϕ is monomorphic, there is a π-homomorphism $\beta : B \to I$ such that $\phi\beta = \alpha$. This is simply to say that a π-homomorphism of a submodule into I can always be extended to the whole module. The reader may compare this definition with that of an injective abelian group (5.3).

We can now state

10.7.5c Theorem. *The system $\{H^*(\pi; A); \tau_*, \phi_*, \nu_*\}$ has the following properties:*

(i) *if $\phi : A \to A$ is the identity, ϕ_* is the identity;*

(ii) *for $\phi_1 : A_1 \to A_2$, $\phi_2 : A_2 \to A_3$, $\phi_{1*}\phi_{2*} = (\phi_1\phi_2)_*$;*

(iii) *for any commutative diagram of left π-modules and π-homomorphisms*

$$\begin{array}{ccccccccc} 0 & \longrightarrow & A' & \xrightarrow{\lambda} & A & \xrightarrow{\mu} & A'' & \longrightarrow & 0 \\ & & \downarrow{\scriptstyle \phi'} & & \downarrow{\scriptstyle \phi} & & \downarrow{\scriptstyle \phi''} & & \\ 0 & \longrightarrow & B' & \xrightarrow{\rho} & B & \xrightarrow{\sigma} & B'' & \longrightarrow & 0 \end{array}$$

in which each row is exact, the diagram

$$\begin{array}{ccc} H^q(\pi; A'') & \xrightarrow{\nu_*} & H^{q+1}(\pi; A') \\ \downarrow{\scriptstyle \phi''_*} & & \downarrow{\scriptstyle \phi'_*} \\ H^q(\pi; B'') & \xrightarrow{\nu_*} & H^{q+1}(\pi; B') \end{array}$$

is commutative;

(iv) *for any exact sequence $0 \to A' \xrightarrow{\lambda} A \xrightarrow{\mu} A'' \to 0$ of left π-modules and π-homomorphisms, the sequence*

$$\ldots \to H^q(\pi; A') \xrightarrow{\lambda_*} H^q(\pi; A) \xrightarrow{\mu_*} H^q(\pi; A'') \xrightarrow{\nu_*} H^{q+1}(\pi; A') \to \ldots$$

is exact;

(v) *for each left π-module A, τ_* is an isomorphism and, if $\phi^\pi : A^\pi \to B^\pi$ is the homomorphism induced by the π-homomorphism $\phi : A \to B$, the diagram*

$$\begin{array}{ccc} H^0(\pi; A) & \xrightarrow{\tau_*} & A^\pi \\ \downarrow \phi_* & & \downarrow \phi^\pi \\ H^0(\pi; B) & \xrightarrow{\tau_*} & B^\pi \end{array}$$

is commutative;

(vi) *if A is π-injective, $H^q(\pi; A) = 0$ for all $q \neq 0$.* ❚

Contrahomology analogues of other results can be easily proved and, in particular,

10.7.14c Theorem. *If A is a left π-module and $X(\pi)$ is as in* 10.7.14, *then $H^*(\pi; A) \cong H^*(X(\pi); \{\breve{A}_x\})$, where $\{\breve{A}_x\}$ is the local system of groups of* 10.7.14. ❚

10.7.15c Corollary. *Under the same hypotheses on $X(\pi)$,*

$$H^*(\pi; G) \cong H^*(X(\pi); G),$$

where G is any abelian group. ❚

10.7.16c Theorem. *If A is a left π-module and if X is a p.c., l.p.c. and l.s.c. space such that $\pi_n(X) = 0$ for $1 < n < k$ and $\pi_1(X) \cong \pi$, then*

$$H^n(\pi; A) \cong H^n(X; \{\breve{A}_x\}), \quad n < k. \text{ ❚}$$

We can show how a homomorphism $f : \pi \to \varpi$ induces

$$f^* : H^*(\varpi; A) \to H^*(\pi; A)$$

and can prove that any contrahomology system for modules satisfying 10.7.5c is isomorphic to the system constructed; this can be proved from a modification of the homology argument. We establish the existence and uniqueness of

$$f^* : \bar{H}^*(\varpi; A) \to \bar{H}^*(\pi; A)$$

satisfying the analogues of (vii), (viii) and (ix) of 10.7.6. Here we use the fact that any left π-module can be embedded in a π-injective module; the proof of this fact is not entirely obvious‡. As in homology theory, the uniqueness of f^* establishes the uniqueness of the contrahomology of modules.

Contrahomology theory is more informative than homology theory when it is given a multiplicative structure. We can do this for the contrahomology theory of a left π-ring R, that is, a π-module in which the operations respect also the multiplication in R. For this case and for any $c \in C_p(\pi) \pitchfork_\pi R$, $d \in C_q(\pi) \pitchfork_\pi R$, we define

$$c \cup d \in C_{p+q}(\pi) \pitchfork_\pi R$$

‡ See B. Eckmann and A. Schopf, *Arch. Math.* **4** (1953), 75.

by $\quad ((\xi_0, \xi_1, \ldots, \xi_{p+q}), c \cup d) = ((\xi_0, \ldots, \xi_p), c) \times ((\xi_p, \ldots, \xi_{p+q}), d),$

the multiplication on the right-hand side being that of R.

We observe that

$$\begin{aligned}(\xi . (\xi_0, \ldots, \xi_{p+q}), c \cup d) &= ((\xi\xi_0, \ldots, \xi\xi_{p+q}), c \cup d) \\ &= ((\xi\xi_0, \ldots, \xi\xi_p), c) \times ((\xi\xi_p, \ldots, \xi\xi_{p+q}), d) \\ &= (\xi((\xi_0, \ldots, \xi_p), c)) \times (\xi((\xi_p, \ldots, \xi_{p+q}), d)) \\ &= \xi . ((\xi_0, \ldots, \xi_{p+q}), c \cup d),\end{aligned}$$

so that $c \cup d$ is a π-homomorphism. As in chapter 6 we verify the contraboundary formula

10.7.19 $$\delta(c \cup d) = (\delta c) \cup d + (-1)^p (c \cup \delta d).$$

Hence $C^{\cdot}(\pi; R)$ is a contrachain ring and the contrahomology group has a product structure in terms of which it forms a contrahomology ring $R^*(\pi; R)$. If R is a ring with trivial operations from π,

$$R^*(\pi; R) \cong R^*(X(\pi); R).$$

We finish this section with a number of results that are of value in studying the homology groups of modules, particularly of trivial modules.

10.7.20 Theorem. *If G is an abelian group (and so a trivial π-module)*

$$H_n(\pi; G) \cong (H_n(\pi) \otimes G) \oplus (H_{n-1}(\pi) * G),$$
$$H^n(\pi; G) \cong (H_n(\pi) \pitchfork G) \oplus (H_{n-1}(\pi) \dagger G),$$

where $H_n(\pi) = H_n(\pi; J)$.

If C is a π-complex we define C_π to be the chain complex for which $(C_\pi)_n = (C_n)_\pi$ with boundary induced from that in C. The proof of 10.7.20 is exactly as for ordinary chain complexes; for, since G and J have trivial operations, $C \otimes_\pi G \cong (C_\pi) \otimes G$ and $(C(\pi))_\pi$ is free. ∎

We remark that $H_0(\pi) = J$ and, moreover, $H_1(\pi) = \pi/[\pi, \pi]$, the abelian group associated with π. The latter follows from 8.8.3 if we know that

$$H_*(\pi) \cong H_*(X(\pi)),$$

since $$\pi_1(X(\pi)) \cong \pi.$$

To verify the result without appealing to $X(\pi)$ we observe that $(C_0)_\pi$ is freely generated by (1); that $(C_1)_\pi$ is freely generated (as an abelian group) by the elements $(1, \xi_1)$, each of which is a cycle of C_π; and that $(C_2)_\pi$ is freely generated by elements $(1, \xi_1, \xi_2)$ with boundaries

$$(\xi_1, \xi_2) - (1, \xi_2) + (1, \xi_1) = (1, \xi_2\xi_1^{-1}) - (1, \xi_2) + (1, \xi_1) \in (C_1)_\pi.$$

Hence $(H_1)_\pi$ is the abelian group generated by elements $(1, \xi)$ subject to the relations

$$(1, \xi_2\xi_1^{-1}) + (1, \xi_1) = (1, \xi_2)$$

or $$(1, \xi) + (1, \xi') = (1, \xi\xi').$$

It follows at once that

10.7.21 $$H_1(\pi) \cong \pi/[\pi, \pi].$$

10.7.22 Theorem. *If $\bar{\pi} = \pi \times \pi'$, the direct product of π and π',*

$$H_n(\bar{\pi}; J) \cong \sum_{p+q=n} H_p(\pi; J) \otimes H_q(\pi'; J) + \sum_{p+q=n-1} H_p(\pi; J) * H_q(\pi'; J).$$

If we had proved the existence of $X(\pi)$ for all π we could have proved this by using the homotopy result that $X(\pi \times \pi') = X(\pi) \times X(\pi')$ and appealing to the Künneth formula for the homology of a topological product. In the absence of such a proof of existence we proceed rather differently. We define $\bar{C}(\pi, \pi')$ to be $C(\pi) \otimes C(\pi')$; here the tensor product of right complexes is defined, as far as its chain complex structure is concerned, in the usual way for tensor products of chain complexes; the operation by $(\xi, \xi') \in \bar{\pi}$ is given by

$$(c \otimes c') . (\xi, \xi') = (c . \xi) \otimes (c' . \xi'), \quad c, c' \in C(\pi), C(\pi').$$

In $\bar{C}$ an augmentation $\bar{\epsilon}$ is given by

$$(c \otimes c') \bar{\epsilon} = (c\epsilon)(c'\epsilon') \in J,$$

where $c, c' \in C_0(\pi)$, $C_0(\pi')$ and ϵ, ϵ' are the augmentations in $C(\pi), C(\pi')$. It is easy to verify that $(\bar{C}; \bar{\partial}, \bar{\epsilon})$ is a right $\bar{\pi}$-complex. Moreover $(\bar{C}; \bar{\partial}, \bar{\epsilon})$ is $\bar{\pi}$-free; the tensor product of a π-free basis for $C(\pi)$ and a π'-free basis for $C(\pi')$ gives a $\bar{\pi}$-free basis for $\bar{C}$. Further, $\bar{C}$ is acyclic; for, $C(\pi)$ being an acyclic geometric chain complex, 5.7.8 shows that $H_q(\bar{C}) = 0$ for $q > 0$, and the argument of 5.7.10 readily establishes that $\bar{\epsilon}_* : H_0(\bar{C}) \cong J$.

We now apply 10.7.9 to deduce that, for any left $\bar{\pi}$-module A

$$H_*(\bar{\pi}; A) \cong H_*(\bar{C}; A).$$

Now $H_*(\bar{C}; J) \cong H_*(\bar{C}_{\bar{\pi}}) = H_*(C_\pi \otimes C'_{\pi'})$, where $C' = C(\pi')$. Since C_π and $C'_{\pi'}$ are free chain complexes we may apply the Künneth formula, which follows 5.7.17, to obtain the result. ∎

For the next theorem we use an acyclic π-complex different from $C(\pi)$ to compute the homology groups.

10.7.23 Theorem. *If G is an abelian group and π is cyclic of order m,*

$$H_{2n-1}(\pi; G) \cong G_m \quad \textit{and} \quad H_{2n}(\pi; G) \cong {}_mG \quad \textit{for} \quad n \geqslant 1.$$

Let ξ be a generator of π and let us define a right π-complex D in which, for all $k \geqslant 0$, D_k is $P_m = J(\pi)$; P_m is the additive group of integral polynomials in ξ of degree $< m$; its elements may be written in the form $\sum_{r=0}^{m-1} a_r \xi^r$, where $a_0, \ldots, a_{m-1}$ are integers. In P_m we pick out two elements, $\sigma = 1 + \xi + \ldots + \xi^{m-1}$ and $\tau = 1 - \xi$. We define $\partial_{2n} : D_{2n} \to D_{2n-1}$ (where $D_{2n} = D_{2n-1} = P_m$) to be given by multiplication by σ and $\partial_{2n-1} : D_{2n-1} \to D_{2n-2}$ by multiplication by τ, for all $n \geqslant 1$; we also define $\epsilon : D_0 \to J$ by the rule that it maps $\Sigma a_r \xi^r$ to Σa_r. Since each D_k is $J(\pi)$ each is a right π-module under the product in $J(\pi)$; since $\sigma\tau = \tau\sigma = 0, \partial\partial = 0$; further $\partial_1 \epsilon = 0$, so that we have defined a right π-complex D.

D is π-free since the element 1 constitutes a π-basis for each D_k. It is also acyclic; for an element becomes zero on multiplication by σ only if it is itself a multiple of τ and vice versa; and $\Sigma a_r = 0$ if and only if $\Sigma a_r \xi^r$ is a multiple of τ.

Consider now $D \otimes A$ for a left π-module A; since $D_k \cong J(\pi)$, $D_k \otimes_\pi A \cong \check{A}$ under the correspondence $\xi^r \otimes a \to \xi^r . a$. Then, identifying $D_k \otimes_\pi A$ with $\check{A}$,

$$a\partial_{2n} = \sigma . a, \quad a\partial_{2n-1} = \tau . a \quad \text{for} \quad n \geqslant 1.$$

Writing $\sigma.A$, $\tau.A$ for the set of all elements $\sigma.a$, $\tau.a$, $a \in A$, and ${}_\sigma A$, ${}_\tau A$ for the set of elements a such that $\sigma a = 0$, $\tau a = 0$, we deduce that

$$H_{2n-1}(\pi; A) \cong {}_\tau A/\sigma.A \quad \text{and} \quad H_{2n}(\pi; A) \cong {}_\sigma A/\tau.A \quad \text{for} \quad n \geqslant 1.$$

In particular, if we consider the case of an abelian group G with trivial operations

$$\sigma.G = mG, \quad \tau.G = 0; \qquad {}_\sigma G = {}_m G, \quad {}_\tau G = G.$$

Hence the theorem follows. ∎

10.7.24 Corollary. *If π is finitely generated and commutative, and if G is an abelian group, $H_*(\pi; G)$ can be calculated.*

By 5.1.1 π is the direct product of a finite set of subgroups each of which is J or J_m for some m. Then 10.7.18, 10.7.20, 10.7.22 and 10.7.23 enable $H_*(\pi; G)$ to be determined. ∎

The following theorem has important applications:

10.7.25 Theorem. *If π is of order m, then $mH_q(\pi; A) = 0$, $q \geqslant 1$.*

Let the elements of π be enumerated as ${}_1\xi, \ldots, {}_m\xi$ and let

$$\Lambda = \Lambda_q : C_q(\pi) \to C_{q+1}(\pi), \quad q \geqslant 0,$$

be given by

$$(\xi_0, \ldots, \xi_q)\,\Lambda = \sum_{r=1}^{m} ({}_r\xi, \xi_0, \ldots, \xi_q).$$

Then Λ is a π-homomorphism and a straightforward computation shows that

$$(\xi_0, \ldots, \xi_q)\,(\Lambda_q \partial_{q+1} + \partial_q \Lambda_{q-1}) = m(\xi_0, \ldots, \xi_q), \quad q \geqslant 1.$$

Thus, if Λ induces $\bar{\Lambda} : C_q(\pi; A) \to C_{q+1}(\pi; A)$, then

$$c(\bar{\Lambda}_q \bar{\partial}_{q+1} + \bar{\partial}_q \bar{\Lambda}_{q-1}) = mc, \quad c \in C_q(\pi; A),\ q \geqslant 1.$$

It follows that if z is a cycle of $C_q(\pi; A)$, $q \geqslant 1$, then $mz = z\bar{\Lambda}\bar{\partial}$, so that mz is a boundary. ∎

10.7.25c Theorem. *If π is of order m, then $mH^q(\pi; A) = 0$, $q \geqslant 1$.* ∎

10.8 The spectral sequence associated with a covering

In this section we consider a topological space X and a regular covering $\tilde{X}$; the quotient group of $\pi_1(X)$ by the projection of $\pi_1(\tilde{X})$ is here written as π. In‡ 6.7.4 it was shown how π may be regarded as a group of left operators on $\tilde{X}$. If $\tilde{\Delta}$ stands for the singular chain complex $\Delta(\tilde{X})$, then π acts as a group of left operators also on $\tilde{\Delta}$, which may thus be regarded as a π-free left π-complex, the augmentation being defined in the obvious way. The projection $p : \tilde{X} \to X$ induces $p : \tilde{\Delta} \to \Delta(X)$ and establishes a canonical isomorphism between $\tilde{\Delta}_\pi$ and $\Delta(X)$, since the kernel of p is generated by elements $\tilde{f} - \xi.\tilde{f}$,

‡ The group here written as π appears in chapter 6 as π/π_0.

$\tilde{f} \in \tilde{\Delta}$, $\xi \in \pi$. We use this algebraic apparatus to study homology relations between $\tilde{X}$ and X.

In 10.7 we made extensive use of the tensor product of a right π-complex and a left π-module; here we shall consider also the tensor product of two π-complexes. If C is a right π-complex and D is a left π-complex we define the chain complex $C \otimes_\pi D$ by defining $(C \otimes_\pi D)_n$ to be $\sum_{p+q=n} (C_p \otimes_\pi D_q)$ and by defining the boundary operator in the usual way (see 10.1.9); it is easy to verify that this operator respects the identifications in $C \otimes_\pi D$ and is therefore well-defined.

If ϵ is the augmentation in C, we extend it to a π-homomorphism of C into J by defining $C_n \epsilon$ to be zero for $n > 0$. With this understanding ϵ induces a chain map

$$\beta : C \otimes_\pi D \to J \otimes_\pi D.$$

Using the canonical isomorphism between $J \otimes_\pi D_n$ and $(D_n)_\pi$, we shall identify $J \otimes_\pi D$ with D_π.

10.8.1 Theorem. *If C is acyclic and D is π-free, β induces an isomorphism $\beta_* : H_*(C \otimes_\pi D) \cong H_*(D_\pi)$.*

Since β is a chain map it is a d.g. homomorphism. We now filter $C \otimes_\pi D$ by subgroups $A_p = \sum_{n \leqslant p} C \otimes_\pi D_n$ and we filter D_π by subgroups $A'_p = \sum_{n \leqslant p} (D_\pi)_n$. Each is now a d.f.g.-group and β is a d.f.g. homomorphism. If $(\mathscr{E}^r, \partial^r)$, $(\mathscr{E}'^r, \partial'^r)$ are the spectral sequences of these two d.f.g.-groups, then $\mathscr{E}^0_{p,q} = C_q \otimes_\pi D_p$ and $(c \otimes d)\,\partial^0 = c\partial \otimes d$. Since D is π-free, $H_{p,q}(\mathscr{E}^0) = H_q(C) \otimes_\pi D_p$; but C is acyclic so that $H_{p,q}(\mathscr{E}^0) = 0$ for $q > 0$. Also $\mathscr{E}'^0_p \cong (D_\pi)_p$ so that ${}_n\mathscr{E}'^0_p = 0$ unless $n = p$ and $H_{p,q}(\mathscr{E}'^0) = 0$ for $q > 0$.

We now examine the homomorphism $H_{p,0}(\mathscr{E}^0) \to H_{p,0}(\mathscr{E}'^0)$ induced by β. It is determined by

$$(c_0 \otimes d)\,\beta = (c_0 \epsilon) \otimes d,$$

and it is therefore the homomorphism $H_0(C) \otimes_\pi D_p \to J \otimes_\pi D_p$ induced by ϵ; since C is acyclic this is an isomorphism. These results prove that β induces an isomorphism $H_{p,q}(\mathscr{E}^0) \cong H_{p,q}(\mathscr{E}'^0)$ and hence a g.-isomorphism $\mathscr{E}^1 \cong \mathscr{E}'^1$. We now invoke 10.3.13 to deduce that β induces an isomorphism $\beta_* : H_*(C \otimes_\pi D) \cong H_*(D_\pi)$. ∎

If M is a chain complex we shall write $H_*(M; G)$ for $H_*(M \otimes G)$, when convenient.

10.8.2 Corollary. *If C is also π-free and G is any abelian group, β induces*

$$\beta_* : H_*(C \otimes_\pi D; G) \cong H_*(D_\pi; G).$$

For then $C \otimes_\pi D$ and D_π are both geometric‡ so that we may apply 3.10.1. ∎

10.8.3 Corollary. *If $\tilde{\Delta}$ is the singular π-complex of a regular covering $\tilde{X}$ of the space X with π as cover-transformation group and if C is a π-free acyclic π-complex, the augmentation in C induces an isomorphism*

$$\beta_* : H_*(C \otimes_\pi \tilde{\Delta}; G) \cong H_*(X; G).$$

For $\tilde{\Delta}$ is a π-free left π-complex and there is a canonical isomorphism between $\tilde{\Delta}_\pi$ and $\Delta(X)$. ∎

10.8.4 Proposition. *If C is a π-free acyclic right π-complex and D is a left π-complex, the filtration of* 10.1.9 *of $C \otimes_\pi D$ gives a spectral sequence (E^r, ∂^r) in which*

$$E^2_{p,q} = H_p(\pi; H_q(D)).$$

In this filtration (compare 10.3.12)

$$E^0_p = C_p \otimes_\pi D \quad \text{and} \quad (c_p \otimes d)\,\partial^0 = (-1)^p\, c_p \otimes d\partial.$$

So, since C_p is π-free,

$$E^1_p \cong C_p \otimes_\pi H_*(D) \quad \text{and} \quad (c \otimes \alpha)\,\partial^1 = c\partial \otimes \alpha,$$

for $c \in C_p$, $\alpha \in H_*(D)$. Hence

$$E^2_{p,q} = H_p(C \otimes_\pi H_q(D)) = H_p(\pi; H_q(D)),$$

since C is π-free and acyclic. ∎

10.8.5 Corollary. *If G is any abelian group, the corresponding filtration of $C \otimes_\pi D \otimes G$ gives a spectral sequence for which*

$$E^2_{p,q} = H_p(\pi; H_q(D; G)).$$

Here $E^0_p = (C_p \otimes_\pi D) \otimes G = C_p \otimes_\pi (D \otimes G)$, with the obvious operation of π on $D \otimes G$. Then $E^1_p = C_p \otimes_\pi H_*(D; G)$ and

$$E^2_{p,q} = H_p(C \otimes_\pi H_q(D; G)) = H_p(\pi; H_q(D; G)). \; ∎$$

10.8.6 Theorem. *If $\tilde{X}$, X are as before and G is any abelian group, there is a spectral sequence (E^r, ∂^r) in which*

$$E^2_{p,q} \cong H_p(\pi; H_q(\tilde{X}; G))$$

‡ We remind the reader that a geometric complex is one that is free and has a bottom dimension.

and E^∞ is isomorphic to the bigraded group associated with a suitable filtration of $H_*(X; G)$. Moreover, in $H_p(\pi; H_q(\tilde{X}; G))$ the operations of π on $H_q(\tilde{X}; G)$ are induced by the operations of π on $\Delta(\tilde{X}; G)$.

The first assertion follows from 10.8.5 with D replaced by $\tilde{\Delta} = \Delta(\tilde{X})$. The E^∞ term of the spectral sequence of 10.8.5 is the graded group associated with the filtration of $H_*(C \otimes_\pi D \otimes G)$; but by 10.8.3 this last group is isomorphic with $H_*(X; G)$. The final part of the statement of 10.8.6 follows from the fact that the operations on the homology group of a π-complex are induced by those on the π-complex itself. ▌

We remark that if X is a topological group then the operations of π on $H_*(\tilde{X}; G)$ are trivial. For then (see 6.9.8) $\tilde{X}$ is a topological group and π may be identified with $p^{-1}(e)$ which is a subgroup of $\tilde{X}$; moreover, the operation of π on $\tilde{X}$ is just the restriction to π of left multiplication by elements of $\tilde{X}$. Now let $\tilde{x}_0$ be a given point of $\tilde{X}$; since $\tilde{X}$ is path-connected, the map $\tilde{x} \to \tilde{x}_0\tilde{x}$, $\tilde{x} \in \tilde{X}$, is homotopic to the identity map. This implies that the operations of $\tilde{X}$ on $H_*(\tilde{X}; G)$ are trivial, and so therefore are the operations of π.

Theorem 10.8.6 is the main result of this section; since it establishes the fact that the E^2 term involves $H_*(\tilde{X})$ and that the E^∞ term involves $H_*(X)$, it provides a connexion between these two homology groups. The theorem is a generalization of 10.7.15. For, under the conditions of 10.7.15, $\tilde{X}$ is acyclic so that $E^2_{p,q}$ is null if $q > 0$. Now ∂^r raises complementary dimension by $r-1$, so that

$$E^2_{p,0} \cong E^3_{p,0} \cong E^4_{p,0} \cong \dots \cong E^\infty_{p,0} \cong H_p(\pi; G),$$

and $E^\infty_{p,q} = 0$ if $q > 0$. Thus the g.-group associated with $H_p(X; G)$ has just one term, whence

$$H_p(X; G) \cong H_p(\pi; G).$$

We now make some applications of 10.8.6.

10.8.7 Theorem. *If π is infinite cyclic and operates trivially on $H_*(\tilde{X}; G)$, then $H_n(X; G)$ is an extension of $H_{n-1}(\tilde{X}; G)$ by $H_n(\tilde{X}; G)$.*

By 10.7.18 in this case $H_p(\pi; H_q(\tilde{X}; G)) = 0$ if $p \geqslant 2$, whereas, since we may take $X(\pi)$ to be S^1, $H_p(\pi; H_q(\tilde{X}; G)) \cong H_q(\tilde{X}; G)$ for $p = 0, 1$; hence by 10.8.6, $E^2_{p,q} = 0$ if $p \geqslant 2$, $E^2_{p,q} \cong H_q(\tilde{X}; G)$ if $p = 0, 1$. This means that, for given total degree n, ${}_nE^2$ contains at most two non-zero terms $H_n(\tilde{X}; G)$ and $H_{n-1}(\tilde{X}; G)$. Since ∂^r lowers filtering degree by r, it follows that ${}_nE^2 \cong {}_nE^3 \cong \dots \cong {}_nE^\infty$. The graded group, therefore, associated with the filtration of $H_n(X; G)$ contains at most two non-zero terms, ${}_nE^\infty_0 \cong H_n(\tilde{X}; G)$ and ${}_nE^\infty_1 \cong H_{n-1}(\tilde{X}; G)$; hence $H_n(X; G)$ contains a subgroup isomorphic with $H_n(\tilde{X}; G)$, the quotient by which is isomorphic with $H_{n-1}(\tilde{X}; G)$. ▮ In fact the monomorphism

$$H_n(\tilde{X}; G) \to H_n(X; G)$$

whose cokernel is $H_{n-1}(\tilde{X}; G)$ is just p_*.

The case in which π is infinite cyclic and is not assumed to operate trivially on $H_*(\tilde{X}; G)$ can be analysed from a refinement of 10.7.18; in this case, as may readily be verified

$$H_0(\pi; A) \cong A_\pi \quad \text{and} \quad H_1(\pi; A) \cong A^\pi.$$

If this is applied to the situation of 10.8.7, we see that

$${}_nE_0^\infty \cong (H_n(\tilde{X}; G))_\pi \quad \text{and} \quad {}_nE_1^\infty \cong (H_{n-1}(\tilde{X}; G))^\pi,$$

so that $H_n(X; G)$ is an extension of $H_{n-1}(\tilde{X}; G)^\pi$ by $(H_n(\tilde{X}; G))_\pi$. This generalization of 10.8.7 has also been proved by Serre‡ by more direct methods.

10.8.8 Theorem. *If π is of order m and operates trivially on $H_*(\tilde{X}; G)$ and if each element of $H_*(\tilde{X}; G)$ is uniquely divisible by m, then*

$$H_*(\tilde{X}; G) \cong H_*(X; G).$$

Notice that the divisibility condition is satisfied, for instance, if G is a field whose characteristic does not divide m; hence if $G = J_p$, where p is a prime not dividing m or if G is the field of rationals, the conclusion holds provided that the condition of trivial operations is satisfied.

Since π operates trivially on $H_*(\tilde{X}; G)$, we have by 10.7.20

$$H_p(\pi; H_q(\tilde{X}; G)) \cong (H_p(\pi) \otimes H_q(\tilde{X}; G)) \oplus (H_{p-1}(\pi) * H_q(\tilde{X}; G)).$$

Suppose $p \geqslant 1$; then $mH_p(\pi) = 0$ by 10.7.25 whence, since every element of $H_q(\tilde{X}; G)$ is divisible by m, $H_p(\pi) \otimes H_q(\tilde{X}; G) = 0$. We next observe that, since every element of $H_q(\tilde{X}; G)$ is *uniquely* divisible by m, ${}_m(H_q(\tilde{X}; G)) = 0$. If $p = 1$, $H_{p-1}(\pi) = J$ and $J * H_q(\tilde{X}; G) = 0$; if $p > 1$, then, by 10.7.25, $mH_{p-1}(\pi) = 0$ and so, by 5.4.22

$$H_{p-1}(\pi) * H_q(\tilde{X}; G) = 0.$$

We have thus proved that $H_p(\pi; H_*(\tilde{X}; G)) = 0$ for $p \geqslant 1$. Theorem 10.8.6 in this case asserts that $E^2_{p,q} = 0$ if $p \geqslant 1$; the argument used in 10.8.7 (and elsewhere) shows that ${}_nE^2 \cong {}_nE^\infty$, so that ${}_nE^\infty$ contains at most one non-zero term $E^\infty_{0,n} \cong H_n(\tilde{X}; G)$. Hence the graded group of $H_n(X; G)$ with its filtration has but this single non-zero term, to which $H_n(X; G)$ must therefore be isomorphic. ▮ The isomorphism is just the homomorphism p_*.

‡ J.-P. Serre, loc. cit. p. 503.

10.9 Appendix: Application to simplex blocks

Let K be a simplicial complex; we shall call a sequence $K_0, K_1, \ldots, K_p, \ldots$ of subcomplexes of K a *block filtration* of K if

(i) it is an increasing filtration; namely, $K_0 \subseteq K_1 \subseteq \ldots \subseteq K_p \subseteq \ldots$ and $\bigcup_p K_p = K$;

(ii) $H_n(K_p, K_{p-1}) = 0$ if $n \neq p$, where K_{-1} is taken to be empty.

10.9.1 Proposition. *If* $\{e_p^i\}$ *is a block dissection of* K (*see* 3.8.1, 3.8.2), *the notion being extended to include the case of* K *infinite, then* $B^{(0)}, B^{(1)}, \ldots, B^{(p)}, \ldots$ *is a block filtration of* K.

Condition (i) for a block filtration is obviously satisfied; condition (ii) follows from 3.8.1 (ii) and the evident isomorphism

$$H_n(B^{(p)}, B^{(p-1)}) \cong \sum_i H_n(\bar{e}_p^i, \dot{e}_p^i). \blacksquare$$

This proposition guarantees that any result about block filtrations has a consequence for block dissections.

Let $[K_p]$ be a block filtration of K; put $\Gamma_n = H_n(K_n, K_{n-1})$ and let $\partial_n^\Gamma : \Gamma_n \to \Gamma_{n-1}$ be the homomorphism ν_* of the exact homology sequence of the triple K_n, K_{n-1}, K_{n-2}. Then $(\Gamma, \partial^\Gamma)$ is plainly a chain complex and we prove

10.9.2 Theorem. $H_*(\Gamma) \cong H_*(K)$.

$C(K)$ is graduated by $\{C_n(K)\}$ and filtered by $A_p = C(K_p)$; moreover, ${}_nA_p = C_n(K_p)$ and $A_p = \sum_n {}_nA_p$; it is not, however, in general true that, for each n, ${}_nA_n = C_n(K)$. The differential ∂ has the usual properties so that $C(K)$ is a d.u.f.g.-group (see 10.3.2, 10.3.3).

The familiar convergence property for d.f.g.-groups is replaced by the property that E^∞ is the direct limit of the sequence $E^2, E^3, \ldots$ and we shall apply 10.3.11 with $r_0 = 2$. As in the usual case

$${}_nE_p^0 = {}_nA_p/{}_nA_{p-1} = C_n(K_p, K_{p-1})$$

and ∂^0 is the boundary homomorphism for the pair (K_p, K_{p-1}). Thus

$$\begin{aligned} {}_nE_p^1 = H_n(K_p, K_{p-1}) &= 0 \quad \text{if} \quad n \neq p \\ &= \Gamma_n \quad \text{if} \quad n = p. \end{aligned}$$

Moreover $\partial^1 = \partial^\Gamma$, as observed in 10.2, so that

10.9.3
$$\begin{cases} {}_nE_p^2 = 0 \quad \text{if} \quad n \neq p, \\ {}_nE_n^2 = H_n(\Gamma). \end{cases}$$

Since ∂^2 maps ${}_nE^2_p$ to ${}_{n-1}E^2_{p-2}$, it follows that $\partial^2 = 0$, and similarly $\partial^r = 0$ for $r \geqslant 2$. Hence $E^2 = E^3 = \ldots$ and by 10.3.11 each is equal to the limit group E^∞. Hence from 10.9.3

10.9.4
$$\begin{cases} {}_nE^\infty_p = 0 \quad \text{if} \quad n \neq p, \\ {}_nE^\infty_n = H_n(\Gamma). \end{cases}$$

Since $[{}_nD_p]$ provides a filtration of $H_n(K)$ it follows from 10.3.7 and 10.9.4 that

$${}_nD_0 = {}_nD_1 = \ldots = {}_nD_{n-1} = 0, \quad {}_nD_n = {}_nD_{n+1} = \ldots = H_n(K)$$

and
$${}_nD_n/{}_nD_{n-1} = {}_nE^\infty_n = H_n(\Gamma).$$

Hence $H_n(K) \cong H_n(\Gamma)$. ∎

Theorem 10.9.2 generalizes 3.8.8; for in the case of a block dissection ∂^Γ coincides with the boundary in the chain complex $C(B(K))$ (see 3.8). The generalization is substantial in three ways: in 3.8, (i) we only discuss finite complexes; (ii) we impose the condition that

$$H_n(\bar{e}_n, \dot{e}_n) = J;$$

and (iii) we insist that the block e_n contains no simplex of dimension greater than n. On the other hand, the restrictions (ii) and (iii) are of practical value in geometrical examples; for (ii) focuses attention on generators (of cyclic infinite groups) and (iii) ensures that it is easy to recognize suitable generators of the groups $H_n(\bar{e}_n, \dot{e}_n)$ and so of $C(B(K))$.

10.10 Appendix: The spectral sequence associated with group, normal subgroup and quotient group

Let M be a group, N a normal subgroup and $\pi = M/N$; in this section we study the relation between the homology groups‡ of M, N and π that arise from a left M-module A.

Let $C(M)$ be the standard M-complex of 10.7.3; it can also be regarded as an N-complex. As such it is of course acyclic and it is also N-free; one may choose as N-basis of $C_q(M)$ the set of arrays $(\mu_0, \ldots, \mu_q)$ where μ_0 ranges over a set of representatives of the cosets with respect to N and the other μ_i $(i > 0)$ range over M. Further, A may be regarded as an N-module and the group $H_*(N; A)$ may be calculated from the complex $\Gamma = C(M) \otimes_N A$.

‡ See R. C. Lyndon, *Duke Math. J.* **15** (1948), 271; G. Hochschild and J.-P. Serre, *Trans. Amer. Math. Soc.* **74** (1953), 110.

The special feature exploited in this section is that π operates on Γ on the left by

10.10.1 $\xi\{(\mu_0, \mu_1, \ldots, \mu_q)\otimes_N a\} = (\mu_0\mu^{-1}, \ldots, \mu_q\mu^{-1})\otimes_N \mu a,$

where $\xi \in \pi$ is the coset containing $\mu \in M$; this operation respects the identities associated with $\otimes_N$ and is independent of the choice of μ in ξ. This operation induces an operation by π on $H_*(N; A)$.

10.10.2 Proposition. *If M operates trivially on G and N is in the centre of M, then π operates trivially on $H_*(N; G)$.*

Let $\rho^\mu : C(M) \to C(M)$ be given by

$$(\mu_0, \mu_1, \ldots, \mu_q)\,\rho^\mu = (\mu_0\mu^{-1}, \ldots, \mu_q\mu^{-1}).$$

Since N is in the centre of M it follows that ρ^μ is an N-map. It thus follows from 10.7.8, 10.7.9 that $\rho^\mu_* = 1 : H_*(N; A) \cong H_*(N; A)$. Now, since M operates trivially on G, reference to 10.10.1 shows that

$$\xi((\mu_0, \mu_1, \ldots, \mu_q)\otimes_N g) = (\mu_0\mu^{-1}, \ldots, \mu_q\mu^{-1})\otimes_N g.$$

Thus the operation of ξ on $H_*(N; G)$ coincides with ρ_* and is therefore the identity. ▮

The main theorem of this appendix, in which we do not make the special assumptions of 10.10.2 is

10.10.3 Theorem. *If M, N, π, A are as at the beginning of this appendix, there is a spectral sequence (E^r, ∂^r) such that $E^2_{p,q} = H_p(\pi; H_q(N; A))$ and E^∞ is the bigraded group associated with a suitable filtration of $H_*(M; A)$.*

The proof is very like that of 10.8.6. We construct the chain complex $C(\pi)\otimes_\pi \Gamma$ and use the augmentation of $C(\pi)$ to define

$$\beta : C(\pi)\otimes_\pi \Gamma \to \Gamma_\pi.$$

We now prove, in analogy with 10.8.1, that β induces a homology isomorphism. Since Γ is not in general π-free we need a special argument to show that, in the notation of 10.8.1, $H_{p,q}(\mathscr{E}^0) = 0$ for $q > 0$; the rest of the proof that β_* is isomorphic goes through with only formal changes.

We wish therefore to show that, if p is fixed but arbitrary and $q > 0$, $H_q(\pi; \Gamma_p) = 0$. To do this we define a chain homotopy

$$\Lambda : C(\pi)\otimes_\pi \Gamma_p \to C(\pi)\otimes_\pi \Gamma_p$$

by the rule that

$$((\xi_0, \ldots, \xi_q)\otimes_\pi y)\,\Lambda = (\xi, \xi_0, \ldots, \xi_q)\otimes_\pi y,$$

where $y = (\mu_0, \ldots, \mu_p)\otimes_N a$ and ξ is the coset containing μ_0. The reader should verify that Λ respects the defining relations of $\otimes_N$ and $\otimes_\pi$.

An elementary computation shows that

$$\Lambda\partial + \partial\Lambda = 1 \quad \text{on} \quad C_q(\pi)\otimes_\pi \Gamma_p \quad \text{for} \quad q \geqslant 1,$$

so that $H_q(\pi; \Gamma_p) = 0$ for $q \geqslant 1$. We thus conclude as in 10.8.1 that

$$\beta_* : H_*(C(\pi)\otimes_\pi \Gamma) \cong H_*(\Gamma_\pi).$$

Now $\Gamma_\pi = (C(M)\otimes_N A)_\pi$, so that, from 10.10.1, $\Gamma_\pi = C(M)\otimes_M A$. We may thus regard β_* as an isomorphism

$$\beta_* : H_*(C(\pi)\otimes_\pi \Gamma) \cong H_*(M; A).$$

Now $H_*(C(\pi)\otimes_\pi \Gamma)$ acquires a filtration from the usual tensor product filtration of $C(\pi)\otimes_\pi \Gamma$; we use β_* to transfer this filtration to $H_*(M; A)$. Turning, on the other hand, to the spectral sequence of this filtration, we may (compare the proof of 10.8.4) identify $E^0_{p,q}$ with $C_p(\pi)\otimes_\pi \Gamma_q$, $E^1_{p,q}$ with $C_p(\pi)\otimes_\pi H_q(N; A)$ and $E^2_{p,q}$ with $H_p(\pi; H_q(N; A))$. ∎

We apply this to prove‡

10.10.4 Theorem. *Let M be a group of order p^n, where p is prime; then the order of $H_2(M; J)$ is p^k, where $k \leqslant \frac{1}{2}n(n-1)$.*

We shall write $|Q|$ for the order of the group Q.

We remark first that since $C_2(M)$ is finitely generated, so is $H_2(M; J)$. Also by 10.7.25 $p^n H_2(M; J) = 0$, whence $H_2(M; J)$ is a finite abelian group whose order is a power of p.

We next quote the essential fact§ that the centre of M must contain a subgroup J_p; we take $N = J_p$ and apply 10.10.3, writing π for M/J_p. We invoke 10.10.2 for the fact that π operates trivially on

$$H_*(N; J) = H_*(J_p; J).$$

Consider the filtration ${}_2D_0 \subseteq {}_2D_1 \subseteq {}_2D_2 = H_2(M; J)$ of 10.10.3. Then ${}_2D_2/{}_2D_1 = E^\infty_{2,0}$, ${}_2D_1/{}_2D_0 = E^\infty_{1,1}$ and ${}_2D_0 = E^\infty_{0,2}$.

Now $E^\infty_{2,0}$ is a subgroup of $E^2_{2,0} = H_2(\pi; H_0(J_p; J)) = H_2(\pi; J)$; hence

10.10.5 $$|E^\infty_{2,0}| \leqslant |H_2(\pi; J)|.$$

Next $E^\infty_{1,1}$ is a quotient group of

$$E^2_{1,1} = H_1(\pi; H_1(J_p; J)) \cong H_1(\pi; J)\otimes H_1(J_p; J);$$

‡ See J. A. Green, *Proc. Roy. Soc.* A, **237** (1956), 574. The group $H_2(M; J)$ is in fact the *multiplicator*, originally defined by I. Schur, *J. reine u. angew. Math.* **127** (1904), 20.

§ Ledermann [20], Th. 2, p. 99.

here we use 10.7.20 and the fact that $H_0(\pi; J) * H_1(J_p; J) = 0$ as $H_0(\pi; J) = J$. By 10.7.21 $H_1(\pi; J)$ is π abelianized and $H_1(J_p; J) = J_p$. Hence $E^2_{1,1}$ is π abelianized and reduced mod p so that

10.10.6 $$|E^{\infty}_{1,1}| \leqslant |E^2_{1,1}| \leqslant |\pi| = p^{n-1}.$$

Finally, $E^{\infty}_{0,2}$ is a quotient group of

$$E^2_{0,2} = H_0(\pi; H_2(J_p; J)) = H_2(J_p; J) = 0$$

by 10.7.23; therefore

10.10.7 $$|E^{\infty}_{0,2}| = 1.$$

Now combining 10.10.5, 10.10.6 and 10.10.7

$$|H_2(M; J)| = |E^{\infty}_{2,0}| \cdot |E^{\infty}_{1,1}| \cdot |E^{\infty}_{0,2}| \leqslant p^{n-1} |H_2(\pi; J)|.$$

Induction on n now completes the argument. ∎

We remark that this result is best possible, since if M is the direct product of n copies of J_p it follows from 10.7.22 that

$$|H_2(M; J)| = p^{\frac{1}{2}n(n-1)}.$$

EXERCISES

Differential filtered groups and spectral sequences

1. Let $(G; A_p)$ be an f.-group, let $G' = \mathrm{Gr}[A_p]$ and let $G'_p = \sum_{n \leqslant p} \mathrm{Gr}[A_p]_n$. Show that $(G; A_p) \cong (G'; A'_p)$ if and only if each A_{p-1} is a direct factor in A_p.

Now let $(G; A_p, \partial)$ be a d.f.-group. Then ∂ induces a differential in each A_p/A_{p-1} and hence a differential ∂' in G' such that $(G'; A'_p, \partial')$ is a d.f.-group. Show that we may have $(G'; A'_p, \partial') \not\cong (G; A_p, \partial)$, although $(G'; A'_p) \cong (G; A_p)$.

2. Let $(G; G_n, A_p, \partial)$ be a d.f.g.-group. Appropriate inclusion maps induce homomorphisms

$${}_n\rho^r_p : H_n(A_p/A_{p-r}) \rightarrow H_n(A_{p+r-1}/A_{p-1}).$$

Show that there are natural isomorphisms

$${}_nE^r_p \cong \text{Image of } {}_n\rho^r_p, \quad r > 0,$$

and identify the image of d^r under these isomorphisms.

3. Let $(G; G_n, A_p, \partial)$, $(G'; G'_n, A'_p, \partial')$ be d.f.g.-groups over a field F; that is, all groups concerned are vector spaces over the field F and all homomorphisms preserve vector space structure. Define

$$\overline{G} = G \otimes_F G', \overline{G}_n = \sum_{r+s=n} G_r \otimes_F G'_s, \quad \overline{A}_p = \sum_{t+u=p} A_t \otimes_F A'_u, \quad \overline{\partial} = \partial \otimes_F 1 + \epsilon \otimes_F \partial',$$

where $g_n \epsilon = (-1)^n g_n$, $g_n \in G_n$. Show that $(\overline{G}; \overline{G}_n, \overline{A}_p, \overline{\partial})$ is a d.f.g.-group. Show, moreover, that the spectral sequences of $G, G', \overline{G}$ are related by

$$\overline{E}^r_{p,q} = \sum_{\substack{a+b=p \\ c+d=q}} E^r_{a,c} \otimes_F E'^r_{b,d}, \quad \overline{d}^r = d^r \otimes_F 1 + \epsilon \otimes_F d'^r,$$

where $$x\epsilon = (-1)^{a+c} x, \quad x \epsilon E^r_{a,c}.$$

4. Let $\phi, \psi : (G; G_n, A_p, \partial) \to (G'; G'_n, A'_p, \partial')$ be d.f.g.-maps and let $\Lambda : G \to G'$ be a chain homotopy from ϕ to ψ, regarded as d.g.-maps. Show that if $A_p \Lambda \subseteq A'_{p+s}$ for all p and fixed s, then Λ induces a homomorphism $\Lambda^s : E^s \to E'^s$ such that

$$E^s_{p,q}\Lambda^s \subseteq E'^s_{p+s,\,q-s+1}$$

and

$$\phi^s \overset{\Lambda^s}{\simeq} \psi^s.$$

Deduce that $\phi^r = \psi^r$, $r > s$, and $\phi_* = \psi_* : H_*(G) \to H_*(G')$.

Maps of fibrations

Let f be a map $f : X, x_* \to Y, y_*$ and let $F = f^{-1}(y_*)$; we write then

$$f : X, F, x_* \to Y, y_*;$$

if also $f' : X', F', x'_* \to Y', y'_*$, then a *map* from f to f' is a pair of maps

$$g : X, F, x_* \to X', F', x'_*, \quad h : Y, y_* \to Y', y'_*$$

such that $fh = gf'$. We suppose F, F' path-connected.

5. Show that a map (g, h) from f to f' induces a map γ of the d.f.g.-group of f into that of f' and hence a map of the spectral homology sequence of f into that of f'. Show also that (g, h) induces a map $\gamma^{\otimes}$ of the d.f.g.-group $C(Y) \otimes C(F)$ into the d.f.g.-group $C(Y') \otimes C(F')$. Let ϕ^0 be the d.g.-homomorphism of 10.4.3. Show that $\phi^0 \gamma^{\otimes 0} = \gamma^0 \phi^0$.

6. Suppose now that $p : X, F, x_* \to B, b_*$, $p' : X', F', x'_* \to B', b'_*$ are fibre maps and that (g, h) is a map from p to p'. Use the result of Q. 5 and Theorem 10.4.11 to deduce that if B, B' are 1-connected and if (g, h) induces isomorphisms $H_*(B) \cong H_*(B')$, $H_*(F) \cong H_*(F')$, then it induces isomorphisms $H_*(X) \cong H_*(X')$. [Remark: one may indeed prove by these methods that if (g, h) induces homology isomorphisms for any two of fibre space, fibre, and base space, then it induces homology isomorphisms of the third. Given Q. 5 and 10.4.11 the argument is purely algebraic: Let $\{E^r\}$ be a spectral sequence in which $E^2_{p,q}$ may be identified with $E^2_{p,0} \otimes E^2_{0,q} \oplus E^2_{p,0} * E^2_{0,q-1}$; let $\{E'^r\}$ be another such spectral sequence and let $\gamma : \{E^r\} \to \{E'^r\}$ be a map of spectral sequences (respecting the given identifications). Consider in particular the maps

$$\gamma^\infty_{p,q} : E^\infty_{p,q} \to E'^\infty_{p,q}, \quad \gamma^2_{p,0} : E^2_{p,0} \to E'^2_{p,0}, \quad \gamma^2_{0,q} : E^2_{0,q} \to E'^2_{0,q}.$$

Then of the three statements

(i) $\gamma^\infty_{p,q}$ is an isomorphism for all p, q,
(ii) $\gamma^2_{p,0}$ is an isomorphism for all p,
(iii) $\gamma^2_{0,q}$ is an isomorphism for all q,

any two imply the third. This is essentially the comparison theorem for spectral sequences, see E. C. Zeeman, *Proc. Camb. Phil. Soc.* **53** (1957), 57.]

Spectral sequence of a fibre map: relative theory

7. Let $p : X, F \to B, b_*$ be a fibre map, let $B_0 \subseteq B$ and let $X_0 = p^{-1}(B_0)$. Apply the Serre technique of 10.4 to deduce a spectral sequence E^r in which

$$E^2_{p,q} = H_p(B, B_0; \{H_q(F)\}),$$

and E^∞ is the graded group associated with $H_*(X, X_0)$, suitably filtered. (See, for example, J. C. Moore, *Ann. Math.* **58** (1953), 325.)

Let B, B_0 be 1-connected, let E be the space of paths on B ending at b_*, let $p: E \to B$ project each path on to its initial point and let $E_0 = p^{-1}(B_0)$. Assume that $H_r(B, B_0) = 0$, $r < n$, and deduce that $p_*: H_n(E, E_0) \cong H_n(B, B_0)$. Hence infer the *relative Hurewicz isomorphism theorem* (8.8.8).

Spectral sequence of a fibre map: applications (see 10.6)

8. Consider the fibre map $p: EX \to X$ with fibre ΩX. Use 10.6.2 to prove that if $\pi_r(X) = 0$, $0 \leqslant r < n$, then the transgression t is an isomorphism

$$t: H_m(X) \cong H_{m-1}(\Omega X),$$

$0 < m \leqslant 2n-2$.

9. Under the assumptions operating in 10.6, prove that, for the values of Π listed below, if the homology groups of any two of X, F, B belong to Π in positive dimensions, so do the homology groups of the third.

(i) Π = collection of *fg* groups,
(ii) Π = collection of finite groups,
(iii) Π = collection of torsion groups,
(iv) Π = collection of those torsion groups whose elements have orders involving only the primes from a given set $\mathscr{P}$.

10. Obtain $R^*(\Omega S^m)$ when m is even (see Example 10.6.20).

Homology suspension

The scheme $H_k(B) \leftarrow H_k(X, F) \to H_{k-1}(F)$ has been used to define a *transgression* from a subgroup of $H_k(B)$ to a quotient group of $H_{k-1}(F)$; it may just as well serve to define the *homology suspension* s from a subgroup of $H_{k-1}(F)$ to a quotient group of $H_k(B)$. Similarly, the *contrahomology suspension* $s^\cdot$ is a homomorphism from a subgroup of $H^k(B)$ to a quotient group of $H^{k-1}(F)$.

11. Show that the homology suspension $s: H_{m-1}(\Omega X) \to H_m(X)$ is always defined and is inverse to t for $0 < m \leqslant 2n-2$ if X is $(n-1)$-connected.

12. Show that if $\alpha \in H^p(X)$ and $\beta \in H^q(X)$, $p, q > 0$, then $s^\cdot(\alpha\beta) = 0 \in H^{p+q-1}(\Omega X)$.

Homology groups of groups and modules

When π is commutative, we permit ourselves to regard π as abelian and so it may be regarded as belonging to a collection Π (see Q. 13, 14).

13. Regard J as a trivial π-module. Show that, in each of the four cases listed in Q. 9, if $\pi \in \Pi$ then $H_k(\pi; J) \in \Pi$ for $k > 0$. Hence deduce that if $K = K(\pi, n)$, $n \geqslant 1$, then $H_k(K) \in \Pi$, $k > 0$. (Note that $H_k(\pi; J) = H_k(K(\pi, 1))$.)

14. Use Q. 9, Q. 13, and the technique of killing homotopy groups (see Example 10.6.22) to prove that if X is 1-connected and its homology groups (in positive dimension) belong to Π, then so do its homotopy groups. By considering the case $X = S^1 \vee S^2$ (and passing to its universal covering space) show that the condition of 1-connectedness cannot be dispensed with.

15. Show that $H_k(\pi; A) = 0$, $k > 1$, if π is locally free.

16. Show that $H^2(\pi; G)$ may be interpreted as the group of central extensions of π by G; that is, to each class of extensions

$$1 \to G \xrightarrow{\lambda} E \xrightarrow{\mu} \pi \to 1,$$

where G is in the centre of E we associate an element of $H^2(\pi; G)$ (compare 5.9.1). Generalize this to $H^2(\pi; A)$. [Hint: the procedure in the first case is to choose representatives e_ξ of $\mu^{-1}\xi$ for each $\xi \in \pi$ and then to consider, for each pair ξ, η in π the element $\lambda^{-1}(e_\xi e_\eta e_{\xi\eta}^{-1})$ of G. In the second case note that if $1 \to A \xrightarrow{\lambda} E \xrightarrow{\mu} \pi \to 1$ with A abelian then π acts on A by $(e\mu)a = eae^{-1}$ (suppressing λ).]

Covering spaces

17. Relativize 10.8.6 in the following form: If, under the hypotheses of 10.8.6, A is a closed subspace of X and if $\tilde{A} = p^{-1}(A)$, then there is a spectral sequence in which $E^2_{p,q} \cong H_p(\pi; H_q(\tilde{X}, \tilde{A}; G))$ and E^∞ is the bigraded group associated with a suitable filtration of $H_*(X, A; G)$. Investigate the sequence when $X = P^7$, $A = P^3 \subseteq P^7$, $\pi = J_2$ and $G = J$, P^n being real projective n-space.

18. Let $f: X, x_* \to Y, y_*$ induce isomorphisms of homology groups and fundamental group and let $\pi_1(X)$, $\pi_1(Y)$ operate trivially on the homology groups of $\tilde{X}$, $\tilde{Y}$, the universal covering spaces of X, Y. Prove that if $\tilde{f}: \tilde{X}, \tilde{x}_* \to \tilde{Y}, \tilde{y}_*$ lifts f then $\tilde{f}$ induces homology isomorphisms. [Hint: use the Zeeman comparison theorem (see Q. 6).]

19. Let X be the polyhedron of an n-dimensional complex, let $\tilde{X}$ be its universal covering space, let G be a coefficient group for homology, and suppose that $\pi = \pi(X)$ operates trivially on $H_*(\tilde{X}; G)$. Suppose further that $\pi_i(X) = 0$, $1 < i < n$. Use the spectral sequence of 10.8 to show that

$$H_i(\pi; G) \cong H_i(X; G), \quad i < n;$$

that there is an exact sequence

$$0 \to H_{n+1}(\pi; G) \to H_n(\tilde{X}; G) \to H_n(X; G) \to H_n(\pi; G) \to 0;$$

and that
$$H_i(\pi; H_n(\tilde{X}; G)) \cong H_{i+n+1}(\pi; G), \quad i > 0.$$

Verify the hypotheses if $\tilde{X} = S^n$, n odd, π is a finite group of fixpoint-free homeomorphisms of S^n, and $X = S^n/\pi$. If π is of order m deduce that

$$H_n(\pi; G) = G_m,$$

$$H_{n+1}(\pi; G) = {}_mG,$$

$$H_i(\pi; G) \cong H_{i+n+1}(\pi; G), \quad i > 0.$$

[Hint: the covering map $p: \tilde{X} \to X$ has degree m.]

BIBLIOGRAPHY

I. GENERAL LIST

[1] ALEXANDROFF, P. and HOPF, H., *Topologie*, Springer (1935).
[2] LEFSCHETZ, S., *Introduction to Topology*, Princeton (1949).
[3] PATTERSON, E. M., *Topology*, Oliver and Boyd (1956).
[4] PONTRYAGIN, L. S., *Foundations of Combinatorial Topology*, Graylock (1952).
[5] REIDEMEISTER, K., *Einführung in die kombinatorische Topologie*, New York (1950).
[6] SEIFERT, H. and THRELFALL, W., *Lehrbuch der Topologie*, New York (1947).
[7] WALLACE, A. H., *Introduction to Algebraic Topology*, Pergamon (1957).
[8] WILDER, R. L., *Topology of Manifolds*, New York (1949).

II. BOOKS ON HOMOTOPY THEORY

[9] HILTON, P. J., *Introduction to Homotopy Theory*, C.U.P. (1953).
[10] HU, S. T., *Homotopy Theory*, New York, (1959).
[11] WHITEHEAD, G. W., *Homotopy Theory* (1953) (lecture notes available from Technology Bookshop. Cambridge, Mass.).

III. FURTHER READING

[12] Séminaire H. CARTAN
1948/49: *Topologie algébrique.*
1949/50: *Espaces fibrés et homotopie.*
1950/51: *Cohomologie des groupes, suite spectrale, faisceaux.*
1954/55: *Algèbres d'Eilenberg–MacLane et homotopie.*
1958/59: *Invariant de Hopf et opérations cohomologiques sécondaires.*
(Available from the Secretariat mathématique, Institut Henri Poincaré, Paris.)
[13] CARTAN, H. and EILENBERG, S., *Homological Algebra*, Princeton (1956).
[14] EILENBERG, S. and STEENROD, N. E., *Foundations of Algebraic Topology*, Princeton (1952).
[15] GODEMENT, R., *Topologie algébrique et theorie des faisceaux*, Hermann, Paris (1958). (A standard text on sheaf theory and the associated contrahomology theory.)
[16] HIRZEBRUCH, F., *Neue topologische Methoden in der algebraischen Geometrie*, Berlin (1956). (Application of contrahomology theory to algebraic geometry.)
[17] NORTHCOTT, D. G., *Introduction to Homological Algebra*, C.U.P. (1960).
[18] STEENROD, N. E., *Topology of Fibre Bundles*, Princeton (1951). (Part 3 is devoted to obstruction theory in fibre bundles.)
[19] STEENROD, N. E., *Cohomology Operations and Obstructions to Extending Continuous Functions.* (Colloquium notes available from Princeton University.)

IV. BACKGROUND BOOKS ON GROUP THEORY, ANALYTIC TOPOLOGY

[20] LEDERMANN, W., *Introduction to the Theory of Finite Groups*, 3rd edition, Oliver and Boyd (1957).
[21] NEWMAN, M. H. A., *Topology of Plane Sets of Points*, C.U.P. (1951).

INDEX